What do you need to learn NOW?

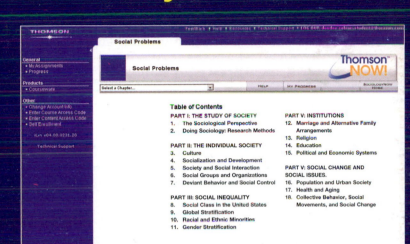

Take charge of your learning with **ThomsonNOW**™, (formerly SociologyNow) the first assessment-centered student learning tool for sociology.

This new online diagnostic tool could be your key to success in your sociology course!

ThomsonNOW™ **will help you:**

◆ create a personalized study plan for each chapter of your sociology text
◆ understand key concepts in the course
◆ better prepare for exams—and increase your chances of success

Lift the page for more information.

THOMSON

™

WADSWORTH

How can you access ThomsonNOW?

◆ **Your instructor may have chosen to package the access code card with your new text.** In this case, you'll find the access card within this text, which contains your free, four-month pass code, allowing you anytime access to ThomsonNOW.

◆ **If your instructor did not order the free access code card to be packaged with your text—or if you have a used copy of the text—you can still obtain an access code for a nominal fee.** Just visit the Thomson Wadsworth E-Commerce site at www.thomsonedu.com/sociology, where easy-to-follow instructions help you purchase your access code.

Are you ready to get started?
Help is just a click away with ThomsonNOW.

The exciting program that is firmly grounded in sociology and lets you take diagnostic quizzes, review chapter content, conduct online research, think critically about sociological concepts, watch videos of well-known sociologists as they discuss important concepts, and ultimately improve your success in the course, all on one, easy-to-use web-based program.

www.thomsonedu.com/thomsonnow

Log on today!

Thomson Wadsworth ◆ P.O. Box 6904 ◆ Florence, KY 41022
800-423-0563 ◆ Fax 859-647-5020 ◆ Email: review@kdc.com

Source code 7TPSOM01

www.wadsworth.com

www.wadsworth.com is the World Wide Web site for Thomson Wadsworth and is your direct source to dozens of online resources.

At *www.wadsworth.com* you can find out about supplements, demonstration software, and student resources. You can also send e-mail to many of our authors and preview new publications and exciting new technologies.

www.wadsworth.com
Changing the way the world learns®

Introduction to Sociology

NINTH EDITION

Henry L. Tischler

Montclair State University

THOMSON

WADSWORTH

Australia • Brazil • Canada • Mexico • Singapore • Spain
United Kingdom • United States

THOMSON

WADSWORTH

Introduction to Sociology, **Ninth Edition**
Henry L. Tischler

Senior Sociology Editor: Robert Jucha
Assistant Editor: Kristin Marrs
Editorial Assistant: Katia Krukowski
Technology Project Manager: Dee Dee Zobian
Marketing Manager: Michelle Williams
Marketing Assistant: Jaren Boland
Marketing Communications Manager: Linda Yip
Project Manager, Editorial Production: Cheri Palmer
Creative Director: Rob Hugel
Print Buyer: Karen Hunt

Permissions Editor: Sarah Harkrader
Production Service: Lisa Royse, Graphic World Inc.
Text Designer: Stuart Paterson
Photo Researcher: Myrna Engler
Illustrator: Graphic World Illustration Studio
Cover Designer: Yvo Riezebos
Cover Image: © Masterfile
Compositor: Graphic World Inc.
Printer: Quebecor World/Versailles

Library of Congress Control Number: 2005936961

ISBN 0-495-09334-3

Thomson Higher Education
10 Davis Drive
Belmont, CA 94002-3098
USA

For more information about our products, contact us at:
Thomson Learning Academic Resource Center
1-800-423-0563

For permission to use material from this text or product, submit a request online at
http://www.thomsonrights.com.
Any additional questions about permissions can be submitted by e-mail to
thomsonrights@thomson.com.

What I know about society could fill a book.
What I don't would fill the world.
Dedicated to my fellow travelers in the journey of life—
Linda, Melissa, and Ben.

Contents in Brief

Contents

Chapter

4 Socialization and Development 80

Chapter

5 Society and Social Interaction 110

Chapter

6 Social Groups and Organizations 134

Features Contents

For Further Thinking

News You Can Use

Preface

As a freshman at Temple University, my first experience with a college textbook was in my sociology course. I dutifully read the assigned chapter during my first week of class hoping to become familiar with the subject matter of this required course. The only problem was that I had no idea what the author was saying. The writing level was advanced, the style dense, and the book downright threatening without photos or illustrations. After several hours of reading I felt frustrated and stupid, and I knew no more about sociology than when I started.

If this was what college was going to be like, I was not going to make it, I thought. I remember admitting reluctantly that I was probably not what guidance counselors in that day referred to as "college material." I pictured myself dropping out after the first semester and looking for a job selling furniture or driving a cab. My family would be disappointed, but my father was a factory worker, and there was no family history of college attendance to live up to. I continued to struggle with the book and earned a D on the mid-term exam. After much effort, I managed to finish the course with a C and a burning disinterest in the field of sociology. I did not take another sociology course for two years, and when I did it was "Marriage and the Family," considered the easiest course on campus.

I often wonder how I came from this inauspicious beginning to become a sociology professor, let alone the author of a widely used introductory sociology textbook. Then again, maybe it is not all that unusual, because that experience continues to have an effect on me each day. Those 15 weeks helped develop my view that little can be gained by presenting knowledge in an incomprehensible or unnecessarily complicated way or by making yourself unapproachable. Pompous instructors and intimidating books are a disservice to education. Learning should be an exciting, challenging, and eye-opening experience, not a threatening one.

One of the real benefits of writing nine editions of this textbook is that I have periodically examined every concept and theory presented in an introductory course. In doing so I have approached the subject matter through a new set of eyes and have consistently tried to find better ways of presenting the material. As instructors, we rarely venture into each other's classrooms and hardly ever do we receive honest, highly detailed, and constructive criticism of how well we are transmitting the subject matter. In the writing of a textbook, we receive this type of information, and we can radically restructure or simply fine-tune our presentation. It is quite an education for those of us who have devoted our careers to teaching sociology.

Student-Oriented Edition

Before revising this edition of *Introduction to Sociology*, we surveyed dozens of instructors to find out what they wanted in a textbook and what would assist them in the teaching of sociology, as well as satisfy student needs. This revised text reflects their significant input. In the surveys for this and past editions, we learned that both students and instructors continue to be concerned about the cost of textbooks. Introductory textbooks have become very attractive and expensive during the last decade because publishers have added hundreds of color photos to the typical volume. This trend has caused the price of textbooks to increase, making them a substantial purchase for the typical student. We did something about the high cost to students—in response to this concern, we broke ranks with textbooks with which we typically competed and went back to the basics. A textbook, after all, is meant to be comprehensive and up-to-date and should serve as an important supplement to a course. It makes no sense to make a book so colorful, and therefore so expensive, that students often forgo purchasing it.

To give students the best value for the dollar, we use black and white photos instead of color and a soft rather than a hard cover. In this way, students will be getting far greater value because nothing of educational content is sacrificed to produce this savings. We are not, however, content to merely provide a better value. We also want to provide a better book.

We, therefore, include a full, built-in study guide with this book that is as extensive, if not more so, than those typically sold separately. In this way, students will be able to purchase the combined textbook and study guide for considerably less than the price of a typical textbook. In fact, the price for our textbook/study guide combination will most likely be lower than the used copy price of a typical hardcover introductory sociology textbook.

Presentation

Even though I began my college career as one of the less-capable students, I was fascinated by what college had to offer. Where else could you be exposed to so much about a world that is so interesting? Belatedly, I began to realize that a great deal of what is interesting falls into the field of sociology. My goal in this book is to demonstrate the vitality, interest, and utility associated with the study of sociology. Examining society and trying to understand how it works is an exciting and absorbing process. I have not set out to make sociologists of my readers (although if that happens I will be delighted), but rather to show how sociology applies to many areas of life and how it is used in day-to-day activities. In meeting this objective, I have focused on two basic ideas: that sociology is a rigorous scientific discipline and that a basic knowledge of sociology is essential for understanding social interaction in many different settings, whether they be work or social. To understand society, we need to understand how it shapes people and how people in turn shape society.

Each chapter progresses from a specific to a general analysis of society. Each part introduces increasingly more comprehensive factors necessary for a broad-based understanding of social organization.

The material is presented through consistently applied learning aids. Each chapter begins with a chapter outline. Then a thought-provoking opening vignette offers a real-life story of the concepts being covered. Key terms are presented in boldfaced type in the text. Key concepts are presented in italicized type in the text. A chapter summary concludes each chapter, and an integrated study guide follows each chapter. A full glossary is in the back of the book for further reference.

Great care has been taken to structure the book in such a way as to permit flexibility in the presentation of the material. Each chapter is self-contained and therefore may be taught in any order.

It has taken nearly two years to produce this revision. Every aspect of this book has been updated, and a great deal has been changed. The information is as current and up-to-date as possible, and there are hundreds of references from 2000 to 2005 throughout the book.

A Comparative and Cross-Cultural Perspective

Sociology is a highly organized discipline shaped by several theoretical perspectives or schools of thought. It is not merely the study of social problems or the random voicing of opinions. In this book, no single perspective is given greater emphasis; a balanced presentation of both functionalist theory and conflict theory is supplemented whenever possible by the symbolic interactionist viewpoint.

The book has received a great deal of praise for being cross-cultural in approach and for bringing in examples from a wide variety of societies. Sociology is concerned with the interactions of people wherever and whenever they occur. It would be shortsighted, therefore, to concentrate on only our own society. Often the best way to appreciate our own situation is through comparison with other societies. We use our cross-cultural focus as a basis for comparison and contrast with U.S. society.

New to This Edition

- Material on social interaction, group behavior, and social organization has been expanded and, where it was previously covered in one chapter, is now presented in two chapters: Chapter 5 ("Society and Social Interaction") and Chapter 6 ("Social Groups and Organizations").
- Chapter 18 has been expanded to now cover "Collective Behavior and Social Change."
- Expanded coverage of global issues throughout the book provides students with an understanding of the interconnectedness of our world.
- Chapter 16 has a redirected focus on "Population and Urban Society," addressing the demographic changes taking place throughout the world.
- Chapter 2 contains useful features on "How to Read a Table" and "How to Spot a Bogus Poll," skill development and critical-thinking features not usually found in other introductory texts.
- One of the hallmarks of this book is interesting theme boxes on relevant sociological issues. Approximately 25 new boxes have been added to this edition.
- A new "News You Can Use" theme box has been added to enable students to see the applicability of sociological issues to their lives.
- New "Social Change" boxes have been added throughout the text to provide examples of major forces that are shaping modern society.
- A wealth of new "Technology and Society" features introduces students to sociological issues dealing with the Internet, media advances, and technology in general.

- Coverage of aging has been expanded in Chapter 17 ("Health and Aging").
- New "For Further Thinking" essays appear at the end of selected chapters to highlight a contemporary issue or application of a sociological concept.

Features

Opening Vignettes

Each chapter begins with a lively vignette that introduces students to the subject matter of the chapter. Many of these are from real-life events to which students can relate. Examples include the scientific validity of the claim that there is an epidemic of missing and abducted children (Chapter 1), whether school bullies are a serious problem (Chapter 2), the cultural adjustment of an American woman in Egypt (Chapter 3), socialization during Marine Corps basic training (Chapter 4), the role names play in our identity (Chapter 6), people who take classes on how to marry a wealthy spouse (Chapter 8), and the personal impact of prenatal screening (Chapter 17). Others deal with unusual circumstances that remind students that there is a wide range of events to which sociology applies. Examples include the eccentric soprano Florence Foster Jenkins (Chapter 7), whites who claim to be black (Chapter 10), a transsexual who believes there are dozens of genders (Chapter 11), and the one-child population control policy in China (Chapter 16).

Theme Boxes

Thought-provoking boxed features bring sociological concepts to life for students. This effective learning tool presents sociological concepts in interesting real-life contexts. You will find six types of boxes in this edition—"News You Can Use," "Technology and Society," "Our Diverse Society," "Controversies in Sociology," "Social Change," and "Global Sociology." In addition, "For Further Thinking" essays appear at the end of selected chapters to provide a more in-depth presentation of a topic.

News You Can Use

The "News You Can Use" boxes examine trends or interesting sociological research that have a connection to students' lives. The instructor will be able to discuss these with an eye toward showing the relevance of sociology to everyday life. Included in this section are such topics as "Is There an Epidemic of College Student Suicides?" "Laugh and the World Laughs with You," "The Strength of Weak Ties in Job Hunting," "Are Peaceful Pot Smokers Being Sent to Prison?" "Are Urban Poverty Ghettos Shrinking?"

"Hispanics: Racial Group? Ethnic Group? Neither?" "Are We Biased against Assertive Women?" "Marriage and Divorce Quiz," "A Nation of Believers," "Do Men without Women Become Violent?" and "Binge Drinking as a Health Problem."

Technology and Society

Social research and technological change often go hand-in-hand. In particular, the Internet has had an enormous impact on society and students' lives. Recognizing the importance of the social impact of technology, we explore such topics as "Does Television Reduce Social Interaction?" "The Girls Who Will Not Be Born," "Hate Sites on the Web," "Is There Gender in Cyberspace?" "Seeking God on the Web," "College Students and the Internet," "Selling Human Life," and "Dispersed Collective Behavior on the Internet."

Our Diverse Society

Anyone studying sociology will quickly become aware of the enormous amount of social diversity. The United States, with its extensive history of immigration, has become one of the most diverse countries in the world. How has this diversity expressed itself in American society? In the "Our Diverse Society" boxes we explore this question when we look at such topics as "Racial Integration in the Military," "Women Who Did Not Want Women to Vote," "Public Heroes, Private Felons: Athletes and Sexual Assault," "Deborah Tannen: Communication between Women and Men," "Should Same-Sex Marriages Be Permitted?" "Who Is God?" "Comparing the Political and Moral Values of the 1960s with Today," "Disorderly Behavior and Community Decay," and "Stereotypes about the Elderly."

Controversies in Sociology

The special "Controversies in Sociology" boxes are designed to show students two sides of an issue. The topics featured will help students realize that most social events require close analysis and that hastily drawn conclusions are often wrong. The students will see that to be a good sociologist one must be knowledgeable about disparate positions and willing to question the validity of all statements and engage in critical thinking.

Included in these boxes are such controversies as "Is There a Difference between Sociology and Journalism?" "Truth in the Courtroom versus Truth in the Social Sciences," "Famous Research Studies You Cannot Do Today," "Is Day Care Harmful to Children?" "Is the Income Gap between the Rich and the Poor a Problem?" "Is the Debate on Race and Intelligence Worthwhile?" "Can Gender Identity Be Changed?" and "What Produces Homelessness?"

Social Change

"Social Change" boxes investigate trends or puzzling developments in society. These boxes allow instructors and students to examine a specific issue to see what answers sociology can offer to help understand the trend. Included are such topics as "Is There a Culture Clash between the United States and Saudi Arabia?" "Limiting Technology to Save the Community," "Serial Murderers and Mass Murderers," "What Causes Poverty?" "Reluctant to Marry: The Men Who Want to Stay Single," "Religion Is Constantly Changing," and "Jonathan Kozol on *The Shame of the Nation*."

Global Sociology

To highlight the cross-cultural nature of this book, many chapters include a "Global Sociology" box. These boxed features encourage students to think about sociological issues in a larger context and explore the global diversity present in the world. Included among these boxes are such topics as "Is McDonald's Practicing Cultural Imperialism or Cultural Accommodation?" "An American Success Story Does Not Translate into Japanese," "Cross-Cultural Social Interaction Quiz," "The United States Is a World Leader in Homicide," "Rich Countries with Poor Children," "HIV/AIDS: Worldwide Facts," "Arranged Marriage in India," "College Graduates: A Worldwide Comparison," "The Worst Offenders of Religious Freedom," "What if the Population Problem Is Not Enough People?" "Women Live Longer than Men throughout the World," and "Global Aging Quiz."

For Further Thinking

At the end of selected chapters, longer boxes appear that discuss a chapter-related topic in greater detail. These boxes appear at the end of the chapter so they do not interrupt the flow of material. Included in this section are "If You Are Thinking about Sociology as a Career, Read This," "The Conflict between Being a Researcher and a Human Being," "The Continuing Debate over Capital Punishment: Does It Deter Murderers?" "How Easy Is It to Change Social Class?" "How Much Are Children Hurt by Their Parents' Divorce?" and "Are College Admissions Tests Fair?"

Built-in Study Guide and Practice Tests

The interactive workbook study guide, by Jay Livingston of Montclair State University, is fully integrated into the book. Each chapter is followed by a study guide section, so students can review the material immediately without having to search for it elsewhere in the book. This encourages students to see the study guide as an integral part of the learning process.

The study guide provides for ample opportunity to review the material with a variety of styles of review questions. All key terms and key sociologists are reviewed with matching questions. Key concepts are revisited with fill-in questions. Critical Thought Exercises help students understand concepts covered in the chapter. Often website URLs are provided for students to expand on their exploration of the topic, and a matching question answer key is provided to allow students immediate review of their answers.

Practice tests are placed at the end of the book to provide students with additional preparation for testing. Whereas other practice tests are limited to recognition and recall items, these questions lead students to engage in such higher-level cognitive skills as analysis, application, and synthesis. The tests encourage students to think critically and apply the material to their experiences. Again, an answer key is provided to allow students full review and preparation.

All of these tools will be very useful for students preparing for essay exams and research papers. The textbook also includes the important section, "How to Get the Most Out of Sociology," which discusses how to use the study guide, practice tests, and lecture material in preparing for exams and getting the most out of the introductory sociology course.

The Ancillary Package

Instructor's Resource Manual and Test Bank

Debra Heath-Thornton of Messiah College prepared the revision of the Instructor's Manual and Test Bank. Both a new and experienced instructor will find plenty of ideas in this Instructor's Manual, which is closely correlated to the textbook and the student study guide. Each chapter of the manual includes teaching objectives, key terms, lecture suggestions, activities, discussion questions, and formatted handouts for many topics. The Instructor's Manual also contains an annotated list of resources for students for reference or as a handout. Instructors will be able to download the Instructor's Manual from the Wadsworth Sociology website.

Consult your sales representative for access information or how to secure the printed version. The Test Bank contains multiple choice, true/false, and essay questions keyed to each learning objective. These test items are page referenced to the textbook and include significant numbers of application as well as knowledge questions. Story problems use names drawn from a variety of cultures, reflecting the diversity of U.S. society. Instructors requested that the questions be tied to the practice tests, and we followed that suggestion.

ExamView® Computerized Testing

Quickly create customized tests that can be delivered in print or online. ExamView's simple "what you see is what you get" interface allows you to easily generate tests of up to 250 items. (Contains all the Test Bank questions electronically.)

ThomsonNOW™

Empower your students with the first assessment-centered student tutorial system for *Introduction to Sociology*. Seamlessly tied to the new edition of this text, this powerful and interactive web-based learning tool helps students gauge their unique study needs, then gives them a Personalized Learning Plan that focuses their study time on the concepts they most need to master. By providing students with a better understanding of exactly that on which they need to focus, ThomsonNOW helps students make the optimum use of their study time, bringing them closer to success! Visit http://www.thomsonedu.com.

Multimedia Manager: A Microsoft® PowerPoint® Tool

This one-stop lecture tool makes it easy for you to assemble, edit, publish, and present custom lectures for your course using Microsoft PowerPoint. The Multimedia Manager lets you bring together text-specific lecture outlines and art from Thomson Wadsworth texts, along with video and animations from the web or your own materials—culminating in a powerful, personalized, media-enhanced presentation.

WebTutor™ ToolBox for WebCT® and Blackboard®

WebTutor Toolbox combines easy-to-use course management tools with content from this text's rich companion website. Ready to use as soon as you log on, you can customize WebTutor Toolbox with weblinks, images, and other resources.

vMentor™

You can experience live, one-on-one tutoring! When you adopt this text packaged with vMentor, you give your students access to virtual office hours—one-on-one, online tutoring help from a subject-area expert, at no additional cost with the text. In vMentor's virtual classroom, students interact with the tutor and other students using two-way audio, an interactive whiteboard for illustrating the problem, and instant messaging. To ask a question, students simply click to raise a "hand." For additional information, please consult your local Thomson representative. (For proprietary, college, and university adopters only.)

Turnitin™

This proven online plagiarism-prevention software promotes fairness in the classroom by helping students learn to correctly cite sources and allowing instructors to check for originality before reading and grading papers. Turnitin quickly checks student papers against billions of pages of Internet content, millions of published works, and millions of student papers and within seconds generates a comprehensive originality report.

Thomson InSite for Writing and Research™

This all-in-one, online writing and research tool includes electronic peer review, an originality checker, an assignment library, help with common errors, and access to InfoTrac® College Edition. InSite makes course management practically automatic! Visit http://insite.thomson.com.

InfoTrac College Edition with InfoMarks®

Four months' access to this online database—featuring reliable, full-length articles from thousands of academic journals and periodicals—is available with this text at no additional charge! Now features stable, topically bookmarked InfoMarks URLs to assist in research, plus InfoWrite critical thinking and writing tools. This fully searchable database offers 20 years' worth of full-text articles from almost 5,000 diverse sources, such as academic journals, newsletters, and up-to-the-minute periodicals, including *Time, Newsweek, Science, Forbes,* and *USA Today.* This incredible depth and breadth of material—available 24 hours a day from any computer with Internet access—makes conducting research so easy that your students will want to use it to enhance their work in every course!

Extension: Wadsworth's Sociology Readings Collection

Create your own customized reader for your introductory class, drawing from dozens of classic and contemporary articles found in the exclusive Thomas Wadsworth TextChoice2 collection. Create a customized reader just for your class, containing as few as two or three seminal articles to more than a dozen edited selections. With Extension, you can preview articles online, make selections, and add original material of your own, to create your own printed reader for your class. To build your own custom reader, visit Thomson's digital library at http://www.textchoice2.com. This site allows you to preview content and create a project online.

Website: http://sociology.wadsworth.com/tischler9e

The book's companion website includes chapter-specific resources for instructors and students. For instructors the site offers a password-protected instructor's manual, Microsoft PowerPoint presentation slides, and more. For students there is a multitude of text-specific study aids, including the following: tutorial practice quizzing that can be scored and emailed to the instructor, weblinks, InfoTrac College Edition exercises, flash cards, MicroCase Online data exercises, crossword puzzles, Virtual Explorations, and much more!

ABC Videos

Launch your lectures with exciting video clips from the award-winning news coverage of ABC. Addressing topics covered in a typical course, these videos are divided into short segments and are perfect for introducing key concepts in contexts relevant to students' lives.

CNN® Today Videos

Volumes V–VII available. Launch your lectures with riveting footage from CNN, the world's leading 24-hour global news television network. Organized by topics covered in a typical course, these videos are divided into short segments and are perfect for introducing key concepts in contexts relevant to students' lives. High-interest clips are followed by questions designed to spark class discussion.

Wadsworth's Lecture Launchers for Introductory Sociology Video/DVD

An exclusive offering jointly created by Thomson Wadsworth and Dallas TeleLearning, this video contains a collection of video highlights taken from the Exploring Society: An Introduction to Sociology Telecourse (formerly The Sociological Imagination). Each 3- to 6-minute video segment has been specially chosen to enhance and enliven class lectures and discussions of 20 key topics covered in the Introduction to Sociology course. Accompanying the video is a brief written description of each clip, along with suggested discussion questions to help effectively incorporate the material into the classroom.

Sociology: Core Concepts Video/DVD

An exclusive offering jointly created by Thomson Wadsworth and Dallas TeleLearning, this video contains a collection of video highlights taken from the Exploring Society: An Introduction to Sociology Telecourse (formerly The Sociological Imagination). Each 15- to 20-minute video segment will enhance student learning of the essential concepts in the introductory course and can be used to initiate class lectures, discussion, and review. The video covers topics such as the sociological imagination, stratification, race and ethnic relations, social change, and more.

Introduction to Sociology 2007 Transparency Masters

A set of black and white transparency masters consisting of tables and figures from Wadworth's introductory sociology texts is available to help prepare lecture presentations. Free to qualified adopters.

Acknowledgments

I am grateful for the thoughtful contributions of the following people who served as official reviewers for this new Ninth Edition:

Laura Dowd
University of Georgia

Nancy Feather
West Virginia University

Hubert Anthony Kleinpeter
Florida A&M University

Steven Patrick
Boise State University

Craig T. Robertson
University of North Alabama

Laurie Smith
East Texas Baptist University

I also wish to thank the many colleagues and reviewers of previous editions of *Introduction to Sociology* for their many contributions and suggestions. I am grateful for the thoughtful contributions of the following people: Patrick Ashton, Indiana University–Purdue University; Froud Stephen Burns, Floyd Junior College; Peter Chroman, College of San Mateo; Mary A. Cook, Vincennes University; William D. Curran II, South Suburban College; Ione Y. Deollos, Ball State University; Stanley Deviney, University of Maryland–Eastern Shore; Brad Elmore, Trinity Valley Community College; Cindy Epperson, St. Louis Community College–Meramac; Larry Frye, St. Petersburg College; Richard Garnett, Marshall University; David A. Gay, University of Central Florida; Daniel T. Gleason, Southern State College;

Charlotte K. Gotwald, York College of Pennsylvania; Richard L. Hair, Longview Community College; Selwyn Hollingsworth, University of Alabama; Sharon E. Hogan, Longview Community College; Bill Howard, Lincoln Memorial University; Sidney J. Jackson, Lakewood Community College; Michael C. Kanan, Northern Arizona University; Ed Kick, Middle Tennessee State University; Louis Kontos, Long Island University; Steve Liebowitz, University of Texas, Pan American; Thomas Ralph Peters, Floyd College; David Phillips, Arkansas State University; Kanwal D. Prashar, Rock Valley Community College; Charles A. Pressler, Purdue University, North Central; Stephen Reif, Kilgore College; Richard Rosell, Westchester Community College; Catherine A. Stathakis, Goldey Beacom College; Doris Stevens, McLennan Community College; Gary Stokley, Louisiana Tech University; Elena Stone, Brandeis University; Judith C. Stull, La Salle University; Lorene Taylor, Valencia Community College; Paul Thompson, Polk Community College; Brian S. Vargus, Indiana University–Purdue University Indianapolis; Steven Vassar, Minnesota State University–Mankato; Peter Venturelli, Valparaiso University; J. Russell Willis, Grambling State University; and Bobbie Wright, Thomas Nelson Community College.

A project of this magnitude becomes a team effort, with many people devoting enormous amounts of time to ensure that the final product is as good as it can possibly be. At Thomson Wadsworth, Robert Jucha, the acquisitions editor, ushered this project through its many stages. Elise Smith, and later Kristin Marrs, assistant editors, served ably on the book development and ancillary package. Michelle Williams led the marketing efforts. Cheri Palmer provided guidance throughout the production process, which resulted in the book you now see. Lisa Royse was also responsible for the smooth production process. I am grateful to all those students and instructors who have shared with me their thoughts about the book over the years. Please continue to let me know how you feel about this book.

Henry L. Tischler
htischl@frc.mass.edu

About the Author

Henry L. Tischler grew up in Philadelphia and received his bachelor's degree from Temple University and his masters and doctorate degrees from Northeastern University. He pursued post-doctoral studies at Harvard University. His first venture into textbook publishing took place while he was still a graduate student in sociology when he wrote the fourth edition of *Race and Ethnic Relations* with Brewton Berry. The success of that book led to his authorship of the eight editions of *Introduction to Sociology.*

Tischler has been a professor at Framingham State College in Framingham, Massachusetts, for more than two decades. He has also taught at Northeastern University, Tufts University, and Montclair State University. He continues to teach introductory sociology every year and has been instrumental in encouraging many students to major in the field. His other areas of interest are race and ethnicity, and crime and deviant behavior.

Professor Tischler has been active in making sociology accessible to the general population and has been the host of an author interview program on National Public Radio. He has also written a weekly newspaper column called "Society Today," which dealt with a wide variety of sociological topics.

Tischler and his wife Linda divide their time between Boston and New York City. Linda Tischler is a senior writer at a national magazine. The Tischlers have a daughter, Melissa, who is a management consultant, and a son, Ben, who is an account executive in advertising.

Effective Study: An Introduction

Why should you read this essay? If you think you have an A in your back pocket, perhaps you shouldn't. Maybe you are just not interested in sociology or about learning ways to become a really successful student. Maybe you're just here because an advisor told you that you need a social science course. Maybe you feel, "Hey, a C is good. I'll never need this stuff." If so, you can stop reading now.

But if you want to ace sociology—thereby becoming a more effective participant in society and social life—and if you want to learn some techniques to help you in other classes, too, this is for you. It's filled with the little things no one ever seems to tell you that improve grades, make for better understanding of classes, and may even make classes enjoyable for you. The choice is yours: to read, or not to read.

Be forewarned. These contents may challenge the habits of a lifetime—habits that have gotten you this far but ones that may endanger your future success.

This essay contains ways to help you locate major ideas in your textbook. It contains many techniques that will be of help in reading your other course textbooks. If you learn these techniques early in your college career, you will have a head start on most other college students. You will be able to locate important information, understand lectures better, and probably do better on tests. By understanding the material better, you will not only gain a better understanding of sociology but also find that you are able to enjoy your class more.

The Problem: Passive Reading

Do you believe reading is one-way communication? Do you expect the author's facts will become apparent if you only read hard enough or long enough? (Many students feel this way.) Do you believe the writer has buried critical material in the text somewhere and that you need only find and highlight it to get all that's important? And do you believe that if you can memorize these highlighted details you will do well on tests? If so, then you are probably a passive reader.

The problem with passive reading is that it makes even potentially interesting writing boring. Passive reading reduces a chapter to individual, frequently unrelated facts instead of providing understanding of important concepts. It seldom digs beneath the surface, relying on literal meaning rather than sensing implications. Since most college testing relies on understanding of key concepts rather than simple factual recall, passive reading fails to significantly help students do well in courses.

Key Features of the Study Guide

For each chapter you will find the following:

Key concepts matching exercise
Includes every major term defined in the chapter
Promotes association of major thinkers with their key ideas or findings
Provides correct answers

Key thinkers/researchers matching exercise (where relevant)
Includes most important theorists or researchers discussed in the text
Promotes association of major thinkers with their key ideas or findings
Provides correct answers

Critical thinking questions
Promotes depth in reflecting on the material
Encourages creative application of the important concepts to everyday life
Presented in increasing levels of complexity, abstraction, and difficulty
Provides help in preparing for essay exams and papers

Comprehensive practice test
Includes questions on all major points in the chapter
Includes true/false, multiple-choice, and essay questions
Provides correct answers

The Solution: Active Reading

Active reading is recognizing that a textbook should provide two-way communication. It involves knowing what aids are available to help understand the text and then using them to find the meaning. It involves prereading and questioning. It includes recording of questions, vocabulary learning, and summarizing. Still, with all these techniques, it frequently takes less time and produces significantly better results than passive reading.

This textbook—especially the Study Guide—is designed to help you become an active reader. For your convenience, the Study Guide material related to each chapter appears right after that chapter. The corners of the Study Guide pages are edged in color for easy reference. In the Study Guide, you will find a variety of learning aids based on the latest research on study skills. If you get into the habit of using the aids presented here, you can apply similar techniques to your other textbooks and become a more successful learner.

Effective Reading: Your Textbook

How should you approach your textbook as an active reader? Here are some techniques for reading text chapters that you should consider.

1. Think first about what you know. Read the title of your chapter, then ask yourself what experiences you have had that relate to that title. For example, if the title is "Social Interaction and Social Groups," ask yourself, "In what ways have I interacted with others in social situations? Have I ever been part of a social group? If so, what do I remember about the experience?" Answers to these questions personalize the chapter by making it relate to your experiences. They provide a background for the chapter, which experts say improves your chances of understanding the reading. Your answers show that you do know something about the chapter, so that its content won't be so alien.

2. Review the learning objectives. Not all textbooks provide learning objectives like this one does, but where available, they can be a valuable study aid. Learning objectives are stated in behavioral terms—they tell you what you should be able to do when you finish the chapter. Ask yourself questions about the tasks suggested in each learning objective, and then read to find the information needed to accomplish that task. For instance, if a learning objective states, "Explain how variations in the size of groups affect what goes on within them," then you'll want to ask yourself something like, "How do groups vary in size?" and "How does each variation affect interaction within the group?"

3. Prior to reading the textbook chapter, read the chapter summary as an index to important terms and ideas. The summary includes all the points you need to find items in the chapter you know already. You may be able to read more quickly through sections covering these items. Some items you may not know anything about. This tells you where to spend your reading time. A good rule is to study most what you know least. Wherever it is, the summary is often your best guide to important material.

4. Pay attention to your chapter outline. This textbook, like most other introductory college textbooks, has an outline at the beginning of each chapter. If you do nothing else besides reading the summary and going through this outline before reading the chapter, you will be far ahead of most students because you will be clued in on what is important. The outline indicates the way ideas are organized in the chapter and how those ideas relate to one another. Certain ideas are indented to show that they are subsets or parts of a broader concept or topic. Knowing this can help you organize information as you read.

5. Question as you read. Turn your chapter title into a question, then read up to the first heading to find your answer. The answer to your question will be the main idea for the entire chapter. In forming your question, be sure it contains the chapter title. For example, if the chapter title is "Doing Sociology: Research Method," your question might be "What research methods does sociology use?" or "Why do you need research methods to do sociology?" As you go through the chapter, turn each heading into a question, and then read to find the answer. Most experts say that turning chapter headings into questions is a most valuable step in focusing reading on important information. You may also want to use the learning objectives as questions because you know that these objectives will point you toward the most important material in a section. However, it is also a good idea to form your own questions to get into practice for books that do not contain this helpful aid. A good technique might be to make your own question, then check it against the appropriate objective before reading. In any case, use a question, then highlight your answer in the text. This will be the most important information

under each heading. Don't read as if every word is important; focus on finding answers.

6. Pay attention to graphic aids. As you read, note those important vocabulary words appearing in bold type. Find the definitions for these words (in this book, definitions appear in italics right next to key words) and highlight them. These terms will be important to remember. Your Study Guide identifies all these important terms in the section headed "Key Concepts." A "Key Thinkers/Researchers" section, if applicable, identifies the sociologists and other important thinkers in the chapter worth remembering. Both the "Key Concepts" and "Key Thinkers/ Researchers" sections are organized as matching exercises. Testing yourself after you read a text chapter (the answer key is at the end of the Study Guide chapter) will let you know whether you recognize the main concepts and researchers.

Pay attention to photos and photo captions. They make reading easier because they provide a visualization of important points in the textbook. If you can visualize what you read, you will ordinarily retain material better than people who don't use this technique. Special boxed sections usually give detailed research information about one or more studies related to a chapter heading. For in-depth knowledge, read these sections, but only after completing the section to which they refer. The main text will provide the background for a better understanding of the research, and the visualization provided by the boxed information will help illuminate the text discussion.

7. When in doubt, use clues to find main ideas. It is possible that, even using the questioning technique, there could be places where you are uncertain whether you're getting the important information. You have clues both in the text and in the Study Guide to help you through such places. In the text, it helps to know that main ideas in paragraphs occur more frequently at the beginning and end. Watch for repeated words or ideas—these are clues to important information. Check examples; any point that the author uses examples to document is important. Be alert for key words (such as "first," "second," "clearly," "however," "although," and so on); these also point to important information. Names of researchers (except for those named only within parentheses) will almost always be important. For those chapters in which important social scientists are discussed, you will find a "Key Thinkers/Researchers" section in your Study Guide.

Guidelines for Effective Reading of Your Textbook

1. Think first about what you know.
2. Review the learning objectives.
3. Prior to reading the textbook chapter, read the chapter summary as an index to important terms and ideas.
4. Pay attention to your chapter outline.
5. Question as you read.
6. Pay attention to graphic aids.
7. When in doubt, use clues to find main ideas.
8. Do the exercises in the Study Guide.
9. Review right after reading.

8. Do the exercises in the Study Guide. The exercises in the Study Guide are designed as both an encouragement and a model of active learning. The exercises are not about mere regurgitation of material. Rather, you are asked to analyze, evaluate, and apply what you read in the text. By completing these exercises you are following two of the most important principles articulated in this essay: You are actively processing the material, and you are applying it to your own life and relating it to your own experiences. This is a guaranteed recipe for learning.

9. Review right after reading. Most forgetting takes place in the first day after reading. A review right after reading is your best way to hold text material in your memory. A strong aid in doing this review is your Study Guide. If a brief review is all you have time for, return to the Learning Objectives at the beginning of the chapter. Can you do the things listed in the objectives? If so, you probably know your material. If not, check the objective and reread the related chapter section to get a better understanding.

An even better review technique is to complete—if you haven't already done so—the exercises. Writing makes for a more active review, and if you do the exercises, you will have the information you need from the chapter. If there are blanks in your knowledge, you can check the appropriate section of text and write the information you find in your Study Guide. This technique is especially valuable in classes requiring essay exams or papers because it gives you a comprehensive understanding of the material, as well as a sense of how it can be applied to real-world situations.

For a slightly longer but more complete review, do the "Key Concepts" and "Key Thinkers/Researchers" matching tests. These will assure that you have mastered the key vocabulary and know the contributions of the most important researchers mentioned in the chapter. Since a majority of test questions are based on understanding of vocabulary, research findings, and major theories, you will be assuring yourself of a testing benefit during your review.

It is also a good idea to review the "Critical Thinking" questions in the Study Guide. One key objective of sociology—indeed, of all college courses—is to help you develop critical thinking skills. Though basic information may change from year to year as new scientific discoveries are made, the ability to think critically in any field is important. If you get in the habit of going beyond surface knowledge in sociology, you can transfer these skills to other areas. This can be a great benefit not only while you're in school but afterward as well. As with the exercises section, these questions provide the kind of background that is extremely useful for essay exams.

What other methods would an active student use to improve understanding and test scores in sociology? The next several sections present a variety of techniques.

Functioning Effectively in Class

To function effectively in class, you must, of course, be there. While no one may take attendance or force you to be present, studies show that you have a significantly greater chance of succeeding in your class if you attend regularly. Lecture material is generally important—and it is given only once. If you miss a lecture, in-class discussion, game, or simulation, there is no really effective way to make it up.

Guidelines for Effective Functioning in Class

1. Begin each class period with a question.
2. Ask questions frequently.
3. Join in classroom discussion.

Assuming you are present, there are two ways of participating in your sociology class: actively and passively. Passive participation involves sitting there, not contributing, waiting for the instructor to tell you what is important. Passive participation takes little effort, but it is unlikely to result in much learning. Unless you are actively looking for what is significant, the likelihood of finding the important material or separating it effectively from what is less meaningful is not great. The passive student runs the risk of taking several pages of unneeded notes or missing key details altogether.

Active students begin each class period with a question. "What is this class going to be about today?" They find an answer to that question, usually in the first minute, and use this as the key to important material throughout the lecture or other activity. When there is a point they don't understand, they ask questions. Active students know that many other students probably have similar questions but are afraid to ask. Asking questions allows you to help others while helping yourself.

Active students also know that what seems a small point today may be critical to understanding a future lecture. Such items also have a way of turning up on tests. If classroom discussion is called for, active students are quick to join in. And the funny thing is, they frequently wind up enjoying their sociology class as they learn.

Effective Studying

As you study your sociology text and notes, both the method you use and the time picked for study will have effects on comprehension. Establishing an effective study routine is important. Without a routine, it is easy to put off study—and put it off, and put it off . . . until it is too late. To be most effective, follow the few simple steps listed below.

Guidelines for Effective Studying

1. When possible, study at the same time and place each day.
2. Study in half-hour blocks with five-minute breaks.
3. Review frequently.
4. Don't mix study subjects.
5. Reward yourself when you're finished.

1. When possible, study at the same time and place each day. Doing this makes use of psychological conditioning to improve study results. "Because it is 7:00 P.M. and I am sitting at my bedroom desk, I realize it is time to begin studying sociology."
2. Study in half-hour blocks with five-minute breaks. Long periods of study without breaks

frequently reduce comprehension to the 40% level. That is most inefficient. By using short periods (about 30 minutes) followed by short breaks, you can move that comprehension rate into the 70% range. Note that if 30 minutes end while you are still in the middle of a text section, you should go on to the end of that section before stopping.

3. For even more efficient study, review frequently. Take about a minute at the end of each study session to mentally review what you've studied so far. When you start the next study session, spend the first minute or two rehearsing in your mind what you studied in the previous session. This weaves a tight webbing in which to catch new associations. Long-term retention of material is aided by frequent review. A 10-minute review planned on a regular basis (about every two weeks) saves on study time for exams and ensures that you will remember needed material.

   Another useful way to review is to try to explain difficult concepts or the chapter learning objectives to someone else. One problem students often have is that after studying and reviewing the material by themselves they think they know it, only to have that knowledge desert them at the time of the exam. Trying to explain something to someone else forces us to be clear about key points and to discover and articulate the relationship among the components of an idea. Ask your friends or family to bear with you as you try to explain the material. After all, they will learn something as well!

4. Don't mix study subjects. Do all of your sociology work before moving on to another course. Otherwise, your study can result in confusion of ideas and relationships within materials studied.

5. Finally, reward yourself for study well done. Think of something you like to do and do it when you finish studying for the day. This provides positive reinforcement, which makes for continued good study.

Successfully Taking Tests

Of course, tests are a payoff for you as a student. Tests are where you can demonstrate to yourself and to the instructor that you really know the material. The trouble is, few people have learned how to take tests effectively, and knowing how to take tests effectively makes a significant difference in exam scores. Here are a few tips to improve your test-taking skills.

Taking the Test

1. Don't come early; don't come late.
2. Be sure you understand all the directions before you start answering.
3. Read through the test, carefully answering only items you know.
4. Now that you've answered what you know, look carefully at the other questions.
5. If you finish early, stay to check answers.
6. Don't be distracted by other test takers.
7. When you get your test back, use it as a learning experience.

Studying for Tests

1. Think before you study. All material is not of equal value. What did the instructor emphasize in class? What was covered in a week? A day? A few minutes? Were any chapters emphasized more than others? Which learning objectives did your instructor stress? Review the "Key Thinkers/Researchers" and "Key Concepts" sections in your Study Guide for important people and terms. Which of these were given more emphasis by your instructor? Use these clues to decide where to spend most of your study time.

2. Begin studying a week early. When you start early, if you encounter material you don't know, you have time to find answers. If you see that you know blocks of material already, you have saved yourself time in future study sessions. You also avoid much of the forgetting that occurs with last-minute cramming.

3. Put notes and related chapters together for study. Integrate the material as much as possible, perhaps by writing it out in a single, comprehensive format. A related technique is to visualize the material on the pages of the text and in your notes. You may even want to think of a visual metaphor for some of the key ideas. This way you can see and remember the connections between similar subjects or similar treatments of the same subject. Grouping the material will also make your studying much more efficient.

   As you study, don't stop for unknown material. Study what you know. Once you know it, go back and look at what you don't know yet. There is no need to study again what you already know. Put it aside, and concentrate on the unknown.

Studying for the Test

1. Think before you study.
2. Begin studying a week early.
3. Put notes and related chapters together for study.
4. Take practice tests.

4. Take practice tests. When you have completed your studying, take the appropriate practice test for each chapter. These tests are grouped together at the back of the book. Tests include true/false and multiple choice questions, with comprehensive or thematic essays at the end. Each test is divided into sections by major headings in the chapter. Within each section, questions are presented in scrambled order, as they are likely to be on the actual test. Taking the practice test contains a double benefit. First, if you get a good score on this test, you know that you understand the material. Second, the format of the practice test is very similar to that of real tests. For this reason, you should develop confidence in your ability to succeed in course tests by doing well on the practice tests. If your course tests include essay questions, you should, in addition to the practice test essays, use the "Critical Thinking" sections to prepare and practice focused, in-depth answers.

Taking the Test

1. Don't come early; don't come late. Early people tend to develop anxieties; late people lose test time. Studies show that people who discuss test material with others just before a test may forget that material on the test. This is another reason that arriving too early puts students in jeopardy. Get there about two or three minutes early. Relax and visualize yourself doing well on the test. After all, if you followed the study guidelines discussed above, you can't help but do well! Be confident; repeat to yourself as you get ready for the test, "I can do it! I will do it." This will set a positive mental tone.
2. Be sure you understand all the directions before you start answering. Not following directions is the biggest cause of lost points on tests. Ask about whatever you don't understand. The points you save will be your own.
3. Read through the test, carefully answering only items you know. Be sure you read every word and every answer choice as you go. Use a piece of paper or a card to cover the text below the line you are reading. This can help you focus on

each line individually—and increase your test score.

Speed creates a serious problem in testing. The mind is moving so fast that it is easy to overlook key words such as "except," "but," "best example," and so on. Frequently, multiple choice questions will contain two close options, one is correct, while the other is partly correct. Moving too fast without carefully reading items causes people to make wrong choices in these situations. Slowing your reading speed makes for higher test scores.

The mind tends to work subconsciously on questions you've read but left unanswered. As you're doing questions later in the test, you may suddenly have the answer for an earlier question. In such cases, answer the question right away. These sudden insights quickly disappear and may not come again while you are taking the test.

4. Now that you've answered what you know, look carefully at the other questions. Eliminate alternatives you know are wrong and then guess. Never leave a blank on a test. You may have a 25% chance when you guess on a four-item multiple choice question, but you have a chance. A chance is better than no chance.
5. If you finish early, stay to check answers. Speed causes many people to give answers that a moment's hesitation would show to be wrong. Read over your choices, especially those for questions that caused you trouble. Don't change answers because you suddenly feel one choice is better than others. Studies show that this is usually a bad strategy. However, if you see a mistake or have genuinely remembered new information, change your answer.
6. Don't be distracted by other test takers. Some people become very anxious because of the noise and movement of other test takers. This is most apparent when several people begin to leave the room after finishing their tests. Try to sit where you will be least apt to see or interact with other test takers. Usually this means sitting toward the front of the room and close to the wall farthest from the door. Turn your chair slightly toward the wall if possible. The more you insulate yourself from distractions during the test, the better off you will be.

Don't panic when other students finish their exam before you do. Accuracy is always more important than speed. Work at your own pace and budget your time appropriately. For a timed test, always be aware of the time remaining. This means that if a clock is not visible in the classroom, you need to have your own wristwatch. Take as much of the available time as you need to do an accurate and complete job. Remember, your grade will be based on the answers you give, not on whether you were the first—or the last—to turn in your exam.

7. When you get your test back, use it as a learning experience. Diagnosing a test after it is returned to you is one of the most effective strategies for improving your performance in a course. What kind of material was on the test: theories, problems, straight facts? Where did the material come from: book, lecture, or both? The same kind of material taken from the same source(s) will almost certainly be on future tests.

    Look at each item you got wrong. Why is it wrong? If you know why you made mistakes, you are unlikely to make the same ones in the future. Look at the overall pattern of your errors. Did you make most of your mistakes on material from the lectures? Perhaps you need to improve your note-taking technique. Did your errors occur mostly on material from the readings? Perhaps you need to pay more attention to main idea clues and highlight text material more effectively. Were the questions you got wrong evenly distributed between in-class and reading material? Perhaps you need to learn to study more effectively and/or to take steps to reduce test anxiety. Following these steps can make for more efficient use of textbooks, better note taking, higher test scores, and better course grades.

A Final Word

As you can see, the key to success lies in becoming an active student. Managing time, questioning at the start of lectures, planning effective measures to increase test scores, and using all aids available to make reading and studying easier are all elements in becoming an active student. The Study Guide and Practice Tests for this textbook have been specially designed to help you be that active student. Being passive may seem easier, but it is not. Passive students spend relatively similar amounts of time but learn less. Their review time is likely to be inefficient. Their test scores are more frequently lower—and they usually have less fun in their classes.

Active students are more effective than passive ones. The benefit in becoming an active student is that activity is contagious; if you become an active student in sociology, it is hard not to practice the same active learning techniques in English and math as well. Once you start asking questions in your textbook and using your Study Guide, you may find that you start asking questions in class. As you acquire a greater understanding of your subject, you may find that you enjoy your class more, as well as learn more and do better on tests. That is the real benefit in becoming an active learner. It is a challenge we strongly encourage you to accept.

1

The Sociological Perspective

AP/Wide World Photos

ThomsonNOW™

Reviewing is as easy as ❶ ❷ ❸
Use ThomsonNOW to help you make the grade on your
next exam. When you are finished reading this chapter,
go to the chapter review for instructions on how to make
ThomsonNOW work for you.

Learning Objectives

After studying this chapter, you should be able to do the following:

- Understand the sociological point of view and how it differs from that of journalists and talk-show hosts.
- Compare and contrast sociology with the other major social sciences.
- Describe the early development of sociology from its origins in nineteenth-century Europe.

- Know the contributions of sociology's pioneers: Comte, Martineau, Spencer, Marx, Durkheim, and Weber.
- Describe the early development of sociology in the United States.
- Understand the functionalist, conflict theory, and the interactionist perspectives.
- Realize the relationship between theory and practice.

T he sweet, smiling faces of America's missing children are on fliers and billboards. Their disappearance launches Amber Alerts and search parties of neighbors and law officials. And newscasters like Katie Couric, of the *Today Show,* eagerly jump into the fray, leading the hour with the latest abduction story and advocating for mandatory education for children in how to avoid being kidnapped. The problem, she warned in one broadcast, was at epidemic levels—58,000 American children annually snatched by strangers.

Over at Fox News, Bill O'Reilly added to the frenzy with even more alarming numbers. Calling the latest incident "the tip of the iceberg," he insisted that there were "more than 100,000 abductions of children by strangers every year in the United States" (VitalStats, 2002).

Numbers like these, cited in congressional testimony, are what eventually led to legislation designed to facilitate recovery of missing kids, in part by centralizing the reporting and tracking of children.

Parents, naturally, are worried sick. Many eagerly trot their youngsters off to places like Wal-Mart or Home Depot or Blockbuster to be documented in case police someday need identifying information. A study of parents' worries by the Mayo Clinic in Rochester, Minnesota, found that nearly three-quarters of parents said they feared their child being abducted.

It's a horrific, national problem.

Or is it?

Losing even one child to a kidnapper is, of course, a tragedy. But the chances of that happening are far less than what news reports would have us believe.

According to a Justice Department study, only about 115 cases annually now qualify as "stereotypical kidnappings," in which a child is taken by a stranger and either held for ransom, abused, or killed. About 43% of the kids in the *Today Show*'s tally were missing for less than an hour, and many of the ones who did disappear know the folks who took them. Indeed, most abductions are related to child custody suits.

In addition, many of the children who are kidnapped by "nonfamily" members are taken by friends, romantic partners, or acquaintances. And a large number of the "missing children" are actually teenagers, particularly teenage girls who run off with their boyfriends.

With 50 million children under the age of 13 in the United States, the actual chance of having a child abducted and murdered is about 1 in 450,000 (Hammer et al., 2002; Cooper, 2005).

Despite the varying reasons for child abduction, no one is denying that the issue is a legitimate social problem with very serious consequences. Is this sociology? Or, for that matter, is this what sociologists do when they study society? The answer would have to be no.

The *New York Times,* at the top of every issue, has the statement "all the news that's fit to print." If this statement were true, each issue would be so large that few would attempt to read it. News is brought to us by people who make choices. Some of their choices, inevitably, are better than others and represent the perceptions of the reporters and editors who produce the papers or news broadcasts (Murray, Schwartz, & Lichter, 2001).

Far too infrequently do we realize that people often use data to persuade and that statistics can be used as part of a strategy to promote concern about a social problem. Much of the information we read every day and mistake for sociology is actually an attempt by one group or another to influence social policy. Other information mistaken for sociology is really an attempt to sell a book, or the efforts of television producers to present entertaining programs.

With the constant bombardment of information about social issues, we could come to believe that nearly everyone is engaged in the study of sociology to some extent and that everyone has not only the right but also the ability to put forth valid information about society. This is not the case. Some people have no interest in putting forth objective information and are instead interested in getting us to support their position or point of view. On other occasions the "researchers" do not have the ability or training to disseminate accurate information about drug abuse, homelessness, welfare, high school dropout rates, white-collar crime, or a host of other sociological topics.

Sociologists have very different goals in mind when they investigate a problem than do journalists or talk-show hosts. A television talk-show host needs to make the program entertaining and maintain high ratings, or the show may be canceled. A journalist is writing for a specific readership. This will certainly limit the choice of topics, as well as the manner in which an issue is investigated. On the other hand, a sociologist must answer to the scientific community as she or he tries to further our understanding of a topic. This means that the goal is not high ratings, but an accurate and scientific approach to the issue being studied.

In this book we will ask you to go beyond popular sociology and investigate society more scientifically than you did before. You will learn to look at major events, as well as everyday occurrences, a little differently and start to notice patterns you may have never seen before. After you are equipped with the tools of research, you should be able to evaluate critically popular presentations of sociology. You will see that sociology represents both a body of knowledge and a scientific approach to the study of social issues.

Vast differences exist within the same society.

Sociology as a Point of View

Sociology is *the scientific study of human society and social interactions.* As sociologists our main goal is to *understand* social situations and look for repeating patterns in society. We do not use facts selectively to create a lively talk show, sell newspapers, or support one particular point of view. Instead, sociologists are engaged in a rigorous scientific endeavor, which requires objectivity and detachment.

The main focus of sociology is the group, not the individual. Sociologists attempt to understand the forces that operate throughout society—forces that mold individuals, shape their behavior, and thus determine social events.

When you walk into an introductory physics class, you may know very little about the subject and hold very few opinions about the various topics within the field. On the other hand, when you enter your introductory sociology class for the first time, you will feel quite familiar with the subject matter. You have the advantage of coming to sociology with a substantial amount of information, which you have gained simply by being a member of society. Ironically, this knowledge also can leave you at a disadvantage,

Figure 1–1 Levels of Social Understanding: Domestic Violence

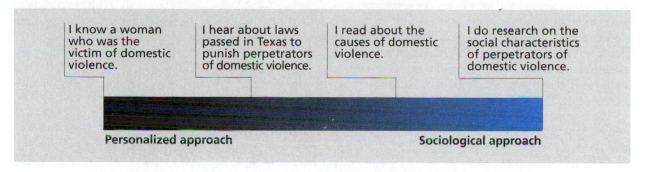

I know a woman who was the victim of domestic violence.

I hear about laws passed in Texas to punish perpetrators of domestic violence.

I read about the causes of domestic violence.

I do research on the social characteristics of perpetrators of domestic violence.

Personalized approach

Sociological approach

because these views have not been gathered in a scientific fashion and may not be accurate.

Over the years and through a variety of experiences we develop a set of ideas about the world and how it operates. This point of view influences how we look at the world and guides our attempts to understand the actions and reactions of others. Even though we accept the premise that individuals are unique, we tend to categorize or even stereotype people to interpret and predict behavior and events.

Is this personalized approach adequate for bringing about an understanding of ourselves and society? Although it may serve us quite well in our day-to-day lives, a sociologist would answer that it does not give us enough accurate information to develop an understanding of the broader social picture. This picture becomes clear only when we know something about the society in which we live, the social processes that affect us, and the patterns of interaction that characterize our lives.

Let us take the issue of domestic violence. Figure 1–1 shows that we could examine the issue in a variety of ways. If we knew a woman who was the victim of domestic violence, we would have personal information about the experience. If she were willing to discuss her experience with us, we would know more about domestic violence on a specific case level. Although this information is important, it is not yet sociology and is closer to the personalized commonsense approach to understanding society. Sociology tries to move beyond that level of understanding.

If we rely on our own experiences, we are like the blind men of Hindu legend trying to describe an elephant: the first man, feeling its trunk, asserts, "It is like a snake"; the second, trying to reach around the beast's leg, argues, "No, it is like a tree"; and the third, feeling its solid side, disagrees, saying, "It is more like a wall." In a small way, each man is right, but not one of them is able to understand or describe the whole elephant.

If we were to look for recurring patterns in domestic violence, we would now be doing what sociologists do. A sociologist examining the issue might

be interested in the age, socioeconomic level, and ethnic characteristics of the victims of domestic violence. A sociologist might want to compare these characteristics with the characteristics of victims of other types of violence: "Are there differences?" he or she asks. "If so, what kinds and why?"

While studying sociology, you will be asked to look at the world a little differently from the way you usually do. Because you will be looking at the world through other people's eyes—using new points of view—you will start to notice things you may never have noticed before. When you look at life in a middle-class suburb, for instance, what do you see? How does your view differ from that of a poor inner-city resident? How does the suburb appear to a recent immigrant from Russia or Cuba or India? How does it appear to a burglar? Finally, what does the sociologist see?

Sociology asks you to broaden your perspective on the world. You will start to see that the reason people act in markedly different ways is not because one person is "sane" and another is "crazy." Rather, it is because they all have different ways of making sense out of what is going on in the world around them.

These unique perceptions of reality produce varying lifestyles, which in turn produce different perceptions of reality. To understand other people, we must stop looking at the world from a perspective based solely on our own individual experiences.

The Sociological Imagination

Although most people interpret social events on the basis of their individual experiences, sociologists step back and view society more as an outsider than as a personally involved and possibly biased participant. For example, whereas we assume that most people in the United States marry because of love, sociologists remind us that the decision to marry—or not to marry—is influenced by a variety of social values taught to us since early childhood.

That is, we select our mates based on the social values we internalize from family, peers, neighbors, community leaders, and even our television heroes. As a result, we are less likely to marry someone from a different socioeconomic class, from a different race or religion, or from a markedly different educational background. Thus, as we pair off, we follow somewhat predictable patterns: In most cases the man is older, earns more money, and has a higher occupational status than the woman.

These patterns may not be evident to the two people who are in love with each other; indeed, they may not be aware that anything other than romance played a role in their choice of a mate. As sociologists we begin to discern marriage patterns.

We may note that marriage rates vary in different parts of the country, that the average age of marriage is related to educational level, and that social class is related to marital stability. These patterns (discussed in Chapter 12) show us that there are forces at work that influence marriage and that may not be evident to the individuals who fall in love and marry.

C. Wright Mills (1959) pointed out different levels on which social events can be perceived and interpreted. He used the term the **sociological imagination** to refer to *this relationship between individual experiences and forces in the larger society that shape our actions.*

The sociological imagination is the process of looking at all types of human behavior patterns and discerning previously unseen connections among them, noting similarities in the actions of individuals with no direct knowledge of one another, and finding subtle forces that mold people's actions. Like a museum-goer who draws back from a painting to see how the separate strokes and colors form subtly shaded images, sociologists stand back from individual events to see why and how they occurred. In so doing, they discover patterns that govern our social existence.

The sociological imagination focuses on every aspect of society and every relationship among individuals. It studies the behavior of crowds at ball games and racetracks; shifts in styles of dress and popular music; changing patterns of courtship and marriage; the emergence and fading of different lifestyles, political movements, and religious sects; the distribution of income and access to resources and services; decisions made by the Supreme Court, by congressional committees, and by local zoning boards; and so on. Every detail of social existence is food for sociological thought and relevant to sociological analysis.

The potential for sociology to be put to use—applied to the solution of real-world problems—is enormous. Proponents of applied sociology believe the work of sociologists can and should be used to help bring about an understanding of, and perhaps even guidelines for changing, the complexities of modern society.

The demand for applied sociology is growing, and many sociologists work directly with government agencies or private businesses in an effort to apply sociological knowledge to real-world problems.

For example, they might investigate such questions as how the building of a dam will affect the residents of the area, how jury makeup affects the outcome of a case, why voters select one candidate over another, how a company can boost employee morale, and how relationships among administrators, doctors, nurses, and patients affect hospital care. The answers to these questions have practical applications. The growing demand for sociological information provides many new career choices for sociologists (see "For Further Thinking: If You Are Thinking about Sociology as a Career, Read This").

Is Sociology Common Sense?

Common sense is what people develop through everyday life experiences. In a very real sense, it is the set of expectations about society and people's behavior that guides our own behavior. Unfortunately, these expectations are not always reliable or accurate because without further investigation, we tend to believe what we want to believe, to see what we want to see, and to accept as fact whatever appears to be logical. Whereas common sense is often vague, oversimplified, and frequently contradictory, sociology as a science attempts to be specific, to qualify its statements, and to prove its assertions.

Upon closer inspection, we find that the proverbial words of wisdom rooted in common sense are often illogical. Why, for example, should you "look before you leap" if "he who hesitates is lost"? How can "absence make the heart grow fonder" when "out of sight, out of mind"? Why should "opposites attract" when "birds of a feather flock together"? The "common sense" approach to sociology is one of the major dangers the new student encounters. Common sense often makes sense after the fact. It is more useful for describing events than for predicting them. It deludes us into thinking we knew the outcome all along (Hawkins & Hastie, 1990).

One researcher (Teigen, 1986) asked students to evaluate actual proverbs and their opposites. When given the actual proverb "Fear is stronger than love," most students agreed that it was true. But so did students who were given the reverse statement, "Love is stronger than fear." The same was true for the statements "Wise men make proverbs and fools repeat them" (actual proverb) and its reversal, "Fools make proverbs and wise men repeat them."

Although common sense gleaned from personal experience may help us in certain types of interactions, it will not help us understand why and under what conditions these interactions are taking place.

FOR FURTHER THINKING

If You Are Thinking about Sociology as a Career, Read This

Speaking from this side of the career-decision hurdle, I can say that being a sociologist has opened many doors for me. It gave me the credentials to teach at the college level and to become an author of a widely used sociology text. It also enabled me to be a newspaper columnist and a talk-show host. Would I recommend this field to anyone else? I would, but not blindly. Realize before you begin that sociology can be an extremely demanding discipline and, at times, an extremely frustrating one.

As in many other fields, the competition for jobs in sociology can be fierce. If you really want this work, do not let the herd stop you. Anyone with motivation, talent, and a determined approach to finding a job will do well. However, be prepared for the long haul: To get ahead in many areas you will need to spend more than four years in college. Consider your bachelor's degree as just the beginning.

Fields like teaching at the college level and advanced research often require a PhD, which means at least four to six years of school beyond the BA.

Now for the job possibilities. As you read through these careers, remember that right now your exposure to sociology is limited (you are only on Chapter 1 in your first college sociology text), so do not eliminate any possibilities right at the start. Spend some time thinking about each one as the semester progresses and you learn more about this fascinating discipline.

Most people who go into sociology become teachers. You will need a PhD to teach in college, but often a master's degree will open the door for you at the two-year college or high school level.

Second in popularity to teaching are nonacademic research jobs in government agencies, private research institutions, and the research departments of private corporations. Researchers carry on many different functions, including conducting market research, public opinion surveys, and impact assessments. Evaluation research, as the latter field is known, has become more popular in recent years because the federal government now requires environmental impact studies on all large-scale federal projects. For example, before a new interstate highway is built, evaluation researchers attempt to determine the effect the highway will have on communities along the proposed route.

This is only one of many opportunities available in government work. Federal, state, and local governments in policy-making and administrative functions also hire sociologists. For example, a sociologist employed by a community hospital provides needed data on the population groups being served and on the health care needs of the community. Another example: Sociologists working in a prison system can devise plans to deal with the social problems that are inevitable when people are put behind bars. Here are a few additional opportunities in government work: community planner, correction officer, environmental analyst, equal opportunity specialist, probation officer, rehabilitation counselor, resident director, and social worker.

A growing number of opportunities also exist in corporate America, including market researchers, pollsters, human resource managers, affirmative action coordinators, employee assistance program counselors, labor relations specialists, and public information officers, just to name a few. These jobs are available in nearly every field from advertising to banking, from insurance to publishing.

Although your corporate title will not be "sociologist," your educational background will give you the tools you need to do the job—and do it well, which, to corporations, is the bottom line.

Whether you choose government or corporate work, you will have the best chance of finding the job you want by specializing in a particular field of sociology while you are still in school. You can become an urban or family specialist or become knowledgeable in organizational behavior before you enter the job market. For example, many demographers, who compile and analyze population data, have specialized in this aspect of sociology. Similarly, human ecologists, who investigate the structure and organization of a community in relation to its environment, have specialized educational backgrounds as well. Keep in mind that many positions require a minor or some course work in other fields such as political science, psychology, ecology, law, or business. By combining sociology with these fields, you will be well prepared for the job market.

What next? Be optimistic and start planning. As the American Sociological Association observed, few fields are as relevant to today and as broadly based as sociology. Yet, ironically, its career potential is just beginning to be tapped. Start planning by reading the *Occupational Outlook Quarterly,* published by the U.S. Bureau of Labor Statistics, as well as academic journals to keep abreast of career trends. Then study hard and choose your specialty. With this preparation, when the time comes to find a job, you will be well prepared.

SOCIAL CHANGE

Too Smart to Marry?

Many of the subjects that we study in sociology are also popular topics in the media, or concepts that people think of as "common sense." Take the idea that the more education a woman has, the less likely she is to marry. Any brainy girl who has ever heard the taunt, "It's not smart to be too smart," is likely to wonder if a high GPA will sink her chances of ever finding wedded bliss. Stereotypes like these are often given additional credence in the press. Take a piece written by Maureen Dowd, a columnist for the *New York Times*. She wrote, "The rule of thumb seems to be that the more successful the woman, the less likely it is that she will find a husband or bear a child. For men the reverse is true." Most of the letters in response to her column agreed.

As with many stereotypes like this, there is often some nugget of truth behind such thinking. But here is the crucial difference between sociology and popular wisdom: As sociologists we do not automatically accept such easy pronouncements as fact. Like scientists—and sociology is, after all, a social *science*—we want proof, and we cultivate a healthy degree of skepticism until we get it. In a case like this, we would look at research data to determine whether these views are really true. Were they accurate at a certain point in time but not at another? Do they describe certain women and not others?

A review of marriage data for the past few decades would show us that although the stereotype once was true, in the past 25 years, the marriage gap between more and less educated women has narrowed significantly. In 1980, a woman who did not have a high school degree was more likely to be married than a woman with a college or graduate degree. Today the reverse is true. College-educated women are now more likely to be married than high school dropouts.

The profile of men most likely to marry has also changed. Today, the person most likely to end up without a wedding ring is the poorly educated man. The real truth now? Smart is sexy—for both sexes.

Percentage of White Males 40–44 Who Are Married

Education	1980	1990	2000
11th Grade	85.6	75.2	65.0
College	86.2	81.0	78.2
Grad School	85.1	84.5	82.5

Percentage of White Females 40–44 Who Are Married

Education	1980	1990	2000
11th Grade	83.9	77.5	70.1
College	83.4	77.7	76.8
Grad School	66.0	71.3	73.8

United States Census of Population, Public Use Microdata Sample (PUMS) 5% sample.

Sources: Rose, Elaina. (2004, March). Education and Hypergamy in Marriage Markets; Department of Economics, Paper #353330. Seattle: University of Washington; Dowd, Maureen. (2002, April 10). "The Baby Bust." *New York Times*.

Sociologists as scientists attempt to qualify these statements by specifying, for example, under what conditions do "opposites tend to attract" or "birds of a feather flock together." Sociology as a science is oriented toward gaining knowledge about why and under what conditions events take place in order to understand human interactions better. (For a discussion of how sociology is different from common sense see "Social Change: Too Smart to Marry?")

Sociology and Science

Sociology is commonly described as one of the social sciences. Science refers to *a body of systematically arranged knowledge that shows the operation of general laws.*

Sociology also uses the same general methods of investigation that are used in the natural sciences. Like the natural scientists, sociologists use the **scientific method,** *a process by which a body of scientific knowledge is built through observation, experimentation, generalization, and verification.*

The collection of data is an important aspect of the scientific method, but facts alone do not constitute a science. To have any meaning, facts must be ordered in some way, analyzed, generalized, and related to other facts. This is known as theory construction. Theories help organize and interpret facts and relate them to previous findings of other researchers.

Science is only one of the ways in which human beings study the world around them. Take feeling happy as an example. A physiologist might describe joy as

Sociologists and anthropologists share many theories and concepts. Whereas sociologists tend to study groups and institutions within large, modern, industrial societies, anthropologists tend to focus on the cultures of small, preindustrial societies.

a biochemical response to certain events. A poet might describe the experience in beautiful language. A theologian might describe happiness as the outcome of a relationship with God.

Unlike other means of inquiry, science for the most part limits its investigations to empirical entities, things that can be observed directly or that produce directly observable events. Therefore, one of the basic features of science is **empiricism,** *the view that generalizations are valid only if they rely on evidence that can be observed directly or verified through our senses.* For example, theologians might discuss the role of faith in producing "true happiness"; philosophers might deliberate over what happiness actually encompasses; but sociologists would note, analyze, and predict the consequences of such measurable items as job satisfaction, the relationship between income and education, and the role of social class in the incidence of divorce.

Sociology as a Social Science

The **social sciences** consist of all *those disciplines that apply scientific methods to the study of human behavior.* Although there is some overlap, each of the social sciences has its own area of investigation. It is helpful to understand each of the social sciences and to examine sociology's relationship to them.

Cultural Anthropology The social science most closely related to sociology is *cultural anthropology.* The two have many theories and concepts in common and often overlap. The main difference is in the groups they study and the research methods they use. Sociologists tend to study groups and institutions within large, modern, industrial societies, using research methods that enable them rather quickly to gather specific information about large numbers of people. In contrast, cultural anthropologists often immerse themselves in another society for a long time, trying to learn as much as possible about that society and the relationships among its people. Thus, anthropologists tend to focus on the culture of small, preindustrial societies because they are less complex and more manageable using this method of study.

Psychology The study of individual behavior and mental processes is part of *psychology;* the field is concerned with such issues as motivation, perception, cognition, creativity, mental disorders, and personality. More than any other social science, psychology uses laboratory experiments.

Psychology and sociology overlap in a subdivision of each field known as *social psychology*—the study of how human behavior is influenced and shaped by various social situations. Social psychologists study

© Frans Lemmens/Lineair/Peter Arnold, Inc.

such issues as how individuals in a group solve problems and reach a consensus, or what factors might produce nonconformity in a group situation. For the most part, however, psychology studies the individual, and sociology studies groups of individuals, as well as society's institutions.

The sociologist's perspective on social issues is broader than that of the psychologist. Take the case of alcoholism, for example. The psychologist might view alcoholism as a personal problem that has the potential to destroy an individual's physical and emotional health, as well as marriage, career, and friendships. The sociologist, on the other hand, would look for patterns in alcoholism. Although each alcoholic makes the decision to take each drink—and each suffers the pain of addiction—the sociologist would remind us to look beyond the personal and to consider the broader aspects of alcoholism, such as its social causes. Sociologists want to know who drinks excessively, when they drink, where they drink, and under what conditions they drink. They are also interested in the social costs of chronic drinking—costs in terms of families torn apart, jobs lost, children severely abused and neglected; costs in terms of highway accidents and deaths; costs in terms of drunken quarrels leading to violence and to murder. Noting the startling increase in heavy alcohol use by adolescents over the past 10 years and the rapid rise of chronic alcoholism among women, sociologists ask, what forces are at work to account for these patterns?

Economics Economists have developed techniques for measuring such things as prices, supply and demand, money supplies, rates of inflation, and employment. This study of the creation, distribution, and consumption of goods and services is known as *economics*. The economy, however, is just one part of society. It is each individual in society who decides whether to buy an American car or a Japanese import, whether she or he is able to handle the mortgage payment on a dream house, and so on. Whereas economists study price and availability factors, sociologists are interested in the social factors that influence the resulting economic behavior. Is it peer pressure that results in the buying of the large flashy car, or is it concern about gas mileage that leads to the purchase of a small, fuel-efficient, modest vehicle? What social and cultural factors contribute to the differences in the portion of income saved by the average wage earner in different societies? What effect does the unequal allocation of resources have on social interaction?

These are examples of the questions sociologists seek to answer.

History Although not exactly a social science, history shares certain attributes with sociology. The study of *history* involves looking at the past in an attempt to learn what happened, when it happened, and why it happened. Sociology also looks at historical events within their social contexts to discover why things happened and, more important, to assess what their social significance was and is. Historians provide a narrative of the sequence of events during a certain period and may use sociological research methods to try to learn how social forces have shaped historical events. Sociologists, on the other hand, examine historical events to see how they influenced later social situations.

Historians focus on individual events—the American Revolution or slavery—and sociologists generally focus on phenomena such as revolutions or the patterns of dominance and subordination that exist in slavery. They try to understand the common conditions that contribute to revolutions or slavery wherever they occur.

Let us consider the subject of slavery in the United States. Traditionally, historians might focus on when the first slaves arrived, on how many slaves existed in 1700 or 1850, and the conditions under which they lived. Sociologists and modern social historians would use these data to ask many questions: What social and economic forces shaped the institution of slavery in the United States? How did the Industrial Revolution affect slavery? How has the experience of slavery affected the black family? Although history and sociology have been moving toward each other over the past 20 years, each discipline still retains a somewhat different focus: sociology on the present, history on the past.

Political Science Concentrating on three major areas, *political science* is the study of political theory, the actual operation of government, and, in recent years, political behavior. This emphasis on political behavior overlaps with sociology. The primary distinction between the two disciplines is that sociology focuses on how the political system affects other institutions in society, whereas political science devotes more attention to the forces that shape political systems and the theories for understanding these forces. However, both disciplines share an interest in why people vote the way they do, why they join political movements, how the mass media are changing political parties and processes, and so on.

Social Work Much of the theory and research methods of social work are drawn from sociology and psychology, but social work focuses to a much greater degree on application and problem solving.

The disciplines of sociology and social work are often confused with each other. In the early days of sociology, women were often unable to attend graduate sociology programs and chose social work studies instead.

The main goal of *social work* is to help people solve their problems, whereas the aim of sociology

CONTROVERSIES IN SOCIOLOGY

Is There a Difference between Sociology and Journalism?

It often seems as if sociologists and journalists are engaged in the same activities. Journalists examine and write about social issues. They interview people. They often conduct polls. They make predictions. They offer recommendations for correcting social problems. If journalists do all this, why would someone need to become trained as a sociologist?

This is a sociology textbook, so needless to say we are going to make the case that there is a difference between sociologists and journalists.

P. J. Baker, L. E. Anderson, and D. S. Dorn (1993) analyze the difference between sociology and journalism, explaining that daily newspapers and weekly news magazines are written for the general public, which wants an overview of local and world events. One of the fundamental features of these media is the timely coverage of recent events. "Nothing is more stale and uninteresting to readers than an old story. Nothing is more enjoyable for reporters and editors than to scoop the competing paper or magazine with a late-breaking story their competitor has missed." Three types of journalists do the writing: "reporters, who write stories;" editors, who generate ideas for stories and review the copy; "and commentators, who interpret events. Professional experts on social problems also are invited to write letters to the editor or commentaries for the editorial page." Journalists usually have a college degree in any of a wide variety of areas, or they may have an advanced degree from a professional journalism program. Jargon is kept to a minimum and elaborate explanations must be presented in manageable terms so that the average reader can understand them.

Baker, Anderson, and Dorn go on to explain that sociologists engage in the study of society with:

The primary intent of sharing their work with other sociologists, not with the general public. . . .

They pay special attention to their methods of investigation, their theories of explanation, and their claims of originality. When sociologists publish their results, they are fully aware that other sociologists may dispute the soundness of their findings or the logic of their explanations. The public has little interest in sociological disputes about methods, theories, or claims of originality.

Sociologists usually publish their writings as either articles in scholarly journals, chapters in books, or full-length books. These writings are screened by editors and critics hired to evaluate the merits of the work. Sociologists aim to have their colleagues recognize their work as truly significant.

"Journalists" however, "are always thinking about tomorrow's headlines or next week's cover story." Sociologists' work can never be completed in such short time frames. According to Baker, Anderson, and Dorn, "a major sociological study may take three to five years," though most take one to two years. Sociologists also have the freedom to study historical materials. "Sociologists are not totally indifferent to the times in which they live; many hope that their work will be relevant to contemporary debates about current issues." Essentially, the two fields represent different approaches to social issues. Journalists get a multifaceted overview of an issue, whereas sociologists have the luxury of exploring a topic in depth and contemplating the ramifications of their findings.

Source: *Social Problems: A Critical Thinking Approach*, 2d ed. (pp. 20–22), by P. J. Baker, L. E. Anderson, & D. S. Dorn, 1993, Belmont, CA: Wadsworth.

is to understand why the problems exist. Social workers provide help for individuals and families who have emotional and psychological problems or who experience difficulties that stem from poverty or other ongoing problems rooted in the structure of society. Social workers also organize community groups to tackle local issues such as housing problems and try to influence policy-making bodies and legislation. Sociologists provide many of the theories and ideas used to help others. Although sociology is not social work, it is a useful area of academic concentration for those interested in entering the helping professions.

The Development of Sociology

It is hardly an accident that sociology emerged as a separate field of study in Europe during the nineteenth century. That was a time of turmoil, a period in which the existing social order was being shaken by the growing Industrial Revolution and by violent uprisings against established rulers (the American and French revolutions). People also were affected by the impact of discovering, through world exploration and colonialization, how others lived. At the same time, a void was left by the declining power of the church to impose its views of right and wrong. New social classes of industrialists and businesspeople emerged to challenge the rule of the feudal aristocracies.

Tightly knit communities, held together by centuries of tradition and well-defined social relationships, were strained by dramatic changes in the social environment. Factory cities began to replace the rural estates of nobles as the centers for society at large. People with different backgrounds were brought together under the same factory roof to work for wages instead of exchanging their services for land and protection. Families now had to protect themselves, to buy food rather than grow it, and to pay rent for their homes. These new living and working conditions led to the development of an industrial, urban lifestyle, which, in turn, produced new social problems.

Many people were frightened by what was going on and wanted to find some way of understanding and dealing with the changes taking place. The need for a systematic analysis of society coupled with acceptance of the scientific method resulted in the emergence of sociology. Henri Saint-Simon (1760–1825) and Auguste Comte (1798–1857) were among the pioneers in the science of sociology.

Auguste Comte (1798–1857)

Born in the French city of Montpellier on January 19, 1798, Auguste Comte grew up in the period of great political turmoil that followed the French Revolution of 1789–1799. In August 1817, Comte met Henri Saint-Simon and became his secretary and eventually his close collaborator. Under Saint-Simon's influence, Comte converted from an ardent advocate of liberty and equality to a supporter of an elitist conception of society.

Saint-Simon and Comte rejected the lack of empiricism in the social philosophy of the day. Instead they turned for inspiration to the methods and intellectual framework of the natural sciences, which they perceived as having led to the spectacular successes of industrial progress. They set out to develop a "science of man" that would reveal the underlying principles of society much as the sciences of physics and chemistry explained nature and guided industrial progress. During their association the two men collaborated on a number of essays, most of which contained the seeds of Comte's major ideas. Their alliance came to a bitter end in 1824 when Comte broke with Saint-Simon for both financial and intellectual reasons.

Financial problems, lack of academic recognition, and marital difficulties combined to force Comte into a shell. Eventually, for reasons of "cerebral hygiene," he no longer read any scientific work related to the fields about which he was writing.

Living in isolation at the periphery of the academic world, Comte concentrated his efforts between 1830 and 1842 on writing his major work,

August Comte coined the term *sociology*. He wanted to develop a "science of man" that would reveal the underlying principles of society, much as the sciences of physics and chemistry explained nature and guided industrial progress.

Cours de Philosophie Positive, in which he coined the term *sociology.* Comte devoted a great deal of his writing to describing the contributions he expected sociology would make in the future. He was much less concerned with defining sociology's subject matter than with showing how it would improve society. Although Comte was reluctant to specify subdivisions of sociology, he identified two major areas that sociology should concentrate on.

These were *social statics,* the study of how various institutions of society are interrelated, focusing on order, stability, and harmony; and *social dynamics,* the study of complete societies and how they develop and change over time. Comte believed all societies move through certain fixed stages of development, eventually reaching perfection (as exemplified in his mind by industrial Europe). Sociologists, however, no longer accept the idea of a perfect society.

Harriet Martineau (1802–1876)

Harriet Martineau, an Englishwoman, was an early and significant contributor to the development of sociology. In 1837 she published *Theory and Practice of Society in America,* in which she analyzed the customs and lifestyle present in the nineteenth-century United States. Her book was based on traveling throughout the United States and observing day-to-day life in all its forms, from that which took place in prisons, mental hospitals, and factories to family

Harriet Martineau

1833.

Boston Filmworks

Harriet Martineau was an early and significant contributor to the development of sociology. She believed that scholars should not simply offer observations but should also use their research to bring about social reform.

Herbert Spencer when 38

Boston Filmworks

Herbert Spencer helped to define the subject matter of sociology. Spencer also became a proponent of a doctrine known as social Darwinism.

gatherings, slave auctions, and even proceedings of the Supreme Court and Senate. The book helped map out what a sociological work dealt with by examining the impact of immigration, family issues, politics, and religion, as well as race and gender issues. In her book she also compared social stratification systems in Europe with those in the United States.

Martineau's work also demonstrated the level of objectivity she thought was necessary for an analysis of society when she noted, "It is hard to tell which is worse, the wide diffusion of things that are not true or the suppression of things that are." Later in her career, she came to the conclusion that scholars should not just offer observations, but should also use their research to bring about social reform for the benefit of society. She asked her readers to "judge for themselves . . . how far the people of the United States lived up to" their stated ideals (Hoecker-Drysdale, 1992).

Martineau's second important contribution to sociology was translating into English August Comte's six-volume *Positive Philosophy*. Her two-volume edition of this book introduced the field of sociology to England and influenced people such as Herbert Spencer, as well as early American sociologists.

Other women also produced important works during the early days of sociology. They included Helen Hunt Jackson and her 1881 *Century of Dishonor*, an exposé of the government's treatment of Native Americans; Frances Kellor's 1901 book, *Experimental Sociology*; Dorothy Swaine Thomas's 1925 *Social Aspects of the Business Cycle*; and Helen Macgill Hughes who in 1944 began a 17-year term as manager of the *American Journal of Sociology*.

Herbert Spencer (1820–1903)

A largely self-educated Briton, Herbert Spencer had a talent for synthesizing information. In 1860 he started work toward the goal of organizing human knowledge into one system. The result was his *Principles of Sociology* (1876, 1882), the first sociology textbook.

Unlike Comte, Spencer was precise in defining the subject matter of sociology. He declared the field of sociology included the study of the family, politics, religion, social control, work, and stratification.

Spencer believed society was similar to a living organism. Just as the individual organs of the body are interdependent and make their specialized contributions to the living whole, so, too, are the various segments of society interdependent. Every part of society serves a specialized function necessary to ensure society's survival as a whole.

Spencer became a proponent of a doctrine known as social Darwinism. **Social Darwinism** *applied to society Charles Darwin's notion of "survival of the fittest," in which those species of animals best adapted to the environment survived and prospered, while those poorly adapted died out.*

Spencer reasoned that people who could not successfully compete in modern society were poorly adapted to their environment and therefore inferior. Lack of success was viewed as an individual failing, and that failure was in no way related to barriers (such as prejudice or racism) created by society. In this view, to help the poor and needy was to intervene vainly in a natural evolutionary process.

Social Darwinism had a significant effect on those who believed in the inequality of races. They now claimed that those who had difficulty succeeding in the white world were really members of inferior races. The fact that they lost out in the competition for status was proof of their poor adaptability to the environment. The survivors were clearly of superior stock (Berry & Tischler, 1978).

Many whites accepted social Darwinism because it served as a justification for their control over institutions. It enabled them to oppose reforms or social welfare programs, which they viewed as interfering with nature's plan to do away with the unfit. Social Darwinism thus became a justification for the repression and neglect of African Americans following the Civil War. It was also used to justify policies that resulted in the decimation of Native American populations and the complete eradication of the native people of Tasmania (near Australia) between 1803 and 1876 by white settlers (Fredrickson, 1971; Parrillo, 1997).

Spencer's ties to social Darwinism have led many scholars to disregard his original contributions to the discipline of sociology. However, Spencer originally formulated many of the standard concepts and terms still current in sociology, and their use derives directly from his works.

During the nineteenth century, sociology developed rapidly under the influence of three scholars of highly divergent temperaments and orientations. Despite their differences, however, Karl Marx, Émile Durkheim, and Max Weber were responsible for shaping sociology into a relatively coherent discipline.

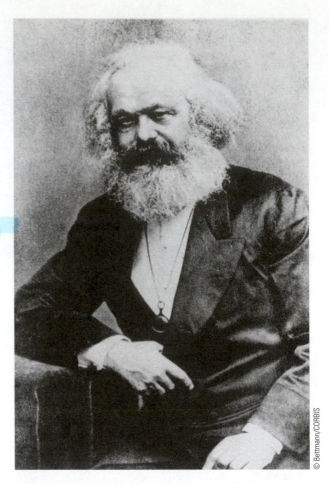

Karl Marx's views on class conflict were shaped by the Industrial Revolution. He believed that capitalist societies produce greater conflict because of the deep divisions between the social classes.

Karl Marx (1818–1883)

Those who are unfamiliar with his writings often think of Karl Marx as a revolutionary proponent of the political and social system seen in countries once labeled communist. Marx lived in Europe during the early period of industrialization, when the overwhelming majority of people in such societies were poor. The rural poor moved to cities where employment was available in the factories and workshops of the new industrial economies. Those who owned and controlled the factories exploited the masses who worked for them. Even children, some as young as 5 or 6 years old, worked 12-hour days, six and seven days a week (Lipsey & Steiner, 1975), and received only a subsistence wage. "The iron law of wages"—the philosophy that justified paying workers only enough money to keep them alive—prevailed during this early period of industrialization.

In this way, the rural poor were converted into an urban poor. Meanwhile, those who owned the means of production possessed great wealth, power, and prestige. Marx tried to understand the societal forces that produced such inequities and looked for a means to change them in order to improve the human condition.

Marx believed the entire history of human societies could be seen as the history of class conflict: the conflict between the bourgeoisie, who own and control the means of production (capitalists), and the proletariat, who make up the mass of workers—the exploiters and the exploited. He believed the capitalists determined the distribution of wealth, power, and even ideas in that society. The wealthy get their power not just from their control of the economy, but also from their control of the political, educational, and religious institutions in their society as well. According to Marx, capitalists make and enforce laws that serve their interests and act against the interests of workers. Their control over all institutions enables them to create common beliefs that make the workers accept their status. These economic, political, and religious ideologies make the masses loyal to the very institutions that are the source of their exploitation and that also are the source of the wealth, power, and prestige of the ruling class. Thus, the prevailing beliefs of any society are those of its dominant group.

Marx predicted that capitalist society eventually would be polarized into two broad classes: the capitalists and the increasingly impoverished workers. Intellectuals like him would show the workers that the capitalist institutions were the source of exploitation and poverty. Gradually, the workers would become unified and organized, and then through revolution they would take over control of the economy.

The means of production would then be owned and controlled by the people in a workers' socialist state. Once the capitalist elements of all societies had been eliminated, the governments would wither away.

New societies would develop in which people could work according to their abilities and take according to their needs. The seeds of societal conflict and social change would then come to an end as the means of production were no longer privately owned.

In many capitalist societies today, regulatory mechanisms have been introduced to prevent some of the excesses of capitalism. Unions have been integrated into the capitalist economy and the political system, giving workers a legal, legitimate means through which they can benefit from the capitalist system.

Marx was not a sociologist, but his considerable influence on the field can be traced to his contributions to the development of *conflict theory,* which will be discussed more fully in this chapter.

Émile Durkheim (1858–1917)

A student of law, philosophy, and social science, Émile Durkheim was the first professor of sociology at the University of Bordeaux, France. Whereas

Émile Durkheim produced the first true sociological study. Durkheim's work moved sociology fully out of the realm of social philosophy and chartered the discipline's course as a social science.

© Bettmann/CORBIS

Spencer wrote the first textbook of sociology, it was Durkheim who produced the first true sociological study. Durkheim's work moved sociology fully out of the realm of social philosophy and helped chart the discipline's course as a social science.

Durkheim believed that individuals were exclusively the products of their social environment and that society shapes people in every possible way. To prove his point, Durkheim studied suicide. He believed that if he could take what was perceived to be a totally personal act and show that it is patterned by social factors rather than exclusively by individual mental disturbances, he would provide support for his point of view.

Émile Durkheim studied suicide because he believed people were the product of their social environments. Differences in these environments were responsible for variations in social behavior.

Durkheim began with the theory that the industrialization of Western society was undermining the social control and support that communities had historically provided for individuals. Industrialization forced or induced individuals to leave rural communities for

Figure I–2   Suicide Rates of People 15–24 in Various Countries

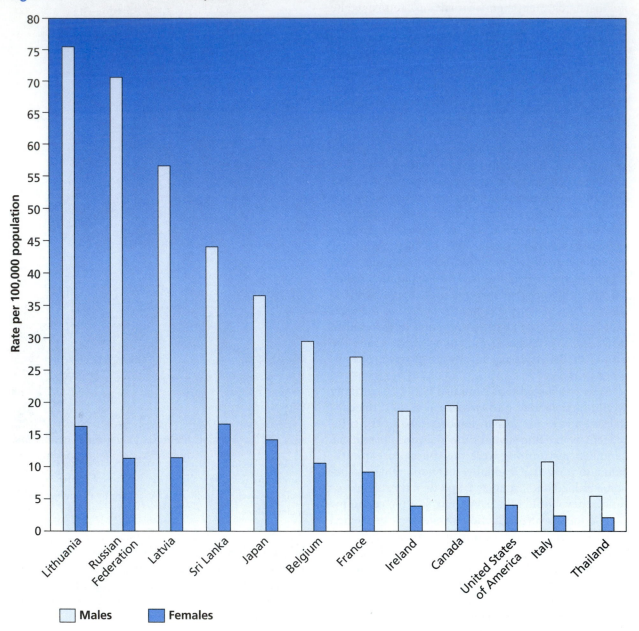

Source: World Health Organization "Suicide Rates per 100,000 by Country, Year, and Sex," available at www.who.int/mental
_health/prevention/suicide_rates/en/index.html. Accessed January 2, 2006.

urban areas, where there were usually greater economic opportunities. The anonymity and impersonality that they encountered in these urban areas, however, caused many people to become isolated from both family and friends. In addition, in industrial societies people are frequently encouraged to aspire to goals that are difficult to attain. Durkheim refined his theory to state that suicide rates are influenced by group solidarity and societal stability. He believed that low levels of solidarity—which involve more individual choice, more reliance on oneself, and less adherence to group standards—would mean high rates of suicide.

To test his idea, Durkheim decided to study the suicide rates of Catholic versus Protestant countries. He assumed the suicide rate in Catholic countries would be lower than in Protestant countries because Protestantism emphasized the individual's relationship to God over community ties. The comparison of suicide records in Catholic and Protestant countries in Europe supported his theory by showing the probability of suicide was indeed higher in Protestant countries. (See Figure 1–2 for suicide rates in various countries.)

Recognizing the fact that lower suicide rates among Catholics could be based on factors other than group solidarity, Durkheim proceeded to test other groups. Reasoning that married people would be more integrated into a group than single people, or people with children more than people without children, or non–college-educated people more than college-educated people (because college tends to break group ties and encourage individualism), or Jews more than non-Jews, Durkheim tested each of these groups, and in each case his theory held. Then, characteristic of the scientist that he was, Durkheim extended his theory by identifying three types of suicide—egoistic, altruistic, and anomic—that take place under different types of conditions.

Egoistic suicide comes from low group solidarity, an underinvolvement with others. Durkheim argued that loneliness and a commitment to personal beliefs rather than group values can lead to egoistic suicide. He found that single and divorced people had higher suicide rates than did married people and that Protestants, who tend to stress individualism, had higher rates of suicide than did Catholics.

Altruistic suicide derives from a very high level of group solidarity, an overinvolvement with others. The individual is so tied to a certain set of goals that he or she is willing to die for the sake of the community.

This type of suicide, Durkheim noted, still exists in the military as well as in societies based on ancient codes of honor and obedience. Perhaps the best-known historical examples of altruistic suicide come from Japan: the ceremonial rite of *seppuku*, in which a disgraced person rips open his own belly, and the kamikaze attacks by Japanese pilots toward the end of World War II.

The Japanese pilots, instead of being morose before their bombing missions that would produce certain death, were often reported to be cheerful and serene. One 23-year-old kamikaze in a letter to his parents voiced the feelings of thousands of his fellows when he wrote: "I shall be a shield for His Majesty and die cleanly along with my squadron leader and other friends." There were, he noted, 16 members in his squadron, and he added: "May our deaths be as sudden and clean as the shattering of crystal" (Axell, 2002).

Today we often see examples of altruistic suicide with the terrorists who flew the planes into the World Trade Towers and the Palestinian suicide bombers. These individuals are willing to sacrifice their lives for their cause as they blow up a building, plane, or restaurant. In addition to destroying property, the terrorists often want to kill as many people as possible. (For a discussion of a contemporary suicide issue see "News You Can Use: Is There an Epidemic of College Student Suicides?")

Anomic suicide results from a sense of feeling disconnected from society's values. A person may know what goals to strive for but not have the means of

Much of Max Weber's work was an attempt to clarify, criticize, and modify the words of Karl Marx. He also showed how religion contributed to the creation of new economic conditions and institutions.

attaining them, or a person may not know what goals to pursue. Durkheim found that times of rapid social change or economic crisis are associated with high rates of anomic suicide.

Durkheim's study was noteworthy not only because it proved that the most personal of all acts, suicide, is in fact a product of social forces, but also because it was one of the first examples of a scientifically conducted sociological study. Durkheim systematically posed theories, tested them, and drew conclusions that led to further theories. He also published his results for everyone to see and criticize.

Durkheim's interests were not limited to suicide. His mind ranged the entire spectrum of social activities. He published studies on *The Division of Labor in Society* (1893) and *The Elementary Forms of the Religious Life* (1917). In both works he drew on what was known about nonliterate societies, following the lead of Comte and Spencer in viewing them as evolutionary precursors of the contemporary industrial societies of Europe.

Durkheim focused on the forces that hold society together—that is, on the functions of various social structures. This point of view, often called the *functionalist theory* or *functionalist perspective*, remains one of the dominant approaches to the modern study of society.

Max Weber (1864–1920)

Much of Weber's work attempted to clarify, criticize, and modify the works of Marx. For that reason we shall discuss Weber's ideas as they relate to and contrast with those of Marx. Unlike Marx, who

NEWS YOU CAN USE

Is There an Epidemic of College Student Suicides?

When Joanne Michelle Leavy, a 23-year-old graduate student in the arts, jumped 12 stories to her death just before the beginning of the fall semester in 2004, the tragedy earned NYU a distinction no college wants: a reputation as one of the nation's premiere suicide universities. Like MIT, which had suffered 11 suicides in 11 years, and Cornell, whose famous gorges were supposedly a magnet for depressed students looking to end their own lives, NYU had experienced a terrible string of student deaths, five in one year.

Soon the papers had marshaled pundits to talk about "suicide contagion," and to speculate on the idea of suicide as a fad, leading readers to believe that college students were at greater risk than others to die by their own hands.

A careful analysis of research data, however, comes to precisely the opposite conclusion. Although suicide is, admittedly, the second leading cause of death among college students (after accidents), their nonstudent peers are actually at far greater risk. The decade-long Big Ten Student Suicide Study, which tracked nearly 350,000 students, found that students average 7.5 suicides per 100,000 annually, compared with the national average of 15 per 100,000. That sample was matched for age, race, and gender. Older students, between the ages of 20 and 24, accounted for 46% of the suicides, and graduate students comprised 32% of the total.

The potential for the egoistic type of suicide that Émile Durkheim described becomes less likely because of the possible attachments a campus provides. A college or university typically provides support and community for its members—whether it be orientation programs, resident advisor assistance, or counseling. Although college mental health programs are often understaffed, their very existence is a resource less easily available to nonstudents.

Another reason why college students are less likely to commit suicide than others in their age group is that although guns are used in 60% of all suicides, college campuses for the most part forbid the ownership of guns.

This is not to say that suicide on campus is not a very serious problem. Suicide rates among the young in general have been rising steadily. Advances in medications and treatments have made it possible for students with major depression, bipolar disorder, and even schizophrenia to attend college who would have not been able to do so in the past. The risk is that some of these students may be overwhelmed by the college experience and have serious problems adjusting.

Sources: Ellen, Elizabeth Fried. (2002, August). "Identifying and Treating Suicidal College Students." *Psychiatric Times* 19(8); Silverman Morton, M., Meyer, P. M., Sloane, F. et al. (1997). "The Big Ten Student Suicide Study: A 10-Year Study of Suicides on Midwestern University Campuses." *Suicide Life Threat Behavior* 27(3), pp. 285–303.

---

was not only an intellectual striving to understand society but also a revolutionary conspiring to overturn the capitalist social system, Weber was essentially a German academic attempting to understand human behavior. Weber believed the role of intellectuals was simply to describe and explain truth, whereas Marx believed the scholar should also tell people what to do.

Marx believed that ownership of the means of production resulted in control of wealth, power, and ideas. Weber showed that economic control does not necessarily result in prestige and power. For example, the wealthy president of a chemical company whose toxic wastes have been responsible for the pollution of a local water supply might have little prestige in the community. Moreover, the company's board of directors might deprive the president of any real power.

Although Marx maintained that control of production inevitably results in control of ideologies, Weber stated that the opposite may happen: Ideologies sometimes influence the economic system.

When Marx called religion an "opium of the people," he was referring to the ability of those in control to create an ideology that would justify the exploitation of the masses. Weber, however, showed that religion could be a belief system that contributed to the creation of new economic conditions and institutions. In *The Protestant Ethic and the Spirit of Capitalism* (1904–1905), Weber tried to demonstrate how the Protestant Reformation of the seventeenth century provided an ideology that gave religious justification to the pursuit of economic success through rational, disciplined, hard work. This ideology, called the Protestant ethic, ultimately helped transform northern European societies from feudal agricultural communities to industrial capitalist societies.

Yet Weber also predicted that science—the systematic, rational description, explanation, and manipulation of the observable world—would lead to a gradual turning away from religion. The apparent decline in the influence of organized religion in highly industrialized societies seems to support Weber's prediction.

Understanding the development of bureaucracy interested Weber. Whereas Marx saw capitalism as the source of control, exploitation, and alienation of human beings and believed that socialism and communism would ultimately bring an end to this exploitation, Weber believed bureaucracy would characterize both socialist and capitalist societies. He anticipated and feared the domination of individuals by large bureaucratic structures. As he foresaw, bureaucracies now rule our modern industrial world, both capitalist and socialist—economic, political, military, educational, and religious. Given the existing situation, it is easy to appreciate Weber's anxiety. As he put it,

> Each man becomes a little cog in the machine and, aware of this, his one preoccupation is whether he can become a bigger cog. . . . The problem which besets us now is not: how can this evolution be changed?—for that is impossible, but what will become of it? (Quoted in Coser, 1977)

The Development of Sociology in the United States

Sociology had its roots in Europe and did not become widely recognized in the United States until almost the beginning of the twentieth century. The early growth of American sociology took place at the University of Chicago. That setting provided a context in which a large number of scholars and their students could work closely to refine their views of the discipline. It was there that the first graduate department of sociology in the United States was founded in the 1890s. From the 1920s to the 1940s, the so-called Chicago school of sociologists led American sociology in the study of communities, with particular emphasis on urban neighborhoods and ethnic areas.

Many of America's leading sociologists from this period were members of the Chicago school, including Robert E. Park, W. I. Thomas, and Ernest W. Burgess. Most of these individuals were Protestant ministers or sons of ministers, and as a group they were deeply concerned with social reform.

Also in Chicago, but not directly part of the university, Jane Addams (1860–1935) was also deeply committed to social reform. Jane Addams was born in 1860 to a prosperous Quaker family dedicated to the antislavery cause. Her father John Addams was a politician and friend of Abraham Lincoln. Jane Addams was part of the first generation of middle-class women to go to college. Addams graduated as valedictorian from Rockford Female Seminary (Illinois) in 1881. There were few professions open to educated women and after graduation Addams returned home and was expected to wait for a marriage proposal (Elshtain, 2001).

During the next few years, Addams traveled through Europe and observed the poverty that existed in the city's slums. She also studied ways in which various organizations attempted to alleviate the problem. During her stay in London, she visited a settlement house run by Oxford University students where they helped the poor. She used this settlement house, called Toynbee Hall, as a model for a program she would later develop in Chicago to assist the poor.

Jane Addams and Ellen Gates Star finally opened the doors to their own version of Toynbee Hall in September of 1889. They called it Hull House. It was designed to serve the immigrant population of Chicago's nineteenth ward. For 40 years, Hull House successfully served the community by offering a wide variety of clubs and activities.

During this time, Hull House and Jane Addams became known internationally for championing the rights of immigrants and fighting for child labor laws. She was also known for actions toward industrial safety, juvenile courts, labor unions, women's suffrage, and world peace.

Addams wrote extensively about Hull House activities. She published 11 books and numerous articles, and she spoke often at venues throughout the United States and the world. She used her inheritance and the proceeds from her writing and speaking engagements to live on, because she did not receive a salary from Hull House. She also used her income to underwrite various social causes throughout her life.

In 1907 she published *Newer Ideals of Peace,* from which she became known internationally as a pacifist. This brought her much ridicule when the United States entered World War I. But in time, the public began to embrace her ideals. By 1931 her reputation as a peacemaker was firmly established and she was awarded the Nobel Peace Prize, shared with Nicholas Murray Butler. After that, people from all over the world began to write her letters and extolling her work. She received pleas for intervention around the world to help alleviate hunger, poverty, and oppression (Swarthmore College).

W. E. B. DuBois (1868–1963) became the first African American to receive a PhD from Harvard in 1896 with his dissertation *The Suppression of the African Slave-Trade to the United States.* DuBois then went on to Atlanta University, where he was in charge of the sociology program until 1910. At that

© The Art Archive/Culver Pictures

W. E. B. DuBois was the first African American to receive a PhD from Harvard University. He wrote dozens of articles and books on the history and sociology of African Americans.

point he left to become editor of *The Crisis,* the journal of the National Association for the Advancement of Colored People. By that time DuBois had written dozens of articles and books on the history and sociology of African Americans and was the country's leading African American sociologist.

At the time when DuBois came of age, racism was very much a part of the American landscape on both a popular and academic level. Politicians and writers were openly declaring that blacks belong to an inferior race that contributes nothing to society. DuBois believed that doctrines and theories had a powerful effect on social conditions. Slavery and the disenfranchisement of blacks were rooted in the notion of the inferiority of the race. It was important, he felt, to change these beliefs in order to change the status of African Americans. Much of his scholarly work was governed by his view that sociological studies of African Americans would have a positive effect on public opinion (Brotz, 1966).

DuBois argued for the acceptance of African Americans into all areas of society and advocated militant resistance to white racism. He believed that it was not solely the responsibility of blacks, nor was it in their capacity, to alter their collective place in American society, but that it was primarily the responsibility of whites, who held the power to effect such change.

In 1903, he published *The Souls of Black Folk,* a collection of eloquent, well-reasoned essays on race relations. Blending sociology and economics he described the injustices that had scarred the black experience in the United States. "The problem of the 20th century is the problem of the color line," he declared (Lewis, 2000).

Throughout his life DuBois considered himself torn between being a black man and being an American. This conflict led him to feel like an exile in the United States. As DuBois noted in his autobiography:

> Had it not been for the race problem early thrust upon me and enveloping me, I should have probably been an unquestioning worshipper at the shrine of the established social order into which I was born. But just that part of this order which seemed to most of my fellows nearest perfection seemed to me most inequitable and wrong: and starting from that critique, I gradually, as the years went by, found other things to question in my environment. (DuBois, 1968)

DuBois died in 1963 at the age of 95, just before the famous march on Washington took place where Martin Luther King, Jr. made his "I Have a Dream" speech. It was ironic that America's preeminent black intellectual died on the eve of this great civil rights gathering, which had gained so much energy from his ideas against segregation. DuBois had long ago concluded that the possibility of racial equality was a receding mirage for people of color. At the time of his death he was leading the life of a political exile in Ghana.

Talcott Parsons (1902–1979) was the sociologist most responsible for developing theories of structural functionalism in the United States. He presided over the Department of Social Relations at Harvard College from the 1930s until he retired in 1973. Parsons's early research was quite empirical, but he later turned to the philosophical and theoretical side of sociology. In *The Structure of Social Action* (1937), Parsons presented English translations of the writings of European thinkers, most notably Weber and Durkheim. In his best-known work, *The Social System* (1951), Parsons portrayed society as a stable system of well-ordered, interrelated parts. His viewpoint elaborated on Durkheim's perspective.

Robert K. Merton also has been an influential proponent of functionalist theory. In his classic work, *Social Theory and Social Structure* (1968), first published in 1949, Merton spelled out the functionalist view of society. One of his main contributions to sociology was to distinguish between two forms of social functions—manifest functions and latent functions. By **social functions** Merton meant *those social processes that contribute to the ongoing operation or maintenance of society.* **Manifest functions** are *the intended and recognized consequences of those processes.* For example, one of the manifest functions of going to college is to obtain knowledge, training, and a degree in a specific area. **Latent functions** are *the unintended or not readily recognized consequences of such processes.* Therefore, college may also offer the opportunity of making lasting friendships and finding potential marriage partners.

Under the leadership of Parsons and Merton, sociology in the United States moved away from a concern with social reform and adopted a so-called value-free perspective. This perspective, which Max Weber advocated, requires description and explanation rather than prescription; it holds that people should be told what is, not what should be.

As critics of Parsons and Merton have pointed out, however, interpretations of what exists may differ depending on the perspective from which reality is viewed and on the values of the viewer (Gouldner, 1970; Lee, 1978; Mills, 1959).

Theoretical Perspectives

Scientists need a set of working assumptions to guide them in their work. These assumptions suggest which problems are worth investigating and offer a framework for interpreting the results of studies. Such sets of assumptions are known as paradigms.

Paradigms are *models or frameworks for questions that generate and guide research.* Of course, not all paradigms are equally valid, even though at first they seem to be. Sooner or later, some will be found rooted in fact, whereas others will remain abstract and unusable, finally to be discarded. We shall examine those paradigms that have withstood the scrutiny of major sociologists.

Functionalism

Functionalism—or structural functionalism, as it is often called—is rooted in the writings of Spencer and Durkheim and the work of such scholars as Parsons and Merton. **Functionalism** *views society as a system of highly interrelated structures or parts that function or operate together harmoniously.*

Functionalists analyze society by asking what each different part contributes to the smooth functioning of the whole. For example, we may assume the education system serves to teach students specific subject matter. However, functionalists might note that it acts as a system for the socialization of the young and as a means for producing conformity. The education system serves as a gatekeeper to the rewards society offers to those who follow its rules.

From the functionalist perspective, society appears quite stable and self-regulating. Much like a biological organism, society is normally in a state of equilibrium or balance. Most members of a society share a value system and know what to expect from one another.

The best-known proponent of the structural-functionalist perspective was Talcott Parsons. His theory centered on the view that there were certain social systems involving major areas of social life, such as the family, religion, education, politics, and economics. Parsons evaluated these systems according to functions they performed both for society as a whole and for one another.

Functionalism is a very broad theory in that it attempts to account for the complicated interrelationships of all the elements that make up human societies, including the complex societies of the industrialized (and industrializing) world. In a way it is impossible to be a sociologist and not be a functionalist, because most parts of society serve some stated or unstated purpose. Functionalism is limited in one regard, however: The preconception that societies are normally in balance or harmony makes it difficult for proponents of this view to account for how social change comes about.

A major point of criticism of functionalist theory is its conservative bias. That is, if all the parts of society fit together smoothly, we can assume that the social system is working well. Conflict is then seen as something that disrupts the essential orderliness of the social structure and produces imbalance between the parts and the whole.

Conflict Theory

Conflict theory is rooted in the work of Marx and other social critics of the nineteenth century. **Conflict theory** *sees society as constantly changing in response to social inequality and social conflict.*

For the conflict theorists, social change pushed forward by social conflict is the normal state of affairs. Static periods are merely temporary way stations along the road. Conflict theorists believe social order results from dominant groups making sure that subordinate groups are loyal to the institutions that are the dominant groups' sources of wealth, power, and prestige. The dominant groups will use coercion, constraint, and even force to help control those people who are not voluntarily loyal to the laws and rules they have made. When this order cannot be maintained and the subordinate groups rebel, change comes about.

Conflict theorists are concerned with the issue of who benefits from particular social arrangements and how those in power maintain their positions and continue to reap benefits from them. The ruling class is seen as a group that spreads certain values, beliefs, and social arrangements to enhance its power and wealth. The social order then reflects the outcome

Table I–I

Major Theoretical Perspectives in Sociology

Perspective	Scope of Analysis	Point of View	Focus of Analysis
Structural-Functional	Macro level	The various parts of society are interdependent and functionally related.	The functional and dysfunctional aspects of institutions and society.
		Social systems are highly stable.	
		Social life is governed by consensus and cooperation.	
Social Conflict	Macro level	Society is a system of accommodations among competing interest groups.	How social inequalities produce conflict.
		Social systems are unstable and are likely to change rapidly.	Who benefits from particular social arrangements.
		Social life involves conflict because of differing goals.	
Interactionist	Micro level	Most of what people do has meaning beyond the concrete act.	How people make sense of the world in which they participate.
		The meanings that people place on their own and one another's behavior can vary.	

of a struggle among those with unequal power and resources.

Conflict perspectives are often criticized as concentrating too much on conflict and change and too little on what produces stability in society.

They also are criticized for being too ideologically based and making little use of research methods or objective statistical evidence. The conflict theorists counter that the complexities of modern social life cannot be reduced to statistical analysis, and that doing so has caused sociologists to become detached from their object of study and removed from the real causes of human problems.

Both functionalist and conflict theories are descriptive and predictive of social life. Each has its strengths and weaknesses, and each emphasizes an important aspect of society and social life.

The Interactionist Perspective

Functionalism and conflict theory can be thought of as the opposite sides of the same coin. Although quite different from one another, they share certain similarities. Both approaches focus on major structural features of entire societies and attempt to give us an understanding of how societies survive and change. Social life, however, also occurs on an intimate scale between individuals. The **interactionist perspective** *focuses on how individuals make sense of—or interpret—the social world in which they participate.* As such, this approach is primarily concerned with human behavior on a person-to-person level. Interactionists criticize functionalists and conflict theorists for implicitly assuming that social processes and social institutions somehow have a life of their own apart from the participants.

Interactionists remind us that the educational system, the family, the political system, and indeed all of society's institutions are ultimately created, maintained, and changed by people interacting with one another.

The interactionist perspective includes a number of loosely linked approaches. George Herbert Mead devised a *symbolic interactionist* approach that focuses on signs, gestures, shared rules, and written and spoken language. Harold Garfinkel used *ethnomethodology* to show how people create and share their understandings of social life. Erving Goffman took a *dramaturgical* approach in which he saw social life as a form of theater. (We will discuss ethnomethodology and dramaturgy in Chapter 5.) Of these three approaches, the symbolic interactionist approach has received the widest attention and presents us with a well-formulated theory. (Table 1–1 compares the functionalist, conflict theory, and interactionist approaches with sociology.)

Symbolic Interactionism As developed by George Herbert Mead (1863–1931), **symbolic interactionism** is *concerned with the meanings that people place on their own and one another's behavior.*

Human beings are unique in that most of what they do with one another has meaning beyond the concrete act. According to Mead, people do not act or react automatically, but carefully consider and even rehearse what they are going to do. They take into account the other people involved and the situation in which they find themselves. The expectations and reactions of other people greatly affect each individual's actions. In addition, people give things meaning and act or react on the basis of these meanings. For example, when the flag of the United States is raised, people stand because they see the flag as representing their country.

Because most human activity takes place in social situations—in the presence of other people—we must fit what we as individuals do with what other people in the same situation are doing. We go about our lives with the assumption that most people share our definitions of basic social situations. This agreement on definitions and meanings is the key to human interactions in general, according to symbolic interactionists. For example, a staff nurse in a mental hospital unlocking a door for an inpatient is doing more than simply enabling the patient to pass from one ward to another. He or she also is communicating a position of social dominance over the patient (within the hospital) and is carrying a powerful symbol of that dominance—the key. The same holds true for a professor writing on a blackboard or a company vice-president dictating to a secretary.

Such interactions, therefore, although they appear to be simple social actions, also are laden with highly symbolic social meanings. These symbolic meanings are intimately connected with our understanding of what it is to be and to behave as a human being. This includes our sense of self; how we experience others and their views of us; the joys and pains we feel at home, at school, at work, and among friends and colleagues; and so on.

Symbolic interaction and its various offshoots have been criticized for paying too little attention to the larger elements of society. Interactionists respond that societies and institutions are made up of individuals who interact with one another and do not exist apart from these basic units. They believe that an understanding of the process of social interaction will lead to an understanding of the rest of society. In actual fact, interactionists still must bridge the gap between their studies of social interaction and those of the broader social structures. Nevertheless, symbolic interactionism does complement functionalism and conflict theory in important ways and gives us important insights into how people interact.

Contemporary Sociology

Contemporary sociological theory continues to build on the original ideas proposed in the interactionist perspective, functionalism, and conflict theory. It would be difficult to see contemporary sociological theory as either conflict theory or functionalism in the original sense. Much of it has been modified to include important aspects of each theory. Even symbolic interactionism has not been wholeheartedly embraced and aspects of it have instead been absorbed into general sociological writing.

Very little contemporary sociological theory still can be identified as true functionalism. Part of this is due to the fact sociologists today have abandoned trying to develop all-inclusive theories and instead opt for what Merton (1968) referred to as middle-range theories. **Middle-range theories** *are concerned with explaining specific issues or aspects of society instead of trying to explain how all of society operates.* A middle-range theory might be one that explains why divorce rates rise and fall with certain economic conditions or how crime rates are related to residential patterns.

Modern conflict theory was initially refined by such sociologists as C. Wright Mills (1959), Ralf Dahrendorf (1958), Randall Collins (1975, 1979), and Lewis Coser (1956) to reflect the realities of contemporary society. Mills and Dahrendorf did not see conflict as confined to class struggle. Rather, they viewed it as applicable to the inevitable tensions that arise between groups: parents and children, producers and consumers, professionals and their clients, unions and employers, the poor and the materially comfortable, and minority and majority ethnic groups. Members of these groups have both overlapping and competing interests, and their shared needs keep all parties locked together within one society. At the same time, the groups actively pursue their own ends, thus constantly pushing the society to change in order to accommodate them.

Coser incorporated aspects of both functionalism and conflict theory, seeing conflict as an inevitable element of all societies and as both functional and dysfunctional for society. Conflict between two groups tends to increase their internal cohesion. For example, competition between two divisions of two computer companies to be the first to produce a new product may draw the members closer to one another as they strive to reach the desired goal. This feeling might not have occurred had it not been for the sense of competition, as the conflict itself becomes a form of social interaction.

Conflict also could lead to cohesion by causing two or more groups to form alliances against a common enemy. For example, a political contest may cause several groups to unite in order to defeat a common opponent.

During the past 35 years, conflict theory has been influenced by a generation of neo-Marxists. These people have helped produce a more complex and sophisticated version of conflict theory that goes beyond the original emphasis on class conflict and instead shows that conflict exists within almost every aspect of society (Gouldner, 1970, 1980; Skocpol, 1979; Wallerstein, 1974, 1979, 1980, 1991; Tilly, 1978, 1981; Starr, 1982, 1992).

Theory and Practice

Sociological theory gives meaning to sociological practice. Merely assembling countless descriptions of social facts is inadequate for understanding society as a whole. Only when data are collected within the conceptual framework of a theory—in order to answer the specific questions growing out of that theory—is it possible to draw conclusions and make valid generalizations. This pursuit is the ultimate purpose of all science.

Theory without practice (research to test it) is at best poor philosophy and at worst unscientific, and practice uninformed by theory is at best trivial and at worst a tremendous waste of time and resources. Therefore, in the next chapter we shall move from theory to practice—to the methods and techniques of social research.

SUMMARY

- A great deal of social issues information comes from sources that have an interest in getting people to support a particular point of view.
- Sociology, by contrast, is the scientific study of human society and social interactions.
- Sociology seeks an accurate and scientific understanding of society and social life.
- The main focus of sociology is the group and not the individual.
- A sociologist tries to understand the forces that operate throughout the society—forces that mold individuals, shape their behavior, and thus determine social events.
- The social sciences consist of all those disciplines that apply scientific methods to the study of human behavior. Though there is some overlap, each of the social sciences has its own area of investigation.
- Cultural anthropology, psychology, economics, history, political science, and social work all have some things in common with sociology, but each

has its own distinct focus, objectives, theories, and methods.
- Sociology emerged as a separate field of study in Europe during the nineteenth century. It was a time of turmoil and a period of rapid and dramatic social change. Industrialization, political revolution, urbanization, and the growth of a market economy undermined traditional ways of doing things.
- The need for a systematic analysis of society coupled with the acceptance of the scientific method resulted in the emergence of sociology.
- In the United States, sociology developed in the early twentieth century. Its early growth took place at the University of Chicago, where the first graduate department of sociology in the United States was founded in 1890.
- The so-called Chicago school of sociology focused on the study of urban neighborhoods and ethnic areas, and included many of America's leading sociologists of the period.
- Scientists need a set of working assumptions to guide them in their professional activities.
- These models or frameworks for questions that generate and guide research are known as paradigms.
- Sociologists have developed several paradigms to help them investigate social processes.
- Functionalism views society as a system of highly interrelated structures that function or operate together harmoniously. Functionalists analyze society by asking what each part contributes to the smooth functioning of the whole. From the functionalist perspective, society appears quite stable and self-regulating. Critics have attacked the conservative bias inherent in this assumption.
- Conflict theory sees society as constantly changing in response to social inequality and social conflict. For these theorists, social conflict is the normal state of affairs. Social order is maintained by coercion. Conflict theorists are concerned with the issue of who benefits from particular social arrangements and how those in power maintain their positions.
- The interactionist perspective focuses on how individuals make sense of, or interpret, the social world in which they participate. This perspective consists of a number of loosely linked approaches.
- Contemporary sociology has built on and modified the insights of these three theoretical perspectives.

Media Resources

The Companion Website for *Introduction to Sociology*, Ninth Edition

http://sociology.wadsworth.com/tischler9e

Supplement your review of this chapter by going to the companion website to take one of the Tutorial Quizzes, use the flash cards to master key terms, and check out the many other study aids you will find there. You will also find special features such as Wadsworth's Sociology Online Resources and Writing Companion, GSS data, and Census 2000 information at your fingertips to help you complete that special project or do some research on your own.

CHAPTER ONE STUDY GUIDE

KEY CONCEPTS AND THINKERS

Match each concept with its definition, illustration, or explanation presented below.

a. Scientific method
b. Middle-range theory
c. Empiricism
d. Paradigms
e. Functionalism
f. Conflict theory

g. Interactionist perspective
h. Macro level
i. Social solidarity
j. Altruistic suicide
k. Manifest function

l. Latent function
m. Symbolic interactionism
n. Anomic suicide
o. Sociological imagination
p. Egoistic suicide

_m_ 1. The branch of sociology that studies how people assign meaning to people's behavior, including their own.

_i_ 2. The degree to which people are bonded to groups and to the society as a whole.

_h_ 3. Large-scale social phenomena such as culture, class systems, population shifts, and so on.

_c_ 4. The view that generalizations are valid only if they rely on evidence that can be observed directly or verified through our senses.

_o_ 5. The ability to see the link between personal experiences and social forces.

_f_ 6. The paradigm that emphasizes the conflict between different sectors of a society, and how groups use resources to secure their own particular interests.

_d_ 7. General views of the world that determine the questions to be asked and the important things to look at in answering them.

_g_ 8. The paradigm that focuses on how people interpret and attempt to influence the social world.

_k_ 9. Intended outcomes of an institution.

_a_ 10. A process by which a body of scientific knowledge is built through observation, experimentation, generalization, and verification.

_b_ 11. Explanations that focus on specific issues rather than society as a whole.

_e_ 12. The paradigm that emphasizes how elements of a society do (or do not) work toward accomplishing necessary functions.

_n_ 13. Suicide caused by feelings of normlessness and confusion, the feeling that the rules of the game no longer make sense.

_l_ 14. Unintended, unrecognized, but often useful consequence of an institution.

_j_ 15. Suicide that results from the willingness to sacrifice one's own life for the good of the social group.

_p_ 16. Suicide related to lack of involvement with others.

Match the thinkers with their main idea or contribution.

a. Auguste Comte
b. Harriet Martineau
c. C. Wright Mills

d. Herbert Spencer
e. Émile Durkheim
f. Karl Marx

g. Max Weber
h. Jane Addams
i. W. E. B. DuBois

_d_ 1. Saw society as an organism; applied Darwin's idea of "survival of the fittest" to explain and justify social conditions of different individuals and groups.

_i_ 2. African American sociologist, early twentieth-century; militant opponent of racism and keen observer of its effects (*The Souls of Black Folk*).

_c_ 3. American sociologist; developed concept of the sociological imagination.

_a_ 4. Coined the term *sociology;* emphasized empiricism; thought society was evolving toward perfection.

_b_ 5. Wrote observations of institutions (prisons, factories, and so on); compared American and European class systems.

_e_ 6. Emphasized social solidarity; studied *rates* of behavior in groups rather than individual behavior.

_h_ 7. American social reformer; founded Hull House, a settlement house for immigrants in Chicago.

f **8.** Viewed social change as resulting from the conflicts between social classes trying to secure their interests. Thought that eventually the workers would overthrow the capitalist-run system.

q **9.** Thought power, wealth, and status were separate aspects of social class. Saw bureaucratization as a dominant trend with far-reaching social consequences. Contradicted Marx in arguing that religious ideas influenced economics, specifically that Protestantism brought the rise of capitalism.

CENTRAL IDEA COMPLETIONS

Following the instructions, fill in the appropriate concepts and descriptions for each of the questions posed in the following section.

1. Recently, child abduction has been a subject of public discussion and legislation. (a) How does using the *sociological imagination* help us better understand the phenomenon of child abductions? (b) What kinds of questions would a sociologist ask when studying a social problem with very serious social consequences?

    a. _____

    b. _____

2. Tischler cites data on suicide among college students. From your own knowledge, identify factors that might promote and inhibit each of the types of suicide in Durkheim's typology.

    a. Egoistic suicide _____

    b. Altruistic suicide _____

    c. Anomic suicide _____

3. Tischler says that some people who use sociological data and ideas "have no interest in putting forth objective information." Imagine that there has been a suicide on campus. How might the interests of the following affect the way they treat information?

    a. The local newspaper _____

    b. The school administration _____

4. What questions and research strategies might each of the major sociological paradigms use in looking at the issue of domestic violence?

    a. Structural-Functionalist _____

    b. Conflict Theory _____

    c. Interactionist Perspective _____

5. Think of some institution or organization—a university or elementary school, a court, a church—and list its manifest functions and latent functions.

    a. Manifest functions _____

    b. Latent functions _____

CRITICAL THOUGHT EXERCISES

1. Women (white, 40–44 years old) with advanced education are much more likely to be married today than were their counterparts in 1980. High school dropouts (white, 40–44 years old) of both sexes are less likely to be married than were their counterparts in 1980. What might account for these changes?

2. In debates over current political and social issues, which arguments and ideas are most compatible with social Darwinism? Which current ideas are most at odds with social Darwinism?

3. Reread Tischler's insert on sociology as a career. Using the occupational handbooks located in your library as well as fliers from your campus employment service, conduct a preliminary career inventory of specific occupations for which a major in sociology would be helpful.

INTERNET ACTIVITIES

1. The American Sociological Association (ASA) website has a section for students at all levels (http://www.asanet.org/apap/student.html). The ASA website also has the ASA booklet on careers in sociology (http://www.asanet.org/student/career/homepage.html).

2. If the sociological theorists seem too imposing, take a look at the Dead Sociologists' Society (http://www2.pfeiffer.edu/~lridener/dss/deadsoc.html). Larry Ridener founded the site after seeing the 1989 Robin Williams movie *Dead Poets Society*. As the Williams character tried to get his students interested in poetry, Ridener wanted his students interested in sociological ideas. The site has excellent links on a wide variety of sociology topics.

3. Look at suicide rates among the different states in the United States (http://www.suicidology.org/associations/1045/files/2002statedatapg.pdf). Can you think of social factors that can account for these differences?

4. Where did all these Madisons come from, and what happened to Mildred? Sociology shows us that decisions that seem highly personal and individual (like the decision to commit suicide) fall into patterns. Another such decision is what to name a baby. You can see these patterns and check your own name at the U.S. Census website (http://www.ssa.gov/OACT/babynames/).

ANSWERS TO KEY CONCEPTS

1.m 2.i 3.h 4.c 5.o 6.f 7.d 8.g 9.k 10.a 11.b 12.e 13.n 14.l 15.j 16.p

ANSWERS TO KEY THINKERS

1.d 2.i 3.c 4.a 5.b 6.e 7.h 8.f 9.g

ThomsonNOW™

Reviewing is as easy as ❶ ❷ ❸

1. Before you do your final exam, take the ThomsonNOW diagnostic quiz to help you identify the areas on which you should concentrate. You will find information on ThomsonNOW and instructions on how to access all of its great resources on the foldout at the beginning of the text.
2. As you review, take advantage of ThomsonNOW's study videos and interactive Map the Stats exercises to help you master the chapter topics.
3. When you are finished with your review, take ThomsonNOW's posttest to confirm you are ready to move on to the next chapter.

2

Doing Sociology: Research Methods

© Taxi/Getty Images

Learning Objectives

After studying this chapter, you should be able to do the following:

- Explain the steps in the sociological research process.
- Analyze the strengths and weaknesses of the various research designs.
- Know what independent and dependent variables are.
- Know what sampling is and how to create a representative sample.
- Recognize researcher bias and how it can invalidate a study.
- Explain the strengths and weaknesses of the various measures of central tendency.
- Read and understand the contents of a table.
- Explain the concepts of reliability and validity.
- Understand the problems of objectivity and ethical issues that arise in sociological research.

On nearly a daily basis we see and hear news reports of social trends. For example, let us consider the problem of schoolyard bullying. In 2001 the prestigious *Journal of the American Medical Association* (JAMA) reported on a study of students in grades 6 through 10. The major finding in the study was that nearly 30% of the students "reported moderate or frequent involvement in bullying." The authors concluded that "bullying is a serious problem for U.S. youth," and that "the prevalence of bullying observed in this study suggests the importance of preventive intervention research targeting bullying behaviors." The information was sent out in a press release to hundreds of media outlets, who then reported it to the public.

Why would people be interested in the problem of bullying? Some of the students who have taken weapons to schools and shot and killed fellow classmates have claimed they were victims of bullying. So it might seem that we could reduce school shooting rampages if we could reduce bullying, but we do not have any evidence to prove the connection.

What exactly is bullying? According to the JAMA researchers "a student is being bullied when another student, or group of students, say or do nasty and unpleasant things to him or her." When a student is teased repeatedly it was also considered bullying.

Two students fighting or quarreling would not count as an example of bullying.

The crucial part of the study revolved around how frequently the students were bullied or bullied others. The researchers decided that bullying incidents that occurred at least weekly were defined as "frequent." Those that took place sometimes were defined as "moderate." Based on these definitions the authors came up with the fact that 30% of the students reported frequent or moderate involvement in bullying.

A closer look at the study shows that the authors group those who were bullied with those who did the bullying. If they had only reported on the victims of bullying the figure would have been 17%. Second, the authors also grouped frequent and moderate bullying together. It turns out that only 8% of students were victims of frequent bullying. In addition, some of these students were victims of bullying outside of school (Best, 2002).

When the media reported on this study in a prestigious journal they thought they were making the public aware of a troubling social issue. The media do not have the time or ability to carefully evaluate the validity of every press release and whether the findings are correct. As you will see in this chapter, the research process involves a number of specific steps that must be followed to produce a valid study. Only when this is done faithfully can we have any confidence in the results of the study. In this chapter we shall examine some of the methods used by scientists in general—and sociologists in particular—to collect data to test their ideas.

The Research Process

How should you conduct a research study? After reading Chapter 1, you know that you should not approach a study and draw conclusions on the basis of your personal experience and perceptions; rather, you should approach the study scientifically.

To approach a study scientifically, you should keep in mind that science has two main goals: (1) to describe in detail particular things or events and (2) to propose and test theories that help us understand these things or events.

There is a great deal of similarity between what a detective does in attempting to solve a crime and what a sociologist does in answering a research problem. In the course of their work, both detectives and sociologists must gather and analyze information.

For detectives, the object is to identify and locate criminals and collect enough evidence to ensure their identification is correct. Sociologists, on the other hand, develop hypotheses, collect data, and develop theories to help them understand social behavior. Although their specific goals differ, both sociologists and detectives try to answer two general questions: Why did it happen, and under what circumstances is it likely to happen again? That is, sociologists seek to explain and predict.

All research problems require their own special emphasis and approach. The research procedure is usually custom-tailored to the research problem.

Nonetheless, there is a sequence of steps called the research process that is followed when designing a research project. In short, the **research process** *involves defining the problem, reviewing previous research on the topic, developing one or more hypotheses, determining the research design, defining the sample and collecting data, analyzing and interpreting the data, and finally preparing the research report.* The sequence of steps in this process and the typical questions asked at each step are illustrated in Table 2–1. If there are any terms in this table that you are not familiar with, do not become concerned. We will define them as we examine each of the various steps.

Define the Problem

"Love leads to marriage." Suppose you were given this statement as a subject for sociological research. How would you proceed to gather data to prove or disprove it? You must begin by defining love, a task

that William Shakespeare himself tried to do in his play *Twelfth Night* when he asked, "What is love?" We would know that we have a problem here, as to this day people are still grappling with the question "How do you know when you are in love?"

Concepts of love vary over time and from one culture to another. Sharon Brehm (1992) noted some of the views of love that have been put forth:

1. Love is insanity.
2. Love is not possible in marriage.
3. Love happens only between people of the same sex.
4. Love should not involve sexual contact.
5. Love is a game.
6. Love is a noble quest.
7. Love is doomed.
8. Love leads to happiness.
9. Love and marriage go together.

We could try a different approach and define love by using a definition that a researcher used in another study. For example, Hatfield (1988) defined love as "a state of intense longing for union with another." You would now have to find some way of determining whether this condition exists. You also must decide whether both people have to be in love in order for marriage to take place. You may already notice that it may be difficult to achieve the level of precision necessary for a useful research project.

Once you accurately define your terms and provide details to clarify your descriptions, you can begin to test the statement we proposed. Even after arriving at a careful definition of your terms and a detailed description of love, you may still have trouble answering the question empirically.

An **empirical question** *can be answered by observing and analyzing the world as it is known.* Examples: How many students in this class have an A average? How many millionaires are there in the United States? Scientists pose empirical questions to collect information, to add to what is already known, and to test hypotheses. To turn the statement about love into an empirical question, you must ask: How do we measure the existence of love?

In trying to define and measure love, one researcher (Rubin, 1970, 1973) used an interesting approach. He prepared a large number of self-descriptive statements that considered various aspects of loving relationships as mentioned by writers, philosophers, and social scientists. After administering these statements to a variety of subjects, he was able to isolate nine items that best reflected feelings of love for another. Three of these items are cited in the following paragraph. In each sentence, the person is to fill in the blank with the name of a particular person and indicate the degree to which the item describes the relationship.

The following statements reflect three components of love. The first is attachment-dependency: "If I

Table 2–1

The Research Process

Steps in the Process	Typical Questions
Define the problem	What is the purpose of the study? What information is needed? How can we operationalize the terms? How will the information be used?
Review previous research	What studies have already been done on this topic? Do we need additional information before we begin? From what perspective should we approach this issue?
Develop one or more hypotheses	What are the independent and dependent variables? What is the relationship among the variables? What types of questions do we need to answer?
Determine the research design	Can we use existing data? What will we measure or observe? What research methods should we use?
Define the sample and collect data	Are we interested in a specific population? How large should the sample be? Who will gather the data? How long will it take?
Analyze the data and draw conclusions	What statistical techniques will we use? Have our hypotheses been proved or disproved? Is our information valid and reliable? What are the implications of our study?
Prepare the research report	Who will read the report? What is their level of familiarity with the subject? How should we structure the report?

were lonely, my first thought would be to seek (blank) out." The second component is caring: "If (blank) were feeling badly, my first duty would be to cheer (him or her) up." The final component is intimacy: "I feel that I can confide in (blank) about virtually everything." These three statements show the strong aspect of mutuality in love relationships.

Using Rubin's scale, you can begin to make some headway toward clarifying an important component of your research problem. In the language of science you have operationalized your definition of love. An **operational definition** is *a definition of an abstract concept in terms of the observable features that describe the thing being investigated.* Attachment-dependency, caring, and intimacy can be three features of an operational definition of love and can indicate the presence of love in a research study.

Review Previous Research

Which questions are the "right" questions? Although there are no inherently correct questions, some are better suited to investigation than are others. To decide what to ask, researchers must first learn as much as possible about the subject. We would want to familiarize ourselves with as many of the previous

studies on the topic as possible, particularly those closely related to what we want to do. By knowing as much as possible about previous research, we avoid duplicating a previous study and are able to build on contributions others have made to our understanding of the topic.

After reviewing the research we might find out that the early anthropologist Ralph Linton thought love was a form of insanity and assuming that it should lead to marriage was absurd. As he noted:

> All societies recognize that there are occasional violent emotional attachments between persons of the opposite sex, but our present American culture is practically the only one which has attempted to capitalize these and make them the basis for marriage. Most groups regard them as unfortunate and point out the victims of such attachments as horrible examples. . . . The percentage of individuals with a capacity for romantic love of the Hollywood type [is] about as large as that of persons able to throw genuine epileptic fits. (Linton, 1936)

Needless to say, Linton would not have thought much of our potential research project.

If you were doing a study of whether love leads to marriage, you would find that it would be quite difficult to define what love means.

Develop One or More Hypotheses

Our original statement "Love leads to marriage" is presented in the form of a hypothesis. A **hypothesis** is *a testable statement about the relationships between two or more empirical variables*. A **variable** is *anything that can change (vary)*. The number of highway deaths on Labor Day weekends, the number of divorces that occur each year in the United States, the amount of energy the average American family consumes in the course of a year, the daily temperature in Dallas, the number of marathoners in Boston or in Knoxville, Tennessee—all these are variables. The following are not variables: the distance from Los Angeles to Las Vegas, the altitude of Denver, or the number of marriages in Ohio in 2006. These are fixed, unchangeable facts.

As we review the previous research on the topic of love, we find we can develop additional hypotheses that help us investigate the issue further. For example, our reading might show that a common stereotype people hold is the notion that women are more romantic than men. After all, it appears that women enjoy movies about love and romantic novels more than men do.

But wait a minute. We may begin to suspect that common stereotypes may be all wrong, that they are related to traditional gender-role models. We note that in most traditional societies, the male is the breadwinner, while the female is dependent on him for economic support, status, and financial security. Therefore it would seem that when a man marries, he chooses a companion and perhaps a helpmate, whereas a woman chooses a companion as well as a standard of living. This leads us to hypothesize that in traditional societies men are more likely to marry for love, but women are more likely to marry for economic security.

There is support for this hypothesis. One study designed a scale to measure belief in a romantic ideal in marriage. Males were more likely than females to agree with such statements as "A person should marry whomever he loves regardless of social position" and "As long as they love one another, two people should have no difficulty getting along together in marriage." Men were more likely to disagree than women with the statement "Economic security should be carefully considered before selecting a marriage partner" (Rubin, 1973).

Contrary to popular opinion, men also tend to be more romantic than women (Sprecher & Metts, 1989). This fact should not be that hard to understand when we see that historically men, with their control over resources, have had the luxury to be romantic. Women, who often have not been in charge of their economic destiny, have had to think of men as providers more so than as lovers. We could hypothesize that as gender-role stereotyping declines in the United States and as more and more families come to depend on the income of both spouses, one of two things could happen: Either the importance of romantic love as a basis for marriage will begin to fade, or it will become stronger as the couple now comes together on the basis of mutual attraction as opposed to economic considerations.

Hypotheses involve statements of causality or association. A **statement of causality** says that *something brings about, influences, or changes something else.* "Love between a man and a woman always produces marriage" is a statement of causality.

A **statement of association,** on the other hand, says that *changes in one thing are related to changes in another but that one does not necessarily cause the other.* Therefore, if we propose that "the greater the love relationship between a man and a woman, then the more likely it is they will marry," we are making a statement of association. We are noting a connection between love and marriage, but also that one does not necessarily cause the other.

Often hypotheses propose relationships between two different kinds of variables. An **independent variable** *causes or changes another variable.* A **dependent variable** *is influenced by the independent variable.* For example, we might propose the following hypothesis: Men who live in cities are more likely to marry young than are men who live in the

In one study, males were more likely than females to agree with the statement "A person should marry whomever he or she loves regardless of the person's social position."

country. In this hypothesis the independent variable is the location: Some men live in the city, some live in the country, but presumably their choice of where to live is not influenced by whether they marry young. The age of marriage is the dependent variable because it is possible that the age of marriage depends on where the men live. If research shows that the age of marriage (a dependent variable) is indeed younger among urban men than among rural men, the hypothesis probably is correct.

If there is no difference in the age of marriage among urban and rural men—or if it is earlier among rural men—then the hypothesis is not supported by the data. Keep in mind that proving a hypothesis false can be scientifically useful: It eliminates unproductive avenues of thought and suggests other, more productive approaches to understanding a problem.

Even if research shows that a hypothesis is correct, it does not mean the independent variable necessarily produces or causes the dependent variable.

For example, if it turns out that we can show that love leads to marriage, we still may not know why. In principle, at least, it is possible to be in love without getting married. However, we still do not know what causes people to take the next step.

Determine the Research Design

Once we have developed our hypotheses, we must design a project in which they can be tested. This is a difficult task that frequently causes researchers a great deal of trouble. If a research design is faulty, it may be impossible to conclude whether the hypotheses are true or false, and the whole project will have been a waste of time, resources, and effort.

A research design must provide for the collection of all necessary and sufficient data to test the stated hypotheses. The important word here is *test*. The researcher must not try to prove a point; rather, the goal is to test the validity of the hypotheses. Although it is important to gather as much information as needed, research designs must guard against the collection of unnecessary information, which can lead to a waste of time and money.

When we design our research project we must also decide which of several research approaches to use. There are four main methods of research used by sociologists: surveys, participant observation, experiments, and secondary analysis. Each has advantages and limitations. Therefore, the choice of methods depends on the questions the researcher hopes to answer.

Surveys   A **survey** is *a research method in which a population, or a portion thereof, is questioned in order to reveal specific facts about itself.* Surveys are used to discover the distribution and interrelationship of certain variables among large numbers of people.

The largest survey in the United States takes place every 10 years when the government takes its census. Many of you may have participated in the 2000 census. The U.S. Constitution requires this census to determine the apportionment of members to the House of Representatives. In theory, at least, a representative of every family and every unmarried adult responds to a series of questions about his or her circumstances.

From these answers it is possible to construct a picture of the social and economic facts that characterize the American public at one point in time. *Such a study, which cuts across a population at a given time,* is called a **cross-sectional study.** Surveys, by their nature, usually are cross-sectional. If the same population is surveyed two or more times at certain intervals, a comparison of cross-sectional research can give a picture of changes in variables over time. *Research that investigates a population over a period of time is called* **longitudinal research.**

Survey research usually deals with large numbers of subjects in a relatively short time. One of the shortcomings resulting from this method is that investigators are not able to capture the full richness of feelings, attitudes, and motives underlying people's responses. Some surveys are designed to gather this kind of information through interviewing. An **interview** consists of *a conversation between two (or occasionally more) individuals in which one party attempts to gain information from the other(s) by asking a series of questions.*

It would, of course, be ideal to gather exactly the same kinds of information from each research subject. One way researchers attempt to achieve this is through interviews in which all questions are carefully worked out to get at precisely the information wanted (What is your income? How many years of schooling have you had?). Sometimes research participants are forced to choose among a limited number of responses to questions (as in multiple-choice tests). This process results in very uniform data easily subjected to statistical analysis.

A research interview entirely predetermined by a questionnaire (or so-called interview schedule) that is followed rigidly is called a **structured interview.**

Structured interviews tend to produce uniform or replicable data that can be elicited time after time by different interviewers.

The use of this method, however, also may allow useful information to slip into "cracks" between the predetermined questions. For example, a questionnaire being administered to married individuals might ask about their age, family background, and what role

One of the shortcomings of survey research is that investigators are not able to capture the full richness of feelings, attitudes, and motives underlying people's responses.

love played in their reasons for getting married. If, however, we do not ask about social class or ethnicity, we may not find out that these characteristics are very important for our study. If such questions are not built into the questionnaire from the beginning, it is impossible to recover this lost information later in the process when its importance may become apparent.

One technique that can prevent this kind of information loss is the **semistructured,** *or* **open-ended interview,** *in which the investigator asks a list of questions but is free to vary them or even to make up new questions on topics that take on importance in the course of the interview.* This means that each interview will cover those topics important to the research project but, in addition, will yield additional data somewhat different for each subject. Analyzing such diverse and complex data is difficult, but the results are often rewarding.

Interviewing, although it may produce valuable information, is a complex, time-consuming art. Some research studies try to get similar information by distributing questionnaires directly to the respondents and asking them to complete and return them. This is the way the federal government obtains much of its census data. Although it is perhaps the least expensive way of doing social research, it is often difficult to assess the quality of data obtained in this manner. For example, people may not answer honestly or seriously for a variety of reasons: They may not understand the questions, they may fear the information will be used against them, and so on. But even data gained from personal interviews may be unreliable. In one study, student interviewers were embarrassed to ask preassigned questions on sexual habits, so they left these questions out of the interviews and filled the answers in themselves afterward. In another study, follow-up research found participants had consistently lied to interviewers.

Participant Observation *Researchers entering into a group's activities and observing the group members* are engaged in **participant observation.** Unlike sociologists employing survey research, participant observers do not try to make sure they are studying a carefully chosen sample. Rather, they attempt to get to know all members of the group being studied to whatever degree possible.

This research method is generally used to study relatively small groups over an extended period. The goal is to obtain a detailed portrait of the group's day-to-day activities, to observe individual and group behavior, and to interview selected informants.

Participant observation depends for its success on the relationship that develops between the researchers and research participants. The closer and more trusting the relationship, the more information will be revealed to the researcher—especially the kind of personal information often crucial for successful research.

One of the first and most famous studies employing the technique of participant observation was a study of Cornerville, a lower-class Italian neighborhood in Boston. William Foote Whyte moved into the neighborhood and lived for three years with an Italian family. He published his results in a book called *Street Corner Society* (1943).

All the information for the book came from his field notes, which described the behavior and attitudes of the people whom he came to know.

Nearly two decades after Whyte's study, Herbert Gans conducted a participant observation study, published as *The Urban Villagers* (1962), of another Italian neighborhood in Boston. The picture Gans drew of the West End was broader than Whyte's study of Cornerville. Gans included descriptions of the family, work experience, education, medical care, relationships with social workers, and other aspects of life in the West End. Although he covered a wider range of activities than Whyte, his observations were not as detailed.

© Antonio Mari

In participant observation, the researcher tries to know personally as many members of the group in question as possible.

On rare occasions, participant observers hide their identities while doing research and join groups under false pretenses. Leon Festinger and his students hid their identities when they studied a religious group preaching the end of the world and the arrival of flying saucers to save the righteous, a group with beliefs similar to the Heaven's Gate group of 1997 (Festinger, Rieken, & Schacter, 1956). However, most sociologists consider this deception unethical. They believe it is better for participant observers to be honest about their intentions and work together with their subjects to create a mutually satisfactory situation. By declaring their positions at the outset, sociologists can then ask appropriate questions, take notes, and carry out research tasks without encountering unnecessary and unethical risks to their study.

Participant observation is a highly subjective research approach. In fact, some scholars reject it outright because the results often cannot be duplicated by another researcher. This method, however, has the benefit of revealing the social life of a group in far more depth and detail than surveys or interviews alone. The participant observer who is able to establish good rapport with the subjects is likely to uncover information that would never be revealed to a survey taker.

The participant observer is in a difficult position, however. He or she will be torn between the need to become trusted (therefore emotionally involved in the group's life) and the need to remain a somewhat detached observer striving for scientific objectivity.

Experiments   The most precise research method available to sociologists is the controlled **experiment,** *an investigation in which the variables being studied are controlled and the researcher obtains the results through precise observation and measurement.*

Because of their precision, experiments are an attractive means of doing research. Experiments have been used to study patterns of interaction in small groups under a variety of conditions such as stress, fatigue, or limited access to information.

Although experimentation is appropriate for small-group research, most of the issues that interest sociologists cannot be investigated in totally controlled situations. Social events usually cannot be studied in controlled experiments because they simply cannot be controlled. For these reasons, experiments remain the least used research method in sociology.

Secondary Analysis   **Secondary analysis** is *the process of making use of data that has been collected by others.* Often the original investigator gathered the data for a specific study. Other times it was merely collected as part of the process of keeping records. The researcher engaged in secondary analysis may use this same data for a new study and a very different purpose. For example, Émile Durkheim, in his classic study of suicide in France in the 1890s, engaged in a secondary analysis of official records and developed his theories based on that research.

The enormous amount of material the federal government collects is often used for secondary analysis. The U.S. Bureau of the Census has data on income, birthrates, migration, marriage, divorce, and education levels in the United States that are invaluable for doing social research. Other agencies that provide data that sociologists use for secondary analysis include the Federal Bureau of Investigation, the Department of Labor, the National Center for Health Statistics, and many others.

An advantage of secondary analysis is that it is useful for collecting or analyzing historical and longitudinal data. It also saves the time and money involved in doing a new study. There are some disadvantages, however. The data may be flawed. You may not know if the original researchers had some biases that are present in the data. Or possibly they were not qualified or knowledgeable enough to collect the data. In addition, the data may not really be suitable for your current study. If you are trying to do a study of economic well-being at different points in history, you may decide to gather that

information from certain questions on the U.S. Census of 1950, 1970, and 2000. But which questions should you use? Should you use those on income, those on poverty, those on net worth, those on satisfaction with the political situation, or some other questions? Different types of questions may produce different results and your study may not turn out to be valid because of the choices you make.

Therefore the use of secondary analysis requires a thorough understanding of the research process and the problems that can arise from a poorly conceived study. (Table 2–2 compares the advantages and disadvantages of the various research methods.)

Define the Sample and Collect Data

After determining how the needed information will be collected, the researchers must decide what group will be observed or questioned. Depending on the study, this group might be college students, Texans, or baseball players. The particular subset of the population chosen for study is known as a **sample.**

Sampling is *a research technique through which investigators study a manageable number of people, known as the* sample, *selected from a larger population or group.* If the procedures are carried out correctly, the sample can be called a **representative sample,** or *one that shows, in equivalent proportion, the significant variables that characterize the population as a whole.* In other words, the sample will be representative of the larger population, and the findings from the research will tell us something about the larger group.

The failure to achieve a representative sample is called **sampling error.** Suppose you wanted to sample the attitudes of the American public on some issue such as military spending or federal aid for abortions. You could not limit your sample to only New Yorkers or Republicans or Catholics or African Americans or home owners. These groups do not represent the nation as a whole, and any findings you came up with would contain a sampling error.

How do researchers make sure their samples are representative? The basic technique is to use a **random sample**—*to select subjects so that each individual in the population has an equal chance of being chosen.* For example, if we wanted a random sample of all college students in the United States, we might choose every fifth or tenth or hundredth person from a comprehensive list of all registered college students in this country. Or we might assign each student a number and have a computer pick a sample randomly. However, there is a possibility that simply by chance, a small segment of the total college student population would fail to be represented adequately. This might happen with Native American students, for instance, who make up less than 1% of

Table 2–2

The Advantages and Disadvantages of Various Research Methods

Research Method	Advantages	Disadvantages
Social survey	Large numbers of people can be surveyed with questionnaires. Data can be quantified and comparisons among groups can be made. Measures can be taken at different points.	Respondents may give false information or responses they think the researcher wants to hear. Surveys do not leave room for answers that may not fit the standardized categories. Response rate may be low.
Participant observation	Allows people to be observed in their "natural" environments. Provides a more in-depth understanding of the people being studied. Hypotheses and theories can be developed and changed as the research progresses.	Findings are open to interpretation and subject to researcher bias. The researcher may have an unintended influence on the subjects. It is time-consuming. The results may be difficult to replicate.
Experiment	Variables can be isolated and controlled. A cause-and-effect relationship can be found. Easy to replicate.	The laboratory setting creates an artificial social environment. The study has to be limited to a few variables. Most things sociologists investigate cannot be studied in a laboratory.
Secondary analysis	Useful for collecting or analyzing historical and longitudinal data. Saves time and money involved in doing a new study.	The data may be flawed. Data may not be suitable for the current study.

college students in the United States. For some research purposes this might not matter, but if ethnicity is an important aspect of the research, it would be important to make sure Native American students were included.

The method used to prevent certain groups from being under- or overrepresented in a sample is to choose a **stratified random sample.** With this technique the population being studied is first divided into two or more groups (or strata) by characteristics such as age, sex, or ethnicity. A simple random sample is then taken within each group. Finally, the subsamples are combined (in proportion to their numbers in the population) to form a total sample. In our example of college students, the researcher would identify all ethnic groups represented among college students in the United States. Then the researchers would calculate the proportion of the total number of college students represented by each group and draw a random sample separately from each ethnic group. The number chosen from each group should be proportional to its representation in the entire college student population. The sample would still be random, but it would be stratified for ethnicity.

For a study to be accurate, it is crucial that a sample be chosen with care. The most famous example of sampling error occurred in 1936, when *Literary Digest* magazine incorrectly predicted that Alfred E.

Landon would win the presidential election. Using telephone directories and automobile registration lists to recruit subjects, *Literary Digest* pollsters sent out more than 10 million straw vote ballots and received 2.3 million completed responses. The survey gave the Republican candidate Landon 55% of the vote and Franklin D. Roosevelt only 41% (the remaining 4% went to a third candidate). Based on this poll, the *Digest* confidently predicted Landon's victory. Instead Landon has become known as the candidate who was buried in a landslide vote for Roosevelt (Squire, 1988).

How could this happen? Two major flaws in the sample accounted for the mistake. First, although the *Literary Digest* sample was large, it was not representative of the nation's voting population because it contained a major sampling error. During the Depression years, only the well-to-do could afford telephones and automobiles, and these people were likely to vote Republican.

The second problem with the study was the response rate. Interestingly, of those who claimed to have received a *Literary Digest* ballot, 55% claimed they would have voted for Roosevelt and 44% for Landon. If these people had actually voted in the poll, the *Digest* would have predicted the correct winner. As it turned out, there was a low response rate, and those who did respond were generally better educated,

A stratified random sample is used to prevent certain groups from being underrepresented or overrepresented in a sample.

wealthier people who could afford cars and telephones and who tended to be Landon supporters (Squire, 1988).

The outcome of the election was not entirely a surprise to everyone. A young pollster named George Gallup forecast the results accurately. He realized that the majority of Americans supported the New Deal policies proposed by the Democrats. Gallup's sample was much smaller, but far more representative of the American public than that of the *Literary Digest*. This points out that the representativeness of the sample is more important than its size.

A similar problem occurred in January of 2000 when the *Kansas City Star* reported that "Hundreds of Roman Catholic priests across the United States have died of AIDS-related illnesses, and hundreds more are living with HIV, the virus that causes the disease."

With great fanfare, the paper broke the first of a sure-to-be-controversial three-part series on "AIDS in the Priesthood," a nationwide story that challenged the Roman Catholic Church's handling of what the *Star* characterized as the "strikingly high incidence of AIDS-related disease and death among the church's clerics." Heavy nationwide coverage of this story followed. The study was based on surveys sent to

"3,013 priests randomly selected from an alphabetical listing in the *Official Catholic Directory* of 1999, included all 50 states and all Catholic dioceses except one" (Stats, March 2000). Eight hundred and one priests (27%) responded to the survey. This means that nearly three of four priests sent the survey failed to answer it.

Usually such a low response rate would warrant a follow-up survey to increase the number of surveys answered or to determine if the minority who did respond are representative of the whole group (Stats, Feb. 4 2000). No follow-up interviews were conducted because of the need for confidentiality.

The *Star* noted that of those who responded, "15% said they were homosexual and 5% bisexual. An additional 8% identified themselves as bisexual or 'other,' leaving only 78% self-identified heterosexuals" (Stats, Feb. 4 2000). But these statistics do not tell us some important information. We do not know if the minority who did respond are particularly concerned about AIDS, unusually open about personal sexuality, or more inclined toward homosexuality than the 2,212 priests who did not respond (Stats, Feb. 1 2000).

A closer look at the above percentages shows that the *Star* received a total of four responses from priests who said that they definitely had AIDS and three more who feared they might have AIDS. As an article critical of this study noted:

> If we simply scale up this result from four priests as though it were a representative sample projected onto 46,000 priests, the result is roughly 60 priests, nationwide, who would have AIDS and know it. . . . If we can further presume that the rate of death from AIDS is going down among priests just as it is in the general population, and if we can further judge that all 60 priests with AIDS today are not likely to die during a single year, we should quickly realize that while 60 priests who will someday die of AIDS is a tragedy, it is also nowhere near 1,000 priests with AIDS, [as the *Star* noted]. (Stats, March 2000)

The fact that the sample was not representative means that the *Star* had no solid data on the true number.

A major problem with this survey was that the priests were able to decide to participate or not in the survey. This meant that the sample was unrepresentative, and the *Star* should not project its results onto the nationwide population of priests.

Second, the finding of only four priests who said they have AIDS and three more who said they might have AIDS is a slender thread upon which to base national projections.

What is really disturbing, however, is that the *Kansas City Star* presented its study as valid research and that this survey—which has little, if any, real value—is taken seriously by large numbers of people. (For a further discussion of deceptive research, see "News You Can Use: How to Spot a Bogus Poll.")

Researcher Bias One of the most serious problems in data collection is **researcher bias,** *the tendency for researchers to select data that support, and to ignore data that seem to go against, their hypotheses.* We see this quite often in mass media publications.

They may structure their study to produce the results they wish to obtain, or they may only publicize information that supports their viewpoint. Researcher bias often takes the form of a self-fulfilling prophecy. A researcher who is strongly inclined toward one point of view may communicate that attitude through questions and reactions in such a way that the subject fulfills the researcher's expectations. For example, a researcher who is trying to prove an association between poverty and antisocial behavior might question low-income subjects in a way that would indicate a low regard for their social attitudes. The subjects, perceiving the researcher's bias, might react with hostility and thus fulfill the researcher's expectations.

Researcher bias might have been the cause of the problems with the data presented by sociologist Lenore Weitzman in her book *The Divorce Revolution* (1985). Weitzman theorized that changes in divorce laws resulting in an increase of no-fault divorces had produced a systematic impoverishment of divorced women and children. Weitzman argued that men usually make more money than women and that in divorce women usually get the children; therefore, treating men and women equally in divorce cheats the wife and children. Weitzman presented data from her study to prove it. She noted that, on the average, divorced women experience a 73% decline in their standard of living in the first year of divorce, whereas divorced men experience a 42% rise in their standard of living.

Weitzman's study received a great deal of attention and was cited in 175 newspaper and magazine articles. It was also cited in 348 social science articles, 250 law review articles, 24 state appellate court cases, and one Supreme Court case (Peterson, 1996). The statistic even appeared in President Clinton's 1996 budget. Anne Colby, director of the Murray Research Center at Radcliffe, where the original data are held, called it "one of the most widely quoted statistics in recent history."

Part of the reason the inaccurate data in the study was cited so often may have been that it could be used for both conservative and liberal political purposes. Conservative groups could claim that no-fault divorce should be abolished because it made the breakup of marriage too easy and produced the impoverishment of women and children. Liberal groups could use to data to show that women needed greater protection and more rights after divorce to prevent the inequities in divorce.

Greg Duncan and Saul Hoffman (1988), two researchers who had been tracking the effect of divorce on income for two decades, found the changes following divorce nowhere near as dramatic as Weitzman described. Although they did find a 30% decline in women's living standards and a 10% to 15% improvement in men's living standards, they also found that five years after divorce, the average woman's living standard was actually higher, mainly because many of them had remarried.

What baffled Duncan and Hoffman most was that Weitzman had used their methods to arrive at her numbers. They were not able to duplicate her findings and noted as much. A Census Bureau study eventually confirmed Duncan and Hoffman's study (Bianchi, 1991).

Even though questions continued to be raised about the research, Weitzman refused to allow researchers access to her data, claiming that she wanted to correct some errors in the master computer

NEWS YOU CAN USE

How to Spot a Bogus Poll

Opinion surveys can look convincing and be completely worthless. But asking four simple questions of any poll can separate the good numbers from the trash.

Politicians use opinion polls as verbal weapons in campaign ads. Journalists use them as props to liven up infotainment shows. Executives are more likely to pay attention to polls when the numbers support their decisions. But this isn't how polls are meant to be used. Opinion polls can be a good way to learn about the views Americans hold on important subjects, but only if you know how to cut through the contradictions and confusion.

Conducting surveys is difficult. It is especially difficult to take a meaningful survey of public opinion, because opinion is a subjective thing that can change rapidly from day to day.

Poll questions sometimes produce conflicting or meaningless results, even when they are carefully written and presented by professional interviewers to scientifically chosen samples.

That's why the best pollsters sweat the details on the order and wording of questions, and the way data are coded, analyzed, and tabulated. So the next time a poker-faced person tries to give you the latest news about how Americans feel, ask some pointed questions of your own:

Did You Ask the Right People?

Even when you start out with a representative sample, you could end up with a biased one. This is a risk all pollsters take, but some particular methods lend themselves to greater error. For example, readers of women's magazines are frequently asked to fill out surveys on weighty subjects like crime and sexual behavior. Not only do such polls ignore the opinions of nonreaders, they are biased toward readers who take the trouble to fill out and return the questionnaire, usually at their own expense.

Likewise, many have criticized online polling because Internet users tend to be wealthier, more educated, and more likely to be male than the larger population.

Television news and entertainment shows get into the act by posting toll-free or even toll numbers that viewers can call to "vote" on an issue. These samples are not only biased, they are prone to "ballot-stuffing" by enthusiasts. In other words, viewers who call 12 times get 12 votes.

Conflicts make news. When journalists are trying to liven up a boring political story, they need angry, well-informed citizens like a fish needs water. This is one reason why older men may be quoted more often than other groups. Those aged 50 and older are more likely than younger adults to follow news stories "very closely," according to the Pew Research Center for The People & The Press in Washington, D.C. Men are more likely than women to follow stories about war, business, sports, and politics.

In the last decade, angry white men have dominated media programs designed to give ordinary people a chance to speak out in public. Two-thirds of regular listeners to political talk-radio programs are men. Republicans outnumber Democrats three to one in the talk-radio audience, and 89% of listeners are white,

---

file before doing so. The refusal to share the data went on for years. Richard R. Peterson of the Social Science Research Council wanted to duplicate the study, but Weitzman would not allow him access to the data either. Peterson appealed to the National Science Foundation (NSF), the organization that had funded Weitzman's research. The NSF threatened to declare Weitzman ineligible for federal grants in the future if she did allow Peterson to examine the data.

A year and a half later, Weitzman sent her data to the Murray Research Center at Radcliffe College, which made the research available to others. After examining the data, Peterson (1996) came up with figures more in line with other national studies, with women's standard of living decreasing 27% in the first year after divorce and men's increasing by 10%. This led Peterson to conclude that Weitzman's erroneous results had lent support to those seeking to restrict access to no-fault divorce. Whether or not one agrees with those restrictions, policy arguments about no-fault divorce legislation should not be clouded by the use of the erroneous figures reported in *The Divorce Revolution*. Although social scientists cannot control how results are reported or used in policy debates, we need to take responsibility for correcting errors that may lead to faulty conclusions or to misguided policies (Peterson, 1996).

Felice Levine, a former officer of the American Sociological Association, has pointed out that when sociologists communicate with the public, they need to make it clear how reliable their results are. It is important to let the public know whether the sociologist's fellow researchers have been able to duplicate the data (Cohen, 2000).

compared with the national average for voters of 83%. Three in five regular listeners to political talk radio perceive a liberal bias in the mainstream media, compared with one in five nonlisteners.

What Is the Margin of Error?

No matter how carefully a survey sample is chosen, there will still be some margin of error. If you selected 10 different sets of 1,000 people using the same rules and asked each group the same question, the results would not be identical. The difference between the results is sampling error. Statisticians know that the error is equally likely to be above or below the true mark, and that larger samples have smaller margins of error if they are properly drawn. They are also able to estimate the margin of error, or the amount by which the result could be above or below the truth. Sampling error will always exist unless you survey every member of a population. If you do that, you have conducted a census.

Sampling error is one reason why two professionally conducted polls can show different results and both be correct. Reputable surveys report a margin of error—usually of 3 or 4 percentage points—at a particular confidence level—typically 95%. This means that 5% of the time, or 1 time in 20, the poll's results will not be reliable. The other 95% of the time, it is within 3 or 4 percentage points of the "truth."

Which Came First?

The order in which questions are asked can have a big effect on the results. Most people want to appear consistent to others and to be consistent in their own minds. When a pollster asks a series of related questions, this desire can lead people to take positions they might not have taken if they were asked only one question. Neither way produces an obviously "correct" response, but the results are different.

One way to handle this problem is to rotate the order of questions. Then the degree of differences due to question order can be described and interpreted. But not everyone pays heed to such fine distinctions.

What Was the Question?

"Do you want union officials, in effect, to decide how many municipal employees you, the taxpayer, must support?" Well, do you? This question, taken from an actual survey, is obviously biased. The results might make good propaganda for an antiunion group, but they are totally bogus as a poll. So before you pass a survey finding on to others, or even believe it yourself, be sure to look at the actual question.

When you're presented with a new survey or a used car, it helps to ask a few key questions before you buy. But for all their flaws, surveys are essential to the work of politicians, journalists, and businesspeople.

Source: Adapted from Brad Edmondson, "How to Spot a Bogus Poll," *American Demographics* 18(10) October 1996, pp. 10, 12–15. Used with permission from Primedia Specialty Group, Inc.

One of the standard means for dealing with research bias is to use **blind investigators,** or *investigators who do not know whether a specific subject belongs to the group of actual cases being investigated or to a comparison group.* For example, in a study on the causes of child abuse, the investigator looking at the children's family background would not be told which children had been abused and which were in the nonabused comparison group.

Sometimes double-blind investigators are used. **Double-blind investigators** *are kept uninformed not only of the kinds of subjects (case subjects or comparison group subjects) they are studying but also of the hypotheses being tested.* This eliminates any tendency on their part to find cases that support—or disprove—the research hypothesis.

(For a discussion of the different definitions of truth, see "Controversies in Sociology: Truth in the Courtroom versus Truth in the Social Sciences.")

Analyze the Data and Draw Conclusions

In its most basic sense, **analysis** is *the process through which large and complicated collections of scientific data are organized so that comparisons can be made and conclusions drawn.* It is not unusual for a sociological research project to result in hundreds of thousands of individual pieces of information. By itself this vast array of data has no particular meaning.

The analyst must find ways to organize such data into useful categories so that the relationships that exist can be determined. In this way the hypotheses forming the core of the research can be tested, and

CONTROVERSIES IN SOCIOLOGY

Truth in the Courtroom versus Truth in the Social Sciences

Lawyers and district attorneys often announce in their opening statements that the upcoming trial will be a search for the "truth." One of the goals of sociology specifically and science in general is to uncover the "truth" and bring about a fuller understanding of an issue. Is there a difference between the search for truth in the courtroom and the search for truth in the social and natural sciences?

The sociologist seeks to discover the truth by interviewing subjects, examining data, and piecing together records. Using this information, generalizations are made about what is likely to occur in the future and new hypotheses are proposed and tested. Even though it may seem objective, this search for the truth is still influenced by personal biases and political orientations.

Sometimes a scientist is looking for one thing and discovers something totally different and unexpected. Other times a discovery may be made by accident or because the original experiment was done incorrectly. None of this invalidates the truth of what was discovered. For example, in 1895 Wilhelm Roentgen accidentally discovered X-rays, a type of radiation that can pass through material where ordinary light cannot. X-rays revolutionized the fields of physics and medicine, and for this discovery Roentgen received the Nobel Prize in 1901. Penicillin, too, was an accidental discovery, as were many others that have changed our lives.

Even if a scientist cheats, exaggerates, or even plagiarizes his or her findings, ultimately the truth or falsity of the information is what remains. In the courtroom there is something known as the exclusionary rule, which throws out any evidence that has not been obtained legally. There is no such thing as an exclusionary rule in science, and truth achieved by unfair means is more likely to endure than falsity achieved by fair means.

In the courtroom, truth is entrusted to a jury selected randomly for their lack of information about the facts. Efforts are made to make them representative of the general population in terms of race, gender, or social class. "The task of discovering such truths in science is entrusted largely to trained experts who have studied the subject for years and are intimately familiar with the relevant facts and theories" (Dershowitz, 1996). They may differ from the general population in many ways.

Finally, scientifically discovered "truths" can always be reconsidered based on new evidence. There is no such thing as a statute of limitation on gathering evidence. Scientific research involves a continuous process of building on and refining other people's findings. That is not the case in the courtroom. You cannot be tried for the same crime twice even if new evidence suggests guilt. This is known as the prohibition against double jeopardy, and courts strive for "finality" in a trial and do not want a case to go on indefinitely. The courts do not want cases to be reopened and retried.

Law professor Alan Dershowitz (1996) writes:

Scientific inquiry is basically a search for objective truth. . . . Objective truth [as] validated by accepted, verifiable, and, if possible, replicable historical and scientific tests. . . . Truth in a criminal trial is quite different. . . . Truth, although *one* important goal of the criminal trial, is not the *only* goal. If it were, judges would not instruct jurors to acquit a defendant who they believe "probably" did it. The requirement is that guilt must be proved "beyond a reasonable doubt." But that is inconsistent with the quest for objective truth, because it explicitly prefers one kind of truth to another. The preferred truth is that the defendant did *not* do it, and we demand that the jurors err on the side of that truth, even in cases where it is probable that he did do it.

The next time you feel frustrated by a highly publicized trial where a defendant who may seem guilty is acquitted, remember that you are applying a standard of truth that is more typical of science than the courtroom. The courts must be as concerned about constitutional issues and the fairness of the process as they must be about the search for the truth. Many times, preserving the integrity of the court system requires a decision that goes against what may be the "truth" in the outside world.

Sources: *Reasonable Doubts* (pp. 34–48), by A. M. Dershowitz, 1996, New York: Simon and Schuster; and an interview with Alan M. Dershowitz, March 1996.

new hypotheses can be formulated for further investigation. (One important device to aid in the analysis of data is the table, which is explained in "News You Can Use: How to Read a Table.")

Sociologists often summarize their data by calculating central tendencies or averages. Actually, sociologists use three different types of averages: the mean, the median, and the mode. Each type is calculated differently, and each can result in a different figure.

Suppose you are studying a group of 10 college students whose verbal SAT scores are as follows:

450	690	280	450	760
540	520	450	430	530

Although you can report the information in this form, a more meaningful presentation would give some indication of the central tendency of the 10 SAT scores. The three measures of central tendency, like the three different types of averages mentioned earlier, are the mean, median, and mode.

The *mean* is what is commonly called the *average*. To calculate the mean, you add up all the figures and divide by the number of items. In our example, the SAT scores add up to 5,100. Dividing by 10 gives a mean of 510.

The *median* is the figure that falls midway in a series of numbers—there are as many numbers above it as below it. Because we have 10 scores—an even number—in our example, the median is the mean (the average) of the fifth and sixth figures, the two numbers in the middle. To calculate the median, rearrange the data in order from the lowest to highest (or vice versa). In our example, you would list the scores as follows: 280, 430, 450, 450, 450, 520, 530, 540, 690, 760. The median is 485—midway between the fifth score (450) and the sixth score (520).

The *mode* is the number that occurs most often in the data. In our example the mode is 450.

The three different measures are used for different reasons, and each has its advantages and disadvantages. The mean is most useful when a narrow range of figures exists, as it has the advantage of including all the data. It can be misleading, however, when one or two scores are much higher or lower than the rest. The median deals with this problem by not allowing extreme figures to distort the central tendency. The mode enables researchers to show which number occurs most often. Its disadvantage is that it does not give any idea of the entire range of data. Realizing the problems inherent in each average, sociologists often state the central tendency in more than one form.

Scientists usually are careful in drawing conclusions from their research. One of the purposes of drawing conclusions from data compiled in the course of research is the ability to apply the information gathered to other, similar situations. Problems thus may develop if there are faults in the research design. For example, the study must show **validity**—that is, *the study must actually test what it was intended to test.* If you want to say one event is the cause of another, you must be able to rule out other explanations to show that your research is valid.

Suppose you conclude that marijuana use leads to heroin use. You must show that it is marijuana use and not some other factor, such as peer pressure or emotional problems, that leads to heroin use.

The study must also demonstrate **reliability**—that is, *the findings of the study must be repeatable.* To demonstrate reliability we must show that research can be replicated—repeated for the purpose of determining whether initial results can be duplicated. Suppose you conclude from a study that whites living in racially integrated housing projects, who have contact with African Americans in the same projects, have more favorable attitudes toward blacks than do whites living in racially segregated housing projects. If you or other researchers carry out the same study in housing projects in various cities throughout the country and get the same results, the study is reliable.

It is highly unlikely that any single piece of research will provide all the answers to a given question. In fact, good research frequently leads to the discovery of unanticipated information requiring further research. One of the pleasures of research is that ongoing studies keep opening up new perspectives and posing further questions.

Prepare the Research Report

Research that goes unreported is wasted. Scientific progress is made through the accumulation of research that tests hypotheses and contributes to the ongoing process of bettering our understanding of the world. Therefore, it is usual for agencies that fund research to insist that scientists agree to share their findings.

Researchers generally publish their findings in scientific journals. If the information is relevant to the public, many popular and semiscientific publications will report these findings as well. It is especially important that research in sociology and other social sciences be made available to the public, because much of this research has a bearing on social issues and public policies.

Unfortunately, the general public is not always cautious in interpreting research findings. Special interest groups, politicians, and others who have a cause to plead are often too quick to generalize from specific research results, frequently distorting them beyond recognition. This happens most often when the research focuses on something of national or emotional concern. It is therefore important to double-check reports of sociological research appearing in popular magazines with the original research.

NEWS YOU CAN USE

How to Read a Table

Statistical tables are used frequently by sociologists both to present the findings of their own research and to study the data of others. We will use Table 2–3 to outline the steps to follow in reading and interpreting a table.

1. *Read the title.* The title tells you the subject of the table. Table 2–3 presents data on life expectancy at birth in various countries for people of both sexes.
2. *Check the source.* At the bottom of a table you will find its source. In this case, the source is the *Statistical Abstract of the United States: 2004-2005.* Knowing the source of a table can help you decide whether the information it contains is reliable. It also tells you where to look to find the original data and how recent the information is. In our example, the source is both reliable and recent. If the source were the 1958 *Abstract,* it would be of limited value in telling you about life expectancy in those countries today. Improvements in health care, control of epidemic diseases, or national birth control programs all are factors that may have altered life expectancy drastically in several countries since 1958. Likewise, consider a table of data about AIDS cases in Thailand. If its source were a government agency (which might be trying to alleviate the fears of tourists about the rampant spread of the disease in that country),

Table 2–3

Life Expectancy at Birth for Selected Countries
(Covers countries with 12 million or more population in 2003)

Country	Expectation of Life at Birth, 2003 (years)
Afghanistan	42.0
Brazil	71.1
Canada	79.8
Ghana	56.5
India	63.6
Japan	80.9
Mozambique	38.2
Poland	73.4
Spain	79.2
Sri Lanka	72.6
United States	77.1
Zimbabwe	39.0

Source: *Statistical Abstract of the United States: 2004-2005,* 124th ed. (p. 835), U.S. Bureau of the Census, Washington, DC, 2005.

you might well be skeptical about the reliability of the information in the table.

3. *Look for headnotes.* Many tables contain headnotes directly below the title. These may explain how the data were collected, why certain variables

Objectivity in Sociological Research

Max Weber believed that the social scientist should describe and explain what is, rather than prescribe what should be. His goal was a value-free approach to sociology. More and more sociologists today, however, are admitting that completely value-free research may not be possible. In fact, one of the trends in sociology today that could ultimately harm the discipline is that some sociologists who are more interested in social reform than social research have abandoned all claims to objectivity.

Sociology, like any other science, is molded by factors that impose values on research. Gunnar Myrdal (1969) lists three such influential factors: (1) the scientific tradition within which the scientist is educated; (2) the cultural, social, economic, and political environment within which the scientist is

trained and engages in research; and (3) the scientist's own temperament, inclinations, interests, concerns, and experiences. These factors are especially strong in sociological research because the researcher usually is part of the society being studied.

Does this mean that all science—sociology in particular—is hopelessly subjective? Is objectivity in sociological research an impossible goal? There are no simple answers to these questions. The best sociologists can do is strive to become aware of the ways in which these factors influence them and to make such biases explicit when sharing the results of their research. We may think of this as disciplined, or "objective," subjectivity, and it is a reasonable goal for sociological research.

Another problem of bias in sociological research relates to the people being studied rather than to the researchers themselves. The mere presence of investigators or researchers may distort the situation and produce unusual reactions from subjects who now feel special because of their selection for study.

(and not others) were studied, why the data are presented in a particular way, whether some data were collected at different times, and so on. In our table on life expectancy, the headnote explains that the numbers in the table refer to the average number of additional years a person can expect to live at birth.

4. *Look for footnotes.* Many tables contain footnotes that explain limitations or unusual circumstances surrounding certain data.

5. *Read the labels or headings for each row and column.* The labels will tell you exactly what information is contained in the table. It is essential that you understand the labels—both the row headings on the left and the column headings at the top. Here the row headings tell you the names of the countries being compared for life expectancy.

   For each group, life expectancy is given. Note the units being used in the table. In this case the units are years. Often the figures represent percentages or rates. Many population and crime statistics are given in rates per 100,000 people.

6. *Examine the data.* Suppose you want to find the life expectancy of a newborn baby in the United States. First look down the row at the left until you come to "United States." Then look across the columns until you come to the number. Reading across, you discover that on average, a baby born

in the United States in 2003 can expect to live another 77.1 years.

7. *Compare the data.* Compare the data in the table both horizontally and vertically. Suppose you want to find in which country people can expect to live longest from birth. Looking down the life expectancy column, we find that people born in Japan can expect to live longest—80.9 years. Among these 12 nations the United States ranks fourth, behind Japan, Canada, and Spain. A baby born in Mozambique or Zimbabwe has the shortest life expectancy—only 38.2 and 39.0 years, respectively.

8. *Draw conclusions.* Draw conclusions about the information in the table. After examining the data in the table, you might conclude that a person born in a relatively developed country (Canada, Japan, Spain, United States) is likely to live much longer than is someone born in a poorer nation (Afghanistan, India, Mozambique, Sri Lanka, and Zimbabwe).

9. *Pose new questions.* The conclusions you reach might well lead to new questions that could prompt further research. Why, you might want to know, do people in Ghana have shorter life expectancies than people in Japan? What causes the gap in life expectancy between the rich and poor nations?

Ethical Issues in Sociological Research

All research projects raise fundamental questions. Whose interests are served by the research? Who will benefit from it? How might people be hurt? To what degree do subjects have the right to be told about the research design, its purposes, and possible applications? Who should have access to and control over research data after a study is completed—the agency that funded the study, the scientists, the subjects? Should research subjects have the right to participate in the planning of projects? Is it ethical to manipulate people without their knowledge to control research variables? To what degree do researchers owe it to their subjects not to invade their privacy and to keep secret (and, therefore, not report anywhere) things that were told in strict confidence? What obligations do researchers

have to the society in which they are working? What commitments do researchers have in supporting or subverting a political order? Should researchers report to legal authorities any illegal behavior discovered in the course of their investigations? Is it ethical to expose subjects to such risk by asking them to participate in a study?

In the 1960s the federal government began to prescribe regulations for "the protection of human subjects." According to Herbert Gans (1979), these regulations are designed to force scientists to consider one central issue: "how to judge and balance the intellectual and [societal] benefits of scientific research against the actual or possible physical and emotional costs paid by the people who are being studied." Gans discusses three potential dilemmas for the researcher.

The first situation involves the degree of "permissible risk, pain, or harm." Gans writes, "Suppose a study which temporarily induces [severe emotional

CONTROVERSIES IN SOCIOLOGY

Famous Research Studies You Cannot Do Today

The proper procedures for research have changed considerably over the last few decades. There are a number of older famous studies that have been referred to hundreds of times in textbooks, other studies, and the classroom that do not for a variety of reasons meet contemporary standards of ethical research.

Zimbardo's Prison Environment—In 1972 Philip Zimbardo took Stanford University undergrads and tried to re-create a prison environment. Some of the students became guards, others prisoners. The experiment, which was to last two weeks, had to be cancelled after six days because some student guards became sadistic, and some student prisoners became distraught and depressed. The subjects had not been properly informed about what might happen in this experiment and had not given informed consent. Today these safeguards would have to be followed.

Tearoom Trade Observation—In 1970 sociologist Laud Humphreys wrote about his research observing homosexual behavior in public restrooms. After each observation of an older man having sex with a younger male, Humphreys would note the license plate number of the older man's vehicle. Through a friend in the police department he would trace the addresses of these men. He would then go to the homes of these individuals and pressure them into being subjects in his study.

Milgram's Obedience to Authority—This was the name of a book by psychologist Stanley Milgram who in 1974 wanted to see how susceptible people were to pressure from authority figures. He used a fake machine that caused the subjects to think they were giving electrical shocks to people in the next room. He wanted to see if the subjects would continue to give increasing shocks at the request of the researcher even though the person behind the wall was pleading to be released. The experiment produced great stress in the subjects and today would violate contemporary research ethics.

Cyril Burt's Twin Studies—Cyril Burt spent a lifetime trying to prove that intelligence is primarily an inherited characteristic. Between 1943 and 1966 he published numerous studies comparing the intelligence of identical twins who had been raised in different homes. Each study came to the same conclusion—that twins' intelligence test scores were very close. After Burt's death critics pointed out that the results were just too consistent. Questions also started to arise about whether his research assistants actually existed. On October 24, 1976, Oliver Gillie, the medical correspondent for the *London Sunday Times*, wrote, "Leading scientists are convinced that Burt published false data and invented crucial facts to support his controversial theory that intelligence is largely inherited" (cited in Plucker, 2003).

The Tuskegee Syphilis Study—Between 1932 and 1974 the U.S. government was interested in studying the long-term effects of syphilis. The experiment involved the withholding of penicillin from black male sharecroppers. These subjects did not know they were part of this unethical study, and needless to say, their conditions deteriorated.

distress] promises significant benefits. Experimental researchers may justify the study." However, we may wonder "whether the promised benefits can be realized, and whether they justify the potential dangers to the subjects, even if these are volunteers who know what to expect, and when all possible protective measures are taken."

A second dilemma is the extent to which subjects should be deceived in a study. It is now necessary for researchers to obtain the "informed consent," in writing, of the people they study. Questions still arise, however, about whether subjects are informed about the true nature of the study and "whether, once informed, they can freely decline to participate."

A third problem in research studies concerns the "disclosure of confidential or personally harmful information." Is the researcher entitled to delve into personal lives? What if the researcher uncovers some information that should be brought to the attention of the authorities? Should confidential information be included in a published study?

Rik Scarce, a doctoral student in sociology, had a personal confrontation with the issues of research ethics and confidentiality. He had been doing research on radical environmental groups when a group known as the Animal Liberation Front broke into a research facility at Washington State University, released animals, destroyed computer files, and did general damage to the lab.

A federal grand jury was set up to investigate the break-in and Scarce, although not a suspect, was summoned to appear before the group. Law enforcement officials thought Scarce, while doing his research, might have come across information that would lead to the guilty parties, and they wanted him to testify about his confidential sources.

Scarce answered those questions that he thought would not violate his agreement of confidentiality with his subjects. When he was asked to disclose the names of the actual subjects, he refused. He pointed out that the American Sociological Association's Code of Ethics states that confidentiality *must* be maintained, even at the risk of being jailed. A federal judge did not accept Scarce's ethical obligation and ordered him held in contempt of court until he revealed his confidential reports.

Scarce spent 159 days in the Spokane County Jail and was not released until the court realized that he would not reveal his sources. No other researcher has ever been incarcerated for this length of time, and Scarce's commitment to his ethical obligation caused him to pay a substantial price (Monaghan, 1993; Scarce, 1994).

Every sociologist must grapple with these questions and find answers that apply to particular situations. However, two general points are worth noting. The first is that social research rarely benefits the research subject directly. Benefits to subjects tend to be indirect and delayed by many years—as when new government policies are developed to correct problems discovered by researchers. Second, most subjects of sociological research belong to groups with little or no power, because they are easier to find and study. It is hardly an accident that poor people are the most studied, rich people the least. Therefore, research subjects typically have little control over how research findings are used, even though such applications may affect them greatly.

This means that sociologists must accept responsibility for the fact that they have recruited research subjects who may be made vulnerable as a result of their cooperation. It is important that researchers establish safeguards limiting the use of their findings, protecting the anonymity of their data, and honoring all commitments to confidentiality made in the course of their research.

The ideal relationship between scientist and research participant is characterized by openness and honesty. Deliberately lying to manipulate the participant's perceptions and actions goes directly against this ideal. Yet often researchers must choose between deception and abandoning the research. With few, if any, exceptions, social scientists regard deception of research participants as a questionable practice to be avoided if at all possible. It diminishes the respect due to others and violates the expectations of mutual trust on which organized society is based. When the deceiver is a respected scientist, it may have the undesirable effect of modeling deceit as an acceptable practice. Conceivably, it may contribute to the growing climate of cynicism and mistrust bred by widespread use of deception

by important public figures. (For some example of questionable research see "Controversies in Sociology: Famous Research Studies You Cannot Do Today.")

It is useful for human beings to seek to understand themselves and the social world in which they live. Sociology has a great contribution to make to this endeavor, both in promoting understanding for its own sake and in providing social planners with scientific information with which well-founded decisions can be made and sound plans for future development adopted. However, sociologists must also shoulder the burden of self-reflection—of seeking to understand the role they play in contemporary social processes while at the same time assessing how these social processes affect their findings.

SUMMARY

- To produce a valid study of the social world, we cannot approach it solely on the basis of personal experience and perceptions; rather, we must do so scientifically.
- Science has two main goals:
  1. to describe in detail particular things or events and
  2. to propose and test theories that help us understand these things or events.
- Like detectives, sociologists want to know "Why did it happen?" and "Under what circumstances is it likely to happen again?"
- The scientific research process involves a sequence of steps that must be followed to produce a valid study.
- The research process begins by defining the problem.
- Next the researcher attempts to find out as much as possible about previous studies on the same topic.
- The next step is to develop a hypothesis, or a testable statement about the relationship between two or more empirical variables.
- Hypotheses are tested by constructing a research design, or a strategy for collecting appropriate data.
- Sociologists use four main research designs:
  1. surveys
  2. participant observation
  3. secondary analysis
  4. occasional experiments
- The particular subset of the population chosen for study is known as a sample.
- Sampling is a research technique through which investigators study a manageable number of people selected from a larger population.

- If the procedures are carried out correctly, the sample will be a representative sample, or one that shows, in equivalent proportion, the significant variables that characterize the population as a whole.
- Failure to achieve a representative sample is known as sampling error.
- Sociology, like any other science, is molded by factors that impose values on research. Thus completely value-free research may not be possible.
- Nevertheless, objectivity, or a kind of disciplined subjectivity, is a reasonable goal for sociological research.
- The central ethical concern in research on human participants is how to judge and balance the intellectual and societal benefits of scientific research against the actual or possible physical and emotional costs to the research participants.

Media Resources

The Companion Website for *Introduction to Sociology*, Ninth Edition

http://sociology.wadsworth.com/tischler9e

Supplement your review of this chapter by going to the companion website to take one of the Tutorial Quizzes, use the flash cards to master key terms, and check out the many other study aids you will find there. You will also find special features such as Wadsworth's Sociology Online Resources and Writing Companion, GSS data, and Census 2000 information at your fingertips to help you complete that special project or do some research on your own.

CHAPTER TWO STUDY GUIDE

KEY CONCEPTS

Match each concept with its definition, illustration, or explanation presented below.

a. Operational definition
b. Hypothesis
c. Empirical
d. Variable
e. Independent variable
f. Dependent variable
g. Association
h. Survey

i. Cross-sectional research
j. Longitudinal research
k. Structured interview
l. Open-ended interview
m. Participant observation
n. Secondary analysis
o. Sample
p. Representative sample

q. Random sample
r. Sampling error
s. Stratified random sample
t. Researcher bias
u. Blind investigators
v. Double-blind investigators
w. Validity
x. Reliability

_j_ 1. A study that observes a population over a period of time.
_c_ 2. Based on, or capable of being based on, observed evidence.
_l_ 3. Research where the researcher follows a set of questions but can add follow-up questions on his or her own.
_h_ 4. A study that asks short-answer questions of a fairly large number of people.
_g_ 5. The simultaneous changing of two variables without one necessarily causing the change in the other.
_a_ 6. The conversion of abstract ideas into specific, observable things or events.
_w_ 7. The relation between what a study is supposed to test and what it actually tests.
_t_ 8. The influence, deliberate or not, a researcher exerts to get the preferred result.
_o_ 9. The population that a researcher gathers data on to assess the entire population.
_b_ 10. A testable statement about the relation between variables.
_x_ 11. The degree to which the results of a study would be repeated in other similar studies.
_q_ 12. A sample in which each individual in the population has an equal chance of being selected.
_d_ 13. Anything that can change or that can be sorted into more than one category or value.
_i_ 14. A study that looks at a population at a single point in time.
_k_ 15. Research where the researcher strictly follows a given set of questions.
_n_ 16. The use of available data gathered by another researcher or agency such as the Census Bureau.
_f_ 17. A variable that changes in response to changes in the independent variable.
_e_ 18. The variable that influences another variable without being influenced by that other variable.
_r_ 19. The failure to achieve a representative sample.
_m_ 20. Research in which the researchers hang out with the people they are doing research on.
_p_ 21. A sample in which the relevant variables are distributed in the same proportions as in the entire population.
_s_ 22. A sample in which subgroups are sampled separately to ensure that no group is disproportionately represented.
_v_ 23. Researchers who do not know which category a subject is in.
_j_ 24. Researchers who are unaware of both the category of the subject and of the hypotheses being tested.

CENTRAL IDEA COMPLETIONS

Following the instructions, fill in the appropriate concepts and descriptions for each of the questions posed in the following section.

1. A researcher wanted to see if the amount of time a student spent studying was associated with the student's grades. In this study, what was the

   a. independent variable? _____

   b. dependent variable? _____

2. Suppose that researchers wanted to find out about cheating in college. Explain the advantages and disadvantages of each of the following four research methods that might be used to assess the extent of cheating and the factors that influence it.

    a. social survey

        advantages _____

        disadvantages _____

    b. participant observation

        advantages _____

        disadvantages _____

    c. experiment

        advantages _____

        disadvantages _____

    d. secondary analysis

        advantages _____

        disadvantages _____

3. How does research bias take the form of a self-fulfilling prophecy?

4. What precautions might you use to detect a bogus poll?

5. Why do inaccurate statistics, such as those in Weitzman's study on divorce, become among the most widely cited?

CRITICAL THOUGHT EXERCISES

1. Find a poll result in a newspaper (*USA Today* often has them) or popular magazine. From the information given, which of the critical questions for detecting a bogus poll can you answer? If the information is not provided for a question, try to imagine what might have been done to make the poll bogus on that question.

2. Find a statement in a letter to the editor, an advertisement, a song, anywhere, and try to turn it into a hypothesis about the relation between variables. (For example, there used to be an ad that said, "Blondes have more fun.") How would you create an operational definition to turn abstract ideas like "fun" into something you could measure?

3. There has been much public discussion (and some proposed legislation) about video games and their effects. Companies that make the games and groups that express concern about the games both make conflicting claims. What steps would you take to assess the validity of the research claims coming from people on either side of the debate?

4. There are two main sources of data on crime in the United States: the FBI's Uniform Crime Reports (UCR) and the National Crime Victim Survey (NCVS). The UCR is the total of all crimes reported to the police. The NCVS is based on a sample of people who are interviewed about whether they have been victims of crime. What differences will there be in the picture of crime that each method presents?

INTERNET ACTIVITY

1. The data from the U.S. census are among the most important social science data used throughout the United States. Because the census is taken only once every 10 years, the data can become outdated. To correct this problem, the U.S. Census Bureau conducts the American Community Survey (ACS). Go to the U.S. Census website (http://www.census.gov) and examine the bureau's discussion of the methods it uses to collect these data.

ANSWERS TO KEY CONCEPTS

1.j 2.c 3.l 4.h 5.g 6.a 7.w 8.t 9.o 10.b 11.x 12.q 13.d 14.i 15.k 16.n 17.f 18.e 19.r 20.m 21.p 22.s 23.u 24.v

ThomsonNOW™

Reviewing is as easy as ❶ ❷ ❸

1. Before you do your final exam, take the ThomsonNOW diagnostic quiz to help you identify the areas on which you should concentrate. You will find information on ThomsonNOW and instructions on how to access all of its great resources on the foldout at the beginning of the text.
2. As you review, take advantage of ThomsonNOW's study videos and interactive Map the Stats exercises to help you master the chapter topics.
3. When you are finished with your review, take ThomsonNOW's posttest to confirm you are ready to move on to the next chapter.

3

Culture

© Paul Sauders/Stone/Getty Images

Learning Objectives

After studying this chapter, you should be able to do the following:

- Understand how culture makes possible the variation in human societies.
- Distinguish between ethnocentrism and cultural relativism.
- Know the difference between material and non-material culture.
- Understand the importance of language in shaping our perception and classification of the world.
- Discuss whether animals have language.
- Understand the roles of innovation, diffusion, and cultural lag in cultural change.
- Explain what subcultures are.
- Describe cultural universals.

In 2003, G. Willow Wilson, a light-haired Colorado woman, moved to Cairo to work as a teacher and a journalist. Wilson had recently converted to Islam, and was eager to be closer to the source and culture of her new religion. As she noted . . .

. . . The transition between life in red-state America and life in the Arab capital was at times overwhelming because of the traditional segregation of men and women in many public and private settings. Especially difficult to navigate at first was the Cairene metro, where choosing to ride in the wrong car could result in serious awkwardness.

Commuting women learn, however, to look on the first car—jokingly referred to as the hareem, or women's quarters—as a safe haven from the persistent scrutiny of men, who still dominate public life in Egypt. The first car, off limits to males above the age of 12 or so, is self-policing; should a man wander on, a quiet word is usually enough to send him out the door again. Few men risk so blatant a violation of a woman's first right in Egyptian society: privacy . . .

One night a few months ago, I took the metro downtown to meet a friend. I rode in the women's car, as usual . . . I noticed that my head scarf had begun to slip. I reached up to unpin it. . . . As I rewrapped my scarf, however, I heard a chorus of hisses. I looked up in alarm. A boy of 16 or 17 was making his way through the car, selling boxes of tissues. I blushed, feeling certain that the other women were reprimanding me for taking off my scarf in the presence of a man . . .

"What are you thinking? Don't you have shame?"

"You're too old to be in the women's car, Son."

. . . The tissue seller went red, muttered something in response and turned into the doorway, trying to appear casual . . . retreated down the train at the next stop.

At that moment, I was grateful to be part of the floating world of the women's car . . . Regardless of the many factors that might separate us on the street, in the women's car my fellow passengers felt I had the same right to privacy as they did. (From "The Comfort of Strangers," by G. Willow Wilson, *New York Times Magazine,* May 29, 2005. Copyright © 2005 by the New York Times Co. Reprinted by permission.)

All human societies have complex ways of life that differ greatly from one to the other. These ways have come to be known as *culture.* In 1871, Edward Tylor gave us the first definition of this concept. Culture, he noted, "is that complex whole which includes knowledge, belief, art, law, morals, custom, and other capabilities and habits acquired by man as a member of society" (Tylor, 1958).

Robert Bierstadt simplified Tylor's definition by stating, "Culture is the complex whole that consists of all the ways we think and do and everything we have as members of society" (Bierstadt, 1974).

Most definitions of culture emphasize certain features. Namely, culture is shared; it is acquired, not inborn; the elements make up a complex whole; and it is transmitted from one generation to the next.

The Concept of Culture

We will define **culture** as *all that human beings learn to do, to use, to produce, to know, and to believe as they grow to maturity and live out their lives in the social groups to which they belong.*

Culture is basically a blueprint for living in a particular society. In common speech, people often refer to a "cultured person" as someone with an interest in the arts, literature, or music, suggesting that the individual has a highly developed sense of style or aesthetic appreciation of the "finer things." To sociologists, however, every human being is "cultured." All human beings participate in a culture, whether they are Harvard educated and upper class, or illiterate and living in a primitive society. Culture is crucial to human existence.

When sociologists speak of culture, they are referring to the general phenomenon that is a characteristic of all human groups. However, when they refer to a culture, they are pointing to the specific culture of a particular group. In other words, all human groups have a culture, but it often varies considerably from one group to the next.

Take the concept of time, which we accept as entirely natural. To Westerners, "time marches on" steadily and predictably, with past, present, and future divided into units of precise duration (minutes, hours, days, months, years, and so on). In the Native American culture of the Sioux, however, the concept of time simply does not exist apart from ongoing events: Nothing can be early or late—things just happen when they happen. For the Navajo, the future is a meaningless concept—immediate obligations are what count. For natives of the Pacific island of Truk, however, the past has no independent meaning—it is a living part of the present (Hall, 1981). These examples of cultural differences in the perception of

Native Americans often wish to follow the traditional practices and customs of their culture. At the same time, the culture of the larger society urges them to adopt the conventions of mainstream American society.

time point to a basic sociological fact: Each culture must be investigated and understood on its own terms before it is possible to make valid cross-cultural comparisons.

In every social group, culture is transmitted from one generation to the next. Unlike other creatures, human beings do not pass on many behavioral patterns through their genes. Rather, culture is taught and learned through social interaction.

Culture and Biology

Human beings, like all other creatures, have basic biological needs. We must eat, sleep, protect ourselves from the environment, reproduce, and nurture our young—or else we could not survive as a species. In most other animals, such basic biological needs are met in more or less identical ways by all the members of a species through inherited behavior patterns or instincts. These instincts are specific for a given species as well as universal for all members of that species.

© Tony Freeman/PhotoEdit

Thus, instinctual behaviors, such as the web spinning of specific species of spiders, are constant and do not vary significantly from one individual member of a species to another.

This is not true of humans, whose behaviors are highly variable and changeable, both individually and culturally. It is through culture that human beings acquire the means to meet their needs. For example, the young, or larvae, of hornets or yellow jackets are housed in paper-walled, hexagonal chambers that they scrape against with their heads when hungry. This is a signal to workers, who immediately feed the young tiny bits of undigested insect parts (Wilson, 1975). Neither the larvae nor the workers learn these patterns of behavior: They are instinctual.

In contrast, although human infants cry when hungry or uncomfortable, the responses to those cries vary from group to group and even from person to person. In some groups, infants are breast-fed; in others, they are fed prepared milk formulas from bottles; and in still others, they are fed according to the mother's preference. Some groups breast-feed children for as long as five or six years, others for no more than 10 to 12 months.

Some mothers feed their infants on demand—whenever they seem to be hungry; other mothers hold their infants to a rigid feeding schedule. In some groups, infants are picked up and soothed when they seem unhappy or uncomfortable. Other groups believe that infants should be left "to cry it out." In the United States, parents differ in their approaches to feeding and handling their infants, but most are influenced by the practices they have observed among members of their families and their social groups. Such habits, shared by the members of each group, express the group's culture. They are learned by group members and are kept more or less uniform by social expectations and pressures.

Culture Shock

Every social group has its own specific culture, its own way of seeing, doing, and making things, its own traditions. Some cultures are quite similar to one another; others are very different. When individuals travel abroad to countries with cultures that are very different from their own, the experience can be quite upsetting. Meals are scheduled at different times of day, "strange" or even "repulsive" foods are eaten, and the traveler never quite knows what to expect from others or what others in turn may expect. Local customs may seem charming or brutal. Sometimes travelers are unable to adjust easily to a foreign culture; they may become anxious, lose their appetites, or even feel sick. Sociologists use the term **culture shock** to describe *the difficulty people have adjusting to a new culture that differs markedly from their own.*

Jonah Blank (1992) experienced culture shock often as he traveled throughout India. One day, he observed three bulls walking in the village:

> [They] ambled lazily by the storefront, leaving three steaming piles of dung in their wake. A few minutes later an old woman waddled along, dropped to her knees, and scooped up the fresh patties with her clapped hands. She slapped them onto an already laden tin plate, and shuffled down the alley. . . . Around the corner from the manure collector, another old woman hung a string of dried cow patties outside her door for luck. A large mound of dung sat at the step, stuck each day with newly plucked flower stems. Had I been rude enough to tell her that the custom was unhygienic, she would assuredly have laughed at my science.

Culture shock can also be experienced within a person's own society. Picture the army recruit having to adapt to a whole new set of behaviors, rules, and expectations in basic training—a new cultural setting.

Ethnocentrism and Cultural Relativism

People often make judgments about other cultures according to the customs and values of their own, a practice sociologists call **ethnocentrism.** Thus, an American might call a Guatemalan peasant's home filthy because the floor is made of packed dirt or believe that the family organization of the Watusi (of East Africa) is immoral because a husband may have several wives. Ethnocentrism can lead to prejudice and discrimination and often results in the repression or domination of one group by another.

Immigrants, for instance, often encounter hostility when their manners, dress, eating habits, or religious beliefs differ markedly from those of their new neighbors. Because of this hostility and because of their own ethnocentrism, immigrants often establish their own communities in their adopted country. Many Cuban Americans, for example, have settled in Miami where they have built a power base through strength in numbers. In Dade County alone, which includes Miami, there are about 850,000 Cuban Americans.

To avoid ethnocentrism in their own research, sociologists are guided by the concept of **cultural relativism,** *the recognition that social groups and cultures must be studied and understood on their own terms before valid comparisons can be made.* Cultural relativism frequently is taken to mean that social scientists never should judge the relative merits of any

© Gianluigi Guercia/AFP/Getty Images

As of 2005, Kind Mswati III of Swaziland had II wives and 25 children. His father had 70 wives. Cultural relativism asks us to not judge these facts from our culture alone.

group or culture. This is not the case. Cultural relativism is an approach to doing objective cross-cultural research. It does not require researchers to abdicate their personal standards. In fact, good social scientists will take the trouble to spell out exactly what their standards are so that both researchers and readers will be alert to possible bias in their studies.

American Moshe Rubinstein encountered the contrasting values between American and Arab culture after a traditional Arabic dinner. Rubinstein was presented with a parable by his host Ahmed.

"Moshe," he said as he put his fable in the form of a question, "imagine that you, your mother, your wife, and your child are in a boat, and it capsizes. You can save yourself and only one of the remaining three. Whom will you save?" For a moment I froze, thoughts raced through my mind. . . . No matter what I might say, it would not be right from someone's point of view, and if I refused to answer I might be even worse off. I was stuck. So I tried to answer by thinking aloud as I progressed to a conclusion, hoping for salvation before I said what came to my mind as soon as the question was posed, namely, save the child.

Ahmed was very surprised. I flunked the test. As he saw it, there was only one correct answer. . . .

"You see," he said, "you can have more than one wife in a lifetime, you can have more than one child, but you have only one mother. You must save your mother!" (Rubinstein, 1975)

This example shows us how the value of individuals such as children, spouses, and mothers can vary greatly from one culture to the next. We can see how what we might consider to be a natural way of thinking is not the case at all in another culture.

Cultural relativism requires that behaviors and customs be viewed and analyzed within the context in which they occur. The packed-dirt floor of the Guatemalan house should be noted in terms of the culture of the Guatemalan peasant, not in terms of suburban America (see "Global Sociology: Is McDonald's Practicing Cultural Imperialism or Cultural Accommodation?"). Researchers, however, may find that dirt floors contribute to the incidents of parasites in young children and may therefore judge such construction to be less desirable than wood or tile floors.

In 2005 King Mswati III of Swaziland married his eleventh wife. She was pregnant with his twenty-fifth child. There are two more young women waiting to marry the king in future ceremonies. The king's 11 wives is still far less than the 70 wives his father had (Mthethwa, 2004; "Swazi King," 2005). The mores of our society make us frown on this type of behavior. Cultural relativism would require that we not judge this practice from the standpoint of our culture alone.

Components of Culture

The concept of culture is not easy to understand, perhaps because every aspect of our social lives is an expression of it and also because familiarity produces a kind of nearsightedness toward our own culture, making it difficult for us to take an analytical perspective toward our everyday social lives. Sociologists find it helpful to break down culture into separate components: material culture (objects), nonmaterial culture (rules and shared beliefs), and language (Hall & Hall, 1990).

Material Culture

Material culture *consists of human technology—all the things human beings make and use, from small handheld tools to skyscrapers*. Without material culture our species could not long survive, for material culture provides a buffer between humans and their environment. Using it, human beings can protect themselves from environmental stresses, as when

GLOBAL SOCIOLOGY

Is McDonald's Practicing Cultural Imperialism or Cultural Accommodation?

McDonald's may be the most obvious example of American culture abroad. There are approximately 30,000 McDonald's restaurants in more than 119 countries serving 47 million customers a day.

In many of these countries McDonald's stands as a symbol for the Americanization of the world, with deep-seated political and cultural implications (Kincheloe, 2002). Like Coke, it is easy to denigrate McDonald's as the symbol of the crass, unhealthy, commercial side of American culture. For example, some Japanese critics have blamed junk food for juvenile crime. In France, a few McDonald's have been fire-bombed. After the terrorist attacks of September 11, 2001, all of McDonald's regional offices were closed.

To succeed, McDonald's has had to be responsive to the local cultures and eating habits. The restaurant chain offers *ayran* (a popular chilled yogurt drink) in Turkey, McLaks (a grilled salmon sandwich) in Norway, and in Japan a chicken sandwich flavored with soy sauce and ginger is popular. In Thailand you can get a Samurai Pork Burger, flavored with teriyaki sauce. Throughout the world you will find items such as the Kiwi Burger, the McHuevo, the McNifica, and the McAfrika on the McDonald's menu.

In New Delhi, India, where Hindus shun beef and Muslims refuse pork, the burgers are made of mutton and are called Maharaja Macs. For the many strict Hindus who are vegetarians, there is the McAloo Tikki burger, a spicy vegetarian patty made of potatoes and peas. McDonald's has even figured out how to make a vegetarian mayonnaise without eggs. Workers with green aprons handle only vegetarian food, and those with black aprons handle nonvegetarian food. The restaurants even separate the two menus—realizing that vegetarians do not want to read about meat dishes. What this has meant is that mixed groups of people, with different tastes and customs, have found a place where they can all eat together.

Compared with American customers, most diners in other countries spend a good deal of time in the restaurants. Far from being a place where they eat and run, in many countries people see McDonald's as a spot where they can linger. In cities where space is at a premium, like Hong Kong, teenagers choose it as an escape from cramped apartments where they can meet their friends and even do their homework. Women in traditional cultures meet at McDonald's because there is no alcohol served, and they see it as a safe, socially acceptable place for women.

That the staff is made up of local people means that the restaurants, though obviously foreign, are not immediately perceived as being American. Even though the golden arches may be conspicuous, whether it is New Delhi, Brazil, or Manila, the food is being cooked and served by people who live nearby and speak the local dialect.

Sources: Watson, James L., ed., *Golden Arches East: McDonald's in East Asia,* 1997, Berkeley: University of California Press; Kincheloe, Joe L. (2002), *The Sign of the Burger: McDonald's and the Culture of Power.* Philadelphia, PA: Temple University Press; McDonald's Worldwide, Corporate Responsibility Report 2004. Oak Brook, IL: McDonald's Corporation.

When people of one society come in contact with people of another society, cultural diffusion takes place, as displayed by this McDonald's in Saudi Arabia.

© Lynsey Addario/CORBIS

Housing is an aspect of material culture that can vary widely, as can be seen by comparing this elaborate home in Saudi Arabia and these thatched huts in Nigeria.

they build shelters and wear clothing to protect them from the cold or from strong sunlight.

Even more important, humans use material culture to modify and exploit the environment. They build dams and irrigation canals, plant fields and forests, convert coal and oil into energy, and transform ores into versatile metals. Using material culture, our species has learned to cope with the most extreme environments and to survive and even to thrive on all continents and in all climates. Human beings have walked on the floor of the ocean and on the surface of the moon. No other creature can do this: None has our flexibility. Material culture has made human beings the dominant life form on earth.

Nonmaterial Culture

Every society also has a **nonmaterial culture,** which consists of *the totality of knowledge, beliefs, values, and rules for appropriate behavior.* The nonmaterial culture is structured by such institutions as the family, religion, education, economy, and government.

Whereas material culture is made up of things that have a physical existence (they can be seen, touched, and so on), the elements of nonmaterial culture are the ideas associated with their use. Although engagement rings and birthday flowers have a material existence, they also reflect attitudes, beliefs, and values that are part of American culture. There are rules for their appropriate use in specified situations.

Norms are central elements of nonmaterial culture. **Norms** are *the rules of behavior that are agreed upon and shared within a culture and that prescribe limits of acceptable behavior.* They define "normal" expected behavior and help people achieve predictability in their lives. For example, one of the few truly universal gestures is the kiss. Anthropologists have speculated that it evolved from the time when mothers would pass food, mouth to mouth, to their infants. Think for a moment how the kiss has permeated our lives. Mothers kiss bruises to "make them better." Athletes kiss their trophies. The French sculptor Auguste Rodin created one—the famous sculpture, *The Kiss.*

Yet most cultures follow unwritten norms concerning kissing in public. In some cultures cheek kissing is a normal way of greeting another person. In Russia, you actually kiss the cheek. In other places, such as France, Italy, or Latin America, you kiss the air—that is, cheeks touch and lips make the sound of kissing, but the lips do not actually press against the cheek. In Latin America, only one cheek is kissed. In France, each cheek is kissed. In Belgium and Russia, you kiss one cheek, then the other, and back to the first.

In some countries, kissing the hand is the acceptable form of greeting. French etiquette suggests that the man kisses the woman's hand without actually touching it with his lips. Kissing your own fingertips is a European gesture that conveys the message

According to the norms of American culture, a common way for men to greet each other is to shake hands. In France, men greet each other with a kiss on each cheek.

"That's great! That's beautiful." The origin for this gesture probably stems from the custom of ancient Greeks and Romans who, when entering or leaving the temple, threw a kiss to sacred objects. In Mexico a kissing sound is used to summon a waiter in a restaurant. In the Philippines, street vendors use it to attract the attention of customers.

In the United States kissing between men and women in public is common. Presidential candidates have even given extended mouth-to-mouth kisses to their spouses during prime-time broadcasts.

At the other extreme, in certain Asian countries such as Japan, kissing is considered an intimate sexual act and not permissible in public, even as a social greeting (Axtell, 1998).

Mores (pronounced more-ays) are *strongly held norms that usually have a moral connotation and are based on the central values of the culture.* Violations of mores produce strong negative reactions, which are often supported by the law. Desecration of a church or temple, sexual molestation of a child, rape, murder, incest, and child beating all are violations of American mores.

Not all norms command such absolute conformity. Much of day-to-day life is governed by traditions, or **folkways,** which are *norms that permit a wide degree of individual interpretation as long as certain limits are not overstepped.* People who violate folkways are seen as peculiar or possibly eccentric, but rarely do they elicit strong public response.

For example, a wide range of dress is now acceptable in most theaters and restaurants. Men and women may wear clothes ranging from business attire to jeans, an open-necked shirt, or a sweater. However, extremes in either direction will cause a reaction. Many establishments limit the extent of informal dress: Signs may specify that no one with bare feet or without a shirt may enter. On the other hand, a person in extremely formal attire might well attract attention and elicit amused comments in a fast-food restaurant.

Good manners in our culture also show a range of acceptable behavior. A man may or may not open a door or hold a coat for a woman, who may also choose to open a door or hold a coat for a man—all four options are acceptable behavior and cause neither comment nor negative reactions from people.

These two examples illustrate another aspect of folkways: They change with time. Not too long ago a man was *always* expected to hold a door open for a woman, and a woman was *never* expected to hold a coat for a man.

Folkways also vary from one culture to another. In the United States, for example, it is customary to thank someone for a gift. To fail to do so is to be ungrateful and ill mannered. Subtle cultural differences can make international gift giving, however, a source of anxiety or embarrassment to well-meaning business travelers. For example, if you give a gift on first meeting an Arab businessman, it may be interpreted as a bribe. If you give a clock in China, it is considered bad luck. In fact, the Mandarin word for *clock* is similar to the one for *death*. In Latin America, you will have a problem if you give knives, letter openers, or handkerchiefs. The first two indicate the end of a friendship; the latter is associated with sadness.

Norms are specific expectations about social behavior, but it is important to add that they are not absolute. Even though we learn what is expected in our culture, there is room for variation in individual interpretations of these norms that deviate from the ideal norm. (For an example different regional norms in the United States, see "Controversies in Sociology: Is There a Culture of Violence in the South?")

Ideal norms are *expectations of what people should do under perfect conditions.* These are the norms we first teach our children. They tend to be simple, making few distinctions and allowing for no exceptions. In reality, however, nothing about human beings is ever that dependable.

Real norms are *norms that are expressed with qualifications and allowances for differences in individual behavior.* They specify how people actually behave. They reflect the fact that a person's behavior is guided by norms as well as unique situations.

In Kazahkistan, for example, although bribery is frowned upon, professors are so poorly paid that over the years a norm system has developed in which they supplement their salaries by taking money or other items from their students in return for higher grades. The cultural norms that have evolved in the higher education system in Kazahkistan include an

CONTROVERSIES IN SOCIOLOGY

Is There a Culture of Violence in the South?

No matter what region of the country you hail from, chances are you have heard a stereotype about the culture and personalities of the people from your area. New Yorkers are pushy. Californians are flaky. New Englanders are standoffish. Texans are arrogant. Oregonians are tree-huggers. But if you tell Southerners that they are more violent than their neighbors to the north, you might just have a fight on your hands.

Still, there is evidence that the characterization might be more than just a regional slur. In their book, *Culture of Honor: The Psychology of Violence in the South,* authors Richard E. Nisbett and Dov Cohen argue that the difference in the two regions' propensity toward violence may stem from culturally acquired beliefs about personal honor that are more embedded in the residents of the South than in those of their northern peers. Southerners fiercely believe that a person's reputation is important and worth defending. As a result, a comment that might be tolerated in Trenton may escalate to lethal violence in Tuscaloosa.

Nisbett and Cohen marshal some impressive evidence to support their hypothesis. They cite the fact that in the South, murder rates are higher than in the North for arguments among friends and acquaintances—where someone's honor is at stake—but not for killings committed in the course of other felonies, like knocking off a convenience store, where there is no relationship with the victim. Neither per capita income, hot weather, nor a history of slavery can explain these differences.

The authors also tested to see what kind of triggers lead to violent responses by asking Northerners and Southerners to read vignettes in which a man's honor was challenged. Southern respondents were much more likely to justify a violent response to an insult, saying a guy would not "be much of a man" if he was not willing to fight.

Nisbett and Cohen have an interesting theory about the origins of such attitudes. Many of the South's early settlers were Scots-Irish livestock herders. Because it is easy to steal sheep or cows in sparsely populated rural areas, people in herding cultures often cultivate a reputation for being trigger-happy hotheads as a deterrent to theft. This theory gains credence from the fact that southern white homicide rates are high in poor, rural regions, but not in more affluent, densely populated areas. The Scots-Irish code of honor, unfortunately, is less useful when it involves two guys trading insults in a bar than two men fighting over a heifer.

Source: *Culture of Honor: The Psychology of Violence in the South,* by Richard E. Nisbett and Dov Cohen, Boulder, CO: Westview Press.

---

entire vocabulary describing the various types of transactions. For example:

Razvodit: To get exams in advance by bribing the professor.

Psevdorepetitorstvo: When 10 or more students pay for extra classes to pass the course. They are not required to attend these classes.

Roga I kopyta: Bribes involving payment of sheep or other livestock. Usually given by students from rural villages.

Real'niy" diplom: A diploma with all necessary seals and signatures. Cost between $800–$1,500.

Chay-kofe-potansuem: A bribe offered by female students "without complexes."
(KIMEP Times, 2002)

The concepts of ideal and real norms are useful for distinguishing between mores and folkways. For mores, the ideal and the real norms tend to be very close, whereas folkways can be much more loosely connected: Our mores say thou shalt not kill and really mean it, but we might violate a folkway by neglecting to say thank you, for example, without provoking general outrage. More important, the very fact that a culture legitimizes the difference between ideal and real expectations allows us room to interpret norms to a greater or lesser degree according to our own personal dispositions.

Values are *a culture's general orientations toward life—its notions of what is good and bad, what is desirable and undesirable.* For example, each year the University of California at Los Angeles surveys college students to get an idea of what values are important to them. In the fall of 2003, the CIRP surveyed 276,449 freshman students at 413 of the nation's baccalaureate institutions.

Most college freshmen feel that abortion should be a legal right. Although support for legalized abortion has declined since 1992, with a low in 1999 of 53%, this view has been steadily gaining ground with over 55% of all entering freshmen expressing support in 2003.

Over 59% of entering freshmen support marriage for same-sex couples, up from 51% in 1997. A wide majority have typically favored exerting greater control over the sale of handguns. Only about one-third of freshmen support abolishing the death penalty, although student support for this view has been growing steadily in the past decade (Higher Education Research Institute, 2004).

Values can also be understood by looking at patterns of behavior. Sociologist Seymour Martin Lipset (1996) believes the United States has unique cultural values that set it apart from the rest of the world. However, these values can produce both positive and negative outcomes. For example, the United States is the most religious, optimistic, productive, well-educated, and individualistic country in the world. It is also one of the most crime-ridden and litigious nations, with a wide gap in income distribution, and some of the lowest levels of welfare benefits.

How can widely held cultural values produce both good and bad outcomes? Part of the answer lies in how people attempt to fulfill these values. Take, for example, the American emphasis on individualism and individual happiness. It may produce technological innovation in a wide variety of areas, but it is also responsible for the country's high divorce rate. Americans believe you should be satisfied with your life, be happy with your work, and like your spouse. If that is not the case, you are expected to make changes to correct the situation. If there are marital problems we want to know why you are staying in the marriage. The Japanese do not automatically assume that if you are dissatisfied with your circumstances you must change them.

The United States is a high-achievement–oriented society. At the same time it also leads the world with many types of crime. In the United States, a lack of success causes the individual to feel much more dissatisfied than in less achievement-oriented societies. Hence, people will try to get ahead by whatever means necessary. For some, this may mean crime.

In American society there is a disdain for authority stemming from the country's revolutionary past. The early founders rejected the control of England and produced a sharp break with the authority of the English powers. Americans do not show the kind of deference to authority as is commonly the case in countries such as Canada or Britain that have not had the same kind of revolutionary history.

The Origin of Language

Language enables humans to organize the world around them into labeled cognitive categories and use these labels to communicate with one another.

Language, therefore, makes possible the teaching and sharing of the values, norms, and nonmaterial culture we just discussed. It provides the principal means through which culture is transmitted and the foundation on which the complexity of human thought and experience rests.

Language allows humans to transcend the limitations imposed by their environment and biological evolution. It has taken tens of millions of years of biological evolution to produce the human species. On the other hand, in a matter of decades, cultural evolution has made it possible for us to travel to the moon. Biological evolution had to work slowly through genetic changes, but cultural evolution works quickly through the transmission of information from one generation to the next. In terms of knowledge and information, each human generation, because of language, is able to begin where the previous one left off. Each generation does not have to begin anew, as is the case in the animal world.

Over the past 75 years, sociologists and anthropologists have formed a standard view of the interplay between language and culture. It can be summarized as follows: Whereas animals are rigidly controlled by their biology, human behavior is determined by culture and language. Free from biological constraints, human cultures can vary from one another in an infinite number of ways.

Human infants are born with nothing more than a few reflexes and an ability to learn. Children learn their culture through their culture's language, socialization, and role models.

Some scientists (Pinker, 1994) believe that the human capacity to use language is one of the most distinctive human attributes and that this critical step in cultural development has a biological basis as well as a cultural one.

The study of the genomes of people and chimpanzees has yielded some insight into the origin of language. It appears that language is a relatively recent development, having evolved only in the last 100,000 years. Some believe the emergence of behaviorally modern humans about 50,000 years ago was set off by a major genetic change that made modern language possible.

In 2001 the first human gene involved specifically in language was discovered. Known as FOXP2, the gene is known to switch on other genes that are important for speech and language.

The discovery took place as part of research on three generations of the KE family, half of whom had speech and language disorders. The thinking at first was that the gene merely caused low intelligence and made speech unintelligible. Testing suggested that the disorder was more complex. Some of the family members did score lower on average in IQ tests. Others, however, scored in the normal range but still had a problem with language. The language problems were not just due to motor control. The family members also had trouble understanding sentences or grammar rules. They even had trouble performing tasks that the average 4-year-old could perform.

The thinking is that the FOXP2 gene has remained largely unaltered during the evolution of mammals, but suddenly changed in humans after they split off from the chimpanzee line of descent. By conferring the ability for rapid articulation, the improved gene may have provided the finishing touch to the acquisition of language and human cultural development (Enard et al., 2002).

Language and Culture

All people are shaped by the **selectivity** of their culture, *a process by which some aspects of the world are viewed as important while others are virtually neglected.* The language of a culture reflects this selectivity in its vocabulary and even its grammar. Therefore, as children learn a language, they are being molded to think and even to experience the world in terms of one particular cultural perspective.

This view of language and culture, known as the **Sapir-Whorf hypothesis,** *argues that the language a person uses determines his or her perception of reality* (Whorf, 1956; Sapir, 1961). This idea caused some alarm among social scientists at first, for it implied that people from different cultures never quite experience the same reality. Although more recent research has modified this extreme view, it remains true that different languages classify experiences differently—that language is the lens through which we experience the world. The prominent anthropologist Ruth Benedict (1961) pointed out, "We do not see the lens through which we look."

The category corresponding to one word and one thought in language A may be regarded by language B as two or more categories corresponding to two or more words or thoughts. For example, we have only one word for water, but the Hopi Indians have two words—*pahe* (for water in a natural state) and *keyi* (for water in a container). Yet the Hopi have only one word to cover every thing or being that flies, except birds. Strange as it may seem to us, they call a flying insect, an airplane, and a pilot by the same word. Verbs also are treated differently in different cultures. In English we have one verb *to go.* In New Guinea, however, the Manus language has three verbs—depending on direction, distance, and whether the going is up or down.

Language helps define our view of the world and others. The Indians of the North American Plains are willing to accept someone's unusual behavior rather than label the person as mentally ill. In fact, in American Indian culture, such people may be considered gifted, and thus very spiritual. Some American Indian cultures believe people with special needs are considered "waken," or holy, and belong to the creator. They are therefore treated with great respect.

Many Native American groups do not have the concept of "mental illness." They view mental illness as a white person's disease defined by mainstream society as shameful and unnatural. As such American Indians are likely to see mental health as a mainstream concept that does not really apply to them. Mental and emotional problems are thought to be brought on by biological, social, or cultural violations or taboos, such as excessive drinking for example. Wellness takes place when there is harmony, among body, mind, and spirit.

The American Indian conception of respect is also different from that of mainstream culture. It is inappropriate to pry into the innermost thoughts and feelings of another person, as is done by mental health professionals. One lives in harmony with all other beings because it is spiritually necessary to do so. All parts of life are interrelated and thus worthy of respect. To be in a state of conflict with people or to offend them is to be in disharmony and thus in a dangerous and vulnerable state.

In the Northern Plains Indians' view of communication, asking direct questions about mental illness may actually produce the behavior. In American Indian culture such questioning can allow spirits to enter the person's essence, producing "ghost illness" (National Institute of Justice, June 2005).

A little bit closer to home, consider the number of words and expressions pertaining to technology that have entered the English language. These include *cyberspace, virtual reality, hackers, phishing, spamming, morphing, wired,* and *zapped.*

These words reflect the preoccupation of American culture with technology. In contrast, many Americans are at a loss for words when they are asked to describe nature: varieties of snow, wind, or rain; kinds of forests; rock formations; earth colors and textures; or vegetation zones. Why?

These things are not of great importance in urban American culture. The translation of one language into another often presents problems. Direct translations are often impossible because (1) words may have a variety of meanings and (2) many words and ideas are culture bound. An extreme example of the first type of these translation problems occurred near the end of World War II. After Germany surrendered, the Allies sent Japan a surrender ultimatum. Japan's premier responded that his government would *mokusatsu* the ultimatum to surrender. *Mokusatsu* has two possible meanings in English: "to consider" or "to take notice of." The premier meant that the government would consider the surrender ultimatum.

SOCIAL CHANGE

Is There a Culture Clash between the United States and Saudi Arabia?

In a provocative book, *The Clash of Civilizations* (1996), Samuel J. Huntington, a Harvard professor and former advisor to President Lyndon Johnson, argued that in the future the main source of international conflict will be cultural differences. In developing his views, Huntington paid particular attention to the cultural gulf between Western and Islamic cultures. According to Huntington, people of different ethnic backgrounds and religions are likely to see their relationships with other groups in terms of "us" versus "them." Many people disagreed with Huntington and criticized him for making what they thought were exaggerated generalizations.

In the wake of September 11 however, Huntington's thesis began receiving renewed attention. "Do clearly delineated 'civilizations' exist or not? If they do, are they really in conflict with one another? And what light can be shed on the way Western and Islamic societies view one another?" (Miller & Feinberg, 2002)

The Roper Polling Service conducted a worldwide survey of 30,000 people in 30 countries to see whether there really are profound cultural differences. In one series of questions people were asked "how close they felt to the cultures and ways of life of various countries— very close, somewhat close, somewhat distant, or very distant." Almost all Americans (93%) felt "very or somewhat close to the culture and way of life of the United States." In addition, fewer than half of the people felt "close to any other specific culture, suggesting that even a country as ethnically diverse as America views the world in an "us" and "them" fashion.

Although compared with most countries the United States was relatively open to other cultures, Americans felt particularly distant from Arab culture. "Very few (4%) felt very or somewhat close to Arab culture." In fact, there was no other culture in the survey from which Americans felt as distant. This suggests that the Islamic presence in the United States has done little to encourage any feeling of closeness.

Like Americans, people from Saudi Arabia were also "nearly unanimous in their feeling of kinship to their own culture (98% felt very or somewhat close to Arab culture"), and few (7%) felt as close to American culture." Of all the countries surveyed, Saudis felt most distant from the American way of life.

The Roper Poll also asked about certain cultural values. "Whereas Americans rank freedom as a top ten value, Saudis do not." Americans put self-esteem on their top-ten list, Saudis put modesty on theirs.

When Saudis were asked to compare themselves to other cultures, they viewed themselves as part of a culture that has little in common with the rest of the world. Furthermore, the Saudis believed they had the "least in common with the United States when it came to personal values." What were these differences? The key American values included freedom, individuality, and self-reliance. For the Saudis they were faith, modesty, obedience, and tradition.

These findings do not, of course, definitively prove Huntington's thesis of American and Islamic culture being in fundamental conflict. They do, however, demonstrate that within the United States and Saudi Arabia there exists a great deal of perceived distance between the cultures.

Source: Adapted from Thomas A. W. Miller and Geoffrey D. Feinberg, "Culture Clash: Personal Values Are Shaping Our Times," *The Public Perspective*, March/April 2002, pp. 6–9.

The English translators, however, used the second interpretation, "to take notice of," and assumed that Japan had rejected the ultimatum. This belief that Japan was unwilling to surrender was a factor in the atomic bombing of Hiroshima and Nagasaki (Samovar, Porter, & Jain, 1981). Most likely the bombing would still have taken place even with the other interpretation, but this example does demonstrate the problems in translating words and ideas from one language into another.

These examples demonstrate the uniqueness of language. No two cultures represent the world in exactly the same manner, and this cultural selectivity, or bias, is expressed in the form and content of a culture's language.

The Symbolic Nature of Culture

All human beings respond to the world around them. They may decorate their bodies, make drawings on cave walls or canvases, or mold likenesses in clay. These all act as symbolic representations of their society. All complex behavior is derived from the ability to use symbols for people, events, or places. Without the ability to use symbols to create language, culture could not exist.

Symbols and Culture

What does it mean to say that culture is symbolic? A **symbol** is *anything that represents something else*

The "thumbs up" gesture is appropriate in American society but may be an insult in other countries.

People who live in Venice, Italy, have adjusted how they build their homes and live their lives to fit their watery environment.

and carries a particular meaning recognized by members of a culture. Symbols need not share any quality at all with whatever they represent. Symbols stand for things simply because people agree that they do.

Thus, when two or more individuals agree about the things a particular object represents, that object becomes a symbol by virtue of its shared meaning for those individuals. When Betsy Ross sewed the first American flag, she was creating a symbol.

The important point about the meanings of symbols is that they are entirely arbitrary, a matter of cultural convention. Each culture attaches its own meaning to things. Thus, in the United States, mourners wear black to symbolize their sadness at a funeral. In the Far East, people wear white. In this case, the symbol is different but the meaning is the same. On the other hand, the same object can have different meanings in different cultures. Among the Sioux Indians, the swastika (a cross made with ends bent at right angles to its arms) was a religious symbol; in Nazi Germany, its meaning was political.

In recent years e-mail messages have produced a whole host of symbols for commonly used expressions. (See "News You Can Use: Symbols in Cyberspace" for examples of some of these symbols.)

Few travelers would think of going abroad without taking along a dictionary or phrase book to help them communicate with the people in the countries they visit. Although most people are aware that symbolic gestures are the most common form of cross-cultural communication, they do not realize that the language of gestures can be just as different, just as regional, and just as likely to cause misunderstanding as the spoken word can.

After a good meal in Naples, a well-meaning American tourist expressed his appreciation to the waiter by making the "A-OK" gesture with his thumb and forefinger. The waiter was shocked. He headed for the manager. The two seriously discussed calling the police and having the hapless tourist arrested for obscene behavior in a public place.

What had happened? How could such a seemingly innocent and flattering gesture have been so misunderstood? In American culture, everyone from astronauts to politicians, to signify that everything is fine, uses the sign confidently in public. In France and Belgium, however, it means "You're worth zero"; while in Greece and Turkey, it is an insulting or vulgar sexual invitation. In parts of southern Italy, it is an offensive and graphic reference to a part

NEWS YOU CAN USE

Symbols in Cyberspace

Communication involves the display of many symbols. When we communicate in e-mail many of the gestures and facial expression that help clarify a message are missing. In response people have developed a whole host of symbols to help make the message clear. They include:

:-)	Smile	:-((	I am very sad
:-(	Sad	I-{	Good grief
:-0	Wow	I-I	Asleep
:-X	My lips are sealed	I^o	Snoring
LOL	Laughing out loud	:-@	Screaming
:-II	I am angry	~ :-(	Steaming mad
}:[	Angry, frustrated	%-)	Dazed or silly
		%*]	Drunk
		%-\	Hungover
		%-{	Ironic
		%-(	Confused
		:-C	Astonished

of the anatomy. Small wonder that the waiter was shocked.

In fact, dozens of gestures take on totally different meanings from one country to another. Is thumbs-up always a positive gesture? It is in the United States and in most of western Europe. When it was displayed by the emperor of Rome, the upright thumb gesture spared the lives of gladiators in the Coliseum. However, do not try it in Sardinia and northern Greece. There the gesture means the insulting phrase, "Up yours." The same naiveté that can lead Americans into trouble in foreign countries also may work in reverse.

After paying a call on Richard Nixon, Soviet leader Leonid Brezhnev stood on a balcony at the White House and saluted the American public with his hands clasped together in a gesture many people interpreted as meaning "I am the champ," or "I won." What many Americans perceived as a belligerent gesture was really just the Russian gesture for friendship (Axtell, 1991).

Looking at culture from this point of view, we would have to say that all aspects of culture—nonmaterial and material—are symbolic.

Culture and Adaptation

Culture probably has been part of human evolution since the time, some 15 million years ago, when our ancestors first began to live on the ground. As we have

stressed throughout this chapter, humans are extraordinarily flexible and adaptable.

Adaptation is *the process by which human beings adjust to changes in their environment.* This adaptability, however, is not the result of being biologically fitted to the environment; in fact, human beings are remarkably unspecialized.

We do not run very fast, jump very high, climb very well, or swim very far. However, we are specialized in one area: We are culture producing, culture transmitting, and culture dependent. This unique specialization is rooted in the size and structure of the human brain and in our physical ability both to speak and to use tools.

Culture, then, is the primary means by which human beings adapt to the challenges of their environment. Thus, using enormous machines we strip away layers of the earth to extract minerals, and using other machines we transport these minerals to yet more machines, where they are converted to a staggering number of different products.

Take away all our technology and American society would cease to exist. Take away all culture and the human species would perish. Culture is as much a part of us as our skin, muscles, bones, and brains.

Mechanisms of Cultural Change

Cultural change takes place at many different levels within a society. Some of the radical changes that have taken place often become obvious only in hindsight.

When people of one society come in contact with people of another society, cultural diffusion takes place, as evidenced by the Dairy Queen in Japan.

When the airplane was invented, few people could visualize the changes it would produce. Not only did it markedly decrease the impact of distance on cultural contact, but also it had enormous impact on such areas as economics and warfare.

It is generally assumed that the number of cultural items in a society (including everything from toothpicks to structures as complex as government agencies) has a direct relation to the rate of social change. A society that has few such items will tend to have few **innovations,** *any new practice or tool that becomes widely accepted in a society.* As the number of cultural items increases, so do the innovations, as well as the rate of social change. For example, an inventory of the cultural items—from tools to religious practices—among the hunting and gathering Shoshone Indians totals a mere 3,000. Modern Americans who also inhabit the same territory in Nevada and Utah are part of a culture with items numbering well into the millions. Social change in American society is proceeding rapidly, whereas Shoshone culture, as revealed by archaeological excavations, appears to have changed scarcely at all for thousands of years.

Two simple mechanisms are responsible for cultural evolution: innovation and diffusion. Innovation takes place in several different ways, including recombining in a new way elements already available to a society (invention), discovering new concepts, finding new solutions to old problems, and devising and making new material objects.

Diffusion is *the movement of cultural traits from one culture to another.* It almost inevitably results when people from one group or society come into contact with another, as when immigrant groups take on the dress or manners of already established groups and in turn contribute new foods or art forms to the dominant culture. Rarely does a trait diffuse directly from one culture into another.

Rather, diffusion is marked by **reformulation,** *in which a trait is modified in some way so that it fits better in its new context.* This process of reformulation can be seen in the transformation of black folk blues into commercial music such as rhythm and blues and rock 'n' roll. Or consider moccasins—the machine-made, chemically waterproofed, soft-soled cowhide shoes—which today differ from the Native American originals and usually are worn for recreation rather than as part of basic dress, as they originally were. Sociologists would say, therefore, that moccasins are an example of a cultural trait that was reformulated when it diffused from Native American culture to industrial America.

Cultural Lag

Although the diverse elements of a culture are inter-related, some may change rapidly while others lag behind. William F. Ogburn (1964) coined the term **cultural lag** to describe *the phenomenon through which new patterns of behavior may emerge, even though they conflict with traditional values*. Ogburn observed that technological change (material culture) is typically faster than change in nonmaterial culture—a culture's norms and values—and technological change often results in cultural lag.

Consequently, stresses and strains among elements of a culture are more or less inevitable. For example, even though the Internet in general and the World Wide Web in particular offer vast educational opportunities, teachers have been slow to incorporate these technologies into the classroom.

Traditional school values may be in conflict with use of the web. Schools often assume that education is best carried out in isolation from the rest of society and that the teacher is the main guide for the students along a path to learning. Education has changed little from 100 years ago and we still expect teachers to talk and groups of students to listen.

The web makes it possible for the student to connect to countless sites outside of the classroom and makes it possible for the student to pursue individual educational goals. The teacher's role and influence becomes less clear with the introduction of this technology. The teacher, instead of being in charge, must now be ready to collaborate with the student and serve as a partner in the exploration of the resources (Maddux, 1997). Traditional teacher-student roles and values are challenged in the process.

Other instances of cultural lag have considerably greater and more widespread negative effects. Advances in medicine have led to lower infant mortality and greater life expectancy, but there has been no corresponding rapid worldwide acceptance of methods of birth control. The result is a potentially disastrous population explosion.

Animals and Culture

Do animals have culture? Many social scientists would say no. Language often is cited as the major behavioral difference between humans and animals. Humans possess language, whereas it is said animals do not. Language is the crucial ingredient in the ability to transmit culture from one generation to the next. Animals may have traits that may be socially transmitted. But they lack the ability to benefit from the accumulation of knowledge and the ability to improve things over time. In human cultures, things change and improve from one generation to the next. People create things that are useful for survival and

these things evolve and get better causing human culture to flourish. No single individual created something as complex and useful as a computer, but the history of advances led to its development.

Others disagree and think animals use language in unique ways that we have overlooked. A number of experiments—the earliest dating back to the mid-1950s—have shown that apes are able to master some of the most fundamental aspects of language. Apes, of course, cannot talk. Their mouths and throats simply are not built to produce speech, and no ape has been able to approximate more than four human words. However, efforts to teach apes to communicate by other means have met with a fair amount of success.

The first and most widely known experiment in ape language research began in 1966, under the direction of Allen and Beatrix Gardner of the University of Nevada, with a chimpanzee named Washoe. This experiment consisted of teaching the chimp American Sign Language (ASL), the hand-gesture language used by deaf people. Washoe learned more than 200 distinct signs and was able to ask for food, name objects, and make reference to her environment. The Gardners replicated their results with four other chimpanzees.

Another experiment involves a female gorilla named Koko. Francine Patterson has been working with Koko since 1972. Koko uses approximately 400 signs regularly and another 300 occasionally. She also understands several hundred spoken words (so much so that Patterson has to spell such words as *candy* in her presence). In addition, Koko invents signs or creates sign combinations to describe new things. She tells Patterson when she is happy or sad, refers to past and future events, defines objects, and insults her human companions by calling them such things as "dirty toilet," "nut," and "rotten stink." Once when Patterson was drilling Koko on body parts, the gorilla signed, "Think eye ear nose boring" (Hawes, 1995).

Koko has taken several IQ tests and has recorded scores just below average for a human child—averaging between 70 and 95 points. However, as Patterson has pointed out, the IQ tests have a cultural bias toward humans, and the gorilla may be more intelligent than the test indicates.

For example, one item instructs the child, "Point to two things that are good to eat." The choices are a block, an apple, a shoe, a flower, and an ice cream sundae. Reflecting her tastes, Koko pointed to the apple and the flower. She likes to eat flowers and has never seen an ice cream sundae. Although this answer is correct for Koko, it is only half right for humans and therefore was scored incorrect.

Some interesting work with apes has been done by Sue Savage-Rumbaugh and her colleagues, who taught a form of computer language to chimpanzees. Using special symbols, they managed to teach apes to name objects and "converse" with each other.

An unexpected turn of events produced some interesting results with a bonobo, or pygmy chimp, named Kanzi. Rumbaugh was having no luck trying to teach Kanzi's adopted mother Matata to use the keyboard. During the lessons while the mother was trying in vain to figure out what the experimenters wanted, Kanzi was spontaneously picking up on the tasks while crawling about and generally being more of a distraction than a participant.

When Rumbaugh finally gave up on Matata and turned her attention to Kanzi, assuming that at age 2 he might now be old enough to learn, she was shocked to find that he already knew most of what she wanted to teach him and more. Kanzi was far more adept at the language tasks than any previous chimp and developed these abilities further over the years.

Kanzi may have learned so well because he was immature at the time. It suggests the possibility that there might be in chimps, as in humans, a critical period when some special language-learning mechanism is activated. Yet the problem with the critical learning period approach is that chimps in the wild do not learn a language. Why should a chimp whose ancestors never spoke (and who himself cannot speak) demonstrate a critical period for language learning (Deacon, 1997)?

Steven Pinker, a cognitive scientist at MIT, believes the various animal language experiments are "exercises in wishful thinking" (Johnson, 1995). He states: "In my mind this kind of research is more analogous to the bears in the Moscow circus who are trained to ride unicycles" Johnson notes: "[Pinker] is not convinced that the chimps have learned anything more sophisticated than how to press the right buttons to get the hairless apes on the other side of the console to cough up M & M's, bananas and other tidbits of food."

Language and the production and use of tools are central elements of nonmaterial and material culture. So does it make sense to say that culture is limited to human beings? Although scientists disagree in their answers to this question, they do agree that humans have refined culture to a far greater degree than have any other animals and also that humans depend on culture for their existence much more completely than do any other creatures.

Subcultures

To function, every social group must have a culture of its own—its own goals, norms, values, and ways of doing things. As Thomas Lasswell (1965) pointed out, such group culture is not just a "partial or miniature" culture. It is a full-blown, complete culture in its own right. Every family, clique, shop, community, ethnic group, and society has its own culture. Hence, every individual participates in a number of different cultures in the course of a day. Meeting social expectations of various cultures is often a source of considerable stress for individuals in complex, heterogeneous societies like ours. Many college students, for example, find that the culture of the campus varies significantly from the culture of their family or neighborhood.

At home they may be criticized for their musical taste, their clothing, their antiestablishment ideas, and for spending too little time with the family. On campus they may be pressured to open up their minds and experiment a little or to reject old-fashioned values.

Sociologists use the term **subculture** to refer to *the distinctive lifestyles, values, norms, and beliefs of certain segments of the population within a society.*

The concept of subculture originated in studies of juvenile delinquency and criminality (Sutherland, 1924), and in some contexts the *sub* in *subculture* still has the meaning of inferior. However, sociologists increasingly use subculture to refer to the cultures of discrete population segments within a society. The term is primarily applied to the culture of ethnic groups (Italian Americans, Jews, Native Americans, and so on) as well as to social classes (lower or working, middle, upper, and so on). Certain sociologists reserve the term *subculture* for marginal groups—that is, for groups that differ significantly from the so-called dominant culture.

Types of Subcultures

Several groups have been studied at one time or another by sociologists as examples of subcultures. These can be classified roughly as follows.

Ethnic Subcultures Many immigrant groups have maintained their group identities and sustained their traditions while at the same time adjusting to the demands of the wider society. Though originally distinct and separate cultures, they have become American subcultures. America's newest immigrants, Asians from Vietnam, Korea, Japan, the Philippines, Taiwan, India, and Cambodia, have maintained their values by living together in tightly knit communities in New York, Los Angeles, and other large cities while at the same time encouraging their children to achieve success by American terms.

Occupational Subcultures Certain occupations seem to involve people in a distinctive lifestyle even beyond their work. For example, New York's Wall Street is not only the financial capital of the world, it is identified with certain values such as materialism, greed, or power. Construction workers, police, entertainers, and many other occupational groups involve people in distinctive subcultures.

Religious Subcultures Certain religious groups, though continuing to participate in the wider society, nevertheless practice lifestyles that set them apart. These include Christian evangelical groups, Mormons, Muslims, Jews, and many religious splinter groups. Sometimes the lifestyle may separate the group from the culture as a whole as well as the subculture of its immediate community. In a drug-ridden area of Brooklyn, New York, for example, a group of Muslims follows an antidrug creed in a community filled with addicts, drug dealers, and crack houses. Their religious beliefs set them apart from the general society while their attitude toward drugs separates them from many other community members.

Political Subcultures Small, marginal political groups may so involve their members that their entire way of life is an expression of their political convictions. Often these are so-called left-wing and right-wing groups may reject much of what they see in American society but remain engaged in society through their constant efforts to change it to their liking.

Geographic Subcultures Large societies often show regional variations in culture. The United States has several geographical areas known for their distinctive subcultures. For instance, the South is known for its leisurely approach to life, its broad dialect, and its hospitality. The North is noted for "Yankee ingenuity," commercial cunning, and a crusty standoffishness. California is known for its trendy and ultra-relaxed or "laid-back" lifestyle. And New York City stands as much for an anxious, elitist, arts-and-literature–oriented subculture as for a city.

Social Class Subcultures Although social classes cut horizontally across geographical, ethnic, and other subdivisions of society, to some degree it is possible to discern cultural differences among the classes. Sociologists have documented that linguistic styles, family and household forms, and values and norms applied to childrearing are patterned in terms of social class subcultures. (See Chapter 8 for a discussion of social class in the United States.)

Deviant Subcultures As we mentioned earlier, sociologists first began to study subcultures as a way of explaining juvenile delinquency and criminality. This interest expanded to include the study of a wide variety of groups that are marginal to society in one way or another and whose lifestyles clash with that of the wider society in important ways.

Some of the deviant subcultural groups studied by sociologists include prostitutes, strippers, hackers, pool hustlers, pickpockets, drug users, and a variety of criminal groups.

Universals of Culture

Despite their individual and cultural diversity, their many subcultures and countercultures, human beings are members of one species with a common evolutionary heritage. Therefore, people everywhere must confront and resolve certain common, basic problems such as maintaining group organization and overcoming difficulties originating in their social and natural environments. **Cultural universals** are *certain models or patterns that have developed in all cultures to resolve these problems.*

Among those universals that fulfill basic human needs are the division of labor, the incest taboo, marriage, family organization, rites of passage, and ideology. It is important to keep in mind that although these forms are universal, their specific contents are particular to each culture.

The Division of Labor

Many primates live in social groups in which it is typical for each adult group member to meet most of his or her own needs. The adults find their own food, prepare their own sleeping places, and, with the exception of infant care, mutual grooming, and some defense-related activities, generally fend for themselves.

This is not true of human groups. In all societies—from the simplest bands to the most complex industrial nations—groups divide the responsibility for completing necessary tasks among their members.

This means that humans constantly must rely on one another; hence they are the most cooperative of all primates.

The Incest Taboo, Marriage, and the Family

All human societies regulate sexual behavior. Sexual mores vary enormously from one culture to another, but all cultures apparently share one basic value: sexual relations between parents and their children are to be avoided. (There is evidence that some primates also avoid sexual relations between males and their mothers.) In most societies it is also wrong

© Warwick Kent/Photolibrary/PictureQuest

Certain patterns of behavior, such as marriage, are found in every culture but take various forms. The marriage ceremony, for example, varies greatly among different cultures.

for brothers and sisters to have sexual contact (notable exceptions being the brother-sister marriages among royal families in ancient Egypt and Hawaii, and among the Incas of Peru). *Sexual relations between family members* is called **incest,** and because in most cultures very strong feelings of horror and revulsion are attached to incest, it is said to be forbidden by taboo. A **taboo** is *the prohibition of a specific action.*

The presence of the incest taboo means that individuals must seek socially acceptable sexual relationships outside their families. All cultures provide definitions of who is or is not an acceptable candidate for sexual contact. They also provide for institutionalized marriages—ritualized means of publicly legitimizing sexual partnerships and the resulting children. Thus, the presence of the incest taboo and the institution of marriage result in the creation of families. Depending on who is allowed to marry—and how many spouses each person is allowed to have—the family will differ from one culture to another.

However, the basic family unit consisting of husband, wife, and children (called the nuclear family) seems to be a recognized unit in almost every culture, and sexual relations among its members (other than between husband and wife) are almost universally taboo. For one thing, this helps keep sexual jealousy under control. For another, it prevents the confusion of authority relationships within the family. Perhaps most important, the incest taboo ensures that family offspring will marry into other families, thus recreating in every generation a network of social bonds among families that knits them together into larger, more stable social groupings.

Rites of Passage

All cultures recognize stages through which individuals pass in the course of their lifetimes. Some of these stages are marked by biological events, such as the start of menstruation in girls. However, most of these stages are quite arbitrary and culturally defined. All such stages—whether or not corresponding to biological events—are meaningful only in terms of each group's culture.

Rarely do individuals drift from one such stage to another; every culture has established **rites of passage,** or *standardized rituals marking major life transitions.*

The most widespread—if not universal—rites of passage are those marking the arrival of puberty (often resulting in the individual's taking on adult status), marriage, and death. Typical rites of passage celebrated in American society include baptisms, bar and bat mitzvahs, confirmations, major birthdays, graduation, wedding showers, bachelor parties, wedding ceremonies, major anniversaries, retirement parties, and funerals and wakes. Such rites accomplish several important functions, including helping the individual achieve a sense of social identity, mapping out the individual's life course, and aiding the individual in making appropriate life plans. Finally, rites of passage provide people with a context in which to share common emotions, particularly with regard to events that are sources of stress and intense feelings, such as marriage and death.

Ideology

A central challenge that every group faces is how to maintain its identity as a social unit. One of the most important ways that groups accomplish this is by promoting beliefs and values to which group members are firmly committed. Such **ideologies,** or *strongly held beliefs and values,* are the cement of social structure.

Ideologies are found in every culture. Some are religious, referring to things and events beyond the perception of the human senses. Others are more secular—that is, nonreligious and concerned with the everyday world. In the end, all ideologies rest on untestable ideas rooted in the basic values and assumptions of each culture.

Even though ideologies rest on untestable assumptions, their consequences are very real. They give direction and thrust to our social existence and meaning to our lives. The power of ideologies to mold passion and behavior is well known. History is filled with both horrors and noble deeds people have performed in the name of some ideology: thirteenth-century Crusaders, fifteenth-century Inquisitors, pro-states rights and pro-union forces in nineteenth-century America, abolitionists, prohibitionists, trade unionists, Nazis and fascists, communists, segregationists, civil rights activists, feminists, consumer activists, environmentalists.

These and countless other groups have marched behind their ideological banners, and in the name of their ideologies they have changed the world, often in major ways.

Culture and Individual Choice

Very little human behavior is instinctual or biologically programmed. In the course of human evolution, genetic programming gradually was replaced by culture as the source of instructions about what to do, how to do it, and when it should be done. This means that humans have a great deal of individual freedom of action—more than any other creature.

However, as we have seen, individual choices are not entirely free. Simply by being born into a particular society with a particular culture, every human being is presented with a limited number of recognized or socially valued choices. Every society has means of training and of social control that are brought to bear on each person, making it difficult for individuals to act or even think in ways that deviate too far from their culture's norms. To get along in society, people must keep their impulses under some control and express feelings and gratify needs in a socially approved manner at a socially approved time (see "For Further Thinking: The Conflict between Being a Researcher and a Human Being"). This means that humans inevitably feel somewhat dissatisfied, no matter to which group they belong. Coming to terms with this central truth about human existence is one of the great tasks of living.

SUMMARY

- All human societies have complex ways of life that differ greatly from one to the other.
- Each society has its own unique blueprint for living, or culture.
- Culture consists of all that human beings learn to do, to use, to produce, to know, and to believe as they grow to maturity and live out their lives in the social groups to which they belong.
- Humans are remarkably unspecialized; culture allows us to adapt quickly and flexibly to the challenges of our environment.
- Sociologists view culture as having three major components: material culture, nonmaterial culture, and language.
- Language and the production of tools are central elements of culture. Evidence exists that animals engage, or can be taught to engage, in both of these activities. This does not mean that they have culture.
- Scientists disagree about how to interpret the evidence. Without question, however, it can be said that humans have refined culture to a far greater degree than other animals and are far more dependent on it for their existence.
- Every social group has its own complete culture.
- Sociologists use the term *subculture* to refer to the distinctive lifestyles, values, norms, and beliefs associated with certain segments of the population within a society. Types of subcultures include ethnic, occupational, religious, political, geographical, social class, and deviant subcultures.
- People in all societies must confront and resolve certain common, basic problems. Cultural universals are certain models or patterns that have developed in all cultures to resolve those problems. Among them are the division of labor, the incest taboo, marriage, family organization, rites of passage, and ideology. Though the forms are universal, the content is unique to each culture.
- By dividing the responsibility for completing necessary tasks among their members, societies create a division of labor.
- Every culture has established rites of passage, or standardized rituals marking major life transitions. Ideologies, or strongly held beliefs and values, are the cement of social structure in that they help a group maintain its identity as a social unit.
- Due to a lack of instinctual or biological programming, humans have a great deal of flexibility and choice in their activities. Individual freedom of action is limited, however, by the existing culture.

FOR FURTHER THINKING

The Conflict between Being a Researcher and a Human Being

Sociologists and anthropologists are supposed to understand the importance of cultural relativism and realize that cultures must be studied and understood in their own terms before valid comparisons can be made. A researcher should avoid imposing his or her values on a people or interfering with a culture to such an extent that it moves away from its origins. Is this a realistic goal, or are we being overly idealistic when we think this can be accomplished?

Kenneth Good (1991) had to deal with these issues often when he studied the Yanomama. The Yanomama are approximately 10,000 South American Indians who live in 125 villages in southern Venezuela and northern Brazil.

In terms of their material and technological culture, the Yanomama stand out in their primitiveness. They have no system of numbers . . . they have not invented the wheel. They know nothing of the art of metallurgy. Until recently they made fire . . . [by] rubbing two sticks together (Good 1991).

Traditionally, the Yanomama do not wear clothing, but they paint red designs on their bodies. "Girls and women adorn their faces by inserting slender sticks through holes in the lower lip at either side of the mouth and in the middle, and through the pierced nasal septum." Their lives are characterized by persistent aggression among village members and perpetual warfare with other groups. They engage in club fighting, gang rape, and murder.

Good encountered many events that went against his cultural value system. One day he saw two groups of tribespeople having a tug of war.

But instead of pulling on a thick vine, they were pulling on a woman. . . . Her assailants on one side were three of the wilder teenage [boys]. Trying to pull her away from them were three elderly women. The tug of war went on for 10 minutes or so while I watched, my blood rising as instinct to me to put a stop to it. . . . "What are they doing to her?" I asked. . . . "They're going to rape her," came the answer as casually as if she had said, "They're going to have a picnic."

Kenneth Good and his Yanomama wife, Yarima.

© Kenneth Good

With a concerted heave, the teenagers pulled her free. . . . Howling in victory, they ran down the trail, yanking her along. As they ran, they were joined by more shouting teenagers. I followed behind as the stampede bore her into the jungle.

I stood there, my heart pounding. I had no doubt I could scare these kids away. On the other hand, I was an anthropologist. I wasn't supposed to take sides and make value judgments. This kind of thing went on. If a woman showed up somewhere unattached, chances were she'd be raped. She knew it, they knew it. It was expected behavior. What was I supposed to do, I thought, try to inject my standards of morality? I hadn't come down here to change these people or because I thought I'd love everything they did: I'd come to study them.

So why was I standing there shaking with anger? Why was I thinking, Come on, Ken, what's wrong with you? Are you going to stand around with your notebook in your hand and observe a gang rape in the name of anthropological science? . . . How could I live with a group of human beings and not be involved with them as a fellow human being?

That afternoon was a turning point for Good. About a month later there was another woman-dragging episode.

After half an hour . . . one of the other men got fed up with the noise. . . . "That's enough," he said, picking up his arrows. "This is really annoying. I'm going to stab her, that will put a stop to it." I watched as he walked up to the three of them with his arrows. He was really going to do it.

When I saw this I yelled, "Don't do that!" [He] stopped and looked around, surprised. Our eyes met, then he walked back to his hammock and put his arrows down. . . . I knew that I was not the same detached observer I had been before.

Good was scheduled to spend 15 months with the Yanomama. After this time passed, however, he did not leave. He stayed and learned to speak their language and to hunt fish and gather as they did. He learned what it meant to be a nomad. Later he was adopted into the lineage of the village and given a wife, Yarima, according to Yanomama custom and in keeping with the wishes of the tribal leader. Eventually, Yarima had a profound effect on him.

I'm in love. Unbelievable, intense emotion, almost all the time. In the morning when she gets up to start the day, when I see her come in from the gardens with a basket of plantains, especially when we make love. Sure it's universal, except that being in love in Yanomama culture with a Yanomama girl is different, a different game, different rules.

In my wildest dreams it had never occurred to me to marry an Indian woman in the Amazon jungle. I was from suburban Philadelphia. I had no intention of going native. . . . That was what I had come to, after all these years of struggling to fit into the Yanomama world, to speak their language fluently, to grasp their way of life from the inside. My original purpose to observe and analyze this people . . . had slowly merged with something far more personal.

Eventually, Kenneth Good brought Yarima out of the Amazon jungle to the United States, where she lived from 1988 to 1993. Yarima had never seen flat cleared land, much less houses and cars. She had never worn clothes or walked in shoes. They had three children. On a trip back to the Yanomama as part of a National Geographic documentary, Yarima decided to remain. Good was horrified and went back to the United States.

Good's last attempt to persuade his wife to rejoin him in the West nearly succeeded. Lengthy negotiations were held with Yarima's brother and she eventually agreed to accompany Good to a jungle landing strip where a plane was waiting to fly them out. At the last minute, she decided she could not do it and ran back into the jungle. Kenneth Good returned to New Jersey, where he lives with their three children. Was Good doing valuable research, or did he impose his values on another culture?

Sources: *Into the Heart,* by Kenneth Good, with D. Chanoff, 1991, New York: Simon & Schuster, and an interview with David Chanoff, July 1994.

Media Resources

The Companion Website for *Introduction to Sociology,* Ninth Edition

http://sociology.wadsworth.com/tischler9e

Supplement your review of this chapter by going to the companion website to take one of the Tutorial Quizzes, use the flash cards to master key terms, and check out the many other study aids you'll find there. You'll also find special features such as Wadsworth's Sociology Online Resources and Writing Companion, GSS data, and Census 2000 information at your fingertips to help you complete that special project or do some research on your own.

CHAPTER THREE STUDY GUIDE

KEY CONCEPTS

Match each concept with its definition, illustration, or explanation presented below.

a. Culture
b. Culture shock
c. Ethnocentrism
d. Cultural relativism
e. Material culture
f. Nonmaterial culture
g. Norms
h. Mores

i. Folkways
j. Ideal norms
k. Real norms
l. Values
m. Sapir-Whorf hypothesis
n. Diffusion
o. Reformulation
p. Cultural lag

q. Subculture
r. Rites of passage
s. Innovation
t. Selectivity
u. Symbol
v. Cultural universals
w. Taboo
x. Ideology

_h_ 1. Strongly held norms that usually have important moral implications.
_p_ 2. The conflict between cultural ideas and new patterns of behavior, especially those that arise because of technological innovation.
_j_ 3. Expectations of what people should do under perfect conditions.
_a_ 4. A pattern for living shared by the members of a society.
_e_ 5. The physical objects human beings make and use.
_g_ 6. Rules, often unwritten, for everyday behavior.
_n_ 7. The movement of cultural traits from one culture to another.
_r_ 8. Standardized rituals marking major life transitions.
_c_ 9. Judging another culture by standards of one's own culture.
_f_ 10. The totality of knowledge, beliefs, values, and rules for appropriate behavior—shared by members of society.
_o_ 11. The modification of a cultural trait adopted from another culture so that it fits better with one's own.
_d_ 12. Withholding judgment and seeking to understand other societies on their own terms.
_i_ 13. Norms or customs that permit a wide degree of individual interpretation.
_b_ 14. The difficulty people have adjusting to a new culture that differs markedly from their own.
_q_ 15. The distinctive lifestyle or culture of certain segments of the population within a society.
_l_ 16. A culture's general orientations toward life.
_k_ 17. Norms that are expressed with qualifications and allowances for differences in individual behavior.
_m_ 18. The language a person uses shapes his or her perception of reality.
_x_ 19. Strongly held but untestable ideas shared by members of a society that uphold the basis of that society.
_u_ 20. Something that represents something else and whose meaning is understood by the members of a culture.
_s_ 21. A tool or practice that becomes widely accepted in society.
_r_ 22. The process by which a culture emphasizes some aspects of reality and ignores others.
_w_ 23. A very strongly held prohibition on some specific action.
_v_ 24. Patterns or models that all cultures have developed.

CENTRAL IDEA COMPLETIONS

Following the instructions, fill in the appropriate concepts and descriptions for each of the questions posed in the following section.

1. Cell phones are a fairly recent phenomenon. The friction over their use indicates some cultural lag, as people try to apply precellular norms. Do norms of cell phone use differ among groups? In some particular situations, what values does cell phone use contradict? What values support cell phone

use? How do you think norms will change in response to cellular technology with its capacity for conversation, pictures, text messages, and so on?

a. norm difference _____

b. values that conflict with cell phone use _____

c. values that support cell phone use _____

d. likely change in norms _____

2. Is there a "youth culture" in the United States? What differences in values and norms are there between younger people (14–25) and people 30 or more years older?

a. differences in norms _____

b. differences in values _____

3. What subcultures exist at your school? How do norms and values differ among these subcultures?

4. In recent years, some African cultures have been criticized for practicing "female circumcision" (the surgical removal of some of the labia or clitoris). Describe both the ethnocentric and cultural relativist reactions to this practice.

a. ethnocentrism _____

b. cultural relativism _____

CRITICAL THOUGHT EXERCISES

1. Tischler discusses the concepts of ideal and real norms in education in Kazahkistan. How do the concepts of ideal and real norms apply at your college or university?

2. Refer to the reading on *Social Change* and discuss the question: Is there a culture clash between the United States and Saudi Arabia?

3. In American politics, people often refer to "culture wars," and cultural differences between "red states" and "blue states." What values seem to be in dispute in these "wars"? How do these values affect what people actually do?

INTERNET ACTIVITIES

1. Visit the Cyberculture section at the ASA Culture Site (http://www.ncf.edu/culture/cyber.htm). Develop a short report on the emerging importance of cyberculture to the traditional study of culture.

2. *The Economist* magazine did an article on "U.S. Exceptionalism," the ways in which Americans differ from their counterparts in other advanced countries. You can see the results here: http://www.economist.com/displaystory.cfm?story_id=1511812.

3. Some of the data for the article in *The Economist* came from the World Values Survey. You can find some of the results here (http://www.worldvaluessurvey.org/statistics/index.html) and elsewhere at their website.

ANSWERS TO KEY CONCEPTS

1.h 2.p 3.j 4.a 5.e 6.g 7.n 8.r 9.c 10.f 11.o 12.d 13.i 14.b 15.q 16.l 17.k 18.m
19.x 20.u 21.s 22.t 23.w 24.v

ThomsonNOW™

Reviewing is as easy as ❶ ❷ ❸

1. Before you do your final exam, take the ThomsonNOW diagnostic quiz to help you identify the areas on which you should concentrate. You will find information on ThomsonNOW and instructions on how to access all of its great resources on the foldout at the beginning of the text.
2. As you review, take advantage of ThomsonNOW's study videos and interactive Map the Stats exercises to help you master the chapter topics.
3. When you are finished with your review, take ThomsonNOW's posttest to confirm you are ready to move on to the next chapter.

4

Socialization and Development

© SuperStock/PictureQuest

Learning Objectives

After studying this chapter, you should be able to do the following:

- Discuss how biology and socialization contribute to the formation of the individual.
- Know what sociobiology is.
- Explain how extreme social deprivation affects early childhood development.
- Identify the stages of cognitive and moral development.

- Explain the views of the self developed by Cooley, Mead, and Freud.
- Understand Erikson's and Levinson's stage models of lifelong socialization.
- Explain how family, schools, peer groups, and the mass media contribute to childhood socialization.
- Know how adult socialization and resocialization differ from primary socialization.

It had been only 11 weeks since the recruits started marine boot camp training at Parris Island, South Carolina, but now, as many revisited civilian society, they felt alienated from their past lives. To Patrick Bayton, "Everything feels different. I can't stand half my friends." Frank Demarco attended a street fair in Bayonne, New Jersey. "It was crowded. Trash everywhere. People were drinking, getting into fights. No politeness whatsoever. This is the way civilian life is—nasty."

As the son of a Wall Street executive, Daniel Keane came from a privileged background. When he came home to spend some time with his family he said, "I didn't know how to act. I didn't know how to carry on a conversation." He found it even more difficult being with his friends. They were "drinking, acting stupid and loud." He was particularly disappointed when two old friends refused to postpone smoking marijuana for a few minutes until he was away from them. "I was disappointed in them doing that. It made me want to be at SOI [the Marines' School of Infantry]."

Keane felt like he had joined a cult or a religion. "People don't understand," he said, "and I'm not going to waste my breath trying to explain when the only thing that really impresses them is how much beer you can chug down in 30 seconds." Military ideology disapproves of the lack of order and respect for authority that it believes characterizes civilian society.

As Sergeant Major James Moore pointed out, "it is difficult to go back into a society of 'What's in it for me?' when a Marine has been taught the opposite for so long" (Ricks, 1997).

During boot camp, the Marine Corps attempts to sever a recruit's ties to his or her private life in order to facilitate a process of socialization to the military culture. By the time it is over, the marines come to see themselves as different from society morally and culturally. The boot camp experience has modified their previous years of socialization enough that they now feel most comfortable with others who are also part of that culture.

A marine's first instruction on becoming a member of society began in childhood. This *process of social interaction that teaches the child the intellectual, physical, and social skills needed to function as a member of society* is called **socialization**. Through socialization experiences children learn the culture of the society into which they have been born. In the course of this process, each child slowly acquires a **personality**—that is, *the patterns of behavior and ways of thinking and feeling that are distinctive for each individual*. Contrary to popular wisdom, nobody is a born salesperson, criminal, or military officer. These things all are learned and modified as part of the socialization process.

Becoming a Person: Biology and Culture

Every human being is born with a set of **genes,** *inherited units of biological material.* Half are inherited from the mother, half from the father. No two people have exactly the same genes, except for identical twins. Genes are made up of complicated chemical substances, and a full set of genes is found in every body cell. The Human Genome Project has found that humans have about 30,000 genes—barely twice the number of a fruit fly.

Genes influence the chemical processes in our bodies and even control some of these processes completely. For example, such things as blood type, the ability to taste the presence of certain chemicals, and some people's inability to distinguish certain shades of green and red are completely under the control of genes. Most of our body processes, however, are not controlled solely by genes but are the result of the interaction of genes and the environment (physical, social, and cultural). Thus, how tall you are depends on the genes that control the growth of your legs, trunk, neck, and head and also on the amount of protein, vitamins, and minerals in your diet. Genes help determine your blood pressure, but so do the amount of salt in your diet, the frequency with which you exercise, and the amount of stress under which you live.

© Big CheesePhoto LLC/Alamy

The resemblance between this mother and son is quite clear because half his genes were inherited from her.

Nature versus Nurture: A False Debate

For more than a century, sociologists, educators, and psychologists have argued about which is more important in determining a person's qualities: inherited characteristics (nature) or socialization experiences (nurture). After Charles Darwin (1809–1882) published *On the Origin of Species* in 1859, human beings were seen to be a species similar to all the others in the animal kingdom. Because most animal behavior seemed to the scholars of that time to be governed by inherited factors, they reasoned that human behavior similarly must be determined by instincts—biologically inherited patterns of complex behavior. Instincts were seen to lie at the base of all aspects of human behavior, and eventually researchers cataloged more than 10,000 human instincts (Bernard, 1924).

Then, at the turn of the century, a Russian scientist named Ivan Pavlov (1849–1936) made a startling discovery. He found that if a bell were rung just before dogs were fed, eventually they would begin to salivate at the ringing of the bell itself, even when no food was served. The conclusion was inescapable: So-called instinctive behavior could be molded or, as Pavlov (1927) put it, conditioned, through a series of repeated experiences linking a desired reaction with a particular object or event. Dogs could be taught to salivate.

Pavlov's work quickly became the foundation on which a new view of human beings was built—one that stressed their infinite capacity to learn and be molded. The American psychologist John B. Watson (1878–1958) taught a little boy to be afraid of a rabbit by startling him with a loud noise every time he was allowed to see the rabbit. What he had done was to link a certain reaction (fear) with an object (the rabbit) through the repetition of the experience. Watson eventually claimed that if he were given complete control over the environment of a dozen healthy infants, he could train each one to be whatever he wished—"doctor, lawyer, artist, merchant, even beggar or thief" (Watson, 1925). Among certain psychologists, conditioning became the means through which they explained human behavior.

Sociologists believe humans are unique because they are the only life form able to accumulate knowledge, improve on it, and pass it on to the next generation. Human society would not be able to advance without this ability. This fact has made it possible for humans to develop societies throughout the world in a wide variety of climates and settings. Genes alone do not make this possible, but the accumulated knowledge from others in the community does (Richerson & Boyd, 2004).

Sociobiology

The debate over nature versus nurture took a different turn when Edward O. Wilson published his book *Sociobiology: The New Synthesis* (1975). The discipline of sociobiology also known as evolutionary psychology or neo-Darwinian theory uses biological principles to explain the behavior of all social beings, both animal and human.

The study of the biological basis of social behavior in animals had long been accepted by biologists. When Wilson published his book, in the twenty-seventh and final chapter, he applied the theories of sociobiology to human beings.

According to Wilson (1975), human beings inherit a propensity for certain behaviors and social structures. These traits commonly are called human nature and include the division of labor between the sexes, bonding between parents and children, heightened altruism toward closest kin, incest avoidance, some forms of ethical behavior, suspicion of strangers, tribalism, dominance orders within groups, male dominance overall, and territorial aggression over limited resources. Although people have free will, our genes make certain behaviors more likely than others. Although cultures may vary, they all begin to lean in certain common directions.

When an especially harsh and prolonged winter leaves an Inuit (Eskimo) family without food supplies, the family must break camp and quickly find a new site in order to survive. Frequently an elderly member of the family, often a grandmother, who may slow down the others and require some of the scarce food, will stay behind and face certain death. Wilson (1975, 1979) saw this as an example of altruism, which ultimately might have a biological component. In sacrificing her own life, the grandmother is improving her kin's chances of survival. She already has made her productive contribution to the family. Now the younger members of the family must survive to ensure the continuation of the family and its genes into future generations.

Wilson concluded that behavior can be explained in terms of the ways in which individuals act to increase the probability that their genes and the genes of their close relatives will be passed on to the next generation. For example, studies of mating strategies in 37 countries show that men and women consistently respond differently to what they look for in their romantic and sexual partners. Men value physical appearance more than women do, and women weigh status and income more than men do. Men's ideal mates are a few years younger than they are on average, and women's a few years older (Buss, 1994). The men are looking for healthy women to bear their children whereas women are looking for mature, responsible men who will not abandon them.

Up until the publication of Wilson's book, the prevailing view in the social sciences had been that there was no biologically based human nature, that human behavior is almost entirely sociocultural in origin, and therefore genes play little or no role in social behavior. Wilson took the opposite position, proposing that human behavior cannot be understood without biology (1994). Proponents of sociobiology believe that social science one day will be a mere subdivision of biology.

Critics saw sociobiology as not just intellectually flawed but also morally wrong. If human nature is rooted in heredity, they suggested, then some forms of social behavior probably cannot be changed. Tribalism and gender differences could be judged as unavoidable and class differences and war in some manner as natural. People could also argue that some racial or ethnic groups differ irreversibly in personal abilities and emotional attributes. Some people could have inborn mathematical genius, others a bent toward criminal behavior (Wilson, 1994).

The furor over sociobiology was so strong that the American Anthropological Association attempted to pass a resolution to censure sociobiology. Only a passionate speech by noted anthropologist Margaret Mead defeated the motion (Fisher, 1994).

Within Wilson's own department at Harvard University, his chair, Richard Lewontin, and another department member, Stephen Jay Gould, formed the Sociobiology Study Group to publicly denounce Wilson's ideas. Wilson needed bodyguards and was publicly attacked at academic conferences.

Stephen Jay Gould (1976) proposed another, equally plausible scenario, one that discounts the existence of a particular gene programmed for altruism. He perceived the grandmother's sacrifice as an adaptive cultural trait. (It is widely acknowledged that culture is a major adaptive mechanism for humans.) Gould suggested that the elders remain behind because they have been socially conditioned from earliest childhood to the possibility and the appropriateness of this choice. They grew up hearing the songs and stories that praised the elders who stayed behind. Such self-sacrificers were the greatest heroes of the clan. Families whose elders rose to such an occasion survived to celebrate the self-sacrifice, but those families without self-sacrificing elders died out.

Wilson made several major concessions to Gould's viewpoint, acknowledging that among human beings, "the intensity and form of altruistic acts are to a large extent culturally determined" and that "human social evolution is obviously more cultural than genetic." He also left the door open to free will, admitting that even though our genetic coding may have a major influence, we still have the ability to choose an appropriate course of action (Wilson,

1978). However, Wilson insisted, "history did not begin 10,000 years ago. . . . [B]iological history made us what we are no less than culture" (1994).

Gould conceded that human behavior has a biological, or genetic, base. He distinguished, however, between genetic determinism (the sociobiological viewpoint) and genetic potential. What the genes prescribe is not necessarily a particular behavior but rather the capacity for developing certain behaviors. Although the total array of human possibilities is inherited, which of these numerous possibilities a particular person displays depends on his or her experience in the culture.

In later years, two similar research fields grew out of sociobiology: human behavioral ecology and evolutionary psychology. Both disciplines try to understand contemporary human behavior by referring to evolutionary developments. As with sociobiology, both approaches are still the cause of much debate.

Some of the opposition to these fields is related to the nature versus nurture debate. Biology is about nature; culture is about nurture. Critics of evolutionary theories note they can explain genetically determined behaviors, but not behaviors that are culturally learned. As most human behavior is learned, they assert little is to be gained from sociobiology, or evolutionary psychology (Richerson & Boyd, 2004).

The debate over the relative contribution of nature and nurture continues. Probably the best way to think about the influence of genes is to compare them to a recipe, one in which the ingredients are influenced by the environment. Some traits are more sensitive to environmental influence than others. However, just as a winter snowfall is the result of both the temperature and the moisture in the air, so must the human organism and human behavior be understood in terms of both genetic inheritance and the effects of environment. Nurture—that is, the entire socialization experience—is as essential a part of human nature as are our genes. It is from the interplay between genes and the environment that each human being emerges (Fisher, 1994).

Deprivation and Development

Some unusual events and interesting research indicate that human infants need more than just food and shelter if they are to function effectively as social creatures.

Extreme Childhood Deprivation   There are only a few recorded cases of human beings who have grown up without any real contact with other humans. One such case was found in January of 1800, when hunters in Aveyron, in southern France, captured a boy who was running naked through the forest. He seemed to be about 11 years old and apparently had

Jean-Marc Itard attempted to establish normal human relationships and communication with the "Wild Boy of Aveyron," whom he nicknamed "Victor."

been living alone in the forest for at least five or six years. He appeared to be thoroughly wild and subsequently was exhibited in a cage, from which he managed to escape several times. Finally, he was examined by "experts" who found him to be an incurable "idiot" and placed him in an institute for deaf-mutes.

A young doctor, Jean-Marc Itard, thought differently however. After close observation, he discovered that the boy was neither deaf nor mute nor an idiot. Itard believed that the boy's wild behavior, lack of speech, highly developed sense of smell, and poor visual attention span all were the result of having been deprived of human contact. It appeared that the crucial socialization provided by a family had been denied him. Though human, the boy had learned little about how to live with other people. Itard took the boy into his house, named him Victor, and tried to socialize him. He had little success. Although Victor slowly learned to wear clothes, to speak and write a few simple words, and to eat with a knife and fork, he ignored human voices unless they were associated with food and developed no relationships with people other than Itard and the housekeeper who cared for him. He died at the age of 40 (Itard, 1932; Shattuck, 1980).

Another sad case concerns a girl named Anna, who grew up in the 1930s and had the misfortune of being born illegitimately to the daughter of an extremely disapproving family. Her mother tried to place Anna with foster parents, was unable to do so, and therefore brought her home. To quiet the family's harsh criticisms, the young mother hid Anna away in a room in the attic, where she could be out of sight and even forgotten by the family. Anna remained there for almost six years, ignored by the whole family, including her mother, who did the very minimum to keep her alive. Finally, Anna was discovered by social workers. The 6-year-old girl was unable to sit up, to walk, or to talk. In fact, she was so withdrawn from human beings that at first she appeared to be deaf, mute, and brain damaged. However, after she was placed in a special school, Anna did learn to communicate somewhat, to walk (awkwardly), to care for herself, and even to play with other children. Unfortunately, she died at the age of 10 (Davis, 1940).

Another case of extreme childhood isolation was that of a girl named Genie, who came to the attention of authorities in California in 1970. Genie's nearly blind mother went to the California social welfare offices looking for help for herself, not Genie. The social worker noticed that the child was small, looked "withered," and had "a halting gait and an unnaturally stooped posture." The worker alerted her supervisor to what she thought was an unreported case of autism in a child estimated to be 6 or 7 years old.

Genie was actually 13-1/2 years old, weighed only 59 pounds, and was only 54 inches tall. She could not focus her eyes beyond 12 feet, could not chew solid food, showed no perception of hot and cold, and could not talk. Her condition was due to her father, who throughout her whole life had confined Genie to a small bedroom, harnessed to an infant's potty seat. Genie was left to sit in the harness, unable to move anything except her fingers and hands, feet and toes, hour after hour, day after day, month after month, year after year (Rhymer, 1993).

When Genie was hospitalized she was unsocialized, severely malnourished, unable to speak or even stand upright. After four years in a caring environment, Genie had learned some social skills, was able to take a bus to school, had begun to express some feelings toward others, and had achieved the intellectual development of a 9-year-old. There were still, however, serious problems with her language development that could not be corrected, no matter how involved the instruction (Curtiss, 1977).

Another example of a child being deprived of proper care and socialization is Oxana Malaya. Oxana spent much of her childhood between the ages of 3 and 8 living in a kennel with dogs in the back garden of the family home in the Ukraine. On some occasions she would be allowed in the house with her alcoholic parents, but most of her time was spent with the dogs. With dogs as her constant companions, she growled, barked, and crouched like a wild dog, smelled her food before she ate it, and was found to have acquired extremely acute senses of hearing, smell, and sight.

When her situation came to light in 1991 at the age of 8 she could hardly speak. She has acquired some language, but her communication with others is limited. It appears she missed some of the critical periods for normal language development. She lives in a home for the mentally disabled (Feralchildren.com).

These examples of extreme childhood isolation point to the fact that none of the behavior we think of as typically human arises spontaneously. Humans must be taught to stand up, to walk, to talk, even to think. Human infants must develop **social attachments;** *they must learn to have meaningful interactions and affectionate bonds with others.* This seems to be a basic need of all primates, as the research by Harry F. Harlow shows.

In a series of experiments with rhesus monkeys, Harlow and his coworkers demonstrated the importance of body contact in social development (Harlow, 1959; Harlow & Harlow, 1962). In one experiment, infant monkeys were taken from their mothers and placed in cages where they were raised in isolation from other monkeys. Each cage contained two substitute mothers: One was made of hard wire and contained a feeding bottle; the other was covered with soft terry cloth but did not have a bottle. Surprisingly, the baby monkeys spent much more time clinging to the cloth mothers than to the wire mothers, even though they received no food at all from the cloth mothers. Apparently, the need to cling to and to cuddle against even this meager substitute for a real mother was more important to them than being fed.

Other experiments with monkeys have confirmed the importance of social contact in behavior. Monkeys raised in isolation *never* learn how to interact with other monkeys or even how to mate. If placed in a cage with other monkeys, those who were raised in isolation either withdraw or become violent and aggressive—threatening, biting, and scratching the others.

Female monkeys who are raised without affection make wretched mothers themselves. After being impregnated artificially and giving birth, such monkeys either ignore their infants or display a pattern of behavior described by Harlow as "ghastly."

As with all animal studies, we must be very cautious in drawing inferences for human behavior. After all, we are not monkeys. Yet Harlow's experiments show that without socialization, monkeys do not develop normal social, emotional, sexual, or maternal behavior. Because human beings rely on learning even more than monkeys do, it is likely that the same is true of us.

It is obvious that the human organism needs to acquire culture to be complete; it is very difficult, if not impossible, for children who have been isolated from other people from infancy onward to catch up. They apparently suffer permanent damage, although human beings do seem to be somewhat more adaptable than were the rhesus monkeys Harlow studied.

Infants in Institutions Studies of infants and young children in institutions confirm the view that human beings' developmental needs include more than the mere provision of food and shelter. Psychologist Rene Spitz (1945) visited orphanages in Europe and found that in those dormitories where children were given routine care but otherwise were ignored, they were slow to develop and were withdrawn and sickly. These children's needs for social attachment were not met.

With the fall of the former Soviet Union, people in the West became aware of the conditions in eastern European orphanages. In many of these orphanages, no consistent responsive caregiving was provided. Children had several caregivers each, which prevented close individual attention from someone who knew the child. The children received little contact due to understaffing and uninformed or disinterested caregivers. Often the rooms had no toys or other objects for mental stimulation. Because malnutrition was common in eastern European culture, many times the malnourished staff would consume food meant for the children. Frequently children were not spoken to or called by their names, and were left in bed for hours without any attention. There was very little attempt to provide physical, mental, or emotional stimulation for these children.

Children from these experiences usually displayed an **attachment disorder**—they were *unable to trust people and to form relationships with others.* Many people who adopted children from these settings and thought that "love was all the children needed" to overcome these early experiences discovered that extensive treatment was necessary for these children to ever become normally functioning adults (Doolittle, 1995).

It appears clear that human infants need more than just food and shelter if they are to grow and develop properly. Every human infant needs frequent contact with others who demonstrate affection, who respond to attempts to interact, and who themselves initiate interactions with the child. Infants also need contact with people who find ways to interest them in their surroundings and who teach them the physical and social skills and knowledge they need to function. In addition, to develop normally, children need to be taught the culture of their society—to be socialized into the world of social relations and symbols that are the foundation of the human experience.

The Concept of Self

Every individual comes to possess a social identity by occupying **statuses**—*culturally and socially defined positions*—in the course of his or her socialization.

This social identity changes as the person moves through the various stages of childhood and adulthood recognized by the society. New statuses are occupied; old ones are abandoned. Picture a teenage girl who volunteers as a candy striper in a community hospital. She leaves that position to attend college, joins a sorority, becomes a premedical major, and graduates. She goes to medical school, completes a residency, becomes engaged, and then enters a program for specialized training in surgery. Perhaps she marries; possibly she has a child. All along the way she is moving through different social identities, often assuming several at once. When, many years later, she returns to the hospital where she was a teenage volunteer, she will have an entirely new social identity: adult woman, surgeon, wife (perhaps), mother (possibly).

This description of the developing girl is the view from the outside, the way that other members of the society experience her social transitions, or what sociologists would call changes in her social identity. A **social identity** is *the total of all the statuses that define an individual.* But what of the girl herself? How does this human being who is growing and developing physically, emotionally, intellectually, and socially experience these changes? Is there something constant about a person's experience that allows one to say, "I am that changing person—changing, but yet somehow the same individual"? In other words, do all human beings have personal identities separate from their social identities? Most social scientists believe that the answer is yes. *This changing yet enduring personal identity* is called the **self**.

The self develops when the individual becomes aware of his or her feelings, thoughts, and behaviors as separate and distinct from those of other people. This usually happens at a young age, when children begin to realize that they have their own history, habits, and needs and begin to imagine how these might appear to others. By adulthood, the concept of self is developed fully. Most researchers would agree that the concept of self includes (1) an awareness of the existence, appearance, and boundaries of one's own body (you are walking among the other members of the crowd, dressed appropriately for the occasion, and trying to avoid bumping into people as you chat); (2) the ability to refer to one's own being by using language and other symbols ("Hi, as you can see from my name tag, I'm Harry Hernandez from Gonzales, Texas."); (3) knowledge of one's personal history ("Yup, I grew up in Gonzales. My folks own a small farm there, and since I was a small boy I've wanted to study farm management."); (4) knowledge of one's

© Hoby Finn/PhotoDisc/Getty Images, Inc.

The process of socialization involves trying on a variety of roles that will eventually make up the self.

needs and skills ("I'm good with my hands all right, but I need the intellectual stimulation of doing large-scale planning."); (5) the ability to organize one's knowledge and beliefs ("Let me tell you about planning crop rotation."); (6) the ability to organize one's experiences ("I know what I like and what I don't like."); and (7) the ability to take a step back and look at one's being as others do, to evaluate the impressions one is creating, and to understand the feelings and attitudes one stimulates in others ("It might seem a little funny to you that a farmer like me would want to come to a party for the opening of a new art gallery. Well, as far back as I can remember, I always kinda enjoyed art, and now that I can afford to indulge myself, I thought maybe I'd buy something I like.") (see Cooley, 1909; Mead, 1935; Erikson, 1964; Gardner, 1978).

Dimensions of Human Development

Clearly, the development of the self is a complicated process. It involves many interacting factors, including the acquisition of language and the ability to use symbols. Three dimensions of human development are tied to the emergence of the self: cognitive development, moral development, and gender identity.

Cognitive Development   For centuries, most people assumed that a child's mind worked in exactly the same way as an adult's mind. The child was thought of as a miniature adult who simply was lacking information about the world. Swiss philosopher and psychologist Jean Piaget (1896–1980) was instrumental in changing that view through his studies of the development of intelligence in children. His work has been significant to sociologists because the processes of thought are central to the development of identity and, consequently, to the ability to function in society.

Piaget found that children move through a series of predictable stages on their way to logical thought and that some never attain the most advanced stages. From birth to age 2, during the *sensorimotor stage,* the infant relies on touch and the manipulation of objects for information about the world, slowly learning about cause and effect. At about age 2, the child begins to learn that words can be symbols for objects. In this, the *preoperational stage* of development, the child cannot see the world from another person's point of view.

The *operational stage* is next and lasts from age 7 to about age 12. During this period, the child begins to think with some logic and can understand and work with numbers, volume, shapes, and spatial relationships. With the onset of adolescence, the child progresses to the most advanced stage of thinking—*formal, logical thought.* People at this stage are capable of abstract, logical thought and can develop ideas about things that have no concrete reference, such as infinity, death, freedom, and justice. In addition, they are able to anticipate possible consequences of their acts and decisions.

Achieving this stage is crucial to developing an identity and an ability to enter into mature interpersonal relationships (Piaget & Inhelder, 1969).

Moral Development   Every society has a **moral order**—that is, *a shared view of right and wrong.* Without moral order a society soon would fall apart. People would not know what to expect from themselves and one another, and social relationships would be impossible to maintain. Therefore, the process of socialization must include instruction about the moral order of an individual's society.

The research by Lawrence Kohlberg (1969) suggests that not every person is capable of thinking about morality in the same way. Just as our sense of self and our ability to think logically develop in stages, our moral thinking develops in a progression of steps as well. To illustrate this, Kohlberg asked children from a number of different societies (including Turkey, Mexico, China, and the United States) to resolve moral dilemmas such as the following: A man's wife is dying of cancer. A rare drug might save her, but it costs $2,000. The man tries to raise the money but can come up with only $1,000. He asks the druggist to sell him the drug for $1,000, and the druggist refuses. The desperate husband then breaks into the druggist's store to steal the drug. Should he have done so? Why or why not?

Kohlberg was more interested in the reasoning behind the child's judgment than in the answer itself. Based on his analysis of this reasoning, he concluded that changes in moral thinking progress step by step through six qualitatively distinct stages (although most people never go beyond Stages 3 or 4):

Stage 1. Orientation toward punishment.

Stage 2. Orientation toward reward.

Stage 3. Orientation toward possible disapproval by others.

Stage 4. Orientation toward formal laws and fear of personal dishonor.

Stage 5. Orientation toward peer values and democracy.

Stage 6. Orientation toward one's own set of values.

Kohlberg found that although these stages of moral development correspond roughly to other aspects of the developing self, most people never progress to Stages 5 and 6. In fact, Kohlberg subsequently dropped Stage 6 from his scheme because it met with widespread criticism that he could not refute. Critics

Boston Filmworks

If a child's need for meaningful interaction with another is not met, the development of a social identity will be delayed.

felt that Stage 6 was elitist and culturally biased. Kohlberg himself could find no evidence that any of his long-term subjects ever reached this stage (Muson, 1979). At times, people regress from a higher stage to a lower one. For example, when Kohlberg analyzed the explanations that Nazi war criminals of World War II gave for their participation in the systematic murder of millions of people who happened to possess certain religious (Jewish), ethnic (gypsies), or psychological (mentally retarded) traits, he found that none of the reasons was above Stage 3 and most were at Stage 1—"I did what I was told to do; otherwise, I'd have been punished" (Kohlberg, 1967). However, many of these war criminals had been very responsible and successful people in their prewar lives and presumably in those times had reached higher stages of moral development.

Gender Identity   One of the most important elements of our sense of self is our gender identity. Certain aspects of gender identity are rooted in biology. Males tend to be larger and stronger than females, but females tend to have better endurance than males. Females also become pregnant and give birth to infants and (usually) can nurse infants with their own milk. However, gender identity is mostly a matter of cultural definition. There is nothing inherently male or female about a teacher, a pilot, a carpenter, or a pianist other than what our culture tells us. As we shall see in Chapter 11, gender identity and sex roles are far more a matter of nurture than of nature.

Theories of Development

Among the scholars who have devised theories of development, Charles Horton Cooley, George Herbert Mead, Sigmund Freud, Erik Erikson, and Daniel Levinson stand out because of the contributions they have made to the way sociologists today think about socialization. Cooley and Mead saw the individual and society as partners. They were symbolic interactionists (see Chapter 1) and as such believed that the individual develops a self solely through social relationships—that is, through interaction with others. They believed that all our behaviors, our attitudes, even our ideas of self arise from our interactions with other people. Hence, they were pure environmentalists, in that they believed that social forces rather than genetic factors shape the individual.

Freud, on the other hand, tended to picture the individual and society as enemies. He saw the individual as constantly having to yield reluctantly to the greater power of society, to keep internal urges (especially sexual and aggressive ones) under strict control.

Erikson and Levinson presented something of a compromise position. They thought of the individual as progressing through a series of stages of development that express internal urges, yet are greatly influenced by societal and cultural factors.

Charles Horton Cooley (1864–1929)

Cooley believed that the self develops through the process of social interaction with others. This process begins early in life and is influenced by such primary groups as the family. Later on, peer groups become very important as we continue to progress as social beings. Cooley used the phrase **looking-glass self** to describe *the three-stage process through which each of us develops a sense of self.* First, we imagine how our actions appear to others. Second, we imagine how other people judge these actions. Finally, we make some sort of self-judgment based on the presumed judgments of others. In effect, other people become a mirror or looking glass for us (1909).

In Cooley's view, therefore, the self is entirely a social product—that is, a product of social interaction. Each individual acquires a sense of self in the course of being socialized and continues to modify it in each new situation throughout life. Cooley believed that the looking-glass self constructed early in life remains fairly stable and that childhood experiences are very important in determining our sense of self throughout our lives.

One of Cooley's principal contributions to sociology was his observation that although our perceptions are not always correct, what we believe is more important in determining our behavior than is what is real. This same idea was also expressed by sociologist W. I. Thomas (1928) when he noted, "If men define situations as real, they are real in their consequences." If we can understand the ways in which people perceive reality, then we can begin to understand their behavior.

George Herbert Mead (1863–1931)

Mead was a philosopher and a well-known social psychologist at the University of Chicago. His work led to the development of the school of thought called symbolic interactionism (described in Chapter 1). As a student of Cooley's, Mead built on Cooley's ideas, tracing the beginning of a person's awareness of self to the relationships between the caregiver (usually the mother) and the child (1934).

According to Mead, the self becomes the sum total of our beliefs and feelings about ourselves. The self is composed of two parts, the "I" and the "me." The **"I"** *portion of the self wishes to have free expression, to be active and spontaneous.* The "I" wishes to be free of the control of others and to take the initiative in situations. It is the part of the individual that is unique and distinctive. The **"me"** *portion of the self is made up of those things learned through the socialization process from family, peers, school, and so on.* The "me" makes normal social interaction possible, while the "I" prevents it from being mechanical and totally predictable.

Mead used the term **significant others** to refer to *those individuals who are most important in our development, such as parents, friends, and teachers.* As we continue to be socialized, we learn to be aware of the views of the generalized others. These **generalized others** are *the viewpoints, attitudes, and expectations of society as a whole, or of a community of people whom we are aware of and who are important to us.* We may believe it is important to go to college, for example, because significant others have instilled this viewpoint in us. While at college we may be influenced by the views of selected generalized others who represent the community of lawyers that we hope to join one day as we progress with our education.

Mead believed that the self develops in three stages (1934). The first or **preparatory stage** is *characterized by the child's imitating the behavior of others, which prepares the child for learning social-role expectations.* In the second or **play stage,** *the*

George Herbert Mead's writings led to development of the school of thought known as symbolic interactionism.

child has acquired language and begins not only to *imitate behavior, but also to formulate role expectations:* playing house, cops and robbers, and so on. In this stage, the play features many discussions among playmates about the way things "ought" to be. "I'm the boss," a little boy might announce. "The daddy is the boss of the house." "Oh no," his friend might counter, "Mommies are the real bosses." In the third or **game stage,** *the child learns that there are rules that specify the proper and correct relationship among the players.* For example, a baseball game has rules that apply to the game in general as well as to a series of expectations about how each position should be played. During the game stage, according to Mead, we learn the expectations, positions, and rules of society at large. Throughout life, in whatever position we occupy, we must learn the expectations of the various positions with which we interact, as well as the expectations of the general audience, if our performance is to go smoothly.

Thus, for Mead the self is rooted in, and begins to take shape through, the social play of children and is well on its way to being formed by the time the child is 8 or 9 years old. Therefore, like Cooley, Mead regarded childhood experience as very important to charting the course of development.

Sigmund Freud (1856–1939)

Freud was a pioneer in the study of human behavior and the human mind. He was a doctor in Vienna, Austria, who gradually became interested in the problem of understanding mental illness.

In Freud's view, the self has three separately functioning parts: the id, the superego, and the ego. The **id** consists of *the drives and instincts that Freud believed every human being inherits, but which for the most part remain unconscious.* Of these instincts, two are most important: the aggressive drive and the erotic or sexual drive (called libido). Every feeling derives from these two drives. The **superego** represents *society's norms and moral values as learned primarily from our parents.* The superego is the internal censor. It is not inherited biologically, like the id, but is learned in the course of a person's socialization. The superego keeps trying to put the brakes on the id's impulsive attempts to satisfy its drives. So, for instance, the superego must hold back the id's unending drive for sexual expression (Freud, 1920, 1923). The id and superego, then, are eternally at war with each other. Fortunately, there is a third functional part of the self called the **ego,** which *tries not only to mediate in the eternal conflict between the id and the superego, but also to find socially acceptable ways for the id's drives to be expressed.* Unlike the id, the ego constantly evaluates social realities and looks for ways to adjust to them (Freud, 1920, 1923).

Freud pictured the individual as constantly in conflict: The instinctual drives of the id (essentially sex and aggression) push for expression, while at the same time the demands of society set certain limits on the behavior patterns that will be tolerated.

Even though the individual needs society, society's restrictive norms and values are a source of ongoing discontent (Freud, 1930). Freud's theories suggest that society and the individual are enemies, with the latter yielding to the former reluctantly and only out of compulsion.

Erik H. Erikson (1902–1994)

In 1950 Erikson, an artist-turned-psychologist who studied with Freud in Vienna, published an influential book called *Childhood and Society* (1964). In it he built on Freud's theory of development but added two important elements. First, he stressed that development is a lifelong process and that a person continues to pass through new stages even during adulthood. Second, he paid greater attention to the social and cultural forces operating on the individual at each step along the way.

In Erikson's view, human development is accomplished in eight separate stages (Table 4–1). Each stage amounts to a crisis of sorts brought on by two

Table 4–1

Erikson's Eight Stages of Human Development

Stage	Age Period	Characteristic to Be Achieved	Major Hazards to Achievement
Trust vs. mistrust	Birth to 1 year	Sense of trust or security—achieved through parental gratification of needs and affection	Neglect, abuse, or deprivation; inconsistent or inappropriate love in infancy; early or harsh weaning
Autonomy vs. shame and doubt	1 to 4 years	Sense of autonomy—achieved as child begins to see self as individual apart from his/her parents	Conditions that make the child feel inadequate, evil, or dirty
Initiative vs. guilt	4 to 5 years	Sense of initiative—achieved as child begins to imitate adult behavior and extends control of the world around him/her	Guilt produced by overly strict discipline and the internalization of rigid ethical standards that interfere with the child's spontaneity
Industry vs. inferiority	6 to 12 years	Sense of duty and accomplishment—achieved as the child lays aside fantasy and play and begins to undertake tasks and schoolwork	Feelings of inadequacy produced by excessive competition, personal limitations, or other events leading to feelings of inferiority
Identity vs. role confusion	Adolescence	Sense of identity—achieved as one clarifies sense of self and what he/she believes in	Sense of role confusion resulting from the failure of the family or society to provide clear role models
Intimacy vs. isolation	Young adulthood	Sense of intimacy—the ability to establish close personal relationships with others	Problems with earlier stages that make it difficult to get close to others
Generativity vs. stagnation	30s to 50s	Sense of productivity and creativity—resulting from work and parenting activities	Sense of stagnation produced by feeling inadequate as a parent and stifled at work
Integrity vs. despair	Old age	Sense of ego integrity—achieved by acceptance of the life one has lived	Feelings of despair and dissatisfaction with one's role as a senior member of society

Source: Adapted from *Childhood and Society,* 2d. ed., by E. H. Erikson, 1964. Copyright © 1950, 1963 by W. W. Norton & Company, Inc., renewed © 1978, 1991 by Erik H. Erikson. Used by permission of W. W. Norton & Company, Inc.

factors: biological changes in the developing individual and social expectations and stresses.

At each stage the individual is pulled in two opposite directions to resolve the crisis. In normal development, the individual resolves the conflict experienced at each stage somewhere toward the middle of the opposing options. For example, very few people are entirely trusting, and very few trust nobody at all. Most of us are able to trust at least some other people and thereby form enduring relationships, while at the same time staying alert to the possibility of being misled.

Erikson's view of development has proved to be useful to sociologists because it seems to apply to many societies. In a later work (1968), he focused on the social and psychological causes of the "identity crisis" that seems to be so prevalent among American and European youths. Erikson's most valuable contribution to the study of human development has been to show that socialization continues throughout a person's life

© David Young-Wolff/Alamy

According to Erik Erikson, adolescence is a time when the teenager must develop an identity as well as the ability to establish close personal relationships with others.

and does not stop with childhood. People continue to develop after age 30—and after 60 and 70 as well. The task of building the self is lifelong; it can be considered our central task from cradle to grave. We construct the self—our identity—using the materials made available to us by our culture and our society.

Daniel Levinson (1920–1994)

A blending of sociology and psychology has occurred in the area of adult development. Through research in this field, we have come to recognize that there are predictable age-related developmental periods in the adult life cycle, just as there are in the developmental cycles for children and adolescents. These periods are marked by a concerted effort to resolve particular life issues and goals.

Levinson (1978) did research in this area involving a male population. He recruited 40 men, ages 35 to 45, from four occupational groups: factory workers, novelists, business executives, and academic biologists. Each participant was interviewed several times during a two- to three-month period and again, if possible, in a follow-up session two years later. Later, Levinson (1996) repeated his study with 45 women in three broad categories: academics, corporate-financial careers, and homemakers.

From these studies, Levinson developed the foundation of his theory. He proposed that adults periodically are faced with new but predictable developmental tasks throughout their lives and that working through these challenges is the essence of adulthood. Levinson believed that the adult life course is marked by a continual series of building periods, followed by stable periods, and then followed again by periods in which people make attempts to change some of the perceived flaws in the previous design.

Levinson believed that both men and women go through the same periods of adult development, although there are differences due to the external and internal constraints. For example, gender-role expectations in society or the biological demands of childbearing will influence how a woman works on the developmental tasks of each stage. The model is particularly interesting to sociologists because it appears to show us that there is a close relationship between individual development and one's position in society at a particular time.

There are, however, problems with being too quick to embrace this type of theory. Levinson's theory of adult development is what is commonly referred to as a "stage theory." Stage theories describe a series of changes that follow an orderly sequential pattern. These theories have been criticized as being too rigid, for assuming that the changes are always in one direction, and for assuming that the stages are universal. Critics say that we can apply stages of development to children, but that when we do so with adults, we leave little room for individual differences, social change, and specific cohort experiences (Neugarten, 1979).

Early Socialization in American Society

Children are brought up very differently from one society to another. Each culture has its own childrearing values, attitudes, and practices. No matter how children are raised, however, each society must provide certain minimal necessities to ensure normal development. The infant's physical needs must, of course, be addressed, but more than that is required. Children need speaking social partners (some evidence suggests that a child who has received no language stimulation at all in the first five to six years of life will be unable ever to acquire speech [Chomsky, 1975]). They also need physical stimulation; objects that they can manipulate; space and time to explore, to initiate activity, and to be alone; and finally, limits and prohibitions that organize their options and channel development in certain culturally specified directions (Provence, 1972).

Every family socializes its children to its own particular version of the society's culture. The values and worldview of a boy brought up on a dairy farm in Wisconsin are likely to differ from those of a child born and brought up in a city such as Los Angeles or New York City.

© Robert Frerck/Odyssey/Chicago

Every society provides this basic minimum care in its own culturally prescribed ways. A variety of agents, which also vary from culture to culture, are used to mold the child to fit into the society. Here we consider some of the most important agents of socialization in American society.

The Family

For young children in most societies—and certainly in American society—the family is the primary world for the first few years of life. Children also are having significant early experiences in day-care centers with no family members. The values, norms, ideals, and standards presented are accepted by the child uncritically as correct—indeed, as the only way things could possibly be. Even though later experiences lead children to modify much of what they have learned within the family, it is not unusual for individuals to carry into the social relationships of adult life the role expectations that characterized the family of their childhood. Many of our gender-role expectations are based on the models of female and male behavior we witnessed in our families.

Every family, therefore, socializes its children to its own particular version of the society's culture. In addition, however, each family exists within certain subcultures of the larger society: It belongs to a geographical region, a social class, one or two ethnic groups, and possibly a religious group or other subculture. Families differ with regard to how important these factors are in determining their lifestyles and their childrearing practices. For example, some families are very deeply committed to a racial or ethnic identification such as African American, Hispanic, Chinese, Native American, Italian American, Polish American, or Jewish American. Much of family life may revolve around participation in social and religious events of the community and may include speaking a language other than English.

Evidence also shows that social class and parents' occupations influence the ways that children are raised in the United States. Parents who have white-collar occupations are accustomed to dealing with people and solving problems. As a result, white-collar parents value intellectual curiosity and flexibility. Blue-collar parents have jobs that require specific tasks, obeying orders, and being on time. They are likely to reward obedience to authority, punctuality, and physical or mechanical ability in their children (Kohn & Schooler, 1983).

The past four decades have seen major changes in the structure of the American family. High divorce rates, the dramatic increase in the number of single-parent families, and the common phenomenon of two-worker families have meant that the family as the major source of socialization of children is being challenged. Child-care providers have become a major influence in the lives of many young children. (For a discussion of the effects of day care on the socialization of children, see "Controversies in Sociology: Is Day Care Harmful to Children?")

The School

The school is an institution intended to socialize children in selected skills and knowledge. In recent decades, however, the school has been assigned additional tasks. For instance, in poor communities and neighborhoods, school lunch (and breakfast) programs are an important source of balanced nutrition for children. The school must also confront a more basic problem. As an institution, it must resolve the conflicting values of the local community and of the state and regional officials whose job it is to determine what should be taught. For example, in many school systems parents want to reintroduce school prayer or discussions of religion, even though the Supreme Court has ruled that such actions are not permissible. In other instances, education officials

CONTROVERSIES IN SOCIOLOGY

Is Day Care Harmful to Children?

Americans have fairly strong opinions concerning child care and ideal working situations. According to a Gallup poll (May 4, 2001), 41% of Americans believe it is best for one parent to stay at home solely to raise children while the other parent works. Only 13% of Americans say the ideal situation is for both parents to work full time outside the home.

According to the U.S. Bureau of Labor, only 13% of families fit the traditional model of husband as wage-earner and wife as homemaker. In 61% of married couple families, both husband and wife work outside the home. Six of every 10 mothers of children younger than age 6 are in the labor force (Winkler, 1998). More than half (50.6%) of children whose mothers work full time spend at least 35 hours a week in day care (Capizzano & Main, 2005). Small wonder that child care is becoming an ever increasing fact of life.

Approximately two-thirds of all preschool children are in one of three types of day care. About 41% of all children are cared for by a relative, principally by their fathers or a grandparent. In the second type of arrangement, a woman may take several unrelated young children into her home on a regular basis; or a child care provider may come into the child's home, 17.4% of child care is of this type. Another 35.4% of the children are enrolled in day-care centers, and 5.0% of the children

are cared for by a parent (typically the mother) while they are working (Figure 4–1) (Smith, 2002).

With many parents putting their preschoolers in organized child care, people want to know whether these settings are doing a good job socializing children compared with traditional childrearing. Sociologists begin to answer this question by noting that just as all parenting is not uniformly good, neither is all day care mediocre. High-quality day care does exist and can be a suitable substitute for home parenting. Unfortunately, poor-quality care, which can threaten a child with serious physical and psychological harm, is also readily available.

As day care has become a common reality for many children, it may be too simplistic to discuss whether day care is good or bad in itself. What is really important is what the quality of the day care is. Studies show that children who spend time in high-quality child care are just as likely to have secure attachment relationships as children who stay at home. In addition, quality child care can also lead to better mother-child interaction. Children whose home circumstances are less than ideal and who attend high-quality day care have been shown to benefit from the relationship they develop with a sensitive and responsive child-care provider (National Institutes of Health, 1997).

Figure 4–1 **Primary Child-Care Arrangements For Preschool Children**

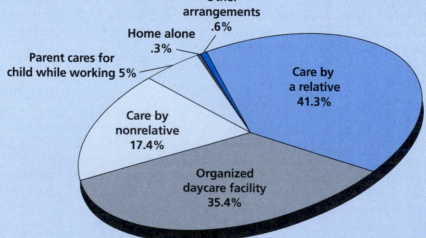

Source: Smith, Kristin, "Who's Minding the Kids? Child Care Arrangements: Spring 1997." *Current Population Reports,* pp. 70–86 (U.S. Census Bureau, Washington, DC, 2002).

Some studies suggest that children in child-care centers do better on tests of verbal fluency, memory, and comprehension. These children also have more advanced language skills and tend to be less timid and fearful, be more outgoing and cooperative, and function more successfully in social relationships. When contrasted with smaller families with only one or two children, day-care children have a chance to interact with their peers at a younger age. When they get to school, such necessities as sharing and making friends do not come as such a shock (Clarke-Stewart & Allhusan, 2005).

At the same time, children in child care tend to be less polite, less agreeable, less respectful of others, more rebellious, loud, and boisterous, and more aggressive. Children suffer when they spend long hours in poor-quality day care, especially when the children are young (Clarke-Stewart & Allhusan, 2005). One study found that children do not do well on school-readiness tests at age 3 if their mothers worked more than 30 hours per week by the time the child was 9 months old (National Institute of Child Health and Human Development, 1998). In addition, "children who experience long hours of child care over the first four years of life are more at risk for showing behavior problems, particularly aggression. Not only were these children more likely to engage in assertive, defiant, and even disobedient activities,

but they were also more likely to bully, fight with, or act mean to other children" (National Institute of Child Health and Human Development, 1998).

High-quality day care depends on the adult-child ratios in the center and the overall group size in which children are cared for. Staffing guidelines recommend that there be a child-care worker-to-child ratio of 1:3 for infants younger than age 2, 1:4 for toddlers, and 1:7 for preschoolers. Unfortunately, many day-care centers fall short in this area. Nearly half of all settings exceed optimal guidelines (Whitebook et al., 2001).

The stability of the caregiver and peers with whom the child spends the day is another important indicator of quality. Staff turnover in the child-care profession is very high. Child-care workers earn considerably less than others in the workforce with similar levels of education. Many hold college degrees, yet the median hourly wage for a child-care worker is $7.86 per hour (Bureau of Labor Statistics, 2005). This is about one-third of what kindergarten teachers earn. Small wonder that there is a 30% yearly turnover rate within the field, because child-care workers see little incentive to remain at the job when other options become available (Whitebrook et al., 2001).

Approximately 21% of working parents place their children in the care of a grandparent. This is often a less

© Owen Franken/CORBIS

Day-care quality is dependent on the adult-child ratios in the center and the overall size of the group within which the children are cared for.

continued

CONTROVERSIES IN SOCIOLOGY

Is Day Care Harmful to Children?   (continued)

ideal option than a day-care center. Although many parents may feel it is safest and most convenient to leave their child with a family member, day-care centers are more likely to plan activities for the children and the workers have more training in the field than relatives.

What we now know about day-care arrangements has been summarized as follows (National Institute of Child Health and Human Development, 1998):

Higher-quality day care is related to
- Better mother-child relationships
- Lower probability of insecure attachment in infants of mothers low in sensitivity
- Fewer reports of children's problem behaviors
- Higher cognitive performance of children
- Higher language ability of children
- Higher level of school readiness

Poor-quality day care produces
- Less harmonious mother-child relationships
- Higher probability of insecure mother-child attachment in infants of mothers already low in sensitivity
- More problem behaviors, lower cognitive and language ability, and lower school-readiness scores

Mothers working outside the home also are providing their children with a different model of what a woman's role in society is. This is particularly important to women who have daughters. Daughters of working women see the world in a less gender-stereotypical fashion than do daughters of nonworking women.

Because day care is now a reality for a large segment of the nation's preschoolers, the issue that needs to be

addressed is how to make it a rewarding socialization experience. Clearly, this requires well-trained staff and a properly supervised environment. If we can create that type of environment, then day care can be a positive experience.

Sources: The Gallup Poll, May 4, 2001, "Few Say It's Ideal for Both Parents to Work Full Time Outside of Home"; Department of Health and Human Services Center for the Child Care Workforce, Washington, DC; National Institute of Child Health and Human Development, 1998, *The NICHD Study of Early Child Care;* National Institutes of Health, April 3, 1997, "Results of NICHD Study of Early Child Care Reported at Society for Research in Child Development Meeting," *NIH News Release;* Kristin Smith, "Who's Minding the Kids? Child Care Arrangements: Spring 1997." *Current Population Reports,* 2002, pp. 70–86, U.S. Census Bureau, Washington DC; Marcy Whitebook, Laura Sakai, Emily Gerber, and Carollee Howes, *Then and Now: Changes in Child Care Staffing 1994–2000,* 2001, Washington, DC, Center for Child Care Workforce; Anne E. Winkler, "Earnings of Husbands and Wives in Dual-Earner Families." *Monthly Labor Review,* April 1998. Bureau of Labor Statistics, U.S. Department of Labor, *Occupational Outlook Handbook, 2004-05 Edition,* Childcare Workers, on the Internet at http://www.bls.gov/oco/ocos170.htm (visited August 09, 2005); Alison Clarke-Stewart and Virginia D. Allhusan, *What We Know about Childcare,* Cambridge, MA: Harvard University Press, 2005; Capizzano, Jeffrey and Regan Main, "Many Young Children Spend Long Hours in Child Care, Washington, DC, The Urban Institute, March 31, 2005.

make curriculum changes in the classroom despite the complaints of parents, whose objections are dismissed as perhaps uninformed or tradition bound.

AIDS education is a vivid example of schools deciding what issues should or should not be presented to children. Some parents have objected to teaching young children about sexuality, condoms, and homosexuality, despite the health risks that ignorance could pose. Many school boards have taken the position that the schools have a responsibility to provide this information, even when large numbers of parents object.

In coming to grips with their multiple responsibilities, many school systems have established a philosophy of education that encompasses socialization as well as academic instruction. Educators often aim

to help students develop to their fullest capacity, not only intellectually, but also emotionally, culturally, morally, socially, and physically. By exposing the student to a variety of ideas, the teachers attempt to guide the development of the whole student in areas of interests and abilities unique to each. Students are expected to learn how to analyze these ideas critically and reach their own conclusions. The ultimate goal of the school is to produce a well-integrated person who will become socially responsible. Two questions arise: Is such an ambitious, all-embracing educational philosophy working? And is it an appropriate goal for our schools?

In a way, the school is a model of much of the adult social world. Interpersonal relationships are not based on individuals' love and affection for one another. Rather, they are impersonal and predefined by the society with little regard for each particular individual who enters into them. Children's process of adjustment to the school's social order is a preview of what will be expected as they mature and attempt to negotiate their way into the institutions of adult society (job, political work, organized recreation, and so on). Of all the socializing functions of the school, this preview of the adult world may be the most important. (The role of the school in socialization will be discussed more extensively in Chapter 14.)

Peer Groups

Peers are *individuals who are social equals.* From early childhood until late adulthood, we encounter a wide variety of peer groups. No one will deny that they play a powerful role in our socialization. Often their influence is greater than that of any other source of socialization.

Within the family and the school, children are in socially inferior positions relative to figures of authority (parents, teachers, principals). As long as the child is small and weak, this social inferiority seems natural, but by adolescence a person is almost fully grown, and arbitrary submission to authority is not so easy to accept. Hence, many adolescents withdraw into the comfort of social groups composed of peers.

Parents may play a major role in the teaching of basic values and the development of the desire to achieve long-term goals, but peers have the greatest influence in lifestyle issues, such as appearance, social activities, and dating. Peer groups also provide valuable social support for adolescents who are moving toward independence from their parents.

As a consequence, their peer-group values often run counter to those of the older generation. New group members quickly are socialized to adopt symbols of group membership such as styles of dress, use and consumption of certain material goods, and stylized patterns of behavior. A number of studies have documented the increasing importance of peer-group socialization in the United States. One reason for this is that parents' life experiences and accumulated wisdom may not be very helpful in preparing young people to meet the requirements of life in a society that is changing constantly. Not infrequently, adolescents are better informed than their parents are about such things as sex, drugs, and technology.

Peer-group influence, for many youths, can lead to wasted lives and violence. For many, gangs—kids banding together for identity, status, petty criminal activity, and mutual protection—often involve drug abuse. In many urban high schools, for example, cocaine is traded, and attempts to emphasize the dangers of drugs fall on deaf ears. (See "Our Diverse

Peer groups provide valuable social support for adolescents.

Society: Win Friends and Lose Your Future: The Costs of Not 'Acting White'" for another aspect of peer-group influence.)

The negative effects of peer pressure are felt on college campuses as well as in poor inner-city neighborhoods. Peer pressure has caused deaths from hazing activities in college fraternities. For example, one month after entering the Massachusetts Institute of Technology, freshman Scott Kruger died after consuming excessive amounts of alcohol as part of a hazing ritual at a fraternity party.

As the authority of the family diminishes under the pressures of social change, peer groups move into the vacuum and substitute their own morality for that of the parents. Peer groups are most effective in molding the behavior of those adolescents whose parents do not provide consistent standards, a principled moral code, guidance, and emotional support. David Elkind (1981) has expressed the view that the power of the peer group is in direct proportion to the extent that the adolescent feels ignored by the parents. In fact, more than a half century ago sociologist David Riesman, in his classic work *The Lonely Crowd* (1950), already thought that the peer group had become the single most powerful molder of many adolescents'

OUR DIVERSE SOCIETY

Win Friends and Lose Your Future: The Costs of Not "Acting White"

"Children can't achieve unless we raise their expectations and turn off the television sets and eradicate the slander that says a black youth with a book is acting white."

—*Senator Barack Obama, 2004 Democratic National Convention Keynote Address*

It's been about 140 years since the end of slavery, yet black students lag significantly behind white students. The average black 17-year-old reads at the level of a white 13-year-old. Black students score one standard deviation below white students on the SAT test. This racial divide exists even when black students attend schools in affluent neighborhoods.

One way to explain this gap is to point to a black student peer culture that encourages students to not "act white." Factors that would be considered examples of acting white could include enrolling in honors or advanced placement courses, raising your hand in class, playing the violin, wearing clothes from Abercrombie and Fitch or the Gap, or speaking Standard English.

Students in high school can attest to the fact that there is a strong desire to be accepted by valued peer groups. Most of the time being accepted by a peer group does not require any major short-term costs. Yes, it is true that being part of one peer group might mean you will be rejected by another peer group. But usually, you will not end up financially poorer or seriously handicapped in some way because you are part of a sports, music, or other social peer group.

Becoming a member of a peer group means you have to decide what proportion of your time you are going to devote to the group and what it values, and what proportion of time to other activities and your studies. For some peer groups, studying and doing well overlaps with the values of the group. Other times these behaviors are in conflict with the group, and members may be ostracized for acting in a way that conflicts with the group values. If a peer group does not value academic achievement, loyalty to the group may result in mediocre or even failing grades.

In many high schools white students who do well actually have more friends than students who do poorly. Often for black students that is not the case. A black student who has a 4.0 average has 1.5 fewer friends than the white student with a 4.0 (Cook & Ludwig, 1997).

Let's look at the Frank W. Ballou High School, in Washington, D.C., as an example. The principal of the school decided that it would be a good idea to give outstanding students cash for their achievements. A student with perfect grades in any of the year's four marking periods would receive $100. With four marking periods, over the course of a year a student could get $400. There was a catch. The straight "A" students had to personally receive each check at an awards assembly. Soon the assemblies turned into forums in which the winners were jeered being called "Nerd!" "Geek!" "Egghead!" and the harshest of all, "Whitey!" The honor students were tormented for months afterwards. With each assembly, fewer students showed up (Susskind, 1998).

In the short term, then, for black students the cost of doing well is not being accepted. The benefit of not doing well is that you will have friends and a more active social life. In the future, however, things are different. Once high school is over, the benefit of peer acceptance rapidly disappears, but the cost of not doing well in school quickly starts to become important. Employers or colleges look at mediocre or poor grades as a sign that this person should be passed over for the job or college admission. The short-term benefit of being accepted by the peer group has now turned into a long-term liability that is difficult to erase.

A peer group socialization process that discourages students from doing well in school because it is considered a sign of "acting white" is exacting a long-term price that most students can ill afford.

Sources: Austen-Smith, David, and Roland G. Fryer, Jr. (2005, May). "An Economic Analysis of 'Acting White.'" *Quarterly Journal of Economics* Vol. 120, Issue 2, pp. 551–583; Fryer, Jr., Roland G., and Paul Torelli. (2005, May 1). "An Empirical Analysis of 'Acting White.'" Harvard University Society of Fellows and NBER, http://post.economics.harvard.edu/faculty/fryer/papers/fryer_torelli.pdf; Cook, Phillip, and Jens Ludwig. (1997). Weighing the "Burden of 'Acting White'": Are There Race Differences in Attitudes Towards Education? *Journal of Public Policy and Analysis* 16, 256–278; Susskind, Ron, *A Hope in the Unseen*, New York: Broadway Books, 1998.

behavior and that striving for peer approval had become the dominant concern of an entire American generation—adults as well as adolescents. He coined the term *other directed* to describe those who are overly concerned with finding social approval.

Television, Movies, and Video Games

It is possible that today Riesman would review his thinking somewhat. Since the late 1960s, the mass media—television, radio, magazines, films,

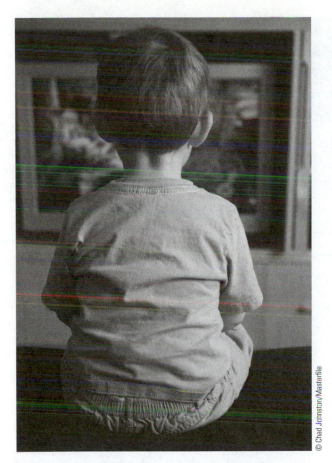

People have become increasingly concerned about the socializing role played by the mass media.

© Chad Johnston/Masterfile

newspapers, and the Internet—have become important agents of socialization in the United States. It is almost impossible in our society to escape from the images and sounds of television or radio; in most homes, especially those with children, the media are constantly visible or audible. With the exception of video games most mass media communication is one-way, creating an audience that is conditioned to receive passively whatever news, messages, programs, or events are brought to them.

Today, 98.2% of all households in the United States have television sets, with an average of 2.4 sets per home. Most children become regular watchers of television between ages 3 and 6 (Bureau of the Census, March 11, 2004).

Schoolchildren watch an average 2.5 hours of television a day on school days and an average of 4 hours and 20 minutes on an average weekend. Their favorite programs are situation comedies, cartoons, music videos, sports, game shows, talk shows, and soap operas. One study concluded that by the time most people reach the age of 18, they will have spent more waking time watching television than doing anything else—talking with parents, spending time with friends, or even going to school (Bureau of the Census, 2002).

Because young children are so impressionable, and because in so many American households the television is used as an unpaid mechanical babysitter, social scientists have become increasingly concerned about the socializing role the mass media play in our society. If children learn from experience, such exposure must certainly have an effect (see "Social Change: Should Television Be Used to Teach Values?"). By the time the average American child leaves elementary school, she or he has seen 8,000 killings and 100,000 violent acts portrayed on television.

Mark Sappenfield (2002) writes:

For much of the past half century, the link between watching violence on television and violent behavior in everyday life has seemed an open question—embraced by one study, rejected by another, and largely left unanswered by years of congressional inquiries. This, however, is rapidly changing. To a growing number of scientists and psychiatrists, the correlation between the two is no longer a point of debate; it is an established fact. . . . Already, six major pediatric, psychiatric, and medical associations have said that the evidence of a link is overwhelming, citing more than 1,000 studies in the past 30 years.

In one study (Centerwall, 1992) an attempt was made to see whether there was a connection between the change in violent crime rates and the introduction of television in the United States. This study found that murder rates in both the United States and Canada increased, almost 93% in the United States and 92% in Canada (adjusted for population increases), between 1945, when widespread commercial television did not exist, and 1970.

In South Africa the apartheid government did not allow television to be viewed for fear of its destabilizing effects. Because of this policy, television was not a part of South African life until many years after its introduction in the United States and Canada. Yet just as in the United States and Canada, the homicide rate rose sharply as the first generation of children who grew up viewing television reached adulthood.

These facts caused the researcher (Centerwall) to conclude that "long term childhood exposure to television is a causal factor behind approximately one half of the homicides committed in the United States, or approximately 10,000 homicides annually." He went on to estimate that up to half of all rapes and assaults in the United States could be directly related to television viewing. Others are more conservative in their views and believe that only 10% of all violence is related to television or movies (Kopel, 1995).

Attempts have been made to warn viewers of violent content in programs. In 1997 the television networks agreed to add content ratings to the existing age-based ratings, using the designation TV-Y, a show appropriate for all children; TV-7, a show

SOCIAL CHANGE

Should Television Be Used to Teach Values?

There have been many complaints about the sex, violence, and pathology that is part of the daily diet of shows on television. At a time when 98% of U.S. households own at least one television set—a set that is turned on for an average of nearly seven hours a day—concerns have been raised about whether people learn from and emulate the behavior of the characters they see on TV.

But if television contributes to poor behavior, as some suggest, might it also be a vehicle for encouraging good behavior? In 1988, Jay Winsten, a professor at the Harvard School of Public Health and the director of the school's Center for Health Communication, conceived a plan to use television to introduce a new social concept—the "designated driver"—to North America. Shows were already dealing with the topic of drinking, Winsten reasoned, so why not add a line of dialogue here and there about not driving drunk? With the assistance of then NBC chairman Grant Tinker, Winsten met with more than 250 writers, producers, and executives over six months, trying to sell them on his designated driver idea.

Winsten's idea worked; the "designated driver" is now common parlance across all segments of American society and in 1991 won entry into a Webster's dictionary for the first time. An evaluation of the campaign in 1994 revealed that the designated driver "message" had aired on 160 prime-time shows in four seasons and had been the main topic of twenty-five 30-minute or 60-minute episodes. More important, these airings appear to have generated tangible results. In 1989, the year after the "designated driver" was invented, a Gallup poll found that 67% of adults had noted its appearance on network television. What's more, the campaign seems to have influenced adult behavior: Polls conducted by the Roper Organization in 1989 and 1991 found significantly increasing awareness and use of designated drivers. By 1991, 37% of all U.S. adults claimed to have refrained from drinking at least once in order to serve as a designated driver, up from 29% in 1989. In 1991, 52% of adults younger than 30 had served as designated drivers, suggesting that the campaign was having greatest success with its target audience.

In 1988 there were 23,626 drunk driving fatalities. By 1997 the number was 16,189. Although the Harvard Alcohol Project acknowledges that some of this decline is due to new laws, stricter anti–drunk driving enforcement, and other factors, it claims that many of the 50,000 lives saved by the end of 1998 were saved because of the designated driver campaign. (The television campaign was only a part of the overall campaign; there were strong community-level and public service components as well.)

Of course, making television an explicit vehicle for manipulating behavior has its dangers. My idea of the good may not be yours; if my ideas have access to the airwaves but yours don't, what I'm doing will seem to you like unwanted social engineering. We can all agree that minimizing drunk driving is a good thing—but not everyone agrees on the messages we want to be sending to, say, teenage girls about abstaining from sex versus using condoms, about having an abortion, or about whether interfaith marriages are okay. Television's power to mold viewers' understanding of the world is strong enough that we need to be aware that embedding messages about moral values or social behavior can have potent effects—for good or for ill.

Source: Rosenzweig, Jane, "Can TV Improve Us." Reprinted with permission from *The American Prospect*, Volume 10, Number 45: July 1, 1999, The American Prospect, 11 Beacon Street, Suite 1120, Boston, MA 02108. All rights reserved.

appropriate for children age 7 and up; TV-G, a show is suitable for all ages; TV-PG, parental guidance is suggested (this rating may also include a V for violence, S for sexual situations, L for language, or D for suggestive dialogue); TV-14, the show may be unsuitable for children under 14; and TV-MA, the show is for mature audiences. The ratings are designed to work in conjunction with the V-chip, a device in all new television sets produced after 1997. The V-chip allows parents to program their sets to screen out any program rated beyond a desired level (Federal Communications Commission, 2005).

This may not really help the problem. When boys, especially those aged 10 to 14, saw that a program or movie had a "parental discretion" advisory, they found that show more attractive and wanted to watch it. In addition, no evidence was found that antiviolence public service announcements altered adolescents' attitudes toward the appropriateness of using violence to resolve conflict (Mediascope, 1996).

Proving cause and effect in sociology is never easy, and the relationship between television and violence may be more complex than is generally acknowledged. Yet, one study (Johnson, 2002) that followed 707 subjects for 17 years uncovered some disturbing findings. According to the study, adolescents who watched more than one hour a day of television—regardless of content—were roughly four times more likely to commit aggressive acts toward other people later in their lives than those who watched less than one hour. Of those who watched more than three hours, 28.8% were later involved in assaults, robberies, fights, and other aggressive behavior (Sappenfield, 2002).

As the popularity of video games has increased, new concerns have been raised about the potential socialization aspects of this medium. Many video games have violent themes and the same concerns about the impact of television on aggressive behavior are now being raised about this entertainment medium. Do games such as Doom, Mortal Kombat, or Wolfenstein increase a person's aggressive thoughts and behavior? One study (Anderson & Dill, 2000) of 227 college students found that those who had played violent video games in junior high school were more prone to aggressive behavior in college. The thinking is that playing video games provides a way of "practicing aggressive solutions to conflict situations." The concern is that exposure to violent video games is more dangerous than exposure to violent television programs or movies because of the interactive nature of the games. The player is more involved in carrying out the violence than the passive viewer of a program.

Repeated exposure to video game violence increases the potential for aggressive behavior because (1) it produces more positive attitudes and expectations regarding the use of aggression; (2) it leads to rehearsing more aggressive solutions to problems; (3) it decreases consideration of nonviolent alternatives; and (4) it decreases the likelihood of thinking of conflict, aggression, and violence as unacceptable alternatives (Anderson, 2003).

Clearly, many other factors are involved in the relationship between television, movies, and video games and violence. The relationship between violent acts and antisocial behavior is much more complicated than was originally thought. For both adults and children, the social context, peer influence, values, and attitudes also play an important a role in determining behavior.

Adult Socialization

A person's primary socialization is completed when he or she reaches adulthood. **Primary socialization** means that *individuals have mastered the basic information and skills required of members of a society.* He or she has (1) learned a language and can think logically to some degree, (2) accepted the basic norms and values of the culture, (3) developed the ability to pattern behavior in terms of these norms and values, and (4) assumed a culturally appropriate social identity.

There is still much for a person to learn, however, and many new social identities to explore. Socialization, therefore, continues during the adult years. **Adult socialization** is *the process by which adults learn new statuses and roles.* It differs from primary socialization in two ways. First, adults are much more aware than young people are of the processes through which they are being socialized. In fact, adults deliberately engage in programs such as advanced education or on-the-job training in which socialization is an explicit goal. Second, adults often have more control over how they wish to be socialized and therefore can generate more enthusiasm for the process. Whether going to business school, taking up a new hobby, or signing up for the Peace Corps, adults can decide to channel their energy into making the most effective use of an opportunity to learn new skills or knowledge.

An important aspect of adult socialization is **resocialization,** which involves *exposure to ideas or values that in one way or another conflict with what was learned in childhood.* This is a common experience for college students who leave their homes for the first time and encounter a new environment, in which many of their family's cherished beliefs and values are held up to critical examination. Changes in religious and political values are not uncommon during the college years, which often lead to a time of stress for students and their parents.

Erving Goffman (1971) discussed the major resocialization that occurs in **total institutions—** *environments such as prisons or mental hospitals in which the participants are physically and socially isolated from the outside world.* Goffman noted several factors that produce effective resocialization. These include (1) isolation from the outside world, (2) spending all of one's time in the same place with the same people, (3) shedding individual identity by giving up old clothes and possessions for standard uniforms, (4) a clean break with the past, and (5) loss of freedom of action. Under these circumstances, an individual usually changes in a major way along the lines prescribed by those doing the resocialization.

During the 1970s and 1980s, a number of religious cults gained notoriety because they attracted thousands of followers. Hundreds of cults continue to exist today, but we notice them only when they have trouble with authorities. The methods various cults use to indoctrinate their members can be seen as a conscious attempt at resocialization. New members are swept up in the communal spirit of the cult. Group pressure eventually can produce major personality changes in the recruits, and new value systems replace the ones learned previously. Consequently, friends and family members may no longer recognize the person who has been resocialized. The tactics that some religious cults use have been criticized widely.

The cults defend their programs as simply a means through which they encourage people to rid themselves of old ideas and replace them with new ones.

It is not unusual for those undergoing resocialization experiences to become confused and depressed and to question whether they have chosen the right course. Some drop out; others eventually stop resisting and accept the values of their instructors.

In the following sections, we will discuss four events in adult socialization: marriage, parenthood, work, and aging.

Marriage and Responsibility

As Ruth Benedict (1938) noted in a now classic article on socialization in America, "our culture goes to great extremes in emphasizing contrasts between the child and the adult." We think of childhood as a time without cares, a time for play. Adulthood, by contrast, is marked by work and taking up the burden of responsibility. One of the great adult responsibilities in our society is marriage.

Indeed, today's young adults no longer accept uncritically many of the traditional role expectations of marriage. For both men and women, choices loom large: How much of oneself should one devote to a career? How much to self-improvement and personal growth? How much to a spouse? Ours is a time of uncertainty and experimentation. Even so, marriage still retains its primacy as a life choice for adults. Although divorce has become acceptable in most circles, marriage still is treated seriously as a public statement that both partners are committed to each other and to stability and responsibility. (We will discuss marriage and family arrangements in greater detail in Chapter 12.)

Once married, the new partners must define their relationships to each other and in respect to the demands of society. This is not as easy today as it used to be when these choices largely were determined by tradition. Although friends, parents, and relatives usually are only too ready to instruct the young couple in the "shoulds" and "should nots" of married life, such attempts at socialization are often resented by young people who wish to chart their own courses. One choice they must make is whether to become parents.

Parenthood

Once a couple has a child, their responsibilities increase enormously. They must find ways to provide the care and nurturing necessary to the healthy development of their baby, and at the same time they must work hard to keep their own relationship intact, because the arrival of an infant inevitably is accompanied by stress. This requires a reexamination of the role expectations each partner has of the other, both as a parent and as a spouse.

Of course, most parents anticipate some stresses and try to resolve them before the baby is born. They make financial plans, create a living space, and study baby care. They ask friends and relatives for advice and secure their future babysitting services. However, not all the stresses of parenthood are so obvious. One that is overlooked frequently is the fact that parenthood is itself a new developmental phase.

The psychology of being and becoming a parent is extremely complicated. During the pregnancy, both parents experience intense feelings—some expected, others quite surprising. Some of these feelings may even be very upsetting: for instance, the fear that one will not be an adequate parent or that one might even harm the child. Sometimes such feelings lead people to reconsider their decision to become parents.

The birth of the child brings forth new feelings in the parents, many of which can be traced to the parents' own experiences as infants. As their child grows and passes through all the stages of development we have described, parents relive their own development. In psychological terms, parenthood can be viewed as a second chance: Adults can bring to bear all that they have learned in order to resolve the conflicts that were not resolved when they were children. For example, it might be possible for some parents to develop a more trusting approach to life while observing their infants grappling with the conflict of basic trust and mistrust (Erikson's first stage).

Career Development: Vocation and Identity

Taking a job involves more than finding a place to work. It means stepping into a new social context with its own statuses and roles, and it requires that a person be socialized to meet the needs of the situation. These may even include learning how to dress appropriately. For example, a young management trainee in a major corporation was criticized for wearing his keys on a ring snapped to his belt. "Janitors wear their keys," his supervisor told him. "Executives keep them in their pockets." The keys disappeared from the trainee's belt.

Aspiring climbers of the occupational ladder even may have to adjust their personalities to fit the job. In the 1950s and 1960s, corporations looked for quiet, loyal, tradition-oriented men to fill their management positions—men who would not upset the status quo (Whyte, 1956)—and most certainly not women. Today, especially in high-tech industries, the trend has been toward recruiting men and women who show drive and initiative and a capacity for creative thinking and problem solving.

Some occupations require extensive resocialization. Individuals wishing to become doctors or nurses, for example, must overcome their squeamishness about blood, body wastes, genitals, and the inside of the body. They also must accept the undemocratic fact that they will receive much of their training while caring for poor patients (usually ethnic minorities). Wealthier patients are more likely to receive care from fully trained personnel.

The armed forces use basic training to socialize recruits to obey orders without hesitating and to accept killing as a necessary part of their work. For many people, such resocialization can be quite confusing and painful. (For an example of career resocialization for an American in Japan, see "Global Sociology: An American Success Story Does Not Translate into Japanese.")

For some individuals, career and identity are so intertwined that job loss can lead to personal crisis. This occurs for many people who are downsized or "encouraged" to retire. For many, losing a job means reevaluation and a new direction. For others, it means spending months looking for a new job and feeling a profound loss of self-identity.

Aging and Society

In many societies, such as Japan and China, age brings respect and honor. Older people are turned to for advice, and their opinions are valued because they reflect a full measure of experience. Often, older people are not required to stop their productive work simply because they have reached a certain age. Rather, they work as long as they are able to, and their tasks may be modified to allow them to continue to work virtually until they die. In this way, people maintain their social identities as they grow old—and their feelings of self-esteem as well.

This is not the case in the United States. Most employers expect their employees to retire well before they have reached age 70, and Social Security regulations restrict the amount of nontaxable income that retired people may earn.

Perhaps the biggest concern of the elderly is where they will live and who will take care of them when they get sick. The American nuclear family ordinarily is not prepared to accommodate an aging parent who is sick or whose spouse has recently died. In addition, with the increasing life span many elderly who may be in their late 70s or 80s have sons and daughters who themselves may be in their 50s or even 60s. As a result, those older people who have trouble moving around or caring for themselves often have no choice but to live in protected environments.

This means that late in life many people are forced to acquire another social identity. Sadly, it is not a valued one, but rather one of being less valued and less important. This change can be very damaging to older people's self-esteem, and it may even hasten

In many societies, older people are turned to for advice, and their opinions are valued because they are based on life experiences.

them to their graves. The past two decades has seen some attempts at reform to address these issues. Age discrimination in hiring is illegal, and some companies have extended or eliminated arbitrary retirement ages. However, the problem will not be resolved until elderly people achieve a position of respect and value in American culture equal to that of younger adults.

Even though aging is a biological process, becoming old is a social and cultural one: Only society can create a senior citizen. From infancy to old age, both biology and society play important parts in determining how people develop over the course of their lives.

SUMMARY

- From infancy to old age, both biology and society play important parts in determining how people develop.
- Unlike other animal species, human offspring have a long period of dependency. During this time, parents and society work together to make children social beings.
- The process of social interaction that teaches children the intellectual, physical, and social skills and the cultural knowledge they need to function as members of society is called socialization. In the course of this process each child acquires a personality, or patterns of behavior and ways of thinking that are distinctive for each individual.
- Every individual comes to possess a social identity by occupying culturally and socially defined positions.
- In addition, individuals acquire a changing yet enduring personal identity called the self, which develops when the individual becomes aware of

GLOBAL SOCIOLOGY

An American Success Story Does Not Translate into Japanese

In American society, hard work is seen as admirable and individual accomplishments are rewarded. In other societies, work success depends on the relationship the individual develops with superiors and others in the work group. Hampden-Turner and Trompenaars (2000) were presented with the following case of opposing cultural views and asked to give their feedback and recommendations.

Jeff was a 28-year-old [American] in Japan who worked for Motorola Nippon, a largely autonomous Japanese subsidiary of Motorola, the American electronics corporation. Jeff was part of a four-person sales force. [He] worked under the supervision of Muneo, his Japanese manager . . . who had almost complete authority over him.

Jeff felt from the beginning that Muneo did not like him. At his first interview he was told his salary was too high. When Jeff met and became engaged to a Japanese woman, Muneo made clear his disapproval. Muneo also refused Jeff's request for a lower sales quota because of Jeff's difficulty with the Japanese language. However, thanks to his fiancée, Jeff's Japanese language skills were improving rapidly.

Jeff believed that he had to prove himself by working harder. He got up at 6:30 A.M. and scheduled four sales visits a day. After nine months in the job, Jeff was outselling the three other sales executives in his unit, all Japanese. After 18 months he had outsold all three of them put together. This was, he felt, a triumph of hard work and persistence. He was out making sales calls 95% of the time and rarely bothered his boss.

It was therefore a considerable shock when Jeff received an "average" rating from Muneo at his annual appraisal meeting. This was the lowest mark possible, short of "unacceptable," which would have led to being fired. It was also the worst appraisal in the department. Jeff was too angry to argue with Muneo, but appealed his "flagrantly unfair" appraisal to Motorola's international human resources function.

After some months, international HR came down on Jeff's side and his appraisal was revised upward. Muneo not only refused to speak to Jeff, he refused to look at him. Finally Jeff asked for a meeting. Muneo was so angry he could hardly speak. "You shoot me, I shoot you," he repeated.

Jeff was outraged at the seeming injustice of the situation. After interviewing Jeff and hearing his story, Hampden-Turner and Trompenaars gave him the following diagnosis and recommendations.

Given specific American values, you have performed extremely well. Your sales record speaks

---

his or her feelings, thoughts, and behaviors as distinct from those of other people.

- The development of the self is a complex process that has at least three dimensions: cognitive development, moral development, and gender identity.
- Agents of socialization vary from culture to culture.
- In American society, the family is the most important socializing influence in early childhood development.
- As the child grows older and moves into society, other agents of socialization come into play.
- Schools are increasingly expected to meet a variety of social and emotional needs as well as to pass on knowledge and help children develop skills.
- From school age to early adulthood, peers powerfully influence lifestyle orientations and, in some cases, values.

- The mass media present today's children with an enormous amount of information, both for better and for worse.
- Primary socialization ends when individuals reach adulthood. By this time they have (1) learned a language, (2) accepted the basic norms and values of the culture, (3) developed the ability to pattern their behavior in terms of those norms and values, and (4) assumed a culturally appropriate social identity.
- Although socialization continues throughout one's life, adult socialization differs from primary socialization in that adults are much more aware of the processes through which they are socialized, and they often have more control over the process.
- One important form of adult socialization is resocialization, which involves exposure to ideas or values that conflict with what was learned in childhood.

for itself, as do the long hours you have worked, and the fluency of your Japanese. It is even to your credit that you did this all by yourself and did not bother your boss.

But you are living and working in Japan, not the United States, and you must expect to be judged by Japanese values, which are [quite different] from the values to which you are accustomed. We have tried to recreate Muneo's objections to your conduct from what we know of Japanese management culture.

Muneo is angry with you because you began by asking for favors rather than concentrating on how you could help him and the team. When your success began, you did not inform him, solicit his advice, or invite him to share that success. You did not inform other team members about the information and approaches underlying your record sales, so that they could benefit from your knowledge.

"Not bothering him," was seen by Muneo as a snub, not a favor. As [a Japanese] boss, he was formally responsible for your successes and wanted to play a genuine part in them. He probably feels you cut him out of participating in your triumphs.

Appealing over his head to the foreign owners of the company was experienced by Muneo as an insult to his authority and undermining the local autonomy of Japanese management. That you did not warn him of your appeal or discuss it with him first is a rejection of . . . the ideal of mutual respect between you. To have a local decision reversed by U.S. headquarters was a matter of shame for him. He loses face before other Japanese colleagues by provoking interference in domestic affairs.

We recommend bimonthly meetings with Muneo and the Japanese sales team in which you seek their advice and share the background information on your successes. For Muneo to be pleased, the whole team must succeed and Muneo himself must lead that success.

In this scenario, Jeff should have sought advice every step of the way. He needed to be more modest than he would have been in an American work environment and understand that shared knowledge is vital in Japanese organizations. All of this is not easy when one is used to American workplace cultural values.

Source: From *Building Cross-Cultural Competence* (pp. 175–177), by C. Hampden-Turner & F. Trompenaars, 2000, New Haven, CT: Yale University Press. Used by permission of Yale University Press.

- Often, resocialization occurs in a total institution, such as a prison or mental hospital, where participants are physically, socially, and psychically isolated from the outside world.

Media Resources

The Companion Website for *Introduction to Sociology*, Ninth Edition

http://sociology.wadsworth.com/tischler9e

Supplement your review of this chapter by going to the companion website to take one of the Tutorial Quizzes, use the flash cards to master key terms, and check out the many other study aids you'll find there. You'll also find special features such as Wadsworth's Sociology Online Resources and Writing Companion, GSS data, and Census 2000 information at your fingertips to help you complete that special project or do some research on your own.

CHAPTER FOUR STUDY GUIDE

KEY CONCEPTS AND THINKERS

Match each concept with its definition, illustration, or explanation presented below.

a. Sociobiology   **f.** Primary socialization   **k.** Adult socialization
b. The "I" and the "me"   **g.** Self   **l.** Personality
c. Attachment disorder   **h.** Generalized other   **m.** Significant others
d. Superego   **i.** Cognitive development   **n.** Resocialization
e. Looking-glass self   **j.** Moral development

f   **1.** The child's assimilation of the basic elements of culture—language, norms, behavior—and adoption of a culturally appropriate identity.

a   **2.** The use of Darwinian principles of evolution to explain the social behavior of animals and humans.

j   **3.** A series of stages of increasingly complex thinking about what is right and wrong in specific situations.

c   **4.** The inability to form relationships or trust other people, often found in children who grew up having minimal interaction with adults.

i   **5.** A set of stages in a person's ability to think logically and abstractly about how the world works.

e   **6.** A sense of who you are based on how you think other people would judge you.

m   **7.** Term used by Mead to include those individuals who are most important in our development, for example, parents, friends.

n   **8.** The viewpoints, attitudes, and expectations of society as a whole or of a community of people of whom we are aware and who are important to us.

b   **9.** The parts of the self, according to Mead; one more expressive, the other more a product of socialization.

d   **10.** According to Freud, this is the part of the self that represents society's norms and moral values learned primarily from parents.

k   **11.** The process by which adults learn new statuses and roles.

n   **12.** Exposure to ideas or values that in one way or another conflict with what was learned in childhood.

g   **13.** An individual's changing yet enduring personal identity.

l   **14.** The patterns of behavior and ways of thinking and feeling that are distinctive for each individual.

Match the thinkers with their main idea or contribution.

a. Harry Harlow   **e.** Daniel Levinson   **i.** George Herbert Mead
b. Jean Piaget   **f.** Stephen Jay Gould   **j.** Erik Erikson
c. Lawrence Kohlberg   **g.** Ivan Pavlov   **k.** Charles Horton Cooley
d. Sigmund Freud   **h.** Edward O. Wilson

g   **1.** Through experiments with dogs, he demonstrated that behavior could be conditioned.

i   **2.** Proposed a theory of socialization based on the development of the "me"; saw children's relation to rules as moving through three stages—preparatory, play, and game.

h   **3.** Coined the term *sociobiology* and was its major advocate as an explanation of human behavior.

f   **4.** Biologist who criticized sociobiology, offering instead explanations based on culture rather than genetics and evolution.

a   **5.** Illustrated the harmful effects of social isolation through his experiments with rhesus monkeys.

j   **6.** Offered a theory of childhood development based on developmental problems rooted both in biological changes in the individual and in social expectations in the culture.

c   **7.** Maintained that moral thinking developed through five to six distinctive stages.

e   **8.** Did in-depth interviews with middle-aged men to document how adults, like children, go through certain predictable developmental tasks.

9. Argued that society's demand for civilized behavior constantly conflicted with the individual's basic instincts of sex and aggression.

10. Studied the stages of cognitive development that children go through in learning to think logically about the world.

11. Offered a theory of childhood development based on the "looking-glass" self—a person's sense of other people's evaluations.

CENTRAL IDEA COMPLETIONS

Following the instructions, fill in the appropriate concepts and descriptions for each of the questions posed below.

1. Discuss the roles biology and socialization play in the formation of the individual.

_____

_____

_____

_____

_____

2. Use the case histories presented in Chapter 4 to discuss how extreme social isolation and deprivation affect a human's early childhood development.

_____

_____

_____

_____

_____

3. Briefly discuss each of the following sources of influence on the socialization of children:

a. Family _____

b. School _____

c. Peer groups _____

d. Mass media _____

4. Define and discuss resocialization.

_____

_____

_____

5. Explain the key features of the developmental stage models of Erikson and Levinson.

a. Erikson

_____

_____

_____

b. Levinson

_____

_____

_____

CRITICAL THOUGHT EXERCISES

1. The nature versus nurture debate has been an ongoing controversy in the social sciences. Develop a pro and con list of evidence for both positions, as one might for a debate situation. Discuss and support your positions on the debate, providing examples to support your positions.

2. Visit the children's section of your local library. After examining several books aimed at different age levels, discuss how these books might reflect Piaget's work of the developmental stages of childhood. Find examples of books you feel to be ideally suited to appeal to children at different levels of development. Select three titles you would want to read to your own children someday and explain why you think they are examples of valuable children's books.

3. As you think about the material you have just read, consider the role that day care plays in the lives of our children. How does the time spent in day care shift the locus of childhood socialization? What might be the strengths and weaknesses of a childhood spent in day-care situations?

4. How important are peer groups compared with families as agents of socialization? After rereading the box on "Acting White," reflect on your own high school experiences and those of people you knew. Did the values and norms of the peer group conflict with those of parents or others? If so, how did these differing worlds affect the reactions, thoughts, and feelings of you and your friends?

INTERNET ACTIVITIES

1. Visit http://mentalhelp.net/psyhelp/chap3/chap3h.htm. This website begins with a discussion and critique of Kohlberg's ideas on moral development. It then takes you to pages that allow you to "write your own philosophy of life" by answering a list of questions. It also shows you how college students 40 years ago answered these questions. Check your answers against theirs. Most important, ask yourself to what extent your answers can be seen as the product of socialization in a particular time and place (America around the turn of the twenty-first century).

2. Visit http://www.anthro.palomar.edu/social/soc_1.htm. Examine this site, which discusses socialization in three settings: the Yanomama, the Semai tribesman, and in Iran. Follow the links to each site and develop a short essay comparing and contrasting the socialization experiences of infants across these three cultures.

ANSWERS TO KEY CONCEPTS

1.f 2.a 3.j 4.c 5.i 6.e 7.m 8.h 9.b 10.d 11.k 12.n 13.g 14.l

ANSWERS TO KEY THINKERS

1.g 2.i 3.h 4.f 5.a 6.j 7.c 8.e 9.d 10.b 11.k

ThomsonNOW™

Reviewing is as easy as ❶ ❷ ❸

1. Before you do your final exam, take the ThomsonNOW diagnostic quiz to help you identify the areas on which you should concentrate. You will find information on ThomsonNOW and instructions on how to access all of its great resources on the foldout at the beginning of the text.

2. As you review, take advantage of ThomsonNOW's study videos and interactive Map the Stats exercises to help you master the chapter topics.

3. When you are finished with your review, take ThomsonNOW's posttest to confirm you are ready to move on to the next chapter.

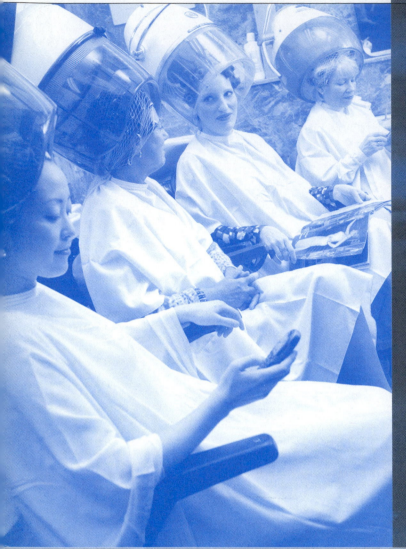

5

Society and Social Interaction

© Alexander Walter/Taxi/Getty Images

Learning Objectives

After studying this chapter, you should be able to do the following:

- Know what the major types of social interaction are.
- Understand the influence of contexts and norms in social interaction.
- Know what ethnomethodology is.
- Be familiar with the different types of social interaction.

- Understand the concepts of status and role.
- Know the difference between role strain and role conflict.
- Understand the differences between the different types of societies: hunting and food-gathering societies, horticultural societies, pastoral societies, agricultural societies, industrials societies, and postindustrial societies.

Wayne Farrell had arrived in New York City only two days before, but already he had his fill of strange experiences. It was almost as if panhandlers, con artists, and a host of individuals pushing unusual causes could sense that he was an easy mark. No sooner would he walk down the street and notice someone distributing leaflets then the person would pick him out of the crowd and engage him in conversation. He tried to be polite and listen, but he repeatedly found himself in the midst of uncomfortable situations. Wayne was starting to wonder whether he was doing anything to contribute to these encounters or if this was just part of living in a large urban environment.

Without realizing it, Wayne was engaging in a pattern of social interaction with these strangers. Coming from a small town in Wyoming, he was unaware of the fact that he was looking at passersby for a second or two longer than the typical person walking down the street. But this was just enough to signal to those looking for such a cue that he was ripe for their approach.

Most Americans use four main distance zones in their business and social relations: intimate, personal, social, and public. Intimate distance varies from direct physical contact with another person to a distance of 6 to 18 inches and is used for private

activities with another close acquaintance. Personal distance varies from 2 to 4 feet and is the most common spacing that people use in conversation. The third zone—social distance—is employed during business transactions or interactions with a clerk or salesperson. As they work with each other, people tend to use this distance, which is usually between 4 and 12 feet. Public distance, used, for instance, by teachers in classrooms or speakers at public gatherings, can be anywhere from 12 to 25 feet or more (Hall, 1969).

As Wayne walks down the busy urban street, he is within the fourth zone. Eye behavior in this environment has some specific rules attached to it. For urban whites, once they are within definite recognition distance (usually 16 to 32 feet), there is mutual avoidance of eye contact—unless they want something specific, such as to make a proposition, receive a handout, or obtain information of some kind. In small towns, however, people are much more likely to look at each other, even if they are strangers (Hall, 1969).

Panhandlers and others seeking to establish contact exploit the unwritten, unspoken conventions of eye contact. They take advantage of the fact that once explicit eye contact is established, it is rude to look away because to do so means brusquely dismissing the other person and his needs. Once having

Social interaction is a central concept to understanding the nature of social life.

© Paul Damien

caught the eye of his mark, the individual locks on, not letting go until he moves through the public zone, the social zone, the personal zone, and, finally, the intimate zone, where people are most vulnerable.

Wayne's experiences were not just a coincidence. He actually was engaging in a social interaction with others in his environment that he was not aware of. He started to make a conscious attempt to avoid prolonged eye contact with strangers, and the unwanted encounters began to subside.

As Wayne discovered, there is no way not to interact with others. Social interaction has no opposite. If we accept the fact that all social interaction has a message value, then it follows that no matter how we may try, we cannot not send a message. Activity or inactivity, words or silence—all contain a message. They influence others, and others respond to these messages. The mere absence of talking or taking notice of each other is no exception, because there is a message in that behavior also.

When sociologists study human behavior, they are interested primarily in how people affect each other through their actions. They look at the overt behaviors that produce responses from others, as well the subtle cues that may result in unintended consequences. Human social interaction is very flexible and quite unlike that of the social animals.

Understanding Social Interaction

Max Weber (1922) was one of the first sociologists to stress the importance of social interaction in the study of sociology. He argued that the main goal of sociology is to explain what he called **social action,** a term he used to refer to *anything people are conscious of doing because of other people.* Weber claimed that to interpret social actions, we have to put ourselves in

the positions of the people we are studying and try to understand their thoughts and motives. The German word Weber used for this is *verstehen,* which can be translated as "sympathetic understanding."

Weber's use of the term *social action* identifies only half of the puzzle, because it deals only with one individual taking others into account before acting. A **social interaction** involves *two or more people taking one another into account.* It is the interplay between the actions of these individuals. In this respect, social interaction is a central concept to understanding the nature of social life.

In this chapter, we will explain how sociologists investigate social interaction. We will start with the basic types of social interaction: verbal and nonverbal. Next, we will examine how social interaction affects those involved in it. We then will broaden our focus a bit and move on to groups and social interactions within them. Finally, we will look at the large groupings of people that make social life possible and that ultimately make up the social structure. In other words, we will start with social behavior at the most basic level and move outward to ever more complicated levels of social interaction.

Social interaction is a central concept to understanding the nature of social life.

Contexts

Where a social interaction occurs makes a difference in what it means. Edward T. Hall (1974) identified *three elements that, taken together, define the* **context** *of a social interaction: (1) the physical setting or place, (2) the social environment, and (3) the activities surrounding the interaction—preceding it, happening simultaneously with it, and coming after it.*

The context of an interaction consists of many elements. Without knowledge of these elements it is

The location and context in which a social interaction takes place make a difference in what it means.

impossible to know the meaning of even the simplest interaction. For example, Germans and Americans treat space very differently. Hall (1969) noted that in many ways, the difference between German and American doors gives us a clue about the space perceptions of these two cultures. In Germany, public and private buildings usually have double doors that create a soundproof environment. Germans feel that American doors, in contrast, are flimsy and light, inadequate for providing the privacy that Germans require.

In American offices, doors usually are kept open; in German offices, they are kept closed. In Germany, the closed door does not mean that the individual wants to be left alone or that the people inside are planning something that should not be seen by others. Germans simply think that open doors are sloppy and disorderly. As Hall explained it:

> I was once called in to advise a firm that has operations all over the world. One of the first questions asked was, "How do you get the Germans to keep their doors open?" In this company the open doors were making the Germans feel exposed and gave the whole operation an unusually relaxed and unbusinesslike

air. Closed doors, on the other hand, gave the Americans the feeling that there was a conspiratorial air about the place and that they were being left out. The point is that whether the door is open or shut, it is not going to mean the same thing in the two countries. (Hall, 1969)

In Japan there is a third view on the issue. Most Japanese executives prefer to share offices to ensure that information can flow easily and each person knows what is happening in the other's area of responsibility. The Japanese executive does not want to risk being unaware of events as they are developing. Japanese firms have ceremonial rooms for receiving visitors, but few other work areas afford any privacy (Hall & Hall, 1987). (See "Global Sociology: Cross-Cultural Social Interaction Quiz.")

Norms

Human behavior is not random. It is patterned and, for the most part, quite predictable. What makes human beings act predictably in certain situations? For one thing, there is the presence of **norms**—*specific rules of behavior, agreed upon and shared, that prescribe limits of acceptable behavior.*

Norms tell us the things we should both do and not do. In fact, our society's norms are so much a part of us that we often are not aware of them until they are violated. Take the unfortunate circumstances that happened in Suzanne Berger's life, for example. One day she bent down to pick up her child only to discover she could not straighten up again. In the process of bending she had injured her back so severely that for years thereafter she could neither walk nor sit for more than a few minutes. Traveling anywhere meant that she had to take along a mat and immediately lie down. In effect, she had to violate the norms that assume that you will not stretch out on the floor in a department store, a train station, a classroom, or at public events. As she described it:

> Strangers try not to stare. . . . At airports and train stations people have thought I was a derelict or crazy or maybe homeless; only the dispossessed lie on floors, children lie on floors, dogs lie on floors . . . but adults? *What's that woman doing over there?* a security guard said at the airport. *Dunno, leave her alone. Must be drunk.* With friends inside my house, being down here upsets a balance of conviviality, of the *whereness* that grounds a conversation. I am always looking up, as though younger or subservient. Outside I lie down with mother-dirt, grass, the asphalt of the city. Wherever I go, I lie down with my mat. *Hey, lady, what the hell you doing down there?* says a child on a city playground. *You sick? You tired?* (Berger, 1996)

GLOBAL SOCIOLOGY

Cross-Cultural Social Interaction Quiz

Brazil

You walk down a busy street and see a man in a group forming two tubes with his hands and looking through them. What is he doing?

1. Trying to read the license plate on a distant car.
2. Making a joking gesture that he is a spy.
3. Letting the other men know that he has spotted a beautiful woman.

CORRECT ANSWER: 3. This is a common gesture men in Brazil use to let on that they admire an attractive woman.

Japan

George is in Japan on a business trip. After dinner he wants to let another businessmen know that he likes him and would like to do business with him. As they leave the restaurant he puts his arm around the man as if he were a college buddy. How will the Japanese man respond?

1. He will like George for being so uninhibited.
2. The Japanese man feels very uncomfortable and will avoid George in the future.
3. The Japanese man will invite George home to meet the rest of the family.

CORRECT ANSWER: 2. Even though the Japanese may permit themselves to be jammed together in subway cars they are not part of a touching society. This type of behavior would make the man feel very uncomfortable.

Morocco

You have been invited to the home of Moroccan family for dinner. The host takes you to a room with low tables and asks you sit on the floor. He asks you to remove you shoes and brings over a basin of water. What should you do?

1. Wash you feet on the water.
2. Wash your hands in the water.
3. Scoop out a handful of water and sprinkle it on your head.

CORRECT ANSWER: 2. It is common to sit cross-legged on the floor and share a meal. You will also notice that the guests will eat from a common platter of food with their fingers.

China

You walk along a street and notice that people spit on the sidewalk or blow their nose without a handkerchief. This is considered:

1. An act of personal hygiene that rids the body of waste.
2. An insult to the foreigner who has just walked past.
3. Rude behavior by ignorant people.

CORRECT ANSWER: 1. Even though the Chinese government is trying to get people to stop doing this, it is still considered appropriate, much like washing your hands.

Source: Roger E. Axtell. (1998). *Gestures: The Do's and Taboos of Body Language around the World.* New York: John Wiley and Sons.

We also have norms that guide us in how we present ourselves to others. We realize that how we dress, how we speak, and the objects we possess relay information about us. In this respect, North Americans are a rather outgoing people. The Japanese have learned that it is a sign of weakness to disclose too much of oneself by overt actions. They are taught very early in life that touching, laughing, crying, or speaking loudly in public are not acceptable ways of interacting.

Not only can the norms for behavior differ considerably from one culture to another, but they also can differ within our own society. Conflicting interpretations of an action can exist among different ethnic groups. Unfamiliarity with such cultural communication can lead to misinterpretations and even unintended insults. Among African Americans, listeners are expected to avert their eyes from the speaker and in doing so are showing respect. Among white Americans, looking at the speaker directly is seen as a sign of respect and looking away may indicate disinterest or boredom.

Let us assume that a white teacher is speaking directly to an African American boy. The youngster may avert his eyes as she is speaking. The teacher may think he is not listening. Worse yet, she may say something like "Look at me when I speak to you," and the boy will be even more confused. He may have been led to believe that looking at the teacher for an extended time may be seen as a challenge to her authority. Sociologists thus need to understand the norms that guide people's behavior, as without this knowledge it is impossible to understand social interaction.

Ethnomethodology

Many of the social actions we engage in every day are commonplace events. They tend to be taken for granted and rarely are examined or considered. Harold Garfinkel (1967) has proposed that it is important to study the commonplace. Those things we take for granted have a tremendous hold over us because we accept their demands without question or conscious consideration. **Ethnomethodology** is *the study of the sets of rules or guidelines that individuals use to initiate behavior, respond to behavior, and modify behavior in social settings.*

For ethnomethodologists, all social interactions are equally important because they provide information about a society's unwritten rules for social behavior—the shared knowledge that is basic to social life.

Garfinkel asked his students to participate in a number of experiments in which the researcher would violate some of the basic understandings among people. For example, when two people hold a conversation, each assumes that certain things are perfectly clear and obvious and do not need further elaboration. Examine the following conversation and notice what happens when one individual violates some of these expectations:

Bob: That was a very interesting sociology class we had yesterday.

John: How was it interesting?

Bob: Well, we had a lively discussion about deviant behavior, and everyone seemed to get involved.

John: I'm not certain I know what you mean. How was the discussion lively? How were people involved?

Bob: You know, they really participated and seemed to get caught up in the discussion.

John: Yes, you said that before, but I want to know what you mean by lively and interesting.

Bob: What's wrong with you? You know what I mean. The class was interesting. I'll see you later.

Bob's response is quite revealing. He is puzzled and does not know whether John is being serious. The normal expectations and understandings around which day-to-day forms of expression occur have been challenged. Still, is it not reasonable to ask for further elaboration of certain statements? Obviously not, when it goes beyond a certain point.

Another example of the confusion brought on by the violation of basic understandings was shown when Garfinkel asked his students to act like boarders in their own homes. They were to ask whether they could use the phone, take a drink of water, have a snack, and so on. The results were quite dramatic:

> Family members were stupefied. They vigorously sought to make the strange actions intelligible and to restore the situation to normal appearances. Reports were filled with accounts of astonishment, shock, anxiety, embarrassment and anger, and with charges by various family members that the student was mean, inconsiderate, selfish, nasty, or impolite. Family members demanded explanations: What's the matter? What's gotten into you? Did you get fired? Are you sick? What are you being so superior about? Why are you mad? Are you out of your mind? Are you stupid? One student acutely embarrassed his mother in front of her friends by asking if she minded if he had a little snack from the refrigerator. "Mind if you have a little snack? You've been eating little snacks around here for years without asking me. What's gotten into you?" (Garfinkel, 1972)

Ethnomethodology seeks to make us more aware of the subtle devices we use in creating the realities to which we respond. These realities are often intrinsic in human nature, rather than imposed from outside influences. Ethnomethodology addresses questions about the nature of social reality and how we participate in its construction.

Dramaturgy

People create impressions, and others respond with their own impressions. Erving Goffman (1959, 1963, 1971) concluded that a central feature of human interaction is impression formation—the attempt to present oneself to others in a particular way. Goffman believed that much human interaction can be studied and analyzed on the basis of principles derived from the theater. This approach, known as **dramaturgy**, states that *in order to create an impression, people play roles, and their performance is judged by others who are alert to any slips that might reveal the actor's true character.* For example, a job applicant at an interview tries to appear composed, self-confident, and capable of handling the position's responsibilities. The interviewer is seeking to find out whether the applicant is really able to work under pressure and perform the necessary functions of the job.

Most interactions require that a person undertake some type of playacting in order to present an image that will bring about the desired behavior from others. Dramaturgy sees these interactions as governed by planned behavior designed to enable an individual to present a particular image to others.

Types of Social Interaction

When two individuals are in each other's presence, they inevitably affect each other. They may do so intentionally, as when one person asks the other for change for a quarter, or they may do so unintentionally, as when two people drift toward opposite sides of the elevator in which they are riding. Whether intentional or unintentional, both behaviors represent types of social interaction.

Nonverbal Behavior

Many researchers have focused attention on how we communicate with one another by using body movements. This study of body movements, known as *kinesics*, attempts to examine how such things as "slight head nods, yawns, postural shifts, and other nonverbal cues, whether spontaneous or deliberate, affect communication (Samovar, Porter, & Jain, 1981). Samovar, Porter, and Jain describe various cultural aspects of nonverbal communication.

> Many of our movements relate to an attitude that our culture has, consciously or unconsciously, taught us to express in a specific manner. In the United States, for example, we show status relationship in a variety of ways. The ritualistic nonverbal movements and gestures in which we engage to see who goes through a door first, or who sits or stands first, are but a few ways our culture uses movement to communicate status. In the Middle East, status is underscored nonverbally by which individual you turn your back to. In Oriental cultures, bowing and backing out of a room are signs of status relationships. Humility might be shown in the United States by a slight downward bending of the head, but in many European countries this same attitude is manifested by dropping one's arms and sighing. In Samoa humility is communicated by bending the body downward.

The use of hand and arm movements as a means of communicating also varies among cultures. We all are aware of the different gestures for derision. For some European cultures, it is a closing fist with the thumb protruding between the index and middle fingers. The Russian expresses this same attitude by moving one index finger horizontally across the other.

In the United States, we can indicate that things are okay by making a circle with one's thumb and index finger while extending the others. If you make this

The use of hand and arm movements, eye contact, and norms of nonverbal behavior are markedly different for Arab Bedouins than they are for Americans.

© Frank Sitman/Stock, Boston

gesture in Japan, you are signifying "money." And in Arab countries, if you bare your teeth while making this gesture, you are displaying "extreme hostility."

In the United States, we say good-bye or farewell by waving the hand and arm up and down. If you wave this way in South America, you may discover that the other person is not leaving but moving toward you. That is because in many countries, the gesture we use as a sign of leaving actually means "come." Eye contact is another area in which some interesting findings have been reported. Samovar, Porter, and Jain explain that in the United States, the following has been noted:

1. We tend to look at our communication partner more when we are listening than when we are talking. The search for words frequently finds us, as speakers, looking into space, as if to find the words imprinted somewhere out there.
2. The more rewarding we find the speaker's message to be the more we will look at him or her.
3. The amount of eye contact we try to establish with other people is determined in part by our perception of their status. . . . When we address someone we regard as having high status we attempt a modest-to-high degree of eye contact. But when we address a person of low status, we make very little effort to maintaining eye contact.
4. We tend to feel discomfort if someone gazes at us for longer than ten seconds at a time.

These notions of eye contact found in the United States differ from those of other societies. In Japan and

AP/Wide World Photos

Spontaneous cooperation that arises from the needs of a particular situation is the oldest and most natural form of cooperation.

China, for example, "it is considered rude to look into another person's eyes during conversation." Arabs, in contrast, use personal space very differently; they stand very close to the person they are talking to and stare directly into the eyes. Arabs believe that the eyes are a "key to a person's being and that looking deeply into another's eyes allows one to see another's soul."

The proscribed relationships between males and females in a culture also influence eye contact. Asian cultures, for example, consider it "taboo for women to look straight into the eyes of males. Most men, out of respect for this cultural characteristic, do not stare directly at women." French men, on the other hand, accept staring as a cultural norm and often stare at women in public (Samovar, Porter, & Jain, 1981; Axtell, 1998).

Exchange

When people do something for each other with the express purpose of receiving a reward or return, they are involved in an exchange *interaction.*

Most employer-employee relationships are exchange relationships. The employee does the job and is rewarded with a salary. The reward in an exchange interaction, however, need not always be material; it can also be based on emotions such as gratitude. For example, if you visit a sick friend, help someone with a heavy package at the supermarket, or help someone solve a problem, you will expect these people to feel grateful to you.

Sociologist Peter Blau (1964) pointed out that exchange is the most basic form of social interaction. He believes social exchange can be observed everywhere once we are sensitized to it.

Cooperation

A **cooperative interaction** occurs *when people act together to promote common interests or achieve shared goals.* The members of a basketball team pass to one another, block off opponents for one another, rebound, and assist one another to achieve a common goal—winning the game. Likewise, family members cooperate to promote their interests as a family—the husband and wife both may hold jobs as well as share in household duties, and the children may help out by mowing the lawn and washing the dishes.

NEWS YOU CAN USE

Laugh and the World Laughs with You

Why do people laugh? It is possible that laughing started out as a community's shared sign of relief after some passing danger. The group would be signaling "we can now let our guard down and relax." You will notice that when people laugh the muscles in their body do in fact relax. And we have all heard stories of people who have laughed so hard they fell over or worse. The relaxation may also make us trust those we are sharing the laugh with.

Laughing is clearly a social event. Studies have shown that people are 30 times more likely to laugh when they are with others than when they are alone. Laughter helps strengthen our social bonds with others. When people are in a group that is laughing they join in so as not to feel left out. The more everyone laughs the stronger the bonds become. Most of this laughter is not really in response to a formal effort at telling a joke. Most of the time the laughter is after a fairly innocuous comment like "wow look what he's wearing" "Gee, tell me how I can be so lucky too."

Laughter is contagious, and from the earliest days of television the comedy shows have known that they seem funnier when they have a laugh track. The discovery was made by accident in 1950 when a show tried to make up for not having a live audience by using recorded laughter. Ever since, the laugh track has been one of the few unchanging aspects of television.

Laughter can change the behavior of others. It helps make threatening or embarrassing situations more comfortable. Laughter can also be a way of deflecting anger. If the threatening person laughs it lowers the risk of a confrontation. The person is saying "this is not as serious as you thought."

Laughter however, is related to power. In the workplace the boss or those with more control use humor more than subordinates. In those types of situations, controlling what is considered funny becomes a way exercising power. Because men often have more power than women, we see women laughing significantly more when the speaker is a man than when it is a woman. In mixed-gender audiences both men and women laugh more when the speaker is a male then a female. This makes it tough for female comedians.

The more we examine the role of laughter in social situations the more complicated we will see it is. In some situations laughter can be an aggressive act, a sign of winning or controlling the situation. It can be used as way to boast about a triumph and belittle a competitor.

Source: Robert A. Provine. (2000). *Laughter: A Scientific Investigation.* New York: Penguin Books.

College students often cooperate by studying together for tests. (For a discussion of how laughter facilitates cooperation, see "News You Can Use: Laugh and the World Laughs with You.")

Conflict

In a cooperative interaction, people join forces to achieve a common goal. By contrast, people in conflict struggle with one another for some commonly prized object or value. In most conflict relationships, only one person can gain at someone else's expense. Conflicts arise when people or groups have incompatible values or when the rewards or resources available to a society or its members are limited. Thus, conflict always involves an attempt to gain or use power.

The fact that conflict often leads to unhappiness and violence causes many people to view it negatively. However, conflict appears to be inevitable in human society. A stable society is not a society without conflicts, but rather one that has developed methods for resolving its conflicts by justly or brutally suppressing them temporarily. For example, Lewis Coser (1956, 1967) pointed out that conflict can be a positive force in society. The American Civil Rights Movement in the 1950s and 1960s may have seemed threatening and disruptive to many people at the time, but it helped bring about important social changes that may have led to greater social stability.

Coercion involves the use of power that is regarded as illegitimate by those on whom it is exerted. The stronger party can impose its will on the weaker, as in the case of a parent using the threat of punishment to impose a curfew on an adolescent.

Coercion rests on force or the threat of force, but usually it operates more subtly.

Competition

The fifth type of social interaction, **competition**, is *a form of conflict in which individuals or groups confine their conflict within agreed upon rules.*

Competition is a common form of interaction in the modern world—not only on the sports field but

TECHNOLOGY AND SOCIETY

Does Television Reduce Social Interaction?

People are much less involved in the social life of their communities than in the past. Membership in such diverse organizations as the PTA, the Elks Club, the League of Women Voters, the Red Cross, and labor unions has declined between 25% and 50% in the past three decades. Time spent on informal socializing and visiting is down by 25%.

Political participation has also declined sharply, including such things as attending a rally or speech (down 36% between 1973 and 1993), attending a meeting on town or school affairs (down 39%), or working for a political party (down 56%). Even church attendance appears to be down by about 30% since the 1960s.

The usual explanations we hear for this trend include that time pressures are greater now than in the past, suburbanization has caused people to be more spread out, and the movement of women into the paid labor force and the stresses that go along with two-career families have resulted in limited free time.

Researcher Robert Putnam does not accept these reasons and believes the real culprit is television. As television's hold on society became greater, people began to give up more of those activities that brought them together in social groups. In 1950 barely 10% of American homes had television sets, but by 1959, 90% did. The viewing hours grew nearly 20% in the 1960s and by an additional 7.5% during the 1970s. By 1998 viewing per TV household was more than 50% higher than it had been in the 1950s.

The average American now watches roughly four hours of television a day. This means that television absorbs more than 40% of the average person's free time. Each hour spent watching television appears to take away time from social interaction.

What is interesting is that other activities do not take time from social interaction. For example, people who listen to a good deal of classical music, read a daily newspaper, or pursue other hobbies are more likely, not less likely, to join social groups. Television watching appears to be the only leisure activity that inhibits participation in social gatherings. Television's negative impact is an outgrowth of both the medium and the message it conveys.

Television, with few exceptions, is a one-way medium. The viewer can control which channel to watch but cannot control the information coming through that channel. Television simply produces a state of passive receptivity.

The message coming from television is one that shows mostly the negative side of people. Violent shows are popular as are talk shows portraying dysfunctional families. Local news focuses on death and destruction. The more television people watch the more they think that the world is a dangerous place they should avoid. People receive the message is that it is safer to watch television than to venture out into the world and engage in social activities. This seems to be a destructive force on social interaction.

Source: Adapted from "Technology and Mass Media," in Robert D. Putnam, *Bowling Alone: The Collapse and Revival of American Community*, New York: Simon & Schuster, 2000, pp. 216–246.

in the marketplace, the education system, and the political system. American presidential elections, for example, are based on competition. Candidates for each party compete throughout the primaries, and eventually one candidate is selected to represent each of the major parties. The competition grows even more intense as the remaining candidates battle directly against each other to persuade a nation of voters they are the best person for the presidency.

One type of relationship may span the entire range of focused interactions: An excellent example is marriage. Husbands and wives cooperate in household chores and responsibilities. They also engage in exchange interactions. Married people often discuss their problems with each other—the partner whose role is listener at one time will expect the spouse to provide a sympathetic ear at another time. Married people also experience conflicts in their relationship. A couple may have a limited amount of money set aside, and each may want to use it for a different purpose. Unless they can agree on a third, mutually desirable use for the money, one spouse will gain at the other's expense, and the marriage may suffer. The husband and wife whose marriage is irreversibly damaged may find themselves in direct competition. If they wish to separate or divorce, their conflict will be regulated according to legal and judicial rules.

Through the course of our lifetimes, we constantly are involved in several types of social interaction because we spend most of our time in some kind of group situation. How we behave in these situations is generally determined by two factors—the statuses we occupy and the roles we play—which together constitute the main components of what sociologists call social organization. (For a look at one of the ways in which television has affected social interaction, see "Technology and Society: Does Television Reduce Social Interaction?")

Elements of Social Interaction

People do not interact with one another as anonymous beings. They come together in the context of specific environments and with specific purposes. Their interactions involve behaviors associated with defined statuses and particular roles. These statuses and roles help to pattern our social interactions and provide predictability.

Statuses

Statuses are *socially defined positions that people occupy.* Common statuses may pertain to religion, education, ethnicity, and occupation: Protestant, college graduate, African American, and teacher, and so on. Statuses exist independently of the specific people who occupy them (Linton, 1936).

For example, our society recognizes the status of politician. Many people occupy that status, including President George W. Bush, Senator Ted Kennedy, Senator Hillary Clinton, and Governor Arnold Schwarzenegger. New politicians appear; others retire or lose popularity or are defeated, but the status, as the culture defines it, remains essentially unchanged. The same is true for all other statuses: occupational statuses such as doctor, computer analyst, bank teller, police officer, butcher, insurance adjuster, thief, and prostitute; and nonoccupational statuses such as son and daughter, jogger, friend, Little League coach, neighbor, gang leader, and mental patient.

It is important to keep in mind that from a sociological point of view, status does not refer—as it does in common usage—to the idea of prestige, even though different statuses often do contain differing degrees of prestige. In the United States, for example, research has shown that the status of Supreme Court justice has more prestige than that of physician, which in turn has more prestige than that of sociologist (Nakao, Keoko, & Treas, 1993).

People generally occupy more than one status at a time. Consider yourself, for example: You are someone's daughter or son, a full-time or part-time college student, perhaps also a worker, a licensed car driver, a member of a church or synagogue, and so forth. Sometimes one of the multiple statuses a person occupies seems to dominate the others in patterning that person's life; such a status is called a *master status.*

For example, George W. Bush has occupied a number of diverse statuses: husband, father, state governor, and presidential candidate. After January 20, 2001, however, his master status was that of president of the United States, as it governed his actions more than did any other status he occupied at the time. A person's master status will change many times in the course of his or her life cycle. Right

Figure 5–1 Status and Master Status

Generally, each individual occupies many statuses at one time. The statuses of a female executive at major television network include author, wife, mother, pianist, and so on. Other statuses could be added to this list. However, one status—vice-president for programming—is most important in patterning this woman's life. Sociologists call such a status a master status.

now, your master status probably is that of college student.

Five years from now it may be graduate student, artist, lawyer, spouse, or parent. Figure 5–1 illustrates the different statuses occupied by a 35-year-old woman who is an executive at a major television network. Although she occupies many statuses at once, her master status is that of vice-president for programming.

In some situations, a person's master status may have a negative influence on the person's life. For example, people who have followed what their culture considers a deviant lifestyle may find that their master status is labeled according to their deviant behavior. Those who have been identified as ex-convicts are likely to be so classified no matter what other statuses they occupy. They will be thought of as ex-convict-painters, ex-convict machinists, ex-convict-writers, and so on. Their master status has a negative effect on their ability to fulfill the roles of the statuses they would like to occupy. Ex-convicts who are good machinists or house painters may find employers unwilling to hire them because of their police records. Because the label *criminal* can stay with individuals throughout their lives, the criminal justice system is reluctant to label juvenile offenders or to open their records to the courts. Juvenile court files are usually kept secret and often permanently sealed when the person reaches age 18.

Some statuses, called **ascribed statuses,** are *conferred on us by virtue of birth or other significant*

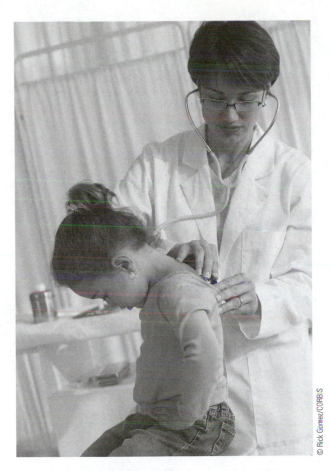

This woman's ascribed status is female; her achieved status is based on her profession.

Figure 5–2   Status and Roles

The status of vice-president for programming at a major television network has several roles attached to it, including attending meetings, making programming decisions, and so on.

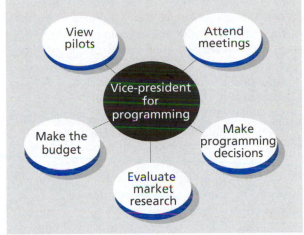

factors not controlled by our own actions or decisions; people occupy them regardless of their intentions. Certain family positions, such as that of daughter or son, are typical ascribed statuses, as are one's gender and ethnic or racial identity. Other statuses, called **achieved statuses**, are *occupied as a result of the individual's actions*—student, professor, garage mechanic, race car driver, artist, prisoner, bus driver, husband, wife, mother, or father.

Roles

Statuses alone are static—nothing more than social categories into which people are put. Roles bring statuses to life, making them dynamic. As Robert Linton (1936) observed, you occupy a status but you play a role. **Roles** are *the culturally defined rules for proper behavior that are associated with every status.*

Roles may be thought of as collections of rights and obligations. For example, to be a race car driver you must become well versed in these rights and obligations, as your life might depend on them. Every driver has the right to expect other drivers not to try to pass when the race has been interrupted by a yellow flag because of danger. Turned around, each

driver has the obligation not to pass other drivers under yellow-flag conditions. A driver also has a right to expect race committee members to enforce the rules and spectators to stay off the raceway. On the other hand, a driver has an obligation to the owner of the car to try hard to win.

In the case of our television executive, she has the right to expect to be paid on time, to be provided with good-quality scripts and staff support, and to make decisions about the use of her budget. On the other hand, she has the obligation to act in the best interest of the network, to meet schedules, to stay within her budget, and to treat her employees fairly. What is important is that all these rights and obligations are part of the roles associated with the status of vice-president for programming. They exist without regard to the particular individuals whose behavior they guide (Figure 5–2).

A status may include a number of roles, and each role will be appropriate to a specific social context. For example, as the child of a military officer, Kay Redfield Jamison found that children had to learn the importance of statuses and roles and the proper behavior to be displayed toward those who occupied those positions.

> [The] Cotillion was where officers' children were supposed to learn the fine points of manners, dancing, white gloves, and other unrealities of life. It also was where children were supposed to learn, as if the preceding fourteen or fifteen years hadn't already made it painfully clear, that generals outrank colonels, who, in turn, outrank majors and captains and lieutenants, and everyone, but everyone, outranks

children. Within the ranks of children, boys always outrank girls.

One way of grinding this particularly irritating pecking order into the young girls was to teach them the old and ridiculous art of curtsying. It is hard to imagine that anyone in her right mind would find curtsying an even vaguely tolerable thing to do. But having been given the benefits of a liberal education by a father with strongly nonconforming views and behaviors, it was beyond belief to me that I would seriously be expected to do this. I saw the line of crisply crinolined girls in front of me and watched each of them curtsying neatly. Sheep, I thought, Sheep. Then it was my turn. Something inside of me came to a complete boil. It was one too many times watching one too many girls being expected to acquiesce; far more infuriating, it was one too many times watching girls willingly go along with the rites of submission. I refused. A slight matter, perhaps, in any other world, but within the world of military custom and protocol—where symbols and obedience were everything, and where a child's misbehavior could jeopardize a father's promotion—it was a declaration of war. Refusing to obey an adult, however absurd the request, simply wasn't done. Miss Courtnay, our dancing teacher, glared. I refused again. She said she was very sure that Colonel Jamison would be terribly upset by this. I was, I said, very sure that Colonel Jamison couldn't care less. I was wrong. (Jamison, 1995)

Role Sets

All the roles attached to a single status are known collectively as a **role set.** However, not every role in a particular role set is enacted all the time. An individual's role behaviors depend on the statuses of the other people with whom he or she is interacting. For example, as a college student you behave one way toward other students and another way toward professors. Similarly, professors behave one way toward other professors, another way toward students, and yet a third way toward deans. So the role behavior we expect in any given situation depends on the pairs of statuses that the interacting individuals occupy. This means that role behavior really is defined by the rights and obligations that are assigned to statuses when they are paired with one another (Figure 5–3).

It would be difficult to describe the wide-ranging, unorganized assortment of role behaviors associated with the status of television vice-president for programming. Sociologists find it more useful to describe the specific behavior expected of a network television vice-president for programming interacting

Figure 5–3   Role Sets

People's role behaviors change according to the statuses of the other people with whom they interact. The female vice-president for programming will adopt somewhat different roles depending on the statuses of the various people with whom she interacts at the station: a writer, a journalist, her assistants, and so on.

with different people. Such a role set would include the following:

Vice-president for programming/network president

Vice-president for programming/other vice-presidents

Vice-president for programming/script writer

Vice-president for programming/administrative assistant

Vice-president for programming/television star

Vice-president for programming/journalist

Vice-president for programming/producer

Vice-president for programming/sponsor

The vice-president's role behavior in each case would be different, meshing with the role behavior of the individual(s) occupying the other status in each pairing (Merton, 1969).

Role Strain

Even though most people try to enact their roles as they are expected to, they sometimes find it difficult. *When a single role has conflicting demands attached to it,* individuals who play that role experience **role strain** (Goode, 1960). For example, the captain of a freighter is expected to be sure the ship sails only when it is in safe condition, but the captain also is expected to meet the company's delivery schedule because a day's delay could cost the company thousands of dollars. These two expectations may exert competing pulls on the captain, especially when some defect is reported, such as a malfunction in the ship's radar system. The stress of these competing pulls is not due to the captain's personality, but rather is built into the nature of the role expectations attached to the captain's status. Therefore, sociologists describe the captain's experience of stress as role strain.

Role Conflict

An individual who is occupying more than one status at a time and who is unable to enact the roles of one status without violating those of another status is encountering **role conflict.** Not long ago, pregnancy was considered "women's work." An expectant father was expected to get his wife to the hospital on time and pace the waiting room anxiously awaiting the nurse's report on the sex of the baby and its health. Today, men are encouraged and even expected to participate fully in the pregnancy and the birth of the child. A role conflict arises, however, in that although the new father is expected to be involved, his involvement is defined along male gender-role lines. He is expected to be helpful, supportive, and essentially a stabilizing force. He really is not allowed to indicate that he is frightened, nervous, or possibly angry about the baby. His role as a male, even in twenty-first-century American society, conflicts with his feelings as a new expectant father (Shapiro, 1987).

As society becomes more complex, individuals occupy increasingly larger numbers of statuses. This increases the chances for role conflict, which is one of the major sources of stress in modern society.

Role-Playing

The roles we play can have a profound influence on both our attitudes and our behavior. Playing a new social role often feels awkward at first, and we may feel we are just acting—pretending to be something that we are not. However, many sociologists feel that the roles a person plays are the person's only true self. Peter Berger's (1963) explanation of role-playing goes further: The roles we play can transform not only our actions but also ourselves.

One feels more ardent by kissing, more humble by kneeling, and more angry by shaking one's fist—that is, the kiss not only expresses ardor but manufactures it. Roles carry with them both certain actions and emotions, and attitudes that belong to these actions. The professor putting on an act that pretends to wisdom comes to feel wise.

Institutions and Social Organization

Anyone who has traveled to foreign countries knows that different societies have different ways of doing things. The basic things that get done actually are quite similar—food is produced and distributed; people get married and have children; and children are raised to take on the responsibilities of adulthood. The vehicle for accomplishing the basic needs of any society is the social institution.

Social Institutions

Sociologists usually speak of five areas of society in which basic needs have to be fulfilled: the family sector, the education sector, the economic sector, the religious sector, and the political sector. For each of these areas, social groups and associations carry out the goals and meet the needs of society.

The behavior of people in these groups and associations is organized or patterned by the relevant **social institutions**—*the ordered social relationships that grow out of the values, norms, statuses, and roles that organize those activities that fulfill society's fundamental needs.* Thus, economic institutions organize the ways in which society produces and distributes the goods and services it needs; educational institutions determine what should be learned and how it should be taught; and so forth.

Of all social institutions, the family is perhaps the most basic. A stable family unit is the main ingredient necessary for the smooth functioning of society. For instance, sexual behavior must be regulated and children must be cared for and raised to fit into society. Hence, the institution of the family provides a system of continuity from one generation to the next.

Using the family as an example, we can see the difference between the concept of group and the concept of institution. A group is a collection of specific, identifiable people. An institution is a system for organizing standardized patterns of social behavior. In other words, a group consists of people, and an institution consists of actions. For example, when sociologists discuss *a* family (say the Smith family), they are referring to a particular group of people. When they discuss *the* family, they are referring to family as an institution—a cluster of statuses, roles, values, and norms that organize the standardized patterns of behavior that we expect to find within family groups. Thus, the family as an American institution typically embodies several master statuses: those of husband, wife, and, possibly, father, mother, and child. It also includes the statuses of son, daughter, brother, and sister. These statuses are organized into well-defined, patterned relationships: Parents

Relatively rapid social change is making less predictable the types of behavior that go along with gender roles.

have authority over their children, spouses have a sexual relationship with each other (but not with the children), and so on.

However, specific family groups may not conform entirely to the ideals of the institution. There are single-parent families, families in which the children appear to be running things, and families in which there is an incestuous parent-child relationship. Although a society's institutions provide what can be thought of as a master plan for human interactions in groups, actual behavior and actual group organization often deviate in varying degrees from this plan.

Social Organization

If we step back from a mosaic, we see the many multicolored stones composing a single, coordinated pattern or picture. Similarly, if we step back and look at society, we see the many actions of all its members fall into a pattern or series of interrelated patterns. These consist of social interactions and relationships expressing individual decisions and choices. These choices, however, are not random; rather, they are an outgrowth of a society's social organization. **Social organization** consists of *the relatively stable pattern of*

social relationships among individuals and groups in society.

These relationships are based on systems of social roles, norms, and shared meanings that provide regularity and predictability in social interaction.

Social organization differs from one society to the next. Thus, Islam allows a man to have up to four wives at once, whereas in American society, with its Judeo-Christian religious tradition, such plural marriage is not an acceptable family form.

Just as statuses and roles exist within ordered relationships to one another, social institutions also exist in patterned relationships with one another in the context of society. All societies have their own patterning for these relationships. For example, a society's economic and political institutions often are closely interrelated. So, too, are the family and religious institutions. Thus, a description of American social organization would indicate the presence of monogamy along with Judeo-Christian values and norms and the institutionalization of economic competition and of democratic political organizations.

A society's social organization tends to be its most stable aspect. The American social organization, however, may not be as static as that of many other societies. American society is experiencing relatively rapid social change because of its complexity and because of the great variety in the types of people that are part of it. This complexity makes life less predictable because new values and norms being introduced from numerous quarters result in changes in social organization. For example, ideas about the behavior that should go along with female gender roles have changed considerably over the past three decades. Traditionally, married women were expected not to work but to stay home and attend to the rearing of children. Today, the majority of American women are working outside the home, and views on what roles mothers should play in the lives of their children are in flux.

Societies

The most complex structures sociologists examine are societies. A **society** is *a grouping of people who share the same territory and participate in a common culture.* Organized and long-lasting societies are rare among most mammals, with the wolf pack, the prairie dog town, and the baboon troop representing some of the notable exceptions. Society, however, is universal among humans. For this to be so, it must have performed major adaptive functions that have increased the chances of human survival. Among certain species of animals, great speed, powerful jaws, or brute strength may be the key adaptive advantage. Among humans, the social organization that results in society has enabled our survival.

Among humans the members of a particular society are so mutually interdependent that often the very survival of each member depends on the behavior of the other members. There have been a variety of different types of societies throughout history. Let us examine these.

Types of Societies

In this section we shall trace the development of the major types of societies. In tracing this development we shall look at hunting and food-gathering societies, horticultural societies, pastoral societies, agricultural societies, and finally, industrial and postindustrial societies.

Hunting and Food-Gathering Societies The members of the earliest human societies—**hunting and food-gathering societies** *survived by foraging for vegetable foods and small game, fishing, collecting shellfish, and hunting larger animals.* They did not plant crops or raise animals for future needs; rather, they subsisted from day to day on whatever was at hand. In modern times a few of the world's simplest societies—in

Australia, Africa, and South America—still subsist using these methods and depend to a large degree on tools made of stone, wood, and bone.

Most hunter-gatherer societies are nomadic and when the food in one area is exhausted they move on. Often this movement is related to the seasons of the year. The group may be able to stay in one area for three or four months and when the climate changes they find a more hospitable climate. Hunter-gatherers tend to be small groups because they need a fairly large area to survive. Farming makes much more efficient use of the land and can support substantially larger populations.

Hunter-gatherer societies have fewer class divisions than other societies. With a nomadic existence there is little opportunity to store surplus food. Therefore, most everyone in the group is on the same level. Of course, greater recognition would be given to the best hunters.

If we were to consider the evolutionary cycle during which humans have been on this earth, we would find that most of it has been marked by hunting and gathering societies. Anthropologists have estimated that humans hunted for at least 1 million years, but that it has been a mere 10,000 years since the first people began to experiment with the possibilities of organized agriculture.

Anthropologist Richard Leakey (Leakey & Lewin, 1977) believes that hunting is the key factor in the development of human social organization. The sharing of food became an integral part of early hunting and gathering groups. This opened up a behavioral gulf between humans and primate groups. Primates are primarily vegetarians. Even though they are social animals, a plant-eating existence tends to make the individual members self-centered and uncooperative. To be a vegetarian is essentially to be solitary. Each member tears a leaf from a branch or plucks fruit from a tree, and promptly eats it. There is little communal eating or sharing of food. Sharing was one of a group of traits acquired through hunting and gathering that helped produce an increasingly adaptable way of life. This adaptability enabled the human species to thrive in practically every corner of the globe.

Horticultural Societies Some 12,000 to 15,000 years ago, coinciding with the retreat of the last glaciers, a drying trend took place in what previously had been rich, subtropical climates, and the giant deserts of Africa, Asia, and the Middle East took shape. Even beyond the deserts' constantly expanding borders, new arid conditions made precarious the hunting and gathering way of life. Some groups continued to eke out an existence using the old methods. Others crowded together in the more abundant regions, harvesting wild grains until population pressures drove them out into less favorable environments.

There they attempted to recreate the rich environments they had left. In doing so, they created a new type of lifestyle—**horticultural societies**—*societies in which human muscle power and handheld tools are used to cultivate gardens and fields* (Flannery, 1965, 1968). It appears that women invented this new and revolutionary form of food production—deliberately planting seeds with the idea of having a sure source of food later on—through their observations of the relationships between seeds and the growth of plants.

The movement from hunting and gathering to farming also represented a dramatic social change. Hunters as a people must have faith in the resources of their physical world and in their ability to make use of them. They live in small, intimate, cooperative groups moving from camp to camp as their food supplies dictate. As they roam the land that they may share with other bands, they may have fights and disputes, but they do not seek them out.

An agricultural existence produces the exact opposite way of life. Because crops must be tended and harvested, farmers are confined to a certain area. A sedentary way of life also offers the possibility of accumulating material possessions. In turn, the land bearing the crops and the farmer's possessions must be defended. A whole new outlook on life thus develops. Agriculture supports many more people, thus giving rise to villages and towns. The possibility of expanding possessions and the ability to have power over others in turn produces the urge to accumulate still more and brings about the necessity of protecting what has already been won. The stage is thus set for conflict and wars.

To hunting and gathering tribes, it made little sense to take the territory of a neighboring band. In contrast, agriculture allowed local populations to grow. If one village decided to take over the crops of another, it would benefit because its own population could then expand with the extra food. In this respect the violence and destruction common in modern times is an outgrowth of the agricultural way of life, and not a biological remnant from our hunting ancestors (Leakey & Lewin, 1977).

Pastoral Societies   **Pastoral societies** *rely on herding and the domestication and breeding of animals for food and clothing to satisfy most of the group's needs.* Animal herds provide food, milk, dung (for fuel), skin, sheared fur, and even blood (which is drunk as a major source of protein in East Africa).

Pastoral societies appear in many regions that are not suitable for agriculture such as semiarid desert regions and the northern tundra plains of Europe and Asia. They are also found in less severe climates, including East African savannas and mountain grasslands. Pastoralism, however, almost never occurs in forest or jungle regions. It is an interesting fact, which scholars have not been able to explain, that no true pastoral societies ever emerged in the Americas before the arrival of the Europeans.

Agricultural Societies   **Agricultural societies** *use the plow in food production.* Agriculture is more efficient than horticulture. Instead of just using human muscle power, plowing turns the topsoil far deeper than hoeing, allowing for better aerating and fertilizing of the ground and therefore producing more food. Interestingly, early agriculture probably did not produce much more than could be harvested in naturally rich environments. However, by about 5500 B.C. farmers in the Middle East were not only using the plow but irrigation (the channeling of water for crops) as well. With irrigation, farmers became capable of producing huge surpluses—enough to feed large numbers of people who did not produce food themselves.

Reliance on agriculture had dramatic consequences for society. Ever-growing populations came together in the broad river valleys like those of the Nile in Egypt, the Tigris and Euphrates in the Middle East, the Huang Ho (Yellow River) in China, and the Danube and Rhine in Europe. This rapidly rising and geographically compressed population density gave rise to cities and to new social arrangements. For the first time society was *not* organized principally in terms of kinship. Rather, occupational diversity and institutional specialization (including political, economic, and religious institutions) became important.

As populations grew and cities developed, the need arose for some central organization that could administer the increasingly complex activities that accompanied the growth. Such a central authority could serve the interests of the leaders by enabling them to collect taxes and create a military to protect the new centers of population. These developments led to the emergence of the centralized state.

Industrial Societies   **Industrialism** *consists of the use of mechanical means (machines and chemical processes) for the production of goods.* The Industrial Revolution at first developed slowly in the mid-eighteenth century and gained momentum by the turn of the nineteenth century. In 1798, Eli Whitney built the first American factory for the mass production of guns near New Haven, Connecticut. By the mid-1800s the Industrial Revolution had swung into high gear with the invention of the steam locomotive and Henry Besemer's development of large-scale production techniques at his steelworks in England in 1858. It is called a revolution because of the enormous changes it brought about in society.

Industrial societies *use mechanical means of production instead of human or animal muscle power.*

They constitute an entirely new form of society that requires an immense, mobile, diversely specialized, highly skilled, and well-coordinated labor force. To meet the demand for people with specialized and complex skills, many members of the labor force must be educated, at least able to read and write. Hence, an educational system open to all is a hallmark of industrial societies—something that was not necessary in preindustrial times. Industrialism also requires the creation of highly organized systems of exchange between the suppliers of raw materials and industrial manufacturers on the one hand, and between the manufacturers and consumers on the other.

Industrial societies are inevitably divided along class lines. The nature of the division varies, depending on whether the society allows private ownership of capital (capitalism) or puts all capital in the hands of the state (socialism). All industrial societies have at least two social classes: (1) a large labor force that produces goods and services but has little or no influence on what is done with them and (2) a much smaller class that determines what will be produced and how it will be distributed.

Postindustrial Society The world has continued to change since the beginning of industrialization. These changes have prompted sociologists to adopt the concept of the postindustrial society (Bell, 1973). Whereas industrial societies depend on inventions and advances made by craftspeople, **postindustrial societies** *depend on specialized knowledge to bring about continuing progress in technology.* The computer and biotech industries are integral parts of postindustrial society. Advances in these fields are made by highly trained specialists who work to create new products for our society. Postindustrial societies are dependent on a well-educated population. Intellectual and technical knowledge and access to large quantities of information are crucial to these types of society. The economies of postindustrial societies are service oriented and more than half of the work in the United States is involved in service occupations. Rapid changes in technology continue to have a major impact on lifestyles in postindustrial society.

Using the concepts we have discussed in this chapter, sociologists study the ways in which individuals interact with one another as individuals and as members of society. In future chapters we will analyze these processes further.

SUMMARY

- Humans are symbolic creatures, and everything they do conveys a message to others.
- Whether we intend it or not, other people take account of our behavior.
- Most Americans distinguish among intimate, personal, social, and public distance.
- People do not interact with each other as anonymous beings.
- People come together in the context of specific environments, with specific purposes, and with specific social characteristics.
- Statuses and roles are some of the most important social characteristics.
- Statuses are socially defined positions that people occupy in a group or society that help determine how they interact with one another.
- Statuses exist independently of the specific people who occupy them.
- Roles are the culturally defined rules for proper behavior that are associated with every status.
- A role is basically a collection of rights and obligations.
- Statuses and roles help define our social interactions and provide predictability.
- The most complex structures sociologists examine are societies.
- A society is a grouping of people who share the same territory and participate in a common culture.
- Social institutions consist of the ordered relationships that grow out of the values, norms, statuses, and roles that organize those activities that fulfill society's fundamental needs. Institutions are systems for organizing standardized patterns of social behavior.
- Social organization consists of the relatively stable pattern of social relationships among individuals and groups in society. It differs from one society to the next.
- The members of the earliest human societies—hunting and food-gathering societies survived by foraging for vegetable foods and small game, fishing, collecting shellfish, and hunting larger animals.
- Horticultural societies are societies in which human muscle power and handheld tools are used to cultivate gardens and fields.
- Pastoral societies rely on herding and the domestication and breeding of animals for food and clothing to satisfy most of the group's needs.
- Agricultural societies use the plow in food production.
- Industrialism consists of the use of mechanical means (machines and chemical processes) for the production of goods.
- Industrial societies use mechanical means of production instead of human or animal muscle power.
- Postindustrial societies depend on specialized knowledge to bring about continuing progress in technology.

Media Resources

The Companion Website for *Introduction to Sociology*, Ninth Edition

http://sociology.wadsworth.com/tischler9e

Supplement your review of this chapter by going to the companion website to take one of the Tutorial Quizzes, use the flash cards to master key terms, and check out the many other study aids you'll find there. You'll also find special features such as Wadsworth's Sociology Online Resources and Writing Companion, GSS data, and Census 2000 information at your fingertips to help you complete that special project or do some research on your own.

CHAPTER FIVE STUDY GUIDE

KEY CONCEPTS AND THINKERS

Match each concept with its definition, illustration, or explanation presented below.

a. Role strain
b. Conflict
c. Norms
d. Exchange interaction
e. Ethnomethodology
f. Role conflict
g. Kinesics

h. Social distance
i. Master status
j. Achieved status
k. Social interaction
l. Dramaturgy
m. Roles
n. Social action

o. Ascribed status
p. Hunter-gatherer society
q. Postindustrial society
r. Agricultural society
s. Horticultural society
t. Pastoral society
u. Industrial society

_c_ **1.** Specific rules of behavior that are agreed upon and shared and that prescribe limits of acceptable behavior.

_d_ **2.** Something done toward another person for the purpose of receiving some reward from that person.

_e_ **3.** The study of the sets of rules or guidelines that individuals use to initiate behavior, respond to behavior, and modify behavior.

_h_ **4.** A distance of between 4 and 12 feet; it is used in more impersonal business interactions.

_b_ **5.** People struggling against one another for some commonly prized object or value.

____ **6.** Anything people are conscious of doing because of other people.

_g_ **7.** The study of how slight nods, yawns, postural shifts, nonverbal cues, and other body movements affect behavior.

_i_ **8.** One of the multiple statuses a person occupies that seems to dominate the others in patterning a person's life.

_j_ **9.** A status occupied as a result of an individual's actions.

_m_ **10.** Culturally defined rules for proper behavior that are associated with every status.

_a_ **11.** Conflicting demands attached to the same role.

_f_ **12.** An inability to enact the roles of one status without violating those of another status.

_l_ **13.** The use of the framework of theater, performance, and role to understand people's behavior.

_o_ **14.** A status conferred on a person because of unchangeable qualities (sex, race, and so on) rather than because of his or her actions.

____ **15.** Two or more people taking one another into account.

____ **16.** Society in which people live by herding animals.

____ **17.** The simplest, oldest type of society.

____ **18.** The simplest farming society.

____ **19.** Society based on exchange of knowledge and service rather than the production of goods.

____ **20.** Society featuring machine-based production.

____ **21.** Society that ramped up farming production by use of the plow.

Match the thinkers with their main idea or contribution.

a. Edward T. Hall   **b.** Erving Goffman   **c.** Harold Garfinkel

____ **1.** Was a pioneer in studying the context of social interaction.

____ **2.** Proposed that it was important to study the commonplace aspects of everyday life.

____ **3.** Developed an approach that focused on how people try to create a favorable impression of themselves and the manner in which others judge their performances.

CENTRAL IDEA COMPLETIONS

Following the instructions, fill in the appropriate concepts and descriptions for each of the questions posed below.

1. Differentiate between **role strain** and **role conflict:** _____
   _____
   _____

   a. What might be **two forms** of role strain facing a female college professor?
   _____

   b. What are the **sources** of these role strains?

   _____

   c. How might the individual **resolve** these strains?

   _____

2. Consider the kinds of interactions in a family.

   a. What activities in a family illustrate **exchange** behaviors?

   _____

   _____

   b. What activities in a family illustrate **conflict** behaviors?

   _____

   _____

   c. What activities in a family illustrate **competitive** behaviors?

   _____

   _____

3. List the four major types of social distance zones outlined in your text.

   (1) _____   (2) _____

   (3) _____   (4) _____

   a. Assuming you were to **violate** each form of social distance zone, which violation would you define
      as most serious? _____

   b. What factors might influence an audience witnessing your violation to regard your behavior as serious?
   _____

4. What is the difference between an achieved status and an ascribed status?
   _____

   In which types of situations is it legitimate to think more in terms of ascribed than achieved status?
   _____

5. What is "master status," and what difficulties can it create in social interaction?
   _____

CRITICAL THOUGHT EXERCISES

1. Following your text's discussion of the four social distances most Americans use in business and social relations (intimate, personal, social, and public), conduct your own ethnomethodological study by adopting a personal distance incongruent with the setting in which you are interacting. For example, with a close acquaintance, begin a conversation while standing face-to-face. As your conversation unfolds, begin backing away and increase your distance from the customary 6 to 18 inches to 2½ to 4 feet, then to more than 4 feet. What happened to the conversation? What was the reaction of your acquaintance? Hypothesize what might have happened if you had elected to use an intimate personal distance within a public distance setting, for example, at a department or grocery store. How might a salesperson or store clerk have reacted if you had begun to move as close as 6 to 18 inches?

2. Attend a meeting of an international students' organization at your college or university. Observe not only the public and personal conversational distances engaged in by the students but also try to determine whether these distance vary by the country the students come from. After the meeting, approach several of the international students and ask them whether they have noticed any differences in the use of personal space among students in their home country and in the United States. If the opportunity presents itself, inquire whether the students have had any initial difficulties adjusting to Americans' use of social space when they first arrived. Finally, ask whether they expect (on the basis of the time they are spending in the United States) to have any adjustment problems when they return to their home country.

3. Make a list of embarrassing incidents. Ask other students or friends to contribute to the list. Analyze the incidents in terms of role, status, and dramaturgy. Are there role requirements that were not being met? Was there role conflict, as when a performance intended for one audience (friends, for example) is seen by someone it was not intended for (a teacher, parents)?

4. Blau says that we are often unaware of the exchange qualities of social interaction. Look at several ordinary interactions and try to find the exchange component. To make the task easier, first look at points of disagreement or conflict, where one person thought another had not done the right thing. Is the complaint really about the person's failure to live up to the expectations of exchange? Then look at more successful interactions to see how each person had contributed to the exchange.

INTERNET ACTIVITIES

1. http://www3.usal.es/~nonverbal/introduction.htm. This resource has links to all sorts of websites with information on nonverbal communication and body language.

2. Dane Archer at the University of California at Santa Cruz has produced several videos on nonverbal communication. Some still shots from them can be seen here: http://zzyx.ucsc.edu/~archer/intro.html.

ANSWERS TO KEY CONCEPTS

1.c 2.d 3.e 4.h 5.b 6.n 7.g 8.i 9.j 10.m 11.a 12.f 13.l 14.o 15.k 16.t 17.p 18.s 19.q 20.u 21.r

ANSWERS TO KEY THINKERS

1.a 2.c 3.b

ThomsonNOW™

Reviewing is as easy as ❶ ❷ ❸

1. Before you do your final exam, take the ThomsonNOW diagnostic quiz to help you identify the areas on which you should concentrate. You will find information on ThomsonNOW and instructions on how to access all of its great resources on the foldout at the beginning of the text.

2. As you review, take advantage of ThomsonNOW's study videos and interactive Map the Stats exercises to help you master the chapter topics.

3. When you are finished with your review, take ThomsonNOW's posttest to confirm you are ready to move on to the next chapter.

6

Social Groups and Organizations

© Ryan McVay/Stone/Getty Images

Learning Objectives

After studying this chapter, you should be able to do the following:

- Distinguish between primary and secondary groups.
- Explain the functions of groups.
- Understand the role of reference groups.

- Know the influence of group size.
- Understand the characteristics of bureaucracy.
- Know what Michels's concept of "the iron law of oligarchy" is.
- Understand why social institutions are important.

W e all feel strongly about our names. We do not like it when people mispronounce our names and we are quick to correct them. Parents give their children names that they think will help their future life prospects. The popularity of names, however, changes with the times. The parents today who are naming their little girls Madison, Olivia, and Hannah would be horrified if someone suggested they name their new baby Bertha, Gertrude, or Myrtle. Yet, these were very popular names in the past. (See Tables 6–1 and 6–2.) Names help identify who we are and what groups we belong to. They can be used to show that we are part of a particular ethnic group, social class, or religious group. People have changed their names to avoid being identified with a certain group. Naming a baby, then, is the parent's first act to show that the child is part of a certain group.

The Nature of Groups

A good deal of social interaction occurs in the context of a group. In common speech the word *group* is often used for almost any occasion when two or more people come together. In sociology, however, we use several terms for various collections of people, not all of which are considered groups. A **social group** consists of *a number of people who have a common identity, some feeling of unity, and certain common goals and shared norms.* In any social group, the

individuals interact with one another according to established statuses and roles.

The members develop expectations of proper behavior for people occupying different positions in the social group. The people have a sense of identity and realize they are different from others who are not members. Social groups have a set of values and norms that may or may not be similar to those of the larger society.

For example, a group of students in a college class may have some common norms, which include taboo subjects, open expression of feelings, interrupting or challenging the professor, avoiding conflict, and length and frequency of contributions. All these are usually hidden or implicit and new members have to learn them quickly. Violations of the norms will involve sanctions (for example, disapproval), which may include comments, disapproving looks, or avoidance of the "deviant."

Our description of a social group contrasts with our definition of a **social aggregate,** which is made up of *people who temporarily happen to be in physical proximity to each other, but share little else.* Consider passengers riding together in one car of a train. They may share a purpose (traveling to Washington, D.C.) but do not interact or even consider their temporary association to have any meaning. It hardly makes sense to call them a group—unless something more happens. If it is a long ride, for instance, and several passengers start a card game, the card players will

Table 6–1

Most Popular Baby Names for 2004

Rank	Male Name	Female Name
1	Jacob	Emily
2	Michael	Emma
3	Joshua	Madison
4	Matthew	Olivia
5	Ethan	Hannah
6	Andrew	Abigail
7	Daniel	Isabella
8	William	Ashley
9	Joseph	Samantha
10	Christopher	Elizabeth

Source: Social Security Administration, http://www.ssa.gov/OACT/babynames/

Table 6–2

Baby Names That Used to be Popular

Male	Female
Cecil	Agnes
Chester	Bertha
Dewey	Bessie
Elmer	Beulah
Floyd	Gertrude
Homer	Myrtle
Mack	Pearl

Source: Social Security Administration, http://www.ssa.gov/OACT/babynames/

have formed a social group: They have a purpose, they share certain role expectations, and they attach importance to what they are doing together. Moreover, if the card players continue to meet one another every day (say, on a commuter train), they may begin to feel special in contrast with the rest of the passengers, who are just "riders." A social group, unlike an aggregate, does not cease to exist when its members are away from one another. Members of social groups carry the fact of their membership with them and see the group as a distinct entity with specific requirements for membership.

A social group has a purpose and is therefore important to its members, who know how to tell an "insider" from an "outsider." It is a social entity that exists for its members apart from any other social relationships that some of them might share. Members of a group interact according to established norms and traditional statuses and roles. As new members are recruited to the group, they move into these traditional statuses and adopt the expected role behavior—if not gladly, then as a result of group pressure.

Consider, for example, a tenants' group that consists of the people who rent apartments in a building.

Most such groups are founded because tenants feel a need for a strong, unified voice in dealing with the landlord on problems with repairs, heat, hot water, and rent increases. Many members of a tenants' group may never have met one another before; others may be related to one another; and some may also belong to other groups such as a neighborhood church, the PTA, a bowling league, or political associations. The group's existence does not depend on these other relationships, nor does it cease to exist when members leave the building to go to work or to go away on vacation. The group remains, even when some tenants move out of the building and others move in. Newcomers are recruited, told of the group's purpose, and informed of its meetings; they are encouraged to join committees, take leadership responsibilities, and participate in the actions the group has planned. Members who fail to support group actions (such as withholding rent) will be pressured and criticized by the group.

People sometimes are defined as being part of a specific group because they share certain characteristics. If these characteristics are unknown or unimportant to those in the category, it is not a social group. Involvement with other people cannot develop unless one is aware of them. People with similar characteristics do not become a social group unless concrete, dynamic interrelations develop among them (Lewin, 1948). For example, although all left-handed people fit into a group, they are not a social group just because they share this common characteristic. A further interrelationship must also exist. They may, for instance, belong to Left-Handers International of Topeka, Kansas, an organization that champions the accomplishments of left-handers. The group has even designated August 13 as International Left-Handers Day. Thousands of left-handers belong to this social group.

Even if people are aware of one another, that is still not enough to make them a social group. We may be classified as Democrats, college students, upper class, or suburbanites. Yet for many of us who fall into these categories, there is no group. We may not be involved with the others in any patterned way that is an outgrowth of that classification. In fact, we personally may not even define ourselves as members of the particular category, even if someone else does.

Social groups can be large or small, temporary or long lasting: Your family is a group, as is your ski club, any association to which you belong, or the clique you hang around with. In fact, it is difficult for you to participate in society without belonging to a number of different groups.

In general, social groups, regardless of their nature, have the following characteristics: (1) permanence beyond the meetings of members, that is, even when members are dispersed; (2) means for identifying members; (3) mechanisms for recruiting new members; (4) goals or purposes; (5) social statuses and

Table 6–3

Relationships in Primary and Secondary Groups

	Primary	Secondary
Physical Conditions	Small number Long duration	Large number Shorter duration
Social Characteristics	Indentification of ends Intrinsic valuation of the relation Intrinsic valuation of other person Inclusive knowledge of other person Feeling of freedom and spontaneity Operation of informal controls	Disparity of ends Extrinsic valuation of the relation Extrinsic valuation of other person Specialized and limited knowledge of other person Feeling of external constraint Operation of formal controls
Sample Relationships	Friend–friend Husband–wife Parent–child Teacher–pupil	Clerk–customer Announcer–listener Performer–spectator Officer–subordinate
Sample Groups	Play group Family Village or neighborhood Work team	Nation Clerical hierarchy Professional association Corporation

Source: *Human Society,* by K. Davis, 1949, New York: Macmillan.

roles, that is, norms for behavior; and (6) means for controlling members' behavior.

The traits we described are features of many groups. A baseball team, a couple about to be married, a work unit, players in a weekly poker game, members of a family, or a town planning board all may be described as groups. Yet being a member of a family is significantly different from being a member of a work unit. The family is a primary group, whereas most work units are secondary groups.

Primary and Secondary Groups

The difference between primary and secondary groups lies in the kinds of relationships their members have with one another. Charles Horton Cooley (1909) defined primary groups as groups that

[A]re characterized by intimate face-to-face association and cooperation. They are primary in several senses, but chiefly in that they are fundamental in forming the social nature and ideas of the individual. The result of intimate association, psychologically, is a certain fusion of individualities in a common whole, so that one's very self, for many purposes at least, is the common life and purpose of the group. Perhaps the simplest way of describing this wholeness is by saying that it is a "we"; it involves the sort of sympathy and mutual identification of which "we" is the natural expression.

Cooley called primary groups the nursery of human nature because they have the earliest and most fundamental effect on the individual's social-

ization and development. He identified three basic primary groups: the family, children's play groups, and neighborhood or community groups.

Primary groups involve *interaction among members who have an emotional investment in one another and in a situation, who know one another intimately, and who interact as total individuals rather than through specialized roles.* For example, members of a family are emotionally involved with one another and know one another well. In addition, they interact with one another in terms of their total personalities, not just in terms of their social identities or statuses as breadwinner, student, athlete, or community leader.

A **secondary group,** in contrast, *is characterized by much less intimacy among its members. It usually has specific goals, is formally organized, and is impersonal.* Secondary groups tend to be larger than primary groups, and their members do not necessarily interact with all other members. In fact, many members often do not know one another at all; to the extent that they do, rarely do they know more about one another than about their respective social identities. Members' feelings about, and behavior toward, one another are patterned mostly by their statuses and roles rather than by personality characteristics. The chair of the General Motors board of directors, for example, is treated respectfully by all General Motors employees—regardless of the chair's gender, age, intelligence, habits of dress, physical fitness, temperament, or qualities as a parent or spouse. In secondary groups, such as political parties, labor unions, and large corporations, people are very much what they do.

Table 6–3 outlines the major differences between primary and secondary groups.

Functions of Groups

To function properly, all groups—both primary and secondary—must (1) define their boundaries, (2) choose leaders, (3) make decisions, (4) set goals, (5) assign tasks, and (6) control members' behavior.

Defining Boundaries

Group members must have ways of knowing who belongs to their group and who does not. Sometimes devices for marking boundaries are obvious symbols, such as the uniforms worn by athletic teams, lapel pins worn by Rotary Club members, rings worn by Masons, and styles of dress. The idea of the British school tie is that, by its pattern and colors, it signals exclusive group membership, has been adopted by businesses ranging from banking to brewing. Other ways by which group boundaries are marked include the use of gestures (special handshakes) and language (dialect differences often mark people's regional origin and social class). In some societies (including our own), skin color also is used to mark boundaries between groups.

Choosing Leaders

All groups must grapple with the issue of leadership. A **leader** is *someone who occupies a central role or position of dominance and influence in a group*. In some groups, such as large corporations, leadership is assigned to individuals by those in positions of authority. In other groups, such as adolescent peer groups, individuals move into positions of leadership through the force of personality or through particular skills such as athletic ability, fighting, or debating. In still other groups, including political organizations, leadership is awarded through the democratic process of nominations and voting. Think of the long primary process the presidential candidates must endure in order to amass enough votes to carry their party's nomination for the November election.

Leadership need not always be held by the same person within a group. It can shift from one individual to another in response to problems or situations that the group encounters. In a group of factory workers, for instance, leadership may fall on different members depending on what the group plans to do—complain to the supervisor, head to a bar after work, or organize a picnic for all members and their families.

Politicians and athletic coaches often like to talk about individuals who are "natural leaders." Although attempts to account for leadership solely in terms of personality traits have failed again and again, personality factors may determine what kinds of leadership functions a person assumes. Researchers (Bales, 1958; Slater, 1966) have identified two types of leadership roles: (1) **instrumental leadership,** *in which a leader actively proposes tasks and plans to guide the group toward achieving its goals,* and (2) **expressive leadership,** *in which a leader works to keep relations among group members harmonious and morale high*. Both kinds of leadership are crucial to the success of a group.

Sometimes one person fulfills both leadership functions, but when that is not the case, those functions are often distributed among several group members. The individual with knowledge of the terrain who leads a group of train crash survivors to safety is providing instrumental leadership. The group members who think of ways to keep the group from giving in to despair are providing expressive leadership. The group needs both kinds of leadership to survive.

Making Decisions

Closely related to the problem of leadership is the way groups make decisions. In many early hunting and food-gathering societies, important group decisions were reached by consensus—talking about an issue until everybody agreed on what to do (Fried, 1967). Today, occasionally, town councils and other small governing bodies operate in this way.

Because this consensus gathering takes a great deal of time and energy, many groups opt for efficiency by taking votes or simply letting one person's decision stand for the group as a whole.

Setting Goals

As we pointed out before, all groups must have a purpose, a goal, or a set of goals. The goal may be very general, such as spreading peace throughout the world, or it may be very specific, such as playing cards on a train. Group goals may change. For example, the card players might discover that they share a concern about the use of nuclear energy and decide to organize a political-action group.

Assigning Tasks

Establishing boundaries, defining leadership, making decisions, and setting goals are not enough to keep a group going. To endure, a group must do something, if nothing more than ensure that its members continue to make contact with one another. Therefore, it is important that group members know what needs to be done and who is going to do it. This assigning of tasks, in itself, can be an important group activity (think of your family discussions about sharing household chores). By taking on group tasks, members not only help the group reach its goals but also show their commitment to one another and to the group as a whole.

This leads members to appreciate one another's importance as individuals and the importance of the group in all their lives—a process that injects life and energy into a group.

Controlling Members' Behavior

If a group cannot control its members' behavior, it will cease to exist. For this reason, failure to conform to group norms is seen as dangerous or threatening, whereas conforming to group norms is rewarded, if only by others' friendly attitudes. Groups not only encourage but often depend for survival on conformity of behavior. A member's failure to conform is met with responses ranging from coolness to criticism or even ejection from the group. Anyone who has tried to introduce changes into the constitution of a club or to ignore long-standing conventions—such as ways of dressing, rituals of greeting, or the assumption of designated responsibilities—probably has experienced group hostility.

Primary groups tend to be more tolerant of members' deviant behavior than secondary groups. For example, families often will conceal the problems of a member who suffers from chronic alcoholism or drug abuse. Even primary groups, however, must draw the line somewhere, and they will invoke negative sanctions (see Chapter 7) if all else fails to get the deviant member to show at least a willingness to try to conform. When primary groups finally do act, their punishments can be far more severe or harsh than those of secondary groups. Thus, an intergenerational conflict in a family can result in the commitment of a teenager to an institution or treatment center. Secondary groups tend to use formal, as opposed to informal, sanctions and are much more likely than primary groups simply to expel, or push out, a member who persists in violating strongly held norms: Corporations fire unsatisfactory employees, the army discharges soldiers who violate regulations, and so on.

Even though primary groups are more tolerant of their members' behavior, people tend to conform more closely to their norms than to those of secondary groups. This is because people value their membership in a primary group, with its strong interpersonal bonds, for its own sake. Secondary group membership is valued mostly for what it will do for the people in the group, not because of any deep emotional ties. Because primary group membership is so desirable, its members are more reluctant to risk expulsion by indulging in behavior that might violate the group's standards or norms than are secondary group members.

Usually, group members will want to conform as long as they experience the group as important. Solomon Asch (1955) showed just how far group members will go to promote group solidarity and con-

Figure 6–1 Group Pressure

In Solomon Asch's experiment on conformity to group pressure, groups of eight students were asked to decide which of the comparison lines (right) was the same length as the standard line (left).

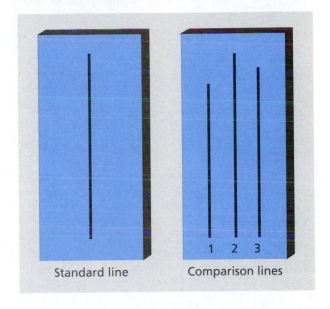

Standard line   Comparison lines

formity. In a series of experiments, he formed groups of eight people and then asked each member to match one line against three other lines of varying lengths (Figure 6–1). Each judgment was announced in the presence of the other group members.

The groups were composed of one real subject and seven of Asch's confederates, whose identities were kept secret from the real subject. The confederates had met previously with Asch and had been instructed to give a unanimous but incorrect answer at certain points throughout the experiment. Asch was interested in finding out how the individual who had been made a minority of one would respond in the presence of a unanimous majority. The subject was placed in a situation in which a group unanimously contradicted the information of his or her senses. Asch repeated the experiment many times. He found that 32% of the answers by the real subjects were identical with, or in the direction of, the inaccurate estimates of the majority. This was quite remarkable, because there were virtually no incorrect answers in the control groups that lacked Asch's accomplices, which rules out the possibility of optical illusion. This illustrates an instance in which individuals are willing to give incorrect answers in order not to appear out of step with the judgments of the other group members. (For further discussion of the impact of group members, see "Controversies in Sociology: Does Society Shape Our Memories?")

Although groups must fulfill certain functions in order to continue to exist, they serve primarily as a point of reference for their members.

CONTROVERSIES IN SOCIOLOGY

Does Society Shape Our Memories?

Any police officer will tell you that eyewitness accounts of events can differ dramatically. Not only are our memories of events shaped by our interaction with other people, but, it turns out, they can be shaped by certain biases we may hold.

Every day we encounter news reports. Most are accurate, but some turn out to be inaccurate. Which ones are we likely to believe? Which ones are we likely to remember?

For example, at the beginning of the Iraq war, there were many reports of the possible discovery of weapons of mass destruction (WMDs). Although these reports were subsequently shown to be false, six months after the invasion, one-third of Americans still believed the WMDs had been found.

In another case, after reading a story about a jewelry theft, people continued to believe the suspect named was guilty, even though his alibi as to his whereabouts during the time of the crime turned out to be true.

Several things seem to be happening here. People tend to remember things that logically go together even if it did not actually happen. Imagine that subjects are given a list of words that include, for example, *thermos, sandwich, beverage*. They may later "remember" that words such *napkin* or *apple* were also on the list, because they all relate to the topic, "lunch."

Second, as we saw with our discussion of the Asch experiment, people's memories of events can also be influenced by others who may hold contradictory positions.

It also turns out that people have "mental maps" of social life and the world in general. Once we have decided to either believe or not believe something, any later change or retraction will merely be filtered through our position on social life or politics.

On the WMD example, Americans were more supportive of the invasion of Iraq than Germans or Australians. Because the Germans and Australians were skeptical of the reasons for going to war in the first place, they tended not to believe events they thought were erroneously reported. Americans, who were more supportive of the invasion, ignored many of the corrections of misinformation. This would help explain why they continued to believe WMDs were found even when that was reported not to be the case.

This willingness to believe false information that fits our views can be particularly troubling in political campaigns. One side can make a false claim about the other side's candidate, allow it to echo through a news cycle, then contritely retract it, knowing full well that receptive audiences are likely to still believe the initial lie.

We have to acknowledge that we have preconceived notions of people or events. We tend to believe what fits in with the social worlds that we are a part of and reject that which does not fit our expectations. Accepting this would be the first step toward becoming more objective about the world around us.

Sources: Stephan Lewandowsky, Werner G.K. Stritzke, Klaus Oberauer, and Michael Morales. "Memory for Fact, Fiction, and Misinformation: The Iraq War 2003." *Psychological Science* Vol 16, No. 3; Sharon Begley. (2005, Feb. 4). "People Believe a 'Fact' That Fits Their Views Even If It's Clearly False." *Wall Street Journal*, p. B1.

Reference Groups

Groups are more than just bridges between the individual and society as a whole. We spend much of our time in one group or another, and the effect that these groups have on us continues even when we are not actually in contact with the other members.

The norms and values of groups we belong to or identify with serve as the basis for evaluating our own and others' behavior.

A **reference group** is *a group or social category that an individual uses to help define beliefs, attitudes, and values and to guide behavior.* It provides a comparison point against which people measure themselves and others. A reference group is often a category we identify with, rather than a specific group we belong to. For example, a communications major may identify with individuals in the media without having any direct contact with them. In this respect, anticipatory socialization is occurring, in that the individual may alter his or her behavior and attitudes toward those he or she perceives to be part of the group he or she plans to join. For example, people who become bankers soon feel themselves part of a group—bankers—and assume ideas and lifestyles that help them identify with that group. They tend to dress in a conservative, "bankerish" fashion, even buying their clothes in shops that other bankers patronize to make sure they have the "right" clothes from the "right" stores. They join organizations such as country clubs and alumni associations so they can mingle with other bankers and clients. Eventually, the norms and values they adopted when they joined the bankers' group become internalized; they see and judge the world around them as bankers.

We can also distinguish between positive and negative reference groups. Positive reference groups are composed of people we want to emulate. Negative reference groups provide a model we do not wish to follow. Therefore, a writer may identify positively with those writers who produce serious fiction but may think of journalists who write for tabloids as a negative reference group.

Even though groups are in fact composed of individuals, individuals are also created to a large degree by the groups they belong to through the process of socialization (see Chapter 4). Of these groups, the small group usually has the strongest direct effect on an individual.

Small Groups

The term **small group** refers to *many kinds of social groups, such as families, peer groups, and work groups, that actually meet together and contain few enough members so that all members know one another.* The smallest group possible is a **dyad,** which *contains only two members.* An engaged couple is a dyad, as are the pilot and copilot of an aircraft.

George Simmel (1950) was the first sociologist to emphasize the importance of the size of a group on the interaction process. He suggested that small groups have distinctive qualities and patterns of interaction that disappear when the group grows larger. For example, dyads resist change in their group size: On the one hand, the loss of one member destroys the group, leaving the other member alone. On the other hand, a **triad,** or *the addition of a third member,* creates uncertainty because it introduces the possibility of a two-against-one alliance.

Triads are more stable in situations when one member can help resolve quarrels between the other two. When three diplomats are negotiating offshore fishing rights, for example, one member of the triad may offer a concession that will break the deadlock between the other two. If that does not work, the third person may try to analyze the arguments of the other two in an effort to bring about a compromise. The formation of shifting pair-offs within triads can help stabilize the group. When it appears that one group member is weakening, one of the two paired members will often break the alliance and form a new one with the individual who had been isolated (Hare, 1976).

This is often seen among groups of children engaged in games. In triads in which alliances do not shift and the configuration constantly breaks down into two against one, the group will become unstable and may eventually break up. In Aldous Huxley's novel *Brave New World,* the political organization of the earth was defined by three eternally warring political powers. As one power seemed to be losing, one of the others would come to its aid in a temporary alliance, thereby ensuring worldwide political stability while also making possible endless warfare. No power

Triads are usually unstable groups because the possibility of two-against-one alliances is always present.

could risk the total defeat of another because the other surviving power might then become the stronger of the surviving dyad.

As a group grows larger, the number of relationships within it increases, which often leads to the formation of **subgroups**—*splinter groups within the larger group.* Once a group has more than five to seven members, spontaneous conversation becomes difficult for the group as a whole. Then two solutions are available: The group can split into subgroups (as happens informally at parties), or it can adopt a formal means of controlling communication (use of *Robert's Rules of Order,* for instance).

For these reasons, small groups tend to resist the addition of new members because increasing size threatens the nature of the group. In addition, there may be a fear that new members will resist socialization to group norms and thereby undermine group traditions and values. On the whole, small groups are much more vulnerable than large groups to disruption by new members, and the introduction of new members often leads to shifts in patterns of interaction and group norms.

Large Groups: Associations

Although all of us probably would be able to identify and describe the various small groups we belong to, we might find it difficult to follow the same process with the large groups that affect us.

As patrons or employees of large organizations and governments, we function as part of large groups all the time. Thus, sociologists must study large groups as well as small groups in order to understand the workings of society.

Much of the activity of a modern society is carried out through large and formally organized groups. Sociologists refer to these groups as **associations,** which are *purposefully created special-interest groups that* *have clearly defined goals and official ways of doing things.* Associations include such organizations as government departments and agencies, businesses and factories, labor unions, schools and colleges, fraternal and service groups, hospitals and clinics, and clubs for various hobbies from gardening to collecting antiques. Their goals may be very broad and general—such as helping the poor, healing the sick, or making a profit—or quite specific and limited—such as manufacturing automobile tires or teaching people to speak Chinese. Although an enormous variety of associations exist, they all are characterized by some degree of formal structure with an underlying informal structure.

Formal Structure

For associations to function, the work that the association must accomplish is assessed and broken down into manageable tasks that are assigned to specific individuals. In other words, associations are run according to a formal organizational structure that consists of planned, highly institutionalized, and clearly defined statuses and role relationships.

The formal organizational structure of large associations in contemporary society is best exemplified by the organizational structure called bureaucracy. For example, when we consider a college or university, fulfilling its main purpose of educating students requires far more than simply bringing together students and teachers. Funds must be raised, buildings constructed, qualified students and professors recruited, programs and classes organized, materials ordered and distributed, grounds kept up, and buildings maintained. Lectures must be given; seminars must be led; and messages need to be typed, copied, and filed. To accomplish all these tasks, the school must create many different positions: president, deans, department heads, registrar, public relations staff, groundskeepers, maintenance personnel, purchasing agents, secretaries, faculty, and students.

Every member of the school has clearly spelled-out tasks that are organized in relation to one another: Students are taught and evaluated by faculty, faculty members are responsible to department heads or deans, deans to the president, and so on. Underlying these clearly defined assignments are procedures that are never written down but are worked out and understood by those who have to get the job done.

Informal Structure

Sociologists recognize that formal associations never operate entirely according to their stated rules and procedures. Every association has an informal structure consisting of networks of people who help out one another by "bending" rules and taking proce-

dural shortcuts. No matter how carefully plans are made, no matter how clearly and rationally roles are defined and tasks assigned, every situation and its variants cannot be anticipated. Sooner or later, then, individuals in associations are confronted with situations in which they must improvise and even persuade others to help them do so.

As every student knows, no school ever runs as smoothly as planned. For instance, going by the book—that is, following all the formal rules—often gets students tied up in long lines and red tape. Enterprising students and instructors find shortcuts. A student who wants to change from Section A of Sociology 100 to Section E might find it very difficult or time-consuming to change sections (add and drop classes) officially. However, it might be possible to work out an informal deal—the student stays registered in Section A but attends, and is evaluated in, Section E. The instructor of Section E then turns the grade over to the instructor of Section A, who hands in that grade with all the other Section A grades—as if the student had attended Section A all along. The formal rules have been bent, but the major purposes of the school (educating and evaluating students) have been served.

In addition, human beings have their own individual needs even when they are on company time, and these needs are not always met by attending single-mindedly to assigned tasks. To accommodate these needs, people often try to find extra break time for personal business by getting jobs done faster than would be possible if they followed all the formal rules and procedures. To accomplish these ends, individuals in associations may cover for one another, look the other way at strategic moments, and offer one another useful information about office politics, people, and procedures. Gradually, the reciprocal relationships among members of these informal networks become institutionalized; "unwritten laws" are established, and a fully functioning informal structure evolves. (For a discussion of applying the informal structure to job hunting, see "News You Can Use: The Strength of Weak Ties in Job Hunting."

Gemeinschaft and Gesellschaft

The Chicago sociologists, in their studies of the city, used some of the concepts developed by Ferdinand Tönnies (1865–1936), a German sociologist. In his book *Gemeinschaft und Gesellschaft*, Tönnies examined the changes in social relations attributable to the transition from rural society (organized around small communities) to urban society (organized around large impersonal structures).

Tönnies noted that in a **Gemeinschaft** *(community), relationships are intimate, cooperative, and personal.* For example, author Philip Roth (1998),

NEWS YOU CAN USE

The Strength of Weak Ties in Job Hunting

Here is a common frustrating job-hunting story: With a great GPA and good references, you send out dozens of resumes to online job postings, newspaper ads, and company recruitment websites, and hear . . . nothing. Then you get news that a less skilled friend whose father's golf buddy works in a top company has landed a plum job. Is the old cliché—"It's not what you know, but *who* you know"—really true?

In some cases, yes. Mark Granovetter, author of a famous study on the strength of what he called "weak ties," found that 56% of all professional job applicants found their job through personal contacts. Only 16% landed jobs through advertisements or employment agencies.

But the nature of those contacts may surprise you. Typically, the person who opened the door was not a close friend or relative, but someone who actually did not know the job seeker very well. Granovetter called these relationships "weak ties." These are the bonds that exist between individuals who see one another infrequently, and whose relationships are casual rather than intimate. Today, we often characterize these clusters of acquaintances as social networks.

But why would weak ties work better than strong ones? Aren't the people who know you well the ones most likely to have your best interests at heart? That may be, but because of their very closeness, they are likely to be exposed to the same sources of information. People outside that tightly knit circle, however, have networks that reach much further afield. By getting to know them, you essentially tap into their networks, just as they, then, have access to yours.

Additionally, people who know us less well are less likely to pigeonhole us based on our past experiences or skills. This is especially true for job seekers who want to branch out into new areas. When you want to reinvent yourself, the people who know you best may hinder rather than help you. They may be supportive, but they could try to preserve the old identities you are trying to shed.

The best strategy for a job hunter, then, is to try and expand your circle of acquaintances. College buddies, friends of friends, professional associations, social clubs, church groups, and civic organizations all can lead to relationships that qualify as "weak tie" networks. (They can also, of course, lead to enduring friendships and personal growth!)

Websites such as Friendster and LinkedIn have been launched to facilitate these kinds of social networks electronically. As people increasingly understand the reciprocal power of social networks, they are more willing to facilitate such relationships. People with contacts in many social networks ultimately do better in the job market.

Sources: Mark Granovetter. (1983). "The Strength of Weak Ties: A Network Theory Revisited." *Sociological Theory*, Vol. 1, pp. 201–233; Herminia Ibarra. (2002). *Working Identity: Unconventional Strategies for Reinventing Your Career.* Cambridge: Harvard Business School Press.

---

in his book *American Pastoral,* describes such a community:

> About one another, we knew who had what kind of lunch in the bag in his locker and who ordered what on his hot dog at Syd's; we knew one another's physical attributes—who walked pigeon-toed and who had breasts, who smelled of hair oil and who oversalivated when he spoke; we knew who among us was belligerent and who was friendly, who was smart and who was dumb; we knew whose mother had an accent and whose father had a mustache, whose mother worked and whose father was dead; somehow we even dimly grasped how every family's different set of circumstances set each family a distinctive difficult human problem.

In a Gemeinschaft the exchange of goods is based on reciprocity and barter, and people look out for the well-being of the group as a whole. Among the Amish, for example, there is such a strong community spirit that, should a barn burn down, members of the community quickly come together to rebuild it. In just a matter of days a new barn will be standing—the work of community members who feel a strong tie and responsibility to another community member who has encountered misfortune. (For a discussion of how the Amish try to maintain a Gemeinschaft, see "Social Change: Limiting Technology to Save the Community.")

In a **Gesellschaft** (society), *relationships are impersonal and independent.* People look out for their own interests, goods are bought and sold, and formal contracts govern economic exchanges. Everyone is seen as an individual who may be in competition with others who happen to share a living space.

Tönnies saw Gesellschaft as the product of mid-nineteenth-century social changes that grew out of

A small town is likely to produce a Gemeinschaft, in which relationships are intimate, cooperative, and personal; a city is likely to produce a Gesellschaft, in which relationships are impersonal and people look out for their own interests.

industrialization, in which people no longer automatically wanted to help one another or to share freely what they had. There is little sense of identification with others in a Gesellschaft, in which each individual strives for advantages and regards the accumulation of goods and possessions as more important than the qualities of personal ties. Modern urban society is, in Tönnies's terms, typically a Gesellschaft, whereas rural areas retain the more intimate qualities of Gemeinschaft.

In small, rural communities and preliterate societies, the family provided the context in which people lived, worked, were socialized, were cared for when ill or infirm, and practiced their religion. In contrast, modern urban society has produced many secondary groups in which these needs are met. It also offers far more options and choices than did the society of Tönnies's Gemeinschaft: educational options, career options, lifestyle options, choice of marriage partner, choice of whether to have children, and choice of where to live. In this sense, the person living in today's urban society is freer.

Mechanical and Organic Solidarity

Tönnies wrote about communities and cities from the standpoint of what we call an ideal type, in that no community or city actually could conform to the definitions he presented. They are basically concepts that help us understand the differences between the two. In the same sense, Émile Durkheim devised ideas about mechanical and organic solidarity.

According to Durkheim, every society has a **collective conscience**—*a system of fundamental beliefs and values*. These beliefs and values define for its members the characteristics of the good society, which is one that meets the needs for individuality, for security, for superiority over others, and for any of a host of other values that could become important to the people in that society. **Social solidarity** *emerges*

SOCIAL CHANGE

Limiting Technology to Save the Community

Some groups go out of their way to preserve a certain lifestyle or a sense of community. Often, decisions about the introduction of technology will either preserve or hinder the development of community.

The Amish, a tight-knit religious community of about 225,000 located in Lancaster County, Pennsylvania, Ohio, and several other states, are widely known for refusing to allow automobiles or technology into their community. They all wear the same identical plain clothing without buttons, live in houses without electricity, and cultivate the fields with horse-drawn machinery. Yet you would probably be surprised to find out that the Amish also use cell phones, gas barbecue grills, and in-line skates. How could we explain this contradiction?

The Amish lifestyle is based on specific decisions made about the impact of technology on community life. The telephone has been the source of intense controversy among the Amish since the 1920s. Eventually the telephone was accepted because it could be used to call doctors, veterinarians, and merchants. But this did not mean that you should have a phone in the home. Rather, the phone should be in a communal location where it could be used by many people. But this is much less convenient than having a phone in the home. Why would you have a phone that you can only use to call people from some communal location? Well, think about the number of times a phone call has interrupted a conversation. Have you ever left a family meal to take a phone call? Now that the cell phone can follow us to most locations, a phone call has the ability to interrupt all kinds of social encounters. Relegating the telephone to some communal spot outside of the home sends the message that the telephone conversations are much less important than those taking place in the community or the home. Keeping the telephone at a distance is a symbolic way of making it your servant, rather than the other way around.

Along the same line, some Amish craftsmen who need electricity in a workshop may use a diesel generator to charge a bank of 12-volt batteries. It would certainly be much easier to have electricity connected to the workshop. But Amos, an Amish craftsman, pointed out that the Bible teaches them to separate themselves from the world: "Connecting to the electric lines would make too many things too easy. Pretty soon people would start plugging in radios and televisions. . . . Batteries and generators only work for a short time and you have to work to keep them going. It is a way of controlling the use of electricity. We try to restrict things that would lead to us losing that sense of being separate, to put the brakes on how fast we change."

Does the cell phone, the beeper, the BlackBerry, call-waiting, or automated voice mail enhance our community life? If we decided that community came first, how would we change our use of technology?

Source: Adapted from Rheingold, Howard, "Look Who's Talking" *Wired*, January, 1999.

---

from the people's commitment and conformity to the society's collective conscience.

A **mechanically integrated society** is one in which *a society's collective conscience is strong and there is a great commitment to that collective conscience.*

In this type of society, members have common goals and values and a deep and personal involvement with the community. A modern-day example of such a society is that of the Tasaday, a food-gathering community in the Philippines. Theirs is a relatively small, simple society, with little division of labor, no separate social classes, and no permanent leadership or power structure.

In contrast, in an **organically integrated society**, *social solidarity depends on the cooperation of individuals in many different positions who perform specialized tasks.* The society can survive only if all the tasks are performed. With organic integration such as is found in the United States, social relationships are more formal and functionally determined than are the close, personal relationships of mechanically integrated societies.

Although we may take for granted the movement from Gemeinschafts to Gesellschafts, or mechanically integrated to organically integrated societies, it is only relatively recently in the course of human history that cities have become the dominant type of living arrangements.

Bureaucracy

Although in ordinary usage the term *bureaucracy* suggests a certain rigidity and red tape, it has a somewhat different meaning to sociologists.

Robert K. Merton (1969) defined **bureaucracy** as "*a formal, rationally organized social structure [with] clearly defined patterns of activity in which, ideally,*

Modern bureaucracies have an organizational structure that makes it clear what rights and duties are attached to the various positions.

every series of actions is functionally related to the purposes of the organization."

The German sociologist Max Weber (1956) provided the first detailed study of the nature and origins of bureaucracy. Although much has changed in society since he developed his theories, Weber's basic description of bureaucracy remains essentially accurate to this day.

Weber's Model of Bureaucracy: An Ideal Type

Weber viewed bureaucracy as the most efficient—although not necessarily the most desirable—form of social organization for the administration of work. He studied examples of bureaucracy throughout history and noted the elements that they had in common. Weber's model of bureaucracy is an **ideal type,** which is *a simplified, exaggerated model of reality used to illustrate a concept.*

When Weber presented his ideal type of bureaucracy, he combined into one those characteristics that could be found in one form or another in a variety of organizations. It is unlikely that we ever would find a bureaucracy that has all the traits presented in Weber's ideal type. However, his presentation can help us understand what is involved in bureaucratic

systems. It is also important to recognize that Weber's ideal type is in no way meant to be ideal in the sense that it presents a desired state of affairs. In short, an ideal type is an exaggeration of a situation that is used to convey a set of ideas. Weber outlined six characteristics of bureaucracies:

1. *A clear-cut division of labor.* The activities of a bureaucracy are broken down into clearly defined, limited tasks, which are attached to formally defined positions (statuses) in the organization. This permits a great deal of specialization and a high degree of expertise.

   For example, a small-town police department might consist of a chief, a lieutenant, a detective, several sergeants, and a dozen officers. The chief issues orders and assigns tasks; the lieutenant is in charge when the chief is not around; the detective does investigative work; the sergeants handle calls at the desk and do the paperwork required for formal booking procedures; and the officers walk or drive through the community, making arrests and responding to emergencies. Each member of the department has a defined status and duty as well as specialized skills appropriate to his or her position.

2. *Hierarchical delegation of power and responsibility.* Each position in the bureaucracy is given

sufficient power so that the individual who occupies it can do assigned work adequately and can also compel subordinates to follow instructions.

Such power must be limited to what is necessary to meet the requirements of the position. For example, a police chief can order an officer to walk a specific beat but cannot insist that the officer join the Lions Club.

3. *Rules and regulations.* The rights and duties attached to various positions are stated clearly in writing and govern the behavior of all individuals who occupy them. In this way, all members of the organizational structure know what is expected of them, and each person can be held accountable for his or her behavior. For example, the regulations of a police department might state, "No member of the department shall drink intoxicating liquors while on duty." Such rules make the activities of bureaucracies predictable and stable.

4. *Impartiality.* The organization's written rules and regulations apply equally to all its members. No exceptions are made because of social or psychological differences among individuals.

Also, people occupy positions in the bureaucracy only because they are assigned according to formal procedures. These positions belong to the organization itself; they cannot become the personal property of those who occupy them. For example, a vice-president of United States Steel Corporation is usually not permitted to pass on that position to his or her children through inheritance.

5. *Employment based on technical qualifications.* People are hired because they have the ability and skills to do the job, not because they have personal contacts within the company. Advancement is based on how well a person does the job. Promotions and job security go to those who are most competent.

6. *Distinction between public and private spheres.* A clear distinction is made between the employees' personal lives and their working lives. It is unusual for employees to be expected to take business calls at home. At the same time, employees' family lives have no place in the work setting.

Although many bureaucracies strive at the organizational level to attain the goals that Weber proposed, most do not achieve them on the practical level.

Bureaucracy Today: The Reality

Just as no building is ever identical to its blueprint, no bureaucratic organization fully embodies all the features of Weber's model. One thing that most bureaucracies do have in common is a structure that separates those whose responsibilities include keeping in mind the overall needs of the entire organization from those whose responsibilities are much more narrow and task oriented. Visualize a modern industrial organization as a pyramid. Management (at the top of the pyramid) plans, organizes, hires, and fires. Workers (in the bottom section) make much smaller decisions limited to carrying out the work assigned to them. A similar division cuts through the hierarchy of the Roman Catholic Church. The pope is at the top, followed by cardinals, archbishops, and bishops; the clergy are below. Only bishops can ordain new priests, and they plan the church's worldwide activities. The priests administer parishes, schools, and missions; their tasks are quite narrow and confined.

Although employees of bureaucracies may enjoy the privileges of their positions and guard them jealously, they may be adversely affected by the system in ways that they do not recognize. Alienation, adherence to unproductive ritual, and acceptance of incompetence are some of the results of a less-than-ideal bureaucracy.

Robert Michels, a colleague of Weber's, also was concerned about the depersonalizing effect of bureaucracy. His views, formulated at the beginning of this century, are still pertinent today.

The Iron Law of Oligarchy

Michels (1911) concluded that the formal organization of bureaucracies inevitably leads to **oligarchy**, *under which organizations that were originally idealistic and democratic eventually come to be dominated by a small self-serving group of people who achieved positions of power and responsibility.* This can occur in large organizations because it becomes physically impossible for everyone to get together every time a decision has to be made. Consequently, a small group is given the responsibility of making decisions. Michels believed that the people in this group would become corrupted by their elite positions and more and more inclined to make decisions to protect their power rather than represent the will of the group they were supposed to serve.

In effect, Michels was saying that bureaucracy and democracy do not mix. Despite any protestations and promises that they will not become like all the rest, those placed in positions of responsibility and power often come to believe that they are indispensable to, and more knowledgeable than, those they serve. As time goes on, they become further removed from the rank and file.

The iron law of oligarchy suggests that organizations that wish to avoid oligarchy should take a number of precautionary steps. They should make sure that the rank and file remain active in the organization and that the leaders not be granted absolute control of a centralized administration. As long as open lines of communication and shared decision making exist between the leaders and the rank and file, an oligarchy cannot develop easily.

Clearly, the problems of oligarchy, of the bureaucratic depersonalization described by Weber, and of personal alienation all are interrelated. If individuals are deprived of the power to make decisions that affect their lives in many or even most of the areas that are important to them, withdrawal into narrow ritualism and apathy are likely responses.

SUMMARY

- A social group consists of a number of people who have a common identity, some feeling of unity, and certain common goals and shared norms.
- Sociologists distinguish between primary groups, which involve intimacy, informality, and emotional investment in one another, and secondary groups, which have specific goals, formal organization, and much less intimacy.
- To function properly, all groups must define their boundaries, choose leaders, make decisions, set goals, assign tasks, and control members' behavior.
- A reference group is a group or social category that an individual uses to help define beliefs, attitudes, and values and to guide behavior.

- When individuals alter their behavior and attitudes toward those in a group they wish to join, they are engaging in anticipatory socialization.
- Associations are purposefully created special-interest groups that have clearly defined goals and official ways of doing things.
- A modern form of large association is bureaucracy, which is a formal, rationally organized social structure with clearly defined patterns of activity that are functionally related to the purposes of the organization.

 Media Resources

The Companion Website for *Introduction to Sociology,* Ninth Edition

http://sociology.wadsworth.com/tischler9e

Supplement your review of this chapter by going to the companion website to take one of the Tutorial Quizzes, use the flash cards to master key terms, and check out the many other study aids you'll find there. You'll also find special features such as Wadsworth's Sociology Online Resources and Writing Companion, GSS data, and Census 2000 information at your fingertips to help you complete that special project or do some research on your own.

CHAPTER SIX STUDY GUIDE

KEY CONCEPTS AND THINKERS

Match each concept with its definition, illustration, or explanation presented below.

a. Social group
b. Social aggregate
c. Primary group
d. Secondary group
e. Instrumental leadership
f. Expressive leadership
g. Reference group

h. Dyad
i. Triad
j. Associations
k. Gemeinschaft
l. Gesellschaft
m. Collective conscience

n. Mechanical integration
o. Organic integration
p. Bureaucracy
q. Ideal type
r. Oligarchy
s. Subgroup

p 1. A formal, rationally organized social structure, divided into offices with specific tasks run on principles of impartiality.
g 2. A group a person uses as a guide to values, beliefs, and behavior.
d 3. A goal-oriented, impersonal, more formal group.
q 4. A simplified model used to illustrate a concept.
s 5. A splinter group, usually created informally, to enable face-to-face interaction.
c 6. A group of people who know one another well and interact as complete individuals rather than in specialized roles.
h 7. A group of two people.
i 8. A group of three people.
n 9. Social solidarity based on similarity among people and strong commitment to the collective conscience.
o 10. Social solidarity based on difference and the fitting together of different specialized tasks.
r 11. Rule by a small group of self-interested people.
k 12. (Community) A group where relations are intimate, personal, and cooperative.
a 13. People who share a common identity, goals, and norms.
j 14. Purposefully created groups with clearly defined goals and procedures.
b 15. People who have little in common but who are in the same place.
m 16. Durkheim's term for the shared fundamental beliefs and values of a group.
l 17. (Society) A group where relations are impersonal and independent.
e 18. Focused on accomplishing concrete tasks.
f 19. Concerning feelings and interpersonal relationships.

Match the thinkers with their main idea or contribution.

a. Charles Horton Cooley
b. Solomon Asch
c. Georg Simmel

d. Ferdinand Tönnies
e. Émile Durkheim

f. Max Weber
g. Robert Michels

f 1. Emphasized the importance of bureaucracy as a social development.
e 2. Proposed the idea that different types of society were held together via different types of solidarity (mechanical and organic).
g 3. Originated the idea of "the iron law of oligarchy."
b 4. Conducted experiments on conformity to group pressure.
d 5. Developed the concepts of Gemeinschaft and Gesellschaft.
c 6. Sociologist who pioneered the idea that group size affects interaction.
a 7. Developed the concepts of primary and secondary groups.

CENTRAL IDEA COMPLETIONS

Following the instructions, fill in the appropriate concepts and descriptions for each of the questions posed in the following section.

1. List and define the six major functions of **social groups:**

   a. _____

   b. _____

   c. _____

   d. _____

   e. _____

   f. _____

2. What are **reference groups?** _____

   What **functions** do reference groups serve? _____

3. Present an example of each of the following:

   a. Aggregate: _____

   b. Social group: _____

   c. Primary group: _____

   d. Secondary group: _____

4. Using the American high school as an organization, pick **three specific characteristics of bureaucracy** outlined by Weber. Describe how informal structures might subvert the three formal characteristics of your example high school.

   a. _____

   b. _____

   c. _____

5. What is Michels' **"iron law of oligarchy"?**

   _____

6. What are the characteristics of bureaucracy? (Can you think of any that are not mentioned in the text?)

   a. _____

   b. _____

   c. _____

7. What are the differences between a social group and an institution?

   _____

   _____

   _____

   _____

CRITICAL THOUGHT EXERCISES

1. Develop a list of the top three (in terms of importance to your life) primary groups to which you belong and the top three secondary groups. What, if anything, do these six groups share in common? What features are unique to each? How do the concepts of primary group and secondary group help you in your examination of the differences among the six groups? Select one of the three secondary groups and discuss what changes would need to occur for that group to become a primary group. Follow the same procedure for one of the three primary groups becoming a secondary group.

2. Consider the functions that all groups must perform (defining boundaries, setting goals, and so on). Compare how two different groups you are familiar with—one primary, one secondary—carry out these functions and how the way in which the group carries out this function affects how the members of the group feel about the group and about one another.

3. To what extent is education at your college or university bureaucratized? How do the various elements of bureaucracy (division of labor, impartiality, and so on) affect the content and quality of education? How might education be different in a less bureaucratically organized school?

4. Observe a group in operation and try to classify each statement as to whether it is instrumental or expressive. See if group members seem to specialize more in one area than the other.

INTERNET ACTIVITIES

1. To gain a sense of the importance of group membership around the world, go to http://degraaff.org/hci, which provides an alphabetized listing of different campus groups and organizations that are concerned with home computer interaction (HCI). What do you see as the commonalties and differences across the groups depicted at these sites? How well do they manage to accomplish the major group tasks discussed in your text?

2. For more information on small groups, look at http://www.abacon.com/commstudies/groups/group.html. This site is designed for entire courses devoted to the topic of small groups, but it has useful information and suggestions for interesting Internet activities related to the various subtopics in small groups.

ANSWERS TO KEY CONCEPTS

1.p   2.g   3.d   4.q   5.s   6.c   7.h   8.i   9.n   10.o   11.r   12.k   13.a   14.j   15.b   16.m   17.l   18.e   19.f

ANSWERS TO KEY THINKERS

1.f   2.e   3.g   4.b   5.d   6.c   7.a

ThomsonNOW™

Reviewing is as easy as ❶ ❷ ❸

1. Before you do your final exam, take the ThomsonNOW diagnostic quiz to help you identify the areas on which you should concentrate. You will find information on ThomsonNOW and instructions on how to access all of its great resources on the foldout at the beginning of the text.
2. As you review, take advantage of ThomsonNOW's study videos and interactive Map the Stats exercises to help you master the chapter topics.
3. When you are finished with your review, take ThomsonNOW's posttest to confirm you are ready to move on to the next chapter.

7

Deviant Behavior and Social Control

Linda Hayes, Boston Filmworks

Learning Objectives

After studying this chapter, you should be able to:

- Understand deviance as culturally relative.
- Explain the functions and dysfunctions of deviance.
- Distinguish between internal and external means of social control.
- Differentiate among the various types of sanctions.
- Describe and critique biological, psychological, and sociological theories of deviance.
- Discuss the concept of anomie and its role in producing deviance.
- Know how the *Uniform Crime Reports* and the *National Crime Victimization Survey* differ as sources of information about crime.
- Describe the major features of the criminal justice system in the United States.

Soprano Florence Foster Jenkins believed she was the "goddess of song." Unfortunately, she had no singing talent. She sang wildly out of tune and her voice was quivering and colorless. She was, however, a wealthy New York socialite and did not hesitate to use her money to let the world know that she should be considered a world-class diva.

Several times a year she would rent the Ritz Carlton Hotel and give stunningly inept renditions of standard opera arias and songs specifically written for her, which she would proceed to mangle with her appalling voice. She would create lavish costumes for these performances, and her pianist, Cosme McMoon, treated her with the utmost respect as he accompanied her with the appropriate music.

Eventually, her reputation produced a following, and tickets to her bizarre performances, which could be purchased only from her directly, were sold out months in advance and were as difficult to get as those for the Metropolitan Opera. Her following continued to build; her final performance took place at Carnegie Hall. When she died a month later, she left the following epitaph on her gravestone: "Some people say I cannot sing, but no one can say that I didn't sing." Florence Foster Jenkins was certainly an eccentric, but was she deviant?

Defining Normal and Deviant Behavior

What determines whether a person's actions are seen as eccentric, creative, or deviant? Why will two men walking hand-in-hand cause raised eyebrows in one place but not in another? Why do Britons waiting to enter a theater stand patiently in line, whereas people from the Middle East jam together at the turnstile? In other words, what makes a given action—men holding hands, cutting into a line—"normal" in one case but "deviant" in another?

The answer is culture—more specifically, the norms and values of each culture (see Chapter 3). Together, norms and values make up the **moral code** of a culture—*the symbolic system in terms of which behavior takes on the quality of being "good" or "bad," "right" or "wrong."* Therefore, to decide whether any specific act is "normal" or "deviant," it is necessary to know more than only what a person did. One also must know who the person is (that is, the person's social identity) and the social and cultural contexts of the act. For example, if Florence Foster Jenkins held her recitals on a street corner in a seedy neighborhood instead of a stage at the Ritz Carlton

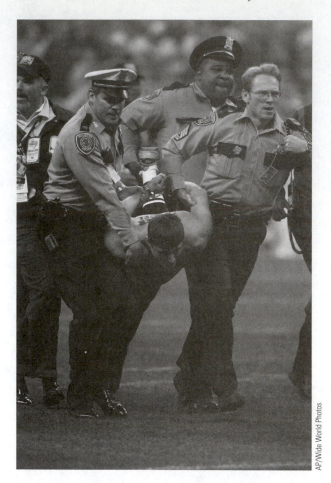

AP/Wide World Photos

When a person violates the norms of society external means of social control may be used.

certain acts are almost universally accepted as being deviant. For example, parent-child incest is severely disapproved of in nearly every society. Genocide, the willful killing of specific groups of people—as occurred in the Nazi extermination camps during World War II—also is considered wrong even if it is sanctioned by a government or an entire society. The Nuremberg trials that were conducted after World War II supported this point. Even though most of the accused individuals tried to claim they were merely following orders when they murdered or arranged for the murder of large numbers of Jews and other groups, many were found guilty. The reasoning was that there is a higher moral order under which certain human actions are wrong, regardless of who endorses them. Thus, despite their desire to view events from a culturally relative standpoint, most sociologists find certain actions wrong, no matter what the context.

The Functions of Deviance

Émile Durkheim observed that deviant behavior is "an integral part of all healthy societies" (1895). Why is this the case? The answer, Durkheim suggested, is that in the presence of deviant behavior, a social group becomes united in its response. In other words, opposition to deviant behavior creates opportunities for cooperation essential to the survival of any group. For example, let us look at the response to a scandal in a small town as Durkheim described it:

> [People] stop each other on the street, they visit each other, they seek to come together to talk of the event and to wax indignant in common. From all the similar impressions which are exchanged, from all the temper that gets itself expressed, there emerges a unique temper . . . which is everybody's without being anybody's in particular. That is the public temper. (1895)

When social life moves along normally, people begin to take for granted one another and the meaning of their social interdependence. A deviant act, however, reawakens their group attachments and loyalties because it represents a threat to the moral order of the group. The deviant act focuses people's attention on the value of the group. Perceiving itself under pressure, the group marshals its forces to protect itself and preserve its existence.

Deviance also offers society's members an opportunity to rededicate themselves to their social controls. In some cases, deviant behavior actually helps teach society's rules by providing illustrations of violation. Knowing what is wrong is a step toward understanding what is right. Deviance, then, may be functional to a group in that it: (1) causes the group's

Hotel, would people still have been as interested in her events? Of course not.

For sociologists, then, **deviant behavior** is *behavior that fails to conform to the rules or norms of the group in question* (Durkheim, 1960/1893). Therefore, when we try to assess an act as being normal or deviant, we must identify the group by whose terms the behavior is judged. Moral codes differ widely from one society to another. For that matter, even within a society, groups and subcultures exist whose moral codes differ considerably. Watching television is normal behavior for most Americans, but it would be seen as deviant behavior among the Amish.

Making Moral Judgments

As we stated, sociologists take a culturally relative view of normalcy and deviance and evaluate behavior according to the values of the culture in which it occurs. Ideally, they do not use their own values to judge the behavior of people from other cultures. Even though social scientists recognize that normal and deviant behavior may vary greatly and that no science can determine what acts are inherently deviant,

The forms of dress and the types of behavior that are considered deviant depend on who is doing the judging and what the context might be.

of deviant behavior acts as a safety valve and actually prevents more serious instances of nonconformity. For example, the Amish, a religious group that does not believe in using such examples of contemporary society as cars, radios, televisions, and fashion-oriented clothing, allows its teenagers a great deal of latitude in their behaviors before they are fully required to follow the dictates of the community. This prevents a confrontation that could result in a major battle of wills.

The Dysfunctions of Deviance

Deviance, of course, has a number of dysfunctions as well, which is why every society attempts to restrain deviant behavior as much as possible. Included among the dysfunctions of deviant behavior are the following: (1) It is a threat to the social order because it makes social life difficult and unpredictable. (2) It causes confusion about the norms and values of a society. People become confused about what is expected, what is right and wrong. The social standards compete with one another, causing tension among the different segments of society. (3) Deviance also undermines trust. Social relationships are based on the premise that people will behave according to certain rules of conduct. When people's actions become unpredictable, the social order is thrown into disarray. (4) Deviance also diverts valuable resources. To control widespread deviance, vast resources must be called on and shifted from other social needs.

Mechanisms of Social Control

In any society or social group, it is necessary to have **mechanisms of social control,** or *ways of directing or influencing members' behavior to conform to the group's values and norms.* Sociologists distinguish between internal and external means of control.

Internal Means of Control

As we already observed in Chapters 3, 5, and 6, people are socialized to accept the norms and values of their culture, especially in the smaller and more personally important social groups to which they belong, such as the family. The word *accept* is important here. Individuals conform to moral standards not just because they know what they are, but also because they have internalized these standards. They experience discomfort, often in the form of guilt, when they violate these norms. In other words, for a group's moral code to work properly, it must be internalized and become part of each individual's emotional life as well as his or her thought processes. As this occurs, individuals begin to pass judgment on their own actions.

members to close ranks; (2) prompts the group to organize to limit future deviant acts; (3) helps clarify for the group what it really does believe in; (4) and teaches normal behavior by providing examples of rule violation. Finally, (5) in some situations, tolerance

In this photo from France at the end of World War II, the woman whose head has been shaven is jeered by the crowd as she is escorted out of town. She had been a Nazi collaborator during the war. Émile Durkheim believed that deviant behavior performs an important function by focusing people's attention on the values of the group. The deviance represents a threat to the group and forces it to protect itself and preserve its existence.

© Robert Capa/Magnum

In this way the moral code of a culture becomes an **internal means of control**—that is, *it operates on the individual even in the absence of reactions by others.*

External Means of Control: Sanctions

External means of control consist of *other people's responses to a person's behavior—that is, rewards and punishments.* They include social forces external to the individual that channel behavior toward the culture's norms and values.

Sanctions are *rewards and penalties that a group's members use to regulate an individual's behavior.* Thus, all external means of control use sanctions of one kind or another. *Actions that encourage the individual to continue acting in a certain way* are called **positive sanctions.** *Actions that discourage the repetition or continuation of the behavior* are **negative sanctions.**

Positive and Negative Sanctions Sanctions take many forms, varying widely from group to group and from society to society. For example, an American audience might clap and whistle enthusiastically to show its appreciation for an excellent artistic or athletic performance, but the same whistling in Europe would be a display of strong disapproval. Or consider the absence of a response. In the United States, a professor would not infer public disapproval because of the absence of applause at the end of a lecture—such applause by students is the rarest of compliments. In many universities in Europe, however, students are expected to applaud after every lecture (if only in a rhythmic, stylized manner). The absence of such applause would be a horrible blow to the professor, a public criticism of the presentation.

Most social sanctions have a symbolic side to them. Such symbolism has a powerful effect on people's self-esteem and sense of identity. Consider the positive feelings experienced by Olympic gold medalists or those elected to Phi Beta Kappa, the national society honoring excellence in undergraduate study. Or imagine the negative experience of being given the "silent treatment," such as that imposed on cadets who violate the honor code at the military academy at West Point (to some, this is so painful that they drop out).

Sanctions often have important material qualities as well as symbolic meanings. Nobel Prize winners receive not only public acclaim but also a hefty check. The threat of loss of employment may accompany public disgrace when an individual's deviant behavior becomes known. In isolated, preliterate societies, social ostracism can be the equivalent of a death sentence.

Both positive and negative sanctions work only to the degree that people can be reasonably sure that they actually will occur as a consequence of a given act. In other words, they work on people's expectations. Whenever such expectations are not met, sanctions lose their ability to mold social conformity.

It is important to recognize a crucial difference between positive and negative sanctions. When society applies a positive sanction, it is a sign that social controls are successful: The desired behavior has occurred and is being rewarded. When a negative sanction is applied, it is because of the failure of social controls: The undesired behavior has not been prevented. Therefore, a society that frequently must punish people is failing in its attempts to promote conformity. A school that must expel large numbers of students or a government that frequently must call out troops to quell protests and riots should begin to look for the weaknesses in its own system of internal means of social control to promote conformity.

Formal and Informal Sanctions **Formal sanctions** *are applied in a public ritual—as in the awarding of a prize or an announcement of expulsion—and are usually under the direct or indirect control of authorities.* For example, to enforce certain standards of behavior and protect members of society, our society creates laws. Behavior that violates these laws may be punished through formal negative sanctions. Not all sanctions are formal, however. Many social responses to a person's behavior involve **informal sanctions**, or *actions by group members that arise spontaneously with little or no formal direction.* Gossip is an informal sanction that is used universally. Congratulations are offered to people whose behavior has approval. In teenage peer groups, ridicule is a powerful, informal, negative sanction. The anonymity and impersonality of urban living, however, decrease the influence of these controls except when we are with members of our friendship and kinship groups.

A Typology of Sanctions Figure 7–1 shows the four main types of social sanctions, produced by combining the two sets of sanctions we have just discussed: informal and formal, positive and negative. Although formal sanctions might appear to be strong influences on behavior, informal sanctions actually have a greater effect on people's self-images and behavior. This is so because informal sanctions usually occur more frequently and come from close, respected associates.

1. **Informal positive sanctions** are *displays people use spontaneously to express their approval of another's behavior.* Smiles, pats on the back, handshakes, congratulations, and hugs are informal positive sanctions.
2. **Informal negative sanctions** are *spontaneous displays of disapproval or displeasure,* such as frowns, damaging gossip, or impolite treatment directed toward the violator of a group norm.
3. **Formal positive sanctions** are *public affairs, rituals, or ceremonies that express social approval of a person's behavior.* These occasions are planned and organized. In our society, they include such events as parades that take place after a team wins the World Series or the Super Bowl, the presentation of awards or degrees, and public declarations of respect or appreciation (banquets, for example). Awards of money are a form of formal positive sanctions.
4. **Formal negative sanctions** are *actions that express institutionalized disapproval of a person's behavior.* They usually are applied within the context of a society's formal organizations—

Figure 7–1   Types of Social Sanctions

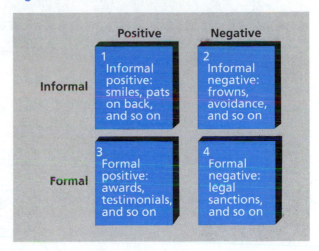

schools, corporations, the legal system, for example—and include expulsion, dismissal, fines, and imprisonment. They flow directly from decisions made by a person or agency of authority, and frequently specialized agencies or personnel (such as a board of directors, a government agency, or a police force) enforce them. (For a discussion of formal sanctions against athletes, see "Our Diverse Society: Public Heroes, Private Felons: Athletes and Sexual Assault.").

Theories of Crime and Deviance

Criminal and deviant behavior has been found throughout history. It has been so troublesome and so persistent that much effort has been devoted to understanding its roots. Many dubious ideas and theories have been developed over the ages. For example, a medieval law specified that "if two persons fell under suspicion of crime, the uglier or more deformed was to be regarded as more probably guilty" (Wilson & Herrnstein, 1985). Modern-day approaches to deviant and criminal behavior can be divided into the general categories of biological, psychological, and sociological explanations.

Biological Theories of Deviance

The first attempts to provide "scientific" explanations for deviant and criminal behavior centered on the importance of inherited factors and downplayed the importance of environmental influences. From this point of view, deviant individuals are born, not made.

Cesare Lombroso (1835–1901) was an Italian doctor who believed that too much emphasis was being put on "free will" as an explanation for deviant behavior. While trying to discover the anatomic differences between deviant and insane men, he came upon what he believed was an important insight. As

OUR DIVERSE SOCIETY

Public Heroes, Private Felons: Athletes and Sexual Assault

Thousand of hours of media attention are devoted to broadcasting sports events or reporting on the professional accomplishments of athletes. Athletes are presented as heroes and role models for the nation's youth. Yet at the same time, stories detailing the violence in the private lives of these public figures also make front-page news.

Does being a celebrity provide an athlete charged with sexual assault certain built-in advantages in the legal system? Jeff Benedict, the former director of research at the Center for the Study of Sport in Society at Northeastern University, and sociologist Alan Klein wanted to answer this question.

They examined 217 criminal complaints of sexual assault against women by athletes filed with the police between 1986 and 1995. They found that when an athlete was arrested for these complaints, only 43 cases resulted in a plea bargain and only 10 resulted in a jury conviction (Figure 7–2).

Why are these numbers so low? Jeff Benedict believes the courts are trying very hard to avoid any accusation of favoritism toward the athletes. Athletes do have other advantages, however. First, when an athlete is charged with a crime, he usually has the ability to pay for first-rate legal representation. Even college athletes, who are supposedly unemployed and not earning any income, get excellent representation. Second, the athlete often has a favorable public image whereas the accuser is unknown. Juries have a hard time believing that the well-dressed, popular athlete sitting at the defense table committed a brutal crime. The defense attorney only has to convince one person on the jury that the athlete did not commit the crime and the case will not result in a conviction. Athletes also have coaches, agents, alumni if they are college athletes, and other respected people who will speak on their behalf.

Community support carries a great deal of weight in court. The prosecution often has to settle for a plea bargain or a reduced sentence because getting a conviction in court is usually difficult.

Figure 7–2   Athletes Accused of Sexual Assault

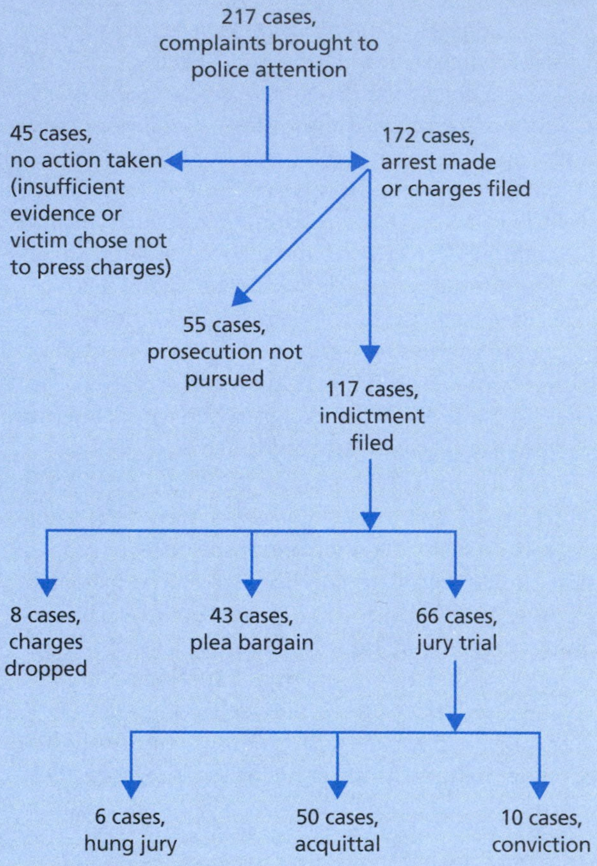

Sources: Based on "Arrest and Conviction Rates for Athletes Accused of Sexual Assault," by J. Benedict & A. Klein, 1997, *Sociology of Sport Journal*, 14, pp. 86–94; *Public Heroes, Private Felons: Athletes and Crimes Against Women*, by J. Benedict, 1997, Boston: Northeastern University Press; and an interview with the author, September 1997.

he was examining the skull of a criminal, he noticed a series of features, recalling an apish past rather than a human present:

> At the sight of that skull, I seemed to see all of a sudden, lighted up as a vast plain under a flaming sky, the problem of the nature of the criminal—an atavistic being who reproduces in his person the ferocious instincts of primitive humanity and the inferior animals. (Quoted in Taylor et al., 1973)

According to Lombroso, criminals are evolutionary throwbacks whose behavior is more apelike than human. They are driven by their instincts to engage in deviant behavior. These people can be identified by certain physical signs that betray their savage nature. Lombroso spent much of his life studying and dissecting dead prisoners in Italy's jails and concluded that their criminality was associated with an animal-like body type that revealed an inherited primitiveness (Lombroso-Ferrero, 1972). He also believed that certain criminal types could be identified by their head size, facial characteristics (size and shape of the nose, for instance), and even hair color. His writings were met with heated criticism from scholars who pointed out that perfectly normal-looking people have committed violent acts. (Modern social scientists would add that by confining his research to the study of prison inmates, Lombroso used a biased sample, thereby limiting the validity of his investigations.)

Shortly before World War II, anthropologist E. A. Hooton argued that the born criminal was a scientific reality. Hooton believed crime was not the product of social conditions but the outgrowth of "organic inferiority."

> [W]hatever the crime may be, it ordinarily arises from a deteriorated organism. . . . You may say that this is tantamount to a declaration that the primary cause of crime is biological inferiority—and that is exactly what I mean. . . . Certainly the penitentiaries of our society are built upon the shifting sands and quaking bogs of inferior human organisms. (Hooton, 1939a)

Hooton went to great lengths to analyze the height, weight, and shape of the body, nose, and ears of criminals. He was convinced that people betrayed their criminal tendencies by the shape of their bodies:

> The nose of the criminal tends to be higher in the root and in the bridge, and more frequently undulating or concave-convex than in our sample of civilians. . . . [B]ootleggers persistently have broad noses and short faces and flaring jaw angles, while rapists monotonously display narrow foreheads and elongated, pinched noses. (Hooton, 1939a)

Following in Hooton's footsteps, William H. Sheldon and his coworkers carried out body measurements of thousands of subjects to determine whether personality traits are associated with particular body types. They found that human shapes could be classified as three particular types: endomorphic (round and soft), ectomorphic (thin and linear), and mesomorphic (ruggedly muscular) (Sheldon & Tucker, 1940). They also claimed that certain psychological orientations are associated with body type. They saw endomorphs as being relaxed creatures of comfort; ectomorphs as being inhibited, secretive, and restrained; and mesomorphs as being assertive, action oriented, and uncaring of others' feelings (Sheldon & Stevens, 1942).

Sheldon did not take a firm position on whether temperamental dispositions are inherited or are the outcome of society's responses to individuals based on their body types. For example, Americans generally expect heavy people to be good-natured and cheerful, skinny people to be timid, and strongly muscled people to be physically active and inclined toward aggressiveness. Anticipating such behaviors, people often encourage them. In a study of delinquent boys, Sheldon and his colleagues (1949) found that mesomorphs were more likely to become delinquents than were boys with other body types. Their explanation of this finding emphasized inherited factors, although they acknowledged social variables. The mesomorph is quick to anger and lacks the ectomorph's restraint, they claimed. Therefore, in situations of stress, the mesomorph is more likely to get into trouble, especially if the individual is both poor and not very smart. Sheldon's bias toward a mainly biological explanation of delinquency was strong enough for him to have proposed a eugenic program of selective breeding to weed out those types he considered predisposed toward criminal behavior.

In the mid-1960s, further biological explanations of deviance appeared, linking a chromosomal anomaly in males, known as XYY, with violent and criminal behavior. Typically, males receive a single X chromosome from their mothers and a Y chromosome from their fathers. Occasionally, a child will receive two Y chromosomes from his father. These individuals will look like normal males; however, based on limited observations, a theory developed that these individuals were prone to commit violent crimes. The simplistic logic behind this theory is that because males are more aggressive than females and possess a Y chromosome that females lack, this Y chromosome must be the cause of aggression, and a double dose means double trouble. One group of researchers noted: "It should come as no surprise that an extra Y chromosome can produce an individual with heightened masculinity, evinced by characteristics such as unusual tallness . . . and powerful aggressive tendencies" (Jarvik, 1973).

Today the XYY chromosome theory has been discounted. It has been estimated (Chorover, 1979; Suzuki & Knudtson, 1989) that 96% of XYY males lead ordinary lives with no criminal involvement. A maximum of 1% of all XYY males in the United States may spend any time in a prison (Pyeritz et al., 1977). No valid theory of deviant and criminal behavior can be devised on the basis of such unconvincing data.

Current biological theorists have gone beyond the simplistic notions of Lombroso, Hooton, Sheldon, and the XYY syndrome. Today such theories focus on technical advances in genetics, brain functioning, neurology, and biochemistry. Contemporary biological theories are based on the notion that behavior, whether conforming or deviant, results from the interaction of physical and social environments.

The best known of these theories is Sarnoff Mednick's theory of inherited criminal tendencies. Mednick proposed that some genetic factors are passed along from parent to child. Criminal behavior is not directly inherited, nor do the genetic factors directly cause the behavior; rather, one inherits a greater susceptibility to criminality or to adapt to normal environments in a criminal way (Mednick, Moffitt, & Stacks, 1987).

Mednick believed that certain individuals inherit an autonomous nervous system that is slow to be aroused or to react to stimuli. Such individuals are then slow to learn control of aggressive or antisocial behavior (Mednick, 1977).

More recent research has focused on trying to show that several brain chemical systems may be involved in sensation-seeking, impulsivity, negative temperament, and other types of antisocial behavior. Some of this research has focused on neurotransmitters, or chemicals in the brain that allow the various regions of the brain to communicate with each other. The neurotransmitter serotonin has been identified with impulsivity and aggression. It is believed that low levels of serotonin produce impulsive and aggressive behavior (Edwards & Kravitz, 1997).

Serotonin levels are related to environmental conditions, however. Studies show that low levels of serotonin are found in individuals who experience high and chronic amounts of stress (Dinan, 1996; Graeff et al., 1996). It also appears that poor parenting lowers serotonin levels and good parenting raises them (Pine et al., 1997; Field et al., 1998).

Critics of biological theories of criminal behavior claim that such theories present an oversimplified view of genetic and biological influences. They also worry that the research may be used to support the view that there are racial differences in the predisposition to crime (Fishbein, 2001).

Psychological Theories of Deviance

Psychological explanations of deviance downplay biological factors and emphasize instead the role of parents and early childhood experiences, or behavioral conditioning, in producing deviant behavior. Although such explanations stress environmental influences, there is a significant distinction between psychological and sociological explanations of deviance.

Psychological orientations assume that the seeds of deviance are planted in childhood and that adult behavior is a manifestation of early experiences rather than an expression of ongoing social or cultural factors. The deviant individual therefore is viewed as a "psychologically sick" person who has experienced emotional deprivation or damage during childhood.

Psychoanalytic Theory Psychoanalytic explanations of deviance are based on the work of Sigmund Freud and his followers. Psychoanalytic theorists believe that the unconscious, the part of us consisting of irrational thoughts and feelings of which we are not aware, causes us to commit deviant acts.

According to Freud, our personality has three parts: the id, our irrational drives and instincts; the superego, our conscience and guide as internalized from our parents and other authority figures; and the ego, the balance among the impulsiveness of the id, the restrictions and demands of the superego, and the requirements of society. Because of the id, all of us have deviant tendencies, though through the socialization process we learn to control our behavior, driving many of these tendencies into the unconscious. In this way, most of us are able to function effectively according to our society's norms and values. For some, however, the socialization process is not what it should be. As a result, the individual's behavior is not adequately controlled by either the ego or superego, and the wishes of the id take over. Consider, for example, a situation in which a man has been driving around congested city streets looking for a parking space. Finally, he spots a car that is leaving and pulls up to wait for the space. Just as he is ready to park his car, another car whips in and takes the space. Most of us would react to the situation with anger. We might even roll down the car window and direct some angry gestures and strong language at the offending driver. There have been cases, however, in which the angry driver has pulled out a gun and shot the offender. Instead of simply saying, "I'm so mad I could kill that guy," the offended party acted out the threat. Psychoanalytic theorists might hypothesize that in this case, the id's aggressive drive took over, because of an inadequately developed conscience.

Psychoanalytic approaches to deviance have been strongly criticized because the concepts are very abstract and cannot easily be tested. For one thing, the unconscious can be neither seen directly nor measured. Also, such approaches tend to overemphasize innate drives at the same time that they underemphasize social and cultural factors that bring about deviant behavior.

Behavioral Theories According to the behavioral view, people adjust and modify their behaviors in response to the rewards and punishments their actions elicit. If we do something that leads to a favorable outcome, we are likely to repeat that action. If our behavior leads to unfavorable consequences, we are not eager to do the same thing again (Bandura, 1969). Those of us who live in a fairly traditional environment are likely to be rewarded for engaging in conformist behavior, such as working hard, dressing in a certain manner, or treating our friends in a certain way. We would receive negative sanctions if our friends found out that we had robbed a liquor store. For some people, however, the situation is reversed. That is, deviant behavior may elicit positive rewards. A 13-year-old who associates with a delinquent gang and is rewarded with praise for shoplifting, stealing, or vandalizing a school is being indoctrinated into a deviant lifestyle. The group may look with contempt at the "straight" kids who study hard, make career plans, and do not go out during the week. According to this approach, deviant behavior is learned by a series of trials and errors. One learns to be a thief in the same way that one learns to be a sociologist.

Crime as Individual Choice James Q. Wilson and Richard Herrnstein (1985) have devised a theory of criminal behavior that is based on an analysis of individual behavior. Sociologists, almost by definition, are suspicious of explanations that emphasize individual behavior, because they believe such theories neglect the setting in which crime occurs and the broad social forces that determine levels of crime. However, Wilson and Herrnstein have argued that whatever factors contribute to crime—the state of the economy, the competence of the police, the nurturance of the family, the availability of drugs, the quality of the schools—they must affect the behavior of individuals before they affect crime. They believe that if crime rates rise or fall, it must be because of changes that have occurred in areas that affect individual behavior.

Wilson and Herrnstein contend that individual behavior is the result of rational choice. A person will choose to do one thing as opposed to another because it appears that the consequences of doing it are more desirable than the consequences of doing something else. At any given moment, a person can choose between committing a crime and not committing it.

The consequences of committing the crime consist of rewards and punishments. The consequences of not committing the crime also entail gains and losses. Crime becomes likely if the rewards for committing the crime are significantly greater than those are for not committing the crime. The net rewards of crime include not only the likely material gain from the crime but also intangible benefits such as obtaining emotional gratification, receiving the approval of peers, or settling an old score against an enemy. Some of the disadvantages of crime include the pangs of conscience, the disapproval of onlookers, and the retaliation of the victim.

The benefits of not committing a crime include avoiding the risk of being caught and punished and not suffering a loss of reputation or the sense of shame afflicting a person later discovered to have broken the law. All of the benefits of not committing a crime lie in the future, whereas many of the benefits of committing a crime are immediate. The consequences of committing a crime gradually lose their ability to control behavior in proportion to how delayed or improbable they are. For example, millions of cigarette smokers ignore the possibility of fatal consequences of smoking because those consequences are distant and uncertain. If smoking one cigarette caused certain death tomorrow, we would expect cigarette smoking to drop dramatically.

Sociological Theories of Deviance

Sociologists have been interested in the issue of deviant behavior since the pioneering efforts of Émile Durkheim in the late nineteenth century. Indeed, one of the major sociological approaches to understanding this problem derives directly from his work. It is called anomie theory.

Anomie Theory Durkheim published *The Division of Labor in Society* in 1893. In it, he argued that deviant behavior can be understood only in relation to the specific moral code it violates: "We must not say that an action shocks the common conscience because it is criminal, but rather that it is criminal because it shocks the common conscience" (1960/1893).

Durkheim recognized that the common conscience, or moral code, has an extremely strong hold on the individual in small, isolated societies where there are few social distinctions among people and everybody more or less performs the same tasks. Such *mechanically integrated* societies, he believed, are organized in terms of shared norms and values: All members are equally committed to the moral code. Therefore, deviant behavior that violates the code is felt by all members of the society to be a personal threat. As society becomes more complex—that is, as work is divided into more numerous and increasingly specialized tasks—social organization is maintained by the interdependence of individuals. In other words, as the division of labor becomes more specialized and differentiated, society becomes more *organically integrated.* It is held together less by moral consensus than by economic interdependence.

A shared moral code continues to exist, of course, but it tends to be broader and less powerful in determining individual behavior. For example, political leaders among the Cheyenne Indians led their people by persuasion and by setting a moral example (Hoebel, 1960). In contrast with the Cheyenne, few modern Americans actually expect exemplary moral behavior from their leaders, despite the public rhetoric calling for it. We express surprise, but not outrage, when less than honorable behavior is revealed about our political leaders. We recognize that political leadership is exercised through formal institutionalized channels and not through model behavior.

In highly complex, rapidly changing societies such as our own, some individuals come to feel that the

Figure 7–3   **Merton's Typology of Individual Modes of Adaptation**

Conformists accept both (a) the goals of the culture and (b) the institutionalized means of achieving them. Deviants reject either or both. Rebels are deviants who may reject the goals or the institutions of the current social order and seek to replace them with new ones that they would then embrace.

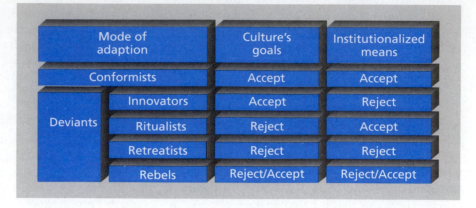

Mode of adaption	Culture's goals	Institutionalized means
Conformists	Accept	Accept
Deviants — Innovators	Accept	Reject
Ritualists	Reject	Accept
Retreatists	Reject	Reject
Rebels	Reject/Accept	Reject/Accept

moral consensus has weakened. Some people lose their sense of belonging, the feeling of participating in a meaningful social whole. Such individuals feel disoriented, frightened, and alone.

Durkheim used the term **anomie** to refer to *the condition of normlessness, in which values and norms have little impact and the culture no longer provides adequate guidelines for behavior.*

Durkheim found that anomie was a major cause of suicide, as we discussed in Chapter 1. Robert Merton built on this concept and developed a general theory of deviance in American society.

Strain Theory   Robert K. Merton (1938, 1969) believed that American society pushes individuals toward deviance by overemphasizing the importance of monetary success while failing to emphasize the importance of using legitimate means to achieve that success. Those individuals who occupy favorable positions in the social-class structure have many legitimate means at their disposal to achieve success. However, those who occupy unfavorable positions lack such means. Thus, the goal of financial success combined with the unequal access to important environmental resources creates deviance.

As Figure 7–3 shows, Merton identified four types of deviance that emerge from this strain. Each type represents a mode of adaptation on the part of the deviant individual; that is, the form of deviance a person engages in depends greatly on the position he or she occupies in the social structure. Specifically, it depends on the availability to the individual of legitimate, institutionalized means for achieving success. Thus, some individuals, called **innovators**, *accept the culturally validated goal of success but find deviant ways of going about reaching it.* Con artists, embezzlers, bank robbers, fraudulent advertisers, drug dealers, corporate criminals, crooked politicians, cops on the take—each is trying to "get ahead" using whatever means are available.

Ritualists are *individuals who reject or deemphasize the importance of success once they realize they*

will never achieve it and instead concentrate on following and enforcing rules more precisely than was ever intended. Because they have a stable job with a predictable income, they remain within the labor force but refuse to take risks that might jeopardize their occupational security. Many ritualists are often tucked away in large institutions such as governmental bureaucracies.

Another group of people also lacks the means to attain success but does not have the institutional security of the ritualists. **Retreatists** are *people who pull back from society altogether and cease to pursue culturally legitimate goals.* They are the drug and alcohol addicts who can no longer function—the panhandlers and street people who live on the fringes of society.

Finally, there are the rebels. **Rebels** *reject both the goals of what to them is an unfair social order and the institutionalized means of achieving them.* Rebels seek to tear down the old social order and build a new one with goals and institutions they can support and accept.

Merton's theory has become quite influential among sociologists. It is useful because it emphasizes external causes of deviant behavior that are within the power of society to correct. The theory's weakness is its inability to account for the presence of certain kinds of deviance that occur among all social strata and within almost all social groups in American society: for example, juvenile alcoholism, drug dependence, and family violence (spouse beating and child abuse).

Control Theory   In control theory, social ties among people are important in determining their behavior. Instead of asking what causes deviance, control theorists ask: What causes conformity? They believe that what causes deviance is the absence of what causes conformity. In their view, conformity is a direct result of control over the individual. Therefore, the absence of social control causes deviance. According to this theory, people are free to violate norms if they lack intimate attachments with parents, teachers, and peers. These attachments help them establish values linked to a conventional lifestyle. Without these attachments and acceptance of conventional norms, the opinions of other people do not matter and the individual is free to violate norms without fear of social disapproval. This theory assumes that the disapproval of others plays a major role in preventing deviant acts and crimes.

According to Travis Hirschi (1969), one of the main proponents of control theory, we all have the potential to commit deviant acts. Most of us never commit these acts because of our strong bond to society. Hirschi's view is that there are four ways in which individuals become bonded to society and conventional behavior:

1. *Attachment to others*. People form intimate attachments to parents, teachers, and peers who display conventional attitudes and behavior.
2. *Commitment to conformity*. Individuals invest their time and energies in conventional types of activities, such as getting an education, holding a job, or developing occupational skills. At the same time, people show a commitment to achievement through these activities.
3. *Involvement in conventional activities*. People spend so much time engaged in conventional activities that they have no time to commit or even think about deviant activities.
4. *Belief in the moral validity of social rules*. Individuals have a strong moral belief that they should obey the rules of conventional society.

If these four elements are strongly developed, the individual is likely to display conventional behavior. If these elements are weak, deviant behavior is likely.

More recently Hirschi and Gottfredson (1993) have proposed a theory of crime based on one type of control only—self-control. They have suggested that people with high self-control will be less likely during all periods of life to engage in criminal acts. Those with low self-control are more likely to commit crime than those with high self-control. The source of low self-control is ineffective parenting. Parents who do not take an active interest in their children and do not socialize them properly produce children with low self-control. Once established in childhood, the level of social control people have acquired will guide them throughout the rest of their lives.

Techniques of Neutralization Most of us think we act logically and rationally most of the time. To violate the norms and moral values of society, we must have **techniques of neutralization,** *a process that makes it possible for us to justify illegal or deviant behavior* (Sykes & Matza, 1957). In the language of control theory, these techniques provide a mechanism by which people can break the ties to the conventional

society that would inhibit them from violating the rules. Techniques of neutralization are learned through the socialization process. According to Sykes and Matza, they can take several forms:

1. *Denial of responsibility*. These individuals argue that they are not responsible for their actions; forces beyond their control drove them to commit the act, such as a troubled family life, poverty, or being drunk at the time of the incident. In any event, the responsibility for what they did lies elsewhere.
2. *Denying the injury*. The individual argues that the action did not really cause any harm. Who really got hurt when the individual illegally copied some computer software and sold it to friends? Who is really hurt in illegal betting on a football game?
3. *Denial of the victim*. The victim is seen as someone who "deserves what he or she got." The man who made an obscene gesture to us on the highway deserved to be assaulted when we caught up with him at the next traffic light. Some athletes, when accused of sexual assault of a woman, have claimed that the woman consented to sex when she agreed to go to the athlete's hotel room.
4. *Condemnation of the authorities*. Deviant or criminal behavior is justified because those who are in positions of power or are responsible for enforcing the rules are dishonest and corrupt themselves. Political corruption and police dishonesty leave us with little respect for these authority figures because they are more dishonest than we are.
5. *Appealing to higher principles or authorities*. We claim our behavior is justified because we are adhering to standards that are more important than abstract laws. Acts of civil disobedience against the government are justified because of the government's misguided policy of supporting a corrupt dictatorship. Our behavior may be technically illegal, but the goal justifies the action.

Using these techniques of neutralization, people are able to break the rules without feeling morally unworthy. They may even be able to put themselves on a higher plane specifically because of their willingness to rebel against rules. They are basically redefining the situation in favor of their actions.

Cultural Transmission Theory This theory relies strongly on the concept of learning, growing out of the work of Clifford Shaw and Henry McKay, who received their training at the University of Chicago. They became interested in the patterning of delinquent behavior in that city when they observed that

Chicago's high-crime areas remained the same over the decades—even though the ethnic groups living in those areas changed. Further, they found that as members of an ethnic group moved out of the high-crime areas, the rate of juvenile delinquency in that group fell; at the same time, the delinquency rate for the newly arriving ethnic group rose. Shaw and McKay (1931, 1942) discovered that delinquent behavior was taught to newcomers in the context of juvenile peer groups. Also, because such behavior occurred mostly in the context of peer-group activities, youngsters gave up their deviant ways when their families left the high-crime areas.

Edwin H. Sutherland and his student Donald R. Cressey (1978) built a more general theory of juvenile delinquency on the foundation laid by Shaw and McKay. This **theory of differential association** *is based on the central notion that criminal behavior is learned in the context of intimate groups* (Table 7–1). When criminal behavior is learned, it includes two components: (1) criminal techniques (such as how to break into houses) and (2) criminal attitudes (rationalizations that justify criminal behavior). In this context, people who become criminals are thought to do so when they associate with the rationalizations for breaking the law more than with the arguments for obeying the law. They acquire these attitudes through long-standing interactions with others who hold these views. Thus, among the estimated 70,000 gang members in Los Angeles County, status is often based on criminal activity and drug use. Even arrests and imprisonment are events worthy of respect. A youngster exposed to and immersed in such a value system will identify with it, if only to survive. In many respects, differential association theory is quite similar to the behavioral theory we discussed earlier. Both emphasize the learning or socialization aspect of deviance. Both also point out that deviant behavior emerges in the same way that conformist behavior emerges; it is merely the result of different experiences and different associations.

Labeling Theory Under **labeling theory** *the focus shifts from the deviant individual to the social process by which a person comes to be labeled as deviant and the consequences of such labeling for the individual.* This view emerged in the 1950s from the writings of Edwin Lemert (1972). Since then, many other sociologists have elaborated on the labeling approach. Labeling theorists note that although we all break rules from time to time, we do not necessarily think of ourselves as deviant—nor are we so labeled by others. However, some individuals, through a series of circumstances, do come to be defined as deviant by others in society. Paradoxically, this labeling process actually helps bring about more deviant behavior.

Table 7–1

Sutherland's Principles of Differential Association

1. Deviant behavior is learned.
2. Deviant behavior is learned in interaction with other people in a process of communication.
3. The principal part of the learning of criminal behavior occurs within intimate personal groups.
4. When deviant behavior is learned, the learning includes (a) techniques of committing the act, which are sometimes very complicated or sometimes very simple, and (b) the specific direction of motives, drives, rationalizations, and attitudes.
5. The specific direction of motives and drives is learned from definitions of the legal codes as favorable or unfavorable. That is, a person learns reasons for both obeying and violating rules.
6. A person becomes deviant because of an excess of definitions favorable to violating the law over definitions unfavorable to violating the law.
7. Differential associations may vary in frequency, duration, priority, and intensity.
8. The process of learning criminal behavior by association with criminal and anticriminal patterns involves all the mechanisms used in any other learning situation.
9. Although criminal behavior is an expression of general needs and values, it is not explained by those general needs and values, because noncriminal behavior is an expression of the same needs and values.

Source: Adapted from *Principles of Criminology,* 10th ed. (pp. 80–82), by E. H. Sutherland & D. R. Cressey, 1978, Chicago: Lippincott.

Being caught in wrongdoing and branded as deviant has important consequences for one's further social participation and self-image. The most important consequence is a drastic change in the individual's public identity. It places the individual in a new status, and he or she may be revealed as a different kind of person than formerly thought to be. Such people may be labeled as thieves, drug addicts, lunatics, or embezzlers and are treated accordingly.

To be labeled as a criminal, one need commit only a single criminal offense. Yet the word carries a number of connotations of other traits characteristic of anyone bearing the label. A man convicted of breaking into a house and thereby labeled criminal is presumed to be a person likely to break into other houses. Police operate on this premise and round up known offenders for questioning after a crime has been committed. In addition, it is assumed that such an individual is likely to commit other kinds of crimes as well, because he or she has been shown to be a person without "respect for the law." Therefore, apprehension for one deviant act increases the likelihood that this person will be regarded as deviant or undesirable in other respects.

Even if no one else discovers the deviance or endorses the rules against it, the individual who has committed it acts as an enforcer. Such individuals may brand themselves as deviant because of what they did and punish themselves in one way or another for the behavior (Becker, 1963).

At least three factors appear to determine whether a person's behavior will set in motion the process by which he or she will be labeled as deviant: (1) the importance of the norms that are violated, (2) the social identity of the individual who violates them, and (3) the social context of the behavior in question. Let us examine these factors more closely.

1. *The importance of the violated norms.* As we noted in Chapter 3, not all norms are equally important to the people who hold them. The most strongly held norms are mores, and their violation is likely to cause the perpetrator to be labeled deviant. The physical assault of an elderly person is an example. For less strongly held norms, however, much more nonconformity is tolerated, even if the behavior is illegal. For example, running red lights is both illegal and potentially very dangerous, but in some American cities it has become so commonplace that even the police are likely to look the other way rather than pursue violators.

2. *The social identity of the individual.* In all societies there are those whose wealth or power (or even force of personality) enables them to ward off being labeled deviant despite behavior that violates local values and norms. Such individuals are buffered against public judgment and even legal sanction. A rich or famous person caught shoplifting or even using narcotics has a fair chance of being treated indulgently as an eccentric and let off with a lecture by the local chief of police. Conversely, there are those marginal or powerless individuals and groups, such as welfare recipients or the chronically unemployed, toward whom society has little tolerance for nonconformity. Such people quickly are labeled deviant when an opportunity presents itself and are much more likely to face criminal charges.

3. *The social context.* The social context within which an action occurs is important. In a certain situation an action might be considered deviant, whereas in another context it will not. Notice that we say social context, not physical location. The nature of the social context can change even when the physical location remains the same. For example, for most of the year the New Orleans police manage to control open displays of sexual behavior, even in the famous French Quarter. However, during the week of Mardi Gras, throngs of people freely engage in what at other times of the year would be called lewd and indecent behavior. During Mardi Gras, the social context invokes norms for evaluating behavior that do not so quickly lead to the assignment of a deviant label.

Labeling theory has led sociologists to distinguish between primary and secondary deviance. **Primary deviance** is *the original behavior that leads to the application of the label to an individual.* **Secondary deviance** is *the behavior that people develop as a result of having been labeled as deviant* (Lemert, 1972). For example, a teenager who has experimented with illegal drugs for the first time and is arrested for it may face ostracism by peers, family, and school authorities. Such negative treatment may cause this person to turn more frequently to using illegal drugs and to associating with other drug users and sellers, possibly resorting to robberies and muggings to get enough money to buy the drugs. Thus, the primary deviant behavior and the labeling resulting from it lead the teenager to slip into an even more deviant lifestyle. This new lifestyle would be an example of secondary deviance.

Labeling theory has proved useful. It explains why society will label certain individuals deviant but not others, even when their behavior is similar. There are, however, several drawbacks to labeling theory. For one thing, it does not explain primary deviance. That is, even though we may understand how labeling may contribute to future, or secondary, acts of deviance, we do not know why the original, or primary, act of deviance occurred. In this respect, labeling theory explains only part of the deviance process. Another problem is that labeling theory ignores the instances when the labeling process may deter a person from engaging in future acts of deviance. It looks at the deviant as a misunderstood individual who really would like to be an accepted, law-abiding citizen. Clearly, this is an overly optimistic view.

It would be unrealistic to expect any single approach to explain deviant behavior fully. In all likelihood, some combination of the various theories discussed is necessary to gain a fuller understanding of the emergence and continuation of deviant behavior.

The Importance of Law

As discussed earlier in this chapter, some interests are so important to a society that folkways and mores are not adequate enough to ensure orderly social interaction. Therefore, laws are passed to give the state the power of enforcement. These laws become a formal system of social control, which is exercised when other informal forms of control are not effective.

Conflict theorists believe that laws are used by the state to promote and protect itself.

It is important not to confuse a society's moral code with its legal code, nor to confuse deviance with crime. Some legal theorists have argued that the legal code is an expression of the moral code, but this is not necessarily the case. For example, although most states and hundreds of municipalities have enacted some sort of antismoking law, smoking is not an offense against morals. Conversely, it is possible to violate American moral sensibilities without breaking the law.

What, then, is the legal code? The **legal code** consists of *the formal rules, called* **laws**, *adopted by a society's political authority*. The code is enforced through the use of formal negative sanctions when rules are broken. Ideally, laws are passed to promote conformity to those rules of conduct that the authorities believe are necessary for the society to function and that will not be followed if left solely to people's internal controls or the use of informal sanctions. Others argue that laws are passed to benefit or protect specific interest groups with political power, rather than society at large (Quinney, 1974; Vago, 1988).

The Emergence of Laws

How is it that laws come into society? How do we reach the point where norms are no longer voluntary and need to be codified and given the power of authority for enforcement? Two major explanatory approaches have been proposed: the consensus approach and the conflict approach.

The **consensus approach** *assumes that laws are merely a formal version of the norms and values of the people*. There is a consensus among the people on these norms and values, and the laws reflect this consensus. For example, people will generally agree that it is wrong to steal from another person. Therefore, laws emerge formally stating this fact and provide penalties for those caught violating the law.

The consensus approach is basically a functionalist model for explaining a society's legal system. It assumes that social cohesion will produce an orderly adjustment in the laws. As the norms and values in society change, so will the laws. Therefore "blue laws," which were enacted in many states during colonial times and which prohibited people from working or opening shops on Sunday, have been changed, and now vast shopping malls do an enormous amount of business on Sundays.

The conflict approach to explaining the emergence of laws sees dissension and conflict between various groups as a basic aspect of society. The conflict is resolved when the groups in power achieve control. The **conflict approach** to law *assumes that the elite use their power to enact and enforce laws that support their own economic interests and go against the* *interests of the lower classes.* As William Chambliss (1973) noted:

> Conventional myths notwithstanding, the history of criminal law is not a history of public opinion or public interest. . . . On the contrary, the history of the criminal law is everywhere the history of legislation and appellate-court decisions which in effect (if not in intent) reflect the interests of the economic elites who control the production and distribution of the major resources of the society.

The conflict approach to law was supported by Richard Quinney (1974) when he noted, "Law serves the powerful over the weak . . . moreover, law is used by the state . . . to promote and protect itself."

Chambliss used the development of vagrancy laws as an example of how the conflict approach to law works. He pointed out that the emergence of such laws paralleled the need of landowners for cheap labor in England during a time when the system of serfdom was breaking down. Later, when cheap labor was no longer needed, vagrancy laws were not enforced. Then, in the sixteenth century, the laws were modified to focus on those who were suspected of being involved in criminal activities and interfering with those engaged in the transportation of goods. Chambliss (1973) noted, "Shifts and changes in the law of vagrancy show a clear pattern of reflecting the interests and needs of the groups who control the economic institutions of the society. The laws change as these institutions change."

Crime in the United States

Crime is *behavior that violates a society's legal code.* In the United States what is criminal is specified in written law, primarily state statutes. Federal, state, and local jurisdictions often vary in their definitions of crimes, though they seldom disagree in their definitions of serious crimes.

A distinction is often made between violent crimes and property crimes. A **violent crime** is *an unlawful event, such as homicide, rape, and assault, that may result in injury to a person.* Robbery is also a violent crime because it involves the use or threat of force against the person. (For a discussion of particularly unusual violent crimes, see "Social Change: Serial Murderers and Mass Murderers.")

A **property crime** is *an unlawful act that is committed with the intent of gaining property but that does not involve the use or threat of force against an individual.* Larceny, burglary, and motor vehicle theft are examples of property crimes. Criminal offenses are also classified according to how the criminal justice system handles them. In this respect, most jurisdictions recognize two classes of offenses: felonies and misdemeanors. Felonies are not distinguished from misdemeanors in the same way in all areas, but most states define **felonies** as *offenses punishable by a year or more in state prison.* Although the same act may be classified as a felony in one jurisdiction and as a misdemeanor in another, the most serious crimes are never misdemeanors, and the most minor offenses are never felonies.

Crime Statistics

It is very difficult to know with any certainty how many crimes are committed in the United States each year. Two major approaches are taken in determining the extent of crime. One measure of crime is provided by the Federal Bureau of Investigation (FBI) through its *Uniform Crime Reports* (UCR). Since 1929, the FBI has been receiving monthly and annual reports from law enforcement agencies throughout the country, currently representing 94% of the national population. The UCR consists of eight crimes: *homicide, forcible rape, robbery, aggravated assault, burglary, larceny-theft, motor vehicle theft,* and *arson.* Arrests are reported for 21 additional crime categories also. Not included are federal offenses—political corruption, tax evasion, bribery, or violation of environmental protection laws, among others.

The UCR has undergone a five-year redesign effort and will soon be converted to a more comprehensive and detailed program known as the National Incident Based Reporting System (NIBRS).

Sociologists and critics in other fields note that for a variety of reasons, these statistics are not always reliable. For example, each police department compiles its own figures, and definitions of the same crime vary from place to place. Other factors affect the accuracy of the crime figures and rates published in the reports—for example, a law enforcement agency or a local government may change its method of reporting crimes so that the new statistics reflect a false increase or decrease in the occurrence of certain crimes. Under some circumstances, UCR data are estimated, because some jurisdictions do not participate or report partial data (Bureau of Justice Statistics, September 2002).

A second measure of crime is provided through the *National Crime Victimization Survey* (NCVS), which began in 1973 to collect information on crimes suffered by individuals and households, whether or not these crimes were reported to the police. The UCR only measures reported crimes. The NCVS collects detailed information on the frequency and nature of the crimes of rape, sexual assault, personal robbery, aggravated and simple assault, household burglary, theft, and motor vehicle theft.

The similarity between these crimes and the UCR categories is obvious and intentional. Some crimes are

SOCIAL CHANGE

Serial Murderers and Mass Murderers

What used to be a rare occurrence in the United States is now becoming increasingly commonplace. There are now two or three mass murders every month in the United States. In fact, 7 of the 10 largest mass killings in American history occurred in the past decade. Many people want to know what makes these individuals kill. Sociologist Jack Levin is one of the nation's best-known authorities on this problem.

Levin points out that, contrary to popular assumptions, mass murderers do not just "snap" or "go crazy." Their killing sprees are methodical and extremely well planned, and the motive usually is to get even. Mass murderers seek revenge against those individuals they feel are responsible for their problems. Levin notes, "The mass killer may be depressed, disillusioned, despondent, or desperate, but not deranged."

There are two types of multiple homicides. First are the mass killings that are often in the news. Here the individual kills a number of people within a short time. The murders could happen at the killer's last place of employment, in a restaurant, or at home. Serial killings differ in that instead of killing in a violent outburst, the murderer kills one victim at a time over a period of days, weeks, years, and even decades.

Mass killers tend to be white, middle-class, middle-aged males. Levin believes it takes a prolonged period of frustration to produce the kind of rage necessary for this type of brutal eruption. Mass killers have seen their lives go downhill for decades. Their relationships with others have fallen apart. Many cannot hold jobs. They are trying to survive with their many problems, but then a catastrophic event pushes them over the edge.

On the surface, serial murderers seem the same in that they are also likely to be white, middle-aged males. The difference is that serial murderers love to kill. Killing becomes a pleasurable end in itself. Most important, killing gives the serial killer a feeling of power. Levin notes that serial murderers have a great need for dominance and control, which they satisfy by taking the last breath from their victims. Very few will use a gun

because they want physical contact with the victim. They are sadistic.

Levin believes it is incorrect to characterize serial killers as insane. These people know what they are doing is wrong; they simply do not care. Levin notes: "They do not have a defect of the mind, they have a defect of character. They are

Jack Levin.

not mad, they are bad. They are not crazy, they are very crafty. They are not sick, they are sickening." They do not feel guilty about their actions. The secret to serial murderers' success lies in the fact that they do not look like the monsters they are. They are thoroughly familiar with the rules of society, but they do not feel that the rules apply to them.

When it comes to the issue of whether the death penalty should apply for these killers, Levin points out that in many ways mass murderers are dead already. They want to die and often kill themselves right after their violent binge. For the serial killers, Levin believes we should "lock them up and throw away the key. These people cannot be rehabilitated."

Mass murders and serial murders are a growing but still rare phenomenon. Levin points out that all told about 500 people die a year at the hand of a mass killer or serial killer. This is quite small compared with the 15,500 single-victim homicides a year. Levin reminds us that "you are still more likely to contract leprosy or malaria than you are to be murdered by a serial killer or mass killer." We cannot suspect everyone around us of being a killer. Our goal should be to understand the basis for mass killings and serial killings so we may one day be able to prevent them.

Source: Jack Levin, presentation at Framingham State College, Framingham, Massachusetts, April 1999.

missing from the NCVS that appear in the UCR. Murder cannot be measured through victim surveys, because obviously the victim is dead. Arson cannot be measured well through such surveys because the victim may in fact have been the criminal. An arson investigator is often needed to determine whether a fire was actually arson. Also, because of problems in questioning the victim, crimes against children younger than age 12 are also excluded.

Whereas the UCR depends on police departments' records of reported crimes, the NCVS attempts to

assess the total number of crimes committed. The NCVS obtains its information by asking a nationally representative sample of 49,000 households (about 101,000 people over the age of 12) about their experiences as victims of crime during the previous six months. The households stay in the sample for three years (Bureau of Justice Statistics, September 2002).

Of the 11.6 million crimes that occurred in 2004, the NCVS estimated that half of violent victimizations and 39% of property crimes were reported to the police. Of the violent crimes in 2004, 35.8% of rapes,

Figure 7–4 **Percentage of Selected Crimes Reported to the Police**

Source: Bureau of Justice Statistics, National Crime Victimization Survey, 2004, September 2005, http://www.ojp.usdoj.gov/bjs/pub/pdf/cv04.pdf, accessed October 28, 2005.

Figure 7–5 **Likelihood That Someone Will Be Arrested for a Known Crime**

Source: U.S. Department of Justice, Federal Bureau of Investigation, Crime in the United States, 2004 http://www.fbi.gov/ucr/cius_04/offenses_cleared/index.html, accessed October 28, 2005.

61.1% of robberies, and 64.2% of aggravated assault with injury were brought to the attention of the police. Motor vehicle theft continued to be the property crime reported to the police the most (84.8%) (Bureau of Justice Statistics, 2004) (Figure 7–4).

The particular reason most frequently mentioned for *not* reporting a crime was that it was not important enough. For violent crimes, the reason most often given for not reporting was that it was a private or personal matter. (See Figure 7–5 for the likelihood that

Figure 7–6   Likelihood That Someone Will Be Sent to Prison for a Known Crime

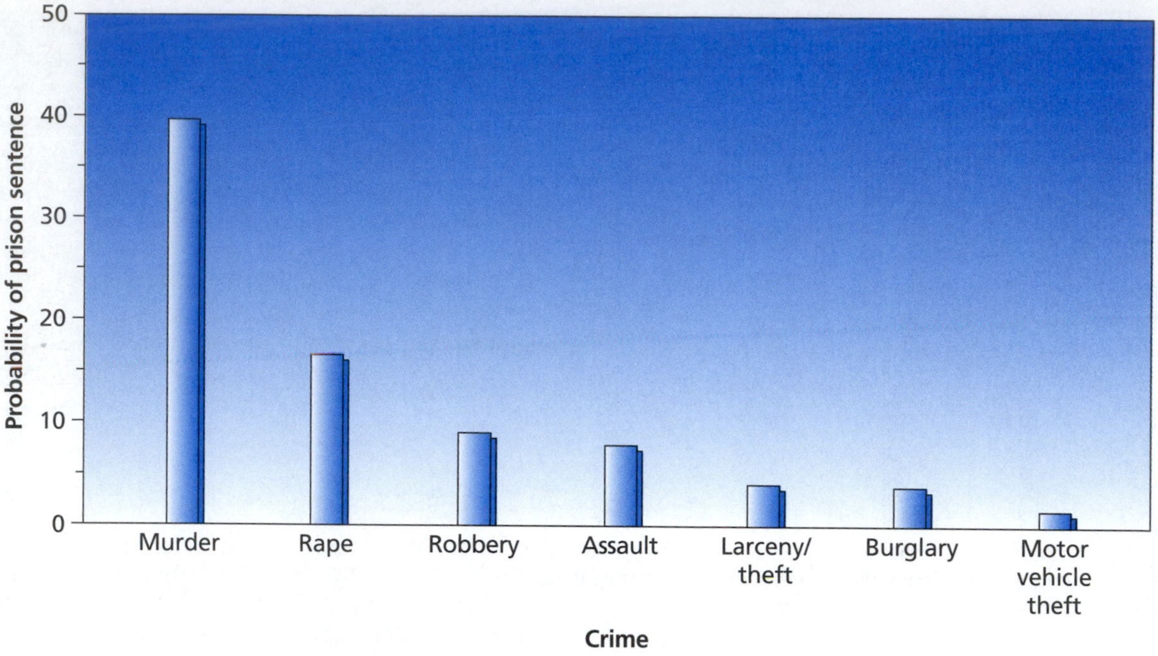

Source: *Crime in the United States,* 1997, Bureau of Justice Statistics, *Felony Defendants Large Urban Counties, 1994,* January 1998, p. 24.

someone will be arrested for a known crime, and Figure 7–6 for the likelihood that someone will be sent to prison for a known crime.)

The Uniform Crime Reports and the National Crime Victimization Survey both contain errors and omissions. Despite their respective drawbacks, they both are valuable sources of data on nationwide crime.

Kinds of Crime in the United States

The crime committed can vary considerably in terms of the effect it has on the victim and on the self-definition of the perpetrator of the crime. White-collar crime is as different from street crime as organized crime is from juvenile crime. In the next section, we will examine these differences.

Juvenile Crime

Juvenile crime refers to *the breaking of criminal laws by individuals younger than age 18.* Regardless of the reliability of specific statistics, one thing is clear: Serious crime among our nation's youth is a matter of great concern. Hard-core youthful offenders—perhaps 10% of all juvenile criminals—are responsible, by some estimates, for two-thirds of

all serious crimes. Although the vast majority of juvenile delinquents commit only minor violations, the juvenile justice system is overwhelmed by these hard-core criminals.

Serious juvenile offenders are predominantly male, disproportionately minority group members (compared with their proportion in the population), and typically disadvantaged economically. They are likely to exhibit interpersonal difficulties and behavioral problems, both in school and on the job. They are also likely to come from one-parent families or families with a high degree of conflict, instability, and inadequate supervision.

Arrest records for 2004 show that youths younger than age 18 accounted for 17.3% of all arrests. The most common crime that juveniles were arrested for was larceny-theft, whereas adults were most often arrested for drug abuse violations (Federal Bureau of Investigation, 2005).

Arrests, however, are only a general indicator of criminal activity. The greater number of arrests among young people may be partly due to their lack of experience in committing crimes and to their involvement in the types of crimes for which apprehension is more likely, for example, theft versus fraud. In addition, because youths often commit crimes in groups, the resolution of a single crime may lead to several arrests. (See Table 7–2 for arrest rates by age.)

Indeed, one of the major differences between juvenile and adult offenders is the importance of gang

Table 7–2

Age Distribution of Arrests, 2004

Age Group	Percentage* of U.S. Population	Percentage* of People Arrested
Age 14 and younger	21.2	5.1
15–19	7.2	20.4
20–24	6.8	19.8
25–29	6.4	12.9
30–34	7.1	10.5
35–39	8.0	9.8
40–44	8.2	9.2
45–49	7.3	6.2
50–54	6.4	3.3
55–59	4.9	1.5
60–64	3.9	0.7
Age 65 and older	12.7	0.6

*Percentages do not equal 100% because of rounding.

Source: Department of Justice Federal Bureau of Investigation, Crime in the United States, 2004, Table 38, http://www.fbi.gov/ucr/cius_04/documents/04tbl38a.xls.

membership and the tendency of youths to engage in group criminal activities. Gang members are more likely than other young criminals to engage in violent crimes, particularly robbery, rape, assault, and weapon violations. Gangs that deal in the sale of crack cocaine have become especially violent in the past decade.

There is conflicting evidence on whether juveniles tend to progress from less serious to more serious crimes. It suggests that violent adult offenders began their careers with violent juvenile crimes; thus they began as, and remained, serious offenders. However, minor offenses of youths are often dealt with informally and may not be recorded in crime statistics.

The juvenile courts—traditionally meant to treat, not punish—have had limited success in coping with such juvenile offenders (Reid, 1991). Strict rules of confidentiality, aimed at protecting juvenile offenders from being labeled as criminals, make it difficult for the police and judges to know the full extent of a youth's criminal record. The result is that violent youthful offenders who have committed numerous crimes often receive little or no punishment.

Defenders of the juvenile courts contend, nonetheless, that there would be even more juvenile crime without them. Others, arguing from learning and labeling perspectives, contend that the system has such a negative effect on children that it actually encourages **recidivism**—that is, *repeated criminal behavior after punishment.* All who are concerned with this issue agree that the juvenile courts are less than efficient, especially in the treatment of repeat offenders. One reason for this is that perhaps two-thirds of juvenile court time is devoted to processing children guilty of what are called **status offenses,** *behavior that is criminal only because the person involved is a minor* (examples are truancy and running away from home). Recognizing that status offenders clog the courts and add greatly to the terrible overcrowding of juvenile detention homes, states have sought ways to deinstitutionalize status offenders. One approach, known as **diversion**—*steering youthful offenders away from the juvenile justice system to nonofficial social agencies*—has been suggested by Edwin Lemert (1972).

Violent Crime

"In general the younger the victim, the higher the rate of fatal and nonfatal violence (rape, robbery, sexual assault, and assault) experienced" (Klaus & Rennison, 2002). In 2004, the violent crime rate in the United States reached the lowest level since the Bureau of Justice Statistics (BJS) started measuring it in 1973. An estimated 1.4 million violent crimes occurred in 2004, compared with 44 million such incidents in 1973 (FBI Uniform Crime Reports, 2005).

Bureau of Justice data show that 54% of all violent crime victims know their attackers. Nearly 70% of the rape and sexual assault victims know the offender as an acquaintance, friend, relative, or intimate. Twenty-eight percent of the rape and sexual assault victimizations are reported to the police (Rennison, 2002). The violent crime rate in the United States also includes the highest homicide rate in the industrialized world. There are more homicides in any one of the cities of New York, Detroit, Los Angeles, or Chicago each year than in all of England and Wales combined. (See "Global Sociology: The United States is a World Leader in Homicide" and "For Further Thinking: The Continuing Debate over Capital Punishment: Does It Deter Murderers?")

In addition to homicide and rape, other violent crimes such as aggravated assault and robbery have an effect on American households. In 2004 there were 850,000 aggravated assaults and about 400,000 robberies (FBI Uniform Crime Reports, 2005).

Property Crime

Seventy-five percent of all crime in the United States is what is referred to as crime against property, as opposed to crime against the person. In all instances of crime against property, the victim is not present and is not confronted by the criminal.

The most significant nonviolent crimes are burglary, auto theft, and larceny-theft. In 2004, more

© Bettmann/CORBIS

Juvenile courts have had limited success in dealing with juvenile offenders. The boot camp approach has been tried with some of these individuals, but the success of this method is also questionable.

than 2.1 million households reported a burglary, 1.2 reported an auto theft, and 10.3 million reported a property crime. Keep in mind that most household thefts are not reported (FBI Uniform Crime Reports, 2005).

Violent and property crime victimizations disproportionately affected urban residents. Urbanites accounted for 29% of the U.S. population and sustained 38% of all violent and property crime victimizations.

White-Collar Crime

White-collar crime is often used to refer to either a certain type of offender (for example, high socioeconomic status or occupation of trust) or a certain type of offense (for example, economic crime).

The term **white-collar crime** was coined by Edwin H. Sutherland (1940) to refer to *the acts of individuals who, while occupying positions of social responsibility or high prestige, break the law in the course of their work for the purpose of illegal personal or organizational gain.* The FBI has decided to approach white-collar crime in terms of the offense, defining it as "those illegal acts which are characterized by deceit, concealment, or violation of trust and

which are not dependent upon the application or threat of physical force or violence" (Barnett, 2002).

White-collar crimes include such illegalities as embezzlement, bribery, fraud, theft of services, kickback schemes, and others in which the violator's position of trust, power, or influence has provided the opportunity for him or her to use lawful institutions for unlawful purposes. White-collar offenses frequently involve deception. Federal prosecutors say that 8,766 defendants were charged with what they call white-collar crimes in the year 2000 alone. The majority of these offenses involved frauds and counterfeiting/forgery (Barnett, 2002).

Although white-collar offenses are often less visible than crimes such as burglary and robbery, the overall economic impact of crimes committed by such individuals are considerably greater. Not only is white-collar crime very expensive, it is also a threat to the fabric of society, causing some to argue that it causes more harm than street crime (Reiman, 1990). Sutherland (1961) has argued that because white-collar crime involves a violation of public trust, it contributes to a disintegration of social morale and threatens the social structure. This problem is compounded by the fact that in the few cases in which white-collar criminals actually are prosecuted and

GLOBAL SOCIOLOGY

The United States Is a World Leader in Homicide

The United States has the dubious distinction of having one of the highest homicide rates of all industrialized and nonindustrialized countries in the world. The nation's murder rate was 5.6 per 100,000 population in 2002, compared with 4.6 per 100,000 in 1950. This number is two to three times that of most European countries. Only five countries—South Africa, Russia, Lithuania, Estonia, and Latvia—have higher rates.

If we look for explanations for this phenomenon, we begin to see that in the United States homicide has become less of a domestic nonstranger event and more of an event that grows out of other criminal situations. This change has also made it more difficult to solve homicides. When homicide was more likely to be a domestic or intimate relationship event, nearly all were solved. For example, 94% of all homicides in 1954 were solved. As of 2002, the solution rate had dropped to 64% as homicides are increasingly likely to be perpetrated by a larger group of individuals as they commit a variety of crimes (Figure 7–7). Homicide has the highest resolution rate of all serious crimes.

Homicide is the sixth leading cause of death among blacks and the twentieth leading cause of death among whites in the United States (Anderson & Smith, 2005). When compared with the death toll from other major causes such as heart disease and cancer, the percentage attributed to homicide seems quite modest. There is another way of looking at it, however. Homicide disproportionately involves young victims without any major diseases, making them more like the victims of fatal automobile accidents than those dying from fatal diseases. In any given year, the median age of death of homicide victims is about 33 (Federal Bureau of Investigation, *Supplementary Homicide Reports, 1976–2002*). The number of years of life lost to a homicide is usually significantly greater than those lost to a disease. Taken as a whole, the years lost to homicide equal nearly 80% of those lost to heart disease and nearly 70% of those lost to cancer (Zimring & Hawkins, 1997).

Sources: *Alcohol and Homicide: A Deadly Combination of Two American Traditions,* by R. N. Parker, 1995, Albany, NY: State University Press; *The Nature of Homicide: Trends and Changes,* U.S. Department of Justice, Office of Justice Programs, 1996, Washington, DC: U.S. Government Printing Office; *Crime Is Not the Problem: Lethal Violence in America,* by F. E. Zimring & G. Hawkins, 1997, New York: Oxford University Press; Robert N. Anderson, and Betty L. Smith, "Deaths: Leading Causes for 2002," National Vital Statistics Reports. Volume 53, Number 17 March 7, 2005; FBI, *Supplementary Homicide Reports, 1976-2002.*

Figure 7–7   **U.S. Homicide Solution Rates**

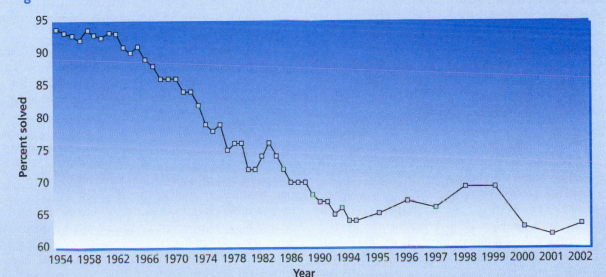

Source: Uniform Crime Reports; FBI Supplement, "Homicide Reports 1976–2002."

convicted, punishment usually is relatively light. Very few people who commit white-collar crimes actually go to prison. Federal Bureau of Prisons statistics for the year 2001 show that only 1,021 prisoners out of a total 156,238 for that year fit the description of a white-collar criminal. More than half of these are held at minimum-security prisons, which some refer to as "Club Feds" (Barnett, 2002).

New forms of white-collar crime involving political and corporate institutions have emerged in the past decade. For example, the dramatic growth in high technology has brought with it sensational accounts of computerized heists by sophisticated criminals seated safely behind computer terminals. The possibility of electronic crime has spurred widespread interest in computer security, by business and government alike. Crimes committed by corporate executives have been highly publicized in recent years.

Victimless Crime

Usually we think of crimes as involving culprits and victims—that is, individuals who suffer some loss or injury as a result of a criminal act. However, a number of crimes do not produce victims in any obvious way, and so some scholars have coined the term *victimless crime* to refer to them.

Basically, **victimless crimes** are *acts that violate those laws meant to enforce the moral code.* Usually they involve the use of narcotics, illegal gambling, public drunkenness, the sale of sexual services, or status offenses by minors. If heroin and crack cocaine addicts can support their illegal addictions legitimately, then who is the victim? If a person bets $10 or $20 per week with the local bookmaker, who is the victim? If someone staggers drunkenly through the streets, who is the victim? If a teenager runs away from home because conditions there are intolerable, who is the victim?

Some legal scholars argue that the perpetrators themselves are victims: Their behavior damages their own lives. This is, of course, a value judgment, but then the concept of deviance depends on the existence of values and norms (Schur & Bedau, 1974). Others note that such offenses against the public order do, in fact, contribute to the creation of victims, if only indirectly: Heroin addicts rarely can hold jobs and eventually are forced to steal to support themselves; prostitutes are used to blackmail people and to rob them; chronic gamblers impoverish themselves and bring ruin on their families; alcoholics drive drunk and cause accidents and may be violent at home; and so on.

Clearly, the problems raised by the existence of victimless crimes are complex. In recent years, American society has begun to recognize that at least some crimes truly are victimless and that they should

therefore be decriminalized. Two major activities that have been decriminalized in many states and municipalities are the smoking of marijuana (though not its sale) and sex between unmarried, consenting adults of the same gender. (For a discussion of sentences for marijuana possession, see "News You Can Use: Are Peaceful Pot Smokers Being Sent to Prison?")

Victims of Crime

We have been discussing crime statistics, the types of crimes committed, and who commits them. But what about the victims of crime? Is there a pattern? Are some people more apt to become crime victims than others are? It seems that this is true; victims of crime are not spread evenly across society. Although, as we have seen, the available crime data are not always reliable, a pattern of victimization can be seen in the reported statistics. A person's race, gender, age, and socioeconomic status have a great deal to do with whether that individual will become a victim of a serious crime.

Statistics show that, overall, males are much more likely to be victims of serious crimes than females are. When we look at crimes of violence and theft separately, however, a more complex picture emerges. Younger people are much more likely than the elderly to be victims of crime. African Americans are more likely to be victims of violent crime than are whites or members of other racial groups. People with low incomes have the highest violent-crime victimization rates. Theft rates are the highest for people with low incomes (less than $7,500 per year) and for those with higher incomes (more than $30,000 per year). Students and the unemployed are more likely than homemakers, retirees, or the employed to be victims of crime. Rural residents are less often crime victims than are people living in cities (Rennison, 2001).

Despite the growing, albeit unfounded, concern about crimes against the elderly, figures show that young people are most likely to be victims of serious crimes. For example, the violent victimization rates for people aged 16 to 19 are 20.3 times higher than for people 65 and older. Similarly, one in eight people murdered are younger than age 18 (Rennison, 2001).

The reason the elderly are less likely to be the victims of violent crime than the young is related in part to differences in lifestyle and income. Younger people may more often be in situations that place them at risk. They may frequent neighborhood hangouts, bars, or events that are likely places for an assault to occur. About 22% of the elderly reported that they never went out at night for entertainment, shopping, or other activities.

The only crime category that affected the elderly at about the same rate as most others (except those ages

NEWS YOU CAN USE

Are Peaceful Pot Smokers Being Sent to Prison?

One of the arguments that advocates for more lenient marijuana laws use to make their case is that thousands of law-abiding citizens are being sent to prison for merely smoking a few cigarettes in the company of their friends, or selling small amounts to others. For example, 24-year-old Donovan James Adams, described as a casual user, was tried in a Montana federal court and sentenced to 66 months in prison for selling 3 ounces of marijuana.

Critics of the nation's drug policies often point out that drug offenders typically serve longer sentences than people convicted of robbery, rape, or assault. Are overzealous police officers going overboard in enforcing marijuana laws or is there more to the story?

The United States Bureau of Justice Statistics (BJS) investigated the matter and came up with a different story than the one put forth by people suggesting our marijuana laws are sending many first-time offenders to prison. Drug offenders can be tried in state or federal courts depending on whether the offense took place locally or crossed state lines. Of those serving time for any drug offense in state prisons, the BJS found that 83% were people who had committed crimes in the past,

with the vast majority having multiple convictions. Only three-tenths of 1% of state inmates were in prison for being first-time marijuana possession offenders. Out of the 1.2 million people in state prisons, that comes down to 3,600 people.

In the federal prison system the story is similar. Of all drug defendants convicted in federal court for marijuana violations, the overwhelming majority were guilty of drug dealing. According to the study only 63 people were serving time in federal prison for simple possession. Yet even these people were not necessarily people smoking marijuana in their apartments or dorm rooms. The median amount of marijuana these 63 people were in possession of was 115 pounds. Using the number of 85 cigarettes to the ounce, that equals approximately 156,400 cigarettes. It would be a bit difficult for one person to smoke all those joints.

Source: Untangling the Statistics: Numbers Don't Lie—But They Can Deceive, http://www.whitehousedrugpolicy.gov/publications/whos _in_prison_for_marij/untangling_the_stats.pdf, accessed August 31, 2005.

12–24) was personal theft, which includes robbery, purse snatching, and pocket picking. Criminals may believe that the elderly are more likely to have large amounts of cash and are less likely to defend themselves, making them particularly vulnerable to these crimes (Rennison, 2001).

Criminal Justice in the United States

Every society that has established a legal code has also set up a **criminal justice system**—*personnel and procedures for arrest, trial, and punishment*—to deal with violations of the law. The three main categories of our criminal justice system are the police, the courts, and the prisons.

The Police

The police system developed in the United States is highly decentralized. It exists on three levels: federal, state, and local. On the federal level, the United States does not have a national police system. Congress, however, does enact federal laws. These laws

govern the District of Columbia and all states when a federal offense has been committed, such as kidnapping, assassination of a president, mail fraud, bank robbery, and so on. The FBI enforces many of these laws and also assists local and state law enforcement authorities in solving local crimes. If a nonfederal crime has been committed, the FBI must be asked by local or state authorities before it can aid in the investigation. If a particular crime is a violation of both state and federal law, state and local police often cooperate with the FBI to avoid unnecessary duplication of effort.

The state police patrol the highways, regulate traffic, and have primary responsibility for the enforcement of some state laws. They provide a variety of other services, such as a system of criminal identification, police training programs, and computer-based records systems to assist local police departments.

The jurisdiction of a police officer at the local level is limited to the state, town, or municipality in which the person is a sworn officer of the law. Some problems inevitably result from such a highly decentralized system. Jurisdictional boundaries sometimes result in overlapping, communication problems, and difficulty in obtaining assistance from another law enforcement agency.

Contrary to some expectations, the public has a great deal of confidence in the police. In a year 2000 Gallup poll, 55% of the public rated the police either "very high" or "high" for honesty and ethics. Only 11% had negative views of the police.

The situation changes, however, once we look at the numbers more closely. The perceptions of police held by African Americans and white Americans differ dramatically. Whereas 63% of whites had a high level of confidence in the police, only 26% of African Americans felt the same way. In fact, 35% of blacks compared with 8% of whites had very little or no confidence in the police (Ludwig, 2000).

People often wonder if we put more police office on the street does it reduce crime. It might seem that it should, but a number of studies have cast doubt on what would seem like a logical assumption. Cities with high crime rates have more police officers. Are they having a effect in deterring crime?

It turns out a good way to determine if the police help to stop crime is to look at police staffing during terror alerts. More police are on the street during terror alerts and fewer after them. Does the crime rate differ during these times? Two researchers (Klick & Tabarrok, 2005) looked at Washington, D.C., between March 12, 2002, and July 2003. During that time, the terror alert level rose and fell four times. On high-alert days, total crimes decreased by an average of 6.6%. On high-alert days the police officers spent an extra four hours on duty after their regular eight-hour shifts.

Were tourists and criminals avoiding the area during terror alert days? No, there were as many tourists as before and crime was down throughout the city. A bigger police presence does not affect all crime levels the same. Murder levels, for example, were unchanged. Street crimes were down considerably. Theft from automobiles and car theft fell 40% during high-alert days. Burglaries dropped 15%.

Klick and Tabarrok estimated that every $1 spent to add police officers would reduce the costs of crime by $4. They suggest that a 10% increase in police nationally would mean about 700,000 fewer property crimes and 213,000 fewer violent crimes.

The Courts

The United States has a dual court system consisting of state and federal courts, with state and federal crimes being prosecuted in the respective courts. Some crimes may violate both state and federal statutes. About 85% of all criminal cases are tried in the state courts.

The state court system varies from one state to another. Lower trial courts exist for the most part to try misdemeanors and petty offenses. Higher trial courts can try felonies and serious misdemeanors. All states have appeal courts. Many have only one court of

Table 7–3

Who Decides?

These criminal justice officials must decide how to proceed with a case:

Police	Enforce specific laws
	Investigate specific crimes
	Search people, vicinities, buildings
	Arrest or detain people
Prosecutors	File charges or petitions for adjudication
	Seek indictments
	Drop cases
	Reduce charges
Judges or magistrates	Set bail or conditions for release
	Accept pleas
	Determine delinquency
	Dismiss charges
	Impose sentences
	Revoke probation
Correctional officials	Assign to type of correctional facility
	Award privileges
	Punish for disciplinary infractions
Paroling authorities	Determine date and conditions of parole
	Revoke parole

appeal, which is often known as the state supreme court. Some states have intermediate appeal courts.

The federal court system consists of three basic levels, excluding such special courts as the U.S. Court of Military Appeals. The U.S. *district courts* are the trial courts. Appeals may be brought from these courts to the *appellate courts*. There are 11 courts at this level, referred to as *circuit courts*. Finally, the highest court is the *Supreme Court,* which is basically an appeals court, although it has original jurisdiction in some cases.

The lower federal courts and the state courts are separate systems. Cases are not appealed from a state court to a lower federal court. A state court is not bound by the decisions of the lower federal court in its district, but it is bound by decisions of the U.S. Supreme Court (Reid, 1991).

How a case progresses through the criminal justice system, or whether it is even addressed at all, depends on the decisions made by people along the way. Table 7–3 lists various ways in which a case can work its way through the criminal justice system.

Prisons

Prisons are a fact of life in the United States. As much as we may wish to conceal them, and no matter how unsatisfactory we think they are, we cannot imagine doing without them. They represent such a fundamental defense against crime and criminals that we now keep a larger portion of our population in prisons than any other nation and for terms that are longer than in many counties. Small wonder that Americans invented prisons as we know them.

Before prisons, serious crimes were redressed by corporal or capital punishment. Jails existed, but mainly for pretrial detention. The closest thing to the modern prison was the workhouse. This was a place of hard labor designed almost exclusively for minor offenders, derelicts, and vagrants. The typical convicted felon was either physically punished or fined, but not incarcerated. Today's system of imprisonment for a felony is a historical newcomer.

Goals of Imprisonment Prisons exist to accomplish at least four goals: (1) separate criminals from society, (2) punish criminal behavior, (3) deter criminal behavior, and (4) rehabilitate criminals.

1. *Separate criminals from society.* Prisons accomplish this purpose once felons reach the prison gates. Inasmuch as it is important to protect society from individuals who seem bent on repeating destructive behavior, prisons are one logical choice among several others, such as exile and capital punishment (execution). The American criminal justice system relies principally on prisons to segregate convicts from society, and in this regard they are quite efficient.
2. *Punish criminal behavior.* There can be no doubt that prisons are extremely unpleasant places in which to spend time. They are crowded, degrading, boring, and dangerous. Not infrequently, prisoners are victims of one another's violence. Inmates are constantly supervised, sometimes harassed by guards, and deprived of normal means of social, emotional, intellectual, and sexual expression. Prison undoubtedly is a severe form of punishment.
3. *Deter criminal behavior.* The general feeling among both the public and the police is that prisons have failed to achieve the goal of deterring criminal behavior. There are good reasons for this. First, by their very nature, prisons are closed to the public. Few people know much about prison life, nor do they often think about it. Inmates who return to society frequently brag to their peers about their prison experiences to recover their self-esteem. For the prison experience to be a deterrent, the very unpleasant aspects of prison life would have to

be constantly brought to the attention of the population at large. To promote this approach, some prisons have allowed inmates to develop programs introducing high school students to the horrors of prison life. From the scanty evidence available to date, it is unclear whether such programs deter people from committing crimes. Another reason that prisons fail to deter crime is the funnel effect, discussed later. No punishment can deter undesired behavior if the likelihood of being punished is minimal. Thus, the argument regarding the relative merits of different types of punishment is pointless until there is a high probability that whatever forms are used will be applied to all (or most) offenders.

4. *Rehabilitate criminals.* Many Americans believe that rehabilitation—the resocialization of criminals to conform to society's values and norms and the teaching of usable work habits and skills—should be the most important goal of imprisonment. It is also the stated goal of almost all corrections officials. Yet there can be no doubt that prisons do not come close to achieving this aim. According to the FBI, 67% of former inmates released from state prisons in 1994 committed at least one serious new crime within the following three years. This rearrest rate was 5% higher than that among prisoners released during 1983 (Bureau of Justice Statistics, June 2002) (see Figure 7–8).

Figure 7–8 **Likelihood of Prisoners Being Arrested Again within Three Years of Release**

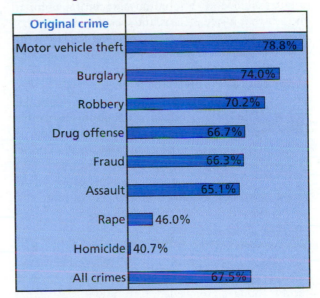

Original crime	
Motor vehicle theft	78.8%
Burglary	74.0%
Robbery	70.2%
Drug offense	66.7%
Fraud	66.3%
Assault	65.1%
Rape	46.0%
Homicide	40.7%
All crimes	67.5%

Source: U.S. Department of Justice, Bureau of Justice Statistics, *Recidivism of Prisoners Released in 1994*, Special Report NCJ 193427. Washington, DC: U.S. Department of Justice, June 2002, p. 9.

Sociological theory helps explain why rehabilitation is often ineffective. Sutherland's ideas on cultural transmission and differential association point to the fact that inside prisons, the society of inmates has a culture of its own, in which obeying the law is not highly valued. New inmates are socialized quickly to this peer culture and adopt its negative attitudes toward the law. Further, labeling theory tells us that once somebody has been designated as deviant, his or her subsequent behavior often conforms to that label. Prison inmates who are released find it difficult to be accepted in the society at large and to find legitimate work. Hence, former inmates quickly take up with their old acquaintances, many of whom are active criminals. It thus becomes only a matter of time before they are once more engaged in criminal activities.

This does not mean that prisons should be torn down and all prisoners set free. As we have indicated, prisons do accomplish important goals, though certain changes are needed. Certainly it is clear that the entire criminal justice system needs to be made more efficient and that prison terms as well as other forms of punishment must follow predictably the commission of a crime. Another idea, which gained some approval in the late 1960s but seems of late to have declined in popularity, is to create "halfway" houses and other institutions in which the inmate population is not so completely locked away from society. This way, they are less likely to be socialized to the prison's criminal subculture. Labeling theory suggests that if the process of delabeling former prisoners were made open, formal, and explicit, released inmates might find it easier to win reentry into society. Finally, just as new prisoners are quickly socialized into a prison's inmate culture, released prisoners must be resocialized into society's culture. This can be accomplished only if means are found to bring ex-inmates into frequent, supportive, and structured contact with stable members of the wider society (again, perhaps, through halfway houses). The simple separation of prisoners from society undermines this goal.

To date, no society has been able to come up with an ideal way of confronting, accommodating, or preventing deviant behavior. Although much attention has been focused on the causes of and remedies for deviant behavior, no theory, law, or social-control mechanism has yet provided a fully satisfying solution to the problem.

A Shortage of Prisons

Today's criminal justice system is in a state of crisis over prison crowding. Even though our national prison capacity has expanded, it has not kept up with demands. The National Institute of Justice estimates that we must add 1,000 prison spaces a week just to keep up with the growth in the criminal population.

Compounding the problem is the fact that many states have mandated prison terms for chronic criminals, drunken drivers, and those who commit gun crimes. Yet nearly every community will have an angry uprising if the legislature suggests building a new prison in their neighborhood. Given state financial pressures, community resistance, and soaring construction costs, people face a difficult choice. They must either build more prisons or let most convicted offenders go back to the community.

A key consideration in sending a person to prison is money. The custodial cost of incarceration in a medium-security prison is $15,000 a year. The cost is closer to $35,000 once you add to this the cost of actually building the prison, and additional payments to dependent families. You can see why judges are quick to use probation as an alternative to imprisonment, particularly when the prisons are already overcrowded.

The other side of the question, however, is how much does it cost us *not* to send this person to prison? Although it is easy to calculate the cost of an offender's year in prison, it is considerably more difficult to figure the cost to society of letting that individual roam the streets. Studies suggest that it is more expensive to release an offender than to incarcerate an offender when you weigh the value of crime prevented through imprisonment.

How much does each crime cost the public? The National Institute of Justice has come up with a figure of $2,300 per crime. This number undoubtedly overestimates the value of petty larcenies and underestimates the cost of rapes, murders, and serious assaults. It is an average, however, and it does give us some way of comparing the costs of incarceration with the costs of freedom. Using the $2,300 per-crime cost, we find that a typical inmate committing 200 crimes (the low estimate) is responsible for $460,000 in crime costs per year. Sending 1,000 additional offenders to prison, instead of putting them on probation, would cost an additional $25 million per year. The crimes averted, however, by taking these individuals out of the community would save society more than $460 million.

This approach merely gives us a dollars-and-cents way of making a comparison. It does not in any way account for the personal anguish and trauma to the victims of crimes that would be averted.

Looking at the issue from this perspective overwhelmingly supports the case for more prison space. It costs communities more in real losses, social damages, and security measures than it does to incarcerate offenders who are crowded out by today's space limitations.

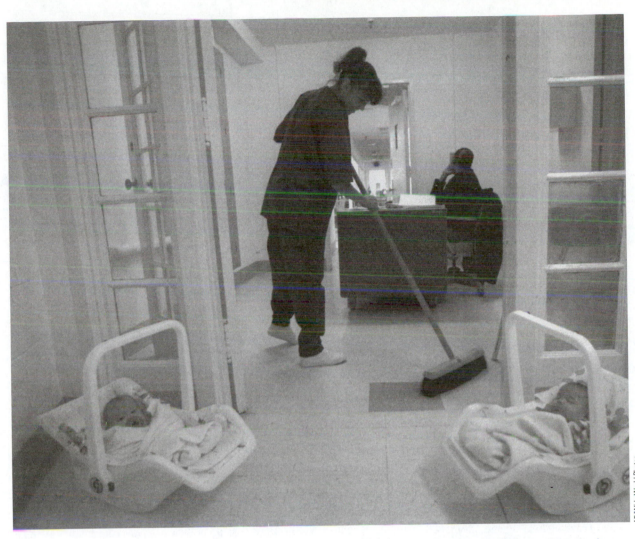

Two-thirds of women in prison are mothers, and the vast majority of their children are younger than age 18. Some prisons, such as the maximum security women's prison in Bedford Hills, New York, allow inmate mothers to keep their babies with them until the babies are 18 months of age.

Women in Prison

As prison reform began in this country, the practice was to segregate women into sections of the existing institutions. There were few women inmates, a fact that was used to justify not providing them with a matron. Vocational training and educational programs were not even considered. In 1873, the first separate prison for women, the Indiana Women's Prison, was opened, with its emphasis on rehabilitation, obedience, and religious education.

In contrast with institutions for adult males, institutions for adult women are generally more aesthetic and less secure. This is an outgrowth of the fact that in the past, women inmates were not considered high security risks, nor have they proved to be as violent as male inmates. Women were more likely to commit property crimes, such as larceny, forgery, and fraud. This trend in crimes has changed, however, and women now commit more violent crimes than property crimes. Still, three-quarters of the violent crimes committed by women are the less serious type known as simple assault. Drug offenses by women have also increased dramatically in recent years (Gowdy et al., 1998).

There are some exceptions, but on the whole, women's institutions are built and maintained with the view that their occupants are not great risks to themselves or to others. Women inmates also usually have more privacy than men do while incarcerated, and women usually have individual rooms. With the relatively smaller number of women in prison, there is a greater opportunity for the inmates to have contact with the staff, and there is also a greater chance for innovation in programming (Reid, 1991).

The number of women in state and federal prisons reached a record of nearly 105,000 in 2004. Since 1995 the annual rate of growth of the female inmate population has been higher than the growth in the number of male inmates. Even though the

Figure 7–9   **Women Prisoners in State and Federal Institutions, 1925–2004**

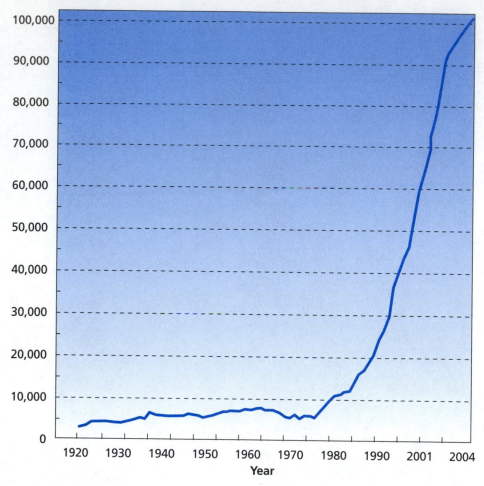

Source: Harrison, Paige M. and Allen J. Beck, "Prisoners in 2004," Bureau of Justice Statistics Bulletin October 2005, http://www.ojp .usdoj.gov/bjs/pub/pdf/p04.pdf Accessed October 28, 2005.

rate of increase in the number of women going to prison has been greater than that for men, females still make up a relatively small segment of the total prison population—7.0% at the end of 2004. Relative to their numbers in the U.S. population, men are 14 times more likely than women to be incarcerated (Harrison & Beck, 2005). (See Figure 7–9.)

Compared with male inmates, female inmates appear to have greater difficulty adjusting to the absence of their families, especially their children. Two-thirds of women in prison are mothers, and the majority (88%) of their children are younger than age 18. Only 25% of these children are cared for by the father while the mother is in prison. Most of the time, a grandparent cares for the children.

The Funnel Effect

One complaint voiced by many of those concerned with our criminal justice system is the existence of the funnel effect, in which many crimes are committed, but few people ever seem to be punished. The funnel effect begins with the fact that of all the crimes committed, only 36.3% are reported to the police (Bureau of Justice Statistics, 2002, *Sourcebook*). Only about 26% lead to an arrest. Further, false arrests, lack of evidence, and plea bargaining (negotiations in which individuals arrested for a crime are allowed to plead guilty to a lesser charge of the crime, thereby saving the criminal justice system the time and money spent on a trial) considerably reduce the number of complaints that actually are brought to trial. To be fair, the situation is not quite as bad as it appears. The number of arrests for serious crimes is considerably higher than it is for crimes in general.

What about punishment? Those who criticize the system's funnel effect seem to regard only a term in prison as an effective punishment. Yet the usual practice is to send to prison only those criminals whose terms of confinement are set at longer than one year. The number of prisoners in federal and state prisons, after declining through the 1960s, rose sharply through the 1990s, reaching more than 1.4 million in 2004 (Harrison & Beck, 2005).

Many thousands of other criminals receive shorter sentences and serve them in municipal and county jails. Thus, if the numbers of people sent to local jails as well as to prison are counted, the funnel effect is seen to be less severe than it often is portrayed. The question then becomes one of philosophy: Is a jail term of less than one year an adequate measure for the deterrence of crime? Or should all convicted criminals have to serve longer sentences in federal or state prisons, with jails used primarily for pretrial detention?

Truth in Sentencing

The amount of time offenders serve in prison is almost always shorter than the time they are sentenced to serve by the court. (See Figure 7–10 for the average time served for selected crimes.) The public has been in favor of longer sentences and uniform punishments for prisoners. Prison crowding and reductions in prison time for good behavior have often resulted in the release of prisoners well before they have served their assigned sentences. In response to complaints that criminals were not paying for their crimes, many states enacted restrictions on the possibility of early release; these laws became known as "truth in sentencing." The truth in sentencing laws require offenders to serve a substantial portion of the prison sentence imposed by the court before being eligible for release. The laws are based on the belief that victims and the public are entitled to know exactly what punishments offenders are receiving (Ditton & Wilson, 1999).

In the 1990s, truth in sentencing laws gained momentum with the help of the U.S. Congress, which authorized grants to expand or build correctional facilities if states would enact such laws. To receive the grants, states had to require people convicted of violent crimes to serve not less than 85% of their prison sentences.

At this point, two-thirds of the states have established truth in sentencing laws. This has limited the powers of parole boards to "set release dates, or of prison managers to award good time, earned time, or both" (Mackenzie, 2000).

Sentencing reforms have also led to more blacks than whites going to prison after arrest. If current trends continue, a black male in the United States would have about a 1 in 3 chance of going to prison during his lifetime, whereas a Hispanic male would have a 1 in 6 chance and a white male would have a 1 in 17 chance of going to prison. In 2001, the chances of going to prison were highest among black

Figure 7–10   Average Time Served for Various Types of Crime

Source: U.S. Department of Justice, Bureau of Justice Statistics, *Trends in State Parole, 1990–2000*, Special Report NCJ 184735. Washington, DC: U.S. Department of Justice, October 2001, p. 5, Table 5.

males (32.2%) and Hispanic males (17.2%) and lowest among white males (5.9%). The lifetime chances of going to prison among black females (5.6%) were nearly as high as for white males. Hispanic females (2.2%) and white females (0.9%) had much lower chances of going to prison (Bonczar, 2003).

SUMMARY

- A culture's norms and values make up its moral code, or the symbolic system by which behavior is viewed as right or wrong, good or bad within that culture.
- Normal behavior is behavior that conforms to the norms of the group in which it occurs.
- Deviant behavior is behavior that fails to conform to the group's norms.
- Criminal and deviant behavior has been found throughout history.
- Scholars have proposed a variety of theories.
- Biological theories such as those propounded by Lombroso and Sheldon stressed the importance of inherited factors in producing deviance.
- Psychological explanations emphasize cognitive or emotional factors within the individual as the cause of deviance.

Text continues on page 184

FOR FURTHER THINKING

The Continuing Debate over Capital Punishment: Does It Deter Murderers?

Many countries throughout the world no longer use the death penalty. Among those that do China stands out with the largest number of executions each year. Amnesty International estimates that China executed more than 3,400 people in 2004. Iran executed 159 people that year and Vietnam executed 64. Next in line is the United States, which executed 59 inmates in 2004, bringing the total number of U.S. executions to 944 since 1976, the year the Supreme Court reinstated the death penalty. Currently, 3,487 prisoners await execution (Figures 7–11 and 7–12). The average person executed in 2004 had been on death row for nearly 11 years. A small number had actually been awaiting execution for more than 20 years. It seems obvious that the vast majority of inmates sentenced to death will not be executed.

The public supports the death penalty. The Gallup poll has been asking Americans about the death penalty

Figure 7–11   Persons Sentenced to Death, 1953–2004

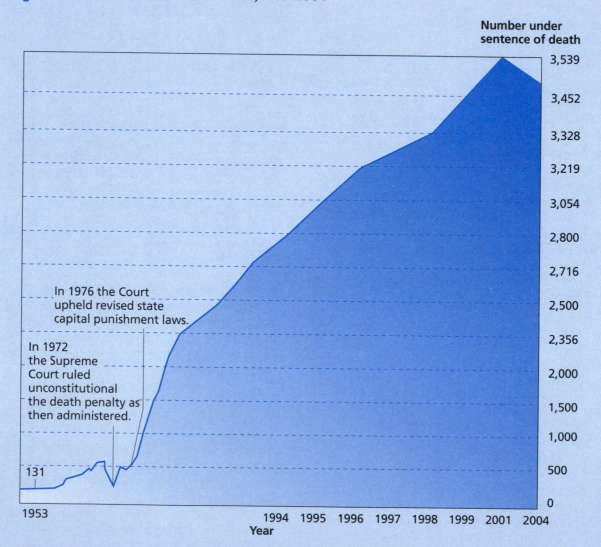

Number under sentence of death

In 1976 the Court upheld revised state capital punishment laws.

In 1972 the Supreme Court ruled unconstitutional the death penalty as then administered.

Sources: "Capital Punishment Statistics 2002, Summary Findings, "U.S. Department of Justice Bureau of Justice Statistics; U.S. Department of Justice, Bureau of Justice Statistics, *Capital Punishment, 2003*, Bulletin NCJ 206627. Washington, DC: U.S. Department of Justice, November, 2004, p. 9, Table 9.

for almost 50 years. As of May 2005, nearly three-quarters (74%) of Americans favor the death penalty in cases of murder, down from its high point of 80% in 1994.

But capital punishment has also been opposed for many years and for many reasons. In the United States, the Quakers were the first to oppose the death penalty and to provide prison sentences instead. Amnesty International, U.S.A., calls capital punishment a "horrifying lottery" in which the penalty is death and the odds of escaping are determined more by politics, money, race, and geography than by the crime committed. The group bases its impression on the fact that black men are more likely to be executed than white men; southern states, including Texas, Virginia, Missouri, Louisiana, and Florida, account for the majority of executions that have occurred since 1977.

It is also no surprise that nearly all death-row inmates are poor. They often had a public defender who might not have been qualified for the task. Even if the inmate's attorney made errors during the defense, the defendant's appellate attorney must demonstrate that the defense counsel's blunders directly affected the jury's verdict and that without those mistakes, the jury would have returned a different verdict (Prejean, 1993). One study (Radelet, Bedau, & Putnam, 1992), for instance, found that between 1900 and 1991, 416 innocent people were convicted of capital crimes, and 23 actually were executed. The two most frequent causes of errors that produced wrongful convictions were perjury by prosecution witnesses and mistaken eyewitness testimony.

Yet the arguments for capital punishment continue to mount, centering mainly on the issue of deterrence. Which brings us back to the age-old question: Does the death penalty deter homicide? Until the 1970s, social scientists continued to argue that they could find no evidence that it did. Sophisticated statistical techniques have been used on capital punishment data. One study (Mocan & Gittings, 2001) concluded that each execution decreased the number of homicides by five or six. Another study claimed that an increase in arrests, sentencings, or executions tends to reduce the murder rate. In particular, each execution results, on average, in 18 fewer murders, according to some economists (Dezhbakhsh, Rubin, & Shepherd, 2002). Another study reported that the unofficial moratorium on executions during most of 1996 in Texas appears to have contributed to additional homicides (Cloninger & Marchesini, 2001).

There may be more involved in deterrence than we think. Plato believed we are deterred from committing

continued

Figure 7–12   **Inmates Executed, 1930–2004**

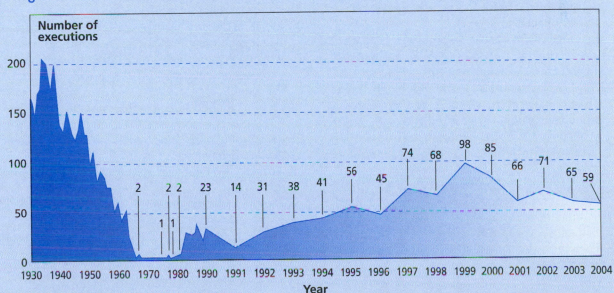

Source: *Capital Punishment, 2003*, Bulletin NCJ 206627. Washington, DC: U.S. Department of Justice, November, 2004, p. 9, Table 9.

FOR FURTHER THINKING

The Continuing Debate over Capital Punishment: Does It Deter Murderers? (continued)

crimes by seeing others punished. He was referring to punishments administered in public, where everyone could see the gory details of torture and execution. Fortunately, today executions are not public, and only a small number of people witness them. In place of actually seeing the execution, we now have mass media reports that become our eyes. Therefore, deterrence should be related to how much an execution is publicized.

People also argue that the death penalty is applied in a racially discriminatory fashion. One extensive study (Baldus, Woodworth, & Pulaski, 1990) concluded that the odds of being condemned to death were 4.3 times greater for defendants who killed whites than for defendants who killed blacks. Opponents of the death penalty have used this information to make the case that it should be abolished entirely on the grounds that racial

bias is an inevitable part of the administration of capital punishment in the United States and that it would be better to have no death penalty than one influenced by prejudice. Others argue that the remedy to the problem is to do what is known as "leveling up"—increasing the number of people executed for murdering blacks.

They point out that if we sentence more murderers of black people to death, we are then eliminating the bias. A third solution is to impose mandatory death sentences for certain types of crimes. In this way we are eliminating discretionary judgments and the potential for bias (Kennedy, 1997).

In recent years the number of executions in the United States has been declining, but with public support for the death penalty continuing, and with no broad legal challenges to capital punishment being waged, we can expect executions to continue.

- Psychoanalytic theory suggests that criminals act on the irrational impulses of the id because they failed to develop a proper superego, or conscience, in the socialization process.
- Behaviorists argue that crime is the product of conditioning.
- Wilson and Herrnstein proposed that criminal activity, like all human behavior, is the product of a rational choice by the individual as a result of weighing the costs and benefits of alternative courses of action.
- Sociological theories of deviance rely on patterns of social interaction and the relationship of the individual to the group as explanations.
- Durkheim argued that, in modern highly differentiated and specialized societies, and particularly under conditions of rapid social change, individuals could become morally disoriented. This condition, which he called anomie, can produce deviance.
- Control theorists like Hirschi have argued that everyone is a potential deviant. The issue, for such theorists, is not what causes deviance, but what causes conformity.
- When individuals have strong bonds to society, their behavior will conform to conventional norms. When any of those bonds are weakened, however, deviance is likely.
- Sykes and Matza have argued that people become deviant as a result of developing techniques of neutralization or rationalizations that make it

possible to justify illegal or deviant behavior. Their view is that these techniques are learned as part of the socialization process.
- Cultural transmission theory, pioneered by Shaw and McKay, emphasizes the cultural context in which deviant behavior patterns are learned.
- Sutherland and Cressey suggested the theory of differential association—that is, that individuals learn criminal techniques and attitudes through intimate contact with deviants.
- Labeling theory shifts the focus of attention from the deviant individual to the social process by which a person comes to be labeled as deviant and the consequences of such labeling for the individual.
- In all likelihood some combination of these various theories is necessary for gaining a fuller understanding of the emergence and continuance of deviant behavior.
- Crime is behavior that violates a society's criminal laws.
- Violent crime may result in injury to a person; property crime is committed with the intent of obtaining property and does not involve the use or threat of force against an individual.
- The most serious crimes are termed felonies; less serious crimes are called misdemeanors.
- The FBI publishes statistics on the frequency of selected crimes in the Uniform Crime Reports. These statistics are not always reliable, however.

- The *National Crime Victimization Survey* shows that only a small fraction of all crimes are reported to the authorities.
- The U.S. violent crime rate includes one of the highest homicide rates in the industrialized world.
- Other violent crimes that affect American households include rape, aggravated assault, and robbery.
- Seventy-five percent of all crime in the United States is crime against property, not against a person.
- The criminal justice system consists of personnel and procedures to facilitate the arrest, trial, and punishment of those who violate the laws.
- The three main aspects of this system are the police, the courts, and prisons.
- The goals of imprisonment include separating the criminal from society, punishing criminal behavior, deterring criminal behavior, and rehabilitating, or resocializing criminals to conform to society's values and norms and teaching them usable work habits and skills.

 Media Resources

The Companion Website for *Introduction to Sociology*, Ninth Edition

http://sociology.wadsworth.com/tischler9e

Supplement your review of this chapter by going to the companion website to take one of the Tutorial Quizzes, use the flash cards to master key terms, and check out the many other study aids you will find there. You will also find special features such as Wadsworth's Sociology Online Resources and Writing Companion, GSS data, and Census 2000 information at your fingertips to help you complete that special project or do some research on your own.

CHAPTER SEVEN STUDY GUIDE

KEY CONCEPTS AND THINKERS

Match each concept with its definition, illustration, or explanation presented below.

a. Informal sanctions
b. Diversion
c. Status offenses
d. Victimless crime
e. Property crime
f. Formal sanctions
g. Mesomorph

h. Anomie
i. Labeling theory
j. Innovators
k. Funnel effect
l. Atavistic beings
m. Techniques of neutralization

n. Rehabilitation
o. Secondary deviance
p. Recidivism
q. White-collar crime
r. Social control
s. Retreatists

r   1. Ways of directing or influencing members to conform to the group's values and norms.
f   2. Acts of approval and disapproval applied in a public ritual, usually under the direct or indirect control of authorities.
a   3. Acts of approval or disapproval applied spontaneously by group members.
i   4. An approach to deviance that emphasizes how some people are defined as deviant and the consequences of being so defined.
e   5. Predatory crimes such as theft, where the criminal does not directly confront the victim.
k   6. The process by which a large number of crimes results in only a small number of offenders being sent to prison.
d   7. Crimes such as drug use and gambling that are not predatory but nevertheless violate the moral code.
n   8. The resocialization of criminals to conform to society's values and norms and the teaching of usable work habits and skills.
o   9. Deviant or criminal behavior that people develop as a result of having been labeled as deviant.
p   10. Crimes committed even after punishment has occurred.
h   11. A state of normlessness, in which values and norms have little effect and the culture no longer provides adequate guidelines for behavior.
j   12. In anomie theory, people who take illegal routes to socially approved goals.
q   13. Acts by individuals who, while occupying positions of social responsibility or high prestige, break the law in the course of their work.
s   14. In anomie theory, people who pull back from society altogether and cease to pursue culturally legitimate goals.
c   15. Offenses that are punishable if committed by a juvenile but not by an adult.
g   16. A ruggedly muscular body type associated with being assertive and action oriented.
l   17. According to Lombroso, evolutionary throwbacks whose behavior is more apelike than human.
m   18. Thought processes that make it possible to justify illegal or deviant behavior.
b   19. Sending offenders, especially juveniles, to agencies outside the justice system.

Match the thinkers with their main idea or contribution.

a. Sigmund Freud
b. James Q. Wilson and Richard Herrnstein

c. Edwin H. Sutherland
d. Émile Durkheim
e. Travis Hirschi

f. Clifford Shaw and Henry McKay
g. Cesare Lombroso

d   1. Argued that deviant behavior is an integral part of all healthy societies; developed the concept of anomie.
g   2. Suggested that criminals were evolutionary throwbacks who could be identified by primitive physical features, particularly with regard to the head.
a   3. Argued that crime is produced by the unconscious impulses of the individual.

___h___   **4.** Argued that crime is the product of a rational choice by an individual as a result of weighing the costs and benefits of alternative courses of action.

___e___   **5.** Developed control theory, in which it is hypothesized that the strength of social bonds keeps most of us from becoming criminals.

___f___   **6.** Suggested that certain neighborhoods generate a culture of crime that is passed on to residents.

___c___   **7.** Developed the theory of differential association to explain why some people and not others become deviant; coined the term *white-collar crime*.

CENTRAL IDEA COMPLETIONS

Following the instructions, fill in the appropriate concepts and descriptions for each of the questions posed below.

1. We usually think that deviance is dysfunctional for the society in which it occurs, but what functions can it serve for the society? The text mentions five. Cite them and give examples.

    a. _____

    Example _____

    b. _____

    Example _____

    c. _____

    Example _____

    d. _____

    Example _____

    e. _____

    Example _____

2. How have rates of crime and imprisonment in the U.S. changed over the course of the three decades between 1973 and 2004?

    _____

3. List, define, and provide an example for each of **Merton's five categories of individual adaptation to anomie.**

    a. _____

    Example _____

    b. _____

    Example _____

    c. _____

    Example _____

    d. _____

    Example _____

    e. _____

    Example _____

4. Apply Sykes and Matza's **five techniques of neutralization** to a situation involving cheating in a college or university community.

   a. _____

   b. _____

   c. _____

   d. _____

   e. _____

CRITICAL THOUGHT EXERCISES

1. Some explanations of crime focus on the individual offender; others look more at the factors in the social environment. The same is true of proposals for policies designed to reduce crime. Compare the ways in which these two different approaches might be applied in thinking about some form of deviance on your own campus (academic dishonesty, excessive drinking, and so on). Describe the possible sanctions, both formal and informal, that might be brought to bear on this form of deviance. How effective is each type of sanction?

2. Are athletes violent? Sociologically, what appears to be the relationship between sports and violent behavior? Apply Sutherland's differential association theory to the issue of violence and the behavior of athletes.

   Are athletes treated differently? Consider the issues examined in the reading "Public Heroes, Private Felons: Athletes and Sexual Assault," as well as any specific cases you know of. Discuss how the factors mentioned in the text (social norms, social identity, social context) affect the labeling or nonlabeling of athletes.

INTERNET ACTIVITIES

1. Part A: Log on to the World Wide Web. Using any available search engine or browser, investigate one or more sites that demonstrate the following:

   a. An example of a behavior you personally would define as both deviant and harmful to society.

   b. An example of a behavior you would define as deviant but not socially harmful to society.

   c. A behavior you would define as deviant but not illegal.

   d. A behavior you believe many people older than you might define as deviant but that you and members of your age cohort would not define as deviant.

   Part B: After you have completed Part A, download an example page for each of the sites you visited and discuss:

   —What are the common features of each form of deviance?

   —What aspects, if any, of these websites made you uncomfortable?

   —Assuming the sites you defined as deviant are defined by others in a similar manner, what is it about the behavior that leads to these definitions?

   —What role might new technologies, such as the Internet, play in a society's shifting definitions of deviance?

2. Crime in your town. Go to the Uniform Crime Reports for 1995 (http://www.fbi.gov/ucr/Cius _97/95CRIME/95crime2.pdf). In "Table 8—Number of Offenses Known to the Police, Cities and Towns 10,000 and over in Population, 1995," select one town or city (your own perhaps) and look at the numbers of crimes in each of the categories (Murder, Rape, Motor Vehicle Theft, and so on). You

can compute the crime rate per 100,000 population by dividing the number of crimes by the town population and multiplying by 100,000. Then go to the UCR main page (http://www.fbi.gov/ucr/ucr.htm#cius) and get the same information for the most recent year. How has crime changed?

ANSWERS TO KEY CONCEPTS

1.r 2.f 3.a 4.i 5.e 6.k 7.d 8.n 9.o 10.p 11.h 12.j 13.q 14.s 15.c 16.g 17.l 18.m 19.b

ANSWERS TO KEY THINKERS

1.d 2.g 3.a 4.b 5.e 6.f 7.c

ThomsonNOW™

Reviewing is as easy as ❶ ❷ ❸

1. Before you do your final exam, take the ThomsonNOW diagnostic quiz to help you identify the areas on which you should concentrate. You will find information on ThomsonNOW and instructions on how to access all of its great resources on the foldout at the beginning of the text.
2. As you review, take advantage of ThomsonNOW's study videos and interactive Map the Stats exercises to help you master the chapter topics.
3. When you are finished with your review, take ThomsonNOW's posttest to confirm you are ready to move on to the next chapter.

© Britt Erlanson/The Image Bank/Getty Images

8

Social Class in the United States

Learning Objectives

After studying this chapter, you should be able to do the following:

- Explain the factors that affect a person's chances of upward social mobility.
- Describe the distribution of wealth and income in the United States.
- Summarize the functionalist and conflict theory views of social stratification.

- Describe the characteristics of each of the social classes in the United States.
- Describe differences in the poverty rate among various groups in American society.
- Compare poverty rates in the United States with those of other industrialized countries.
- Describe some of the personal and social consequences of a person's position in the class structure.

Ginie Sayles went from being a single mother on welfare to a wealthy lifestyle by marrying a rich man. She gives lectures on how others can do the same. "Anyone can learn how to tie the knot with that seven-figure bank account," Sayles advises in a Texas drawl. What you need to do is to meet Mr. or Ms. Rich and then "display the social graces needed for those cozy limousine rides and weekends at country estates . . .I don't tell you to marry the rich," she said. "But I teach you how, so at least if you don't . . . it's your choice."

Jensen, 1999

We wore ties on Sunday, and black wool suits called B-suits, with the school crest on our top pocket. The crest was a dragon, whose head reached out toward the sun. Under that came the motto—*Arduus ad Solem.*

My father had taught me a song to help me do up my tie. It had the tune of "Twinkle, Twinkle, Little Star" and went "Over, under, over, through, pull the little end away from you . . ."

Some Sunday afternoons, the school was like a ghost town. The day before, parents had clogged the school with their Bentleys and Range Rovers and Rollses and driven away their sons. The ones who remained were mostly boys who lived abroad. They weren't foreign. It was just that their parents were working in Singapore or Hong Kong or Bermuda.

—Paul Watkins, *Stand Before Your God: An American Schoolboy in England*

Americans like to think that social stratification and social class are minor issues. After all, we do not have inherited ranks, titles, or honors. We do not have coats of arms or rigid caste rankings, besides, equality among men—and women—is an ideal guaranteed by our Constitution and summoned forth regularly in speeches from podiums and lecterns across the land. Yet lavish displays of wealth and the attempts of many people to obtain power and privilege makes it difficult to ignore social inequality and the uneven distribution of material rewards. In this chapter we will begin to see that social stratification is quite complex and open to many subtle variations. It does not always fit neatly into our stereotypes.

The United States is characterized by an enormous diversity of wealth and power. Once rewards are distributed unequally within a society, then economic, political, and social stratification begin.

The American Class Structure

A **social class** consists of *a category of people who share similar opportunities, similar economic and vocational positions, similar lifestyles, and similar attitudes and behaviors.* A society that has several different social classes and permits social mobility is based on a **class system of stratification.** Class boundaries are maintained by limiting social interaction, intermarriage, and mobility into that class.

Some form of class system is usually present in all industrial societies, whether they be capitalist or communist. Social mobility in a class system is often the result of an occupational structure that opens up higher-level jobs to anyone with the education and experience required. A class society encourages striving and achievement. Here in the United States we should find this concept familiar, for ours is basically a class society.

There is little agreement among sociologists about how many social classes exist in the United States and what their characteristics may be. For our purposes here, however, we will follow a relatively common approach of assuming that there are five social classes in the United States: upper class, upper-middle class, lower-middle class, working class, and lower class (Rossides, 1990). Table 8–1 presents descriptions of each of these social classes. (For a discussion of movement between social classes, see "For Further Thinking: How Easy Is It to Change Social Class?", p. 210.)

The Upper Class

Members of the upper class have great wealth, often going back for many generations. They recognize one another, and are recognized by others, by reputation

Social inequality involves the uneven distribution of privileges, material rewards, and power.

© Christopher Morris/Black Star Publishing/PictureQuest

and lifestyle. They usually have high prestige and a lifestyle that excludes those of other classes. Members of this class often influence society's basic economic and political structures. The upper class usually isolates itself from the rest of society by residential segregation, private clubs, and private schools. Historically, they

Table 8–1

Social Classes in the United States

Class	Occupation	Education	Children's Education
Upper class	Corporate ownership; upper-echelon politics; honorific positions in government and the arts	Liberal arts education at elite schools	College and postcollege
Upper-middle class	Professional and technical fields; managers; officials; proprietors	College and graduate training	College and graduate training
Lower-middle class	Clerical and sales positions; small-business semiprofessionals; farmers	High school; some college	Option of college
Working class	Skilled and semiskilled manual labor; craftspeople; foremen; nonfarm workers	Grade school; some or all of high school	High school; vocational school
Lower class	Unskilled labor and service work; private household work and farm labor	Grade school; semi-illiterate	Little interest in education; high school dropouts

Source: Adapted from *Statistical Abstract of the United States: 1981,* U.S. Bureau of the Census, 1981, Washington, DC: U.S. Government Printing Office.

© Topham/The Image Works

Members of the upper class often engage in activities that exclude those of other classes.

area is a long-term consideration. These people often have a college education, own property, and have a savings reserve. They usually live in comfortable homes in the more exclusive areas of a community, are active in civic groups, and carefully plan for the future. They very likely belong to a church. The most common denominations represented are Presbyterians, Episcopalians, Congregationalists, Jews, and Unitarian Universalists. In the United States, 10% to 15% of the population falls into this category.

A large percentage of the new upper-middle class are two-income couples, both of whom are college-educated and employed as corporate executives, high government officials, business owners, or professionals. These relatively affluent individuals are changing the face of many communities. They are gentrifying rundown city neighborhoods with their presence and their money.

The Lower-Middle Class

The lower-middle class shares many characteristics with the upper-middle class, but its members have not been able to achieve the same kind of lifestyle because of economic or educational shortcomings.

Usually high school graduates with modest incomes, they are semiprofessionals, clerical and sales workers, and upper-level manual laborers.

They emphasize respectability and security, have some savings, and are politically and economically conservative. They often are dissatisfied with their standard of living, jobs, and family incomes. They are likely to be represented among the Protestant denominations such as Baptists, Methodists, and Lutherans, or they might be Catholic or Greek Orthodox. They make up 25% to 30% of the United States population. (For further discussion of the middle class, see "Our Diverse Society: The Black Middle Class: Fact or Fiction?")

The Working Class

The working class is made up of skilled and semi-skilled laborers, factory employees, and other blue-collar workers. These are the people who keep the country's machinery going. They are assembly-line workers, auto mechanics, and repair personnel. They are the most likely to be affected by economic downturns. More than half belong to unions.

Working-class people live adequately but have little left over for luxuries. They are less likely to vote than the higher classes, and they feel politically powerless. Although they have little time to be involved in civic organizations, they are very much involved with their extended families. The families are likely to be patriarchal, with sharply segregated sex roles. They stress obedience and respect for elders. Many

have been Protestant, especially Episcopalian or Presbyterian. This is less true today. It is estimated that in the United States, the upper class consists of 1% to 3% of the population.

Since the 1970s, the upper class also has come to include society's new entrepreneurs—people who have often made many millions, and sometimes billions, of dollars in business. In many respects these people do not resemble the upper class of the past. Included are people like William Gates, the founder of Microsoft Corporation, who has a net worth of $63 billion; Warren Buffett, the founder of Berkshire Hathaway, who has a net worth of $43 billion; and Larry Ellison, the founder of Oracle Corporation, whose net worth exceeds $13.7 billion.

Not all billionaires lead opulent lifestyles, and many in the upper class do not approve of displaying wealth. For many, the money is merely a way of keeping score of how well they are doing at their chosen endeavors.

The Upper-Middle Class

The upper-middle class is made up of successful business and professional people and their families. They are usually just below the top in an organizational hierarchy but still command a reasonably high income. Many aspects of their lives are dominated by their careers, and continued success in this

OUR DIVERSE SOCIETY

The Black Middle Class: Fact or Fiction?

Fifty years ago the number of African Americans who could be considered middle class was extremely small. Figures from that era show that 5% of black men and 6% of black women were engaged in white-collar work of any kind. Only 1 out of 20 blacks was a professional, owner or manager of a business, salesperson, or secretary.

Six out of ten African American women were household workers engaged in cleaning, cooking, or watching children for low wages. The majority of black men were unskilled laborers, sharecroppers, or domestic servants.

Today approximately 40% of all blacks can be considered middle class on the basis of self-identification and income level. This is about as large as the white middle class was at the end of President Dwight Eisenhower's second term, a time when American society as a whole was described as predominantly middle class (Thernstrom & Thernstrom, 1997).

Sociologists Melvin Oliver and Thomas Shapiro believe that it is premature to celebrate the rise of the black middle class. The glass is both half full and half empty, because at the same time that black wealth has grown, it has fallen even further behind that of whites. Oliver and Shapiro believe that, materially, whites and blacks constitute two nations with two middle classes. Blacks achieve middle-class status because of income, not assets such as property, stocks, and other forms of wealth. In contrast, the white middle class possesses these assets. Even with similar levels of income, middle-class whites have nearly five times as much of their wealth in assets as middle-class blacks. A house, stocks, and savings all add security and stability to a family's lifestyle. The middle-class black position is precarious and fragile (Oliver & Shapiro, 1997).

Orlando Patterson believes that Oliver and Shapiro have discredited "the hard work, intelligence, and industriousness that middle-class [African Americans] have put into acquiring the very real status and power they possess and the pride that they justly feel in their achievements" (Patterson, 1997). Patterson notes that

the extraordinary growth of income inequality in America since the mid-1970s is a national rather than a racial issue. All middle-class Americans have been losing ground to the wealthiest 1% of families, who now own more than 42% of all wealth, compared with the 22% they owned in 1975. This is a good example of how a serious nonracial issue is translated into a racial one. If two nations are emerging in America, they are the haves and have-nots, a divide that cuts right across race. Indeed, greater inequality actually exists among the whole of African Americans than between African Americans and whites.

Almost all new middle classes in history have had precarious economic starts, including whites, most of whose families rose to middle-class status only after World War II. There is no reason to assume that African Americans should follow the same paths into middle-class status as whites did 40 or more years ago. In the first place, the American economy has gone through a variety of changes, and the avenues of social class mobility of 50 years ago may no longer be available. For many it may make more sense today to continually reinvest in the improvement of skills than to lock up funds in a mortgage. Second, it may well be that the African American middle class has different lifestyle preferences and that asset accumulation may not be one of them. If this is the case, then such choices are entirely their business. Taking a long perspective, the important thing to note is that the children of the new middle-class African Americans will be second- and third-generation members with all the confidence, educational resources, and, most of all, cultural capital to find a more secure niche in the nation's economy.

Sources: *Black Wealth/White Wealth,* by M. L. Oliver and T. M. Shapiro, 1997, New York: Routledge, pp. 90–125; *The Ordeal of Integration,* by O. Patterson, 1997, Washington, DC: Civitas, pp. 24–25; *America in Black and White,* by S. Thernstrom and A. Thernstrom, 1997, New York: Simon & Schuster, pp. 183–202.

of them have not finished high school. The religious makeup is similar to that of the lower-middle class. They represent 25% to 30% of the United States population.

The Lower Class

The lower class comprises the people at the bottom of the economic ladder. They have little in the way of education or occupational skills and consequently are unemployed or underemployed. Lower-class families often have many problems, including broken homes, illegitimacy, criminal involvement, and alcoholism.

Members of the lower class have little knowledge of world events, are not involved with their communities, and usually do not identify with other poor people. They have low voting rates. Because of a variety of personal and economic problems, they often have no way of improving their lot in life. For them, life is a matter of surviving from one day to the next. Their dropout rate from school is high, and they have the highest rates of illiteracy of any of the groups. The lower class is disproportionately African American and Hispanic, but race and poverty do not define it exclusively. Rather, it is defined by a set of characteristics and conditions that are part of a broader lifestyle. Lower-class people often belong to fundamentalist or revivalist religious sects. About 15% to 20% of the population falls into this class.

Money, power, and prestige are distributed unequally among these classes. However, a desire to advance and achieve success is shared by members of all five classes, which makes them believe that the system is just and that upward mobility is open to all. Therefore, they tend to blame themselves for lack of success and for material need (Vanfossen, 1979).

Income Distribution

The U.S. Census Bureau has published annual estimates of the distribution of family income since 1947. Those figures show a highly unequal distribution of wealth. In 2004, for example, the richest one-fifth of families earned 50.1% of the total income for the year, whereas the poorest one-fifth earned only 3.4% (Current Population Survey, 2005).

Without further elaboration, this information allows us to imagine that the richest one-fifth of families consists of millionaire real estate moguls, Wall Street professionals, and CEOs of major companies. The image is somewhat misleading. In 2004, the richest one-fifth included all families with incomes of $100,000 or more (Figure 8–1). Keep in mind that this is a family income derived from jobs held by husbands, wives, and all other family members. Family incomes for the richest 5% of the population begin at $173,640 (Bureau of the Census, *Current Population Survey, 2005 Annual Social and Economic Supplement*).

This is not to imply, though, that there is not a significant difference in the distribution of wealth in the United States. Income, however, is only part of the picture. Total wealth—in the form of stocks, bonds, real estate, and other holdings—is even more unequally distributed. The richest 20% of American families owns more than three-fourths of all the country's wealth. In fact, the richest 5% of all families owns more than half of America's wealth. There is also evidence to support the old adage that "The rich get richer, and the poor get poorer." The number of people in poverty grew from 24.5 million in 1978 to 37 million in 2004 (Current Population Survey, 2005). (See "Controversies in Sociology: Is the Income Gap between the Rich and the Poor a Problem?")

Figure 8–1   Family Income by Quintile, 2004

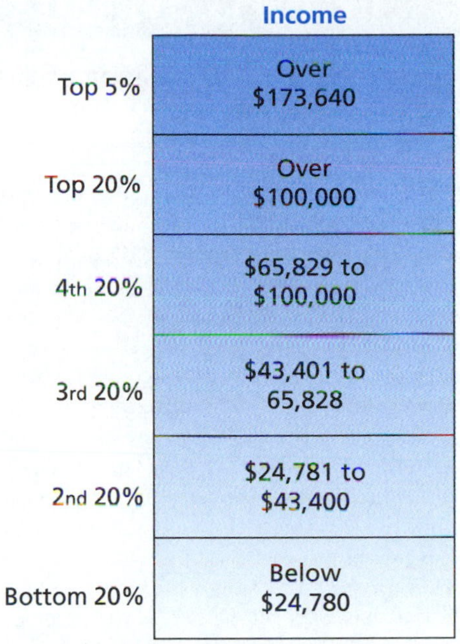

*Richest 5% of all families (included in the fifth quintile)

Source: U.S. Census Bureau, *Current Population Survey, 2005 Annual Social and Economic Supplement*, June 24, 2005. http://pubdb3.census.gov/macro/032005/faminc/new06_000.htm.

Poverty

On a basic level, poverty refers to a condition in which people do not have enough money to maintain a standard of living that includes the basic necessities of life. Depending on which official or quasi-official approach we use, it is possible to document that anywhere from 14 million to 45 million Americans are living in poverty. The fact is, we really do not have an unequivocal way of determining how many poor people there are in the United States.

Poverty seems to be present among certain groups much more than among others. In 2004, 12.7% of all Americans lived below the poverty level. Whereas 8.6% of all whites were living in poverty, 24.7% of all blacks and 21.9% of those of Hispanic origin fell into this group (Bureau of the Census, 2005). (See Figure 8–2 for the poverty rates by race and Hispanic origin.)

People living in certain regions of the United States are much more likely to live in poverty than those living elsewhere. For example, the poverty rates in Louisiana and New Mexico are more than twice those of Maryland (Proctor & Dalaker, 2002).

It is also a fact that the level of poverty in rural areas actually is higher than that in our cities. Thirty percent of the nation's poor live in rural America—a reality that is often overlooked by those who focus only on the problems of the urban poor. Even worse, the economic conditions of the rural poor are expected to deteriorate along with the decline of

CONTROVERSIES IN SOCIOLOGY

Is the Income Gap between the Rich and the Poor a Problem?

Throughout much of U.S. history, there has been a large gap in the amount of wealth held by America's richest and poorest citizens. During the nineteenth and twentieth centuries, the richest 1% of U.S. adults held between 20% and 30% of all private wealth in the country.

Around the start of the 1970s, the income gap began to grow—slowly at first, and then more rapidly during the early 1980s. Today, the gap between rich and poor Americans is at the highest level since the late 1970s. The top wage earners have seen large increases in their real incomes, while those at the bottom have actually experienced losses in real income.

One reason for the increasing wage gap is a decline in low-skill jobs based in the United States as manufacturers move their jobs to other countries where labor is cheaper. It has become harder for people without special skills to land jobs that pay relatively high wages.

Does the U.S. government have an obligation to shrink the income gap? Or is a substantial gap between the highest- and the lowest-paid workers simply a natural part of a capitalist economy?

Some people believe that the income gap is trapping the lowest-paid workers in poverty. Others downplay the importance of the gap in wages. They say it would be harmful to the economy as a whole to interfere with the distribution of income. People who earn higher incomes have worked hard to attain their positions, they say, and their high wages are a just reward for the innovations and expertise they bring to society.

Others believe that the situation is not as dire as some claim because the overall standard of living in the United States has increased in recent decades. Even people at the lowest rung of the wage scale have more material possessions and more money to spend than did their counterparts in earlier generations. In addition, in the United States, they claim, the truly motivated workers can lift themselves into higher income brackets.

It appears the income gap is leading to a society in which people segregate themselves along socioeconomic class lines. People have physically separated themselves from the less fortunate by moving to wealthy areas far away from the problems associated with poverty.

Government approaches to dealing with the income gap include the Earned Income Tax Credit, which provides tax rebates for low-paid workers in nearly 20 million households, and increases in the minimum wage. The next decade will tell us whether these programs are enough to address the problem.

Figure 8–2   **Poverty Rates by Race and Hispanic Origin, 1959 to 2004**

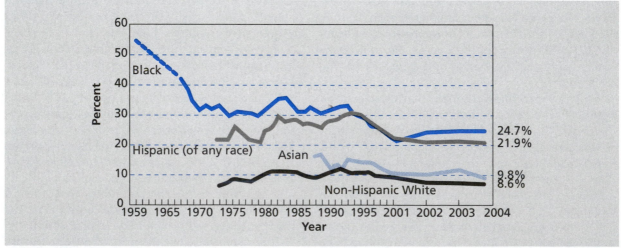

Source: DeNavas-Walt, Carmen, Bernadette D. Poctor, and Cheryl Hill Lee, US. Census Bureau, Current Population Reports, P60-229, *Income, Poverty, and Health Insurance Coverage in the United States: 2004*, U.S. Government Printing Office, Washington, DC, 2005.

unskilled manufacturing jobs and changes in the mining, agricultural, and oil industries. The problem is especially acute for those with little education or marketable job skills.

The Feminization of Poverty

Different types of families also have different earning potentials. In 2004, a family with both a husband and wife present had a median income of $80,410. For a male householder without a wife the figure was $51,745, and for a female householder without a husband it was $35,399. This has caused some sociologists to note the "feminization of poverty," a phrase referring to the disproportionate concentration of poverty among female-headed families.

The real impact of these differences becomes even more striking when we look at single women with children. Whereas 12.7% of all people were below the poverty line in 2004, 28.4% of all single women with children were living in poverty. The percentages are even higher if we limit the pool to just single African American (43.4%) and Hispanic (45.9%) women with children. If present trends continue, 60% of all children born today will spend part of their childhood in a family headed by a mother who is divorced, separated, unwed, or widowed. There is substantial evidence that women in such families are often the victims of poverty. Almost half of all female-headed families with children younger than 18 live below the poverty line (Bureau of the Census, 2005).

© Wally McNamee/CORBIS

Poverty statistics may not adequately count the homeless, who have no permanent address.

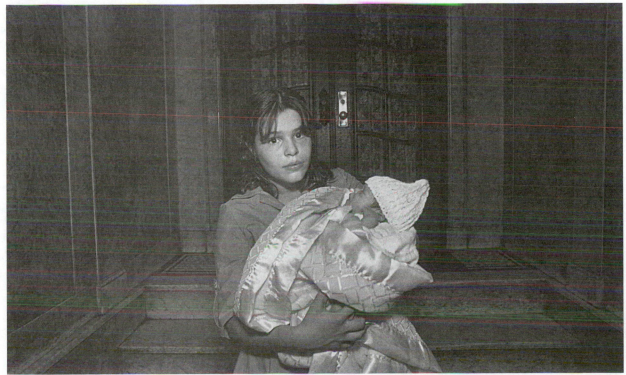

© John Lei/Stock, Boston

Out-of-wedlock births to young women have contributed to the feminization of poverty.

Not all female-headed families are the same, however. The feminization of poverty is both not as bad as, and much worse than, the previous statement suggests. Families headed by divorced mothers are doing better than the 50% figure, whereas families headed by never-married mothers are doing much worse.

What accounts for the fact that never-married mothers are so much poorer than their divorced counterparts? Seventy percent of all out-of-wedlock births occur to young women between the ages of 15 and 24. They are, on average, 10 years younger than divorced mothers. Never-married mothers are also, on average, much less educated. The gender gap in poverty rates is greatest at low educational levels. It is much narrower among those with a high school diploma and practically nonexistent among those with a college education. Single mothers without a high school diploma often have difficulty finding a job that pays enough to cover child-care costs, leading to a dependence on welfare programs (O'Hare, 1996).

How Do We Count the Poor?

To put a dollar amount on what constitutes poverty, the federal government has devised a poverty index of specific income levels, below which people are considered to be living in poverty. Many people use this index to determine how many poor people live in the United States. According to the index, the poverty level for a family of four in 2004 was $19,484 (Table 8–2). The poverty index is based solely on money income and does not reflect the fact that many low-income people receive noncash benefits such as food stamps, Medicaid, and public housing.

The official definition of poverty used for the index was developed by the Social Security Administration in 1964. It was calculated first by estimating the national average dollar cost of a frugal but adequate diet; however, because a 1955 Department of Agriculture study on food consumption found that families of three or more people spent about one-third of their income on food, food costs were multiplied by three to estimate how much total cash income was needed to cover food and other necessities. The poverty index was not originally intended to certify that any individual or family was in need. In fact, the government specifically has warned against using the index for administrative use in any specific program. Despite this warning, people continue to use, or misuse, the poverty index and variations of it for a variety of purposes for which it was not intended. For example, those wanting to show that current government programs are inadequate for the poor will try to inflate the numbers of those living in poverty. Those trying to show that government policies are adequate for meeting the needs of the poor will try to show that the number of poor people is decreasing.

Table 8–2

Income Levels below Which Individuals and Families Are Considered to Be Living in Poverty (2004)

Size of Unit	Income
1 person	$ 9,827
2 people	$12,649
3 people	$14,776
4 people	$19,484
5 people	$23,497
6 people	$27,025
7 people	$31,096
8 people	$34,778
9 people	$31,836
10 people	$36,580

Source: U.S. Bureau of the Census, *Current Population Survey,* January 2005.

Those who think the poverty index overestimates the poor offer three major criticisms. First, when the federal government developed the poverty index in 1964, about one-quarter of federal welfare benefits were in the form of goods and services. Today, noncash benefits account for about two-thirds of welfare assistance. For example, about 23.9 million people received food stamps in 2004, which is not considered income under existing poverty-index rules. Complicating the issue further, the market value of in-kind benefits—such as housing subsidies, school lunch programs, and health care services, among others—has been multiplied by a factor of 40. Some suggest that if the noncash benefits were counted as income, the poverty rate would be 3 percentage points lower.

Second, the poverty measure looks only at income, not assets. If the value of a home or other assets were included, the poverty rate would also be lower.

Third, food typically accounts for a considerably smaller proportion of family expenses today than it did previously. If we were to try to develop a poverty index today, we would probably have to multiply minimal food costs by a factor of five instead of three.

Those who think the poverty figures underestimate the poor have their criticisms also. First, they point out that money used to pay taxes, alimony, child support, health care, or work-related expenses should be excluded when considering assets because these sums cannot be used to buy food or other necessities.

Second, there is no geographic cost-of-living adjustment. The federal government uses the same poverty-level figures for every part of the country. That means the poverty threshold is the same in rural Mississippi as it is in New York City.

Third, many believe that the poverty threshold is unrealistically low. Rather than use an absolute number, poverty status should be determined by

Figure 8–3   **Number in Poverty and Poverty Rates, 1959–2004**

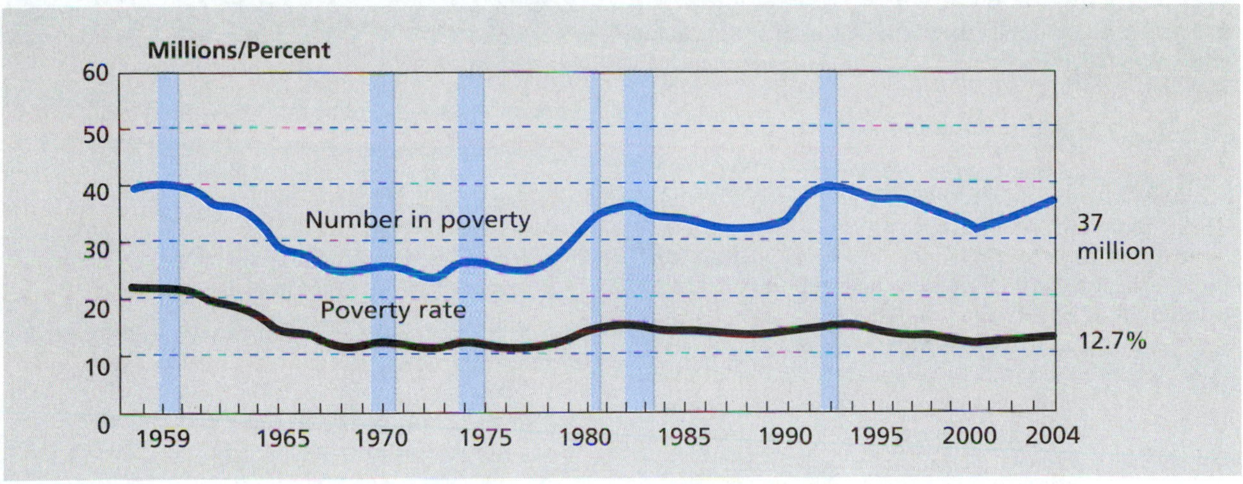

Source: U.S. Census Bureau, Current Population Survey, 1960 to 2005 Annual Social and Economic Supplements.

comparing a person's financial situation with that of the rest of society.

The poverty index has become less and less meaningful. However, its continued existence over all these years has given it somewhat of a sacred character. Few people who cite it know how it is calculated, and they choose to assume it is a fair measure for determining the number of poor in the country. The poverty index has never been a sufficiently precise indicator of need to make it an indisputable test of which individuals and families were poor and which were not.

The number of people living in poverty also is distorted by the fact that the Census Bureau's *Current Population Survey* is derived from households. It excludes all those people who do not live in traditional housing, specifically, the growing numbers of the homeless, estimated at anywhere from 350,000 to 1 million. People in nursing homes and other types of institutions also are not included in the poverty figures because of surveying techniques.

This is not to downplay the number of poor people in the United States. The basic fact is that trying to determine how many poor people there are depends on whom you ask and what type of statistical maneuvering is involved (Figure 8–3).

Myths about the Poor

We are presented with differing views on poverty and what should be done about it. One side argues that more government aid and the creation of jobs are needed to combat changes in the employment needs of the national economy. The other side contends that government assistance programs launched with the War on Poverty in the mid-1960s have encouraged many of the poor to remain poor and should be eliminated for the able-bodied poor of working age (Murray, 1994).

Our perceptions of the poor shape our views of the various government programs available to help them. It is important that we have a clear understanding of who the poor are to direct public policy intelligently. Many Americans believe a number of common myths about the poor. Let us try to clear some of them up.

Myth 1: People Are Poor Because They Are Too Lazy to Work   Half of the poor are not of working age. About 40% are younger than 18; another 10% are older than 65. Most of the able-bodied poor of working age are working or looking for work.

Many of the poor adults who do not work have good reasons for not working; they may be ill or disabled, and many others are going to school (mostly those in their late teens from poor families). Many of the poor work and many work year round.

However, a person working 40 hours a week, every week of the year at minimum wage, will not earn enough to lift a family of three out of poverty. The numbers of the working poor are increasing. There are several reasons for this growth. First, although there are more jobs in the economy than ever before, many of these jobs are in low-paying service industries. A janitor or a cook at a fast-food restaurant earns no more than minimum wage. Second, the better jobs the poor used to hold are no longer part of the U.S. economy. Many companies, seeking sources of cheap labor, have set up manufacturing operations overseas to increase their ability to compete in the world market. Finally, many of the working poor are women or young people with few marketable skills. Often, they are forced to settle for poorly paid, part-time work.

In many ways, the working poor are in worse straits than those on the welfare rolls. For example, a mother on welfare may be eligible for public

NEWS YOU CAN USE

Are Urban Poverty Ghettos Shrinking?

Concentrated poverty ghettos—neighborhoods where at least 40% of the people live in poverty—have been declining. In 1999, about 2.8% of the U.S. population lived in such neighborhoods, down from 4.6% in 1989.

Are we as a society finally solving the poverty dilemma? To answer this question, we need to examine why poverty ghettos developed in the first place. It would be good news because poverty neighborhoods also display high levels of crime, unemployment, female-headed households, welfare dependency, drug use, substandard education, and out-of-wedlock births. Reducing the number of people who live in these environments could also help reduce a great deal of misery.

Looking at poverty historically we find that before the twentieth century the poor usually lived near the rich. Class segregation only became common with improvements in transportation and the appearance of the automobile. After World War II, suburbanization surged dramatically and the poor were left behind in urban poverty ghettos.

Other factors that increased poverty ghettos include:

Federal housing policies—low-income housing projects were built in already poor inner-city neighborhoods.
Federal assistance to highway construction—helped the middle and upper classes move to the suburbs.
Loss of low-skill manufacturing jobs in central cities—produced high unemployment.

Redlining neighborhoods—the tendency for banks to deny loans on the basis of race.
Welfare policies—some policies contributed to less self-reliance and destructive behaviors.

Progress has been made in reversing these factors and recent Census Bureau data show that the number of people living in high-poverty neighborhoods declined by 25% in the 1990s, with the steepest declines among African Americans. Between 1993 and 2000, the poverty rate among African Americans fell from 33.1% to an all-time low of 22.5%.

Poverty ghettos have also been reduced because increasing numbers of minorities have moved to the suburbs. For example, racial and ethnic minorities made up more than one-quarter (27%) of the suburban population in the 100 largest metropolitan areas in 2000, up from 19% in 1990.

As suburbs become more diverse, it is still possible that some suburban areas may gain poverty concentrations as more low-income people and immigrants move to the suburbs rather than central cities.

Source: Iceland, John, "Why Concentrated Poverty Fell in the United States in the 1990s," Population Reference Bureau, August 2005, http://www.prb.org/Template.cfm?Section=PRB&template=/ContentManagement/ContentDisplay.cfm&ContentID=12769, accessed September 12, 2005.

---

housing and a variety of services that a working poor two-parent family may not be able to receive.

It is easy for the government to ignore the plight of the working poor. Scattered throughout the country and with no collective voice to express protest, they are relatively invisible and, therefore, easily forgotten.

Myth 2: Most Poor People Are Minorities, and Most Minorities Are Poor Neither of these statements is true. Most poor people are white, merely because many more whites than minorities live in the United States. The poverty rate, however, remains considerably higher for African Americans and Hispanics than whites (24.7% and 21.9%, respectively, versus 8.6% in 2004) (Bureau of the Census, *Current Population Survey, March 2005*).

One of the reasons that African Americans are associated with the image of poverty is that they make up more than half of the long-term poor. Another reason is that the War on Poverty was motivated in

part, and occurred simultaneously with, the civil rights movement of the 1960s.

Myth 3: Most of the Poor are Single Mothers with Children It is true that a disproportionate share of poor households are headed by women and that the poverty rate for female-headed families is extremely high. For example, about 60% of mothers receiving assistance have never been married. The majority of people in poverty, however, live in other family arrangements. About one-third of the poor live in married-couple families, and nearly one-fourth live alone or with nonrelatives. The remainder lives in a male-headed or other family setting (O'Hare, 1996; Bureau of the Census, *Current Population Survey*, March 2000).

Myth 4: Most People in Poverty Live in the Inner Cities Historically, poverty has been more prevalent in rural areas than in urban areas. In 2002 the poverty rate in rural areas was 14.2% compared with 11.6% in urban areas. Rural residents have

SOCIAL CHANGE

What Causes Poverty?

Americans are divided over what they believe causes poverty. About half the public believes that circumstances beyond their control cause the poor to be poor, whereas the other half notes the poor are not doing enough to help themselves out of poverty.

According to a National Public Radio, Kaiser Family Foundation, and Harvard University Kennedy School of Government School Poll on Poverty in America, "low-income people—that is, those making . . . less than $34,000 per year for a family of four—are only slightly more likely than other Americans to feel it is due to [external] circumstances." The study noted that when low-income people are asked about the specific causes of poverty they "are significantly more likely than other Americans to name drug abuse, medical bills, too few jobs (or too many being part time or low wage), too many single-parent families, and too many immigrants." Black and white Americans differ in their views of how big a problem poverty is and what causes it. Forty-nine percent of whites believe the poor are not doing enough to help themselves out of poverty. Only 36% of blacks agree with this view. Fifty-six percent of whites believe the government cannot eliminate poverty. Only 31% of blacks believe likewise. When asked what is the No. 1 cause of poverty, low-income people are much more likely to name drug abuse, and the poorest Americans—those living below the federal poverty level—are nearly twice as likely as middle- and upper-income people to rank drug abuse so high. Middle-income people are more likely to believe that the No. 1 cause of poverty is poor-quality education, but, as noted in Table 8–3, both groups are equally likely to name poor schools as a major cause of poverty.

Table 8–3

Percentage of Middle-Income and Low-Income People Who Thought a Particular Factor Was a Major Cause of Poverty

Poverty-Causing Factor	Middle-Income	Low-Income
Drug abuse	68	75
Medical bills	54	69
Too many jobs being part-time or low-wage	50	64
Too many single-parent families	52	61
A shortage of jobs	27	52
Too many immigrants	27	39
Poor people lacking motivation	51	56
Poor-quality public schools	47	46

Source: "Poverty in America," a poll conducted by National Public Radio, Kaiser Family Foundation, and Harvard University Kennedy School of Government; available at http://www.npr.org/programs/specials/poll/poverty/summary.html (accessed June 27, 2002). Used by permission.

higher unemployment rates and earn lower wages than urban residents. Rural residents also tend to have below-average educational levels and limited job skills (U.S. Dept. of Agriculture, 2004).

Much of rural poverty is invisible because it occurs in isolated pockets. Poverty rates are exceptionally high in rural counties in Appalachia, the Mississippi Delta, and American Indian reservations. Except for rural Appalachia, which is predominantly white, most rural pockets of poverty are disproportionately composed of African Americans, Hispanic Americans, and Native Americans. (See "News You Can Use: Are Urban Poverty Ghettos Shrinking?")

Myth 5: Welfare Programs for the Poor Are Straining the Federal Budget Since the passage of welfare reform in 1996, the number of families receiving aid has decreased by about 50%, from 3 million to 1.5 million (Lichter & Crowley, 2002). Social assistance programs for low-income people cost the federal government only about one-third as much as other types of social assistance, such as Social Security and Medicare, which mainly go to middle-class Americans, not to the poor (O'Hare, 1996). (For a discussion of what the public sees as the root of poverty, see "Social Change: What Causes Poverty?")

Government Assistance Programs

The public appears to be quite frustrated and upset about the costs of poverty. Much of this frustration, however, stems from a misperception of what programs are behind the escalating government expenditures, a misunderstanding about who is receiving government assistance, and an exaggerated notion of the amount of assistance going to the typical person in poverty. Most government benefits go to the middle class. Many of the people reading this book will be surprised to know that they or their families actually may receive more benefits than those people typically defined as poor. The value of benefits going to the poor actually has fallen in recent years, whereas that going to the middle class actually has risen.

Government programs that provide benefits to families or individuals can be divided into two categories: (1) social insurance and cash benefits going to people of all income levels and (2) means-tested programs and cash assistance going only to the poor.

Social insurance benefits are not means tested, meaning that you do not have to be poor to receive them. They go primarily to the middle class. Many people receiving payments from social insurance programs, such as Social Security retirement and unemployment insurance, feel they are simply getting back the money they put into these programs. They accuse those receiving benefits from means-tested programs of getting something for nothing. This is not exactly true when we recognize that many social insurance recipients receive back far more than they put in, and that the poor, the majority of whom work, pay taxes that contribute to their own means-tested benefits.

Social insurance programs account for the overwhelming majority of federal cash assistance expenditures, and their share has been rising rapidly. Female-headed families in poverty, often portrayed as a heavy drain on the government treasury, account for only 2% of the federal outlays for human resources. In contrast, Social Security and Medicare for the retired elderly, the vast majority of whom are middle class, accounts for 35%.

The Changing Face of Poverty

It appears that economic rewards are distributed more unequally in the United States than elsewhere in the Western industrialized world. In addition, the United States experiences more poverty than other capitalist countries with similar standards of living.

In one international study, the poverty rates for children, working-age adults, and the elderly were tabulated for a variety of countries. The results showed that the United States has been successful in holding down poverty among the elderly. The American elderly experience far less poverty than the elderly in Great Britain, approximately the same as the elderly in Norway and Germany, and far more poverty than the elderly in Canada and Sweden.

The United States has been much less successful in keeping children and working-age adults out of poverty. The U.S. child poverty rate is higher than the rate in Great Britain, and more than double the rate in Norway, Denmark, and France. (See "Global Sociology: Rich Countries with Poor Children.")

How has it happened that the United States has made progress in combating poverty among the elderly but not among other groups? Since 1960, a variety of social policies have been enacted that have improved the standard of living of the elderly relative to that of the younger population. Social Security benefits were increased significantly and protected against the threat of future inflation; Medicare provided the elderly with national health insurance; Supplemental Security Income provided a guaranteed minimum income; special tax benefits for the elderly protected their assets during the later years; and the Older Americans Act supported an array of services specifically for this age group. As a consequence of these measures, poverty among the elderly has declined substantially. Whereas 24.6% of those families 65 and older lived below the poverty level in 1970, only 9.8% did so by 2004 (Bureau of the Census, 2005).

To achieve this dramatic improvement in the conditions of the elderly, it has been necessary to increase greatly the federal money spent on this age group. If these arrangements are maintained, projections show that about 60% of the federal budget will be going to the elderly by the year 2030. A group that has suffered particularly under this shift in expenditures to the elderly is the young. Whereas 14% of children lived in poverty in 1970, 17.8% did so in 2004 (Bureau of the Census, 2005) (Figure 8–4).

It would also surprise many people to learn that not only are the elderly as a group not poor, but that they are actually better off than most Americans. They are more likely than any age group to possess money market accounts, certificates of deposit, U.S. government securities, and municipal and corporate bonds. The median household net worth of those aged 65 to 69 is the highest of any age group, followed by those 70 to 74 years old. They have the highest rate of home ownership of any age group. Seventy-seven percent of those age 65 to 74 own a home, and most of these homes are paid for in full.

Figure 8-4 Poverty Rates for People over 65 and under 18, 1960–2004

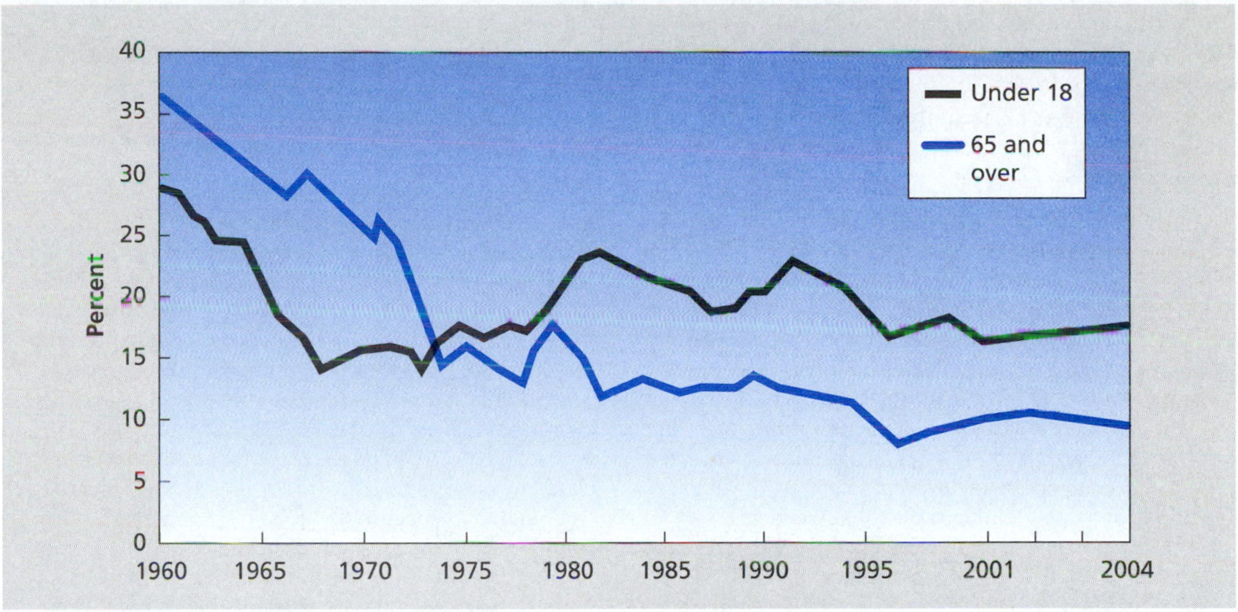

Source: U.S. Bureau of Census, Housing and Household Economic Statistics Division, "Poverty Status of People, By Age, Race, and Hispanic Origin: 1959 to 2004," August 30, 2005, www.census.gov/hhes/www/poverty/histpov/hstpov3.html, accessed August 31, 2005.

Consequences of Social Stratification

Studies of stratification in the United States have shown that social class affects many factors in a person's life. Striking differences in health and life expectancy are apparent among the social classes, especially between the lower-class poor and the other social groups. As might be expected, lower-class people are sick more frequently than are others.

Women living in poverty are more likely to have babies with low birth weights, putting them at higher risk for various cognitive and physical problems. Poor adults are four times as likely to regard themselves in "fair" or "poor" health compared with wealthier adults.

The poor of all races and ethnicities also experience lower life expectancy. The poor are more likely to develop illnesses that shorten the life span, such as heart disease, lung cancer, diabetes, and other degenerative diseases. Poor people are also more likely to die violent deaths, with homicide and suicide rates being substantially higher than among people with higher incomes (Lichter & Crowley, 2002).

Poverty particularly affects the health of the young. Babies born into poverty are significantly more likely to die before their first birthday than those born into families living above the poverty level. One study (Kiely, 1995) found that the infant mortality rate for poor children was more than 50% higher than that for children not born in poverty.

Babies born to African American girls between ages 15 and 19 are more than twice as likely to die as those born to white teenagers. In addition, an African American mother is more than three times as likely to die giving birth than a white mother (Bureau of the Census, 2000). Diet and living conditions also give a distinct advantage to the upper classes because they have access to better and more sanitary housing and can afford more balanced and nutritious food. A direct consequence of this situation is seen in the life-expectancy pattern for each social class. Not surprisingly, lower-class people do not live as long as do those in the upper classes. White males born in 2002 have a life expectancy six years longer than that for African American males, many of whom are concentrated in the lower income brackets (National Center for Health Statistics, 2004).

Family, childbearing, and childrearing patterns also vary according to social class. Women in the higher-income groups, who have more education, tend to have fewer children than do lower-class women with less schooling. Women more often head the family in the lower class, compared with women in the other groups. Middle-class women discipline their children differently than do working-class mothers. The former punish boys and girls alike for the same infraction, whereas the latter often have different standards for sons and daughters. Also,

GLOBAL SOCIOLOGY

Rich Countries with Poor Children

We know that children have a miserable existence in third-world countries, but wealthy countries have trouble keeping children out of poverty also. The United States has been very successful at lifting the elderly out of poverty. It has not been as successful with children. In recent years the U.S. child poverty rate has fluctuated between a high of 22.7% in 1993 to a low of 17.8% in 2004. (Bureau of the Census, 2005). Not only has child poverty remained stubbornly high, but when compared with other countries in the world, the United States appears to do less to improve the living conditions of its poor children than many countries.

Even though 27.7% of children in France start out in poverty, the government's assistance expenditures reduce that number to 7.5% with a variety of programs. In the United States, 26.6% of all children start out in poverty, but the government's programs reduce that number to only 17.8% (Figure 8–5) (UNICEF, 2005).

There are a number of reasons why the United States has such high rates of poverty among children. First, many children are born to unmarried women. Nine percent of children in married-couple families live in poverty, but it rises to 42% of those in female-headed households. Second, American mothers with limited education or skills are less likely than European mothers to return to work quickly after childbirth, because high-quality child care is comparatively expensive and the jobs will not cover the costs. Third, the United States has more poor immigrants than any other country. Many of the children in poverty are born to poor immigrants who have been in the country for only a few years. A fourth factor is an outgrowth of the realities of the American political system, which depends on advocates supporting programs for special constituencies. Children obviously do not vote, and the poor in general have low voting records. The elderly, on the other hand, have high voting records and therefore have substantial political clout. Politicians are usually not voted out of office for cutting benefits to children, whereas they are if they do so for the elderly.

Some have suggested (Bradsher, 1995) that other countries have avoided high levels of child poverty by limiting economic growth and lowering the living standard for everyone. Many European countries with generous social assistance programs also have high rates of unemployment and living standards that do not match those in the United States. These critics charge that instead of bringing everyone up, other countries have brought everyone down. They claim that the high rate of American child poverty appears greater because the gap between the rich and the poor is so much larger in the United States and other countries just have gone further in redistributing income.

Despite the arguments over the data, the fact remains that substantial numbers of poor children live in the United States. Great harm is being done to these children when their poor living conditions are not addressed.

Figure 8–5   Child Poverty Rates in Rich Countries

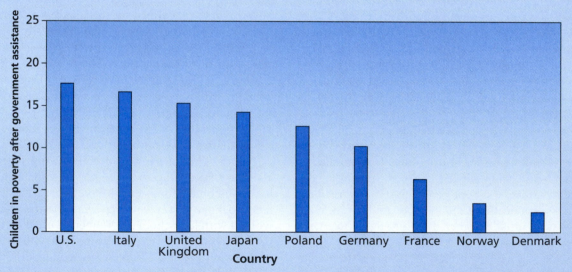

Source: UNICEF, "Child Poverty in Rich Countries, 2005," *Innocenti Report Card* No. 6. UNICEF Innocenti Research Centre, Florence, Italy, 2005; U.S. Bureau of Census, Housing and Household Economic Statistics Division "Poverty Status of People, By Age, Race, and Hispanic Origin: 1959 to 2004, August 30, 2005, www.census.gov/hhes/www/poverty/histpov/hstpov3.html, accessed August 31, 2005.

middle-class mothers judge the misbehaving child's intention, whereas working-class women are more concerned with the effects of the child's action.

There is a direct relationship between a person's social class and the possibility of his or her arrest, conviction, and sentencing if accused of a crime. For the same criminal behavior, the poor are more likely to be arrested; if arrested, they are more likely to be charged; if charged, they are more likely to be convicted; if convicted, they are more likely to be sentenced to prison; and if sentenced, they are more likely to be given longer prison terms than members of the middle and upper classes (Reiman, 1990).

The poor are singled out for harsher treatment at the very beginning of the criminal justice system. Although many surveys show that almost all people admit to having committed a crime for which they could be imprisoned, the police are prone to arrest a poor person and release, with no formal charges, a higher-class person for the same offense. A well-to-do teenager who has been accused of a criminal offense frequently is just held by the police at the station house until the youngster can be released to the custody of the parents; poorer teenagers who have committed the same kind of crime more often are automatically charged and referred to juvenile court.

The poor tend to commit violent crimes and crimes against property—they have little opportunity to commit such white-collar crimes as embezzlement, fraud, or large-scale tax evasion—and they are much more severely punished for their crimes than upper-class criminals are for theirs. Yet white-collar crimes are far more damaging and costly to the public than are the crimes more often committed by poor people. The government has estimated that white-collar crimes cost more than $40 billion a year—more than 10 times the total amount of all reported thefts and more than 250 times the amount taken in all bank robberies.

Even the language used to describe the same crime committed by an upper-class criminal and a poor one reflects the disparity in the treatment they receive. The poor thief who takes $2,000 is accused of stealing and usually receives a stiff prison sentence. The corporate executive who embezzles $200,000 merely has "misappropriated" the funds and is given a lighter sentence or none at all, on the promise to make restitution. A corporation often can avoid criminal prosecution by signing a consent decree, which is in essence a statement that it has done nothing wrong and promises never to do it again. Were this ploy available to ordinary burglars, the police would have no need to arrest them; a burglar would merely need to sign a statement promising never to burgle again and file it with the court.

Once charged, the poor are usually dependent on court-appointed lawyers or public defenders to handle their cases. The better-off rely on private lawyers who have more time, resources, and personal interest in defending their cases.

If convicted of the same kind of crime as a well-to-do offender, the poor criminal is more likely to be sentenced and will generally receive a longer prison term. As for prison terms, the sentence for burglary, a crime of the poor, is generally more than twice as long as that for fraud, whereas a robber will draw an average sentence more than six times longer than that of an embezzler. The result is a prison system heavily populated by the poor.

Another serious consequence of social stratification is mental illness. Studies have shown that at least one-third of all homeless people suffer from schizophrenia, manic-depressive psychosis, or other mental disorders. Such people are the least likely to reach out for help and the most likely to remain on the streets in utter poverty and despair year after year (Torrey, 1988; Jencks, 1994).

Thus, social class has very real and immediate consequences for individuals. In fact, class membership affects the quality of people's lives more than any other single variable.

Why Does Social Inequality Exist?

Sociologists and social philosophers before them have long tried to explain the presence of social inequality—that situation in which the very wealthy and powerful coexist with the poverty-stricken and powerless. Several theories have been put forth to explain this phenomenon.

The Functionalist Theory

Functionalism is based on the assumption that the major social structures contribute to the maintenance of the social system (see Chapter 1). The existence of a specific pattern in society is explained in terms of the benefits that society receives because of that situation. In this sense the function of the family is the socialization of the young, and the function of marriage is to provide a stable family structure.

The functionalist theory of stratification as presented by Kingsley Davis and Wilbert Moore (1945) holds that social stratification is a social necessity. Every society must select individual members to fill a wide variety of social positions (or statuses) and then motivate those people to do what is expected of them in these positions—that is, fulfill their role expectations. For example, our society needs teachers, engineers, janitors, police officers, managers, farmers, crop dusters, assembly-line workers, firefighters, textbook writers, construction workers, sanitation workers, chemists, inventors, artists, bank tellers, athletes,

A direct relationship exists between a person's social class and the possibility of his or her arrest, conviction, and sentencing if accused of a crime.

pilots, secretaries, and so on. To attract the most talented individuals to each occupation, society must set up a system of differential rewards based on the skills needed for each position.

> If the duties associated with the various positions were all equally pleasant . . . all equally important to social survival, and all equally in need of the same ability or talent, it would make no difference who got into which positions. . . . But actually it does make a great deal of difference who gets into which positions, not only because [some] positions are inherently more agreeable than others, but also because some require special talents or training and some are functionally more important than others. Also, it is essential that the duties of the positions be performed with the diligence that their importance requires. Inevitably, then, a society must have, first, some kind of rewards that it can use as inducements, and, second, some way of distributing these rewards differentially according to positions. The rewards and their distribution become part of the social order, and thus give rise to stratification. (Davis & Moore, 1945)

According to Davis and Moore, (1) different positions in society make different levels of contributions to the well-being and preservation of society, (2) filling the more complex and important positions in society often requires talent that is scarce and has a long period of training, and (3) providing unequal rewards ensures that the most talented and best trained individuals will fill the roles of greatest importance. In effect, Davis and Moore believe that those people who are rich and powerful are at the top because they are the best qualified and are making the most significant contributions to the preservation of society (Zeitlin, 1981).

Many scholars, however, disagree with Davis and Moore, and their arguments generally take two forms. The first is philosophical and questions the morality of stratification. The second is scientific and questions its functional usefulness. Both criticisms share the belief that social stratification does more harm than good, that it is dysfunctional.

The Immorality of Social Stratification On what grounds, one might ask, is it morally justifiable to give widely different rewards to different occupations,

© Gloria Karlson

The mentally ill among the homeless are the least likely to reach out for help and the most likely to remain on the streets.

© Neal Preston/CORBIS

Neither functionalist theory nor conflict theory can fully explain why media people earn very large sums of money.

when all occupations contribute to society's ongoing functioning? How can we decide which occupations contribute more? After all, without assembly-line workers, mail carriers, janitors, auto mechanics, nurse's aides, construction laborers, truck drivers, secretaries, shelf stockers, sanitation workers, and so on, our society would grind to a halt. How can the multimillion-dollar-a-year incomes of a select few be justified when the earnings of 12.7% of the American population fall below the poverty level, and many others have trouble making ends meet (Bureau of the Census, 2005)? Why are the enormous resources of our society not more evenly distributed?

Many people find the moral arguments against social stratification convincing enough. However, there are other grounds on which stratification has been attacked—namely, that it is destructive for individuals and society as a whole.

The Neglect of Talent and Merit Regardless of whether social stratification is morally right or wrong, many critics contend that it undermines the very functions that its defenders claim it promotes. A society divided into social classes (with limited mobility among them) is deprived of the potential contributions of many talented individuals born into the lower classes. From this point of view, it is not necessary to do away with differences in rewards for different occupations. Rather, it is crucial to put aside all the obstacles to achievement that currently handicap the children of the poor.

Barriers to Free Competition It can also be claimed that access to important positions in society is not really open. That is, those members of society who occupy privileged positions allow only a small number of people to enter their circle. Thereby, shortages are created artificially. This, in turn, increases the perceived worth of those who are in the important positions. For example, the American Medical Association (AMA) is a wealthy and powerful group that exercises great control over the quality and quantity of physicians available to the American public. Historically, the AMA has directly influenced the number of medical schools in the United States and thereby the number of doctors produced each year, effectively creating a scarcity of physicians. A direct result of this influence is that medical care costs and physicians' salaries have increased more rapidly than has the pace of inflation.

This situation is beginning to change, however. The allure of high earnings that attracts many new doctors, along with changing demographic characteristics, could mean a surplus of physicians in the future. In addition, as more and more doctors fight for the same patient dollars, and as health maintenance organizations (HMOs) try to control costs, earnings may begin to suffer. Thus, although barriers to free competition exist in our society, often the marketplace overrules them in the end.

Functionally Important Jobs When we examine the functional importance of various jobs, we become aware that the rewards attached to jobs do not necessarily reflect the essential nature of the functions. Why should a Hollywood movie star receive an enormous salary for starring in a film and a child-protection worker receive barely a living wage? It is difficult to prove empirically which positions are most important to society or what rewards are necessary to persuade people to want to fill certain positions.

Conflict Theory

As we saw, the functionalist theory of stratification assumes society is a relatively stable system of interdependent parts in which conflict and change are abnormal. Functionalists maintain that stratification is necessary for the smooth functioning of society. Conflict theorists, in contrast, see stratification as the outcome of a struggle for dominance.

Current views of the conflict theory of stratification are based on the writings of Karl Marx. Later, Max Weber developed many of his ideas in response to Marx's writings.

Karl Marx   To understand human societies, Karl Marx believed one must look at the economic conditions centering around producing the necessities of life. Stratification emerged from the power struggles for scarce resources.

> The history of all hitherto existing society is the history of class struggles. [There always has been conflict between] freeman and slave, patrician and plebeian, lord and serf, guild-master and journeyman, in a word, oppressor and oppressed. (Marx & Engels, 1961)

Those groups who own or control the means of production within a society obtain the power to shape or maintain aspects of society that favor their interests. They are determined to maintain their advantage. They do this by setting up political structures and value systems that support their position. In this way the legal system, the schools, and the churches are shaped in ways that benefit the ruling class. As Marx and his collaborator Friedrich Engels put it, "The ruling ideas of each age have always been the ideas of its ruling class" (1961). Thus, the pharaohs of ancient Egypt ruled because they claimed to be gods. In the first third of the twentieth century, America's capitalist class justified its position by misusing Charles Darwin's theory of evolution: The capitalists adhered to the view—called social Darwinism (see Chapter 1)—that those who rule do so because they are the most "fit" to rule, having won the evolutionary struggles that promote the "survival of the fittest."

Marx was most interested in the social impact of the capitalist society that was based on industrial production. In a capitalist society, there are two great classes: the bourgeoisie, or the owners of the means of production or capital, and the proletariat, or the working class. Those in the working class have no resources other than their labor, which they sell to the capitalists. In all class societies, one class exploits another.

Marx believed that the moving force of history was class struggle, or class conflict. This conflict grows out of differing class interests. As capitalism develops, two conflicting trends emerge. On one hand, the capitalists try to maintain and strengthen their position. The exploitative nature of capitalism is seen when the capitalists pay the workers a bare minimum wage, below the value of what the workers actually produce. The remainder is taken by the capitalists as profit and adds to their capital. This capital, which Marx said rightfully belongs to the workers, is then used to build more factories, machines, or anything else to produce more goods. As Marx saw it, "Capital is dead labor, that vampirelike, only lives by sucking living labor, and lives the more, the more labor it sucks" (Marx, 1906).

Eventually, in the face of continuing exploitation, the working classes find it in their interest to overthrow the dominant class and establish a social order more favorable to their interests. Marx believed that with the proletariat in power, class conflict would finally end. The proletariat would have no class below it to exploit. The final stage of advanced communism would include an industrial society of plenty, where all could live in comfort.

Marx was basically a materialist. He believed that people's lives are centered on how they deal with the material world. The key issue is how wealth is distributed among the people. Wealth can be distributed in at least four ways:

1. *To each according to need.* In this kind of system, the basic economic needs of all the people are satisfied. These needs include food, housing, medical care, and education. Extravagant material possessions are not basic needs and have no place in this system.
2. *To each according to want.* Here, wealth will be distributed according to what people desire and request. Material possessions beyond the basic needs are included.
3. *To each according to what is earned.* People who live according to this system become the source of their own wealth. If they earn a great deal of money, they can lavish extravagant possessions on themselves. If they earn little, they must do without.
4. *To each according to what can be obtained—by whatever means.* Under this system, everyone ruthlessly attempts to acquire as much wealth as possible without regard for the hardships that might be brought on others because of these actions. Those who are best at exploiting others become wealthy and powerful, and the others become the exploited and poor (Cuzzort & King, 1980).

In Marxist terms, the first of these four possibilities is what would happen in a socialist society. Although many readers will believe that the third possibility describes U.S. society (according to what is earned), Marxists would say that a capitalist society is characterized by the last choice—the capitalists obtain whatever they can get in any possible way.

Max Weber Weber expanded Marx's ideas about class into a multidimensional view of stratification. Weber agreed with Marx on many issues related to stratification, including the following:

1. Group conflict is a basic ingredient of society.
2. People are motivated by self-interest.
3. Those who do not have property can defend their interests less well than those who have property.
4. Economic institutions are of fundamental importance in shaping the rest of society.
5. Those in power promote ideas and values that help them maintain their dominance.
6. Only when exploitation becomes extremely obvious will the powerless object. (Vanfossen, 1979)

From those areas of agreement, Weber went on to add to and modify many of Marx's basic premises. Weber's view of stratification went beyond the material or economic perspective of Marx. He included status and power as important aspects of stratification as well.

Weber was not interested in society as a whole but in the groups formed by self-interested individuals who compete with one another for power and privilege. He rejected the notion that conflict between the bourgeoisie and proletariat was the only, or even the most important, conflict relationship in society.

Weber believed that there were three sources of stratification: economic class, social status, and political power. Economic classes arise out of the unequal distribution of economic power, a point on which Marx and Weber agreed. Weber went further, however, maintaining that social status is based on prestige or esteem—that is, status groups are shaped by lifestyle, which is in turn affected by income, value system, and education. People recognize others who share a similar lifestyle and develop social bonds with those who are most like them. From this inclination comes an attitude of exclusivity: Others are defined as being not as good as those who are a part of the status group. Weber recognized that there is a relationship between economic stratification and social-status stratification. Typically, those who have a high social status also have great economic power.

Inequality in political power exists when groups are able to influence events in their favor. For example, representatives of large industries lobby at the state and federal levels of government for legislation favorable to their interests and against laws that are unfavorable. Thus, the petroleum industry pushed for lifting restrictions on gasoline prices and the auto industry lobbied for quotas on imported cars. In exchange for "correct" votes, a politician is often promised substantial campaign contributions from wealthy corporate leaders or endorsement and funding by a large labor union whose members' jobs will be affected by the government's decisions. The individual consumer who will pay the price for such political arrangements is powerless to exert any influence over these decisions.

Class, status, and power, though related, are not the same. One can exist without the others. To Weber, they are not always connected in some predictable fashion, nor are they always tied to the economic mode of production. An "aristocratic" southern family may be in a state that is often labeled genteel poverty, but the family name still elicits respect in the community. This kind of status sometimes is denied to the rich, powerful labor leader whose family connections and school ties are not acceptable to the social elite. In addition, status and power are often accorded to those who have no relationship to the mode of production. For example, Nobel Peace Prize winner Mother Teresa, known for her work with the poor in India, controlled no industry, nor did she have any great personal wealth; yet her influence was felt by heads of state the world over.

Whereas Marx was somewhat of an optimist in that he believed that conflict, inequality, and exploitation eventually could be eliminated in future societies, Weber was much more pessimistic about the potential for a more just and humane society.

Modern Conflict Theory

Conflict theorists assume that people act in their own self-interest in a material world in which exploitation and power struggles are prevalent. Modern conflict theory has five aspects:

1. Social inequality emerges through the domination of one or more groups by other groups. Stratification is the outgrowth of a struggle for dominance in which people compete for scarce goods and services. Those who control these items gain power and prestige. Dominance can also result from the control of property, as others become dependent on the landowners.
2. Those who are dominated have the potential to express resistance and hostility toward those in power. Although the potential for resistance is there, it sometimes lies dormant. Opposition may not be organized, because the subordinated groups may not be aware of their mutual interests. They may also be divided because of racial, religious, or ethnic differences.

FOR FURTHER THINKING

How Easy Is It to Change Social Class?

We know that in the United States you can change your educational and financial circumstances through hard work. Is this the same as changing your social class? Or is that a harder job that requires many trade-offs? R. Todd Erkel grew up in a working-class family in Pittsburgh. In the following excerpt, he writes about how the newly middle class often feel uncomfortable, adrift, and profoundly alone as they make their journey to a new culture.

When I decided to go to college, my mother and father offered the only advice they could: "Well," they said, "we hope you know what you're doing." But implicit in their lukewarm endorsement was the obvious truth: I didn't know—any more than they did. So, without a guidance counselor, or parents versed in the calculus of financial aid and college applications, I proceeded arbitrarily, applying to colleges whose names or looks appealed to me. I eventually settled, for no particular reason, on Penn State.

Having applied too late to attend orientation, I arrived in State College, Pennsylvania, with what I soon learned was more than a clothes problem. My credentials—an impressive grade point average and high test scores—gave my application a veneer of promise. But there was no measure for the things I didn't know, nothing to suggest I might have to learn everything about this new world of college from scratch. The system, like me, was blind to the ways in which my working-class background left me unprepared for this new world. It wasn't just poise or spending money that I lacked. Everything from my colloquial speech to my primitive social skills to my wardrobe drew a discreet line between me and my new peer group.

Adrift in this community of 60,000 not-so-kindred souls, I looked to the only place I knew of to place the blame for my ineptitude—inside, with myself.

Overwhelmed by feelings of alienation and worthlessness, I quit. In retrospect, I wonder why nobody—if not my parents, then a teacher or college counselor—could have foreseen my difficult transition to college; not just from one phase of education to another, but from one set of cultural assumptions to another, one entire world to another. I wish, too, that I could have let the full extent of my alienation be known.

Even at the University of Pittsburgh, where I eventually transferred and where the presence of students from working-class backgrounds similar to mine was plain to see, the issue of class remained eerily unspoken. Though part of me knew better, I could not escape the crippling feeling that I remained alone in my bewilderment.

Quiet and hiding in the back corner of a college classroom, I began to make the connection between my feelings of shame and the working-class misgivings toward education I had long ago internalized. For while lip service is paid at every class level in America to the idea that education is the route to a better life, the reality in most working-class homes is that knowledge counts for less than good behavior. The question my parents asked every day when I came home from school was not "What new thing did you learn today?" but "Did you get into any trouble?"

The working-class experience makes the child particularly vulnerable to low self-expectations. Before I could sing the alphabet, I knew something of what it felt like to be my parents: low-achieving, poorly spoken, lacking confidence, afraid to challenge authority, reluctant to ask for help, willing to accept their situation, content to do without.

The message received by children whose parents have battled with the world and come away feeling defeated is that they are better off not even

3. Conflict will most often center on the distribution of property and political power. The ruling classes will be extremely resistant to any attempts to share their advantages in these areas. Economic and political powers are the most important advantages in maintaining a position of dominance.
4. What are thought to be the common values of society are really the values of the dominant groups. The dominant groups establish a value system that justifies their position. They control

the systems of socialization, such as education, and impose their values on the general population. In this way, the subordinate groups come to accept a negative evaluation of themselves and to believe that those in power have a right to that position.

5. Because those in power are engaged in exploitative relationships, they must find mechanisms of social control to keep the masses in line. The most common mechanism of social control is the

trying. A pervasive feeling of helplessness hangs over the working-class house like the secondhand smoke that passes silently from parent to child.

Embracing the promise of an education requires working-class children to construct an inner sense of themselves that is radically different from that of their parents, siblings, and friends, to betray their allegiance to the only source of identity and support they have ever known. At each crossroad, and with every success, I became more aware of the dichotomy—the ways in which my education simultaneously would provide me options and distance me from the life I trusted.

A decade later, the anger I have long felt toward my parents has slowly faded. I realize now that the gifts I so desperately wanted from them (an easy self-confidence and a deep well of optimism) were not theirs to give. Instead, they handed their children a promise, visibly broken, in the hope that we might know better than they did how to make it work. In America, the illusion of free and open passage between classes is preached with religious zeal. But parents who wish for something better for their children must struggle against more than an incomplete education and economic deprivation. They must confront the truth behind the myth of making it in America: The land of opportunity is also the land of persistent class structures and struggles. And though many try, most people never rise very far from the socioeconomic level into which they are born.

Without an awareness of their experience of class discrimination, working-class kids will continue to absorb their parents' shame. They will continue to view the world—through their parents' eyes—as a malevolent force to be approached with rightful caution.

Without some explanation for the feelings present in the home, children internalize their parents' sense of powerlessness. Their unexpressed fears quietly sap the strength of the family. My middle-class friends know enough about the destructive force of class to see me as an exception: a triumph of will over environment. But we sidestep any discussion of a class system in America, afraid maybe of the feelings such awareness might stir. It's easier to assume that we now all occupy one vast, comprehensive middle class.

What I don't mention, and what others don't see, is that I often feel more lost than ever, caught between two widely separated social rungs, never sure whether I should forge ahead or fall back, uncertain whether either option really is mine. I have learned to pose in the middle-class culture, but at a price. I live most of the time on borrowed instincts, afraid to trust that part of the working class I still carry inside.

Today, my family is still wondering what on earth I will one day do for a living and uncomfortable with my willingness to expose the personal details of my life on paper where anyone might read them . . . I am engaged with the world in a way they have never known, and still don't altogether trust. Looking back, the memory of growing apart from the people and habits I know and love stirs a swirl of feelings. I still yearn to believe that my parents know what is best. The adult I've become appreciates why such knowledge eluded them. I understand more clearly the gain of my leaving their world. I'm only now willing to consider the loss.

Source: Excerpted from "The Mighty Wedge of Class" in *Family Therapy Networker,* by R. Todd Erkel, July/August, 1994, pp. 45–47. Used with permission from the author.

---

threat—or the actual use—of force, which can include physical punishment or the deprivation of certain rights. However, more subtle approaches are preferred. By holding out the possibility of a small amount of social mobility for those who are deprived, the power elite will try to induce them to accept the system's basic assumptions. Thus, the subordinate masses will come to believe that by behaving according to the rules, they will gain a better life (Vanfossen, 1979).

The Need for Synthesis

Any empirical investigation will show that neither the functionalist theory nor the conflict theory of stratification is entirely accurate. This does not mean that both are useless in understanding how stratification operates in society. Ralf Dahrendorf (1959) suggested that the two really are complementary rather than opposed. (Table 8–4 compares the two.) We do not need to choose between the two, but instead should see how

Table 8–4

Functionalist and Conflict Views of Social Stratification: A Comparison

The Functionalist View	The Conflict View
1. Stratification is universal, necessary, and inevitable.	1. Stratification may be universal without being necessary or inevitable.
2. Social organization (the social system) shapes the stratification system.	2. The stratification system shapes social organizations (the social system).
3. Stratification arises from the societal need for integration, coordination, and cohesion.	3. Stratification arises from group conquest, competition, and conflict.
4. Stratification facilitates the optimal functioning of society and the individual.	4. Stratification impedes the optimal functioning of society and the individual.
5. Stratification is an expression of commonly shared social values.	5. Stratification is an expression of the values of powerful groups.
6. Power usually is distributed legitimately in society.	6. Power usually is distributed illegitimately in society.
7. Tasks and rewards are allocated equitably.	7. Tasks and rewards are allocated inequitably.
8. The economic dimension is subordinate to other dimensions of society.	8. The economic dimension is paramount in society.
9. Stratification systems generally change through evolutionary processes.	9. Stratification systems often change through revolutionary processes.

Source: Adapted from "Some Empirical Consequences of the Davis-Moore Theory of Stratification," by A. L. Stinchcombe, 1969, in J. L. Roach, L. Gross, & O. R. Gursslin, eds., *Social Stratification in the United States,* Englewood Cliffs, NJ: Prentice-Hall, p. 55. Used by permission.

each is qualified to explain specific situations. For example, functionalism may help explain why differential rewards are needed to serve as an incentive for a person to spend many years training to become a lawyer. Conflict theory would help explain why the offspring of members of the upper classes study at elite institutions and end up as members of prestigious law firms, whereas the sons and daughters of the middle and lower-middle classes study at public institutions and become overworked district attorneys.

SUMMARY

- Despite the American political ideal of the basic equality of all citizens and the lack of inherited ranks and titles, the United States nonetheless has a class structure that is characterized by extremes of wealth and poverty.
- Class distinctions exist in the United States based on race, education, family name, career choice, or wealth.
- Social stratification has shown that social class affects many aspects of people's lives.

- Lower-class people get sick more often and have higher infant mortality rates, shorter life expectancies, and larger families.
- The poor are more likely to be arrested, charged with a crime, convicted, and sentenced to prison and are likely to get longer prison terms than middle- and upper-class criminals.
- The functionalist theory of stratification as presented by Davis and Moore holds that stratification is socially necessary.
- Davis and Moore argue that:
  - Different positions in society make different levels of contributions to the well-being and preservation of society.
  - Filling the more complex and important positions in society often requires talent that is scarce and has a long period of training.
  - Providing unequal rewards ensures that the most talented and best trained individuals will fill the statuses of greatest importance and be motivated to carry out role expectations competently.
- Critics of this view suggest that stratification is immoral because it creates extremes of wealth

and poverty and denigrates the people on the bottom.

- In addition, it is dysfunctional in that it neglects the talents and merits of many people who are stuck in the lower classes.
- It also ignores the ability of the powerful to limit access to important positions, and overlooks the fact that the level of rewards attached to jobs does not necessarily reflect their functional importance.
- Conflict theorists see stratification as the outcome of a struggle for dominance.
- Karl Marx believed that to understand human societies, one must look at the economic conditions surrounding production of the necessities of life.
- Marx believed that those groups that own or control the means of production within a society also have the power to shape or maintain aspects of society to favor their interests.

 *Media Resources*

The Companion Website for *Introduction to Sociology*, Ninth Edition

http://sociology.wadsworth.com/tischler9e

Supplement your review of this chapter by going to the companion website to take one of the Tutorial Quizzes, use the flash cards to master key terms, and check out the many other study aids you will find there. You will also find special features such as Wadsworth's Sociology Online Resources and Writing Companion, GSS data, and Census 2000 information at your fingertips to help you complete that special project or do some research on your own.

CHAPTER EIGHT STUDY GUIDE

KEY CONCEPTS AND THINKERS

Match each concept with its definition, illustration, or explanation presented below.

a. Feminization of poverty
b. Distribution of income
c. Conflict theory
d. Upper-middle class
e. Poverty index

f. Distribution of wealth
g. Poverty
h. Lower class
i. Working class

j. Social class
k. Class system of stratification
l. Functionalist theory
m. Upper class

j   **1.** A category of people who share similar opportunities, similar economic and vocational positions, similar lifestyles, and similar attitudes and behaviors.

k   **2.** A system of stratification that includes several different social classes and permits social mobility.

m   **3.** A U.S. social class characterized by corporate ownership, elite schools, upper-echelon politics, and a higher education.

h   **4.** A U.S. social class characterized by unskilled labor, little interest in education, grade school completion, service work, and farm labor.

d   **5.** A U.S. social class characterized by professional and technical occupations and college and graduate school training.

i   **6.** A U.S. social class made up of skilled and semiskilled laborers.

l   **7.** An explanation for the existence of social classes based on the idea that in order to attract talented individuals to each occupation, society must set up a system of differential rewards.

a   **8.** The phrase referring to the increasing concentration of poverty among female-headed households.

e   **9.** The U.S. government's specification of income levels below which people are considered to be living in poverty.

b   **10.** The degree to which all earnings in the nation are concentrated.

g   **11.** The condition in which people do not have enough money to maintain a standard of living that includes the basic necessities of life.

c   **12.** An explanation that says social class arises and persists because those with more wealth and power use their means to enhance their own position at the expense of others.

f   **13.** The degree of concentration or spreading out of property and other financial assets.

Match the thinkers with their main idea or contribution.

a. Max Weber   b. Karl Marx   c. Harold Kingsley Davis and Wilbert Moore

c   **1.** Developed the functionalist theory of social stratification.

b   **2.** Argued that class was based on ownership and that capitalism required a conflict between owners and workers.

a   **3.** Argued that social stratification was not just a matter of wealth but included prestige and political power as well.

CENTRAL IDEA COMPLETIONS

Following the instructions, fill in the appropriate concepts and descriptions for each of the questions posed below.

1. Assess each of the following ideas about poverty in America:

   a. People are poor because they are too lazy to work: _____

_____

_____

b. Most poor are minorities, and most minorities are poor: _____

_____

_____

c. Most people in poverty live in inner-city areas: _____

_____

_____

d. Welfare programs for the poor are straining the federal budget: _____

_____

_____

2. As your text notes, those arrested, accused, convicted, and imprisoned are disproportionately poor. Yet wealthier people also commit crimes. Compare and contrast the ways the criminal justice system operates for the poor and for the rich in each of the following areas.

   a. Type of crime committed: _____

   _____

   _____

   b. Legal representation: _____

   _____

   _____

   c. Conviction outcomes: _____

   _____

   _____

3. Summarize each of these theories of stratification in one or two sentences:

   a. Functionalist _____

   _____

   b. Conflict _____

   _____

   c. Modern conflict _____

   _____

CRITICAL THOUGHT EXERCISES

1. Assess the performance of the United States, compared with other Western industrialized counties, in reducing poverty among these three groups: children, the elderly, and working-age adults. Where does the United States fare best and where does it do the worst? Describe some of the major factors responsible for the United States' performance with regard to each of the three groups.

2. Try thinking of stratification in your school based on the classes of faculty, students, administrators, and perhaps other groups. How would a conflict theorist look at the actions of each of these groups? How would a functionalist look at them? To what extent do the class distinctions of the wider society come into play on campus and intertwine with the campus categories?

3. Some government policies benefit the wealthy and not the poor—i.e., certain tax policies, lower minimum wage, less spending for welfare, restrictions on food stamps, and so on. How do proponents of these policies justify them? Do these arguments provide examples of Marx and Engels's observation that "The ruling ideas of each age have always been the ideas of its ruling class"?

4. Develop a research paper within which you explore why economic rewards are distributed more unequally in the United States than elsewhere in the Western industrialized world. What are the positive and negative outcomes of this system?

INTERNET ACTIVITIES

1. Visit: http://www.pbs.org/peoplelikeus/film/index.html. This site is connected with a video, "People Like Us," produced by PBS and often presented in sociology courses. For this exercise, you do not need to see the entire video. Rather visit the section that has short video clips with commentaries from the persons interviewed in the complete film. There is an option to e-mail, through PBS, any or all of the persons appearing in the short video clips. After watching the clips, select any three, develop your question, and e-mail each. Report your findings in a short paper for the class.

2. In 2005, the *New York Times* ran a series of articles on social class. The articles are on their website, along with some Flash graphics on social class, income inequality, and social mobility. http://www.nytimes.com/pages/national/class/.

3. Robert E. Wood of Rutgers University has developed an interactive page where you can take a virtual tour of the U.S. class system. http://camden-www.rutgers.edu/~wood/332virtualtour.htm.

ANSWERS TO KEY CONCEPTS

1.j 2.k 3.m 4.h 5.d 6.i 7.l 8.a 9.e 10.b 11.g 12.c 13.f

ANSWERS TO KEY THINKERS

1.c 2.b 3.a

ThomsonNOW™

Reviewing is as easy as ❶ ❷ ❸

1. Before you do your final exam, take the ThomsonNOW diagnostic quiz to help you identify the areas on which you should concentrate. You will find information on ThomsonNOW and instructions on how to access all of its great resources on the foldout at the beginning of the text.
2. As you review, take advantage of ThomsonNOW's study videos and interactive Map the Stats exercises to help you master the chapter topics.
3. When you are finished with your review, take ThomsonNOW's posttest to confirm you are ready to move on to the next chapter.

9

Global Stratification

Sue Cunningham/Alamy

Learning Objectives

After studying this chapter, you should be able to do the following:

- Be able to describe the caste, estate, and class systems of social stratification.
- Know the theories of global stratification.
- Describe the phenomenon of exponential growth.

- Discuss world health trends.
- Know the determinants of fertility and family size.
- Discuss the problems of overpopulation and possible solutions.
- Understand the trends in global aging.

Should there be limits on inequality? Adam Smith certainly thought there had to be limits to deprivation. Smith noted, "No society can be flourishing and happy of which the far greater part of members are poor and miserable." Smith believed all members of society should have enough to enable them to walk down the street "without shame" (Smith, 1776/1976). Nearly all religions note that it is a moral obligation to help the less fortunate. Yet, despite these expressions, global inequality is widespread.

For example, a girl born in Japan today may have a 50% chance of seeing the twenty-second century, whereas a newborn in Afghanistan has a 1 in 4 chance of dying before age 5. And the richest 5% of the world's people have incomes 114 times those of the poorest 5%. Every year about 11 million children die of preventable causes, often for want of simple and easily provided improvements in nutrition, sanitation, and maternal health and education. Every year more than 500,000 women die because of pregnancy and childbirth, with huge regional disparities. By the end of 2002, 15 million children had lost their mother or both parents to the human immunodeficiency virus (HIV), and more than 40 million people were infected with the disease—90% of them in developing countries and 75% in sub-Saharan Africa.

In sub-Saharan Africa, human development has actually regressed in recent years, and the lives of its very poor people are getting worse. More than half of the world's people live on less than $2 a day, the internationally defined poverty line (Figure 9–1). Because of population growth, the number of poor people in the region has increased. As of 2001, 1.1 billion people lacked access to safe water, and 2.4 billion did not have access to any form of improved sanitation services (United Nations Development Report, 2002).

These examples demonstrate the extreme levels of social inequality that exist in the world today. And recent information points to the fact that the situation may be getting worse, not better. "In 73 countries representing 80% of the world's people, 48 have seen inequality increase since the 1950s, 16 have experienced no change, and only nine—with just 4% of the world's people—have seen inequality decline" (United Nations, 2002).

Social inequality is the outgrowth of social stratification, which is the uneven distribution of privileges, material rewards, opportunities, power, prestige, and influence among individuals and groups. Social inequality exists in all societies. The inequality may occur because of wealth, prestige, or power. Societies differ in terms of what is unequal and how the inequality comes about.

219

Figure 9–1   **Percentage of People Who Live on Less Than $2 a Day**

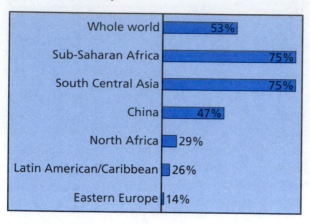

Whole world	53%
Sub-Saharan Africa	75%
South Central Asia	75%
China	47%
North Africa	29%
Latin American/Caribbean	26%
Eastern Europe	14%

Source: World Bank, *World Development Indicators 2005.*

In some Middle Eastern societies, women are expected to cover themselves in public. Such a situation helps perpetuate inequality between men and women.

© Andrew Holbrooke/Black Star

Stratification Systems

Stratification can happen in two ways: (1) People can be assigned to societal roles, using as a basis for the assignment an ascribed status—an easily identifiable characteristic, such as gender, age, family name, or skin color, over which they have no control. This will produce the caste and estate systems of stratification. Or (2) people's positions in the social hierarchy can be based to some degree on their achieved statuses (see Chapter 5), gained through their individual, direct efforts. This is known as the class system.

The Caste System

The **caste system** is *a rigid form of stratification, based on ascribed characteristics such as skin color or family identity, that determines a person's prestige, occupation, residence, and social relationships.* Many people believe that true caste systems are found only in India, although some have suggested that they exist in southwestern Ethiopia and other parts of Africa as well (Pankhurst, 1999).

People are born into and spend their entire lives within a caste, with little chance of leaving it. The caste is a closed group whose members are severely restricted in their choice of occupation and degree of social participation. Marriage outside the caste is prohibited. Social status is determined by the caste of one's birth, and it is unusual for a person to be able to overcome his or her origins.

Contact between castes is minimal and governed by a set of rules or laws. If interaction must take place, it is impersonal, and examples of the participants' superior or inferior status are abundant. Access to valued resources is extremely unequal.

A set of religious beliefs often justifies a caste system. The caste system as it existed for centuries in India before the 1950s is a prime example of how this

kind of inflexible stratification works. The Hindu caste system, in its traditional form in India, consisted of four *varnas* (literally meaning "color"), each of which corresponded to a body part of the mythical Purusa, whose dismemberment was believed to have given rise to the human species. Purusa's mouth issued forth priests *(Brahmans),* and his arms gave rise to warriors *(Kshatriyas).* His thighs produced artisans and merchants *(Vaishyas),* and his feet brought forth menial laborers *(Shudras).* Below the *Shudras* were the untouchables, or *Panchamas* (literally "fifth division"), who were considered to fall outside the caste system and performed the most menial tasks. Hindu scripture holds that each person's varna is inherited directly from his or her parents and cannot change during the person's life (Gould, 1971).

Each varna had clearly defined rights and duties associated with it. Hindus believed in reincarnation of the soul and that, to the extent that an individual followed the norms of behavior of his or her varna, the state of the soul increased in purity and the individual could expect to be born to a higher varna in a subsequent life. (The opposite was also true, in that failure to act appropriately according to the varna resulted in a person being born to a lower varna in the next life.)

This picture of India's caste system is complicated by the presence of thousands of subcastes, or *jatis*. Each of these jatis corresponds in name to a particular occupation (leather worker, shoemaker, cattle herder, barber, potter, and so on). Only a minority within each jati actually performs the work of that subcaste; the rest find employment when and where they can.

Changing the caste system sanctified by Hindu religious texts is difficult. Millions of *Dalits,* traditionally known as members of India's untouchable caste, have "tried to escape the system by converting to Islam, Christianity or Buddhism. But the system is so rooted in South Asian society that caste persists even in Christian and Muslim communities." Doing away with the caste system also threatens upper-caste privilege. "Few upper-caste people are eager to overturn a social order that supplies them with cheap labor and social standing" (Power & Mazumdar, 2000).

Hindus have never placidly accepted the caste system. Since the nineteenth century economic developments have made the caste system less stringent. In the 1930s Mohandas Gandhi tried to change attitudes toward the untouchables, and untouchability was declared illegal in 1949. Officially, the Indian caste system has been outlawed, but evidence of its existence is still present, particularly in rural areas ("Caste," 2002).

The Estate System

The **estate system** is *a closed system of stratification in which a person's social position is defined by law, and membership is determined primarily by inheritance.* An estate is a segment of a society that has legally established rights and duties. The estate system is similar to a caste system, but not as extreme. Some mobility is present, but by no means as much as exists in a class system.

The estate system of medieval Europe is a good example of how this type of stratification system works. The three major estates in Europe during the Middle Ages were the nobility, the clergy, and, at the bottom of the hierarchy, the peasants. A royal landholding family at the top had authority over a group of priests and the secular nobility, who were quite powerful in their own right. The nobility were the warriors; they were expected to give military protection to the other two estates. The clergy not only ministered to the spiritual needs of all the people but also were often powerful landowners as well. The peasants were legally tied to the land, which they worked to provide the nobles with food and wealth. In return, the nobles were supposed to provide social order, not only with their military strength, but also as the legal authorities who held court and acted as judges in disputes concerning the peasants who belonged to their land. The peasants had little freedom or economic standing, low social status, and almost no power.

Just above the peasants was a small but growing group, the merchants and craftsmen. They operated somewhat outside the estate system in that, although they might achieve great wealth and political influence, they had little chance of moving into the estate of nobility or warriors. It is worth noting that it was this marginal group, which was less constricted by norms governing the behavior of the estates, that had the flexibility to gain power when the Industrial Revolution, starting in the eighteenth century, undermined the estate system.

Individuals were born into one of the estates and remained there throughout their lives. Under unusual circumstances people could change their estate, as, for example, when peasants—using produce or livestock saved from their own meager supply or a promise to turn over a bit of land that by some rare fortune belonged to them outright—could buy a position in the church for a son or daughter. For most, however, social mobility was difficult and extremely limited because wealth was permanently concentrated among the landowners. The only solace for the poor was the promise of a better life in the hereafter (Vanfossen, 1979).

The Class System

In Chapter 8 we noted that a society that has several different social classes and permits greater social mobility than a caste or estate system is based on a class system of stratification. Remember that a social class consists of a category of people who share similar opportunities, economic and vocational positions, lifestyles, and attitudes and behaviors. Some form of class system is usually present in all industrial societies, whether they are capitalist or communist. Mobility is greater in a class system than in either a caste or an estate system. This mobility is often the result of an occupational structure that allows higher-level jobs to be available to anyone with the education and experience required. A class society encourages striving and achievement.

Theories of Global Stratification

Rich and poor countries exist throughout the world, and development occurs at varying paces. Are these varying rates of development related to a country's location and the availability of natural resources or to cultural and social factors? Could it be that development is tied to the relationships between the rich and poor nations? Two theories have been proposed to help explain the different rates of development among countries.

Modernization Theory

Modernization theory *assumes that the economic differences among countries are due to technological and cultural differences.* According to this theory, modernization depends on a country having a cultural environment that welcomes innovation and makes it possible for technological advancements to be incorporated into that society.

Strong ties to religious or historical traditions are seen as the greatest barriers to modernization. Economic impoverishment occurs when people are discouraged from adopting technologies that would raise living standards. Other barriers to modernization may include government policies that are not conducive to business, or the lack of money to invest in Western-style industry and agriculture. Modernization theorists have also seen lack of education, low motivation toward achievement, and high birth rates as barriers to development (Rostow, 1960; Huntington, 1968).

Modernization theory assumes that developed countries play an important role in helping the less developed countries to modernize. The developed countries introduce modern technology into these countries, which helps accelerate economic growth. Countries may be given foreign aid to purchase machinery and agricultural products that boost production. The developed countries may help the less developed countries increase food supplies by providing fertilizers, irrigation methods, and insect control. They may also help developing countries control their population (Firebaugh & Sandu, 1998).

Critics of modernization theory see it as a defense of capitalism. They claim that it holds up the developed countries as a model that the developing countries should emulate. Critics also believe it places a disproportionate blame for poverty on the poor societies themselves. In "blaming the victim," they contend that modernization theorists downplay the role rich nations have played in producing the plight poorer countries experience (Wiarda, 1987).

Dependency Theory

Dependency theory *proposes that the economic positions of rich and poor nations are linked and cannot be understood in isolation from each other. Global inequality is due to the exploitation of poor societies by the rich ones.*

Dependency theory is an outgrowth of a Marxist view of the interconnection among world systems. From this perspective capitalism is seen as having a global effect instead of one specific just to the country in which it is practiced. The Western developed nations need to buy raw materials cheaply from the developing countries. At the same time they need to sell manufactured goods at a high price. This situation puts the developing countries in a dependent position that forces them to become debtors. According to this view the developing countries would be able to develop more quickly if their dependence on the developed countries were reduced.

Dependency theorists believe that the prosperity of the more developed countries came about because of the impoverishment of other countries. The rich countries depend on being able to buy inexpensive raw materials from the poor ones. They use these raw materials to produce expensive products that they sell to the poor countries. The control of global resources has fallen into the hands of the rich countries, and the poor countries have become dependent on massive amounts of foreign aid. These countries are then caught in a cycle of dependence in which they must sell their raw materials to pay off the debt they are incurring in their trade with the rich countries.

Critics of dependency theory believe that no systematic empirical support exists for many of the basic claims about the role dependence plays in blocking development. They point out that many of the poorest countries have had little contact with rich nations and that trade has improved the economy of many other countries without making them dependent.

Critics also point out that it is simplistic to believe that only one factor—capitalism—is the cause of global inequality. The theory ignores other factors within a country that could produce poverty, such as corruption, mismanagement, or the suppression of minorities. Critics also see dependency theory as more ideology than theory. They believe that dependency theory is designed to inspire revolutionaries in less developed countries rather than provide an accurate account of economic development.

Global Diversity

World population, 6.477 billion in 2005, has more than doubled since 1960 and is projected to increase to 9.3 billion by 2050 (Population Reference Bureau, *2005 World Population Data Sheet*). Yet the world's richest countries, with 20% of the global population, account for 86% of private consumption; the poorest 20% account for just 1.3%. A child born in an industrialized country will add more to consumption and pollution over his or her lifetime than 30 to 50 children born in developing countries (United Nations, 2001, *State of World Population*).

Throughout the world, some 2 billion people are malnourished. Nearly 60% of people in developing countries lack basic sanitation, one-third do not have access to clean water, one-quarter lack adequate housing, 20% do not have access to modern health services, and 20% of children do not attend school.

In the next section we will look at some of the vast differences in everyday life that exist throughout the world.

World Health Trends

The World Health Organization defines health as "a state of complete mental, physical, and social well-being." This concept may appear straightforward, but it does not easily lend itself to measurement. Consequently, to describe the state of health in the world, we must look at trends. When we look at human

© Peter Johnson/CORBIS

More than 300 million people live in 24 countries where life expectancy is less than 50 years. In these countries, 1 out of 10 newborns die by age 1, and 3 million a year do not survive for one week. In some African villages, death among infants and young children occur 10 times more frequently than deaths among the aged.

health from this perspective, we find that the twentieth century has seen unprecedented gains in health and survival. On a worldwide basis, the average life expectancy for a newborn more than doubled, from 30 years in 1900 to 67 years in 2005 (Population Reference Bureau, *2005 World Population Data Sheet*). For a country like China, this has meant moving from conditions at the turn of the century, when scarcely 60% of newborns reached their 5th birthday, to the present, when more than 60% can expect to reach their 70th birthday.

Health advances in some countries have reached the point at which it appears that the population is approaching the upper limit of average life expectancy. In Japan, where life expectancy is nearly 80 years, a newborn has only a 4 in 1,000 chance of dying before its 1st birthday and less than a 1 in 1,000 risk of dying by age 40.

Unfortunately, the same cannot be said for less developed countries. More than 300 million people live in 24 countries where life expectancy is less than 50 years. In these countries, 1 of 10 newborns die by age 1, and 3 million a year do not survive for one week. In some African villages, deaths among infants and young children occur 10 times more frequently than deaths among the aged.

Currently, 80% of the world's population does not have access to any health care. Malnutrition and parasitic and infectious diseases are the principal causes of death and disability in the poorer nations. The problems these diseases cause are largely preventable, and most of these conditions could be dramatically reduced at a relatively modest cost.

The Health of Infants and Children in Developing Countries

When we look at infant and child health from a global perspective, we find that death among children is overwhelmingly a problem of the developing countries in Africa, Asia, and Latin America. Those countries account for 98% of the world's deaths among children younger than 5. To make matters worse, UNICEF estimates that 95% of these deaths are preventable. Even though child mortality fell by 11%, from 93 to 83 deaths per 1,000 live births during the 1990s, 1 in every 12 children dies before age 5 from preventable diseases. More than 10.5 million children die a year, 150 million are malnourished, and 120 million—mainly girls—never go to school. Let us examine the causes of these deaths.

Child Killers   Every year, millions of children younger than age 5 die from preventable and treatable illnesses, such as diarrheal dehydration, acute respiratory infection, measles, and malaria. In half of the cases, illness is complicated by malnutrition. In addition, 30 million of the world's children are still not routinely vaccinated and continue to die as a result of major childhood killers such as diphtheria, tuberculosis, pertussis, measles, and tetanus (Annan, 2001).

Although infectious diseases account for only 1% of deaths in developed countries, they are the major killers in the developing world, accounting for a staggering 16.3 million deaths, about 41.5% of all deaths. Almost two-thirds of the deaths among children in the developing world each year are caused by just three diseases: pneumonia, diarrhea, and measles. All three could be treated by available and affordable means. Malaria, which is extremely rare in developed countries, accounts for 2 million deaths annually.

Acute respiratory illness (ARI) remains one of the most common causes of child deaths in developing countries. ARI refers to infections of the respiratory tract, including the nose, middle ear, throat, voice box, air passage, and lungs. Pneumonia is the most serious of the acute respiratory diseases, but it can be easily treated with antibiotics. About 60% of ARI deaths could be prevented by a five-day treatment course of antibiotics costing 25 cents. More than half of the children with ARI are never treated.

Dirty drinking water and general unsanitary conditions bring on diarrhea causing the body to lose large quantities of salt and water. In 1990, diarrhea was the number one killer of children under 5. By the year 2000, the number of child deaths from this condition had been cut in half and 1 million deaths were being prevented each year. Progress in addressing this condition was made because of oral rehydration therapy.

In the countries of sub-Saharan Africa and South Asia, measles continues to be a major killer of children under 5 because less than 50% of them are vaccinated against the disease. Measles is a highly contagious disease and to control it at least 90% of the children must be vaccinated. High doses of vitamin A also help with the disease. Those children who get the disease and do not die are often left blind and deaf.

Malaria has reemerged as a major cause of child deaths. The disease causes severe anemia and is one of the main causes of low birth weight in infants. Malaria can be easily prevented by making sure that pregnant women and children sleep under insecticide-treated mosquito nets. This relatively simple action could prevent many deaths. It is estimated that providing safe sleeping environments could prevent 400,000 child deaths in Africa each year. Yet, few children sleep under mosquito nets and those that are used have not been treated.

Neonatal tetanus is a common and preventable illness. In developing countries it occurs when unhygienic birth practices, such as cutting the umbilical cord with a rusty or dirty knife, take place. At least 70% of the young victims die. The highest instances of neonatal tetanus occur in India, Nigeria, and Pakistan. Neonatal tetanus accounts for approximately 14% of all neonatal deaths.

A child deficient in vitamin A is 25 times more likely to die of treatable diseases such as measles, malaria, and diarrhea. Vitamin A deficiency, as people well know, can also lead to irreversible blindness. Vitamin A is crucial in developing resistance to infection and in preventing anemia and night blindness. The problem is that poor families can rarely afford foods rich in vitamin A, such as meat, eggs, fruits, red palm oil, and green leafy vegetables. Many countries now fortify staple foods like flour and sugar with vitamin A and other important micronutrients. Many children between 6 and 59 months are now receiving high-dose vitamin A capsules, which cost only a few cents per year.

Iodine deficiency, which results from not having enough iodized salt in your diet, is the world's single greatest cause of preventable mental retardation. Iodine deficiency can also produce stillbirth and miscarriage in women. During pregnancy, even mild iodine deficiency can hamper learning ability. More than 40 million newborns are still unprotected from learning disabilities, impaired speech, and hearing development problems because of iodine deficiency (Annan, 2001).

Yet there are some success stories. Smallpox has been eradicated (although recently there are fears that terrorists may have access to it), and the World Health Organization believes that polio is on the verge of being eradicated. However, those are the exceptions.

Maternal Health   The survival of infants and children depends critically on the nurturing the mother provides during pregnancy and childhood. When an expectant mother suffers from sexually transmitted illnesses, malaria, or malnutrition during pregnancy, she will be more likely to have a low-birth-weight baby. According to the United Nations, an estimated 18 million low-birth-weight babies are born each year, two-thirds in either South Asia or sub-Saharan Africa. Low birth weight is defined as those babies that weigh less than 5 pounds at birth. Newborns of low birth weight are more likely to die. Those who survive are at risk of having impaired immune systems and being susceptible to chronic illnesses. They are also likely to have lower intelligence and cognitive disabilities leading to learning difficulties.

In South Asia and sub-Saharan Africa, about half of all women between the ages of 15 and 19 are or have been married. Thus, fertility rates are extremely high in these regions.

© David Turnley/CORBIS

increase of three years of education produces 20% to 30% declines in the mortality of children younger than age 5. A mother's education affects her child's health in a number of ways. Better-educated mothers know more about good diet and hygiene. They are also more likely to use maternal and child health services—specifically, prenatal care, delivery care, childhood immunization, and other therapies (Cleland & Ginneken, 1988).

HIV/AIDS

There is a growing gap between infection with HIV in the developed world and in developing countries. In North America, Western Europe, Australia, and New Zealand, antiretroviral drugs have reduced the speed at which HIV-infected people develop AIDS.

For the rest of the world, the news is not as good. Some 4.9 million people worldwide were newly infected with HIV during 2004, bringing the total number of people living with HIV or AIDS to 40 million, up from 34.3 million in 1999. Since the epidemic was identified, about 30 million people have died from AIDS.

The United Nations estimates that without substantially expanded prevention and treatment programs, approximately 68 million people will die of AIDS in the 45 most affected countries between 2000 and 2020. This estimate equals over five times the 13 million deaths in the previous two decades of the epidemic (UNAIDS, 2002a).

Around the world, the epidemic varies. East Asia and the Pacific Rim countries are still keeping the epidemic at bay. Infection rates are lower, but the numbers are still large. More than half the world's population lives in the region, and even though the percentage of the population with HIV may be lower, the absolute number of people is very large. "In China, where almost all cases of HIV/AIDS were previously transmitted through intravenous drug use and unsafe blood practices, the epidemic is now spreading through heterosexual contact." HIV infections rose almost 70% in the first six months of 2001 throughout the country.

The HIV epidemic is growing fastest in the Russian Federation and Eastern Europe. Previously the disease was spread primarily through intravenous drug use, but now nearly 25% of new infections occur through sexual contact between heterosexuals. The disease is rapidly spreading from a select group to the wider population (UNAIDS, 2002b).

In India, infection rates are at about 5.1 million people (UNAIDS, AIDS Epidemic Update 2004). In South and Southeast Asia, approximately 5.8 million people are living with HIV or AIDS. In Indonesia, Malaysia, the Philippines, and Singapore, the infection

In addition to maternal malnutrition, other causes of low-birth-weight babies are the mother's excessively heavy work, malaria infection, severe anemia related to an iron-deficient diet, and hookworm infestation.

Maternal Age An infant is at higher health risk if the mother is in her teens or older than 40, if she has given birth more than seven times, or if the interval between births is less than two years. The interval between births is the most important factor in infant and child mortality. Infants born to mothers at less than two-year intervals between births are 80% more likely to die than children born at birth intervals of two to three years.

This information suggests that family planning programs that discourage early childbearing can substantially reduce infant and child mortality by preventing births to high-risk mothers. When childbearing is spread out, the woman has more time and energy to devote to her own health and the care of her other children.

Maternal Education Children's chances of surviving improve as their mother's education increases. An

GLOBAL SOCIOLOGY

HIV/AIDS: Worldwide Facts

Nearly twenty-five years after the first clinical evidence of acquired immunodeficiency syndrome (AIDS) was reported, it has become one of the most devastating diseases humankind has ever faced. Since the epidemic began, more than 60 million people have been infected with the human immunodeficiency virus (HIV) that causes AIDS. Sixty countries are greatly affected by the disease and AIDS is now the leading cause of death in sub-Saharan Africa. Worldwide, it is the fourth-biggest killer. At the end of 2004, nearly 40 million people globally were living with HIV. In many parts of the developing world, the majority of new infections occur in young adults, with young women especially vulnerable. About one-third of those currently living with HIV/AIDS are aged 15 to 24. Most of them do not know they carry the virus.

Many millions more know nothing or too little about HIV to protect themselves against it. Nearly 30 million people have died from AIDS since the epidemic began. Of all people living with HIV, 90% live in sub-Saharan Africa, Southeast Asia, or Latin America, and most of these do not know they are infected. Figure 9–2 shows AIDS statistics for some regions of the world.

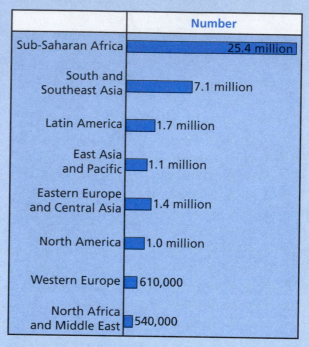

Figure 9–2   People Infected with HIV in Various World Regions

	Number
Sub-Saharan Africa	25.4 million
South and Southeast Asia	7.1 million
Latin America	1.7 million
East Asia and Pacific	1.1 million
Eastern Europe and Central Asia	1.4 million
North America	1.0 million
Western Europe	610,000
North Africa and Middle East	540,000

Sources: UNAIDS, AIDS epidemic update: December 2004, "Global Overview," http://www.unaids.org/wad2004/EPI _1204_pdf_en/Chapter11_maps_n.pdf accessed September 9, 2005, and Population Reference Bureau, "2005 World Population Data Sheet."

rate is well under 1%. In Thailand, a country that has a large group of sex workers and drug users, the infection rate is 2.3%. The picture is also very bleak in Cambodia, with at least 2% of the population infected. The AIDS epidemic arrived late in Asia, and today the region accounts for 20% of all infections worldwide.

In Latin America and the Caribbean, an estimated 1.4 million people now live with HIV/AIDS. Intravenous drug use, as well as unsafe heterosexual and homosexual sex, complicates the eradication of the disease. And though awareness of the epidemic is on the rise in the Caribbean, the region continues to be second only to Africa in the number of infected people.

Prevention efforts in wealthier countries have slowed the spread of the disease dramatically. Yet, in 2004 alone, approximately 64,000 adults in Western and Central Europe and North America were believed to have contracted the disease. In the United

States, AIDS is one the top three causes of death for African American men aged 25–54. The bulk of these new infections have been spread through intravenous drug use. There is also evidence of a decline in safer sex practices in gay communities. This decline will inevitably lead to more infections among gay men, coming after a period of several years in which efforts to stabilize or reduce the epidemic have proved successful (UNAIDS, AIDS Epidemic Update 2004).

The epidemic is moving the fastest in sub-Saharan Africa. The region has about 10% of the world's population but fully 60% of people living with HIV in the world. During 2004, 4.9 million people became infected with HIV in sub-Saharan Africa, bringing the total number of people living with HIV or AIDS in the region to 25.4 million. At the same time, 3.1 million people died in Africa of AIDS in 2004, up from 2.2 million in 1999 (UNAIDS, AIDS Epidemic Update 2004).

The region has reached the unprecedented level of 7.4% of all those between ages 15 and 49 being infected with HIV. Heterosexual transmission accounts for most infections throughout Africa. (See "Global Sociology: HIV/AIDS: Worldwide Facts.")

One of the greatest causes for concern is that over the next few years, the epidemic is bound to get worse before it gets better. Sub-Saharan Africa is faced with a trifold challenge: "providing care for the growing population of people infected with HIV, bringing down the number of new infections through more effective prevention, and coping with the effect that 17 million deaths has had on the continent" (UNAIDS, 2002b).

AIDS is systematically cutting down life expectancy in the countries where the disease is most common. In Southern Africa, life expectancy has fallen from 62 years between 1990 and 1995 to 48 years between 2000 and 2005. It could drop to 43 years in the next decade. Life expectancy in Zimbabwe has fallen by 25 years because of AIDS, and it will be shortened by a third in South Africa. Life expectancy in Botswana rose from 43 years in 1955 to 61 years in 1990. Now, with larger numbers of adults infected with HIV, life expectancy is going back to late 1960 levels. In Botswana, Lesotho, and Swaziland, the population will actually decrease as deaths outnumber births (United Nations, 2005.)

AIDS does not affect just the people with the disease. It also has a major impact on families. As life expectancies fall, the social safety net provided by extended families breaks down, health and education systems falter, and businesses and governments lose their most productive staff.

AIDS is shattering family structures. It is estimated that since the beginning of the epidemic, more than 15 million children have lost one or both parents to AIDS. Increasingly, elderly grandparents and older children are the sole caretakers of scores of younger children (United Nations, 2005).

Even though it appears that in the United States medical science is finding ways to control the spread of the disease, for the rest of the world the worst is still to come. The vast majority of people living with HIV live in the developing world where access to the new antiretroviral drugs is difficult or impossible. By the end of 2001, only 230,000 people in developing countries received antiretroviral drug therapy. This is less than 4% of the 6 million people in need of the drugs. Yet in high-income countries, antiretroviral drug therapy contributed to a decline in the number of AIDS deaths. In these countries only 25,000 people died of AIDS in 2001 out of the 500,000 who were taking the drugs. The disparity is seen in Africa, though, where only 30,000 people were taking the drugs out of the 28.5 million people infected. This low rate of treatment contributed to the 2.2 million people who died of AIDS in Africa in 2001. Unless there is a major cost-effective medical advance, many if not most of the millions currently infected with AIDS in the developing world will die within the next decade (UNAIDS, 2002b).

Population Trends

Global stratification is greatly influenced by population growth. Every minute, 249 babies are born in the world. That means about 358,988 new human beings a day—131.4 million a year—who need to be fed, clothed, sheltered, educated, and employed. The 6 billionth person arrived in the year 2000. Another billion people will be added every 11 to 13 years until the middle of the twenty-first century. It took nearly all of human history for the world's population to reach the first billion.

Population growth has continued throughout the past three decades despite the decline in fertility rates that began in many developing countries in the late 1970s and, surprisingly enough, despite the toll taken by the HIV/AIDS epidemic. Although the rate of increase is slowing, in absolute terms world population growth continues to be substantial. According to U.S. Census Bureau projections, world population will increase to a level of nearly 8 billion people by the end of the next quarter century and will reach 9.3 billion—a number more than half again as large as today's total—by 2050. Ninety-nine percent of global natural increase—the difference between numbers of births and deaths—occurs in the developing world.

The U.S. Census Bureau's projections indicate that early in the next century, death rates will exceed birthrates for the world's more developed countries. As the growth rate in the world's more affluent nations comes to an end, *all* the net annual gain in global population will, in effect, come from the world's developing countries (Bureau of the Census, March 1998).

The issue is underscored by the increasing speed with which the world's population is multiplying. During the first 2 million to 5 million years of human existence, the world population never exceeded 10 million people. The death rate was about as high as the birthrate, so there was no population growth. Population growth began around 8000 BC, when humans began to farm and raise animals. In AD 1650, an estimated 510 million people lived in the entire world. One hundred years later, there were 710 million, an increase of some 39%. By 1900, there were 1.6 billion. Only 100 years later the world population had spiraled to 6.08 billion, with 131.4 million people added each year (Population Reference Bureau, 2000). By the year 2025, the global population will be greater than 8 billion. Of that total, approximately 7 billion will be residents of the poorest and least developed nations.

Table 9–1

Factors Affecting Global Fertility

Number of Children

1. Women's average age at first marriage
2. Breast-feeding
3. Infant mortality

Demand for Children

1. Gender preferences
2. Value of children
   a. Children as insurance against divorce
   b. Children as securers of women's position in family
   c. Children's value for economic gain
   d. Children's value for old-age support
3. Cost of children

Fertility Control

1. Use of contraception
2. Factors influencing fertility decisions
   a. Income level
   b. Education of women
   c. Urban or rural residence

Table 9–2

Countries with the Highest and Lowest Fertility

Lifetime Births per Woman

Highest		Lowest	
Niger	8.0	Belarus	1.2
Guinea-Bissau	7.1	Czech Republic	1.2
Mali	7.1	Poland	1.2
Somalia	7.0	South Korea	1.2
Uganda	6.9	Taiwan	1.2

Source: Population Reference Bureau, "2005 World Population Data Sheet."

Right now, the world population is doubling about every 51 years. If it continues to expand this way, it will quadruple within 102 years and increase eightfold to an incredible 40 billion within 153 years—a situation in which widespread poverty and famine are almost assured. In recent years, however, a small but significant slowing in the rate of world population growth has occurred: Between 1965 and 1970, the annual rate of growth was 2.1%, but by 1999, it had dropped to 1.4% (Population Reference Bureau, 2000). If this slowing trend continues, the world growth rate will be far smaller than previously predicted. This more hopeful pattern is contingent on the average family size being limited to two children. There are also significant differences in fertility in various parts of the world (Table 9–1).

How many children a family has will affect that family's lifestyle in both developed and undeveloped countries. Many factors determine the typical family size in a country. In this section we will examine some of those factors to gain an insight into the variety of economic, social, and psychological issues that have to be addressed when trying to limit population growth. (See Table 9–2 for a summary of some of the determinants of fertility.)

Average Age of Marriage Early marriage provides more years for conception to occur. It also decreases the parents' years of schooling and limits their employment opportunities. In South Asia and sub-Saharan Africa, nearly three-fourths of all women between 15 and 19 are or have been married (Table 9–3). There are even areas of West and Eastern Africa and South Asia where many marriages take place before puberty. In Latin America and Eastern Europe, marriage for girls between 16 and 18 is common. In Afghanistan, the mean age for women at marriage is 17.8 years, and thus the fertility rates in those areas are extremely high. By contrast, the average age of marriage in Sri Lanka is 24.4. The result is that in Afghanistan the average woman has 6.9 children, whereas in Sri Lanka she has 2.3 children (United Nations, 2005; UNICEF, 2005).

To limit fertility some countries have tried to establish a minimum age for marriage. In 1950, China legislated the minimum age for marriage as 18 for women and 20 for men. It also recognized the effect that social controls can have on the marriage age and therefore increased institutional and community pressures for later marriage. In 1980, China again raised the legal minimum marriage ages to 20 for women and 22 for men; it is one of the few countries that has been successful in raising the average age of marriage. When Chinese youth were asked what would be the ideal age for their own marriage, the responses averaged 24.3 years for women and 25.4 for men (Yu, 1995).

Breast-Feeding Breast-feeding delays the resumption of menstruation and therefore offers limited protection against conception. In developing countries, it is much safer for infants to be breast-fed during the first six months of life than to be bottle-fed. Bottle-fed babies are three to six times more likely to experience respiratory infections or diarrhea than breast-fed babies. These risks have produced an outcry against certain large companies that produce powdered milk and that have been trying to encourage bottle-feeding in less developed countries. Bottle-feeding has been introduced into developing countries by hospitals that are trying to follow Western practices as well as by commercial concerns that aggressively promote breast milk substitutes. The World Health Organization and

Table 9–3

Teenage Marriages

	Percentage of 15- to 19-Year-Olds Who Are Married	
Sub-Saharan Africa	**Boys**	**Girls**
Dem. Rep. of Congo	5	74
Niger	4	70
Congo	12	56
Uganda	11	50
Mali	5	50
Asia		
Afghanistan	9	54
Bangladesh	5	51
Nepal	14	42
Middle East		
Iraq	15	28
Syria	4	25
Yemen	5	24
Latin America and Caribbean		
Honduras	7	30
Cuba	7	29
Guatemala	8	24

Source: UN Population Division, Department of Economic and Social Affairs, "World Marriage Patterns 2000."

UNICEF have tried to combat this trend, with limited success.

In Bangladesh, Pakistan, Nepal, and most of sub-Saharan Africa, where few women use contraception, breast-feeding is one of the few controls on fertility rates. In the United States, breast-feeding was relatively unpopular until the late 1960s. As the advantages of breast-feeding became known, however, better-educated women began to use it more. Today, college-educated women are the most likely to start breast-feeding and continue it for the longest periods.

Infant and Child Mortality High infant mortality promotes high fertility. Parents who expect some of their children to die may give birth to more babies than they really want as a way of ensuring a family of a certain size. This sets in motion a pattern of many children born close together, weakening both the mother and babies and producing more infant mortality.

In the short term, the prevention of 10 infant deaths may produce only one to five fewer births. Initially, lower infant and child mortality will lead to somewhat larger families and faster rates of population growth. However, the long-term effects are most important. With improved chances of survival, parents devote greater attention to their children and are willing to spend more on their children's health and education. Eventually, lower mortality rates help parents achieve the desired family size with fewer births and also leads them to want a smaller family.

Education and child health go hand in hand. In Indonesia, the infant mortality rate drops from 131 per 1,000 births for mothers with no schooling to 51 for those with the most education. The mother's education also has an effect on a child's health and nutrition. Children whose mothers complete secondary education are much less likely to be underweight for their age than those whose mothers have less education (Riley, 1997).

Gender Preferences The effect of gender preference on fertility is actually more complicated than might appear at first glance. In most countries, there is a strong preference for male children. Logically, this does not make much sense. In most underdeveloped countries, daughters typically help their mothers with household chores. One would think that this would increase their worth to the family. Clearly, there must be countervailing factors to reduce the desire for daughters.

Three sets of factors influence the desire for male children. They include (1) economic factors—the value assigned to women's work and the ability to contribute to family income; (2) social factors—kinship, marriage patterns, and religion; and (3) psychological factors—influences on parent's decisions about size and composition of the family (Hudson & den Boer, 2004).

The preference for sons, however, may not be all that important in countries with high fertility rates. In Bangladesh, for example, the preference for sons is extremely strong, but because couples desire large families and contraception is relatively rare, it appears to have little effect on fertility (Rob, 1988).

In countries with declining fertility rates, the preference for sons may cause the fertility rate to level off above replacement level. Couples may continue to have children until they have the desired son.

Girls have a better survival rate than boys, and women normally live longer than men. Even with the natural compensation of more male than female births, this still means that most nations have more women than men.

In a number of developing countries, there are far fewer women than men. This is not what would normally be expected and is often due to willful acts or discriminatory customs. One explanation would be lower survival rates for girls—either because of female infanticide or because girls receive less health care than boys. It also is related to the estimated half a million women who die each year of pregnancy and birth complications. Unsafe abortions kill another 100,000 each year. Other factors include widow burnings, in which a woman is burned when her husband dies, and dowry deaths, in which family disputes over the dowry result in the woman being killed. Unfortunately, as appalling as these events sound, they occur regularly in some countries. Figure 9–3 shows some of the countries with significantly fewer women than men.

In China there are 117 boys born for every 100 girls. Worldwide data show that about 105 to 107 boys are born for every 100 girls, implying that 12% of all Chinese baby girls (about 1.7 million) are missing annually (Hudson & den Boer, 2004). This does not mean that they are killed. Some are not reported to the authorities, others are adopted informally by friends or relatives (Riley, 1996).

The major reason for the missing girls seems to be technology. As early as 1979, China began manufacturing ultrasound scanners. The scanners are meant to help doctors check whether fetuses are developing normally, but they can also be used to determine whether a fetus is male or female. By 1990, about 100,000 scanners were available throughout China. Women go to clinics and have an ultrasound exam, and if the fetus appears to be a girl, an abortion is performed in the same clinic. It is estimated that between 500,000 and 750,000 female fetuses are

Figure 9–3   Countries with Fewer Women Than Expected

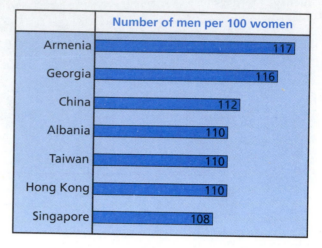

Source: *CIA World Factbook,* March 2005.

aborted in China every year as a result of couples having access to the ultrasound scanner that reveals the sex of a fetus (Albright, 1999). Combination ultrasound/abortion clinics are common in China, South Korea, India, and Afghanistan (Kristof & Wudunn, 1994). (See "Technology and Society: The Girls Who Will Not Be Born.")

We would be wrong to assume that a strong preference for a son is limited just to less developed countries. A recent Gallup poll showed that the American preference for male children is just as strong today as a half century ago. "Forty-two percent of Americans say they would prefer a boy if they could have only one child, whereas 27% said they would prefer a girl. One-quarter of Americans said it would not matter to them either way. In 1941 . . . 38% said they would prefer to have a boy, compared with 24% who said they would choose a girl and 23% who said they did not have a preference."

Younger people between 18 and 29 are actually more likely to prefer a boy than those 65 and older (Gallup Poll, December 2000).

Benefits and Costs of Children   In underdeveloped countries, the benefits for an individual family of having children have generally been greater than the costs. However, those costs and benefits change with the second, third, fourth, and fifth children.

For a rural sample in the Philippines, three-quarters of the costs involved in rearing a third child come from buying goods and services; the other quarter comes from costs in time (or lost wages). Receipts from child earnings, work at home, and old-age support offset 46% of the total. The remaining 54%, the net cost of a child, is equivalent to about 6% of a husband's annual earnings. By contrast, a study of urban

TECHNOLOGY AND SOCIETY

The Girls Who Will Not Be Born

Women are making strides in both India and China, which together are home to a third of humanity. They are living longer and are more likely to be able to read and write than ever before.

Yet, despite this progress, female fetuses are being aborted at startling rates in China and across broad swaths of India. The spread of ultrasound technology in these societies, with their strong preferences for sons, has made it easy to find out the sex of a child before birth and to abort unwanted daughters.

But why is this bias against girls so pronounced in large parts of South and East Asia when it is not in many other regions of the developing world? What is it about China and these areas of India, with their radically different histories and forms of government, that explains this terrible trend? And why, even now, don't the women themselves—more likely than ever before to be literate—insist on giving birth to their girls?

Economists and demographers think the answer lies in a particular form of patriarchal family, common in most parts of both regions. In such families, the daughter's responsibility to care for her parents largely ends at marriage, whereas the son's lasts for life. And in India and China, where there is still no universal, government-sponsored social security system, the question of who supports aging parents is very important, these researchers say.

In most parts of both India and China, where more than 2 billion people live, it is generally a son who carries on the family line, inherits ancestral land, cares for his parents as they age, and performs the most important ceremonial roles when they die. In India, the son lights his parents' funeral pyres, helping set their souls at peace. In China, he cares for his parents' spirits in the afterlife so they do not wander for eternity as hungry ghosts.

In contrast, a daughter in either society typically leaves her natal family after marrying to become part of her husband's family, moving in with them and tending to her in-laws. "There's a traditional expression used in China," said Li Shuzhuo, director of the Population Research Center at Xian Jiaotong University in Xian, China: "Daughters are like water that splashes out of the family and cannot be gotten back after marriage."

Before ultrasound, girls were sometimes victims of infanticide, but much more commonly they were victims of neglect; they were not fed as well as boys or taken to the doctor as quickly. But the advent of sex-selective abortions has added a new and definitive means for acting on a prejudice against girls—and statistics reflect the results, experts say.

In sub-Saharan Africa, Latin America, and Southeast Asia, daughters and sons share the job of caring for their aging parents—and in these regions, the sex ratios are normal. Normally, women around the world give birth to 105 or 106 boys for every 100 girls. But according to China's latest census, there were 117 boys born for every 100 girls in 2000, up from 114 in 1990.

Sources: Excerpted from Celia W. Dugger, "Modern Asia's Anomaly: The Girls Who Don't Get Born," *New York Times*, May 6, 2001, p. 4. Copyright © 2001 by The New York Times Co. Reprinted with permission.

---

areas of the United States showed that almost half the costs of a third child are time costs. Receipts from the child offset only 4% of all costs.

Only economic costs and benefits are taken into account in these calculations. To investigate social and psychological costs, other researchers have examined how individuals perceive children. Economic contributions from children are clearly more important in the Philippines, where fertility is higher than in the United States; concern with the restrictions children impose on parents is clearly greatest in the United States. In many countries, however, couples demonstrate a progression in the values they emphasize as their families grow. The first child is important to cement the marriage and bring the spouses closer together, as well as to have someone to carry on the family name if it is a boy. Thinking of the first child, couples also stress the desire to have someone to love and care for and the child's bringing play and fun into their lives.

In considering a second child, parents emphasize more the desire for a companion for the first child. They also place weight on the desire to have a child of the opposite sex from the first. Similar values are prominent in relation to third, fourth, and fifth children; emphasis is also given to the pleasure derived from watching children grow.

Beyond the fifth child, economic considerations predominate. Parents speak of the sixth child or later children in terms of their helping around the house, contributing to the support of the household, and providing security during old age. For first to third children, the time taken away from work or other pursuits is the main drawback; for fourth and later children, the direct financial burden is more prominent than the time costs (World Bank, 1984).

Contraception Apart from the factors already mentioned, fertility rates eventually are tied to the increasing use of contraception. Contraception is partly a function of a couple's wish to avoid or delay having children and partly related to costs. People have regulated family size for centuries through abortion, abstinence, and even infanticide. In many countries, the costs of preventing a birth, whether economic, social, or psychological, may be greater than the risk of having another child.

Use of contraception varies widely: 18% or fewer for married women in almost all of sub-Saharan Africa, but between 70% and 80% for women in Europe, Asia, and the United States (Population Reference Bureau, 2000).

Contraception is most effective when such programs are publicly subsidized. Not only do the programs address the economic costs of spreading contraception, but they also help communicate the idea that birth control is possible. These programs also offer information about the private and social benefits of smaller families, which helps reduce the desired family size.

The support given to family-planning programs differs dramatically from country to country. At one end of the family-planning spectrum are the governments of India and China, which provide birth control information and devices and actively support abortion. At the other end are predominantly Muslim countries, such as Bangladesh, or predominantly Roman Catholic countries, such as Ireland, whose populations follow the anti–birth control and antiabortion teachings of the religion.

Worldwide, abortion is the most widely used form of birth control, and it is common even when it is illegal. The demand for abortions is rising, and it has been estimated that throughout the world one in three pregnancies ends in abortion. Abortion is legal in the world's three most populous countries (China, India, and the United States) as well as in Japan and all of Europe except Belgium and Ireland (Weeks, 1994). In Russia, where contraceptives are hard to find, more than half of all pregnancies end in abortion (Haub, 1994).

Income Level It is a well-established fact that people with higher incomes want fewer children. Alternative uses of time such as earning money, developing or using skills, and pursuing leisure activities become more attractive, particularly to women. The children's economic contributions become less important to the family welfare, as the family no longer needs to think of children as a form of social security for old age. It is not the higher income itself but, rather, the life change it brings about that lowers fertility.

This relationship between income and fertility holds true only for those with an income greater than a certain minimum level. If people are extremely poor, increases in income will actually increase fertility. In the poorest countries in Africa and Asia, families are often below this threshold. Above the threshold, though, the greatest fertility reduction with rising income occurs among low-income groups.

Education of Women In nearly all societies, the amount of education a woman receives affects the number of children she has. Fertility levels are usually the lowest among the most highly educated women within a country. In Guatemala, for example, women with no formal schooling have an average of 6.9 children, whereas those with a secondary or higher education have only 2.7 children on average. In Zimbabwe, the trend is similar, with women with no schooling having 6.9 children and those with secondary or higher education having 3.8 (Riley, 1997).

Despite this information, in 1999 only 52% of girls in the least developed countries stayed in school past fourth grade. In 22 African nations and 9 Asian nations, school enrollment for girls is less than 80% that for boys. Two-thirds of the world's 876 million illiterate people are women (United Nations, March–May 2002) (Figure 9–4).

Studies also show that women's level of education affects fertility more than does that of men. There are a number of reasons for this. In most instances, children have a greater effect on the lives of women, in terms of time and energy, than they do on the lives of men. The more educated a woman is, the more opportunities she may encounter that conflict with having children. Education also appears to delay marriage, which in itself lowers fertility.

In 10 of 14 developing countries, women with 7 or more years of education marry at least 3.5 years later than do women with no education. Educated women are also more likely to know about and adopt birth control methods. In Mexico, 72% of women with 9 or more years of education are likely to use contraception, whereas only 31% of those with 5 or fewer years of education are likely to do so. A study of 37 countries showed that one additional year of schooling lowers desired family size by more than 0.1 children, so that women with 10 or more years of education want, on average, 1.3 fewer children than women with no formal education (Lutz, 1994).

Urban or Rural Residence Rural lifestyles that involve farming and herding tend to perpetuate high birthrates because children are seen as contributing workers. Urban fertility tends to be lower in developing countries, although it still is at least twice as high as fertility in developed countries (Goliber, 1997). Urban dwellers usually have access to better education and health services, a wider range of jobs,

Figure 9–4   Children of Primary School Age Who Are Not in School

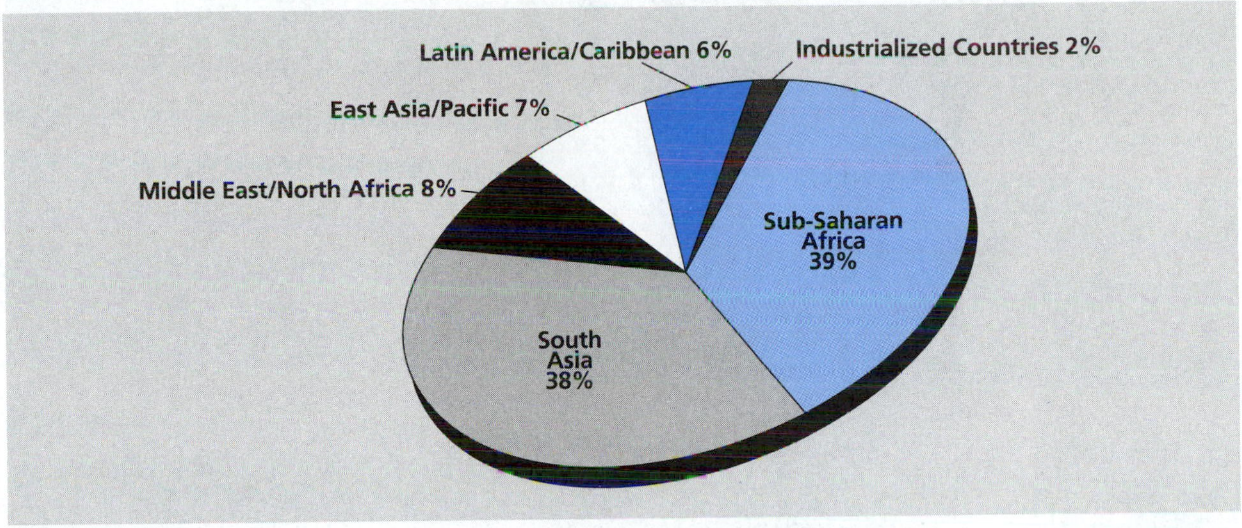

Latin America/Caribbean 6%
Industrialized Countries 2%
East Asia/Pacific 7%
Middle East/North Africa 8%
Sub-Saharan Africa 39%
South Asia 38%

Source: Annan, Kofi A. "We the Children: Meeting the Promises of the World Summit for Children," UNICEF, 2001.

and more avenues for self-improvement and social mobility than do their rural counterparts. They are exposed to new consumer goods and are encouraged to delay or limit childbearing to increase their incomes. They also face higher costs in raising children. As a result, urban fertility rates are usually one to two children lower than rural fertility rates.

The urban woman marries, on average, at least 1.5 years later than the rural woman does. She is more likely to accept the view that fertility should be controlled, and the means for doing so are more likely to be at her disposal.

Global Aging

Population aging is having major consequences and implications in all areas of day-to-day human life, and it will continue to do so. Global aging will affect economic growth, savings, investment and consumption, labor markets, pensions, taxation, and the transfers of wealth, property, and care from one generation to another. The numerical and proportional growth of older populations around the world is indicative of major achievements—decreased fertility rates, reductions in infant and maternal mortality, reductions in infectious and parasitic diseases, and improvements in nutrition and education—that have occurred, although unevenly, on a global scale.

The rapidly expanding numbers of older people represent a social phenomenon without historical precedent. Worldwide, the number of persons aged 60 years or over will nearly triple, increasing from 672 million in 2005 to nearly 1.9 billion by 2050. Today 60% of the elderly live in developing countries. By 2050, the number will rise to 80%. The number of people over 80 will increase from 86 mil-

lion in 2005 to 394 million in 2050 (United Nations, 2005). By 2050, the number of older people (older than 65) in the world will exceed the number of young for the first time in the history of humankind. This historic reversal in relative proportions of young and old occurred by 1998 in the more developed regions of the world (United Nations, April 2002).

In most countries, the elderly population is growing faster than the population as a whole. Almost half of the world's elderly live in China, India, the United States, and the countries of the former Soviet Union. China alone is home to more than 20% of the global total.

The oldest old (85 and older) are the fastest growing segment of the population in many countries worldwide. Unlike the elderly as a whole, the oldest old today are more likely to live in developed than developing countries, although this trend, too, is changing.

The percentage of the elderly population living alone varies widely among nations. In developed countries, percentages range from 9% in Japan to a high 40% in Sweden. In developing countries, few elderly people live alone. In China 3% of the elderly live alone, in South Korea 2%, and in Pakistan 1%. Living alone in those countries is often the result of having a spouse, siblings, and even children who have died.

To date, population aging has been a major issue mainly in the industrialized nations of Europe, Asia, and North America. In at least 30 of these countries, 15% or more of the entire population is 60 and older. Those nations have experienced intense public debate over elder-related issues such as social security costs and health care provisions (Figure 9–5).

© Banana Stock/SuperStock

In most countries the elderly population is growing faster than the population as a whole. This phenomenon represents a social development without historical precedent.

Figure 9–5 World Population 65 and Older, 2000 and 2025

Source: United Nations, Division of the Department of Economic and Social Affairs of the United Nations Secretariat, *World Urbanization Prospects: The 2001 Revision.*

SUMMARY

- Social stratification can happen in two ways:
  - People can be assigned to societal roles, using as a basis for the assignment an ascribed status—an easily identifiable characteristic, such as gender, age, family name, or skin color, over which they have no control. This will produce the caste and estate systems of stratification.
  - People's positions in the social hierarchy can be based to some degree on their achieved statuses gained through their individual, direct efforts. This is known as the class system.
- Modernization theory assumes that the economic differences among countries are due to technological and cultural differences.
- According to this theory, modernization depends on a country having a cultural environment that welcomes innovation and makes it possible for technological advancements to be incorporated into that society.
- Strong ties to religious or historical traditions are seen as the greatest barriers to modernization.
- Economic impoverishment occurs when people are discouraged from adopting technologies that would raise living standards and modernization.
- Dependency theory proposes that the economic positions of rich and poor nations are linked and cannot be understood in isolation from each other.
- Dependency and global inequality is due to the exploitation of poor societies by the rich ones.
- World population, now 6.477 billion, has doubled since 1960 and is projected to grow by half, to 9.3 billion, by 2050.
- The world's richest countries, with 20% of the global population, account for 86% of private consumption; the poorest 20% account for just 1.3%.
- Health advances in some countries have reached the point at which it appears that the population is approaching the upper limit of average life expectancy. That is not true for less developed countries.
- More than 300 million people live in 24 countries where life expectancy is less than 50 years.
- In those countries, 1 in 10 newborns dies by age 1, and 3 million a year do not survive for one week.
- Currently, 80% of the world's population does not have access to any health care.
- Malnutrition and parasitic and infectious diseases are the principal causes of death and disability in the poorer nations.
- When we look at infant and child health from a global perspective, we find that death among children is overwhelmingly a problem of the developing countries in Africa, Asia, and Latin America.
- Those countries account for 98% of the world's deaths among children younger than 5.
- Ninety-five percent of these deaths are preventable.
- There is a growing gap developing between infection with HIV in the developed world and in developing countries.
- One of the greatest causes for concern is that over the next few years, the epidemic is bound to get worse.
- Sub-Saharan Africa faces a triple challenge: providing care for the growing population of people infected with HIV, bringing down the number of new infections through more effective prevention, and coping with the effect that millions of deaths have had on the continent.
- AIDS is systematically cutting down life expectancy in the countries where the disease is most common.
- Population growth has continued throughout the past three decades despite the decline in fertility rates that began in many developing countries in the late 1970s and, in some countries, despite the toll taken by the HIV/AIDS epidemic.
- Ninety-nine percent of global natural increase—the difference between numbers of births and deaths—occurs in the developing world.
- The U.S. Census Bureau's projections indicate that within a decade or two, death rates will exceed birthrates for the world's more developed countries.
- As the growth rate in the world's more affluent nations becomes negative, *all* the net annual gain in global population will, in effect, come from the world's developing countries.
- Factors that determine the typical family size in a country include:
  - average age of marriage
  - breast-feeding
  - infant and child mortality
  - gender preferences
  - benefits and costs of children
  - the availability and use of contraception
  - income level, education, and urban or rural residence of women
- Population aging is having major consequences and implications in all areas of day-to-day human life, and it will continue to do so. Global aging will affect economic growth, savings, investment, and consumption. By 2050, the number of people older than 65 in the world will exceed the number of young.

Media Resources

The Companion Website for *Introduction to Sociology,* Ninth Edition

http://sociology.wadsworth.com/tischler9e

Supplement your review of this chapter by going to the companion website to take one of the Tutorial Quizzes, use the flash cards to master key terms, and check out the many other study aids you will find there. You will also find special features such as Wadsworth's Sociology Online Resources and Writing Companion, GSS data, and Census 2000 information at your fingertips to help you complete that special project or do some research on your own.

CHAPTER NINE STUDY GUIDE

KEY CONCEPTS

Match each concept with its definition, illustration, or explanation presented below.

a. Modernization theory
b. Dependency theory
c. Dehydration, measles, and malaria
d. Maternal education

e. Acute respiratory illness
f. Global stratification
g. Social class

h. Estate system
i. Caste system
j. Varna

_____ 1. The parts of the Hindu god Purusa comprising the Indian caste system.
_____ 2. The *most* rigid form of social stratification, based on ascribed characteristics.
_____ 3. A closed system of social stratification in which a person's social position is defined by law and membership is established mainly by inheritance.
_____ 4. A category of people who share similar opportunities and similar economic and vocational positions, lifestyles, attitudes, and behaviors.
_____ 5. Proposes that the economic positions of the rich and poor nations are linked and cannot be understood apart from each other.
_____ 6. Assumes that the economic differences among countries are due to technological and cultural differences.
_____ 7. Preventable diseases that are child killers in developing nations.
_____ 8. Influences children's chances of survival.
_____ 9. One of the biggest child killers in developing nations.
_____ 10. The degree of difference in standard of living among the countries of the world.

CENTRAL IDEA COMPLETIONS

Following the instructions, fill in the appropriate concepts and descriptions for each of the questions posed below.

1. Briefly describe the degree of inequality among the world's nations. Give three statistics that you think convey the extent of this inequality.

   _____

   _____

   _____

   a. _____

   b. _____

   c. _____

2. Describe the importance, distribution, and impact of HIV/AIDS.

   _____

   _____

   _____

   _____

3. List and explain two costs and two benefits from having children in less developed countries.

   Costs: _____

   Benefits: _____

4. Briefly explain how maternal age, maternal health, and maternal education are related to infant and child mortality.

   Maternal age: _____

   Maternal health: _____

   Maternal education: _____

5. Compare and contrast the process of mobility within the estate system with that of the class system, providing examples in your comparison.

   _____

   _____

6. Briefly define the modernization theory of global stratification.

   _____

   _____

7. What are three factors that help account for the fact that developing countries have far fewer girls than boys?

   a. _____

   b. _____

   c. _____

8. What are two outcomes of women's education that affect a country's fertility rate?

   a. _____

   b. _____

CRITICAL THOUGHT EXERCISES

1. Compare and contrast the health factors that shorten the lives of children in developing countries and then contrast them with those factors that reduce the life expectancy among adults in the same countries.

2. Compare and contrast how gender preferences influence the ratio of boys to girls in Bangladesh and China. What factors affect the preference of girls or boys in each of these countries?

INTERNET ACTIVITIES

1. The Atlas of Global Inequality (http://ucatlas.ucsc.edu/) at the University of California at Santa Cruz has a wealth of maps, charts, graphs, and information on several aspects of global inequality: income, development, trade, and foreign investment; health; gender issues; and so on.

2. The term *globalization* refers to increasing connectedness of all nations of the world, rich and poor. The Global Policy Forum (http://www.globalpolicy.org/globaliz/charts/) has graphs and articles on globalization trends in culture, economy, politics, and so on.

3. The World Bank has information on poverty and other economic indicators on countries around the world. Their website (http://web.worldbank.org/poverty) has information on policies regarding poverty.

ANSWERS TO KEY CONCEPTS

1.j 2.i 3.h 4.g 5.b 6.a 7.c 8.d 9.e 10.f

ThomsonNOW™

Reviewing is as easy as ❶ ❷ ❸

1. Before you do your final exam, take the ThomsonNOW diagnostic quiz to help you identify the areas on which you should concentrate. You will find information on ThomsonNOW and instructions on how to access all of its great resources on the foldout at the beginning of the text.

2. As you review, take advantage of ThomsonNOW's study videos and interactive Map the Stats exercises to help you master the chapter topics.

3. When you are finished with your review, take ThomsonNOW's posttest to confirm you are ready to move on to the next chapter.

10

Racial and Ethnic Minorities

© Patrik Giardino/CORBIS

Learning Objectives

After studying this chapter, you should be able to do the following:

- Describe the genetic, legal, and social approaches to defining race.
- Explain the concept of ethnic group.
- Know how the sociological concept of "minority" is used.
- Understand the relationship between prejudice and discrimination.

- Recognize the effect of institutionalized prejudice and discrimination.
- Discuss the history of immigration to the United States.
- Describe the characteristics of the major racial and ethnic groups in the United States.

Paul and Philip Malone, twin brothers, had always dreamed of becoming firefighters in the Boston Fire Department. The only problem was that the brothers received scores of 69 and 57, respectively, on the written portion of the entry test, when the passing grade was 82. At the time, Boston was under pressure to increase its minority population within the Fire Department and had separate, and lower, passing grades for nonwhites.

Paul and Philip were aware that they had a great-grandmother who was African American. The Malones had not given the issue of race much thought during their original application. They decided to apply again, only this time as African Americans. They retook the test, and this time their scores were considered passing because of the lower passing grades for minority applicants. The Malones were hired and were listed in the records as African American.

Once they were on the job, the issue of race appeared to be relatively unimportant. The Malones did their jobs well for 10 years and then decided to take the civil service examinations for lieutenants. They scored exceptionally high and their names were forwarded to the fire commissioner for promotion. As their applications were being reviewed, questions arose about their race because they did not look African American. They continued to insist that their

claim to being African American was legitimate and produced photos of their great-grandmother.

Their protests were to no avail, however, and they were fired for misrepresenting their race on the original application. They appealed the decision in the courts, but eventually the Massachusetts Supreme Judicial Court ruled against them (Hernandez, 1988, 1989).

If you believe that the Malones misrepresented their race, then what would your view be of Walter White, who was head of the National Association for the Advancement of Colored People (NAACP) from 1931 to 1955? Although he thought of himself as a black man, it has been estimated that he had no more than one-sixty-fourth African American ancestry, considerably less than that of the Malones. White would often go undercover and pass for white while investigating lynchings in the South for the NAACP. In 1923, he even deceived a Ku Klux Klan recruiter into inviting him to Atlanta to advise him on recruitment (Kilker, 1993).

Or what are we to make of Mark Linton Stebbins's claim of being black? In a close contest for a seat on the Stockton, California, city council, Ralph Lee White, an African American, lost his spot on the council to Stebbins, a pale-skinned, blue-eyed man with kinky reddish-brown hair. White claimed that

Throughout history, races have been defined along genetic, legal, and social lines, each presenting its own set of problems.

© Brownie Harris/CORBIS

Stebbins would not represent the minority district because he was white. Stebbins claimed that he was African American. Birth records showed that Stebbins's parents and grandparents were white. He has five sisters and one brother, all of whom are white. Yet, when asked to declare his race he noted: "First, I'm a human being, but I'm black."

Stebbins did not deny that he was raised as a white person. He began to consider himself black only after he moved to Stockton. "As far as a birth certificate goes, then I'm white; but I am black," he maintains. "There is no question about that."

Stebbins now belongs to an African American Baptist church and to the NAACP. Most of his friends are African American. He has been married three times, first to a white woman, and then to two African American women. He has three children from the first two marriages; two from his white wife are being reared as whites, the third from his black wife as African American. He states he considers himself black—"culturally, socially, genetically."

Ralph Lee White remained unconvinced, especially with his former council seat having gone to Stebbins. "Now, his mama's white and his daddy's white," White says, "so how can he be black? If the mama's an elephant and the daddy's an elephant, the baby can't be a lion. He's just a white boy with a permanent."

Stebbins believes the issue of race is tied to identifying with a community in terms of beliefs, aspirations, and concerns. He asserts that a person's racial identity depends on much more than birth records.

Linda Fay McCord would agree with Stebbins's view. McCord was adopted in the 1950s by a black couple. She grew up thinking she was black like her parents. When other children would make fun of her for being so white, she would insist she was black. Then one day in 1998 a stranger called her on the phone claiming to be her aunt. Linda Fay was puzzled because the caller sounded white. After a few questions the caller confirmed she was white, but also noted "so are you."

Linda Fay continued to be skeptical, so the aunt sent a copy of Linda Fay's birth certificate, which stated that she was born in Toast, North Carolina, on November 18, 1946. The birth certificate also clearly noted that her race was "white."

"I don't even know who I am," McCord now says. "I'm caught in the middle of something. My mind says I'm black. Then I look at my skin, and it says I'm white. I've come to the conclusion that color is just a state of mind" (Leland, 2002).

These four examples, which on one hand may seem lighthearted, are at the same time related to some very serious issues. Throughout history, people have gone to great lengths to determine what race a person belonged to. In many states in the United States, laws were devised to determine a person's race if, like the Malones, that person had mixed racial ancestry. Usually these laws existed for the purpose of discriminating against certain minority groups. Our examples also raise the question of whether one can change one's race from white to black, even if one has no black ancestors. Mark

Stebbins seems to think so. In this chapter, we will explore these and other issues related to race and ethnicity as we try to understand how people come to be identified with certain groups and what that membership means.

The Concept of Race

Although the origin of the word is not known, the term *race* has been a highly controversial concept for a long time. Many authorities suspect that it is of Semitic origin, coming from a word that some translations of the Bible render as "race," as in the "race of Abraham," but that is otherwise translated as "seed" or "generation." Other scholars trace the origin to the Czech word *raz*, meaning "artery" or "blood"; others to the Latin *generatio* or the Basque *arraca* or *arraze*, referring to a male stud animal. Some trace it to the Spanish *ras*, itself of Arabic derivation, meaning "head" or "origin." In all these possible sources, the word has a biological significance that implies descent, blood, or relationship.

We shall use the term **race** to refer to *a category of people who are defined as similar because of a number of physical characteristics*. Often the category is based on an arbitrary set of features chosen to suit the labeler's purposes and convenience. As long ago as 1781, German physiologist Johann Blumenbach realized that racial categories did not reflect the actual divisions among human groups. As he put it, "When the matter is thoroughly considered, you see that all [human groups] do so run into one another, and that one variety of mankind does so sensibly pass into the other, that you cannot mark out the limits between them" (Montagu, 1964a). Blumenbach believed that racial differences were superficial and changeable, and modern scientific evidence seems to support this view.

Throughout history, races have been defined along genetic, legal, and social lines, each presenting its own set of problems.

Genetic Definitions

Geneticists define race by noting differences in gene frequencies among selected groups. The number of distinct races that can be defined by this method depends on the particular genetic trait under investigation. Differences in traits, such as hair and nose type, have proved to be of no value in making biological classifications of human beings. In fact, the physiological and mental similarities among groups of people appear to be far greater than any superficial differences in skin color and physical characteristics.

Also, the various so-called racial criteria appear to be independent of one another. For example, any form of hair may occur with any skin color; a narrow nose gives no clue to an individual's skin pigmentation or hair texture. Thus, Australian aborigines have dark skins, broad noses, and an abundance of curly-to-wavy hair; Asiatic Indians also have dark skins but have narrow noses and straight hair. Likewise, if head form is selected as the major criterion for sorting, an equally diverse collection of physical types will appear in each category thus defined. If people are sorted on the basis of skin color, therefore, all kinds of noses, hair, and head forms will appear in each category.

Legal Definitions

By and large, legal definitions of race have not been devised to determine who is black or of another race, but rather who is not white. The laws were to be used in instances in which separation and different treatments were to be applied to members of certain groups. Segregation laws are an excellent example. If railroad conductors had to assign someone to either the black or white cars, they needed fairly precise guidelines for knowing whom to seat where. Most legal definitions of race were devices to prevent blacks from attending white schools, serving on juries, holding certain jobs, or patronizing certain public places. The official guidelines could then be applied to individual cases. The common assumption that "anyone not white was colored," although imperfect, did minimize ambiguity.

There has been, however, very little consistency among the various legal definitions of race that have been devised. The state of Missouri, for example, made "one-eighth or more Negro blood" the criterion for nonwhite status. Georgia was even more rigid in its definition and noted:

> The term "white person" shall include only persons of the white or Caucasian race, who have no ascertainable trace of either Negro, African, West Indian, Asiatic Indian, Mongolian, Japanese, or Chinese blood in their veins. No person, any of whose ancestors [was] . . . a colored person or person of color, shall be deemed to be a white person.

Virginia had a similar law but made exceptions for individuals with one-fourth or more American Indian "blood" and less than one-sixteenth Negro "blood." Those Virginians were regarded as Indians as long as they remained on an Indian reservation, but if they moved, they were regarded as blacks (Berry & Tischler, 1978; Novit-Evans & Welch, 1983).

Most of these laws are artifacts of the segregation era. However, if people think that all vestiges of them have disappeared, they are wrong. As recently as 1982, a dispute arose over Louisiana's law requiring anyone of more than one-thirty-second African descent to be classified as black.

Louisiana's one-thirty-second law is actually of recent vintage, having come into being in 1971. Before this law, racial classification in Louisiana depended on what was referred to as "common repute." The 1971 law was intended to eliminate racial classifications by gossip and inference. In September 1982, Susie Guillory Phipps obtained a copy of her birth certificate so that she could apply for a passport. She was surprised to see that her birth certificate classified her as black. Phipps, who at the time was 49, had lived her entire life as a white person. She requested that her race be noted as white. The state objected and produced an 11-generation family tree with ancestors Phipps knew nothing about, including an early eighteenth-century black slave and a white plantation owner. Phipps responded: "My children are white. My grandchildren are white. Mother and Daddy were buried white." Louisiana was not convinced and calculated that she was three-thirty-seconds black, more than enough to make her black under state law (Novit-Evans & Welch, 1983; Cose, 1997).

Social Definitions

The social definition of race, which is the decisive one in most interactions, pays little attention to an individual's hereditary physical features or to whether his or her percentage of "Negro blood" is one-fourth, one-eighth, or one-sixteenth. According to social definitions of race, if a person presents himself or herself as a member of a certain race and others respond to that person as a member of that race, then it makes little sense to say that he or she is not a member of that race.

In Latin American countries, having African ancestry or African features does not automatically define an individual as black. For example, in Brazil many individuals are listed in the census as white and are considered white by their friends and associates even if they had a grandparent who was of pure African descent. It is much the same in Puerto Rico, where anyone who is not obviously of African descent is classified as either mulatto or white.

The U.S. Census relies on a self-definition system of racial classification and does not apply any legal or genetic rules. In 2000 the Census Bureau noted "Race is a self-identification data item in which respondents choose the race or races with which they most closely identify" (Grieco & Cassidy, 2001).

The 2000 census was also the first that allowed people to check more than one category under race. People were able to declare themselves as members of any one or more of five racial categories: American Indian/Alaskan Native, Asian, African American, Native Hawaiian/Pacific Islander, or white. Those listing themselves as white and a member of a minority were counted as a minority (Holmes, 2000). (See Table 10–1 for a listing of the various racial and ethnicity categories the Census Bureau has used over the years.)

Multiracial Ancestry The new census policy allowing people to choose more than one racial category has ignited debate among various interest groups. Parents of mixed-race children liked the change because they no longer needed to pick one race for their offspring. Others were concerned that the new rules would decrease the official numbers of those considered African American or Native American and decrease their political clout (Holmes, 2000).

In the 2000 census, 6.8 million people (2.4% of the population) identified themselves as belonging to two or more races. The most common combination was "white" and "some other race" (32%). This was followed by "white" and American Indian/Alaska Native (16%), "white" and Asian (13%), and "white" and African American (11%). "There are 63 possible combinations of racial categories, but these four combinations alone account for 72% of those who selected more than one race" (AmeriStat, June 2001).

Interracial Marriage Attitudes toward interracial marriage are telling indicators of the social distance between racial groups. Between the 1660s and the 1960s, dozens of states enacted laws prohibiting interracial marriage. These laws did not just outlaw marriage between whites and blacks, but also between whites and Native Americans, or people of Chinese, Japanese, Korean, and Filipino ancestry (Kennedy, 2003).

In June 1999, the Alabama Senate "voted to repeal the state's constitutional prohibition against interracial marriage." Even though the U.S. Supreme Court struck down prohibitions against interracial marriage 32 years earlier, in parts of rural Alabama, probate judges would still refuse to issue marriage licenses to interracial couples (Greenberg, 1999).

Currently, 65% of Americans say they approve of marriage between blacks and whites, whereas 29% say they disapprove. Blacks are more likely than whites to approve of interracial marriage, by a margin of 78% to 60% (Gallup Poll, June 18, 2002). The interesting paradox is that at the same time that people are becoming more tolerant of interracial marriage, they are increasingly more likely to celebrate their racial and ethnic heritages.

About 5% of U.S. married couples include spouses of a different race. This small percentage masks a remarkable growth in the number of interracial marriages. Since 1970, the number of black and white interracial married couples has increased from 300,000 to 3 million in 2004. Census data also show that the number of children in interracial families increased from 500,000 in 1970 to 3 million in 2004. The

Table 10–1

Race/Ethnicity Categories in the Census, 1860–2000

Census	1860	1890*	1900	1970	2000
Race	White	White	White	White	White
	Black	Black	Black	Negro or	Black
	Mulatto	Mulatto	(Negro Descent)	Black	African American
		Quadroon			or Negro
		Octoroon			
		Chinese	Chinese	Chinese	Chinese
		Japanese	Japanese	Japanese	Japanese
		Indian	Indian	Indian (American)	American Indian
					or Alaska Native
				Filipino	Filipino
				Hawaiian	Native Hawaiian
				Korean	Korean
					Asian Indian
					Vietnamese
					Guamanian or Chamorro
					Samoan
					Other Asian
					Other Pacific Islander
				Other	Some Other Race
Hispanic Ethnicity				Mexican	Mexican
					Mexican American
					or Chicano
				Puerto Rican	Puerto Rican
				Central/South American	
				Cuban	Cuban
				Other Spanish	Other Spanish/ Hispanic/Latino
				(None of these) Hispanic/Latino	Not Spanish

*In 1890, mulatto was defined as a person who was three-eighths to five-eighths black. A quadroon was one-quarter black and an octoroon one-eighth black.

National Center for Health Statistics acknowledges that these numbers are probably undercounts, because the father's race is unspecified in a significant number of births each year.

Native Americans are the most likely to marry outside of their racial group. In fact, they are more likely to marry a white person than another Native American. People of Asian ancestry are also likely to marry interracially. In about 15% of married couples with an Asian American member, the other member is not Asian. Racial intermarriage is much higher among native-born Asians, where the percentage of those who are intermarried reaches 40%.

As more Asian Americans marry interracially, future generations of Asian Americans may increasingly blend with other racial and ethnic groups, mirroring the experience of European ethnic groups in the United States over the past century. Interestingly, the future size of the U.S. Asian population depends in part on whether the children of these marriages identify themselves as Asians. If they do, the Asian American population will increase faster than projected.

African Americans are much less likely to marry outside their race. About 9% of couples with a black spouse include a spouse of another race. The vast majority of these are black-white unions.

Whites are the least likely to marry outside of their race, with only about 3% of married couples including a white person married to a person of another race. Most of these whites are married to an Asian or Native American.

Interracial births are highest for Native Americans —about one-half of all Indian births are biracial. About 20% of births to Asian women are biracial. Only about 5% of births to white and black women are biracial.

Most people with one white and one black parent, when given the opportunity to label themselves, have historically chosen (or been forced to choose, as noted elsewhere in this chapter) one parent's racial

Table 10–2

Facts about Racial Intermarriage

- There are over 3 million racial intermarriages a year in the United States.
- Racial intermarriages represent 5.4% of all married couples. This is up from 1% in 1970.
- The most common types of intermarriages are between white men and Asian or multiple-race women.
- Intermarriage between minority racial groups is much less likely.
- The least common type of intermarriage is between whites and blacks.
- People who intermarry are younger and better educated than average couples.
- Three million children are growing up in interracial families. This is up from 900,000 in 1970.
- Black men are much more likely to intermarry than black women. Ten percent of black men have a nonblack spouse.
- More than 10% of the married couples in Hawaii, California, Oklahoma, Alaska, and Nevada were interracial.

Source: Sharon M. Lee and Barry Edmonston, "New Marriages, New Families: U.S. Racial and Hispanic Intermarriage," *Population Bulletin* 60, No. 2 (Washington, D.C.: Population Reference Bureau, 2005.

identity, and that most often has been and continues to be black. Confusion arises because one can be black and have "white blood," even to the point of having a white parent, but historically one could not be white and have "black blood."

Regional variations can also indicate that different multiracial identities exist in parts of the United States. The most multiracial U.S. states have significantly different racial combinations. In Hawaii, the most multiracial state, the most common combinations are Asian and Native Hawaiian/other Pacific Islander, and white and Asian. In Oklahoma, two-thirds of the multiracial population is white and American Indian. In California, the most frequent combination is white and "some other race" (AmeriStat, November 2001). (See Table 10–2 for some facts about the multiracial population.)

The Concept of Ethnic Group

An **ethnic group** *has a distinct cultural tradition that its own members identify with and that may or may not be recognized by others* (Glazer & Moynihan, 1975). An ethnic group need not necessarily be a numerical minority within a nation (although the term sometimes is used that way).

Many ethnic groups form subcultures (see Chapter 3). They usually possess a high degree of internal loyalty and adherence to basic customs, maintaining a similarity in family patterns, religion, and cultural values. They often possess distinctive folkways and mores; customs of dress, art, and ornamentation; moral codes and value systems; and patterns of recreation. The whole group is usually devoted to something, such as a monarch, religion, language, or territory. Above all, members of the group have a strong feeling of association. The members are aware of a relationship because of a shared loyalty to a cultural tradition.

The folkways may change, the institutions may become radically altered, and the object of allegiance may shift from one trait to another, but loyalty to the group and the consciousness of belonging remain as long as the group exists.

An ethnic group may or may not have its own separate political unit; it may have had one in the past, it may aspire to have one in the future, or its members may be scattered throughout existing countries. Political unification is not an essential feature of this classification. Accordingly, despite the unique cultural features that set them apart as subcultures, many ethnic groups—for example, Arabs, French Canadians, Flemish, Scots, Jews, and Pennsylvania Dutch—are part of larger political parties.

The Concept of Minority

Whenever race and ethnicity are discussed, it is usually assumed that the object of the discussion is a minority group. Technically this is not always true, as we will see shortly. A minority is often thought of as being few in number. The concept of minority, rather than implying a small number, should be thought of as implying differential treatment and exclusion from full social participation by the dominant group in a society. In this sense, we will use Louis Wirth's definition of a **minority** as *a group of people who, because of physical or cultural characteristics, are singled out from others in society for differential and unequal treatment, and who therefore regard themselves as objects of collective discrimination* (Linton, 1936).

In his definition, Wirth speaks of physical and cultural characteristics and not of gender, age, disability, or undesirable behavioral patterns. Clearly he is referring to racial and ethnic groups in his definition of minorities. Some writers have suggested, however, that many other groups are in the same position as those more commonly thought of as minorities and endure the same sociological and psychological problems. In this light, women, homosexuals, adolescents, the aged, the handicapped, the radical right or left, and intellectuals can be thought of as minority groups.

Many ethnic groups form subcultures with a high degree of internal loyalty and adherence to basic customs.

© Ragu Rai/Magnum

Problems in Race and Ethnic Relations

All too often when people with different racial and ethnic identities come together, frictions develop among the groups. People's suspicions and fears are often aroused by those whom they feel to be different.

Prejudice

There are many definitions of prejudice. People, particularly those with a strong sense of identity, often have feelings of prejudice toward others who are not like themselves. For example, in 1945, 56% of the American public said they opposed a law that would require employees to work alongside people of any race or color (Gallup Poll, 1972). In 1966, 3 in 10 Americans (31%) thought houses for sale in their neighborhood should not be offered to

people regardless of race or nationality (Gallup Poll, 1972).

Literally, *prejudice* means a "prejudgment" or "an attitude with an emotional bias" (Wirth, 1944). However, this definition has a problem. All of us, through the process of socialization, acquire attitudes, which may not be in response only to racial and ethnic groups but also to many things in our environment. We come to have attitudes toward cats, roses, blue eyes, chocolate cheesecake, television programs, and even ourselves. These attitudes run the gamut from love to hate, from esteem to contempt, from loyalty to indifference. How have we developed these attitudes? Has it been through the scientific evaluation of information, or by other, less logical means? For our purposes we will define **prejudice** as *an irrationally based negative, or occasionally positive, attitude toward certain groups and their members.*

What is the cause of prejudice? Although pursuing this question is beyond the scope of this book, we can list some of the uses to which prejudice is put and

TECHNOLOGY AND SOCIETY

Hate Sites on the Web

In recent years we have seen the rise of hate sites on the World Wide Web. Hate sites are generally defined as those that "express prejudiced and resentful views about a particular group of people, such as blacks or Jews" ("Hate Speech," 1999). The Southern Poverty Law Center (SPLC), a civil-rights group in Montgomery, Alabama, regularly monitors these sites. The actual number of hate sites is in itself open to debate. Estimates can range anywhere between 400 to 1,200.

In addition to sites that are specifically antiblack, anti-Jewish, or anti-immigrant, there are hate sites known as "Third Position" sites. They express a mix of "left" and "right" ideas with strong neo-fascist overtones. They can be divided into two groups. The first group includes those sites affiliated with the European Liberation Front and its ally, the Liaison Committee for Revolutionary Nationalism. Many of those behind this first group of sites are adherents of Odinism, a neo-Pagan religion popular among skinheads. The second group of sites are affiliated with a key British group, the International Third Position.

Many people would like to ban hate sites. But groups who champion free speech, such as the American Civil Liberties Union, oppose proposals to ban hate sites, noting that hate groups have First Amendment rights also.

In an attempt at a compromise, the Anti-Defamation League has developed HateFilter, a free software product designed to act as a gatekeeper. It blocks access to websites of individuals or groups that, in the judgment of the Anti-Defamation League, advocate hatred, bigotry, or even violence toward groups on the basis of their religion, race, ethnicity, sexual orientation, or other immutable characteristics.

Some observers fear that filters could become subject to manipulation by interest groups who want certain sites blocked. At the request of gay activist groups, CyberNOT software blocks the Internet site of the American Family Association, a group that believes

---

the social functions it serves. First, a prejudice, simply because it is shared, helps draw together those who hold it. It promotes a feeling of "we-ness," of being part of an in-group, and it helps define such group boundaries. Especially in a complex world, belonging to an in-group and consequently feeling special or superior can be an important social identity for many people.

Second, when two or more groups are competing for access to scarce resources (jobs, for example), it makes it easier on the conscience if one can write off his or her competitors as somehow less than human or inherently unworthy. Nations at war consistently characterize each other negatively, using terms that seem to deprive the enemy of any humanity whatsoever.

Third, psychologists suggest that prejudice allows us to project onto others those parts of ourselves that we do not like and therefore try to avoid facing. For example, most of us feel stupid at one time or another. How comforting it is to know that we belong to a group that is inherently more intelligent than another group. Who does not feel lazy sometimes? But how good it is that we do not belong to that group—the one everybody knows is lazy.

Of course, prejudice also has many negative consequences, or dysfunctions, to use the sociological term. For one thing, it limits our vision of the world around us, reducing social complexities and richness to a sterile and empty caricature. Aside from this effect on us as individuals, prejudice also has negative consequences for the whole of society. Most notably, it is the necessary ingredient of discrimination, a problem found in many societies, including our own. (See "Technology and Society: Hate Sites on the Web" for an example of new outlets for prejudice.)

Discrimination

Prejudice is a subjective feeling, whereas discrimination is an overt action. **Discrimination** refers to *differential treatment, usually unequal and injurious, accorded to individuals who are assumed to belong to a particular category or group*. Discrimination against African Americans and other minorities has occurred throughout U.S. history. At the start of World War II, for example, blacks were not allowed in the Marine Corps; blacks could be admitted into the Navy only as mess stewards; and the Army had a 10% quota on black enlistments. (For an example of how the military deals with racial issues, see "Our Diverse Society: Racial Integration in the Military.")

Prejudice does not always result in discrimination. Although our attitudes and our overt behavior are closely related, they are neither identical nor dependent on each other. We may have feelings of

homosexuality is immoral and anti-Christian. The group does not consider itself a hate site and believes that its First Amendment rights are being violated.

Some think that our fear of hate sites is overblown. According to the SPLC, the membership of hate groups has remained about the same over the past few years. Despite the rhetoric about the global reach of the Internet, recruitment has not increased measurably.

David Goldman, executive director of the now defunct Hatewatch, a site that from 1995 to 2001 monitored hate sites, believes that hate groups, which once thrived and created fear in the shadows, wither and hide from the public scrutiny on the Internet. These groups did not count on the fact that forcing their way into people's homes via the Web would have the effect of mobilizing people to join in the fight against them. "Far from persuading a supposed 'silent white majority' of angry Aryans to join their ranks," Goldman says, "these self-proclaimed white warriors made moms and dads into determined anti-hate activists" (Chaudhry, 2000).

Goldman believes that the news is much more encouraging than anyone expected: Hate groups have done an extremely poor job of using the Internet to increase their membership, failing to gain widespread acceptance for their belief that bigotry, hate, and violence are viable responses to human diversity. This is not to say that we no longer have cause for concern. The advent of the "lone wolf" gunman whose hatred may be fed by hate group propaganda, bigoted organizations who use e-commerce to support their hateful enterprises, and the newly emerging racist cyberterrorists all will continue to challenge law enforcement and online civil rights.

Sources: "Hate Speech on the Internet," June 4, 1999, *Issues and Controversies on File;* "Hate Sites Bad Recruiting Tools," by L. Chaudhry, *Wired News,* http://www.wired.com/news/culture/0,1284,36478,00 .html, accessed May 23, 2000.

---

antipathy without expressing them overtly or even giving the slightest indication of their presence. This simple fact—namely, that attitudes and overt behavior vary independently—has been applied by Robert K. Merton (1969/1949) to the classification of racial prejudice and discrimination. There are, he believes, the following four types of people.

Unprejudiced Nondiscriminators
These people are neither prejudiced against the members of other racial and ethnic groups, nor do they practice discrimination. They believe implicitly in the American ideals of justice, freedom, equality of opportunity, and dignity of the individual. Merton recognizes that people of this type are properly motivated to spread the ideals and values of the creed and to fight against those forms of discrimination that make a mockery of them. At the same time, unprejudiced nondiscriminators have their shortcomings. They enjoy talking to one another, engaging in mutual exhortation and thereby giving psychological support to one another. They believe their own spiritual house is in order, thus they do not feel pangs of guilt and accordingly shrink from any collective effort to set things right.

Unprejudiced Discriminators
This type includes those who constantly think of expediency. Though they themselves are free from racial prejudice, they will keep silent when bigots speak out. They will not condemn acts of discrimination but will make concessions to the intolerant and will accept discriminatory practices for fear that to do otherwise would hurt their own position.

Prejudiced Nondiscriminators
This category is for the timid bigots who do not accept the ideal of equality for all but conform to it and give it lip service when the slightest pressure is applied. Those who hesitate to express their prejudices when in the presence of those who are more tolerant belong in this category. Among them are employers who hate certain minorities but hire them rather than run afoul of affirmative action laws and labor leaders who suppress their personal racial bias when the majority of their followers demand an end to discrimination.

Prejudiced Discriminators
These are the bigots, pure and unashamed. They do not believe in equality, nor do they hesitate to give free expression to their intolerance, both in their speech and in their actions. For them, attitudes and behavior do not conflict. They practice discrimination, believing that it is not only proper but in fact their duty to do so (Berry & Tischler, 1978).

OUR DIVERSE SOCIETY

Racial Integration in the Military

When one thinks of institutions in America that have been at the vanguard of social change, the U.S. Army does not spring readily to mind. Yet over the past two decades, the Army has become the most successfully integrated institution in America—from the ranks of the lowliest privates to the highest level of command. Sociologists Charles Moskos and John Sibley Butler studied the military and found that what has made the Army's experience so striking is that this success was achieved without resort to numerical quotas or manipulation of test scores, nor has the promotion of black officers engendered the racial resentment that has become all too common in business, government, and higher education.

A visitor to an Army dining facility (as the old "mess hall" has been renamed) is likely to see a sight rarely encountered elsewhere in American life: blacks and whites commingling and socializing by choice. This stands in stark contrast to the self-imposed racial segregation in most university dining halls today—not to mention within most other locales in our society. In the Army whites and blacks not only inhabit the same barracks but also patronize equally such nonduty facilities as barber shops, post exchanges, libraries, movie theaters, and snack bars. And, in the course of their military duties, blacks and whites work together with little display of racial animosity. Give or take a surly remark here, a bruised sensibility there, the races get along remarkably well.

Even off duty and off post, far more interracial mingling is noticeable around military bases than in civilian life. Most striking, the racial integration of military life has some carry-over into the civilian sphere. The most racially integrated communities in America are towns with large military installations.

One key difference between the way the Army and many civilian organizations reflect the racial climate is that an officer's failure to maintain a bias-free environment is an absolute impediment to advancement in a military career. Most soldiers Moskos and Butler spoke to could not conceive of an officer who expressed racist views being promoted. This is not true in many civilian organizations.

Another, perhaps more important, distinction is that the Army does not lower its standards in order to assure an acceptable racial mix. When necessary, the Army makes an effort to compensate for educational or skill deficiencies by providing specialized, remedial training. Affirmative action exists, but without timetables or quotas governing promotions. Goals that do exist are pegged to the proportion of blacks in the service promotion pool. Even then, these goals can be bypassed if the candidates do not meet standards.

In this regard, compared with most private organizations, the Army has an obvious advantage. The Army can maintain standards while still promoting African Americans at all levels because of the large number of black personnel within the organization. The Army's experience with a plenitude of qualified black personnel illuminates an important lesson. When not marginalized, African American cultural patterns can mesh with and add to the effectiveness of mainstream organizations. The overarching point is that the most effective and fairest way to achieve racial equality of opportunity in the United States is to increase the number of qualified African Americans available to fill positions. Doing so is no small task. But as an objective and basic principle, it is infinitely superior to a system under which blacks in visible positions of authority are presumed to have benefited from relaxed standards.

Source: From *All That We Can Be: Black Leadership and Racial Integration in the Army* (pp. 2, 9–10), by C. C. Moskos and J. S. Butler, 1996, New York: Basic Books. Copyright © 1996 by the Twentieth Century Fund. Reprinted by permission of Basic Books, a member of Perseus Books, L.L.C.

Knowing a person's attitudes does not mean that that person's behavior always can be predicted. Attitudes and behavior are frequently inconsistent because of such factors as the nature and magnitude of the social pressures in a particular situation. The influence of situational factors on behavior can be seen in Figure 10–1.

Institutional Prejudice and Discrimination

Sociologists tend to distinguish between individual and institutional prejudice and discrimination. When individuals display prejudicial attitudes and discriminatory behavior, it is often based on the assumption of the out-group's genetic inferiority. By contrast, **institutionalized prejudice and discrimination** refer to *complex societal arrangements that restrict the life chances and choices of a specifically defined group in comparison with those of the dominant group*. In this way, benefits are given to one group and withheld from another. Society is structured in such a way that people's values and experiences are shaped by a prejudiced social order. Discrimination is seen as a by-product of a purposive attempt to maintain social, political, and economic advantage (Davis, 1979).

Figure 10–1   The Interaction of Prejudice and Discrimination

As this diagram shows, the degree of social pressure being exerted can cause individuals of inherently dissimilar attitudes to exhibit relatively similar behaviors in a given situation.

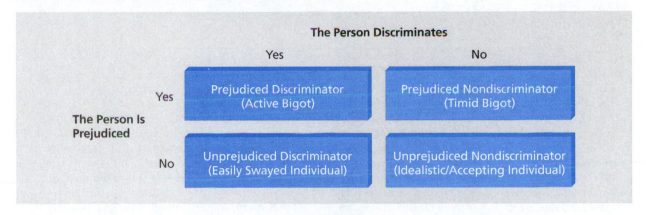

		The Person Discriminates	
---	---	Yes	No
The Person Is Prejudiced	Yes	Prejudiced Discriminator (Active Bigot)	Prejudiced Nondiscriminator (Timid Bigot)
	No	Unprejudiced Discriminator (Easily Swayed Individual)	Unprejudiced Nondiscriminator (Idealistic/Accepting Individual)

Some people argue that institutionalized prejudice and discrimination are responsible for the substandard education that many African Americans receive in the United States. Schools that are predominantly black tend to be inferior at every level to schools that are predominantly white. The facilities for blacks are usually of poorer quality than are those for whites. Many blacks also attend unaccredited black colleges, where the teachers are less likely to hold advanced degrees and are poorly paid. The poorer education that blacks receive is one of the reasons they generally are in lower occupational categories than whites are. In this way, institutionalized prejudice and discrimination combine to maintain blacks in a disadvantaged social and economic position (Kozol, 1992).

Patterns of Racial and Ethnic Relations

Relations among racial and ethnic groups seem to include an infinite variety of human experiences. They run the gamut of emotions and appear to be unpredictable, capricious, and irrational. They range from curiosity and hospitality, at one extreme, to bitter hostility at the other. In this section, we will show that a limited number of outcomes exist when racial and ethnic groups come into contact. These include assimilation, pluralism, subjugation, segregation, expulsion, and annihilation. In some cases, these categories overlap—for instance, segregation can be considered a form of subjugation—but each has distinct traits that make it worth examining separately. (For a discussion of the value of the debate about race and intelligence, see "Controversies in Sociology: Is the Debate on Race and Intelligence Worthwhile?")

Assimilation

In 1753, 23 years before he signed the Declaration of Independence, Benjamin Franklin wondered about the costs and benefits of German immigration. On one hand, he commented that Germans are "the most stupid" and "great disorders may one day arise among us" because of these immigrants. Yet on the other hand he pointed out that they "contribute greatly to the improvement of a country." He finally decided that the benefits of German immigration could outweigh the costs if we "distribute them more equally, mix them with the English, establish English schools where they are now too thick settled" (Borjas, 1999).

Franklin was concerned with the assimilation of the German immigrants. **Assimilation** *is the process whereby groups with different cultures come to have a common culture.* It refers to more than just dress or language, including less tangible items such as values, sentiments, and attitudes. Assimilation is really referring to the fusion of cultural heritages.

Assimilation is the integration of new elements with old ones. The transferring of culture from one group to another is a highly complex process, often involving the rejection of ancient ideologies, habits, customs, attitudes, and language. It also includes the elusive problem of selection. Of the many possibilities presented by a culture, which ones will another culture adopt? Why did some Native Americans, for example, when they were confronted with the white civilization, take avidly to guns, horses, rum, knives, and glass beads, while showing no interest in certain other features to which whites themselves attached the highest value?

In the process of assimilation, one society sets the pattern, for the give and take of culture seems never to operate on a 50-50 basis. Invariably, one group has a much larger role in the process than the other, and various factors interact to make it so. Usually one of the societies enjoys greater power or prestige than the other, giving it an advantage in the assimilation process; one is better suited to the environment than the other; or one has greater numerical strength than

CONTROVERSIES IN SOCIOLOGY

Is the Debate on Race and Intelligence Worthwhile?

In their controversial book *The Bell Curve*, Richard Herrnstein and Charles Murray argued that a transformation was occurring in American society, with a cognitive elite becoming the leaders of the country and another group of people with lesser endowments and abilities doomed to labor on the fringes of society. Herrnstein and Murray also noted that American society was experiencing a crisis because certain groups, such as African Americans, possessed average intelligence levels below those of white Americans. Sociologist Orlando Patterson questions the interest that people have shown in this proposition. He suggests that the discussion would not be important if we were committed to accepting every group as a member of a society that cannot be broken or separated into parts.

It has long been established that significant differences exist in measured average IQ. . . . [Whites] in Tennessee and rural Georgia score significantly lower than [whites] in the Northeast, and, on average, live in far worse conditions than their northern counterparts. No one, however, has ever chosen to call the nation's attention to these differences, except in a sympathetic and appropriately sensitive manner. We do not neglect them, but neither do we make a national issue of them, in the process wantonly insulting and dishonoring these people. No one raises the issue of whether or not these two states in the South are an intellectual drag on the nation; no one bemoans the fact that were they not a part of the nation our ranking in the international IQ parade would be much higher. . . . Why so? Because [whites] in Tennessee and rural Georgia are seen as belonging to the social and moral community that constitutes the American people. Whatever we are incorporates them. If they are different from Northeasterners in one or more respects, the reasoning goes, that is just a fact of our national life, a part of the diverse fabric we take pride in.

The same goes for the elderly. Until very recently, there was a consensus that intellectual functioning declined with age, especially after middle age. . . .

According to recent estimates, by 2025 one in five Americans will be over sixty-five. . . . [There] has also been the growing tendency of those over sixty-five remaining in the workplace, reflected most dramatically in the abolition of compulsory ages of retirement. Hence we now face the prospect that one of the fastest growing and most powerful segments of the population—including the working population—is, according to the IQ specialists, an intellectually impaired group with rapidly declining cognitive competence. This surely ought to provoke great alarm among those who make it their business to guard the intellectual integrity of the nation. After all, one in five is substantially larger than 13 percent, which is the menace posed by intellectually inferior Afro-Americans; and the elderly are growing a lot faster, from all demographic projections. And yet, we have heard almost nothing on this subject from any of those accustomed to warning the nation about impending psychological and social disasters.

Why not? For the same reason that we do not engage in public hand-wringing about the intellectual inferiority of rural [white] Georgians and Tennesseeans. The elderly are an integral part of our community; whatever we are as a people, the reasoning goes—and quite rightly, I might add—they help define it. If an important and essential part of what we are is in intellectual decline, then so are we, and so be it.

A principle is at work here. . . . We may call it the principle of infrangibility. It refers to our commitment to a unity that cannot be broken or separated into parts, a commitment to the elements of a moral order and social fabric that is inviolable and cannot be infringed. Afro-Americans too are an infrangible part of the nation.

Source: Excerpted from *The Ordeal of Integration* (pp. 137–139), by O. Patterson, 1997, Washington: DC: Civitas Counterpoint. Used with permission from the author.

---

the other. Thus, the pattern for the United States was set by the British colonists, and to that pattern the other groups were expected to adapt. This process has often been referred to as **Anglo conformity**—*the renunciation of the ancestral cultures in favor of Anglo-American behavior and values* (Gordon, 1964; Berry & Tischler, 1978).

The Anglo-conformity viewpoint was at its strongest around World War I, as demonstrated by this excerpt from a speech by President Woodrow Wilson:

You cannot dedicate yourself to America unless you become in every respect and with every purpose of your will thorough Americans. You cannot become thorough Americans if you think of yourselves in groups. America does not consist of groups. A man who thinks of himself as belonging to a particular national group in America has not yet become an American, and the man who goes among you to trade upon your nationality is no worthy son to live under the Stars and Stripes. (Wilson, 1915)

Although assimilation frequently has been a professed political goal in the United States, it has seldom been fully achieved. Consider the case of the Native Americans: In 1924, they were granted full U.S. citizenship. Nevertheless, the federal government's policies regarding the integration of Native Americans into American society wavered back and forth until the Hoover Commission Report of 1946 became the guideline for all subsequent administrations. The report stated that

A program for the Indian peoples must include progressive measures for their complete integration into the mass of the population as full, tax-paying [members of the larger society]. . . . Young employable Indians and the better cultured families should be encouraged and assisted to leave the reservations and set themselves up on the land or in business. (Shepardson, 1963)

However, to this day Native American groups have yet to be fully integrated into the mainstream of American life. About 54% live on or near reservations, and most of the rest live in impoverished urban areas. In addition, many Native Americans who left the reservation for greater opportunity in the cities are returning to the reservation. Despite the economic and lifestyle hardships they face on the reservation, their ethnic pride overrides any desire to assimilate.

Other groups, whether or not by choice, also have not assimilated. The Amish, for instance, have steadfastly maintained their subculture in the face of the pressures of Anglo conformity from the larger American society.

China provides an interesting example of what might be called reverse assimilation. Usually defeated minority groups are assimilated into the culture of a politically dominant group. In the seventeenth century, however, Mongol invaders conquered China and installed themselves as rulers. The Mongols were nomadic pastoralists. They were so impressed with the advanced achievements of the Chinese civilization that they gave up their own ways and took on the trappings of Chinese culture: language, manners, dress, and philosophy. During their rule, the Mongols fully assimilated the Chinese culture.

Pluralism

Pluralism, or *the development and coexistence of separate racial and ethnic group identities within a society,* is a philosophical viewpoint that attempts to produce what is considered to be a desirable social situation. When people use the term *pluralism* today, they believe they are describing a condition that seems to be developing in contemporary American society. They often ignore the ideological foundation of pluralism.

The person principally responsible for the development of the theory of cultural pluralism was Horace Kallen, born in Germany. He came to Boston at the age of 5 and was raised in an Orthodox Jewish home. As he progressed through the Boston public schools, he underwent a common second-generation phenomenon: He started to reject his home environment and religion and developed an uncritical enthusiasm for the United States. As he put it, "It seemed to me that the identity of every human being with every other was the important thing, and that the term 'American' should nullify the meaning of every other term in one's personal makeup" (Kallen, 1956).

While Kallen was a student at Harvard, he experienced a number of shocks. Working in a nearby settlement house, he came in contact with liberal and socialist ideas and observed people expressing numerous ethnic goals and aspirations. This exposure caused him to question his definition of what it meant to be an American. This quandary was compounded by his experiences in the American literature class of Barrett Wendell, who believed that Puritan traits and ideals were at the core of the American value structure. The Puritans, in turn, had modeled themselves after the Old Testament prophets. Wendell even suggested that the early Puritans were largely of Jewish descent. These ideas led Kallen to believe that he could be an unassimilated Jew and still belong to the core of the American value system.

After discovering that he could be totally Jewish and still be American, he came to realize that the application could be made to other ethnic groups as well. All ethnic groups, he felt, should preserve their own separate cultures without shame or guilt. As he put it, "Democracy involves not the elimination of differences, but the perfection and conservation of differences."

Pluralism is a reaction against the assumption of assimilation and the idea of America as a "melting pot" in which immigrants from around the world combine into the "new metal" of the American. It is a philosophy that not only assumes that minorities have rights but also considers the lifestyle of the minority group to be a legitimate, and even desirable, way of participating in society. The theory of pluralism celebrates the differences among groups of people. The theory also implies a hostility to existing inequalities in the status and treatment of minority groups. Pluralism has provided a means for minorities to resist the pull of assimilation by allowing them to claim that they constitute the very structure of the social order. From the assimilationist point of view, the minority is seen as a subordinate group that should give up its identity as quickly as possible. Pluralism, on the other hand, assumes that the minority is a primary unit of society and that the unity of the whole depends on the harmony of the various parts.

Switzerland provides an example of balanced pluralism that, so far, has worked exceptionally well (Kohn, 1956). After a short civil war between Catholics and Protestants in 1847, a new constitution —drafted in 1848—established a confederation of cantons (states), and church-state relations were left up to the individual cantons. The three major languages—German, French, and Italian—were declared official languages for the whole nation, and their respective speakers were acknowledged as political equals (Petersen, 1975).

Switzerland's linguistic regions are culturally quite distinctive. Italian-speaking Switzerland has a Mediterranean flavor; in French-speaking Switzerland, one senses the culture of France; and German-speaking Switzerland is distinctly Germanic. However, all three linguistic groups are fiercely pro-Swiss, and the German Swiss especially have strong anti-German sentiments.

Subjugation

In theory, we could assume that two groups may come together and develop an egalitarian relationship. However, racial and ethnic groups rarely have established such a relationship. One of the consequences of the interaction of racial and ethnic groups has been **subjugation**—*the subordination of one group and the assumption of a position of authority, power, and domination by the other.* The members of the subordinate group may for a time accept their lower status and even devise ingenious rationalizations for it.

For the most part this is so because there are few instances in which group contact has been based on the complete equality of power. Differences in power will invariably lead to a situation of superior and inferior positions. The greater the discrepancy in the power of the groups involved, the greater the extent and scope of the subjugation will be.

By the mid-1870s, the various Native American tribes were at the mercy of white Europeans. Native American traditions, beliefs, and ways of life were condemned as backward and immoral. It was thought that the best way to help Native Americans was to ensure that their tribal cultures were destroyed. The Indian peoples would have to be forced to assimilate into the mainstream culture. President Benjamin Harrison's commissioner of Indian affairs, Thomas Jefferson Morgan, expressed this view in 1889 when he noted:

> The logic of events demands the absorption of the Indians into our national life, not as Indians, but as American citizens. . . . The Indians must conform to "the white man's ways," peaceably if they will, forcibly if they must. . . . This civilization may not be the best possible, but it is the best the Indians can get. They cannot escape it, and must either conform to it or be crushed by it. (Josephy, 1994)

To subjugate and "Americanize" the Native Americans, the government banned their religions, rituals, and sacred ceremonies. Medicine men and shamans were either jailed or exiled. Attempts were even made to stop them from speaking their tribal languages. The final step was to send Native American children to boarding schools, where they were taught to become part of the white culture. As a Taos Pueblo youth noted:

> We all wore white man's clothes and ate white man's food and went to white man's churches and spoke white man's talk. And so after a while we also began to say Indians were bad. We laughed at our own people and their blankets and cooking pots and sacred societies and dances. (Embree, 1967/1939)

Why should different levels of power between two groups lead to the domination of one by the other? Gerhard Lenski (1966) proposed that it is because people have a desire to control goods and services. No matter how much they have, they are never satisfied. In addition, high status is often associated with the consumption of goods and services.

Therefore, demand will exceed supply, and as Lenski asserts, a struggle for rewards will be present in every human society. The outcome of this struggle thus will lead to the subjugation of one group by another. When a racial or ethnic group is placed in an inferior position, its people often are eliminated as competitors. In addition, their subordinate position may increase the supply of goods and services available to the dominant group.

Segregation

Segregation, *a form of subjugation, refers to the act, process, or state of being set apart.* It places limits and restrictions on the contact, communication, and social relations among groups. Many people think of

The subjugation of Native Americans is evident in these before and after photos of three Sioux pupils at the Carlisle Indian School in Pennsylvania. At left are three boys upon their arrival at the school. At the right we see them six months later.

segregation as a negative phenomenon—a form of ostracism imposed on a minority by a dominant group—and this is most often the case. However, for some groups, such as the Amish or Chinese in America, who wish to retain their ethnicity, segregation is voluntary.

The practice of segregating people is as old as the human race itself. Examples of it exist in the Bible and in preliterate cultures. American blacks originally were segregated by the institution of slavery and later by both formal sanction and informal discrimination. Although some African Americans formed groups that preached total segregation from whites as an aid to black cultural development, for most it was an involuntary and degrading experience. The word *ghetto* originally refers to the segregated quarter of a city where the Jews in Europe were often forced to live. Native American tribes were often forced to choose segregation on a reservation in preference to annihilation or assimilation. Segregation has operated in a wide range of circumstances.

Expulsion

Expulsion is *the process of forcing a group to leave the territory in which it lives.* This can be accomplished indirectly by making life increasingly unpleasant for a group, as the Germans did for Jews after Adolf Hitler was appointed chancellor in 1933. Over the following six years, Jews were stripped of their citizenship, made ineligible to hold public office, removed from the professions, and forced out of the artistic and intellectual circles to which they belonged. In 1938, Jewish children were barred from public schools. At the same time, the government encouraged acts of violence and vandalism against Jewish communities. These actions culminated in Kristallnacht, November 9, 1938, when the windows in synagogues and Jewish homes and businesses across Germany were shattered and individuals were beaten. Under these conditions, Jews left Germany by the thousands. In 1933, some 500,000 Jews lived in Germany; by 1940, before Hitler began his "final solution"—that is, the murder of all remaining Jews—only 220,000 remained (Robinson, 1976).

Expulsion also can be accomplished through **forced migration,** or *the relocation of a group through direct action.* For example, forced migration was a major aspect of the U.S. government's policies toward Native American groups in the nineteenth century. When the army needed to protect its lines of communication to the West Coast, Colonel Christopher "Kit" Carson was ordered to move the Navajos of Arizona and New Mexico out of the way. He was instructed to kill all the men who resisted and to take

These Jews are arriving at the Nazi concentration camp known as Auschwitz. Most of them were murdered and cremated a short time later as part of the German policy of genocide against the Jews.

everybody else captive. He accomplished this in 1864 by destroying their cornfields and slaughtering their herds of sheep, thereby confronting the Navajos with starvation. After a last showdown in Canyon de Chelly, some 8,000 Navajos were rounded up at Fort Defiance. They then were marched on foot 300 miles to Fort Sumner, where they were to be taught the ways of "civilization" (Spicer, 1962).

Although expulsion is an extreme attempt to eliminate a certain minority from an area, annihilation is the most extreme action one group can take against another.

Annihilation

Annihilation refers to *the deliberate extermination of a racial or ethnic group.* In recent years it has also been referred to as *genocide,* a word coined to describe the crimes committed by the Nazis during World War II—crimes that induced the United Nations to draw up a convention on genocide. Annihilation denies an entire group of people of their right to life, in the same way that homicide denies one person of the right to life.

Sometimes annihilation occurs as an unintended result of new contact between two groups. For example, when the Europeans arrived in the Americas, they brought with them smallpox, a disease that was new to the people they encountered. Native American groups, including the Blackfeet, the Aztecs, and the Incas, who had no immunity against this disease, were nearly wiped out (McNeill, 1976). In most cases, however, the extermination of one group by another has been the result of deliberate action. Thus, the native population of Tasmania, a large island off the coast of Australia, was exterminated by Europeans during the 250 years after the island was discovered in 1642.

The largest, most systematic program of ethnic extermination was the killing of 11 million people, close to 6 million of whom were Jews, by the Nazis before and during World War II. In nearly every country the Germans occupied, the majority of the Jewish population was killed. Relatively few Jews anywhere in Europe escaped (Dwork & van Pelt, 2002). Thus, in the mid-1930s, before the war, about 3.3 million Jews lived in Poland, but at the end of the war in 1945, only 73,955 remained (Baron, 1976). Among them, not a single known family remained intact.

Although attempts have been made to portray this mass murder of Jews as a secret undertaking of the Nazi elite that was not widely supported by the German people, the historical evidence suggests otherwise. For example, during a wave of anti-Semitism (anti-Jewish prejudice, accompanied by violence and repression) in Germany in the 1880s—long before the Nazi regime—only 75 German scholars and other distinguished citizens protested publicly. During the 1930s, the majority of German Protestant churches endorsed the so-called racial principles that the Nazis used to justify first the disenfranchisement of Jews, then their forced deportation, and finally their extermination. (Jews were blamed for a bewildering combination of "crimes," including "polluting the purity of the Aryan race" and causing the rise of communism while at the same time manipulating capitalist economies through their "secret control" of banks.)

It would seem, then, that the majority of Germans supported the Nazi racial policies or at best were apathetic (Goldhagen, 1996). Although in 1943 both the Catholic Church and the anti-Nazi Confessing Church finally condemned the killing of innocent people and pointedly stated that race was no justification for murder, it is fair to say that even this opposition was "mild, vague, and belated" (Goldhagen, 2002). That such objections were raised at all suggests that the Nazis' plan to exterminate all Jews was not a well-kept military secret. The measure of its success is that some 60% of all Jews in Europe—36% of all Jews in the world—were slaughtered (computed from figures in Baron, 1976).

Another "race" also slated for extermination by the Nazis were the Gypsies, small wandering groups who appear to be the descendants of the Aryan invaders of India, the central Eurasian nomads. For the past thousand years or so, Gypsy bands have spread throughout the continents, largely unassimilated (Ulc, 1975/1969). In Europe, they were widely disliked and constantly accused of small thefts and other criminal behavior.

The sheer magnitude and horror of the Nazi attempt to exterminate the Jews provoked outrage and efforts by the nations of the world to prevent such circumstances from arising again. On December 11, 1946, the General Assembly of the United Nations passed by unanimous vote a resolution affirming that genocide was a crime under international law and that both principals and accomplices alike would be held accountable and would be punished. The assembly called for a convention on genocide that would define the offense more precisely and provide enforcement procedures for its repression and punishment. After two years of study and debate, the draft of the convention on genocide was presented to the General Assembly and adopted (United Nations, 1948). Article II of the convention defines genocide as

any of the following acts committed with intent to destroy, in whole or in part, a national, ethnical, racial or religious group as such:

(a) Killing members of the group;

(b) Causing serious bodily or mental harm to members of the group;

(c) Deliberately inflicting on the group conditions of life calculated to bring about its physical destruction in whole or part;

(d) Imposing measures intended to prevent births within the group;

(e) Forcibly transferring children of the group to another group.

The convention further provided that any of the contracting parties could call on the United Nations to take action under its charter for the "prevention and suppression" of acts of genocide. In addition, any of the contracting parties could bring charges before the International Court of Justice.

In the United States, President Harry Truman submitted the resolution to the Senate on June 16, 1949, for ratification. However, the Senate did not act on the measure, and the United States did not sign the document. In 1984, President Ronald Reagan again requested the Senate to hold hearings on the convention so that it could be signed. The United States finally signed the document in 1988.

In the more than 50 years of its existence, the Genocide Convention has never been used to bring charges of genocide against a country. Numerous examples of genocide have occurred during that period, however. It appears to serve more of a symbolic purpose, by asking nations to go on record as being opposed to genocide, than being an effective means of dealing with actual instances of genocide (Berry & Tischler, 1978).

Racial and Ethnic Immigration to the United States

Since the settlement of Jamestown in 1607, well over 45 million people have immigrated to the United States. Up until 1882, the policy of the United States was almost one of free and unrestricted admittance. The country was regarded as the land of the free, a haven for those oppressed by tyrants, and a place of opportunity. The words of Emma Lazarus, inscribed on the Statue of Liberty, were indeed appropriate:

. . . Give me your tired, your poor,
Your huddled masses yearning to breathe free;
The wretched refuse of your teeming shore.
Send these, the homeless, tempest-tost to me.
I lift my lamp beside the golden door!

To be sure, there were those who had misgivings about the immigrants. George Washington wrote to John Adams in 1794, "My opinion with respect to immigration is that except for useful mechanics and some particular descriptions of men or professions, there is no need for encouragement." Thomas Jefferson was even more emphatic in expressing the wish that there might be "an ocean of fire between this country and Europe, so that it would be impossible for any more immigrants to come hither." Such fears, however, were not widely felt. There was the West to be opened, railroads to be built, and canals to be dug; there was land for the asking. People poured across the mountains, and the young nation was eager for population.

Immigration of white ethnics to the United States can be viewed from the perspective of old migration and new migration. The old migration consisted of people from northern Europe who came before the 1880s. The new migration was much larger in numbers and consisted of people from southern and eastern Europe who came between 1880 and 1920. The ethnic groups that made up the old migration included the English, Dutch, French, Germans, Irish, Scandinavians, Scots, and Welsh. The new migration included Poles, Hungarians, Ukrainians, Russians, Italians, Greeks, Portuguese, and Armenians.

Figure 10–2 shows the number of immigrants who came to the United States in each year from 1820 to 2004. The new migration sent far more immigrants to the United States than the old migration. The earlier

Figure 10–2   Immigration to the United States, 1820–2004

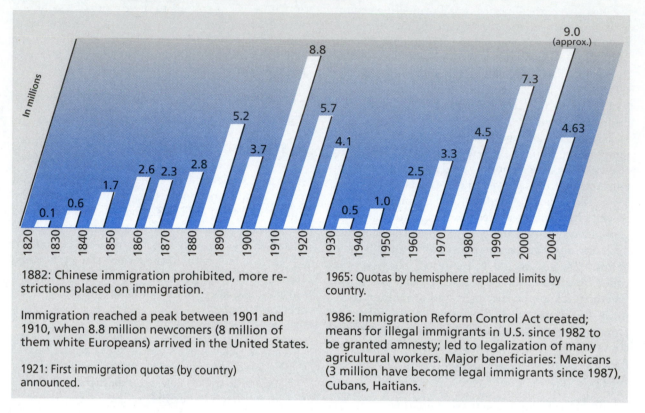

1882: Chinese immigration prohibited, more restrictions placed on immigration.

Immigration reached a peak between 1901 and 1910, when 8.8 million newcomers (8 million of them white Europeans) arrived in the United States.

1921: First immigration quotas (by country) announced.

1965: Quotas by hemisphere replaced limits by country.

1986: Immigration Reform Control Act created; means for illegal immigrants in U.S. since 1982 to be granted amnesty; led to legalization of many agricultural workers. Major beneficiaries: Mexicans (3 million have become legal immigrants since 1987), Cubans, Haitians.

Source: *Yearbook of Immigration Statistics 2004*, http://uscis.gov/graphics/shared/statistics/yearbook/2004/table1.xls, accessed October 11, 2005.

immigrants felt threatened by the waves of unskilled and uneducated newcomers, whose appearance and culture were so different from their own. Public pressure for immigration restriction increased. After 1921, quotas were established limiting the number of people who could arrive from any particular country. The quotas were designed specifically to discriminate against potential immigrants from the southern and eastern European countries. The discriminatory immigration policy remained in effect until 1965, when a new policy was established.

Table 10–3 lists the people who were excluded from immigrating to the United States during each of the periods in its history. As you can see, the United States was much more lenient during the early days of its history. However, even with periods of restrictive immigration, the United States has had one of the most open immigration policies in the world, and it continues to take in more legal immigrants than the rest of the world combined (Kotkin & Kishimoto, 1988).

Immigration Today Compared with the Past

The past 40 years have seen a marked shift in U.S. immigration patterns. From the beginning of the country's birth until the 1960s, most immigrants came primarily from northwestern European countries—Great Britain, Ireland, Germany, Scandinavia, France—and from Canada.

In 1965 a major change in U.S. immigration policy occurred. The national origins quota system, which granted visas mainly to people coming from western European countries, particularly Great Britain and Germany, was repealed. Under the new immigration system, family ties to people already living in the United States became the key factor that determined whether a person was admitted into the country. The number of people allowed to come into the country was also increased.

The change has produced such a dramatic shift in the immigrants coming to the United States that by 2003, 53.3% of the foreign-born population were from Latin America, 25% from Asia, and 13.7% from Europe. Together, Latin America and Asia accounted for 78.3% of the foreign-born population, up from 28.3% in 1970 (Larsen, 2004).

These shifts in immigration patterns have resulted in a much more racially and ethnically diverse foreign-born population. In 1890, only 1.4% of the foreign-born population was nonwhite. By 1970, 27% of this population was nonwhite, and by 2000, 75% of this population was nonwhite.

Table 10–3

United States Immigration Restrictions

1769–1875	No restrictions; open-door policy
1875	No convicts; no prostitutes
1882	No idiots; no lunatics; no people likely to need public care Start of head tax
1882–1943	No Chinese
1885	No gangs of cheap contract laborers
1891	No immigrants with dangerous contagious diseases; no paupers; no polygamists Start of medical inspections
1903	No epileptics; no insane people; no beggars; no anarchists
1907	No feeble-minded; no children under 16 unaccompanied by parents; no immigrants unable to support themselves because of physical or mental defects No immigrants from most of Asia or the Pacific islands; no adults unable to read or write Start of literacy tests
1921	No more than 3% of foreign-born of each nationality in U.S. in 1910; total about 350,000 annually
1924–1927	National Origins Quota Law; no more than 2% of foreign-born of each nationality already in U.S. in 1890; total about 150,000 annually
1940	Alien Registration Act; all aliens must register and be fingerprinted
1950	Exclusion and deportation of aliens dangerous to national security
1952	Codification, nationalization, and minor alterations of previous immigration laws
1965	National Origins Quota system abolished; no more than 20,000 from any one country outside the Western Hemisphere; total about 170,000 annually Start of restrictions on immigrants from other Western Hemisphere countries; no more than 120,000 annually Preference to refugees, aliens with relatives here, and workers with skills needed in the United States
1980	Congress passes the Refugee Act of 1980, repealing ideological and geographical preferences that had favored refugees fleeing communism and Middle Eastern countries
1986	The Immigration Reform and Control Act (IRCA) takes effect; grants amnesty to illegal immigrants living in the United States since 1982; increases sanctions against employers for hiring illegal immigrants
1990	President George Bush increases immigration quotas
1996	Laws enacted to make it easier to deport immigrants who commit crimes Tougher restrictions on ability of immigrants to collect welfare

Source: Smithsonian Institution exhibit, Washington, DC.

With the large number of people immigrating to the United States, nearly 12% were foreign-born in 2003. This is the highest percentage of foreign-born residents since World War II and more than double the 1970 level of 4.7% (Figure 10–3). One-third of the foreign-born population was from Mexico or another Central American country and about one-fourth of this population was from Asia (Larsen, 2004).

Many of these people are recent arrivals. Of the 33.5 million foreign-born people living in the United States in 2003, one-half had arrived after 1990. Sixty percent live in four states: California, New York, Florida, and Texas (Larsen, 2004). Today's immigrants are unique in their ethnic origins, education, and skills. The waves of immigration during earlier periods in U.S. history were mostly from European countries, but today's immigrants are primarily from Latin American and Asian countries.

Their education levels are at two extremes. Whereas most native-born people have completed high school, immigrants are only half as likely to have done so. At the same time, many other immigrants are highly educated, and immigrants are more likely than native-born people to have advanced college degrees. (See Figure 10–3.)

Illegal Immigration

The U.S. Census estimated that in 2003, there were about 9 million undocumented immigrants living in the United States. Of that total, 5.5 million are from Mexico. Another 1.3 million illegal immigrants are from Asia, 1.1 million from Europe, and 625,000 from South America (Bureau of the Census, March 2004).

There are two types of illegal immigrants: those who enter the United States legally but overstay their

Between 1892 and 1924, some 16 million immigrants came through Ellis Island in New York Harbor on the way to their new life in the United States.

visa limits, and those who enter illegally to begin with. Those who migrate over the border each day to work must also be figured into the total.

Since 1970, illegal immigration has figured prominently in the ethnic makeup of several regions of the United States, as illegal immigrants tend to settle in certain states. California is home to 40% of the total, with New York, Florida, Texas, and Illinois accounting for most of the rest.

In 1986, the Immigration Reform and Control Act (IRCA) was passed, a law designed to control the flow of illegal immigrants into the United States. The law makes it a crime for employers, even individuals hiring household help, to knowingly employ an illegal immigrant. Stiff fines and criminal penalties can be imposed if they do so. The law also provided legal status to illegal immigrants who entered the United States before 1982 and who have lived here continuously since. Between 1989 and 1993, 2.7 million people were granted legal resident status under special provisions of the IRCA. These people had lived and worked in the country illegally during the 1980s. Interestingly, only 4% of those who applied for amnesty under the 1986 act worked in farming, fishing, or forestry, running counter to the perception that most illegal immigrants are migrant farmworkers.

America's Ethnic Composition Today

What is America's racial and ethnic composition today? (Figure 10–4). The United States is perhaps the most racially and ethnically diverse country in the world. Unlike many other countries, it has no ethnic group that makes up a numerical majority of the population. In the following discussion, we will examine the major groups in American society.

White Anglo-Saxon Protestants

According to the 2000 census about 41 million people claim some English, Scottish, or Welsh ancestors. These Americans of British origin are often grouped together and called white Anglo-Saxon Protestants (WASPs). Although in numbers they are a minority within the total American population, they have been in America the longest (aside from the Native Americans) and, as a group, have always had the greatest economic and political power in the country. As a result, WASPs often have acted as if they were the ethnic majority in America, encouraging other ethnic groups to assimilate or acculturate to their way of life, the ideal of Anglo conformity.

Figure 10–3 **Where Do Immigrants Come From?**

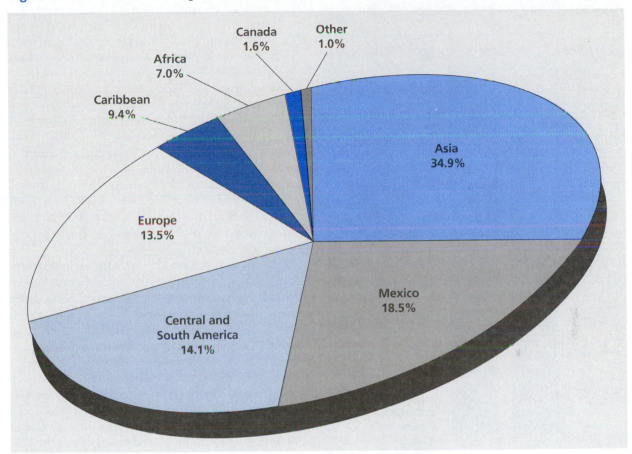

Source: *Yearbook of Immigration Statistics: 2004*, U.S. Department of Homeland Security, 2005.

The Americanization of immigrant groups has been the desired goal of the dominant WASPs during many periods in American history. Contrary to the romantic sentiments expressed on the base of the Statue of Liberty, immigrant groups who came to America after the British Protestants had become established met with considerable hostility and suspicion.

The 1830s and 1840s saw the rise of the "native" American movement, directed against recent immigrant groups (especially Catholics). In 1841, the American Protestant Union was founded in New York City to oppose the "subjugation of our country to control of the Pope of Rome, and his adherents" (Leonard & Parmet, 1972). On the East Coast, Irish Catholics were feared, and in the Midwest, the German "free-thinkers." Protestant religious organizations across America joined forces and urged "native" Americans to organize to offset foreign voting blocs. They also conducted intimidation campaigns against foreigners and tried to persuade Catholics to renounce their religion for Protestantism (Leonard & Parmet, 1972).

As the twentieth century dawned, American sentiments against immigrants from southern and eastern Europe were running high. In Boston, the Immigration Restriction League was formed, which directed its efforts toward keeping out racially "inferior" groups—

who were depicted as inherently criminal, mentally defective, and marginally educable. The league achieved its goal in 1924, when the government adopted a new immigration policy that set quotas on the numbers of immigrants to be admitted from various nations. Because the quotas were designed to reflect (and reestablish) the ethnic composition of America in the 1890s, they heavily favored the admission of immigrants from Britain, Ireland, Germany, Holland, and Scandinavia. This new policy was celebrated as a victory for the "Nordic" race (Krause, 1966).

Another expression of Anglo-conformity pressure was the Americanization Movement, which gained strength from the nationalistic passions brought on by World War I. Its stated purpose was to promote the very rapid acculturation of new immigrants. Thus, "federal agencies, state government, municipalities, and a host of private organizations joined in the effort to persuade the immigrant to learn English, take out naturalization papers, buy war bonds, forget his former origins and culture, and give himself over to patriotic hysteria" (Gordon, 1975/1961). From World War II until the early 1960s, Anglo conformity was essentially an established ideal of the American way of life. Since the upheavals of the 1960s, however, other ethnic groups have organized

Figure 10–4   Racial and Ethnic Makeup of U.S. Population, 2000 and 2050

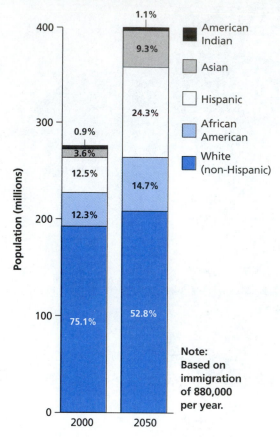

Source: U.S. Bureau of the Census, "Overview of Race and Hispanic Origin," Census 2000 Brief, March 2001; Population Projections Program.

and reacted strongly against Anglo conformity. Strong social political movements, organized along ethnic group lines, have formed. African Americans led the way in the late 1960s with the black power movement, and they were joined by Italian Americans, Mexican Americans (Chicanos), Puerto Ricans, Native Americans, and others. America once again is focusing on its ethnic diversity, and the assumptions of Anglo conformity are being questioned.

African Americans

African Americans represent the third largest race/ethnic group in the United States. According to the Census Bureau, 37.5 million African Americans live in the United States. Blacks now make up 12.8% of the U.S. population, the largest percentage since 1880 (Bureau of the Census, August, 2005).

Roughly 86% of all African Americans live in urban areas, and about 55% live in the southern states (Bureau of the Census, August 2001). This is a significant shift from the 1940s, when roughly 80% of American blacks lived in the South and worked in agriculture.

In three states, African Americans represent more than 30% of the population—Mississippi (37%), Louisiana (33%), and South Carolina (30%). In the District of Columbia (Washington, D.C.), 61% of the population is African American (Bureau of the Census, August, 2005).

Immigrants have been accounting for a greater share of all blacks in the United States as their numbers have increased. These black immigrants have come to the United States from the West Indian countries of Jamaica, Haiti, the Dominican Republic, Barbados, and Trinidad. In addition, a significant number of African blacks are also entering the United States each year. About 2 million African Americans are foreign born. In some cities such as Miami, immigrants have become a clearly distinguishable segment of the black population.

Even though the number of African immigrants from Ethiopia, Ghana, Kenya, and Nigeria has increased dramatically during the past decade, Latin America in general, and the Caribbean basin specifically, is the source of most U.S. immigrants of African descent. Of the African American foreign-born population, about 63% come from Caribbean nations (Bureau of the Census, October 1999).

The African American economic situation has been improving. Although 24.7% of the African American population fell below the poverty rate in 2004, that was lower than most years since the U.S. Census Bureau began collecting poverty data in 1959. The median African American household income of $30,134 was near an all-time high in 2004. The income of a white family has been increasing more rapidly, however, and the ratio of African American to white earnings has actually fallen (Bureau of the Census, September, 2005).

Why have African American families lost ground over the past two decades? A major reason is the growth in female-headed families, who have only one-third the annual income of African American married-couple families. Only 47% of African American families are married-couple families. Another factor is that the average African American family has fewer members in the labor force than white families. Even if African Americans and whites held comparable jobs and earned equal pay, the higher number of wage earners per family for whites would still keep the average African American family income below that of whites (Bureau of the Census, August 2001).

Hispanics (Latinos)

"Hispanic," "Hispano," "Latino," "Latin," "Mejicano," "Spanish," "Spanish-speaking"—which is the right term to use when referring to the Spanish-speaking population in the United States? Political, racial, linguistic, and historical arguments have been

Immigrants of African descent from the Caribbean basin account for an increasingly greater share of all African Americans in the United States.

advanced for using each of these terms. Historically, usage has gone from "Spanish" or "Spanish speaking" to "Latin American," "Latino," or "Hispanic." Geographically, Hispanic is preferred in the Southeast and much of Texas. New Yorkers use both Hispanic and Latino. Some people in New Mexico prefer Hispano.

Politically, the term Hispanic has been identified with more conservative viewpoints, and the term *Latino* has been associated with more liberal politics. This is partially because *Hispanic* is an English word meaning "pertaining to ancient Spain." The U.S. Census Bureau has decided to settle on one term— "Hispanic."

The U.S. Hispanic population increased more than 62% between 1990 and July 2001. This growth rate was significantly faster than that of the U.S. population as a whole. The 41.3 million Hispanics in the United States in 2004 constitute 14.1% of the total population. These totals do not include the 3.8 million people living in Puerto Rico. For the next 50 years, the Hispanic population is projected to add more people to the United States than any single race or ethnic group, reaching 98.2 million in 2050. Results from the 2000 census indicate that Hispanics have surpassed non-Hispanic African Americans as the nation's largest race/ethnic group, behind

non-Hispanic whites. Blacks constitute 12.8% of the U.S. population (Bureau of the Census, June 2005).

The 2000 census found the Hispanic population to be more diverse than it was in 1990. U.S. Hispanics of Mexican origin—the largest Hispanic group— added more than 7 million people over the decade. But the fastest growing group, in percentage terms, was "other" Hispanics, whose population increased from 5 million in 1990 to 10 million in 2000. Many of those who chose "Other Spanish/ Hispanic/Latino" on the census form were immigrants from Central or South America. "Other" Hispanics also include a growing number of people with multiethnic backgrounds who do not identify with a single country or region of origin. (For more on Hispanic diversity, see "News You Can Use: Hispanics: Racial Group? Ethnic Group? Neither?")

Hispanics are not a very well-understood segment of the population. First, no one knows exactly how many Hispanics have crossed the border from Mexico as illegal immigrants. Second, many Americans of Hispanic descent do not identify themselves as "Hispanic" on census forms and are not counted as such.

Nearly two-thirds (66.1%) of the nation's Hispanics are of Mexican origin; 14% are of Central and South

Table 10–4

Cities With Large Hispanic Populations, 2000

Place and State	Percentage Hispanic of Total Population
El Paso, TX	76.6
San Antonio, TX	58.7
Los Angeles, CA	46.5
Houston, TX	37.4
Dallas, TX	35.6
Phoenix, AZ	34.1
San Jose, CA	30.2
New York, NY	27.0
Chicago, IL	26.0
San Diego, CA	25.4

Source: U.S. Bureau of the Census, Census 2000 Summary File 1.

American origin; 9% are of Puerto Rican origin; and 4% are of Cuban origin.

More than three-fourths of Hispanics live in the West or South. The states with the highest concentration of Hispanics are New Mexico (where Hispanics constitute 41% of the total population), California (32%), Texas (30%), Arizona (23%), Nevada (17%), and Florida, Colorado, and New York (15% each). (See Table 10–4 for cities with the largest concentrations of Hispanics.)

People born in Latin America can be found all across the United States, but most live in only a few areas. The difference is the place of birth. For example, about three of four U.S. residents born in the Caribbean live either in the New York or the Miami metro areas. More than half of the Mexican-born population lives in the Los Angeles and Chicago metro areas or the state of Texas (Bureau of the Census, May 2001).

Interestingly, although the vast majority of Hispanics come from rural areas, 90% settle in America's industrial cities and surrounding suburbs. Living together in tightly knit communities, they share a common language and customs.

With the Latin migration north, the United States has experienced the largest migration in its history. Around one-half of Hispanic residents in the United States were born abroad. Newcomers to the United States begin their stay with many economic and educational disadvantages compared with the average American: More than half have not graduated from high school; only 20% of those who came to the United States as adults are bilingual in Spanish and English (*Migration News*, 2000).

Mexican Americans The 27 million Hispanics of Mexican descent make up two-thirds of the Spanish-speaking population in the United States. The main reason is proximity. The 1,936-mile border the United States shares with Mexico is the site of millions of legal and illegal crossings each year.

Some people of Mexican origin call themselves Chicanos, although the term is somewhat controversial. It has long been used as a slang word in Mexico to refer to people of low social class. In Texas, it came to be used for Mexicans who illegally crossed the border in search of work. Many Mexican Americans have taken to using the term themselves to suggest a tough breed of individuals of Mexican ancestry who are committed to achieving success in this country and are willing to fight for it (Madsen, 1973). In his study of Mexican Americans in Texas, William Madsen (1973) noted that among them, three levels of acculturation may be distinguished. Although most Mexican Americans appreciate American technology, one group has retained its Mexican peasant culture, at least in regard to values. Another group consists of people torn between the traditional culture of their parents and grandparents and Anglo-American culture (learned in American schools). Many individuals in this group suffer crises of personal and ethnic identity. Finally, there are those Mexican Americans who have acculturated fully and achieved success in Anglo-American society. Some remain proud of and committed to their ethnic origins, but others would just as soon forget them and assimilate fully into the Anglo-American world.

Mexican Americans have been exploited for many years as a source of cheap agricultural labor. Their median family income of $33,000 (2003) is one of the lowest of any Spanish-speaking group. More than 22% of Mexican Americans live in poverty (Bureau of the Census, 2005). Also, because of their poverty, they have long been willing to do the "stoop labor" (literally bending over and working close to the ground, or menial labor in general) that most Anglo-Americans refuse to do. In addition, 21.1% of all Mexican American families are headed by women, and 37% of all births are out of wedlock (Schmidley, 2001).

Puerto Ricans Also included under the category of Hispanics are Puerto Ricans. In 1898, the United States fought a brief war with Spain and as a result took over the former Spanish colonies in the Pacific (the Philippines and Guam) and the Caribbean (Cuba and Puerto Rico). Puerto Ricans were made full citizens of the United States in 1917. Although government programs improved their education and dramatically lowered the death rate, rapid population growth helped keep the Puerto Rican people poor. American business took advantage of the large supply of cheap, nonunion Puerto Rican labor and built plants there under very favorable tax laws.

More than 3.4 million Puerto Ricans live in the United States. Many have migrated to the American

NEWS YOU CAN USE

Hispanics: Racial Group? Ethnic Group? Neither?

Thirty years ago immigrants from Latin America who settled in the United States were identified in terms of their home nation—as, for example, Cuban Americans or Mexican Americans, just as European newcomers were seen as Italian Americans or Polish Americans. Today the immigrant flow from Central and South America has grown substantially, and the newcomers are known as Hispanics.

Some observers have expressed concern that efforts to make Hispanics a single minority group—for purposes ranging from elections to education to the allocation of public funds—are further dividing American society along racial lines . . .

U.S. race relations have long been understood in terms of black and white. Until recently, many books on the subject did not even mention other races, or did so only as a brief afterthought. Now recognition is growing that Hispanics are replacing blacks as the primary minority . . .

[I]n 1980, "Hispanics" became a distinct statistical and social category in the [U.S.] census, as all households were asked whether they were of "Spanish/Hispanic origin or descent." Had no changes been made in 1980, we might well have continued to think of Hispanics as we do about other white Americans, as several ethnic groups, largely from Mexico and Cuba . . .

Because of their evolving status in the census, Hispanics are now sometimes treated not as a separate ethnic group but as a distinct race. (Race marks sharper lines of division than ethnicity.) Often, for example, when national newspapers and magazines, such as the *Washington Post* and *U.S. News and World Report*, graphically depict racial breakdowns on various subjects, they list Hispanics as a fourth group, next to white, black, and Asian . . .

Should one mind the way the census keeps its statistics— . . . one need not have an advanced degree to realize that the ways we divide people up—or combine them—have social consequences. . . . In short, the census greatly influences the way we see each other and ourselves, individually and as a community . . .

How do Hispanics see themselves—First of all, the vast majority prefer to be classified as a variety of ethnic groups rather than as one. The National Latino Political Survey, for example, found that three out of four respondents chose to be labeled by country of origin, rather than by . . . terms such as "Hispanic" or "Latino." Hispanics are keenly aware of big differences among Hispanic groups, especially between Mexican Americans (the largest group) and Cuban Americans, the latter being regarded as more likely to be conservative, to vote Republican, to become American citizens, and so on . . .

Many Hispanics resist being turned into a separate race or being moved out of the white category. In 1990, the census allowed people to buy out of racial divisions by checking "other" when asked about their racial affiliation. Nearly 10 million people—almost all of them Hispanics—did so . . .

[T]he 2000 census dropped "other" and instead allowed people to claim several races . . . The long list of racial boxes to be checked ended with "some other race," with a space to indicate what that race was. . . . Of those who chose only "some other race," almost all (97 percent) were Hispanic. Among Hispanics, 42.2 percent chose "some other race," 47.9 percent chose white (alone) as their race, 6.3 percent chose two or more races, and 2 percent chose black (alone). In short, the overwhelming majority of Hispanics either chose white or refused racial categorization.

Source: Amitai Etzioni, "Inventing Hispanics: A Diverse Minority Resists Being Labeled," *The Brookings Review,* Winter 2002, Vol. 20, No. 1, pp. 10–13.

mainland seeking better economic opportunities. Most make their homes in the New York City area but return frequently to the island to visit relatives and friends.

Puerto Ricans living in the United States have one of the lowest median incomes among Hispanics. Ironically, the poverty in many Puerto Rican families is in part due to the ease of going back and forth between their homeland and the United States. The desire to one day return permanently to Puerto Rico interferes with a total commitment to assimilate into American culture.

Cuban Americans Currently, 1.25 million people of Cuban ancestry live in the United States. The first immigrants from Cuba arrived in the area now known as Florida as early as the 1500s, but most have come to the United States since the late 1950s. Only in the 1970s did they begin to have a visible cultural and economic effect on the cities where they have settled in sizable numbers.

Many Cubans came to the United States as a result of the 1959 revolution that catapulted Fidel Castro into power. At that time, Castro's rebel forces overthrew the government of Fulgencio Batista.

Castro, a Marxist closely aligned with the former Soviet Union, began to restructure the social order, including appropriating for the state privately owned land and property. Professionals and businesspeople who were part of the established Cuban society felt threatened by these changes and fled to the United States. More than 155,000 Cubans immigrated to the United States between 1959 and 1962. As a whole, these immigrants have done extremely well in American society; they had the distinct advantage of coming to the United States with marketable skills and money.

Another large wave of Cuban immigration occurred in 1980, when Castro allowed people to leave Cuba by way of Mariel Harbor. The result was a flotilla of boats bringing 125,000 refugees to the United States. This wave of immigrants was poorer and less well educated than the first. It also included several thousand prisoners and mental patients, many of whom were imprisoned in the United States as soon as they arrived. Others fled to Miami and other cities to lead a life of crime. Serious friction existed between these two waves of immigrants because of differences in background and social class.

Cubans are relatively recent immigrants, and the first-generation foreign-born predominate. They are exiles who came to the United States not so much because they preferred the U.S. way of life, but because they felt compelled to leave their country. For that reason, most Cuban immigrants fiercely attempt to retain the culture and way of life they knew in Cuba. Of all the Hispanic groups, they are the most likely to speak Spanish in the home: 8 of 10 families do so.

Cubans are largely found in a few major cities. Metropolitan Miami (Dade County, Florida) is the undisputed center, with nearly 65% of all Cubans in the United States living there. In Miami, where Latins make up a larger percentage of the population than Anglos, the city has become distinctively Cuban. Most other Cubans live in New York City, Jersey City and Newark in New Jersey, Los Angeles, and Chicago, all of which are large centers for Hispanics in general.

Acculturation and assimilation have been slow in Miami, in light of the fact that the community is so self-sufficient and has such a large immigrant base. There also appears to be a lack of social and cultural integration between Cubans and other Hispanic groups in U.S. cities with sizable and differentiated Spanish-speaking populations. Of the major Hispanic groups, only the Cubans have come as political exiles, and this has resulted in social, economic, and class differences. In the New York City area, Cubans and Puerto Ricans maintain a distinct social distance. Many Cubans feel or perceive that they have little in common with Puerto Ricans, Mexican Americans, or Dominicans.

The 1.4 million Cubans who live in the United States have fared better than any other Hispanic immigrant group. They are the best educated of any of the Spanish-speaking groups, with 23% having a bachelor's degree compared with 6.9% of Mexicans. The median family income of Cubans is $38,312, nearly a third higher than the earnings of the average Hispanic family (Proctor & Dalaker, 2001). At the current rate of growth, the Cuban income could surpass the national median income within a few years.

Asian Americans

According to the Census Bureau, approximately 12.3 million people of Asian background live in the United States, constituting 4.2% of the total population (Bureau of Census, 2005). Between 1990 and 2000, the Asian and Pacific Islander population increased more quickly than any other race or ethnic group (Barnes & Bennett, 2002). The five largest Asian population groups counted in the 2000 census include the Chinese, Filipino, Asian Indian, Vietnamese, and Korean.

Most Asian Americans are concentrated in the major metropolitan areas. Their percentage of the total population in these cities varies from 29.1% in San Francisco to 0.8% in Detroit. In Honolulu, 70.5% of the population is Asian or Pacific Islander.

The first Asians to settle in America in significant numbers were the Chinese. Some 300,000 Chinese migrated here between 1850 and 1880 to escape the famine and warfare that plagued their homeland. Initially they settled on the West Coast, where they took backbreaking jobs mining and building railroads. However, they were far from welcome and were subjected to a great deal of harassment. In 1882, the government limited further Chinese immigration for 10 years. This limitation was extended in 1892 and again in 1904, finally being repealed in 1943. The state of California imposed special taxes on Chinese miners, and most labor unions fought to keep them out of the mines because they took jobs from white workers. In the late 1800s and early 1900s, numerous riots and strikes were directed against the Chinese, who drew back into their "Chinatowns" for protection. The harassment proved successful. In 1880, 105,465 Chinese lived in the United States; by 1900, the figure had dropped to 89,863 and by 1920, to 61,729. The Chinese population in the United States began to rise again only after the 1950s (Bureau of the Census, 1976). As of 2000, 2.7 million people of Chinese ancestry lived in the United States (Barnes & Bennett, 2002).

Japanese immigrants began arriving in the United States shortly after the Chinese and quickly joined them as victims of prejudice and discrimination. Feelings against the Japanese ran especially high in California, where one political movement attempted

These Japanese Americans were about to be taken to Seattle by a special ferry, which connected with a train to California, as part of their evacuation and internment during World War II.

© Brown Brothers

to have them expelled from the United States. In 1906, the San Francisco Board of Education decreed that all Asian children in that city had to attend a single, segregated school. The Japanese government protested, and after negotiations, the United States and Japan reached what became known as a "gentlemen's agreement." The Japanese agreed to discourage emigration, and President Theodore Roosevelt agreed to prevent the passage of laws discriminating against Japanese in the United States.

Initially, Japanese immigrants were minuscule in number: In 1870, only 55 Japanese lived in America, and in 1880, a mere 148. By 1900, 24,326 lived in this country, and subsequently their numbers have grown steadily. By 1970 they had surpassed the Chinese in number (Bureau of the Census, 1976), but later figures showed that despite a sharp increase since 1970, the number of ethnic Japanese has been far fewer than the number of Chinese. In 2000, approximately 800,000 people of Japanese ancestry lived in the United States (Barnes & Bennett, 2002).

Japanese Americans were subjected to especially vicious mistreatment during World War II. Fearing espionage and sabotage from the ethnic minorities whose home countries were at war with the United States, President Franklin D. Roosevelt signed Executive Order 9066, empowering the military to "remove any and all persons" from certain regions of the country. Although many German Americans actively demonstrated on behalf of Germany before the United States entered World War II, no general action was taken against them as a group. Nor was

any general action taken against Italian Americans. Nonetheless, General John L. DeWitt ordered that all individuals of Japanese descent be evacuated from three West Coast states and moved inland to relocation camps for the duration of the war. In 1942, 120,000 Japanese, including some 77,000 who were American citizens, were moved and imprisoned solely because of their ethnicity—even though not a single act of espionage or sabotage against the United States ever was attributed to one of their number (Simpson & Yinger, 1972). Many lost their homes and possessions in the process. Even with the rampant racism some Japanese Americans volunteered for the 442nd Regimental Combat Team, a fighting group composed solely of Japanese Americans. The 442nd became the most decorated unit in U.S. history.

In 1988, President Reagan signed legislation apologizing for this wartime action. The legislation moved to "right a great wrong" by establishing a $1.25 billion trust fund as reparation for the imprisonment. Each eligible person was to receive a $20,000 tax-free award from the government. The president noted as he signed the legislation: "Yes, the nation was then at war, struggling for its survival. And it's not for us today to pass judgment upon those who may have made mistakes while engaging in the great struggle. Yet we must recognize that the internment of Japanese Americans was just that, a mistake" (Johnson, 1988).

Compared with the earlier group of Asian immigrants who came primarily from China and Japan, since the 1960s many Asians from Vietnam, the

Philippines, Korea, India, Laos, and Cambodia have come to the United States. Many of the Asian immigrants from Vietnam, Cambodia, and Laos were involuntary migrants who were forced to leave their homes because they feared persecution after the United States left Southeast Asia. Two distinct waves of refugees came from these countries. The first began in the 1960s and continued until the end of the Vietnam War in 1975. The first group of Southeast Asian immigrants were mainly middle- and upper-class Vietnamese who found ways to get their families and financial assets out of the country when it became clear that military victory for the United States was not going to be swift. This group numbered about 25,000.

The second wave of refugees was very different. Harsh economic conditions, political persecution within Vietnam, and widespread genocide by the Khmer Rouge government in Cambodia created a flood of refugees desperate to leave the area. Many crowded into unsafe boats and hoped to reach Hong Kong, Malaysia, and other neighboring countries. Some of these "boat people" eventually settled in the United States. Between 1975 and 1994, more than 700,000 Vietnamese refugees and 500,000 Cambodians and Laotians resettled in the United States (U.S. Office of Refugee Resettlement, 1995). Close to half (about 45%) of the nation's Asian-born population live in three metropolitan areas: Los Angeles, New York, and San Francisco.

"Despite their diversity, Asian Americans are often perceived as a single group, not only by other Americans but in their own minds as well" (Jacoby, 2000). What appears to unite them is their newness to this country. Approximately 60 to 70% of Asian Americans are foreign-born, and as many as 30% report that they speak English poorly. Most came to the United States as young adults seeking a better lifestyle (Jacoby, 2000).

Most Asian immigrants are middle class and highly educated. More than a third have a college degree, twice the rate for Americans born here (immigrants from Vietnam, Cambodia, and other Indochinese countries are the exception). The education, occupations, and income attainments of Asian Americans have been far above the national average. In 2004, the median income of Asian Americans was $57,518, compared with $48,977 for white Americans (Bureau of the Census, 2005).

Although Asian Americans make up 4.2% of the total U.S. population, they account for nearly 19% of the freshman class at Harvard and other elite universities. They have achieved stunning success in science and business. As a group, they have the highest percentage of college graduates of any group in the United States. For example, 44% of all Asians older than age 25 have a bachelor's degree. For whites the percentage is 26%. Seven in 10 Asian Americans between 18 and 21 are attending college, versus half of whites. Asian Americans received 10% of the doctorates conferred by the nation's colleges and universities. This included 22% of the doctorates in engineering and 21% of those in computer science (Barnes & Bennett, 2002).

Jews

According to the 2000–2001 National Jewish Population Survey, 5.2 million Jews live in the United States, slightly below the 5.5 million counted in 1990. The Jewish population is substantially older than that of the total U.S. population. In 2000 the median age was 41; for the total U.S. population it is 35. Nineteen percent are older than age 65.

The survey found that "fertility, along with mortality and migration, is one of the major demographic factors determining the size of the Jewish population. Jews are having fewer children than the number required for the population to remain stable." More than half (52%) of Jewish women age 30 to 34 have no children, compared with 27% of all American women, and the Jewish women who do have children aren't having enough to keep the population even, the survey found. Researchers say that the numbers partly reflect that Jewish women, like other women, have been delaying marriage to pursue higher education and careers. "Jewish women who are now approaching the end of their childbearing years, ages 40 to 44, have had approximately 1.8 children, which is below the replacement level of 2.1 (United Jewish Communities, 2002).

"Jews are dispersed across the United States, but their regional distribution differs markedly from that of non-Jews," the survey found. The Jewish population is concentrated in the Northeast (43%). The region containing the fewest Jews is the Midwest (13%), and Jews are also proportionately underrepresented in the South (22%) and West (22%).

The Jewish population is very well educated. "A quarter (24%) of Jewish adults 18 years of age and older has received a graduate degree, and 55% have earned at least a bachelor's degree." This compares with 5% and 28% for the general population, respectively. The additional years of education produce a median household income that is 16% higher than the median for all U.S. households (United Jewish Communities, 2002; Zoll, 2002).

There is no satisfactory answer to the question "What makes the Jews a people?" other than to say that they see themselves—and are seen by others—as a people. Judaism is a religion, of course, but many Jews are nonreligious. Some think of Jews as a race,

About one in five Native American households on reservations lacks complete plumbing facilities or kitchens.

but their physical diversity makes this notion absurd. For more than 2,000 years, Jews have been dispersed around the world. Reflecting this geographic separation, three major Jewish groups have evolved, each with its own distinctive culture: the Ashkenazim, the Jews of eastern and western Europe (excluding Spain); the Sephardim, the Jews of Turkey, Spain, and western North Africa; and the Oriental Jews of Egypt, Ethiopia, the Middle East, and central Asia. Nor are Jews united linguistically. In addition to speaking the language of whatever nation they are living in, many Jews speak one or more of three Jewish languages: Hebrew, the language of ancient and modern Israel; Yiddish, a Germanic language spoken by Ashkenazi Jews; and Ladino, an ancient Romance language spoken by the Sephardim.

The first Jews came to America from Brazil in 1654, but it was not until the mid-1800s that large numbers of Jews began to arrive. These were mostly German Jews, refugees from European anti-Semitism. Then, with especially violent anti-Semitism erupting in eastern Europe in the 1880s, Jewish immigration to America increased considerably, coming in two waves: the last two decades of the nineteenth century and the first two decades of the twentieth.

Jewish immigration was similar to that of other groups in that it consisted overwhelmingly of young people, although the Jewish immigration also had some unique features. First, it was much more a migration of families than that of other European immigrants, who were mostly single males. Second, Jewish immigrants were much more committed to staying here: Two-thirds of all immigrants to the United States between 1908 and 1924 remained, but 94.8% of the Jewish immigrants settled here permanently. Third, the Jewish immigrant groups contained a higher percentage of skilled and urban workers than did other groups. Fourth, especially after the turn of the century, many scholars and intellectuals were among Jewish immigrants, which was not true of other immigrant groups (Howe, 1976).

These differences account for the fact that even though Jews encountered at least as much hostility from white Anglo-Saxon Protestants as did other immigrant groups (and also were subject to intense prejudice from Catholics), they have had relatively more success in pulling themselves up the socioeconomic ladder than other immigrant groups.

Native Americans

The 2000 census made it possible for us to know how many people in the United States think of themselves as only Native American and how many define themselves as Native American in combination with another race. According to the census, 2.8 million people, or 0.9%, reported only American Indian and Alaska Native heritage (Bureau of Census, 2005). In addition, 1.6 million people, or 0.6%, of the U.S. population reported American Indian and Alaska Native as well as one or more other races, for a total of 4.1 million people, or 1.5%, of the population. The most common combinations were "American Indian and White" (66%), followed by "American Indian and African American" (11%).

The 1990 census showed nearly 2 million Native Americans. Using only the Native American population in 2000, this population increased by 26% between 1990 and 2000. If the population of Native Americans alone or in combination is used, an increase of 2.2 million, or 110%, results. Thus, from 1990 to 2000, the range for the increase in the Native American population was 26% to 110%. In comparison, the total population grew by 13%, from 248.7 million in 1990 to 281.4 million in 2000 (Ogunwole, 2002).

This increase cannot be attributed to just natural population growth. Other factors that may have contributed to the higher number include improvements in the way the U.S. Census Bureau counts people on reservations and a great propensity for people (especially those of mixed Native American and white parentage) to report themselves as American Indian.

As early Europeans first stepped ashore in what they considered the New World— they usually were welcomed by the Native Americans. The Indians seemed to regard their lighter-complexioned visitors as something of a marvel, not only for their dress, beards, and "winged" ships but even more for the items they brought that were unusual to their way of life, such as steel knives and swords.

Early European colonists encountered Native American societies that in many ways were as advanced as their own. Especially impressive were their political institutions. For example, the League of the Iroquois, a confederacy that ensured peace among its five member nations and was remarkably successful in warfare against hostile neighbors, was the model on which Benjamin Franklin drew when he was planning the Federation of States (Kennedy, 1961).

Native Americans were surprised at European intolerance for native religious beliefs, sexual and marital arrangements, eating habits, and other customs. At the same time, they became perplexed when Europeans built permanent structures of wood and stone, thus precluding movement. Even village- and town-dwelling Native Americans were used to relocating when local game, fish, and especially firewood gave out.

The colonists and their descendants never questioned the view that the land of the New World was theirs. They took land as they needed it— for agriculture, for mining, and later for industry— and drove off the native groups. Some land was purchased, some acquired through political agreements, some through trickery and deceit, and some through violence. In the end, hundreds of thousands of Native Americans were exterminated by disease, starvation, and deliberate massacre. By 1900, only about 250,000 Indians remained (perhaps one-eighth of their numbers in precolonial times) (McNeill, 1976).

The five states with the largest Native American populations are California, Oklahoma, Arizona, Texas, and New Mexico. Overall, nearly one-half of the nation's American Indians and Alaska Natives live in western states (Ogunwole, 2002).

To make up for past injustices perpetrated against Native Americans, the federal government pays for a number of programs to assist them with education, health care, and housing. The government has also granted Native Americans special rights to govern themselves, so that they are subject only to federal rather than state and local laws. These rights, based on hundreds of treaties made in the eighteenth and nineteenth centuries, provide for Native American sovereignty and give tribes independence from outside governments in return for land. These rights have made it possible for some tribes to open casinos in states that do not allow gambling establishments.

Interestingly, many people informally claim American Indian ancestry. Yet less than one-third of them identify themselves as Native American. Most people who claim Native American ancestry do so in combination with another ancestry group such as the English or Irish.

Native Americans have a median household income that is higher than that of African Americans, similar to that of Hispanics, and lower than that of non-Hispanic whites, Asians, and Pacific Islanders.

The five largest Native American tribes are the Cherokee, Navajo, Sioux, Chippewa, and Choctaw. Considerable differences exist among the various tribes. For instance, the Iroquois and the Creek are much better off economically than the Navajo (Bureau of the Census, 2000).

More than half of all Native Americans live on or near reservations administered fully or partly by the federal Bureau of Indian Affairs. Conditions on some reservations are quite bad, particularly for Native American youths. For example, Indian youths are only half as likely to have both parents at home as rural white teenagers in Minnesota, and they are twice as likely to have experienced the death of a parent. One Native American teenager in six has attempted suicide, a rate four times that of other teenagers (*Society*, March/April 1997).

The physical environment is poor also. About one in five Native American homes on reservations lacks complete plumbing facilities or a complete kitchen. More than half of these homes do not have phones (Bureau of the Census, April 1995).

A Diverse Society

As is evident by now, the many racial and ethnic groups in the United States present a complex and constantly changing picture. Some trends in intergroup relations can be discerned and are likely to continue; new ones may emerge as new groups gain prominence. The resurgence of ethnic-identity movements probably will spread and may be coupled with more collective protest movements among disaffected ethnic and racial minorities who may demand that they be given equal access to the opportunities and benefits of American society.

It is important to realize that the old concept of the United States as a melting pot is both simplistic and idealistic. Many groups have entered the United States. Most encountered prejudice, some severe discrimination, and others the pressures of Anglo conformity.

Contemporary American society is the outcome of all these diverse groups coming together and trying to adjust. Indeed, if these groups are able to interact on the basis of mutual respect, this diversity may offer America strengths and flexibility not available in a homogeneous society.

SUMMARY

- Race refers to a category of people who are defined as similar because of a number of physical characteristics.
- Racial characteristics are arbitrarily chosen to suit the labeler's purposes.
- Races have historically been defined according to genetic, legal, and social criteria, each with its own problems.
- Although genetic definitions center on inherited traits such as hair and nose type, in fact, these traits have been found to vary independently of one another. Moreover, all humans are far more genetically alike than they are different.
- An ethnic group is a group with a distinct cultural tradition with which its own members iden-

tify and that may or may not be recognized by others.
- Many ethnic groups form subcultures, with a high degree of internal loyalty and distinctive folkways, mores, values, customs of dress, and patterns of recreation. Above all, members of the group have a strong feeling of association.
- A minority is a group of people who, because of physical or cultural characteristics, are singled out for differential and unequal treatment and who therefore regard themselves as objects of collective discrimination.
- Prejudice is an irrationally based negative, or occasionally positive, attitude toward certain groups and their members. Prejudice is a subjective feeling; discrimination is an overt action. Discrimination can be defined as differential treatment, usually unequal and injurious, accorded to individuals who are assumed to belong to a particular category or group.
- Assimilation is the process whereby groups with different cultures come to share a common culture. Invariably, one group has a much larger role in the process than the other or others.
- The particular form of assimilation found in the United States is called Anglo conformity—the renunciation of ancestral cultures in favor of Anglo-American behavior and values.
- Pluralism is the development and coexistence of separate racial and ethnic group identities within a society. Pluralism celebrates the differences among groups.
- Historically, the majority of American immigrants have been from Europe.
- The old migration consisted of people from northern Europe who came before 1880.
- The new migration was far larger and consisted of people who came from southern and eastern Europe.
- Discriminatory quotas were set up at the beginning of the twentieth century to restrict the immigration of the latter groups.
- Today, the overwhelming majority of immigrants to the United States come from Latin America, Asia, and the Caribbean.
- Legal immigration to the United States has increased in recent decades.
- There could be as many as 10 million illegal immigrants in the United States.
- Illegal immigrants tend to be young and to settle in California, New York, Florida, Texas, and Illinois.
- The United States is perhaps the most racially and ethnically diverse country in the world; no single ethnic group makes up a majority of the population.

Media Resources

The Companion Website for *Introduction to Sociology,* Ninth Edition

http://sociology.wadsworth.com/tischler9e

Supplement your review of this chapter by going to the companion website to take one of the Tutorial Quizzes, use the flash cards to master key terms, and check out the many other study aids you will find there. You will also find special features such as Wadsworth's Sociology Online Resources and Writing Companion, GSS data, and Census 2000 information at your fingertips to help you complete that special project or do some research on your own.

CHAPTER TEN STUDY GUIDE

KEY CONCEPTS AND THINKERS

Match each concept with its definition, illustration, or explanation presented below.

a. Pluralism
b. Expulsion
c. Americanization movement
d. Discrimination

e. Prejudice
f. Assimilation
g. Anglo conformity
h. Subjugation

i. Segregation
j. Minority group
k. Race
l. Ethnic group

k **1.** A category of people who are defined as similar because of a number of physical characteristics.

h **2.** The domination of one group by another.

l **3.** A group with a distinct cultural tradition with which its own members identify and that may or may not be recognized by others.

j **4.** A group of people who, because of physical or cultural characteristics, are singled out for differential and unequal treatment and who therefore regard themselves as objects of collective discrimination.

e **5.** An irrationally based negative, or occasionally positive, attitude toward certain groups and their members.

d **6.** Differential treatment, usually unequal and injurious, accorded to individuals who are assumed to belong to a particular category or group.

b **7.** The process of forcing a group to leave the territory in which it lives.

f **8.** The process whereby groups with different cultures come to share a common culture.

c **9.** A social movement advocating the complete assimilation of immigrant groups to the United States.

g **10.** The renunciation of other ancestral cultures in favor of Anglo-American behavior and values.

a **11.** The development and coexistence of separate racial and ethnic group identities within a society.

i **12.** The act, process, or state of being set apart.

Match the thinkers with their main idea or contribution.

a. Robert K. Merton
b. Johann Blumenbach
c. Gerhard Lenski

d. Herrnstein and Murray
e. Louis Wirth
f. Benjamin Franklin

g. Hoover Commission
h. Horace Kallen

b **1.** Eighteenth-century German physiologist who realized that racial categories did not reflect the actual divisions among human groups.

a **2.** Showed that there are various ways in which prejudice and discrimination can interact with each other.

d **3.** Argued in *The Bell Curve* that members of a cognitive elite were becoming the leaders of America.

e **4.** Developed a definition of minority group that considers only race and ethnic status.

g **5.** Stated that a program for Native Americans must include progressive measures for their complete integration as tax-paying members of the larger society.

h **6.** Principally responsible for the development of the theory of cultural pluralism.

c **7.** Proposed that dominance of one group over another arises because people have a desire to control goods and services.

f **8.** Wondered in 1753 about the costs and benefits to the United States of German immigration and concluded Germans would "contribute greatly to the improvement of a country."

CENTRAL IDEA COMPLETIONS

Following the instructions, fill in the appropriate concepts and descriptions for each of the questions posed below.

1. Describe the legal and social definitions of race.

    a. Legal: _____

    _____

    b. Social: _____

    _____

2. Define the concepts of minority group, ethnic group, and racial group.

    a. Minority group: _____

    b. Ethnic group: _____

    c. Racial group: _____

3. Compare and contrast the basic characteristics of the following minority groups:

    a. Asian Americans: _____

    _____

    b. Cuban Americans: _____

    _____

    c. Jewish Americans: _____

    _____

    d. Mexican Americans: _____

    _____

4. Explain how the U.S. Army differs from most civilian organizations in the way in which it seeks to create and maintain racial integration in its ranks.

    _____

    _____

5. Which ethnic/racial group in the United States is least likely to marry interracially and which is most likely to do so. Why is this the case?

    Least: _____

    Most: _____

    Reasons: _____

6. Drawing on people you know personally or from news and the media, provide an example for each of Merton's categories of discriminators and nondiscriminators.

    a. Unprejudiced nondiscriminator: _____

    b. Unprejudiced discriminator: _____

    c. Prejudiced nondiscriminator: _____

    d. Prejudiced discriminator: _____

7. Briefly describe the unique features of the minority group experience of Native Americans relative to other minority groups' experiences in the United States.

_____

_____

_____

8. Define Anglo conformity, showing specific examples of its historical expression.

1840s: _____

1920s: _____

1960s: _____

CRITICAL THOUGHT EXERCISES

1. What are the essential elements of race and racial identity? Review the example of Mark Linton Stebbins's claim of being black. Do you feel he made an adequate case for his claim? What criteria would you use if you were faced with the task of determining a person's race? What is the relative importance of biology versus cultural experience in the self-definition people have or do not have regarding the racial category to which they claim membership?

2. Consider the different ways of defining race: genetic markers, ancestry ("blood"), social reaction, self-definition. Is one of these best in all situations? Describe how each definition might be very useful in one circumstance but much less so in others.

3. The U.S. Army has been notably successful in overcoming racism. In what ways might the techniques used by the military be effective in other settings (business, other government agencies, education, and so on), and why might these methods not be transferable to other institutions?

4. Develop a critical thought paper in which you examine the changes in U.S. immigration policy since 1965. How did changes in the immigration system affect the country of origins from which the new immigrants have come? What do the results of 2000 census data tell us about the composition of the United States relative to racial and ethnic group membership? What changes in residence, popular culture, and education do you hypothesize are likely to develop during the next 25 years given the changing face of U.S. immigration that you have just described?

5. How has the rise of the World Wide Web increased the potential for spreading messages of hate and bigotry? Visit at least six sites that have as their goal the promotion of prejudiced and resentful views about a particular group of people. What are the common features of your six sites? Some people see these sites as a threat to American society, whereas others view this issue as overblown. After your examination, what is your assessment of the seriousness of the threat these sites pose? What action, if any, would you propose relative to restricting the dissemination of information across the Internet?

INTERNET ACTIVITIES

1. Visit http://www.pbs.org/race/000_General/000_00-Home.htm. This is the companion site to a PBS series with slide shows, quizzes, a sorting game where you try to classify photos of people by race, and other fun stuff.

2. A psychology project at Harvard offers an interactive exercise to test your "Implicit Associations" to categories of race, age, gender, and so on: https://implicit.harvard.edu/implicit/. Take some of the "tests" and see what they reveal about attitudes you might not have realized you had.

3. Susan A. Vega Garcia at Iowa State University has created a website with links for just about any topic you might want to explore in the general area of race and ethnicity: http://www.public.iastate.edu/~savega/multicul.htm.

ANSWERS TO KEY CONCEPTS

1.k 2.h 3.l 4.j 5.e 6.d 7.b 8.f 9.c 10.g 11.a 12.i

ANSWERS TO KEY THINKERS

1.b 2.a 3.d 4.e 5.g 6.h 7.c 8.f

ThomsonNOW™

Reviewing is as easy as ❶ ❷ ❸

1. Before you do your final exam, take the ThomsonNOW diagnostic quiz to help you identify the areas on which you should concentrate. You will find information on ThomsonNOW and instructions on how to access all of its great resources on the foldout at the beginning of the text.
2. As you review, take advantage of ThomsonNOW's study videos and interactive Map the Stats exercises to help you master the chapter topics.
3. When you are finished with your review, take ThomsonNOW's posttest to confirm you are ready to move on to the next chapter.

Jack Hayes/Boston Filmworks

11

Gender Stratification

Learning Objectives

After studying this chapter, you should be able to do the following:

- Contrast biological and sociological views of sex and gender.
- Describe the concept of patriarchal ideology.
- Understand the functionalist and conflict theory viewpoints on gender stratification.

- Explain the process of gender-role socialization.
- Describe gender differences in the world of work.
- Be aware of the effect of changes in gender roles in U.S. society.

Kate Bornstein is 6 feet tall, she weighs about 190 pounds, and as you are introduced to her you notice the large hands and the powerful handshake. Kate used to be Al Bornstein before she underwent a sex-change operation. Even though Kate acquired the anatomical features of a woman, she still likes to have physical relationships with women. She considers herself a lesbian transsexual.

Why would a man undergo a sex-change operation to become a woman if he/she is physically attracted to women? Kate makes a distinction between gender identity and sexual attraction. She says, "Just because they did a lot of clever surgical manipulation with my genitals, that doesn't change my desire for a romantic partner. I've always been attracted to women. When I grew up in the fifties I knew I wasn't a man, but how could I be a woman if I loved women?"

Kate believes there are many more genders than just male or female. She asserts that her confusion throughout her life stems from being forced to pick one gender over the other. "I know I'm not a man—about that much I'm very clear, and I've come to the conclusion that I'm probably not a woman either" (Bornstein, 1994).

It is difficult for us to imagine that someone could be something other than a man or a woman if we have spent our lives thinking about a world in which there are only men and women. Kate Bornstein forces us to clarify what we mean by gender. What role

does our anatomy play? What does it mean to be socialized as a man or a woman? How much of gender is ascribed and how much is achieved?

In this chapter, we will look at some of the differences between the sexes, examine cross-cultural variations in gender roles, and try to understand how a gender identity is acquired. In the process, we will also focus on the changes that are occurring in gender roles in American society.

Are the Sexes Separate and Unequal?

Sociology makes an important distinction between sex and gender. **Sex** refers to *the physical and biological differences between men and women*. In general, sex differences are made evident by physical distinctions in anatomical, chromosomal, hormonal, and physiological characteristics. At birth, the differences are most evident in the male and female genitalia.

We also need to learn how to be a man or a woman. **Gender** refers to *the social, psychological, and cultural attributes of masculinity and femininity that are based on the previous biological distinctions*. Gender pertains to the socially learned patterns of behavior and the psychological or emotional expressions of attitudes that distinguish males from

279

females. Ideas about masculinity and femininity are culturally derived and pattern the ways in which males and females are treated from birth onward. Gender is an important factor in shaping people's self-images and social identities. Sex is thought of as an *ascribed* status; a person is born either a male or a female (although transsexuals such as Kate Bornstein make us realize that sex can be changed). Gender is learned through the socialization process and thus is an *achieved* status.

In many circles, Kate Bornstein would be dismissed as a "freak of nature" instead of someone who forces us to clarify exactly what we mean by gender. The dominant view in many societies is that gender identities are expressions of what is natural. People tend to assume that acting masculine or feminine is the result of an innate, biologically determined process rather than the result of socialization and social-learning experiences.

To support the view that gender-role differences are innate, people have sought evidence from religion and the biological and social sciences. Whereas most religions tend to support the biological view, biology and the social sciences provide evidence that suggests that what is natural about sex roles expresses both innate and learned characteristics.

Historical Views

We must be careful when we use today's standards to evaluate the statements of people who lived during another time and in another place. Lifting these statements out of context and bringing them into the present has been compared to "trying to plant cut flowers" (Ellis, 1997). Yet it is important to look at these views and see what has changed, because they represented the common thinking.

The ancient Greek philosopher Aristotle, for example, pointed out that

> It is fitting that a woman of a well-ordered life should consider that her husband's wishes are as laws appointed for her by divine will, along with the marriage state and the fortune she shares. (Aristotle, 1908)

The third-century Chinese scholar Fu Hsuan penned these lines about the status of women during his era:

> Bitter indeed it is to be born a woman,
>
> It is difficult to imagine anything so low!
>
> Boys can stand openly at the front gate,
>
> They are treated like gods as soon as they are born. . . . But a girl is reared without joy or love,
>
> And no one in her family really cares for her,
>
> Grown up, she has to hide in the inner rooms,

> Cover her head, be afraid to look others in the face,
>
> And no one sheds a tear when she is married off.
>
> (Quoted in Bullough, 1973)

In traditional Chinese society, women were subordinate to men. Chinese women were often called *Neiren,* or "inside person." To keep women in shackles, Confucian doctrine created what were known as the three obediences and the four virtues. The three obediences were "obedience to the father when yet unmarried, obedience to the husband when married, and obedience to the sons when widowed." Thus, traditionally, Chinese women were placed under the control of men from cradle to grave.

The four virtues were "women's ethics," meaning a woman must know her place and in every way comply with the old ethical code; "women's speech," meaning a woman must not talk too much, taking care not to bore people; "women's appearance," meaning a woman must pay attention to adorning herself with a view to pleasing men; and "women's chores," meaning a woman must willingly do all the chores in the home.

A book of Chinese poetry, *The Book of Songs,* believed to date from 1000–700 B.C., offers this advice to new parents:

> When a son is born
>
> Let him sleep on the bed,
>
> Clothe him with fine clothes.
>
> And give him jade to play with. . . .
>
> When a daughter is born,
>
> Let her sleep on the ground,
>
> Wrap her in common wrappings,
>
> And give her broken tiles for playthings.

Thomas Jefferson believed in supreme personal liberty and the equal creation of all men. Yet when it came to women, he thought very differently. He did not think women should be involved in politics or that they should have the right to vote. He felt strongly that women had a single purpose in life—marriage and the subordination to a husband. When his oldest daughter married, he wrote her: "The happiness of your life now depends on continuing to please a single person. To this all other objects must be secondary" (Nock, 1996). For a further discussion of historical views about women see "Our Diverse Society: Women Who Did Not Want Women to Vote."

In nineteenth-century Europe, attitudes toward women had not improved appreciably. The father of modern sociology, Auguste Comte (1968/1851), in

OUR DIVERSE SOCIETY

Women Who Did Not Want Women to Vote

"It seems to me," Jeannette Gilder wrote in 1894, "that it's a bigger feather in a woman's cap—a brighter jewel in her crown—to be the mother of George Washington than to be a member of Congress from the 32nd District."

Ms. Gilder was arguing that women shouldn't be allowed to vote. In her essay, "Why I Am Opposed to Women Suffrage," Ms. Gilder insisted that women belonged in the home, where they could exert more political influence by nurturing sons, fathers, and brothers than they could ever command with a single ballot. "Politics is too public, too wearing and too unfitted to the nature of women," Ms. Gilder concluded. "It is my opinion that letting women vote would loose the wheels of purgatory."

Until 1920, women—along with paupers, felons, and so-called idiots—couldn't vote in federal elections. At the time, it was believed that women simply couldn't be trusted to take the long, objective view. "The female vote . . . is always more impulsive and less subject to reason, and almost devoid of the sense of responsibility," wrote Francis Parkman, a historian and antisuffragist.

Women were believed to be "too frail for rough usage" and were also beleaguered by their household responsibilities, to the point where many seemed to hover on the verge of constant breakdowns. "The instability of the female mind is beyond the comprehension of the majority of men," declared Edith Melvin, a Concord, Mass., antisuffragist.

Not surprisingly, many men agreed that females should not vote. One of their biggest fears was that women would outlaw drinking, and various breweries supported antisuffrage political candidates. The men's antisuffrage movement even went so far as to produce bogus statistics: "If women achieve the feministic idea and live as men do," wrote a male doctor who opposed female suffrage, "they would incur the risk of 25% more insanity than they have now."

But tens of thousands of women also enlisted in the war against women voting, claiming that it was a slippery slope from the ballot box to depravation. If women got the vote, they would have to serve in the army and on juries. There would be fewer children and more divorce. Men would become less chivalrous and reverent of womanhood. Women would take up men's occupations, and men would take up women's occupations; the result, according to an antisuffrage booster, would be a "race of masculine women and effeminate men and the mating of these would result in the procreation of a race of degenerates."

And if women did run for office, wouldn't they invariably win? When all women can vote, wrote Goldwin Smith, "as the women slightly outnumber the men, and many men, sailors or men employed on railways or itinerants, could not go to the poll, the woman's vote would preponderate, and government would be more female than male."

Here the antisuffragists couldn't have been more wrong. Of the 535 members of the 108th Congress, only 73, or less than 14%, are women. All but 6 of America's 50 governors are men.

Source: From Cynthia Crossen, "Even Women Didn't Want to Give Women the Vote," *Wall Street Journal*, March 5, 2003, p. B1. Reprinted by permission.

---

constructing his views of the perfect society, also dealt with questions about women's proper role in society. Comte saw women as the mental and physical inferiors of men. "In all kinds of force," he wrote, "whether physical, intellectual or practical, it is certain that man surpasses women in accordance with the general law prevailing throughout the animal kingdom." Comte did grant women a slight superiority in the realms of emotion, love, and morality.

Comte believed that women should not be allowed to work outside the home, to own property, or to exercise political power. Their gentle nature, he said, required that they remain in the home as mothers tending to their children and as wives tending to their husbands' emotional, domestic, and sexual needs.

Comte viewed equality as a social and moral danger to women. He felt that progress would result only from making the female's life "more and more domestic; to diminish as far as possible the burden of out-door labour." Women, in short, were to be "the pampered slaves of men."

Religious Views

Many religions have overtly declared that men are superior to women. Men are thought within such religious outlooks to be spiritually superior to women, who are deemed dangerous and untrustworthy. For centuries of Judeo-Christian history, two short passages in scripture helped shape the common view of women's character and proper place: the creation of

Society causes us to expect certain gender-role behaviors from males and females. A changing society, however, produces changes in how these roles are carried out.

© Dana White/PhotoEdit

Eve from Adam's rib, and Eve's transgression and God's subsequent curse on her. The first of these passages presents Eve and women as an afterthought:

> And the Lord God said, It is not good that the man should be alone; I will make a help mate for him. . . . And the rib, which the Lord God had taken from man, made he a woman, and brought her unto the man. (Gen. 2:18, 22)

The second passage relates how, after God made the world a paradise, Eve disobeyed God and helped make the world the imperfect place we know:

> Unto the woman he said, I will greatly multiply thy sorrow and thy conception; in sorrow thou shalt bring forth children; and thy desire shall be to thy husband, and he shall rule over thee. (Gen. 3:16)

War, pestilence, famine, and every sin imaginable were the prices all humanity had to pay for Eve's disobedience. The biblical story of creation presents a God-ordained gender-role hierarchy, with man created in the image of God and woman as a subsequent and secondary act of creation. This account has been used as the theological justification for a **patriarchal ideology,** or *the belief that men are superior to women and should control all important aspects of society.* This kind of legitimization of male superiority is displayed in the following passage from the New Testament:

> But I would have you know, that the head of every man is Christ; and the head of the woman is the man; . . . For the man is not of the woman; but the woman of the man. Neither was the man created for the woman; but the woman for the man. (1 Cor. 11:3, 8–9)

In traditional India, the Hindu religion conceived of women as strongly erotic and thus a threat to male asceticism and spirituality. Women were cut

© The Hutchinson Library

From birth, parents all over the world treat boys and girls differently. Although the Amazon Indian boy is the youngest, he wears the chief's headdress.

off physically from the outside world. They wore veils and voluminous garments and were never seen by men who were not members of the family. Only men were allowed access to and involvement with the outside world.

Women's precarious and inferior position in traditional India is illustrated further by the ancient Manu code, which was drawn up between 200 B.C. and A.D. 200. The code states that if a wife had no children after 8 years of marriage, she would be banished; if all her children were dead after 10 years, she could be dismissed; and if she had produced only girls after 11 years, she could be repudiated.

Stemming from the Hindu patriarchal ideology was the practice of prohibiting women from owning and disposing of property. The prevalent custom in traditional Hindu India was that property acquired by the wife belonged to the husband. Similar restrictions on the ownership of property by women also prevailed in ancient Greece, Rome, Israel, China, and Japan. Such restrictions are still followed by fundamentalist Muslim states such as Saudi Arabia and Iran.

According to traditional Islamic law and tradition, three groups of people are not eligible for legal and religious equality: unbelievers, slaves, and women. Women were in the worst position of these three groups. The slave could become free, the unbeliever could become a believer, but the woman could do nothing to change her status. She was permanently doomed to her second-class status (Lewis, 2002).

Biological Views

Supporters of the belief that the basic differences between males and females are biologically determined have sought evidence from two sources: studies of other animal species, including nonhuman primates (monkeys and apes), and studies of the physiological differences between men and women. We shall examine each.

Animal Studies and Sociobiology Ethology is *the scientific study of animal behavior*. Ethologists have observed that sexual differences in behavior exist throughout much of the nonhuman animal world. Evidence indicates that these differences are biologically determined—that in a given species, members of the same sex behave in much the same way and perform the same tasks and activities. Popularized versions of these ideas, such as those of Desmond Morris in *The Human Zoo* (1970) or Lionel Tiger and Robin Fox in *The Imperial Animal* (1971), generalize from the behavior of nonhuman primates to that of humans. They maintain that in all primate species, including *Homo sapiens*, fundamental differences exist between males and females. They try to explain human male dominance and the traditional sexual division of labor in all human societies on the basis of inherent male or female capacities. They even have extended their analysis to explain other human phenomena, such as war and territoriality, through evolutionary comparisons with other species. A more sophisticated treatment of this same theme is found in the field of sociobiology (see Chapter 1), through the study of the genetic basis for social behavior (Wilson, 1975, 1978, 1994).

Sociobiologists believe that much of human social behavior has a genetic basis. Patterns of social organization such as family systems, organized aggression, male dominance, defense of territory, fear of strangers, the incest taboo, and even religion are seen to be rooted in the genetic structure of our species. The emphasis in sociobiology is on the inborn structure of social traits.

Critics contend that sociobiologists overlook the important role that learning plays among nonhuman primates in their acquisition of social and sexual behavior patterns (Montagu, 1973). They also assert that by generalizing from animal to human behavior, sociobiologists do not take into account fundamental differences between human and nonhuman primates, such as the human use of a complex language system. While freely acknowledging the biological basis for sex differences, these critics assert that among humans, social, and cultural factors overwhelmingly account for the variety in the roles and attitudes of the two sexes. Human expressions of maleness and femaleness, they argue, although influenced by biology, are not determined by it; rather, gender identities acquired through social learning provide the guidelines for appropriate gender-role behavior and expression.

Gender and Physiological Differences Even ardent critics cannot deny that certain genetic and physiological differences exist between the sexes—differences that influence health and physical capacity. Accordingly, the study of gender roles must take those differences into account in such areas as size and muscle development (both usually greater in males), longevity (females, with few exceptions, live longer in nearly every part of the world, sometimes as much as nine years longer on average), and susceptibility to disease and physical disorders (considerable variation exists between men and women). As you can see from Table 11–1, men and women are afflicted by very different chronic conditions.

There has been an explosion of interest in women's health after a highly publicized 1990 government study that raised concerns about the small numbers of women in clinical trials sponsored by the National Institutes of Health (NIH). In 1992, Congress made it illegal to exclude women as subjects in medical research. The National Institutes of Health now has an Office of Research on Women's Health to oversee the representation of women in the institutes' studies. As a result, women's health research has increased substantially.

The medical treatment of women also is getting more attention in the medical school curriculum after a House Appropriations Committee report noted that U.S. physicians are not trained adequately to address the needs of women. Traditionally, medical schools have taught about disease and treatment in terms of the 70-kilogram male. The NIH women's

Table 11–1

Gender and Disease

Heart Attack—Men are more likely to suffer heart attacks. Heart disease is the leading cause of death for women and a major contributor to their morbidity and disability. More women than men die each year from heart disease, and among those who survive a heart attack, nearly twice as many women as men die within the following year.

Cancer—Cancers are the second leading cause of death for women. Lung cancer (attributed to increased cigarette smoking among women) heads the list, with breast cancer coming in second. Women smokers are 20% to 70% more likely to develop lung cancer than male smokers.

HIV/AIDS—HIV incidence is increasing among women, especially in the African American and Hispanic populations. Women are 10 times more likely then men to contract HIV during unprotected sex with an infected partner.

Cardiovascular Disease—Men have a greater prevalence of heart disease, but one in nine women ages 45 to 64 has some cardiovascular disease, rising to one in three at age 65 and older.

Diabetes and Other Chronic Illnesses—Diabetes is a major health problem and is a cause of increased mortality among minority women, especially among middle-aged and older American Indian/Alaska Native, Hispanic, and African American women.

Osteoporosis—Twenty-four million people, primarily women, suffer from this disease. Eighty percent of people with osteoporosis are women; more than half of women older than 65 (mostly white) are afflicted with it.

Immunological Diseases—Autoimmune thyroid diseases have a 15:1 ratio of women to men. Rheumatoid arthritis has a 3:1 ratio of women to men. Multiple sclerosis occurs more often in women.

Mental Disorders—Women are twice as likely as men to be depressed and two to three times more prone to anxiety disorders. In elderly women, the prevalence of depression is 3.64% versus 1% in elderly men.

Alzheimer's Disease—The incidence of this disease is higher among women, and it increases dramatically after age 85.

Visual and Hearing Impairments—Men have nearly a 50% greater likelihood of experiencing these problems.

Source: From National Institutes of Health, "Women's Health," 2001, Health Resources and Services Administration, available at www.hrsa.gov.

health research office and the Health Resources and Services Administration have joined forces to find out exactly where the gaps in medical education are and to recommend a model curriculum to Congress.

A number of medical schools and teaching hospitals have begun to rethink their curricula to address more medical issues affecting women. The new approaches to teaching medicine require students to consider examples of women in all their subjects, not just in obstetrics and gynecology.

The differences between men and women go far beyond the obvious, and findings from research in the emerging field of gender-based biology could one day lead to treatments that vary depending on the sex of the patient. Gender-based biology, a field less than a decade old, identifies the biological and physiological differences between men and women as well as differences in responses to drugs.

A scientific literature review recently presented at the meeting of the Society for the Advancement of Women's Health Research found a growing body of evidence confirming the effect of gender on disease. The review identified 10 diseases, including multiple sclerosis, diabetes, lupus, and rheumatoid arthritis, that disproportionately affect women.

At menopause, women's brains experience a major reduction in estrogen, whereas men, paradoxically, continue producing androgen, which metabolizes into an estrogenlike hormone in the brain. As a result, men maintain high levels of estrogen in their brains as they age. Such findings might explain why aging women experience a higher incidence of cognitive decline such as Alzheimer's disease than men.

Hormone differences between men and women affect how the brain functions. For example, researchers have found that estrogen interacts with both the metabolism and serotonin, a neurotransmitter that helps control moods. Major estrogen receptors are scattered throughout the body but are concentrated in a region of the brain that controls thinking, emotion, sex, and eating.

Researchers will need to determine the significance of some of the gender differences they are starting to discover (Shelton, 1999).

Responses to Stress Gender differences also influence the way men and women react to stress—the "fight-or-flight" reaction that is thought to play a part in heart disease, stroke, and coronary-artery disease, among other ailments. In earlier days, when a primitive man was threatened by wild animals while hunting, testosterone combined with adrenaline enabled him to react quickly to danger. This intense type of reaction is no longer important and may be part of

© Robert Brenner/PhotoEdit

The medical treatment of women is now getting more attention in medical schools, in response to studies that showed that U.S. physicians were not adequately trained in women's health issues.

the reason men suffer more heart attacks than women. Women, it appears, react more slowly to stress, putting less pressure on the blood vessels and the heart. Although learned behavior may play a role in women's response to stress, biology is no less important.

When women are confronted by stress, they tend to respond by seeking out contact with and support from others. The support they seek is usually from other women. This "befriending" behavior has been linked with the hormone oxytocin, which is released by the body during stress. It has been shown to make both rats and humans calmer, less fearful, and more social. Although men also secrete oxytocin, male hormones reduce its effects. Female hormones, on the other hand, amplify its effects (Jablon, 2000).

Although many differences between males and females have a biological basis, other physical conditions may be tied to cultural influences and variations in environment and activity. Men react differently to psychological stress than women do: Each sex develops severe but dissimilar symptoms. Changing cultural standards and patterns of social behavior have had a pronounced effect on other traits that formerly were thought to be linked to sex. For example, the rising incidence of lung cancer among

women—a disease historically associated primarily with men—can be traced directly to changes in social behavior and custom, not biology: Women now smoke as freely as men.

In sum, differing learned behaviors contribute to the relative prevalence of certain diseases and disorders in each sex. However, not all male-female differences in disease rate and susceptibility can be attributed to these factors. In addition to genetically linked defects, differences in some basic physiological processes such as metabolic rates and adult secretions of gonadal hormones may make males more vulnerable than females to certain physical problems.

Gender and Sex

We know that men and women think differently about sex, but how differently? Is it just a stereotype that men seek sex more than women? Some researchers have tried to answer this question. In one study (Buss, 1994), men and women were asked, "How many sexual partners would you like to have in (a) the next month, (b) in the next two years, (c) in your lifetime?" Women averaged 8/10 of a sexual partner in the next month, one in the next two years, and four or five over their lifetimes. Men had very different expectations. Men wanted two sex partners within the next month, eight in the next two years, and 18 over their lifetimes.

Men and women also differed in how well they needed to know a person before having sex. The participants were asked whether they would consider having sex with a desirable partner they had known for (a) one year, (b) one month, or (c) one week. Women preferred to know a person for a year, whereas men had no problem having sex with a woman they had known for a week.

The differences were even more striking in another study (Clark & Hatfield, 1989). Attractive men and women were hired to approach strangers of the opposite sex on college campuses and say to them, "I have been noticing you around campus; I find you very attractive," and then ask one of three questions: (a) "Would you go out with me tonight?" (b) "Would you come over to my apartment tonight?" or (c) "Would you go to bed with me tonight?" Half the men and half the women consented to a date, but that was where the similarity ended. Only 6% of the women consented to go to the confederate's apartment, compared with 69% of the men. None of the women consented to sex; however, 75% of the men consented to sex. Of the remaining 25%, many were apologetic, asking for a rain check or explaining that they could not do it now because they had to meet their girlfriend.

Sociological View: Cross-Cultural Evidence

Most sociologists believe that the way people are socialized has a greater effect on their gender identities than do biological factors. Cross-cultural and historical research offers support for this view, revealing that different societies allocate different tasks and duties to men and women and that males and females have culturally patterned conceptions of themselves and of one another.

Until the pioneering work of anthropologist Margaret Mead (1901–1978) was published in the 1930s, it was widely believed that gender identity (what then was called sex temperament) was a matter of biology alone. It never occurred to Westerners to question their culture's definitions of male and female temperament and behavior, nor did most people doubt that these were innate properties. In 1935, Mead published a refutation of this assumption in *Sex and Temperament in Three Primitive Societies,* which has become a classic. While doing research among isolated tribal groups on the island of New Guinea, she found three societies with widely differing expectations of male and female behavior. The Arapesh were characterized as gentle and home loving, with a belief that men and women were of similar temperament.

Adult men and women subordinated their needs to those of the younger or weaker members of the society. The Mundugamor, by contrast, assumed a natural hostility between members of the same sex and only slightly less hostility between the sexes. Both sexes were expected to be tough, aggressive, and competitive. The third society, the Tchambuli, believed that the sexes were temperamentally different, but the gender roles were reversed relative to the Western pattern.

> I found . . . in one, both men and women act as we expect women to act—in a mild parental responsive way; in the second, both act as we expect men to act—in a fierce initiative fashion; and in the third, the men act according to our stereotype for women—are catty, wear curls and go shopping, while the women are energetic, managerial, unadorned partners. (Mead, 1935)

Although Mead's findings are interesting and suggestive, anthropologists have cautioned against overinterpreting them. They point out that Mead's research was limited to a matter of months and that her then husband and fellow anthropologist, Reo Fortune, rejected her view that the Arapesh did not distinguish between male and female gender roles. Others have suggested (Gewertz, 1981) that the situation with the Tchambuli was short-lived, resulting from tribal wars during which most property was lost and the men had little else to do but work at home.

Most sociologists tend to agree that even in preliterate societies, culture is central to the patterning of gender roles. Nevertheless, biological factors may play a more prominent part in structuring gender roles in societies less technologically developed than our own. Anthropologist Clellan S. Ford (1970) believes that for preindustrial peoples, "the single most important biological fact in determining how men and women live is the differential part they play in reproduction." The woman's life is characterized by a continuing cycle of pregnancy, childbearing, and nursing for periods of up to three years. By the time a child is weaned, the mother is likely to be pregnant again. Not until menopause, which frequently coincides with the end of a woman's life, is her reproductive role over. In those circumstances, it is not surprising that such activities as hunting, fighting, and forest clearing usually are defined as male tasks; whereas gathering and preparing small game, grains, and vegetables; tending gardens; and building shelters are typically female activities, as is caring for the young.

In an early study, George Murdock (1937) provided data on the division of labor by sex in 224 preliterate societies. Men performed such activities as working with metal, wood, and stone; making weapons; hunting and trapping; building houses and boats; and clearing the land for agriculture. Women's activities included grinding grain; gathering and cooking herbs, roots, seeds, fruits, and nuts; weaving baskets; making pottery; and making and repairing clothing. A review of the cross-cultural literature (D'Andrade, 1966) concluded that the division of labor by sex occurs in all societies. Generally, male tasks require vigorous physical activity or travel, whereas female tasks are less physically strenuous and more sedentary.

The almost universal classification of women to secondary status has had a profound effect on their work and family roles. Anthropologist Sherry Ortner (1974) observed that "everywhere, in every known culture, women are considered in some degree inferior to men." One important result of this attitude is the exclusion of women from participation in, or contact with, those areas of the particular society believed to be most powerful, whether they be religious or secular. We see these exclusionary patterns in our own society. There are no female priests in the Catholic Church or women rabbis in Orthodox Judaism. No woman has ever been the presidential candidate of a major political party, and there are very few female chief executive officers of major U.S. corporations.

Another anthropologist, Michelle Rosaldo (1974), believes that "women's status will be lowest in those societies where there is a firm differentiation between domestic and public spheres of activity and where women are isolated from one another and placed under a single man's authority in the home." She believes that the time-consuming and emotionally compelling involvement of a mother with her child is unmatched by any single involvement and commitment by a man. The result is that men are free to form broader associations in the outside world through their involvement in work, politics, and religion. The relative absence of women from these public spheres results in their lack of authority and power. Men's involvements and activities are viewed as important, and the cultural systems accord authority and value to men's activities and roles. Women's work, especially when it is confined to domestic roles and activities, tends to be oppressive and lacking in value and status. Women are seen to gain power and a sense of value only when they can transcend the domestic sphere of activities. This differentiation is most acute in those societies that practice sexual discrimination. Societies in which men value and participate in domestic activities tend to be more egalitarian.

Nevertheless, even though physiological factors tend to play an influential part in gender-role differences, biology does not determine those differences. Rather, people acquire much of their ability to fulfill their gender roles through socialization. (For a discussion of gender in new settings, see "Technology and Society: Is There Gender in Cyberspace?")

What Produces Gender Inequality?

Sociologists have devoted much thought and research to answering this question. They also have tried to explain why males dominate in most societies. Two theoretical approaches have been used to explain dominance and gender inequality: functionalism and conflict theory.

The Functionalist Viewpoint

From Chapter 1 you may recall functionalists (or structural functionalists, to be more precise) believe that society consists of a system of interrelated parts that work together to maintain the smooth operation of society. Functionalists argue that it was quite useful to have men and women fulfill different roles in preindustrial societies. The society was more efficient when tasks and responsibilities were allocated to particular individuals who were socialized to fulfill specific roles.

That the human infant is helpless for such a long time made it necessary that someone look after the child. It was also logical that the mother who gave birth to the child and nursed it was also the one to take care of it. Because women spent their time near the home, they also took on the duties of preparing the food, cleaning clothes, and attending to the other necessities of daily living. To the male fell the duties of hunting, defending the family, and herding. He also became the one to make economic and other decisions important to the family's survival.

This division of labor created a situation in which the female was largely dependent on the male for protection and food, and so he became the dominant partner. This dominance, in turn, caused his activities to be more highly regarded and rewarded. Over time, this pattern came to be seen as natural and was thought to be tied to biological sex differences. Talcott Parsons and Robert Bales (1955) applied functionalist theory to the modern family. They have argued that the division of labor and role differentiation by sex are universal principles of family organization and are functional to the modern family. They believe that the family functions best when the father assumes the *instrumental role,* which focuses on relationships between the family and the outside world. It mainly involves supporting and protecting the family. The mother concentrates her energies on the *expressive role,* which focuses on relationships within the family and requires her to provide the love and emotional support needed to sustain the family. The male is required to be dominant and competent, and the female to be passive and nurturing.

As you can imagine, the functionalist position has been much criticized. The view that gender roles and gender stratification are inevitable does not fit with

TECHNOLOGY AND SOCIETY

Is There Gender in Cyberspace?

Can you have a gender if you do not have a body?

In the day-to-day physical world, what you look and sound like relays information about your gender and produces certain reactions. In cyberspace, it is possible to communicate with others without disclosing any of these characteristics. In essence, you have the ability to engage in the totally new experience of communicating without your gender being a key factor in the interaction. For some, this experience may be quite liberating; for others, very disconcerting. Some women may feel that they no longer have to fear the danger that their comments could cause them if they were made in the physical world. Others may feel constrained communicating with someone if they do not know that person's gender. In cyberspace communication groups, men often impersonate women and women impersonate men.

Yet communicating in cyberspace may not really be as devoid of gender as we may think. Women seem to have recognizably different styles of communicating on the Internet. Susan Herring (1994) analyzed male and female participation styles on several electronic lists. She found that the communication was not equal. Men and women had different expectations about what was and what was not appropriate in online interaction. It also appeared that a small male minority dominated the discourse both in terms of sheer amount of postings and with self-promotional and adversarial comments. Moreover, when women do attempt to participate on a more equal basis, they risk being actively censored by the reactions of the men, or they may be ignored. Women tend to be intimidated by the attacks and avoid participating as a result. Herring believes that the communication styles appear to be recognizably gendered and they perpetuate preexisting interaction patterns in society as a whole.

Herring found that "68% of the messages posted by men made use of an adversarial style in which the poster distanced himself from, criticized, and/or ridiculed other participants, often while promoting his own importance." What is known as "flaming," making offensive or derogatory comments about someone or the person's views, appears to be a male characteristic. The male style is characterized by adversarial comments such as put-downs; strong, often contentious assertions; lengthy or frequent postings; self-promotion; and sarcasm. Men

are also likely to take an authoritative, self-confident stance in which they represent themselves as experts.

The women, in contrast, displayed such features as "hedging, apologizing, asking questions rather than making assertions, and . . . revealing thoughts and feelings and interacting with and supporting others." The female style includes expressions of appreciation, thanking, and community-building activities that make the participants feel accepted and welcomed.

Lisa King (2000) describes Herring's analysis of gender characteristics in online interactions with a comparison between the language women and men use:

Women's Language	Men's Language
Attenuated assertions	Strong assertions
Apologies	Self-promotion
Explicit justifications	Presuppositions
Questions	Rhetorical questions
Personal orientation	Authoritative orientation
Supports others	Challenges others
	Humor/sarcasm

King summarizes Herring's conclusions: ". . . there is no possibility of gender-neutral communication because gender cues are scattered throughout online communications. Therefore, the ideal of free and equal participation is impossible."

Sources: S. Herring, "Gender Differences in CMC: Findings and Implications," *CPSR Newsletter,* *18*(1), Winter 2000; S. Herring, "Gender Differences in Computer-Mediated Communication: Bringing Familiar Baggage to the New Frontier," keynote talk at panel entitled "Making the Net*Work*: Is There a Z39.50 in Gender Communication?", American Library Association annual convention, Miami, June 27, 1994; "Posting in a Different Voice," by S. Herring, 1996, *Philosophical Perspectives on Computer-Mediated Communication,* edited by C. Ess, New York: State University of New York Press; Lisa J. King, "Gender Issues in Online Communities," *CPSR Newsletter,* *18*(1), Winter 2000.

cross-cultural evidence and the changing situation in American society (Crano & Aronoff, 1978). Critics contend that industrial society can be quite flexible in assigning tasks to males and females. Furthermore, they assert that the functionalist model was developed during the 1950s, an era of very traditional family patterns, and that rather than being predictive of family arrangements, it merely is representative of the era during which it became popular.

The Conflict Theory Viewpoint

Although functionalist theory may explain why gender-role differences emerged, it does not explain why they persisted. According to the conflict theory, males dominate females because of their superior power and control over key resources. A major consequence of this domination is the exploitation of women by men. By subordinating women, men gain greater economic, political, and social power. According to conflict theory, as long as the dominant group benefits from the existing relationship, it has little incentive to change it. The resulting inequalities are therefore perpetuated long after they may have served a functional purpose. In this way, gender inequalities resemble race and class inequalities.

Conflict theorists believe that the main source of gender inequality is the economic inequality between men and women. Economic advantage leads to power and prestige. If men have an economic advantage in society, that advantage will produce a superior social position in both society and the family.

Friedrich Engels (1942/1884) linked gender inequalities to capitalism, contending that primitive, noncapitalistic hunting and gathering societies without private property were egalitarian. As these societies developed capitalistic institutions of private property, power came to be concentrated in the hands of a minority of men, who used their power to subordinate women and to create political institutions that maintained their power. Engels also believed that to free women from subordination and exploitation, society must abolish private property and other capitalistic institutions. He believed that socialism was the only solution to gender inequality.

Today, many conflict theorists accept the view that gender inequalities may have evolved because they initially were functional. Many functionalists also agree that gender inequalities are becoming more and more dysfunctional. They agree that the origins for gender inequalities are more social than they are biological.

Gender-Role Socialization

Gender-role socialization is *a lifelong process whereby people learn the values, attitudes, motivations, and behavior considered appropriate to each sex according*

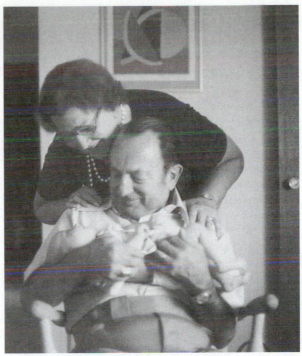

Parents and grandparents respond differently to boys and girls right from the beginning, and they carry in their minds images of what the child should be like, how he or she should behave, and what he or she should be in later life.

to their culture. In our society, as in all others, males and females are socialized differently. In addition, each culture defines gender roles differently. This process is not limited to childhood but continues through adolescence, adulthood, and old age.

Childhood Socialization

Even before a baby is born, its sex is a subject of speculation, and the different gender-role relationships it will form from birth already are being decided. Maccoby and Jacklin (1975) describe this phenomenon using a familiar cultural medium—the musical.

A scene from [the 1940s] musical *Carousel* epitomizes (in somewhat caricatured form) some of the feelings that parents have about bringing up sons as opposed to daughters. A young man discovers he is to be a father. He rhapsodizes about what kind of son he expects to have. The boy will be tall and tough as a tree, and no one will dare to boss him around; it will be all right for his mother to teach him manners, but she mustn't make a sissy out of him. He'll be good at wrestling and will be able to herd cattle, run a riverboat, drive spikes, etc. Then the prospective father realizes, with a start, that the child may be a girl. The music changes to a gentle theme. She will have ribbons in her hair; she will be

CONTROVERSIES IN SOCIOLOGY

Can Gender Identity Be Changed?

In 1963, two healthy twin boys were born. Seven months later, the twin named David Reimer lost his entire penis in a botched circumcision. As a result of the irreparable injury, his parents took him to an expert in sex research at the renowned Johns Hopkins Hospital in Baltimore. There, Dr. John Money persuaded the parents to allow their son to have a surgical sex change. The process involved genital surgery, followed by a 12-year program of social, mental, and hormonal therapy to complete the metamorphosis.

The case was reported in the medical literature as an unqualified success. In 1973, *Time* magazine reported that the case "provides strong support . . . that conventional patterns of masculine and feminine behavior can be altered. It also casts doubt on the theory that major sex differences, psychological as well as anatomical, are immutably set by the genes at conception." For 20 years, this case was cited in numerous sociology, psychology, and human sexuality textbooks. Researchers on these topics used the case as proof that gender is very much the product of socialization and that if intervention occurs early enough, a child may be raised in a gender that is different from that with which he or she was born. As Money noted, "the gender identity gate is open at birth. . . . [I]t stays open at least for something over a year." The case also served as a model of how to treat infants with ambiguous genitalia. Many biologi-

cal boys had operations that turned them into girls, and many girls had procedures that turned them into boys.

The one dissenting voice during this period was Milton Diamond, a young graduate student at the University of Kansas. Diamond was not convinced that there was any reason to believe John Money's theory of psychosexual neutrality in children. He published an article on the topic, and then he contacted Money and offered to do a joint study with him. Money, a respected researcher, saw no reason to pay any attention to the unknown Diamond.

Money's view that gender is the product of socialization was challenged by a case that Dr. William G. Reiner, a faculty member at the University of Oklahoma and Johns Hopkins, encountered in the early 1980s. He had a patient who was a 14-year-old "girl" who said he was a boy not a girl. At birth, the child had ambiguous genitalia. The child had one testicle, but internally, she/he was part female. His parents had raised him female.

At puberty, the one testicle began producing enough male hormones to make her/him seem more male. At which point he started to insist that he was a boy. The child wanted Dr. Reiner to convince his parents that he was a boy and operate on him so that he would have an appropriate penis. With Dr. Reiner's help the parents

---

sweet and petite (just like her mother); and suitors will flock around her. There's a slightly discordant note, introduced for comic relief from sentimentality, when the expectant father brags that she'll be half again as bright as girls are meant to be; but then he returns to the main theme: she must be protected, and he must find enough money to raise her in a setting in which she will meet the right kind of man to marry.

Parents carry in their minds images of what girls and boys are like, how they should behave, and what they should be in later life. Parents respond differently to girls and boys right from the beginning. Girls are caressed more than boys, whereas boys are jostled and roughhoused more. Mothers talk more to their daughters, and fathers interact more with their sons.

These differences are reinforced by other socializing agents—siblings, peers, educational systems, and the mass media. By the first two or three years of life, core gender identity—the sense of maleness or

femaleness—is established as a result of the parents' conviction that their infant's assignment at birth to either the male or female sex is correct.

(For a discussion of whether biology determines gender, see "Controversies in Sociology: Can Gender Identity Be Changed?")

Adolescent Socialization

Most societies have different expectations for adolescent girls and boys. Erik Erikson (1968) believed that the most important task in adolescence is the establishment of a sense of identity. During the adolescent stage, Erikson felt, both boys and girls undergo severe emotional crises centered on questions of who they are and what they will be. If the adolescent crisis is resolved satisfactorily, a sense of identity will be developed; if not, role confusion will persist.

According to Erikson, adolescent boys in our society generally are encouraged to pursue role paths that will prepare them for an occupational commitment, whereas girls generally are encouraged to

agreed that this was the correct course of action and the child was operated on and adjusted well.

The problem for many of the intersex infants is that genetic girls may look like males because they were exposed to testosterone at a critical stage of fetal development. The genetic boys who look like females did not get enough testosterone in utero.

It now appears that Diamond was correct and Money woefully misguided. The view that gender can be changed through socialization and surgery appears to be decidedly wrong. After further investigation, it turned out that the twin David never adjusted to being a girl. He steadfastly refused to grow into a woman and stopped taking the estrogen pills that were prescribed for him at age 12. David also refused to undergo additional surgery that the physicians tried to persuade him he needed to fully become a woman. At 14 he found other physicians who were sympathetic to his plight. He underwent a double mastectomy and a phalloplasty and began a regimen of male hormones. He refused to ever return to Johns Hopkins University where the original diagnosis and surgery had taken place. After David's return to the male gender identity, he felt that his attitudes, behavior, and body were once again united. At 25, David married a woman several years older and adopted her children. He died at an early age.

Many of the people who received these gender-changing operations in early life have experienced difficulties. They have formed support groups, and some have undergone surgery to reverse the process.

In October 1998, the American Academy of Pediatrics invited Diamond to address its prestigious annual meeting. Physicians now reject John Money's views on gender identity and no longer treat infants with ambiguous genitalia the way they did 25 years ago.

All this points to the current view that people are not gender neutral at birth and that they are predisposed to be male or female. Gender is more complex than originally imagined.

Sources: *As Nature Made Him,* by J. Colapinto, 2000, New York: HarperCollins; "Sex Reassignment at Birth: A Long Term Review and Clinical Implications," by M. Diamond & H. K. Sigmundson, March 1997, *Archives of Pediatric and Adolescent Medicine, 151,* pp. 298–304; Claudia Dreifus, "Declaring with Clarity, When Gender Is Ambiguous, *New York Times,* May 31, 2005; *Sexual Signatures: On Being a Man or Woman,* by J. Money & P. Tucker, 1975, Boston: Little, Brown; "Changing Sex Is Hard to Do," by C. Holden, March 1997, *Science,* 275(5307), p. 1745.

---

develop behavior patterns designed to attract a suitable mate. Erikson observed that it is more difficult for girls than for boys to achieve a positive identity in Western society. This is because women are encouraged to be more passive and less achievement oriented than men and to stress the development of interpersonal skills—traits that are not highly valued in our society. Men, by contrast, are encouraged to be competitive, to strive for achievement, and to assert autonomy and independence—characteristics that are held in high esteem in our competitive society.

Nonetheless, female gender roles are changing rapidly, although traditional attitudes toward careers and marriage undoubtedly still remain part of the thinking of many people in our society. Girls are being encouraged not to limit themselves to these stereotyped roles and attitudes. More and more young women expect to pursue careers before and during marriage and childrearing. Marriage no longer is considered the only desirable goal for a woman, nor is it even considered necessary to a woman's success and happiness.

Gender Differences in Social Interaction

There are some interesting differences in how men and women think about the future and solve problems. In one study, researchers asked undergraduates to describe what they expected would happen to them over the next 10 years (Maines & Hardesty, 1987). For both male and female undergraduates, work appeared to be a "universal expectation." Likewise, the kinds of jobs wanted were nearly identical. Both groups mentioned jobs in business, law, hospital administration, and the computer industry. Further education was anticipated by more than half of the men and women. The vast majority (94%) also saw themselves as eventually married with children.

Thus, at first glance there appears to be a striking similarity in men's and women's future plans. Both groups expressed equal desires for marriage, children, work, and higher education. However, there are significant gender differences in expectations of how family, work, and education will be integrated. Men and women have different assumptions about and

In every society boys and girls are socialized differently, and as they mature, a variety of roles are tried out. In this picture, the 6-year-old boy is experimenting with one version of the male role.

tactics for achieving the similarly desired events in their lives.

Men operate in what Maines and Hardesty call a *linear temporal world*. When they try to project what the future might hold for them, they almost always define it in terms of career accomplishments—lawyer, doctor, college professor, business executive, and so on. Education is seen as something that is pursued to attain the desired career.

Men see a family as desirable and not much of an issue in terms of pursuing career goals. They see little problem in coordinating career and family demands. Many expect to have a traditional division of labor in their families, which will provide a support system for their career pursuit. Family problems are seen as easily resolved and very few men surveyed mention career adjustments as an option for addressing the wife's and children's needs.

Young women, in contrast, operate in *contingent temporal worlds*. Work, education, and family all are seen as having to be balanced. Careers are seen as pursuits that may have to be suspended or halted at certain points. The vast majority of women envision problems in their career pursuits, and they see

family responsibilities as a major issue that requires adjustments. Nearly half say they will quit work for a few years as a solution to family/work demands. Instead of having a clear vision of steps needed to accomplish their career goals, women become much more tentative about their future, since they expect it to entail adjustments and compromises.

Young men seem to take their autonomy for granted. That is, they assume that they will be able to accomplish what they set out to do if they have the necessary education, skill, and good fortune. Young women, however, feel much more limited in the control of their future, even with the necessary education and skill. The problems surrounding the integration of family and career lend an element of uncertainty to their ability to accomplish their goals. Women consider flexibility part of their plan when considering career and family needs. This flexibility gives them only partial autonomy in controlling their lives (Maine & Hardesty, 1987).

This element of tentativeness about the future, the willingness to be flexible and to adjust to the needs of others, and the realization that goals cannot be achieved easily without compromise evidently

OUR DIVERSE SOCIETY

Deborah Tannen: Communication between Women and Men

Deborah Tannen, a sociolinguist, is the author of the best-seller *You Just Don't Understand: Women and Men in Conversation* (1990) and *Talking from 9 to 5* (1994). Tannen believes that gender differences are widespread in everyday speech, and that in many ways, men and women are living in different worlds when they try to communicate with each other.

Tannen notes that just about every woman she spoke to recalled saying something at a meeting and being ignored, only to have the same idea taken seriously when it was repeated by a man. Tannen believes this is due to the different ways in which men and women use language. Women use language primarily to create intimacy and connections to other people. For women, language is the glue that holds relationships together.

Men, however, use language mainly to convey information. For men, activities are the things that hold people together, and in the absence of doing something with others, talking about an activity is the next best thing. This involves talking about concrete events or facts. This explains the tendency among men to engage in endless discussions about sporting events, batting averages, and so on. Men are acutely aware of the status differences implied by knowledge, or in one way or another of speaking. For men, language and information are used to attain status, not intimacy.

When people speak at a meeting, they may speak softly or loudly, with or without a disclaimer, in a self-deprecating or declamatory way, briefly or at length, tentatively or with apparent certainty. The first of each of these choices is more likely to be done by a woman than a man, and this plays a role in how seriously the idea is taken.

Tannen also notes that women engage in "troubles talk," or talking about the day's problems. Confronted with this type of talk, a man is likely to offer advice or solutions and get on with it. Women are not looking for this response. More typically, when one woman mentions a problem, someone else will mention a similar problem. Women are looking for agreement and mutual understanding. When someone offers advice or a solution, it creates "one-upmanship." It is no longer a situation in which two people are in similar circumstances but one in which two people are jockeying for status.

Tannen tells a joke about a man whose wife wants to divorce him for not talking to her for two years. At the divorce hearing, the judge asks the man why he has not talked to her for all this time. He answers that he did not want to interrupt her. This joke implies that women talk more than men. Upon closer observation, though, we must distinguish between speaking in public situations and speaking at home. Men seem to be more comfortable with the language used in a public setting, whereas women find the language used in private settings more natural. Essentially, the public language is about information and one person attempting to attain status with knowledge, whereas private language is about personal feelings and sharing them with others. It therefore seems understandable that men talk more in public settings, whereas women talk more in the home. Men feel they do not need to talk much at home, as their status is established. They use language when they are struggling for status. The home is not such a place.

Should people try to change their conversational styles? In business settings, there may be some advantage to certain ways of presenting information. However, even if no one changes, understanding those gender differences improves relationships. Once people realize that their partners and coworkers have different conversational styles, they are inclined to accept differences without blaming themselves, others, or their relationships.

Sources: Interviews with the author, June 1990 and November 1994; *You Just Don't Understand: Women and Men in Conversation,* by D. Tannen, 1990, New York: William Morrow; *Talking from 9 to 5,* by D. Tannen, 1994, New York: William Morrow.

produces a difference in how men and women approach issues. This observation led Carol Gilligan (1982) to believe that men and women think differently when it comes to problem solving.

Some men often think that the highest praise they can bestow on a woman is to compliment her for "thinking like a man." That usually means that the woman has been decisive, rational, firm, and clear. To think "like a woman" in our society always has had negative overtones, being characterized as fuzzy, indecisive, unpredictable, tentative, and softheaded.

Gilligan challenged the value judgments made about male versus female styles of reasoning, especially in the area of moral decision making. She argued that a woman's perspective on things is not inferior to a man's; it is just different.

To illustrate, Gilligan described the different responses that 11-year-old boys and girls made to an example used by psychologist Lawrence Kohlberg. In the example, Heinz, a fictional character, is caught up in a complex moral question. Heinz's wife is dying of cancer. The local pharmacist has discovered a drug that might cure her, but it is very expensive. Heinz has done all he can to raise the money necessary to buy it but can come up with only half the amount, and the druggist demands the full price. The question is: Should Heinz steal the drug?

Boys and girls differed significantly on how they answered this question. Boys often saw the problem as the man's individual moral choice, stating that Heinz should steal the drug, as the right to life supersedes the right to property. Case closed.

The girls Gilligan questioned always seemed to get bogged down in peripheral issues. No, they maintained, Heinz should not steal the drug because stealing is wrong. Heinz should have a long talk with the pharmacist and try to persuade him to do what is right. Besides, they pointed out, if Heinz steals the drug, he might be caught and go to jail. Then what would happen to his wife? What if there were children?

Instead of labeling this tentativeness as a typical example of women's inability to make firm decisions, Gilligan sees it as an attempt by women to deal with the consequences of actions rather than simply with "what is right." For women, moral dilemmas involving people have a greater complexity and therefore a greater ambiguity.

If your morality stresses the importance of not hurting others, as seems to be the case with many women, you will often face failure. As one of the men Gilligan interviewed said, making a moral decision is often a matter of "choosing the victim." There is "violence inherent in choice," Gilligan wrote, and "the injunction not to hurt can paralyze women." If you base your decision on an absolute principle (for example, abortion is murder; therefore, it is wrong), you may act with the decisiveness so admired by both men and women.

If you base your decision on what you imagine is likely to happen (for example, if this child would be born with no father, if the fetus is likely to be seriously defective, or if the mother's life would be endangered by the birth), you often will face uncertainty. Women, whose value systems are usually focused more on people than on principles, consequently find themselves wrestling with the problems that might result from their decisions.

Gilligan hopes that by ceasing to label a man's perspective as right and a woman's as wrong, we can begin to understand that each may be valuable, though different. For this to happen, according to Gilligan, girls and women must gain confidence in their own ethical perspectives. Indeed, Gilligan feels, if society finally accepted a woman's moral view of the interconnectedness of actions and relationships, enormous consequences could result for everything from scholarship to politics to international relations. (For a further discussion of these issues, see

"Our Diverse Society: Deborah Tannen: Communication between Women and Men.")

Stephen Pinker (2002) is not convinced. Gilligan's view that men think about rights and justice and women have feelings of compassion, nurturing, and peaceful accommodation would disqualify women from becoming constitutional lawyers or Supreme Court justices, he notes. Other studies (for instance, Jaffe & Hyde, 2000) have tested Gilligan's hypothesis and found that men and women differ little or not at all in moral reasoning.

Gender Inequality and Work

Women's numerical superiority over men has not enabled them as yet to avoid discrimination in many spheres of American society. In this discussion, we will focus on economic and job-related discrimination, because these data are easily quantified and serve well to highlight the problem. Remember, though, that discrimination against women in America actually is expressed in a far wider range of social contexts and institutions.

Job Discrimination

In 2004 more than 59% of all American women were part of the paid labor force. The median income for men in 2004 was $60,093, whereas for women it was $42,307 (U.S. Department of Labor, 2005). Women and men are concentrated in different occupational groups. Some would argue that the income and earnings differences between them result from pay differences across occupations rather than gender. Within each broad occupational category, women and men tend to work in different specific occupations. In the professional occupations, where women earn the most, they are much less likely than men to be employed in some higher-paying occupations, such as engineers and mathematical and computer scientists.

However, even if we adjust for occupations, we find differences in earnings. Women earn less than men in almost every occupation. For example, male managers and professionals earned $53,976 in 2001, compared with $38,064 for females. Male technical sales workers earned $34,684, whereas females earned $24,596. (See Table 11–2 for the jobs that are disproportionately male and female.)

Men and women differ in terms of educational attainment. Could it be that these salary differences result from educational disparities? It does not seem as if that is the answer either. In 2001, male high school graduates who worked full-time year-round had an average salary of $31,720, whereas women had an average salary of $22,932. For college graduates,

Table II–2

Professions That Are Disproportionately Male and Female

Disproportionately Male	% Male	Disproportionately Female	% Female
Automobile mechanic	98.9	Secretary	98.9
Carpenter	98.3	Dental hygienist	98.5
Airplane pilot	96.3	Receptionist	96.7
Truck driver	95.3	Speech therapist	93.5
Firefighter	96.2	Bank teller	90.0
Mechanical engineer	93.7	Dietitian	89.9
Forestry and logging	91.6	Librarian	85.2
Police detective	87.9	Elementary school teacher	83.3
Architect	76.5	Health aide	82.6
Computer programmer	73.5	Special education teacher	82.6

Source: Bureau of the Census, 2002, *Statistical Abstract of the United States: 2001*, pp. 380–382.

the average salaries were $56,264 for men and $40,768 for women.

A striking fact in the previous information is that not only are the average earnings for women lower than for men, but male high school graduates earn more than women with a two-year college degree. One possible explanation for this disparity is that the less skilled jobs men hold, such as construction, are typically unionized and therefore pay better.

Job discrimination against women is a complicated phenomenon. Generally, women experience discrimination in the business world in three ways: (1) during the hiring process, when women are given jobs with lower occupational prestige than men who have equivalent qualifications receive; (2) through unequal wage policies, by which women receive less pay than men for equivalent work; and (3) in the awarding of promotions, as women find it more difficult than men to move up the career ladder.

Discrimination against women in the economic sector is often quite subtle. Women are more or less channeled away from participation in occupations that are socially defined as appropriate to men. For example, it cannot really be argued that female bank presidents are paid less than men; instead, there are few women bank presidents.

Women and men often do not perform equal work; therefore, the phrase "equal pay for equal work" has little relevance. In some instances, men and women perform similar work, but there may be two job titles and two pay scales. (See "News You Can Use: Are We Biased against Assertive Women?")

Having painted a somewhat pessimistic picture here, we should note that recent years have shown some improvement. Women's share of jobs with high earnings has grown. In 2001, women held nearly

Figure II–I Advanced Degrees Awarded to Women, 1970 and 2002

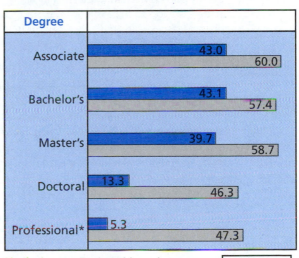

*Includes medical and law degrees.

■ 1970
▢ 2002

Source: U.S. Department of Education, "Digest of Education Statistics" 2003.

49.5% of all executive, administrative, and managerial jobs, up from 34.2% in 1983. Women now represent 51.8% of the professional workforce.

Women also constitute a rising share of people being awarded college and postgraduate degrees. In 2002, they represented 57.4% of people awarded bachelor's degrees, 58.7% of those awarded master's degrees, and 46.3% of the doctorate recipients (U.S. Department of Education, 2003). (See Figure 11–1.) In addition, young women and men (those younger than age 25) had fairly similar earnings;

NEWS YOU CAN USE

Are We Biased against Assertive Women?

In 1953, when Americans were asked whether they preferred a male or female boss, 66% said they preferred a man and 5% favored a woman. Today, people continue to prefer male bosses, but the margin in men's favor is shrinking. In December 2000, "48% of Americans expressed a preference for a male boss and 22% opted for a female one" (Gallup Poll, September 2001). Even though the bias against female bosses is improving, other barriers to women achieving leadership positions still exist.

Among the problems women encounter is that people pay attention to and use information contributed by a man but ignore the identical information presented by a woman (Propp, 1995).

In addition, a woman acting in a competent and knowledgeable manner can interfere with her ability to influence others. Studies show that when women act in the same assertive and competent manner as men, they are less influential and less well liked. For example, men can be equally persuasive, using either assertive or indirect speaking styles. Women, however, are handicapped if they use an assertive speaking style. Even though fellow women may respond favorably to a women speaking in a direct and competent manner, men respond unfavorably to women using an assertive style of speaking. Men reported feeling more threatened by and less inclined to like a competent woman and resisted being influenced by her. Whereas women were equally influenced by competent men and women, men find women most persuasive when they communicate in an indirect and less competent manner (Carli, LaFleur, & Loeber, 1995).

Another study (Matschiner & Murnen, 1999) showed that women in the workplace who adopted a highly traditional and subordinate gender role exerted more influence over men than women who acted in a more assertive manner.

Speaking style also handicaps women in job interviews. Job interviewers preferred to hire men who communicated in a highly competent manner, showing directness and initiative, rather than men who were less assertive and indirect. In contrast, these same interviewers were less likely to hire a woman who presented herself in a highly competent style and preferred women who spoke and acted in a less assertive manner (Buttner & McEnally, 1996).

It appears clear that many subtle factors continue to operate when men and women interact in the workplace.

Sources: E. H. Buttner & M. McEnally, "The Interactive Effect of Influence Tactic, Applicant Gender, and Type of Job on Hiring Recommendations," *Sex Roles,* 34, 1996, pp. 581–591; L. L. Carli, "Gender and Social Influence," *Journal of Social Issues,* 57(4), 2001, p. 725; L. L. Carli, S. J. LaFleur, & C. C. Loeber, "Nonverbal Behavior, Gender, and Influence," *Journal of Personality and Social Psychology,* 68(6), 1995, pp. 1030–1041; Gallup Poll, September 15, 2001, *Gallup Management Journal,* accessed at http://www.gallupjournal.com/GMJarchive/issue3/2001915g.asp; Matschiner & S. K. Murnen, "Hyperfemininity and Influence," *Psychology of Women Quarterly,* 23, 1999, pp. 631–642; K. M. Propp, "An Experimental Examination of Biological Sex as a Status Cue in Decision-Making Groups and Its Influence on Information Use," *Small Group Research,* 26, 1995, pp. 451–474.

---

however, women's earnings were much lower than men's in older age groups.

Despite this progress, women still dominate low-paying fields. Five of the top 10 occupations employing women are secretaries, bookkeepers, cashiers, sales workers, and typists. The two professional positions that women dominate are relatively low paying, namely, nursing and elementary school teaching (Bureau of Labor Statistics, 2002).

SUMMARY

- Sociology distinguishes between sex and gender.
- Sex refers to the physical and biological differences between men and women.

- Gender refers to the social, psychological, and cultural attributes of masculinity and femininity that are based on biological distinctions.
- Ideas about masculinity and femininity are culturally derived and are an important factor in shaping people's self-images and social identities.
- Whereas sex is generally an ascribed status, gender is learned through the socialization process and is thus an achieved status.
- Many people, on the other hand, believe that gender identities and masculine or feminine behavior result from an innate, biologically determined process.
- Historically, societies in both the East and the West have viewed women as inherently inferior to men.

- Not only did intellectuals argue from the point of view of a patriarchal ideology, or a belief that men are superior to women and should control all important aspects of society, but religions in both the East and West supported this view.
- Functionalists argue that the sexual division of labor was necessary for the efficient operation of preindustrial societies. Women, who birthed and nursed infants, remained involved in child care and the necessities of daily living. Men hunted, herded, and defended the family.
- Fuctionalists argue that because the female was largely dependent on the male for protection and food, he became the dominant partner in the relationship.
- Functionalists maintain that this role differentiation is functionally necessary and efficient in modern society as well.
- Critics contend that this view is outdated and that industrial society provides for more flexibility in gender-role assignment.
- According to conflict theory, males dominate females because of their superior power and control over key resources. A major consequence of this domination is the exploitation of women by men.

- Conflict theorists maintain that as long as men benefit from this arrangement, they have little incentive to change it.
- Gender-role socialization is a lifelong process whereby people learn the values, attitudes, motivations, and behavior considered appropriate to each sex according to their culture.

 ## *Media Resources*

The Companion Website for *Introduction to Sociology,* Ninth Edition

http://sociology.wadsworth.com/tischler9e

Supplement your review of this chapter by going to the companion website to take one of the Tutorial Quizzes, use the flash cards to master key terms, and check out the many other study aids you will find there. You will also find special features such as Wadsworth's Sociology Online Resources and Writing Companion, GSS data, and Census 2000 information at your fingertips to help you complete that special project or do some research on your own.

CHAPTER ELEVEN STUDY GUIDE

KEY CONCEPTS AND THINKERS

Match each concept with its definition, illustration, or explanation presented below.

a. Sex
b. Gender
c. Patriarchal ideology
d. Instrumental role

e. Expressive role
f. Contingent temporal world
g. Linear temporal world

h. Sociobiology
i. Gender role socialization
j. Gender identity

h   **1.** A social science theory focusing on the inborn nature of social traits and the role of evolution in shaping human behavioral patterns.

b   **2.** Cultural attributes of masculinity and femininity.

e   **3.** Behaviors that focus on feelings and relationships among members of a group.

i   **4.** The process of acquiring and internalizing the ideas and behaviors deemed appropriate for males or females by their society.

f   **5.** The view that family and career responsibilities·must be balanced.

j   **6.** The individual's sense of maleness or femaleness.

c   **7.** The belief that men are superior to women and should control all important aspects of society.

g   **8.** The view that emphasizes a direct career path with little consideration for the distractions of family and other noninstrumental relationships.

d   **9.** Behaviors that focus on the accomplishment of specific tasks for a group rather than personal feelings.

a   **10.** The physical and biological aspects of male and female.

Match the thinkers with their main idea or contribution.

a. Erik Erikson
b. Friedrich Engels
c. Deborah Tannen

d. Carol Gilligan
e. Auguste Comte
f. Talcott Parsons and Robert Bales

g. Desmond Morris, Robin Fox, and Lionel Tiger
h. Margaret Mead

e   **1.** The founder of sociology, he believed that women should not be allowed to work outside the home, to own property, or to exercise political power.

h   **2.** A pioneering anthropologist who found that gender roles were not innate properties and could vary widely among societies.

f   **3.** Applied the functionalist idea about division of labor to sex roles in the modern family.

b   **4.** Colleague of Karl Marx, theorized that capitalism and private property were the source of gender inequalities and the subordination of women.

a   **5.** Psychologist who pointed out that U.S. society pressures adolescent boys to base their identity on achievement but pressures girls to base their identity on finding a husband.

d   **6.** Challenged the value judgments made by traditional theorists about male and female styles of reasoning, especially in the area of moral decision making.

g   **7.** Writers popularizing the work of ethology and generalizing from the behavior of nonhuman primates to that of humans.

c   **8.** Conducted research on the conversational style and content differences among females and males.

CENTRAL IDEA COMPLETIONS

Following the instructions, fill in the appropriate concepts and descriptions for each of the questions posed below.

1. Compare and contrast the biological and sociological perspectives on sex and gender.

   a. Biological: _____

_____

b. Sociological: _____

_____

2. Following the research of Buss (1994) and by Clark and Hatfield (1989), what can you conclude about the differences in how women and men think about sex?

   a. Women: _____

   _____

   b. Men: _____

   _____

3. Reread the Social Change insert "Are We Biased against Assertive Women?" What are two key points from that reading that would lead you to conclude "yes, we are"?

   a. _____

   b. _____

4. Explain the connection between stress and the release of the hormone oxytocin.

   _____

   _____

5. State the explanations for stratification by gender according to the functional perspective and the conflict theory perspective.

   a. Functionalist: _____

   _____

   b. Conflict: _____

   _____

CRITICAL THOUGHT EXERCISES

1. Your chapter begins with the discussion of Kate Bornstein, who used to be Al Bornstein before she underwent a sex-change operation. The case raises the question of nature and nurture regarding gender orientation and sexual identity. What can this case—both Bornstein's reactions and those of other people—tell us about gender identity among homosexuals, or even among heterosexual men and women?

   How have ideas about sex, gender, and homosexuality changed in the United States over the past few decades? What factors have been responsible for these changes? What will affect whether similar changes happen with ideas about transgendered people?

2. Do research on the friendship patterns of women and men. Write a report in which you briefly analyze your personal friendship patterns, but also develop a design for a "mini-study" of the gender friendships of your close associates. What variables and patterns would you be looking for in your research? How would you gather your data? Develop two hypotheses about what you expect to find.

3. With the agreement of the participants, tape-record a discussion. It might be a class discussion, an informal chat among friends, or a formal public discussion. (If you can't tape it, listen closely and observe.) Keeping in mind the work of Deborah Tannen (read her book, *You Just Don't Understand*) and the idea of Instrumental and Expressive content, describe the conversational styles of men and women both in single-sex groups and in groups with both men and women.

INTERNET ACTIVITIES

1. Deborah Tannen says that men and women have different styles of speech. Do they also write differently? Two scientists have a developed a computer program that supposedly can tell. Find a passage, preferably longer than 500 words, and paste it into the box at http://www.bookblog.net/gender/genie.html. Try it with a few passages and see how well the program does.

2. British psychologist Simon Baron-Cohen has a theory of sex differences in brain "hard wiring" men for "systematizing" and women for "empathy." Go to http://glennrowe.net/BaronCohen/MaleFemale.asp. to take tests to see how you score on each of these.

3. The United Nations is concerned with gender discrimination. Their website (http://un.org/womenwatch) provides information on gender issues in many countries. For information on gender inequality in pay and education, look at http://www.aauw.org/research/statedata/index.cfm, run by the American Association of University Women. Click on your state to see how it ranks on the earnings gap. The National Organization for Women (http://www.now.org/) has information on current political issues related to gender inequality.

ANSWERS TO KEY CONCEPTS

1.h   2.b   3.e   4.i   5.f   6.j   7.c   8.g   9.d   10.a

ANSWERS TO KEY THINKERS

1.e   2.h   3.f   4.b   5.a   6.d   7.g   8.c

ThomsonNOW™

Reviewing is as easy as ❶ ❷ ❸

1. Before you do your final exam, take the ThomsonNOW diagnostic quiz to help you identify the areas on which you should concentrate. You will find information on ThomsonNOW and instructions on how to access all of its great resources on the foldout at the beginning of the text.
2. As you review, take advantage of ThomsonNOW's study videos and interactive Map the Stats exercises to help you master the chapter topics.
3. When you are finished with your review, take ThomsonNOW's posttest to confirm you are ready to move on to the next chapter.

Boston Filmworks

12

Marriage and Alternative Family Arrangements

Learning Objectives

After studying this chapter, you should be able to do the following:

- Explain the functions of the family.
- Describe the major variations in family structure.
- Define marriage and describe its relationship to the phenomenon of romantic love.
- Describe the various rules governing marriage.
- Explain the ways in which mate selection is not random.

- Summarize recent changes in the family as an institution.
- Explain the impact of changes in divorce and child-custody laws.
- Describe the various alternative family arrangements in contemporary American society.

The American family has changed dramatically since 1970. Some of the changes that have taken place include the following:

The marriage rate has fallen more than 40%.

When men and women marry today they are on average four years older than in 1970.

The number of single-parent households has more than doubled.

The proportion of those who have not married by age 35 has more than tripled for both men and women.

Women are nearly twice as likely to be divorced as in 1970.

The divorce rate has increased by nearly 40%.

Unmarried-couple households have increased nearly fivefold.

Twenty-nine percent of all births in 2003 were to unmarried women, compared with one-tenth in 1970.

Half of all children are expected to spend some part of their childhood in a single-parent home (Fields, 2001).

Men and women who grew up in the 1950s and 1960s watched television programs such as *Leave It to Beaver* and *Father Knows Best,* which presented a glorified version of the family of that era. Even though many families did not fit the model, the typical idealized family was one in which mother stayed at home in the suburbs with a brood of children, baking cookies for them and tending to their needs, while father was off working in the city.

Today, many family forms are common: single-parent families (resulting from either unmarried parenthood or divorce), remarried couples, unmarried couples, stepfamilies, and extended or multigenerational families. Mothers are more likely to be full- or part-time workers than they are to be full-time homemakers.

The United States is not alone in experiencing such changes. Family patterns in the United States reflect broad social and demographic trends that are occurring in most industrialized countries around the world.

Has the American family always been in such flux? Although information about family life during the earliest days of our country is not very precise, it does appear quite clear, beginning with the 1790 census, that the American family was quite stable. Divorce and family breakup were not common. If a marriage ended because of desertion, death, or divorce, it was seen as a personal and community tragedy.

The American family has always been quite small. There has never been a strong tradition here of the extended family, in which relatives of several generations lived within the same dwelling. Even in the 1700s, most American families consisted of a husband, wife, and approximately three children. This private, inviolate enclave made it possible for the family to endure severe circumstances and to help build the American frontier.

By the 1970s, radical changes were becoming evident. The marriage rate began to fall; the divorce rate, which had been fairly level, began accelerating upward; and fertility began to decline. The situation today shows that this trend has changed somewhat with the marriage rate continuing to drop, but the divorce rate leveling off and declining slightly.

There also are other signs of change and family instability. Of all children born in 2003, 29% were born to single women. Among African Americans, the percentage of children born out of wedlock was 69.1% (Bureau of the Census, 2005).

The inescapable conclusion that can be drawn from all this is that the American family is in a state of transition. But transition to what? In this chapter, we shall study the institution of the family and look more closely at the current trends and what they forecast for the future.

The American family will continue to evolve and change even further during the next decade. Traditional families will account for only two out of three households, and married couples will make up slightly more than half of all households.

The Nature of Family Life

The U.S. Bureau of the Census distinguishes between a *household* and a *family*. Households consist of people who share the same living space. A household may include one person who lives alone or several people who share a dwelling. A family, on the other hand, has a more precise definition. For example, in his classic study of social organization and the family, anthropologist George P. Murdock (1949) defined the family as:

A social group characterized by common residence, economic cooperation, and reproduction. It includes adults of both sexes, at least two of whom maintain a socially approved sexual relationship, and one or more children, own or adopted, of the sexually cohabiting adults.

This definition has proved to be too limited. For one thing, it excludes many kinds of social groups that seem, on the basis of the functions they serve, to deserve the label of family. For example, in America single-parent families are widely recognized. Despite Murdock's restricted definition of the family, it was his 1949 study of kinship and family in 250 societies, which showed the institution of the family to be present in every one of them, that led social scientists to believe that some form of the family is found in every known human society.

Perhaps a better understanding of the concept of the family may be gained by examining the various functions it performs in society.

Functions of the Family

Social scientists often assign to the fundamental family unit—married parents and their offspring—a number of the basic functions that it serves in most, if not all, societies. Although in many societies the basic family unit serves some of these functions, in no society does it serve all of them completely or exclusively. For example, among the Nayar of India, a child's biological father is socially irrelevant. Generally, another man takes on the social responsibility of parenthood (Gough, 1952). Or consider the Trobrianders, a people living on a small string of islands north of New Guinea. There, as reported by anthropologist Bronislaw Malinowski (1922), the father's role in parenthood was not recognized, and the responsibility for raising children fell to the family of their mother's brother.

The basic family is not always the fundamental unit of economic cooperation. Among artisans in preindustrial Europe, the essential economic unit was not the family but rather the household, typically consisting of the artisan's family plus assorted apprentices and even servants (Laslett, 1965). In some societies, members of the basic family group need not necessarily live in the same household. Among the Ashanti of western Africa, for example, husbands and wives live with their own mothers' relatives (Fortes, Steel, & Ady, 1947). In the United States, a small but growing number of two-career families are finding it necessary to set up separate households in different communities, sometimes hundreds of miles apart. Husband and wife travel back and forth between these households to be with each other and the children on weekends and during vacations.

In all societies, however, the family does serve the basic social functions discussed next.

Regulating Sexual Behavior No society permits random sexual behavior. All societies have an **incest taboo,** which *forbids intercourse among closely related individuals,* although who is considered closely related varies widely. Almost universally, incest rules prohibit sex between parents and their children and between brothers and sisters. However, there are exceptions: The royal families of ancient Egypt, the Inca nation, and early Hawaii did allow sex and marriage between brothers and sisters.

Anthropologist George P. Murdock defined the family as a social group characterized by common residence, economic cooperation, and reproduction.

In the United States, marriage between parents and children, brothers and sisters, grandparents and grandchildren, aunts and nephews, and uncles and nieces is defined as incest and is forbidden. In addition, approximately 30 states prohibit marriage between first cousins. The incest taboo usually applies to members of one's family (however the family is defined culturally) and thus it promotes marriage—and, consequently, social ties—among members of different families.

Patterning Reproduction Every society must replace its members. By regulating where and with whom individuals may enter into sexual relationships, society also patterns sexual reproduction. By permitting or forbidding certain forms of marriage (multiple wives or multiple husbands, for example) a society can encourage or discourage reproduction.

Organizing Production and Consumption In preindustrial societies, the economic system often depended on each family's producing much of what it consumed. In almost all societies, the family consumes food and other necessities as a social unit. Therefore, a society's economic system and family structures often are closely correlated.

Socializing Children Not only must a society reproduce itself biologically by producing children, it also must ensure that its children are encouraged to accept the lifestyle it favors, to master the skills it values, and to perform the work it requires. In other words, a society must provide predictable social situations within which its children are to be socialized (see Chapter 4). The family provides such a context almost universally, at least during the period when the infant is dependent on the constant attention of others. The family is ideally suited to this task because its members know the child from birth and are aware of its special abilities and needs.

Providing Care and Protection Every human being needs food and shelter. In addition, we all need to be among people who care for us emotionally, who help us with the problems that arise in daily life, and who back us up when we come into conflict with others. Although many kinds of social groups are capable of meeting one or more of these needs, the family often is the one group in a society that meets them all.

Providing Social Status Simply by being born into a family, each individual receives both material goods and a socially recognized position defined by ascribed statuses (see Chapter 5). These statuses include social class or caste membership and ethnic identity. Our inherited social position, or family background, probably is the single most important social factor affecting the predictable course of our lives.

Thus, we see that Murdock's definition of the family, based only on specific roles, is indeed too restrictive. A much more productive way of defining the family would be to view it as a universal institution that generally serves the functions discussed previously, although the way it does so may vary greatly from one society to another. The different ways in which various social functions are fulfilled by the institution of the family depend, in some instances, on the form the family takes.

Family Structures

The **nuclear family** is *the most basic family form and is made up of a married couple and their biological or adopted children*. The nuclear family is found in all societies, and it is from this form that all other (composite) family forms are derived.

Two major composite family forms have been defined: polygamous and extended families. **Polygamous families** are *nuclear families linked together by multiple marriage bonds, with one central individual* *married to several spouses. The family is* **polygynous** *when the central person is male and the multiple spouses are female. The family is* **polyandrous** *when the central person is female and the multiple spouses are male.* Polyandry is rare; only one-half of 1% of all societies permit a woman to take several husbands simultaneously, and the women have to be wealthy members of those societies to do so. The Tlingit of southern Alaska are one example. Polyandry also occurs among well-to-do Tibetan families in the highlands of Limi, Nepal (Fisher, 1992).

Extended families *include other relations and generations in addition to the nuclear family, so that along with married parents and their offspring, there may be the parents' parents, siblings of the parents, the siblings' spouses and children, and in-laws.* All the members of the extended family live in one house or in homes close to one another, forming one cooperative unit.

Families, whether nuclear or extended, trace their relationships through the generations in several ways. Under the **patrilineal system,** *the generations are tied together through the males of a family; all members trace their kinship through the father's line.* Under the **matrilineal system,** exactly the opposite is the case: *The generations are tied together through the females of a family.* Under the **bilateral system,** *descent passes through both females and males of a family.* Although in American society descent is bilateral, the majority of the world's societies are either patrilineal or matrilineal (Murdock, 1949).

In patrilineal societies, social, economic, and political affairs usually are organized around the kinship relationships among men, and men tend to dominate public affairs. Polygyny often is permitted, and men also tend to dominate family affairs. Sociologists use the term **patriarchal family** to describe situations in which *most family affairs are dominated by men.* The **matriarchal family,** in which *most family affairs are dominated by women,* is relatively uncommon but does exist. Typically, it emerges in matrilineal societies. The matriarchal family is becoming increasingly common in American society, however, with the rise of single-parent families (most often headed by mothers).

Whatever form the family takes and whatever functions it serves, it generally requires a marriage to exist. Like the family, marriage varies from society to society in its forms.

Defining Marriage

Marriage, an institution found in all societies, is the *socially recognized, legitimized, and supported union of individuals of opposite sexes.* It differs from other unions (such as friendships) in that (1) it takes place in a public (and usually formal) manner; (2) it includes sexual intercourse as an explicit element of the relationship; (3) it provides the essential condition for legitimizing offspring (that is, it provides newborns with socially accepted statuses); and (4) it is intended to be a stable and enduring relationship. Thus, although almost all societies allow for divorce—that is, the breakup of marriage—no society endorses it as an ideal norm.

Romantic Love

What would you think of someone who used the following proposal of marriage:

> It does not appear to me that my hand is unworthy of your acceptance, or that the establishment I can offer would be any other than highly desirable. My situation in life, my connections with the family, and my relationship to your own, are circumstances highly in my favor; and you should take it into further consideration, that in spite of manifold attractions, it is by no means certain that another offer of marriage may even be made to you.

This proposal, which would very likely be rejected by many women today, was made by Mr. Collins to Elizabeth in Jane Austen's novel *Pride and Prejudice* (Austen, 1813). In addition to seeming exceedingly arrogant it also appears to lack any sign that love exists. Americans are strong believers in the power of love. Over half of American adults (52%) said they believed in "love at first sight." Four in 10 Americans said they have actually fallen in love at first sight and almost three-quarters believed that "one true love" existed (Gallup Poll, February 2001).

We also expect to be able to have an exceptionally close relationship with the person we marry. Emotional and spiritual connection rank far above the need for financial security in forming a romantic partnership. Eighty-one percent of women ages 20 to 29 reported that "it is more important to have a husband who can communicate about his deepest feelings than it is to have a husband who makes a good living" (Maybury, 2002).

Americans are also strongly wedded to the view that there is a compatibility between romantic love and the institution of marriage. Not only do we believe they are compatible, but we expect them to coexist. Our culture implies that without the prospect of marriage, romance is immoral and that without romance, marriage is empty. This view underlies most romantic fiction and other media presentations.

When college students from 10 countries were asked if they would marry without love, 86% of American college students said they would not. Only 24% of college students from India said no to marriage without love. Cultures are likely to indulge in

GLOBAL SOCIOLOGY

Arranged Marriage in India

Shona Bose's parents gave her all of the freedoms available to progressive, upper-middle-class Indian youngsters of the 1990s. She attended a co-educational college, was encouraged to get a job with a foreign bank in Calcutta, went to discos with her friends, and was told that if she fell in love, she could pick the husband of her choice.

Yet when the time came for Bose—like most other Westernized young urban professionals in India—she chose the age-old tradition that had been forced on previous generations—arranged marriage.

"When you fall in love as a teenager, you look for different things: looks, glamour, is he fun to go out with? I met a lot of people I liked, but no one who was suitable for marriage," said Bose. "Love is important, but it is not sufficient."

In contrast to some of their parents, who rebelled in the 1960s and 1970s by opting for "love marriages," members of this generation assess their plans for marriage as they might a business plan. Is it financially viable? Are our lifestyles compatible? Is each side likely to uphold his or her end of the deal?

Young people know the divorce rate in India, long dominated by arranged marriages, is less than 5%, while in the United States roughly half of all marriages break up. While India's low rate can be attributed to cultural disapproval of divorce, advocates of arranged marriage insist it is also because research goes into matching mates with optimum compatibility.

Until a few decades ago, nearly every family married their children to eligible mates based on horoscopes, caste, community standing, and pedigree without consulting the children. The bride and groom were often not allowed to meet before the wedding, and even if they did they were not allowed to refuse the match. Most children are still expected to marry the spouse of their family's choice. Recently, newspapers carried a story of a young couple who were beheaded for having eloped.

Among educated Indians, some things have changed. Families who once relied on professional neighborhood matchmakers now often opt for computerized marriage bureaus and pages of highly specific matrimonial ads in the Sunday papers.

Young people are commonly allowed to meet alone a few times before the final decision is made, although relationships are still quite restricted compared to the West. Fewer than half of middle-class urban couples had even held hands before the wedding.

Young people support these arranged marriages. In a survey of urban professionals, 81% said their marriages were arranged, and 94% rated these marriages as "very successful." A poll of college students found that two-thirds wanted an arranged marriage.

The young people's attraction to arranged marriage may be a reaction, in part, to a period in the 1970s and 1980s when many marriages that had been based on love ran into trouble. A 1973 study of upper-middle-class north Indian women found 39% said love was essential for marital happiness. Today, that number has fallen to 11%.

Arranged marriages may also persist because they guarantee that almost everyone will have a marriage partner. Perhaps, most of all, they persist because of the family ties they create. There will always be a great deal of family support if the marriage runs into trouble.

Source: Adapted with permission from "Love No Match for Custom: India's Young Favor Arranged Unions," by I. A. R. Lakshmanan, *The Boston Globe*, August 26, 1997, pp. A1, A10.

---

romantic love if they are wealthy and value individualism over the community (Levine, 1993).

Romantic love can be defined in terms of five dimensions: (1) idealization of the loved one, (2) the notion of a one and only, (3) love at first sight, (4) love winning out over all, and (5) an indulgence of personal emotions (Lantz, 1982).

In some cultures, people expect love to develop after marriage. Hindu children are taught that marital love is the essence of life. Men and women often enter married life enthusiastically expecting a romance to develop. As the Hindu saying goes, "First we marry, then we fall in love" (Fisher, 1992). (See "Global Sociology: Arranged Marriage in India.")

In many of the world's other societies, romantic love is unknown or seen as a strange maladjustment. It may exist, but it has nothing to do with marriage. Marriage in these societies is seen as an institution that organizes or patterns the establishment of economic, social, and even political relationships among families. Three families ultimately are involved: **families of origin** or **families of orientation**, *the two families that produced the two spouses* and their **family of procreation**, *the family created by the spouses' union.*

A belief in romantic love helps make men and women more independent of their relatives. Romantic love weakens the emotional ties that bind people to their families. It makes it natural for partners to be committed to each other and provide mutual support. Romantic love provides a legitimate way to move out of the family without creating tensions and jealousies.

Marriage Rules

In every society, marriage is the binding link that makes possible the existence of the family. All societies have norms or rules governing who may marry whom and where the newlywed couples should live. These rules vary, but certain typical arrangements occur in many societies around the world.

Almost all societies have two kinds of marriage norms or rules: *rules of* **endogamy** *limit the social categories from within which one can choose a marriage partner.* For example, many Americans still attempt to instill in their children the idea that one should marry one's own kind—that is, someone within the ethnic, religious, or economic group of one's family of origin.

Rules of **exogamy**, by contrast, *require an individual to marry someone outside his or her culturally defined group.* For example, in many tribal groups, members must marry outside their lineage. In the United States, there are laws forbidding the marriage of close relatives, although the rules are variable.

These norms vary widely across cultures, but everywhere they serve basic social functions. Rules of exogamy determine the ties and boundaries between social groups, linking them through the institution of marriage and whatever social, economic, and political obligations go along with it. Rules of endogamy, by requiring people to marry within specific groups, reinforce group boundaries and perpetuate them from one generation to the next.

Marriage rules also determine how many spouses a person may have at one time. Among many groups, Europeans and Americans, for example, marriage is **monogamous**—that is, *each person is allowed only one spouse at a time.* However, many societies allow **multiple marriages**, in which *an individual may have more than one spouse* (polygamy). Polygyny, the most common form of polygamy, is found among such diverse peoples as the Swazi of Africa, the Tiwi of Australia, and, formerly, the Blackfeet Indians of the United States. Polyandry, as noted earlier, is extremely rare (Murdock, 1949).

As Marvin Harris (1975) has noted, "Some form of polygamy occurs in 90 percent of the world's cultures." However, within each such society, only a minority of people actually can afford it. In addition, the Industrial Revolution favored monogamy, for reasons we shall discuss shortly. Therefore, monogamy is the most common and widespread form of marriage in the world today.

In India, it is not unusual for girls to be betrothed at an early age; this bride is 4 years old.

Marital Residence

Once two people are married, they must set up housekeeping. Most societies have strongly held norms that influence where a couple lives. **Marital residence rules** *govern where a couple settles down.* **Patrilocal residence** *calls for the new couple to settle down near or within the husband's father's household*—as among Greek villagers and the Swazi of Africa. **Matrilocal residence** *calls for the new couple to settle down near or within the wife's mother's household*—as among the Hopi Indians of the American Southwest.

Some societies, such as the Blackfeet Indians, allow **bilocal residence**, in which *new couples choose whether to live with the husband's or wife's family of origin.* In modern industrial society, newlyweds typically have even more freedom. With **neolocal residence**, *the couple may choose to live virtually anywhere,* even thousands of miles from their families of origin. In practice, however, it is not unusual for

American newlyweds to set up housekeeping near one of their respective families.

Marital residence rules play a major role in determining the compositions of households. With patrilocal residence, groups of men remain in the familiar context of their father's home, and their sisters leave to join their husbands. In other words, after marriage the women leave home to live as strangers among their husband's kinfolk. With matrilocal residence just the opposite is true: Women and their children remain at home, and husbands are the outsiders. In many matrilocal societies this situation often leads to considerable marital stress, with husbands going home to their own mothers' families when domestic conflict becomes intolerable.

Bilocal residence and neolocal residence allow greater flexibility and a wider range of household forms because young couples can move to places in which the social, economic, and political advantages may be greatest. One disadvantage of neolocal residence is that a young couple cannot count on the immediate presence of kinfolk to help out in times of need or with demanding household chores (including the rearing of children). In the United States today, the surrogate or nonkin "family," made up of neighbors, friends, and colleagues at work, may help fill this void. In other societies, polygynous neolocal families help overcome such difficulties, with a number of wives cooperating in household work.

Mate Selection

Like our patterns of family life, America's rules for marriage, which are expressed through mate selection, spring from those of our society's European forebears. Because we have been nourished, through songs and cinema, by the notions of "love at first sight," "love is blind," and "love conquers all," most of us probably believe that in the United States there are no rules for mate selection. Research shows, however, that this is not necessarily true.

If we think statistically about mate selection, we must admit that in no way is it random. Consider for a moment what would happen if it were: Given the population distribution of the United States, blacks would be more likely to marry whites than members of their race; upper-class individuals would have a greater chance of marrying a lower-class person; and various culturally improbable but statistically probable combinations of age, education, and religion would take place. In actual fact, **homogamy**—*the tendency of like to marry like*—is much more the rule.

There are numerous ways in which homogamy can be achieved. One way is to let someone older and wiser, such as a parent or matchmaker, pair up appropriately suited individuals. Throughout history, this has been one of the most common ways by which marriages have taken place. The role of the couple in question can range from having no say about the matter whatsoever to having some sort of veto power. This tradition is quite strong in Islamic countries such as Pakistan. Benazir Bhutto, the two-time Pakistani prime minister, agreed to an arranged marriage despite her Western education and feminist leanings. She submitted to Islamic cultural traditions, under which dating is not considered acceptable behavior for a woman.

In the United States, most people who marry do not use the services of a matchmaker, though the result in terms of similarity of background is so highly patterned that it could seem as if a very conscious homogamous matchmaking effort were involved.

Age In American society, people generally marry within their own age range. Comparatively few marriages have a wide gap between the ages of the two partners. In addition, only 23.2% of American women marry men who are younger than themselves. On the average, in a first marriage for both the man and the woman, the man tends to be about 2 years older than the woman. This is, however, related to age at the time of marriage. For example, 20-year-old men marry women with a median age only 1 month younger. Twenty-five-year-old men marry women with a median age of 11 months younger. For 30- to 34-year-old men, the median age for wives is 2.8 years younger; whereas for men over 65, the median age for wives is nearly 10 years younger (Bureau of the Census, 1991).

In 1890, the estimated median age at first marriage was 26.1 years for men and 22.0 years for women. At that time, a decline in the median age at first marriage began; it did not end until 1956, when the median age reached a low of 20.1 years for women and 22.5 years for men. This 66-year decline has been reversed, and in 2004 the median age at first marriage exceeded the 1890 level, standing at 27.4 years for men and 25.8 for women (Bureau of the Census, June 29, 2005).

For remarriages after a divorce, as might be expected, the average age is older. For men it is 37.4, whereas for women it is 34.2. For widows and widowers, the average age of remarriage is older still, 54.0 for women and 63.1 for men (Bureau of the Census, 1997).

Age homogamy appears to hold for all groups within the population. Studies show that it is true for blacks as well as whites and for professionals as well as laborers. Clearly, the norms of our society are very effective in causing people of similar age to marry each other.

Race Homogamy is most obvious in the area of race. It was only 39 years ago that Thurgood Marshall, who was only months away from appointment to the

Supreme Court, suffered an indignity that seems hard to believe today. Marshall and his wife had found a house in a Virginia suburb of Washington, D.C. There was one problem, however. Virginia did not allow interracial marriages and Marshall was black and his wife was Asian. Just in the nick of time the Supreme Court ruled that these laws were unconstitutional in January of 1967. A Gallop Poll from that time showed that 72% of southern whites and 42% of northern whites thought interracial marriage should be banned (Sailer, 1997).

The laws in 19 states that sought to stop interracial marriage through legislation in the 1960s varied widely, and there was great confusion because of various court interpretations. In Arizona before 1967, it was illegal for a white person to marry a black, Hindu, Malay, or Asian person. The same thing was true in Wyoming, and residents of that state were also prohibited from marrying mulattos.

In 1966, Virginia's Supreme Court of Appeals had to decide on the legality of a marriage that had taken place in Washington, D.C., in 1958 between Richard P. Loving, a white man, and his part-Indian and part-black wife, Mildred Loving. The court unanimously upheld the state's ban on interracial marriages. The couple appealed the case to the United States Supreme Court, which agreed to decide whether state laws prohibiting racial intermarriage were constitutional. Previously, all courts had ruled that the laws were not discriminatory because they applied to both whites and nonwhites. However, on June 12, 1967, the Supreme Court ruled that states could not outlaw racial intermarriage. Today about 5% of U.S. married couples include spouses of a different race. This small percentage masks a remarkable growth in the number of interracial marriages. Since 1970, the number of black and white interracial married couples has increased from 300,000 to 1,400,000 in 1998 (Bureau of the Census, August 2000).

The most common type of interracial marriage is one between a white husband and a wife of a race other than black. The next most common is between a white woman and a man of a race other than black. These types of interracial marriages have been increasing substantially, whereas black-white interracial marriages appear to be leveling off.

Interracial marriages are more likely to involve at least one previously married partner than are marriages between spouses of the same race. In addition, brides and grooms who marry interracially tend to be older than the national average.

The degree of education also differs in interracial marriages. White grooms in interracial marriages are more likely to have completed college than white grooms who marry within their race. In all-white couples, 18% of the grooms hold college degrees. In a marriage involving a white groom and a black bride, 24% of the men finished college.

Among black men married to women of other races only 5% had completed college, whereas 13% of black men married to white women had done so. This compares with 9% of black men whose spouses were also black who held college degrees.

Although the rate of interracial marriage is growing, the proportions vary widely by state. Hawaii has the distinction of having both the highest number and the greatest proportion of interracial marriage. Almost one-fourth of all marriages there are interracial, the majority of which are Asian-Caucasian unions. Florida runs a distant second.

Illinois has the largest number of white brides marrying black grooms of any state. The record for the reverse combination—black brides marrying white grooms—is held by New Jersey. Alaska is second to Hawaii in the proportion of interracial marriages. Thirteen percent of the state's marriages are interracial. None of the other states comes close to these percentages.

Even though their total numbers are still small, racially mixed marriages have become increasingly common in America's melting pot.

Religion Unlike many European nations, none of the American states has ever had legislation restricting interreligious marriage. Religious homogamy is not nearly as widespread as is racial homogamy, though most marriages still involve people of the same religion.

Attitudes toward religious intermarriage vary somewhat from one religious group to another. Almost all religious bodies try to discourage or control marriage outside the religion, but they vary in the extent of their opposition. Before 1970, the Roman Catholic Church would not allow a priest to perform an interreligious marriage ceremony unless the non-Catholic partner promised to rear the children of the union as Catholics, and the Catholic partner promised to encourage the non-Catholic to convert. However, in the late 1960s the Catholic Church softened its policy, and since 1970 the pope has allowed local bishops to permit mixed-religion marriages to be performed without a priest and has also eliminated the requirement that the non-Catholic partner promise to rear the children as Catholics. Nevertheless, the Catholic partner must still promise to try to have the children raised as Catholic.

Protestant denominations and sects also differ with regard to the barriers they place on the intermarriage of their members. At one extreme are the Mennonites, who excommunicate any member who marries outside the faith. At the other are numerous denominations that may encourage their members to

As late as 1966, 19 states sought to stop interracial marriage through legislation.

Boston Filmworks

marry within the faith, but provide no formal penalties for those who do not (Heer, 1980).

Jewish religious bodies also differ in their degree of opposition to religious intermarriage. Orthodox Jews are the most adamantly opposed to intermarriage, whereas Conservative and Reform Jews, though by no means endorsing it, are more tolerant when it does take place. Jewish intermarriage rates have been increasing dramatically in recent years, with about 50% of all Jews marrying someone of another religion.

Religious leaders often are concerned about religious intermarriage, for a variety of reasons. Some claim that one or both intermarrying parties are lost to the religion, and others believe that the potential for marital success is decreased greatly in intermarriage. Complex factors are involved in intermarriages, and simplistic and unequivocal predictions are not warranted.

Most marriages still involve those of the same religion. There is, however, a clear trend of more religious intermarriages in the United States today.

Social Status Level of education and type of occupation are two measures of social status. In these areas, there is usually a great deal of similarity between people who marry each other. Men tend to marry women who are slightly below them in education and social status, though these differences are within a narrow range. Wide-ranging differences between the two people often contain an element of exploitation. One partner may either be trying to make a major leap on the social class ladder or be looking for an easy way of taking advantage of the other partner because of unequal power.

The typical high school environment often plays a major role in maintaining social status homogamy, as it is in high school that students have to start making plans for their future careers. Some may go to college, and others may plan on going to work directly after graduation. This process causes the students to be divided into two groups: the college bound and the workforce bound. Although the lines separating these two groups are by no means impenetrable, in many high schools these two groups maintain separate social activities. In this way, barriers against dating and future marriage between those of unlike social status are set up. After graduation, those who attend college are more likely to associate with other college students and to choose their mates from that pool. Those who have joined the workforce are more likely to choose their mates from that environment. In this way, similarities in education between marriage partners really are not accidental.

As with several of the items we have discussed already, education, social class, and occupation produce a similarity of experience and values among people. Just as growing up in an Italian American family may make one feel comfortable with Italian customs and traits, going to college may make one feel comfortable with those who have experienced that environment. Similarities in social status, then, are as much a result of socialization and culture as of conscious choice. We most likely will marry a person we feel comfortable with—a person who has had experiences similar to our own.

Coming to terms with the constraints on our marriage choices can be a sobering experience: What we thought to be freedom of choice in selecting a mate is revealed instead by various studies to be governed by rules and patterns.

The Transformation of the Family

Most scholars agree that the Industrial Revolution had a strong impact on the family. In his influential study of family patterns around the world, William J. Goode (1963) showed that the modern, relatively isolated nuclear family with weak ties to an extensive kinship network is well adapted to the pressures of industrialism.

First, industrialism demands that workers be geographically mobile so that a workforce is available wherever new industries are built. The modern nuclear family, by having cut many of its ties to extended family networks, is freer to move than its predecessors. It was among laborers' families that extended kinship ties first were weakened. Only in the past few decades have middle- and upper-class families become similarly isolated.

Second, industrialism requires a certain degree of social mobility (see Chapter 8). This is so talented workers may be recruited to positions of greater responsibility (with greater material rewards and increased prestige). A family that is too closely tied to other families in its kinship network will find it difficult to break free and climb into a higher social class. However, if families in the higher social classes are too tightly linked by kinship ties, newly arriving families will find it very difficult to fit into their new social environment. Hence, the isolated nuclear family is well suited to the needed social mobility in an industrial society.

Third, the modern nuclear family allows for inheritance and descent through both sides of the family. Further, material resources and social opportunities are not inherited mainly by the oldest males (or females), as in some societies. This means that all children in a family will have a chance to develop their skills, which in turn means that industry will have a larger, more talented, and flexible labor force from which to hire workers.

By the early twentieth century, then, the nuclear family had evolved fully among the working classes of industrial society. It rested on (1) the child-centered family, (2) **companionate marriage** (that is, *marriage based on romantic love*), (3) increased equality for women, (4) decreased links with extended families or kinship networks, (5) neolocal residence and increased geographical mobility, (6) increased social mobility, and (7) the clear separation between work and leisure. There also was boring and alienating work and an associated expectation that the nuclear family would fulfill the function of providing emotional support for its members.

The World War II years also had a profound effect on the American family, for it was during the war that a process begun in the Great Depression really accelerated. The war made it necessary for hundreds of thousands of women to work outside the home to support their families. They often had to take jobs, vital to the American economy, that had been vacated when their husbands went to fight overseas. After the war, an effort was made to "defeminize" the workforce. Nevertheless, many women remained on the job, and those who left knew what it was like to work for compensation outside the home. Things were never the same again for the American family, and family life began to change.

The initial changes were not all that apparent. Indeed, by the 1950s the United States had entered the most family-oriented period in its history. This was the era of the baby boom, and couples were marrying at the youngest ages in recorded American history. During the 1950s, 96% of women in the childbearing years married (Blumstein & Schwartz, 1983). The war years' experiences also paved the way for secondary groups and formal organization (see Chapter 6) gradually to take over many of the family's traditional activities and functions. As social historian

In 1946 (when this photo was taken), the median family income was $2,800, television was in its infancy, and only 12.5% of people aged 18–22 went to college. In the years following World War II, a radical transformation took place in our society and in the family.

Christopher Lasch (1977) pointed out, this trend was supported by public policy makers, who came to see the family as an obstacle to social progress. Because the family preserved separatist cultural and religious traditions and other "old-fashioned" ideas that stood in the way of "progress," social reformers sought to diminish the family's hold over its children. Thus, the prime task of socializing the young was shifted from the family to centrally administered schools. Social workers intruded into the home, offering constantly expanding welfare services from outside agencies. The juvenile court system expanded, in the belief that deficiencies in the families of youthful offenders caused crime among children.

Thus, the modern period has seen what sociologists refer to as the transfer of functions from the family to outside institutions. This transfer has had a great effect on the family, and it underlies the trends that currently are troubling many people.

There are several problems in trying to assess the prevailing state of the family. Some feel that the family is deteriorating, and they cite the examples of divorce rates and single-parent families to support their view. Others think of the family as an institution that is in transition but just as stable as ever. Was the family of the past a stable extended-family unit, with everyone working for the betterment of the whole?

Or has the family structure changed throughout history in response to the economic and political changes within society? In this next section, we shall explore these views and attempt to clarify the current direction of family life.

The Decline of the Traditional Family

The number of "traditional" families, that is, married couples with children, has been declining in recent years. During the 1990s the percentage fell from 26 to 24%. The 1990s also saw an overall decrease in households headed by married couples among families with children—that percentage fell from 76% in 1990 to 50.2% in 2004 (U.S. Census Bureau, 2004; American Community Survey, 2005). The increasing rarity since the 1950s of the traditional family structure can be attributed to various social factors, such as postponement of marriage, rising divorce rates, and declining birthrates. With the decline of the traditional family, alternative family structures have become more prominent in society, particularly single-parent families and families headed by unmarried couples.

Nontraditional households are on the rise overall in the United States, and recent census data show significant regional variations in the numbers of alternate family structures. In Rhode Island only 21% of the

Figure 12–1 Number of Marriages, in Millions, 1960–2004

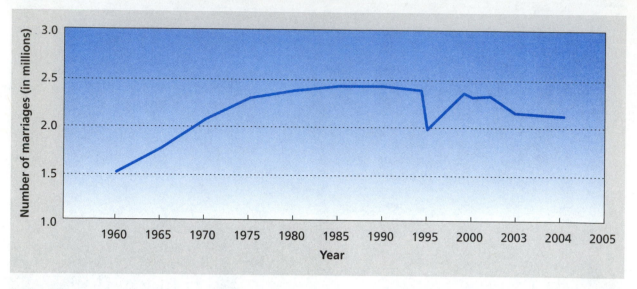

Source: National Center for Health Statistics, 2005

Figure 12–2 Number of Marriages per 1,000 Population, 1960–2003

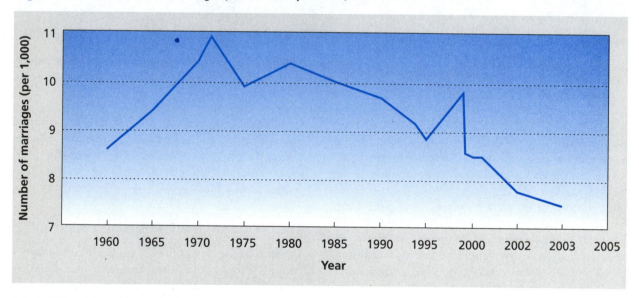

Source: National Center for Health Statistics, 2005

families can be considered traditional, and Mississippi and Louisiana have a high percentage of female-headed households with children (AmeriStat, November 2001).

Changes in the Marriage Rate

Are fewer people marrying now than in the past? The answer depends on how you evaluate the data. One way to look at it is to ask how many marriages there are. In 2004, there were 2.160 million marriages (Figure 12–1). A second way to look at it is to calculate the marriage rate, the proportion of the total population marrying. The 2.160 million marriages in

2004, divided by the total population of million Americans, yields a marriage rate of 7.5 per 1,000 (National Center for Health Statistics, 2005). The number of marriages per 1,000 population reached a peak of 16.4 in 1946, at the end of World War II, and a low of 8.4 in 1958, when an economic recession coincided with a relatively small number of young adults reaching marriageable age in the population. According to the marriage rate, the institution is holding its own with a rate slightly below mid-1960s levels (Figure 12–2).

The third way to get a true picture of the marriage situation in the United States is to take the marriage

Figure 12-3 Number of Marriages per 1,000 Unmarried Women, Ages 15 and Older, 1960-2004

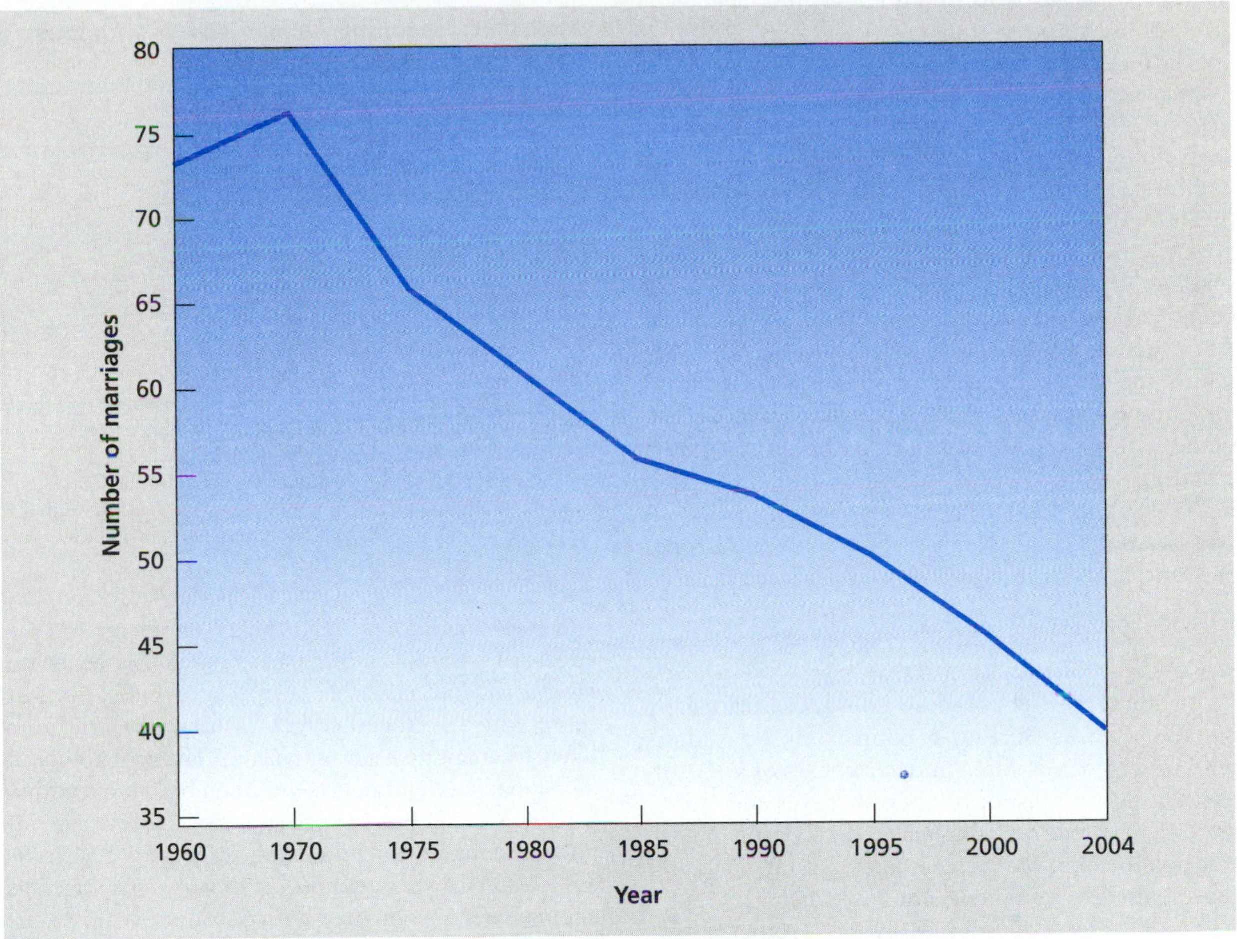

Source: US Department of the Census, Statistical Abstract of the United States, 2001, Page 87, Table 117; and Statistical Abstract of the United States, 1986, Page 79, Table 124. Births, Marriages, Divorces, and Deaths: Provisional Data for 2004, National Vital Statistics Report 53:21, June 26, 2005, Table 3. http://www.cdc.gov/nchs/data/nvsr/nvsr53/nvsr53_21.pdf.

rate per 1,000 never-married women ages 15 and older. This is the proportion of eligible women marrying. The marriage rate for unmarried women began to plummet around 1975, and in 2004 was at an all-time low of 39.9 per 1,000 (Figure 12-3).

Since 1960, the number of marriages and the marriage rate were inflated because the pool of eligible adults kept expanding. Hidden in these numbers, though, is the fact that a rising share of eligible people are choosing not to marry.

Others (Besharov, 1999) think the declines in marriage are exaggerated. The age 15 cutoff point for comparing marriage rates is wrong, because people now just do not wed at such an early age. Since 1960, the average age for women to marry for the first time has increased by 5.6 years—from 20.3 years to 25.9 in 2004.

To measure marriage rates accurately today, we should look at the percentage of women who are "ever married." "Assuming that one should not count teenagers because so few marry, in 1960 the proportion of women 20 and older who had ever been married was 90%." Forty-four years later there has been little change in that number. Young people today marry later, and often plan to make their marriage a partnership of equals (Besharov, 1999).

All this may begin to explain why cohabitation has emerged as a major change in couple relationships. Although the marriage rate may be down, cohabitation is certainly up.

Cohabitation

Cohabitation, or *unmarried couples living together out of wedlock,* has increased dramatically in the past 20 years and is already having a significant impact on the American family. In 1988, fewer than one in five married Americans said they lived with their spouse before marriage. The latest numbers show that percentage has nearly doubled to 37%. The increase is largely accounted for by the behavior of younger Americans (those under age 50), a majority of whom

(51%) say they lived together before marriage. Among those over the age of 50, only 15% say they lived with their spouse before getting married (Jones, 2002).

Although we have a great deal of information on marriages, we have very little data on cohabitation in the United States within the past two decades, and almost no information about it before that time.

Most adults in the United States eventually marry. In 2003, nearly 91% of women ages 45 to 54 had been married at least once. The increasing social acceptance of cohabitation has resulted in definitions of being married or being single becoming less clear-cut. As the personal lives of unmarried couples start to resemble those of their married counterparts, the meaning and permanence of marriage changes also. The number of unmarried-couple households has grown eightfold since 1970, from 523,000 to 5.08 million in 2004 (Table 12–1). There were 5 unmarried couples for every 100 married couples in the United States in 1998, compared with only 1 for every 100 in 1970 (Fields, 2004). Unmarried-couple households are also likely to include children (43%) (Casper & Cohen, 2000).

An unmarried-couple household, as defined by the U.S. Census Bureau, contains only two adults with or without children under 15 years of age present. The adults must be of opposite sexes and not related to one another (Bureau of the Census, 2000). It is possible that the increase in cohabitation figures represents better data collection as much as it represents an increase in couples living together. However, all signs point to an increase in cohabitation.

Even though the percentage of cohabiting couples in the total population is relatively small, many people have lived with a partner outside of marriage at some point. More than 4 in 10 women have been in an unmarried domestic partnership at some point in their lives. The proportion of women in certain age groups cohabiting is quite striking.

For some, cohabitation is a prelude to marriage, for others, an alternative to marriage, and for still others, simply an alternative to living alone. Cohabitation is more common among those of lower educational and income levels. Recent data show that among women in the 19 to 44 age range, 60% of high school dropouts have cohabited compared with 37% of college graduates (Bumpass & Lu, 2000). Cohabitation is also more common among those who are less religious than their peers, those who have been divorced, and those who have experienced parental divorce, fatherlessness, or high levels of marital discord during childhood. A growing percentage of cohabiting couple households, now more than one-third, contain children (Smock, 2000).

Many young people believe that living together before marriage is a good way to "find out whether you really get along," which would prevent a bad marriage

Table 12–1

Number of Cohabiting, Unmarried Adult Couples of the Opposite Sex, 1960–2004

Year	Number
1960	0.439
1970	0.523
1980	1.589
1990	2.856
2000	4.736
2004	5.080

Source: U. S. Bureau of the Census, Current Population Reports, Series P20-537; America's Families and Living Arrangements: March 2000; and U. S. Bureau of the Census, Population Division, Current Population Survey, 2004 Annual Social and Economic Supplement (http://www.census.gov/population/socdemo/hh-fam/cps2004).

and eventual divorce. "But the available data on the effects of cohabitation fail to confirm this belief. In fact, a substantial body of evidence indicates that those who live together before marriage are more likely to break up after marriage." The probability of a first marriage ending in separation or divorce within 5 years is 20%, but the probability of a marriage involving cohabitation breaking up within 5 years is 49%. After 10 years, the probability of a first marriage ending is 33%, compared with 62% for cohabitations (Bramlett & Mosher, 2002). (See Figure 12–4 for high school seniors' attitudes about cohabitation.)

This evidence is controversial, because it is difficult to discount the fact that people who cohabit before marriage have different characteristics from those who do not. It may be these characteristics, and not the experience of cohabitation, that lead to marital instability. There is no clear evidence yet that those who cohabit before marriage have stronger marriages than those who do not (Whitehead & Popenoe, 2002).

Childless Couples

Childlessness among married couples has been increasing in recent years. Many women in the childbearing years see postponement of marriage and childbearing as pathways to a good job and economic independence. For many women this temporary postponement becomes permanent either by choice or by chance. Women with the highest levels of education, those engaged in managerial and professional occupations, and those with the highest family incomes have the highest levels of childlessness.

Childlessness among women 40 to 44 years old (most women have completed childbearing by this age) increased from 10% in 1980 to 19% in 2000.

Figure 12–4 Percentage of High School Seniors Who Thought It Was a Good Idea to Live Together before Getting Married

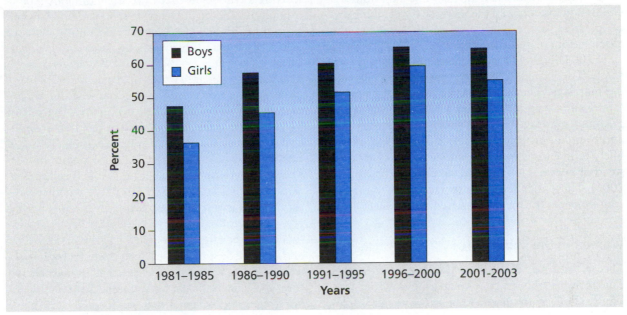

Source: Monitoring the Future surveys conducted by the Survey Research Center at the University of Michigan.

This is the continuation of a trend that began in 1984. That was the year when there were more childless couples in the United States than couples with children younger than 18. This reversal of the ratio of couples with children to childless couples from what was the case throughout most of our history will continue into the foreseeable future (Fields, 2001). Some of these couples are preparents who plan to have children sometime. Over half of all women younger than 35 who are childless expect to have a child at some point. Others are nonparents either by choice or because of fertility problems.

Changes in Household Size

Although changes in household size may be neither positive nor negative, some social scientists use this point to support a negative view of the future of the family. The American household of 1790 had an average of 5.8 members; by 2000, the average number had dropped to 2.62 (Bureau of the Census, 2002). The same trend also has been evident in other parts of the world. The average rural household in Japan in 1660 often had 20 or more members, but by the 1960s the rural Japanese household averaged only 4.5 members. One reason for the reduction in size of the American household may be that today it is very unusual to house unrelated people. Until the 1940s, for a variety of reasons, it was common for people to have nonkin living with them, either as laborers for the fields or as boarders who helped with the rent payment.

The reduction in the number of nonrelatives living with the family explains only part of the continuing reduction in the average household size. Another reason may be a rapid decrease in the number of aging parents living with grown children and their families. Some point to this as evidence of the fragmentation and loss of intimacy present in the contemporary family. At the turn of the century, more than 60% of those 65 or older lived with one or more of their children; today this figure is less than 10%.

How can we account for so many more old people living apart from their families? We might be tempted to say that the family has become so self-centered and so unable to fulfill the needs of its members that the elderly have become the first and most obvious castoffs. However, this trend of the elderly living away from their children can also be seen as a result of the increasing wealth of the population, including the elderly. In the past, many elderly lived with their children because they could not afford to do otherwise.

Another reason for the change in the size of households is the increasing divorce rate. As more families separate legally and move apart physically, the number of people living under one roof has fallen.

A further explanation for the smaller families of today is the tendency of young people to postpone marriage and the increase in the number of working women. As people marry later, they have fewer children. Many couples also are deciding to have no children, as more and more women become involved in work and careers.

All these factors point to the most significant causes of the sharp decline in the average size of the American household: the decrease in the number of children per family and the increase in the number of people living alone. The number of Americans living alone has increased substantially.

In 2004 about 44% of all people 15 and older were single. These single people are not necessarily rejecting marriage. They are just a reflection of changing family and living arrangements. Many young people postpone marriage until they have completed college and found appropriate jobs. Older people are more likely to live in their own homes and apartments instead of nursing homes (U.S. Census, September 12, 2005). These changes ultimately will have important consequences for the entire structure of society.

Women in the Labor Force

The period since World War II has seen a dramatic change in the labor force participation rates of American women. More than 68 million women had paying jobs in 2004, representing more than a 300% increase in 50 years. The number of men in the labor force during this same period increased by only 50%. The change in the labor force is probably the single most important recent change in American society.

This change has come about because of a number of factors. After World War II, many women who took jobs in record numbers to ease labor shortages during the war remained on the job. Their numbers increased the social acceptability of the working woman. Widespread use of contraceptives is a second important factor. Effective contraception gave women the freedom to decide whether and when to have children. As a result, many women postponed childbearing and continued their educations. Baby boomers also had different economic expectations when they entered the workforce. Having two incomes became important to ensure the lifestyle and standard of living they had come to expect.

Family Violence

It appears that we are in greater danger of being the victim of violence when we are with our families than when we are in a public place. Recent years have seen a number of high-profile cases in which spouses have attacked each other, parents have attacked children, and children have attacked and killed parents.

Family violence is greater in poor and minority households, but it occurs in families at all socioeconomic levels. Research has shown the incidence of family violence to be highest among urban lower-class families. It is high among families with more than four children and in those in which the husband is unemployed. Families in which child abuse occurs tend to be socially isolated, living in crowded and otherwise inadequate housing. Research on family violence has tended to focus on lower socioeconomic groups, but scattered data from school counselors and mental health agencies suggest that family violence also is a serious problem among America's more affluent households.

Children who are victims of abuse are more likely to be abusive as adults than children who did not experience family violence. Research shows that about 30% of adults who were abused as children are abusive to their own children (Gelles & Conte, 1990). In essence, they have been socialized to respond to frustrating situations with anger and violent outbursts.

Sociologists are trying to determine whether family violence is rising or decreasing in American society. One national survey (Gelles & Straus, 1988) found that the rate of child abuse had declined by 47% and the rate of spouse abuse had declined by 27% in a decade. The researchers believe that a major reason for the decline is the changing public attitude toward family violence. Behavior that was formerly acceptable is now considered wrong. As a result, people may be less willing to accept violent behavior from a spouse or a parent, and outsiders are more willing to intervene when they suspect abuse.

Changes in society that have affected the family also may have contributed to a decline in family violence. People are getting married later and having fewer children. Many women are in the labor force. These facts lower the risk of family violence. Public policies toward family violence have changed also, and there are many more programs for treating victims and abusers. Shelters for battered women have increased dramatically. Police who used to talk to an abusive husband and try to calm him down are now often required to arrest him. Finally, the media have played a large role in defining family violence as a social problem deserving attention.

Yet other information leads to a less optimistic picture. A Gallup Poll study found that 22% of American women (about 30 million) report having been physically abused at some point in their lives by their spouses or companions. One in five of these women report that the abuse took place within the previous year. In addition, 53% of the people in the study knew of friends or relatives who had been physically abused by their husbands or boyfriends (Gallup Poll, October 1997).

Divorce

Until the past few decades, married couples could divorce only under certain special circumstances. Consequently, the divorce rate (the number of divorces per 1,000 people) was relatively low in the early twentieth century. In 1900, for example, the divorce rate was only

Figure 12–5   **Annual Divorce Rate per 1,000 Population, 1970–2003**

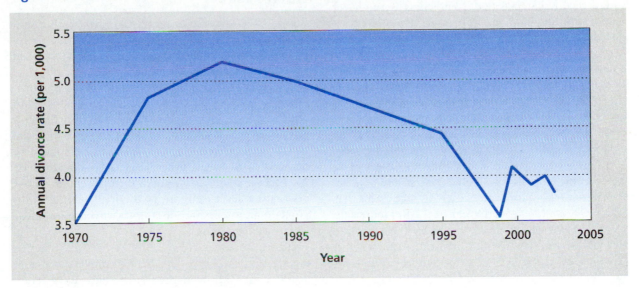

Source: National Center for Health Statistics, 2005

0.7, or less than one-fifth the 2003 divorce rate of 3.8 (National Center for Health Statistics, June 2005).

Rising divorce rates are not unique to the United States. Most industrialized nations have experienced similar patterns. Divorce rates in France and Great Britain have more than doubled in the past two decades. Divorce, however, is far more prevalent in the United States than elsewhere (Bureau of the Census, 2002).

All segments of American society have been affected by the rising divorce rate. Among white and Hispanic women ages 15 to 44, more than one-third have experienced a divorce. Among African American women, the figure is one-half. Religious sanctions against divorce seem to have little influence. Catholics are no less likely to divorce than non-Catholics, despite strong opposition to divorce by the Catholic Church (Bumpass, 1991).

The likelihood of divorce varies considerably with several factors. For example, education levels seem to have a strong effect on divorce rates. The likelihood of a first marriage ending in divorce is nearly 60% for those people with some college education but no bachelor's degree. Those who have a college degree but no graduate-school training have nearly a 40% chance of divorce and are the least divorce prone. We could argue that people with the personality traits and family background that lead them to achieve a college degree are also those most likely to achieve marital stability.

Women who have gone on to graduate school have a greater likelihood of divorce than some less educated women: Approximately 53% of them will divorce. As more and more women earn graduate degrees and as some of the barriers impeding women in combining professional and personal lives are removed, the higher rate of divorce for these women may also decline.

Although the divorce rate rose sharply during the 1970s, it has dropped off in recent years. In 2003, the divorce rate per 1,000 people was 3.8, the lowest it has been since 1972 (National Center for Health Statistics, June 2005) (Figure 12–5).

Even though the rate of increase in divorces has stopped, there is little evidence to suggest that the rate will decline substantially. The United States still has the highest divorce rate in the world. Current divorce rates imply that half of all marriages will end in divorce. Many argue that society cannot tolerate such a high rate of marital disruptions. Thirty years ago, few would have believed society could tolerate even one-third of all married couples divorcing, but that level has already been reached by some marriage cohorts.

The large number of divorces is itself a force that helps to keep the divorce rate high. Divorced people join the pool of available marriage partners, and a large majority remarry. Those remarriages then have a higher overall risk of divorce, and thus an impact on the divorce rate.

Even though divorce rates were lower in 1910, 1930, and 1950 than they are today, can we assume that family life then was happier or more stable? Divorce during those periods was expensive, legally difficult, and socially stigmatized. Many who would have otherwise considered divorce remained married because of those factors. Is it thus accurate to say that it was better for the children and society for the partners to maintain these marriages?

As divorce becomes more common, it also becomes more visible, and such visibility actually can produce more divorces. Others become a model of how difficult marriages are handled. The model of people suffering in an unhappy marriage is being replaced by one in which people start new lives after dissolving a marriage.

Divorce also may be encouraged by the increasing tendency, mentioned earlier, for outside social institutions to assume traditional family functions that once helped hold the family together. Then again, divorce has become a viable option because people can look forward to living longer today, and they may be less willing to endure a bad marriage if they feel there is time to look for a better way of life.

Another reason for today's high divorce rate is that we have come to expect a great deal from marriage. It is no longer enough, as it might have been at the turn of the century, for the husband to be a good provider and the wife to be a good mother and family caretaker. We now look to marriage as a source of emotional support in which each spouse complements the other in a variety of social, occupational, and psychological endeavors.

Divorce rates have also increased because the possibilities for women in the workforce have improved. During earlier eras, divorced women had great problems contending with the financial realities of survival, and many were discouraged from seeking a divorce because they could not envision a realistic way of supporting themselves. With their greater economic independence, many women can now consider divorce as an option.

Today's high divorce rates can also be traced to a number of legal changes that have taken place to make divorce a more realistic possibility for couples who are experiencing difficulties. Many states have instituted no-fault divorce laws, and many others have liberalized the grounds for divorce to include mental cruelty and incompatibility. These are rather vague terms and can be applied to many problem marriages. Even changes by the American Bar Association, which now allows lawyers to advertise, contribute to the increased divorce rate. Advertisements that state that an uncomplicated divorce will cost only $250 put this option within the reach of many couples.

These legal changes are but a reflection of society's attitudinal changes toward divorce. We are a far cry from a few decades ago. Divorce was to be avoided at all costs then, and when it did occur, it became a major source of embarrassment for the entire extended family. The fact that Ronald Reagan was divorced and remarried had little impact on his election to the presidency in 1980 and 1984. With so little public concern being shown, we can be sure that the role of peer-group and public opinion in preventing divorce has been diminished greatly. (See "For Further Thinking: How Much Are Children Hurt by Their Parents' Divorce?", p. 330.)

Divorce Laws

Before 1970, divorces in most American states could be obtained only on fault grounds. Fault grounds are those that assess blame against one of the spouses, and typically consist of adultery, desertion, physical and mental cruelty, long imprisonment for a felony, and drunkenness (Scott, 1990). The fault laws led to wide variations among the states as to how difficult or easy it was to obtain a divorce. New York State had a restrictive law that required one spouse to demonstrate that the other had engaged in adultery. No other grounds for divorce were accepted. Those restrictions led to migratory divorces (getting a divorce in another state) or couples' inventing nonexistent fault grounds.

Those issues led to attempts to change divorce laws. California introduced the Family Law Act of 1969, which allowed divorce if there were irreconcilable differences between the spouses. Thus, California became the first state to enact a no-fault divorce law. Other states quickly followed suit, and by 1985, all 50 states had some form of no-fault divorce law.

No-fault divorce reflects changes in the traditional view of legal marriage. No-fault divorce laws allow couples to dissolve their marriage without either partner having to assume blame for the failure of the marriage. By eliminating the fault-based grounds for divorce and the adversary process, no-fault divorce laws recognize that frequently both parties are responsible for the breakdown of the marriage. Further, these laws recognize that previously, the divorce procedure often worsened the situation by forcing potentially amicable individuals to become antagonists.

The vast majority of states make no-fault divorces possible with virtually no waiting periods. Even with the one-year waiting period that exists in a few states, spouses can divorce much more quickly now than under the previous, fault-based system (Walker & Elrod, 1993). The framers of the no-fault divorce laws did not intend to make divorce easy to obtain, however. Their intent was to permit divorce only after there was considerable evidence that the marriage could not be salvaged (Jacob, 1988).

Critics of no-fault divorce believe it has led to a skyrocketing divorce rate. Within 10 years of no-fault laws being passed the divorce rate jumped from 3.2 to 5.3. Critics believe the laws make divorce too easy. They also lament the fact that most states allow a partner to receive a divorce without the spouse also wanting one ("Covenant Marriage," 1999).

Yet others praise no-fault divorce for easing the pain of divorce for couples by preventing drawn-out and expensive legal proceedings. Defenders of no-fault divorce say it makes it easier for victims of

NEWS YOU CAN USE

Marriage and Divorce Quiz

1. One of the reasons there is more divorce today than in the past is because people live longer and there is more time to get divorced.

   False. Even though people live longer, they also marry later than in the past. About half of all divorces take place before the seventh year of marriage and are not influenced by life span. The length of the typical divorce-free marriage today is similar to what it was 50 years ago.

2. Living together before marriage increases your chance of divorce.

   True. People who live together before marriage are considerably more likely to get divorced than those that do not. It is not clear whether those that live together represent a different population or whether cohabitation introduces an attitude that the relationship is temporary.

3. An unmarried woman is more likely to experience domestic violence than a married woman.

   True. Unmarried women, particularly those who cohabitate with a man are more likely to be the victims of domestic abuse.

4. Now that people are more likely to divorce than in the past, those that stay married are happier than when people stayed married because of the stigma against divorce.

   False. Studies show that the general level of marital satisfaction has not increased. Marriages today have more work-related stress and less marital interaction than in the past

5. Second marriages are more successful than first marriages because people learn from their mistakes.

   False. The divorce rate for second marriages is higher than for first marriages.

6. If your parents divorced your chances of divorcing are increased.

   True. It appears that living in a family that has experienced divorce undermines the view that marriage is a lifelong commitment. The risk nearly triples if both partners come from homes that experienced a divorce.

7. Women are more likely than men to be the ones who initiate a divorce.

   True. Women initiate two-thirds of all divorces. One reason may be that in most states men stand little chance of gaining custody of their children. A divorce usually means that they will have to pay child support, but have limited contact with the children.

8. Teenage marriages are fairly successful if they can get through the first year.

   False. Marrying in your teens increases the likelihood of divorce two to three times over that of couples in their twenties and older.

Source: David Popenoe, "The Top Ten Myths of Divorce" National Marriage Project, http://marriage.rutgers.edu/Publications/Print/Print%20Myths%20of%20Divorce.htm, accessed October 18, 2005.

---

mental and physical abuse to obtain a divorce without having to go through a long, agonizing process.

Despite the criticism, efforts to repeal no-fault divorce laws have been largely unsuccessful. Legislation that would modify or repeal no-fault divorce has failed in at least 22 states ("Covenant Marriage," 1999).

No-fault divorce laws advocate that the financial aspects of marital dissolution be based on equity, equality, and economic need rather than on fault- or gender-based role assignments. Alimony also is to be based on the respective spouses' economic circumstances and on the principle of social equality, not on the basis of guilt or innocence. No longer is alimony automatically awarded to the injured party, regardless of that person's financial needs; no-fault divorce does not recognize an injured party. Instead these laws seek to reflect the changing circumstances of women and their increased participation in the labor force.

Under no-fault divorce law, women are encouraged to become self-supporting and husbands are not automatically expected to continue to support their ex-wives throughout their lives.

Some see no-fault divorce legislation as a redefinition of the traditional marital responsibilities of men and women by the affirmation of a new norm of equality between women and men. Husbands are no longer automatically designated as the head of the household, solely responsible for support, nor are wives alone expected to assume the responsibility of household activities and childrearing. Gender-neutral obligations that fall equally upon the husband and wife have been institutionalized by these new divorce laws. These changes are reflected most clearly in the new considerations for alimony allocation. In addition, property is to be divided equally. Finally, child-support expectations and the standards for child custody also reflect the new egalitarian criteria of no-fault divorce legislation. Today, both father and mother are expected to be equally responsible for the financial support of their children after divorce. In theory, mothers no longer receive custody of the child automatically; rather, a gender-neutral standard instructs judges to award custody in the best interests of the child.

No-fault has worked well for some divorcing couples, yet it has had devastating consequences for many others. Problems often emerge for older homemakers married 35 years or more, lacking any labor-force experience or skills, who may be awarded short-term settlements, ordered to sell the family home, and instructed by the court to pursue job training. Similarly, mothers with toddlers are routinely left with virtually full responsibility for their support.

No-fault divorce laws are based on an idealized picture of women's social, occupational, and economic gains in achieving equality that, in fact, may not reflect their actual conditions and circumstances. This discrepancy between reality and the ideal can have extremely detrimental effects on women's ability to become self-sufficient after divorce.

Child Custody Laws

The view that children would benefit most from living with their mother following divorce was widely accepted and seen in divorce and child custody laws until the 1970s. Fathers often were advised by legal counsel of the futility of contesting custody, and the burden of proof was on the father to document the unfitness of the mother or to prove his ability to be a better parent than the mother. However, there has been an increased recognition of fathers' rights regarding custody, reflecting the changing role of American fathers and the reevaluation of the judicial practice of automatically awarding custody to the mother. Also, "increasing numbers of states are allowing and encouraging joint custody, which involves both parents having decision-making authority for their children even if the child lives predominantly with one parent." California was the pioneer in this area (Fine, 1994). Despite the fact that state laws emphasize joint custody to varying degrees, an overwhelming majority of custody awards are made solely to mothers (Maccoby, et al., 1993).

In a legal sense, joint custody means that parental decision-making authority has been given equally to both parents after a divorce. It implies that neither parent's rights will be considered paramount. Both parents will have an equal voice in the children's education, upbringing, and general welfare. Joint legal custody is not a determinant of physical custody or postdivorce living arrangements. It is, however, often confused with complicated situations in which parents share responsibility for the physical day-to-day care of the children. Such arrangements usually require children to alternate between the parents' residences every few days, weeks, or months.

Although alternating living environments may accompany joint-custody decisions, in most instances they do not. In 90 to 95% of joint-custody awards the living arrangements are exactly the same as those under sole-custody orders; namely, the child physically resides with only one parent. However, both parents make decisions regarding the welfare of the child.

Those who believe joint legal custody is a good idea cite a variety of reasons. They assert that sole-custody arrangements, which almost always involve the child's living with the mother, weaken father-child relationships. They create enormous burdens for the mothers and tend to exacerbate hostilities between the custodial parent and the visiting parent. They continue to perpetuate outmoded gender-role stereotyping. Studies also show that sole-custody arrangements are associated with poverty, antisocial behavior in boys, depression in children, lower academic performance, and juvenile delinquency. Advocates of joint-custody assume that if fathers are given the opportunity to be available as nurturers, to be accessible, they will begin to participate more in the lives of their children. Furthermore, advocates say, such participation will have beneficial effects on children.

Before we too quickly assume that joint custody alleviates problems and produces benefits, we should note that it is far from being a panacea. If couples had trouble communicating and agreeing on things before the divorce, there is no reason to assume that they will have an easier time of it afterward. Most joint-custody orders are vague and do not decide at what point the joint-custodial parent's rights end and those of the parent with the day-to-day care of the child begin. What sorts of responsibilities can one parent require of the other parent? Issues such as these can easily erupt into disputes, particularly when a history of disagreement and distrust has preceded the joint-custody arrangement.

Joint custody does not give either parent the right to prevail over the other. To solve serious disputes the parents must return to court, where they must engage in litigation to prove that one or the other is unfit—the very process that the original decision of joint custody was to have avoided.

Joint custody appears to work best for those parents who have the capacity, desire, and energy to make it work—and for the children whose characteristics and desires allow them to expend the effort necessary to make it work and to thrive under it.

With high divorce rates come high remarriage rates. The resulting blended families produce new relationships that need to develop over time.

Remarriage and Stepfamilies

Throughout history, the reputation of stepfamilies has been surprisingly negative. The general thinking has been that the stepchild suffers in such families and that the suffering is caused by the stepmother. The stepmother has come down to us as a figure of cruelty and evil, constantly plotting to harm her stepchildren in a variety of ways. Just think of the Cinderella or Snow White fairy tales. The French word *maratre* means both "stepmother" and a "cruel and harsh mother." In English literature, "stepmother" is often preceded by "wicked." The stepchild was often a child who did not belong or whose status was similar to that of orphan. In fact, "stepchild" originally meant orphan.

Much of this has changed today, and stepfamilies have become a common sight on the American family landscape. The United States has the highest incidence of stepfamilies in the world. It is estimated that about 17% of married-couple households involve a stepparent. About one child in six is a stepchild.

The stepfamilies of today are different from those of the past in how they have come into being. The vast majority of stepfamilies now come about because of the marriage or cohabitation of mothers and fathers of children whose other parent is still living. Of these, the largest group by far is composed of families formed by the remarriage of divorced men and women. The high divorce rate is the key factor in the rise of the modern stepfamily.

In the past, stepfamilies resulted from quite different conditions and had different implications. Generally, stepfamilies were the product of remarriages after the death of a spouse. Mortality and the frequency of remarriage by the widowed determined the number of stepfamilies. Moreover, unlike today, children who lived in stepfamilies rarely had more than one living parent. In the past, a stepparent replaced a deceased parent.

Today, a stepparent is often an additional parent figure that a child must incorporate. A stepparent must enter a family system that has been created by the custodial parent and her children. The new marriage partners must establish their own relationship in an existing family structure. They must create new rules for how the family is to run. Such changes may produce disadvantages or tensions among the family members.

Remarriage makes parenthood and kinship an achieved status rather than an ascribed status. Traditionally a person became a father or a mother at the birth of their child. They did not have to do anything else to be a parent, nor could they easily resign from the job, especially in a family system that strongly discouraged divorce.

Remarriage after divorce, though, adds a number of other, potential kinship positions. The new parent must now achieve the status of parent or some other role that may resemble it. Working out these arrangements presents many more problems than are usually encountered in a first-marriage family (Cherlin & Furstenberg, 1994).

When people get divorced, they are not rejecting the institution of marriage but rather a specific marriage partner. About 54% of divorced women and 62% of divorced men remarry within five years (Bramlett & Mosher, 2002).

A woman's age, race, income, education, and the presence of children all affect her remarriage rates. Men's remarriage rates are less influenced by these factors, and if they are affected, it is in the opposite direction. Women divorced after age 40 have a lower rate of remarriage than younger women. For men, age appears to be a less significant factor in remarriage. Men with higher incomes and higher levels of education are more likely to remarry than those with less education and lower incomes. Women with higher incomes are less likely to remarry, and the education level of white women appears to make little difference in these rates. For African American women, higher education does increase the likelihood of remarriage (Bramlett & Mosher, 2002).

The presence of children lowers the probability of remarriage for women, but not for men. Nevertheless, about 80% of remarriages involve children, so it should come as little surprise that with the divorce and remarriage rates so high, stepfamilies are becoming a permanent part of the social landscape. Stepfamilies are changing such businesses as the greeting card industry (we now have birthday wishes to stepmothers and thank-you cards to stepfathers). Schools must now ask for information on stepparents as well as biological parents.

Stepfamilies, also known as blended families, are transforming basic family relationships. Where there were once two sets of grandparents, there now may be four; an only child may acquire siblings when his mother remarries a man with children. It used to be that stepfamilies typically followed the death of a spouse. Today's stepfamilies typically follow a divorce. "Nearly two-thirds (65%) of children living in stepfamily situations live with their biological mother. Stepfamilies are more likely to be white than black, and they are more often poor than rich. Stepparents are typically younger than other parents, and they are less likely to have a college education" (Larson, 1992).

Stepparents are taking on roles formerly held by biological parents. They are staying up nights with the spouse's sick children, attending class recitals, and having heated battles over such essential issues of childhood as curfews, TV, homework, and rights to the family car.

Try as they might, in one study of emotionally healthy middle-class families, only one-half the stepfathers and one-quarter of the stepmothers claimed to have "parental feelings" toward their stepchildren, and fewer still claimed to love them.

Stepparenthood is the strongest risk factor for child abuse. A stepparent is 40 to 100 times more likely to kill a young child than a biological parent (Daly & Wilson, 1988). Contributing to this situation is the fact that in the minds of most children, stepparents can never take the place of their real father or mother. Studies have shown that it takes at least four years for children to accept a stepparent in the same way they do their biological parents. Researcher James H. Bray has noted that this acceptance is harder for girls than boys: "Although divorce appears harder on elementary school-age boys, remarriage appears harder on girls" (Kutner, 1988).

At the heart of most stepfamily relationships are children who, like their parents, are casualties of divorce. In the best stepfamily relationships, all the adults work together to meet the needs of the children, realizing that, all too often, no matter what they do, there still will be problems. In reality, stepfamilies are torn apart by many of the same pressures that divide intact families. Financial problems can be especially acute when parents must support children from different marriages. Resentment builds quickly when stepparents feel they have little power or authority in their own homes. With all the conflicting interests and emotions, it's little wonder that about half of stepfamilies end in divorce (Larson, 1992).

The most difficult stage occurs during the early years, when stepparents want everything to go right. Once stepparents realize that relationships with stepchildren build over time, and that their potential network of allies includes all the other adults in the stepfamily relationship, the adjustment for all is faster and healthier.

Family Diversity

A number of options are increasingly available to people who, for various reasons, find the traditional form of marriage impractical or incompatible with their lifestyles. More young people are selecting

SOCIAL CHANGE

Reluctant to Marry: The Men Who Want to Stay Single

Most men marry by about age 30, but there is a significant group of about 20% of young men who are reluctant to commit to marriage. Who are they and how do they differ from men who marry earlier? There seem to be a number of key beliefs that are connected to a significantly more negative attitude toward marriage. Compared to men who marry earlier these men are more likely to:

- Worry about the risks of divorce
- Not want children
- Believe women cannot be trusted to tell the truth about past relationships
- Think single men have better sex lives than married men
- Believe marriage will reduce their personal freedom

These men are also:

- Less likely to come from traditional families
- Likely to have had fathers who were not involved in their lives very much
- More likely to question the value of marriage

For many of these men there are attractive alternatives to marriage. The urban single lifestyle provides the potential for sexual partners. Today there is little social disapproval of being a single man in your 30s or 40s. The increase in cohabitation rates has also made it an attractive alternative to the long-term commitment of marriage. These men do not need to worry about the "biological clock" that influences women's lives. It is not clear that these men will never marry. For the moment they are happy to postpone it indefinitely.

Source: David Popenoe and Barbara Dafoe Whitehead, "The State of Our Unions 2004" The National Marriage Project, June 2004 http://marriage.rutgers.edu/Publications/SOOU/SOOU2004.pdf accessed October 14, 2005.

---

cohabitation as a permanent alternative to marriage (although many consider it a prelude to marriage). In addition, some older men and women are opting to live together in a permanent relationship without getting married. These people choose cohabitation primarily for economic reasons—many would lose sources of income or control of their assets if they entered into a legal marriage. Several other trends are discussed next.

The Growing Single Population

Americans traditionally have been the marrying kind. In 2000, 61.5% of American men and nearly 58% of American women over the age of 18 were married. Younger people, however, may be rejecting this tradition. In 1970, only 10.5% of the women and 19.1% of the men between the ages of 25 and 39 had never been married. In 2003, 40.3% of women and 54.6% of men that age had never been married (Fields, 2004).

Even though the number of people living alone more than doubled between 1970 and 2003, this trend may mean only that more young people are postponing marriage. It also could mean, however, that a growing proportion of adults are staying single permanently. In fact, as we mentioned earlier in this chapter, some studies show that the marriage rate is declining.

There are a number of reasons why people are choosing not to marry. Working women do not need the financial security that a traditional marriage brings, and sex outside of marriage has become much more widespread. Moreover, many singles view marriage as merely a prelude to divorce and are unwilling to invest in a relationship that is likely to fail. As sociologist Frank Furstenburg notes, "Men who weren't married by their late 20's in the 60s were oddballs. Now they're just successful 29-year-olds." (See "Social Change: Reluctant to Marry: The Men Who Want to Stay Single.")

Clearly, many singles would gladly change their marital status if the right person came along. The ratio of unmarried men to unmarried women varies by age group. There are more unmarried men than women at the younger ages, reflecting the fact that men have a later age at first marriage than women do. After age 40 there are fewer unmarried men than unmarried women because women tend to live longer and are less likely to remarry after divorce and widowhood.

The elderly make up a significant proportion of single-person households. Currently, 39% of one-person households are maintained by people ages 65 or older, and fully 80% of those are elderly women. Nearly 40% of women 65 or older live alone, but only 17% of men in that age group do so (Bureau of the Census, 2002).

Figure 12–6   Percentage of Children under Age 18 Living with a Single Parent

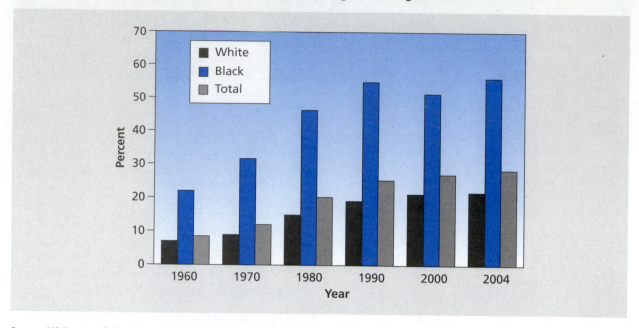

Source: US Bureau of the Census, Current Population Reports, Series P20-537; and U. S. Bureau of the Census, Population Division, Current Population Survey, 2004 Annual Social and Economic Supplement http://www.census.gov/population/socdemo/hh-fam/cps2004.

Single-Parent Families

In the early years of the twentieth century, it was not uncommon for children to live with only one parent because of the high death rate. As the death rate started to fall and the number of single widowed parents declined, it was not long before the divorce rate started to increase and single parents because of divorce became more common. Yet even by 1960, nearly one-third of all single mothers with children under 18 were widows. As divorce rates started to rise dramatically in the 1970s, the situation started to change and most single mothers were divorced or separated. By 1980, only 11% of single mothers were widowed and two-thirds were divorced or separated. The path to single motherhood changed again when many women bypassed marriage altogether. By 2003, 43.5% of single mothers had never been married (Fields, 2004). (See Figure 12–6 for children living with a single parent.)

There has been a significant increase in the number of single-parent families in the United States. In 2003 they constituted 27% of family households with children, up from 24% in 1990 and 11% in 1970 (Popenoe & Whitehead, 2004). A child in a one-parent family was just as likely to be living with a divorced parent in 2003 as with a never-married parent. A decade ago, a child living with one parent was almost twice as likely to be living with a divorced parent as with a never-married parent. The increase in the divorce rate is a major reason for the increase in single-parent families. Most divorced parents now set up new households, whereas in earlier times many would have returned to their own parents' households. The dramatic shift in family structure indicated by these measures has been generated mainly by three trends: divorce, unmarried births, and unmarried cohabitation. The incidence of divorce began to increase rapidly during the 1960s. There are about one million children under age 18 who lose a resident father each year because of divorce.

The second reason for the shift in family structure is an increase in the percentage of babies born to unwed mothers, which suddenly and unexpectedly began to increase rapidly in the 1970s. Since 1960, the percentage of babies born to unwed mothers has increased more than sixfold. The number of births to unmarried women is the highest ever recorded. About a third of all births and more than two-thirds of black births are out of wedlock, although the percentage of unwed black births declined slightly in the late 1990s.

A third and still more recent family trend that has affected family structure is the rapid growth of unmarried cohabitation. Cohabitation has become common among those previously married as well as the young and not-yet-married. An estimated 40% of all children today are expected to spend some time in a cohabiting household during their growing-up years (Bumpass & Lu, 2000).

Cohabitation is often seen as a prelude to marriage. Single people see it as a way to make sure that the couple is compatible before getting married. At least one of the partners expects the arrangement to result

Table 12–2

Percentage of Births to Unmarried Women

Iceland	64%
Sweden	54%
Norway	49%
Denmark	45%
France	40%
United Kingdom	38%
United States	33%
Canada	28%
Germany	14%
Italy	9%
Japan	1%

Source: Ventura SJ, Bachrach CA. Nonmarital Childbearing in the United States, 1940–99. National Vital Statistics Reports; Vol 48 No 16. Hyattsville, Maryland: National Center for Health Statistics, 2000.

Nearly 1.2 million people in the United States identify themselves as part of a gay or lesbian couple.

in marriage in 90% of cohabitations. The reality does not bear out these overly optimistic expectations, however. Only 58% of women who cohabitate end up marrying the man (Bramlett & Mosher, 2002).

Children living with cohabiting couples tend to be disadvantaged compared with those living with married couples. Cohabiting couples have a much higher breakup rate than married couples, a lower level of household income, and a much higher level of child abuse and domestic violence (Whitehead & Popenoe, 2002).

Most single mothers are not poor, but they tend to be younger, earn lower incomes, and are less educated than married mothers. Although the number of single mothers seems to have peaked, the number of single fathers is growing substantially. Men now constitute one-sixth of the nation's 11.9 million single parents (Fields, 2001).

People have become alarmed at the high percentage of births to unwed mothers in the United States. After rising dramatically between 1940 and 1990, out-of-wedlock births have leveled off at about 33% of all births. The number of births to unmarried women reached an annual total of 1.3 million in 2002.

The same thing, however, has also been taking place in a number of other industrialized countries. In 1998, half or more of births in Norway and Sweden were out of wedlock. Denmark, France, and the United Kingdom also have high percentages of out-of-wedlock births. However, levels in the United States are much higher than in some industrialized countries, such as Germany, Italy, Greece, and Japan, where less than 15% of births are out of wedlock (Ventura & Bachrach, 2000). (See Table 12–2.)

Gay and Lesbian Couples

A phenomenon that is not new but one that has become more and more visible is the household consisting of a gay or lesbian couple. Before 1970, almost all gay people wished to avoid the risks that would come with disclosure of their sexual orientation. It was not until 1990 that the U.S. Census attempted to calculate the number of gay and lesbian households. Through a self-identification process, nearly 1.2 million people say they are part of gay and lesbian couples in the United States, and though most live in four metropolitan areas, San Francisco, New York, Los Angeles, and Washington, D.C., nearly one in six live in a rural community. There has also been an increase in the number of recorded lesbian couples. Many more male couples than lesbian couples were recorded in the 2000 census, but new data show that the numbers are beginning to even out. There has also been an increase in social awareness about issues related to homosexuality.

It is also important to note that census data include only that portion of the gay population who are part of a couple, which studies show is only about one-third of gay men and lesbians. Furthermore, the total number of gay and lesbian couples is underreported in the census figures because some are hesitant about sharing information about personal relationships with the government, even though the forms are confidential (Cohn, 2001).

Gay and lesbian couples often are not eligible for the same benefits as traditional couples. In 1989, Denmark became the first country to recognize homosexual marriages. In the United States, only one of the 50 states permits gay or lesbian marriages. This causes many of the gay and lesbian partners to be ineligible for health care or other benefits that might be provided to the married employees and their families. A number of states and municipalities, however, have passed "domestic partnership" statutes to make such

OUR DIVERSE SOCIETY

Should Same-Sex Marriages Be Permitted?

On Nov. 18, 2003 the Massachusetts Supreme Judicial Court ruled that the state Constitution guaranteed equal marriage rights for same-sex couples. The ruling produced a shocked reaction and an outpouring of favorable and unfavorable responses from around the country. Undeterred by the opposition, on May 17, 2004 Massachusetts became the first and only state to issue marriage licenses to same-sex couples.

The Massachusetts ruling produced swift action and nearly half the states now outlaw or are poised to ban same-sex marriages. Only one state, Washington is on the verge of joining Massachusetts in legalizing same-sex marriage.

Unknown to most Americans, European countries had already tackled the same-sex marriage issue with much less commotion. The Netherlands became the first country to legalize gay marriages outright in 2001. Belgium followed in 2002. Other countries such as Denmark, France, and Germany, allow civil unions with basically the same rights as marriage (CBS News).

Some have noted that the increased recognition of domestic partnerships in many jurisdictions makes the movement to legalize same-sex marriages unnecessary. Marriage, however, is an important issue for lesbians and gay men because if domestic partnership is the only vehicle of legal recognition for same-sex relationships, such relationships are relegated to second-class status. Legal recognition of same-sex marriages, they point out, is an important step closer to full equality.

Lesbians and gays would gain many legal rights from the right to marry. Joint tax returns, insurance policies, social security and pension benefits, property inheritance, and veteran's discounts on medical care and on educational and home loans are all examples of the economic benefits that accompany marriage. Next-of-kin status for hospital visits and medical decisions, as well as adoption and guardianship rights, also go along with the right to marry. Gays and lesbians point out that interracial marriage was once illegal, yet many of the same arguments used to fight the legalization of inter-racial marriage are used by today's opponents of same-sex marriage.

Supporters of same-sex marriage point out that few would argue that gays and lesbians are capable of the

benefits available. In addition, 19 states, the District of Columbia, and 119 cities and counties have adopted policies that provide varying degrees of civil rights protection for gays and lesbians (*Harvard Law Review,* 1993).

Many gay and lesbian families include children (22% of lesbian families and 5% of gay families). Many of these children were part of a mother-father family and continued to live with the parent as she or he made the transition to same-sex relationships. Seventeen percent of gays and 29% of lesbians had previously been in a heterosexual marriage.

Gays and lesbians in couple relationships have higher educational levels than men and women in heterosexual marriages. Cohabiting gay men 25 to 34 years old are twice as likely to have a postgraduate degree as married men. The differences are even greater for lesbian couples in this age group, who are three times as likely to have a postgraduate education compared with married women (Black et al., 2000).

Traditionally, researchers and the media have concentrated on the ways in which gays and lesbians are different from heterosexuals. Yet gay and lesbian couples are similar to heterosexual couples in many

ways. They form long-term relationships and have problems similar to those of heterosexual couples. Gay and lesbian couples are concerned with having family benefits such as health insurance, life insurance, and family leave offered to gay and lesbian couples as well as married couples. Many are also involved in extending adoption rights and same-sex marriage recognition to gay and lesbian couples. (See "Our Diverse Society: Should Same-Sex Marriages Be Permitted?")

The Future: Bright or Dismal?

Given all these changes in the American family, should we be concerned that marriage and family life as we know them will one day disappear? Probably not. The divorce rate is high and will continue to be high during the next decade. It is important, however, to keep things in perspective. Divorce is just as much a social universal as is marriage. Anthropologist Margaret Mead noted that "no matter how free divorce" or "how frequently marriages break up," most societies have assumed that marriages would be

sacrifice, commitment, and responsibilities of marriage. If that is the case and they are denied equal legal standing not because of anything about the relationship itself, but purely because of the involuntary nature of homosexuality itself, then it is a clear example of bias and discrimination.

Same-sex marriage would also provide role models for gay youth. It would bring the gay couple into the heart of the traditional family and do more to heal the gay-straight rift than any gay rights legislation (Sullivan, 1995).

Critics of same-sex marriage note that marriage is a sacred union that unites a man and woman together for life. It is central to every faith and every modern society has embraced this view and rejected same-sex marriages.

Those opposed to same-sex marriage argue that the fundamental purpose of marriage is procreation, and because lesbians and gay men cannot procreate, they should not be allowed to marry. Gays and lesbians counter that if we follow this line of reasoning, then infertile heterosexual couples should not be allowed to marry, nor should those straight couples who choose not to have children for any number of reasons.

Opponents also wonder why same-sex marriage should be recognized when the high divorce rates show that we are having enough trouble maintaining the institution of marriage at all. Critics claim support for same-sex marriage would strike most people as a parody of marriage that could further weaken an already strained institution. Critics also argue that extending marriage to gay people could lead the way to a further redefinition of marriage. They fear that people might begin to demand to be allowed to marry more than one person at a time (Wilson, 1996).

Sources: *Virtually Normal: An Argument about Homosexuality,* by A. Sullivan, 1995, New York: Alfred A. Knopf; "Against Homosexual Marriage," by J. Q. Wilson, March 1996, *Commentary;* "Same-Sex Partnerships," February 18, 2000, *Issues and Controversies on File,* pp. 49–56; CBS News, *The Global View Of Gay Marriage* March 4, 2004 http://www.cbsnews.com/stories/2004/03/04/world/main604084.shtml. Accessed November 29, 2005.

permanent. Despite this belief, societies have also recognized that some marriages are incapable of lasting a lifetime and have provided mechanisms for dissolution (Riley, 1991).

Even though the divorce rate is high, the remarriage rate is also very high. The vast majority of people who divorce remarry, usually within a short time after they divorce. The high divorce rate does not necessarily mean that people are giving up on marriage. It just means there is a growing belief that marriage can be better. The high remarriage rate indicates that people are willing to continue trying until they reach their expectations. Obtaining a divorce does not mean the person believes that the idea of marriage is a mistake, only that a particular marriage was a mistake.

Despite claims to the contrary, there is little evidence that the family as an institution is in decline, or any weaker today than a generation ago. Nor is there any indication that people place less value on their own family relationships, or on the role of the family within society, than they once did.

The traditional family is being replaced by family arrangements that better suit today's lifestyles: There are fewer full-time homemakers because more

women are in the workforce. Nonfamily households have increased substantially. The traditional family with a working dad, homemaker mom, and two or more children is now a distinct minority.

The institutions of marriage and the family have proved to be extremely flexible and durable and have flourished in all human societies under almost every imaginable condition. As we have seen, these institutions take on different forms in differing social and economic contexts, and there is no reason to suspect that they will not continue to do so. Therefore, to make predictions about the future of the American family is equivalent to making predictions about the future of American society in particular and industrial society in general. This is extremely difficult to do, given the social, economic, political, and ecological problems facing us. However, for the foreseeable future, it seems reasonable to assume that the forces of industrialism and public policy that helped shape the current nuclear family in its one-parent and two-parent forms will persist. Therefore, the contemporary nuclear family will continue to provide the basic context within which American society will reproduce itself for several generations to come.

FOR FURTHER THINKING

How Much Are Children Hurt by Their Parents' Divorce?

Since 1970, at least a million children a year have seen their parents divorce—building a generation of Americans that is now coming of age. For the past 30 years, Judith Wallerstein and Mavis Hetherington have been studying the children of divorce. Now that the children have grown to adulthood, they provide important information about the life course of individuals whose parents separated when they were young children.

Judith Wallerstein notes:

When I began studying the effects of divorce on children and parents in the early 1970s, I, like everyone else, expected them to rally. But as time progressed, I grew increasingly worried that divorce is a long-term crisis that was affecting the psychological profile of an entire generation. The long-term effects on the children of divorce began to appear in late adolescence and early adulthood, but it was not until the children were fully grown that the whole picture emerged. Divorce is a life-transforming experience. After divorce, childhood is different. Adolescence is different. Adulthood—with the decision to marry or not and have children or not—is different. Whether the final outcome is good or bad, the whole trajectory of an individual's life is profoundly altered by the divorce experience.

Contrary to what we have long thought, the major impact of divorce does not occur during childhood or adolescence. Rather, it rises in adulthood as serious romantic relationships move center stage. When it comes time to choose a life mate and build a new family, the effects of divorce crescendo. A central finding to my research is that children identify not only with their mother and father as separate individuals but with the relationship between them. They carry the template of this relationship into adulthood and use it to seek the image of their new family. The absence of a good image negatively influences their search for love, intimacy, and commitment. Anxiety leads many into making bad choices in relationships, giving up hastily when problems arise, or avoiding relationships altogether.

The divorced family is not a truncated version of the two-parent family. It is a different kind of family in which children feel less protected and less certain about their future than children in reasonably good intact families. Mothers and fathers who share their beds with different people are not the same as mothers and fathers living under the same roof. The divorced family has an entirely new cast of characters and relationships featuring stepparents and stepsiblings, second marriages and second divorces, and often a series of live-in lovers. The child who grows up in a postdivorce family often experiences not one loss—that of the intact family—but a series of losses as people come and go. This new kind of

SUMMARY

- The American family has changed dramatically in the past 35 years.
- The marriage rate is down, the divorce rate is up, and more children are being born to single women.
- A household consists of people who share the same living space.
- Sociologists generally agree that some form of family is found in every known human society.
- The family serves several important functions: regulating sexual behavior, patterning reproduction, organizing production and consumption, socializing children, providing care and protection, and providing social status.

- No society permits random mating; all societies have an incest taboo, which forbids sexual intercourse among closely related individuals.
- Sex between parents and their children is universally prohibited, but who else is considered to be closely related varies widely among societies.
- Every society must replace its members. By regulating where and with whom individuals may enter into sexual relationships, society patterns sexual reproduction.
- Although almost all societies allow for divorce, or the breakup of marriage, none endorses it as an ideal norm.
- American culture is relatively unique in linking romantic love and marriage.

family puts very different demands on each parent, each child, and each of the many new adults who enter the family orbit.

Children are not passive vessels but rather active participants who help shape their own destiny and that of their family. They make gallant efforts to fit into the new requirements of the postdivorce family. . . . Because they are in their formative years, the new roles they assume in the family are built on their character. The roles they adopt to adjust to the new circumstances in the divorced family are likely to endure into adulthood and are frequently reinstalled in their adult relationships.

Mavis Hetherington disagrees with Wallerstein and believes divorce experiences are varied. She notes that initially, especially in marriages involving children, divorce is miserable for most couples.

In the early years, ex-spouses typically must cope with lingering attachments; with resentment and anger, self-doubts, guilt, depression, and loneliness; with the stress of separation from children or of raising them alone; and with the loss of social networks and, for women, of economic security. Nonetheless, she found that a gradual recovery usually begins by the end of the second year. And by six years after divorce, 80 percent of both men and women have moved on to build reasonably or exceptionally fulfilling lives.

Indeed, about 20 percent of the women she observed eventually emerged from divorce enhanced and exhibiting competencies they never would have developed in an unhappy or constraining marriage. They had gone back to school or work to ensure the economic stability of their families, they had built new social networks, and they had become involved and effective parents and socially responsible citizens. Often they had happy second marriages. Divorce had offered them an opportunity to build new and more satisfying relationships and the freedom they needed for personal growth. This was especially true for women moving from . . . disengaged marriage, or from one in which a contemptuous or belligerent husband undermined their self-esteem and child-rearing practices. Divorced men, she found, were less likely to undergo such remarkable personal growth; still, the vast majority of the men in her study did construct reasonably happy new lives for themselves.

Sources: J. S. Wallerstein, J. M. Lewis, and S. Blakeslee, *The Unexpected Legacy of Divorce* (pp. xxvi–xxvii, xxix–xxx), 2000, New York: Hyperion; E. Mavis Hetherington, "Marriage and Divorce American Style," *The American Prospect,* April 8, 2002; E. Mavis Heatherington and John Kelly, *For Better or For Worse: Divorce Reconsidered,* New York: W. W. Norton and Company, 2002.

- Romantic love—which involves the idealization of the loved one, the notion of a one and only, love at first sight, love winning out over all, and an indulgence of personal emotions—has nothing to do with marriage throughout most of the world.
- Marriage in many societies establishes social, economic, and even political relationships among families.
- Three families are ultimately involved in a marriage: the two spouses' families of origin or families of orientation, the families in which they were born or raised, and the family of procreation, which is created by the union of the spouses.

 ## Media Resources

The Companion Website for *Introduction to Sociology,* Ninth Edition

http://sociology.wadsworth.com/tischler9e

Supplement your review of this chapter by going to the companion website to take one of the Tutorial Quizzes, use the flash cards to master key terms, and check out the many other study aids you'll find there. You'll also find special features such as Wadsworth's Sociology Online Resources and Writing Companion, GSS data, and Census 2000 information at your fingertips to help you complete that special project or do some research on your own.

CHAPTER TWELVE STUDY GUIDE

KEY CONCEPTS

Match each concept with its definition, illustration, or explanation below.

a. Incest taboo
b. Exogamy
c. Matriarchal family
d. Family of procreation
e. Nuclear family
f. Matrilineal descent

g. Extended family
h. Endogamy
i. Homogamy
j. No-fault divorce
k. Polygamous family

l. Patrilocal residence
m. Neolocal residence
n. Family of orientation
o. Polyandry
p. Companionate marriage

_____ 1. A norm that forbids sexual intercourse among closely related individuals.
_____ 2. A marriage based on romantic love.
_____ 3. A set of nuclear families linked by multiple marriage bonds, with one central individual married to several spouses.
_____ 4. A family in which the central individual is a female and the multiple spouses are males.
_____ 5. Two spouses and their children.
_____ 6. Divorce granted without one partner having to prove adultery, desertion, abuse, or other faults on the part of the other.
_____ 7. The family in which a person was born and raised.
_____ 8. Marriage within one's own social group.
_____ 9. The family created by marriage.
_____ 10. Family consisting of relatives in addition to the spouses and their children.
_____ 11. Marriage outside one's own culturally defined group.
_____ 12. A situation in which a newly married couple may choose to live virtually anywhere.
_____ 13. Families where the female is the dominant figure in decision making.
_____ 14. Kinship traced through the mother's family (for example, if your last name were your mother's maiden name).
_____ 15. A requirement that a new couple settle down near or within the husband's father's household.
_____ 16. Marrying someone with similar social characteristics.

CENTRAL IDEA COMPLETIONS

Following the instructions, fill in the appropriate concepts and descriptions for each of the questions posed in the following section.

1. How have divorce rates changed over the last half-century, and what factors have affected those rates? _____

   _____

   _____

2. Identify at least three aspects of the Industrial Revolution that promoted the shift from extended families to nuclear families.

   a. _____

   _____

   b. _____

   _____

   c. _____

   _____

3. What are two advantages of arranged marriages over marriages based on romantic love?

a. _____

_____

b. _____

_____

4. What are three conclusions one can draw from the fact that there is a growing single population?

_____

_____

_____

5. Identify three major trends in the American family that have developed over the last few decades.

a. _____

_____

b. _____

_____

c. _____

_____

6. List five functions usually fulfilled by the family; in each case, describe how these functions have been fulfilled by your own family.

a. _____

_____

b. _____

_____

c. _____

_____

d. _____

_____

e. _____

_____

7. Common-sense arguments often suggest "falling in love" and "selecting a mate" are matters of chance or random events; however, your text demonstrates that mate selection is not at all random. Explain two of the ways in which mate selection is not random.

a. _____

_____

_____

b. _____

_____

_____

8. Contemporary American society includes many people with alternative lifestyles. Providing examples, discuss two of these alternative lifestyles as they relate to the family as a social institution.

    a. _____

    _____

    b. _____

    _____

9. Family violence is a major concern. What sociodemographic variables are associated with family violence? What changes in society have affected rates of domestic violence in the last 20 years?

    a. Sociodemographic variables: _____

    _____

    _____

    _____

    b. Factors affecting changes in violence: _____

    _____

    _____

    _____

CRITICAL THOUGHT EXERCISES

1. Present an in-depth discussion of how mate selection takes place in the United States. Be sure to devote attention to the effects of such variables of age, race, religion, and social status on American mate selection.

2. Before reading this chapter, what was your thinking about how parental divorce *hurts* children? Your author devoted considerable attention to the question: *How much are children hurt by their parents' divorce? What were his conclusions?* What was meant by the statement: *The divorced family is not a truncated version of the two-parent family*? What factors appear to affect the extent of adjustment or nonadjustment by children to parental divorce?

3. For Americans, who overwhelmingly favor marriage based on romantic love, it's easy to see the problems that can arise in a system of arranged marriage. You could easily make a list of all the advantages of romantic-love marriages and the difficulties in arranged marriage. But imagine that you are from an arranged-marriage society and make a list of the advantages of arranged marriage and the problems that can easily arise in a system of marriage based on love. Imagine the things that you personally would find unpleasant about an arranged marriage. Do you think people in arranged-marriage societies have similar feelings, and if so, how might they deal with them? What institutions in American society are there to deal with the problems of our system of marriage? (For example, finding a mate, which is not problematic in arranged-marriage societies, has in our own society given rise to the institution of dating as well as a variety of more formal and sometimes commercial solutions to this problem.)

4. Your text describes "the decline of the traditional family." Yet despite this decline, Americans still claim to live in families and to value family. Obviously, there has been a change in ideas about what constitutes a family and family values. Will we see further changes in the definition of family? If so, what will these changes be? What social factors will affect these changes and the speed at which they take place?

INTERNET ACTIVITIES

1. The National Marriage Project (http://marriage.rutgers.edu/) has a wealth of long and short reports on marriage, including the "myths about marriage" cited in this chapter (there's also a link to myths about divorce) as well as articles with titles like "Why Men Don't Commit" based on solid sociological research.

2. The world of the family is strongly affected by the world of work. Families need the income they get from work, but they suffer if work demands too much of their time and energy. The Work and Family Equity Index is a measure developed to assess the impact of work on family, especially in regard to government policies such as requiring employers to provide paid maternity leave. Find out more about it at http://www.hsph.harvard.edu/globalworkingfamilies and see how the United States compares with other countries (not too well).

3. For more on changes in the American family over the last century, see http://www.pbs.org/fmc/book/pdf/ch4.pdf for easily understood charts.

ANSWERS TO KEY CONCEPTS

1.a 2.p 3.k 4.o 5.e 6.j 7.n 8.h 9.d 10.g 11.b 12.m 13.c 14.f 15.l 16.i

ThomsonNOW™

Reviewing is as easy as ❶ ❷ ❸

1. Before you do your final exam, take the ThomsonNOW diagnostic quiz to help you identify the areas on which you should concentrate. You will find information on ThomsonNOW and instructions on how to access all of its great resources on the foldout at the beginning of the text.
2. As you review, take advantage of ThomsonNOW's study videos and interactive Map the Stats exercises to help you master the chapter topics.
3. When you are finished with your review, take ThomsonNOW's posttest to confirm you are ready to move on to the next chapter.

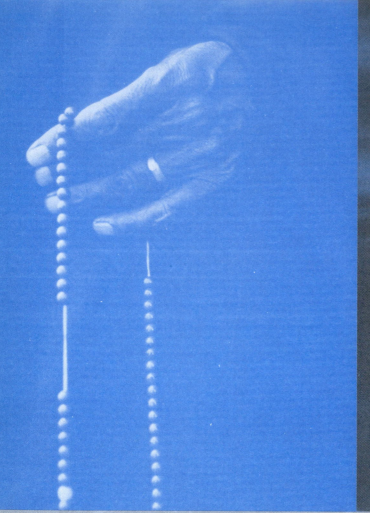

13

Religion

© Eric van den Brulle

Learning Objectives

After studying this chapter, you should be able to do the following:

- Define the basic elements of religion.
- Differentiate among the major types of religion.
- Describe the functions of religion according to the functionalist perspective.

- Explain the conflict theory perspective on religion.
- Describe the basic types of religious organization.
- Describe important aspects of contemporary American religion.
- Describe the major religions in the United States.

The belief in one God is a basic tenet of Christianity and the cornerstone of Judaism. Islam also proclaims, "There is no God but Allah, and Muhammad is His Prophet." Hinduism, however, expresses a very different concept: the idea of the "manyness" of God. Hinduism does not have one creed, one founder, one prophet, or one central moment of revelation. It is an expression of 5,000 years of religious and cultural development in Asia. During this development, Hinduism has assimilated many ideas and ideologies. In Hinduism, whether there is one god or many gods is unimportant. Rather than worshipping a single god, Hinduism involves worshipping one god at a time. Each Hindu is free to choose his or her own god or goddess.

In his book *All Religions Are True*, Mohandas Karamchand Gandhi wrote:

> It has been a humble but persistent effort on my part to understand the truth of all the religions of the world, and adopt and assimilate in my own thought, word, and deed all that I have found to be best in those religions.

Gandhi's acceptance of diversity, however, has a downside. If all religions are true, then they all must be imperfect. After all, if any one religion were perfect, it would be better than all others.

The idea of the manyness of God is alien not only to Western religion, but also to Western culture. Our monotheism leads us to believe in one Truth. In our minds, the ultimate or best can stand out in many areas, whether it be the best car, the best university, or the ultimate religion (Goldman, 1991).

The important thing to realize is that although religion assumes many different forms, it is a universal human institution. To appreciate the many possible kinds of religious experiences, from the belief in one God to the belief in the manyness of God, requires an understanding of the nature and functions of religion in human life and society.

The Nature of Religion

Religion is *a system of beliefs, practices, and philosophical values shared by a group of people; it defines the sacred, helps explain life, and offers salvation from the problems of human existence.* It is recognized as one of society's important institutions.

In his classic study *The Elementary Form of Religious Life*, first published in 1915, Émile Durkheim observed that all religions divide the universe into two mutually exclusive categories: the profane and the sacred. The **profane** consists of *all empirically observable things—that is, things that are knowable through common, everyday experiences.* In contrast, the **sacred** consists of *things that are awe inspiring and knowable only through extraordinary experiences.*

The sacred may consist of almost anything: objects fashioned just for religious purposes (such as a cross),

One of the functions of ritual and prayer is to produce an appropriate emotional state.

© Kim Kyung-Hoon/Reuters/CORBIS

object that had little social value in itself. Today, however, one of Babe Ruth's bats is enshrined in baseball's Hall of Fame. It is no longer used in a profane way but instead is seen as an object that represents the values, sentiments, power, and beliefs of the baseball community. The bat has gained some of the qualities of a sacred object, thus changing from a private object to a public object.

In addition to sacred symbols and a system of beliefs, religion also includes specific rituals. **Rituals** are *patterns of behavior or practices that are related to the sacred*. For example, the Christian ritual of Holy Communion is much more than the eating of wafers and the drinking of wine. To many participants, these substances are the body and blood of Jesus Christ. Similarly, the Sun Dance of the Plains Indians was not merely a group of men dancing around a pole to which they were attached by leather thongs that pierced their skin and chest muscles. It was a religious ritual in which the participants were seeking a personal communion.

The Elements of Religion

All religions contain certain shared elements, including ritual and prayer, emotion, belief, and organization.

Ritual and Prayer All religions have formalized social rituals, but many also feature private rituals such as prayer. Of course, the particular events that make up rituals vary widely from culture to culture and from religion to religion.

All religions include a belief in the existence of beings or forces that human beings cannot experience. In other words, all religions include a belief in the supernatural. Hence, they also include **prayer,** or *a means for individuals to address or communicate with supernatural beings or forces,* typically by speaking aloud while holding the body in a prescribed posture or making stylized movements or gestures.

In the United States 8 out of 10 Americans pray at least weekly. Less than 1 in 10 claim to never pray. When they pray these people are seeking guidance (62%), giving thanks (54%), and seeking forgiveness (47%) or healing (45%). Interestingly, a substantial number of those that pray claim no religion (Holt, 2004).

a geographical location (Mount Sinai), a place constructed for religious observance (a temple), a word or phrase ("Our Father, who art in heaven . . ."), or even an animal (the cow to Hindus, for example). To devout Muslims, the Sabbath, which falls on Friday, is a sacred day. To Hindus, the cow is holy, not to be killed or eaten. These are not ideas to be debated; they simply exist as unchallengeable truths. Similarly, to Christians, Jesus was the Messiah; to Muslims, Jesus was a prophet; but to sociologists, the person of Jesus is a religious symbol. Religious symbols acquire their particular sacred meanings through the religious belief system of which they are a part.

Durkheim believed that every society must distinguish between the sacred and the profane. This distinction is essentially between the social and nonsocial. What is considered sacred has the capacity to represent shared values, sentiments, power, or beliefs. The profane is not supported in this manner; it may have utility for one or more individuals, but it has little public relevance.

We can look at the famous baseball player Babe Ruth's bat as an example of the transformation of the profane to the sacred. At first, it was merely a profane

Emotion One of the functions of ritual and prayer is to produce an appropriate emotional state. This may be done in many ways. In some religions, participants in rituals deliberately attempt to alter their state of consciousness through the use of drugs, fasting, sleep deprivation, and induction of physical pain. Thus, Scandinavian groups ate mushrooms that caused euphoria, as did many native Siberian tribes. Various

Native American religions feature the use of peyote, a button-like mushroom that contains a hallucinogen.

There are approximately 250,000 members of the Native American church who believe that the use of peyote brings them closer to God. Even though it is illegal to use peyote in most states, Congress made it possible for federally recognized tribes to practice traditional Indian ceremonies even if it violates local laws (Gehrke, 2001).

Although not every religion tries to induce altered states of consciousness in believers, all religions do recognize that such states may happen and believe that they may be the result of divine or sacred intervention in human affairs. Prophets, of course, are thought to receive divine inspiration. Religions differ in the degree of importance they attach to such happenings.

Belief All religions endorse a belief system that usually includes a supernatural order and also often a set of values to be applied to daily life. Belief systems can vary widely. Some religions believe that a valuable quality can flow from a sacred object—animate or inanimate, part or whole—to a lesser object. Numerous Christian sects, for instance, practice the laying on of hands, whereby a healer channels divine energy into afflicted people and thus heals them.

Some Christians also believe in the power of relics to work miracles simply because those objects once were associated physically with Jesus or one of the saints. Such beliefs are quite common among the world's religions: Native Australians have their sacred stones, and shamans among African, Asian, and North American societies try to heal through sympathetic touching. In some religions, the source of the valued quality is a personalized deity. In others, it is a reservoir of supernatural force that is tapped.

People's belief can be very strong. In Milton, Massachusetts, in June 2003, people believed they saw an image of the Virgin Mary and Child in a window at Milton Hospital. The event gained international publicity and large crowds started to show up. In April 2005 in Chicago, a steady stream of people flocked to the Kennedy Expressway underpass to view a yellow and white stain on a concrete wall that many believed was an image of the Virgin Mary. Police had to patrol the area under the Kennedy Expressway after hundreds of people gathered to see the image. Some women knelt with rosary beads while other people stood praying (CBS News, 2005).

One of the best-known sightings of the Virgin Mary took place in office windows in Clearwater, Florida. Within weeks, a half-million people had traveled to the site. Many of those who came hoped for a miracle that would cure their ills. Others wanted to receive their own message from Mary, to see visions of Jesus, or to strengthen their faith.

Organization Many religions have an organizational structure through which specialists can be recruited and trained, religious meetings conducted, and interaction facilitated between society and the members of the religion.

The organization also promotes interaction among the members of the religion to foster a sense of unity and group solidarity. Rituals may be performed in the presence of other members. They may be limited to certain locations such as temples, or they may be processions from one place to another. Although some religious behavior may be carried out by individuals in private, all religions demand some public, shared participation. (See "Our Diverse Society: Who is God?")

Magic

In some societies, magic serves some of the functions of religion, though there are essential differences between the two. **Magic** is *an active attempt to coerce spirits or to control supernatural forces. It differs from other types of religious beliefs in that god or gods are not worshipped.* Magic is used to manipulate and control matters that seem to be beyond human control and that may involve danger and uncertainty. It is usually a means to an end, whereas religion is usually an end in itself, although prayer may be seen as utilitarian when a believer asks for a personal benefit. In most instances, religion serves to unify a group of believers, whereas magic is designed to help the individual who uses it. Bronislaw Malinowski (1954) explains:

> We find magic wherever the elements of chance and accident, and the emotional play between hope and fear have a wide and extensive range. We do not find magic wherever the pursuit is certain, reliable, and well under the control of rational methods and technological processes. Further we find magic where the element of danger is conspicuous.

During the Middle Ages, when most of the population was illiterate, the belief in magic was quite extensive. Almost everyone believed in sorcery, werewolves, witchcraft, and black magic. If a noblewoman died, her servants ran around the house emptying all containers of water so her soul would not drown. Bloodletting to cure illnesses was popular. Plagues were believed to be the result of an unfortunate conjuncture of the stars and planets. The air was believed to be infested with such soulless spirits as unbaptized infants, ghouls who pulled out cadavers in graveyards and gnawed on their bones, and vampires who sucked the blood of stray children. For the medieval mind, magic provided an understanding of how the

OUR DIVERSE SOCIETY

Who Is God?

Why does God exist? How have the three dominant monotheistic religions—Judaism, Christianity, and Islam—shaped and altered the conception of God? Karen Armstrong, one of Britain's foremost commentators on religious affairs, traced the history of how men and women have perceived and experienced God from the time of Abraham to the present.

Armstong points out that what the idea of God means depends on the time, the place, and the people who are making the interpretation. How God is viewed by one group may be meaningless to another group. Even the statement "I believe in God" only means something when it is understood in the context of the time period and who is making the statement.

Armstrong notes that what people imply when they refer to "God" may have many meanings that can often be contradictory. In fact, she believes that this flexibility is necessary to keep the notion of God alive. It makes it possible to discard and replace conceptions of God when they cease to have meaning or relevance. She points out that Christianity, Judaism, and Islam have each had to re-create the image of God as times have changed.

As strange as it may seem the same is true for atheism also. Stating, "I do not believe in God" has also had a variety of meanings throughout history. Atheists are denying a particular conception of God that is rooted to a particular time and place. As Armstrong notes "Is the 'God' whom atheists reject today the God of the patriarchs, the God of the prophets, the God of the philosophers, the God of the mystics, or the God of the eighteenth century?" These ideas of God are all very different from one another. In addition "Jews, Christians, and Muslims were all called "atheists" by their pagan contemporaries because they had adopted a revolutionary notion of divinity and transcendence." The atheist of today is denying the existence of a God adequate to address contemporary issues.

Armstrong believes religion has had to be highly practical. As soon as particular views about God ceased to be effective they were adjusted to accommodate changing times. People had less trouble accepting this view in the past than they do today. She points out that our ancestors knew that ideas about God were provisional and open to reinterpretation.

Sources: Adapted from *A History of God* (pp. xx–xxi), by Karen Armstrong, 1994, New York: Ballantine Books; and interviews with the author, October 1994 and 1996.

---

world worked. It helped relieve anxiety and allowed people to blame events on bad luck or evil spirits, and it permitted them to cast blame on curses and witchcraft. Astrology was the most popular science of that time. Only religion could rival astrology as an all-embracing explanation for the unpredictability of life (Shermer, 2000).

Rodney Stark and William Bainbridge (1985) have noted that a belief in magic has always been a major part of Christian faith. A common theme throughout the centuries has been the effort of organized religion to prohibit unorthodox practices and practitioners and to monopolize magic. Nonchurch magic was identified as superstition. Serious efforts to root out magic once and for all emerged in the fifteenth century. Eventually, as many as 500,000 people may have been executed for witchcraft. "In order to monopolize religion," Stark and Bainbridge wrote,

[A] church must monopolize all access to the supernatural. . . . But if the church is to deny others access to the supernatural, it must remain in the magic business. The demand for magic is too great to be ignored. . . . Thus the Catholic

Church remained deeply involved in dispensing magic. Immense numbers of magical rites and procedures were developed. . . . Saints and shrines that performed specialized miracles proliferated, and new procedures for seeking saintly intercession abounded. Many forms of illness, especially mental illness, were defined as cases of possession, and legions of official exorcists appeared to treat them.

Stark and Bainbridge noted that magic's respectability has decreased as more scientific attitudes have proliferated. Magic, especially magical healing, is now found mostly among sectarians and cultists. This fact makes the religious beliefs of sects and cults particularly vulnerable to criticism and refutation (Beckwith, 1986).

Major Types of Religions

The earliest evidence of religious practice comes from the Middle East. In Shanidar Cave in Iraq, archaeologist Ralph Solecki (1971) found remains of burials of Neanderthals—early members of our own species, *Homo sapiens*—dating from between 60,000 and 45,000 years ago. Bodies were tied into a fetal position, buried on their sides, provided with morsels of food placed at their heads, and covered with red powder and sometimes with flower petals. Those practices—the food and the ritual care with which the dead were buried—point to a belief in some kind of existence after death.

Using studies of present-day cultures, as well as historical records, sociologists have devised a number of ways of classifying religions. One of the simplest and most broadly inclusive schemes recognizes four types of religion: supernaturalism, animism, theism, and abstract ideals.

Supernaturalism

Supernaturalism *postulates the existence of nonpersonalized supernatural forces that can, and often do, influence human events.* These forces are thought to inhabit animate and inanimate objects alike—people, trees, rocks, places, even spirits or ghosts—and to come and go at will. The Melanesian/Polynesian concept of *mana* is a good example of the belief in an impersonal supernatural power.

Mana is *a diffuse, nonpersonalized force that acts through anything that lives or moves,* although inanimate objects such as an unusually shaped rock also may possess mana. The proof that a person or thing possesses mana lies in its observable effects. A great chief, merely by virtue of his position of power, must possess mana, as does the oddly shaped stone placed in a garden plot that then unexpectedly yields huge crops. Although it is considered dangerous because of its power, mana is neither harmful nor beneficial in itself, but it may be used by its possessors for either good or evil purposes. An analogy in our culture might be nuclear power, a natural force that intrinsically is neither good nor evil but can be turned to either end by its possessors. We must not carry the analogy too far, however, because we can account for nuclear power according to natural, scientific principles and can predict its effects reliably without resorting to supernatural explanations. A narrower, less comprehensive, but more appropriate analogy in Western society is our idea of luck, which can be good or bad and over which we feel we have little control.

Although certain objects possess mana, taboos may exist in relation to other situations. A **religious taboo** is *a sacred prohibition against touching, mentioning, or looking at certain objects, acts, or people.* Violating a taboo results in some form of pollution. Taboos may exist in reference to foods not to be eaten, places not to be entered, objects and people not to be touched, and so on. Even a person who

becomes a victim of some misfortune may be accused of having violated a taboo and may also become stigmatized.

Taboos exist in a wide variety of religions. Polynesian peoples believed that their chiefs and noble families were imbued with powerful mana that could be deadly to commoners. Hence, elaborate precautions were taken to prevent physical contact between commoners and nobles. The families of the nobility intermarried (a chief often would marry his own sister), and chiefs actually were carried everywhere to prevent them from touching the ground and thereby killing the crops. Many religions forbid the eating of selected foods. Jews and Muslims have taboos against eating pork at any time, and up until the early 1960s, Catholics were forbidden to eat meat on Fridays. Most cultures forbid sexual relations between parents and children and between siblings (the incest taboo) (see Chapter 3).

Supernatural beings fall into two broad categories: those of nonhuman origin, such as gods and spirits, and those of human origin, such as ghosts and ancestral spirits. Chief among those of nonhuman origin are the gods who are believed to have created themselves and may have created or given birth to other gods. Although gods may create, not all religions attribute the creation of the world to them.

Many of those gods thought to have participated in creation have retired, so to speak. Having set the world in motion, they no longer take part in day-to-day activities. Other creator gods remain involved in ordinary human activities. Whether or not a society has creator gods, many other affairs are left to lesser gods. For example, the Maori of New Zealand have three important gods: a god of the sea, a god of the forest, and a god of agriculture. They call upon each god for help in the appropriate area.

Below the gods in prestige, but often closer to the people, are the unnamed spirits. Some of these can offer constructive assistance, and others take pleasure in deliberately working evil for people.

Ghosts and ancestor spirits represent the supernatural beings of human origin. Many cultures believe that everyone has a soul, or several souls, which survive after death. Some of those souls remain near the living and continue to be interested in the welfare of their kin (Ember & Ember, 1981).

Animism

Animism is *the belief in inanimate, personalized spirits or ghosts of ancestors that take an interest in, and actively work to influence, human affairs.* Spirits may inhabit the bodies of people and animals as well as inanimate phenomena such as winds, rivers, or mountains. They are unique beings with feelings, motives, and a will of their own. Unlike mana, spirits may be

© Tim Hall/PhotoDisc/Getty Images

Only three religions are known to be monotheistic: Judaism, Christianity, and Islam.

Table 13–1

Major Religions of the World—2000

Total World Population   6,050,000,000*

		Percentage of Total
Christians	1,999,564,000	33.0
Roman Catholic	1,057,328,000	17.5
Protestants	342,002,000	5.6
Orthodox	215,129,000	3.7
Anglican	79,650,000	1.3
Unaffiliated Christians	111,125,000	1.8
Baha'	7,106,000	0.1
Buddhists	353,794,000	6.0
Chinese folk-religionists	359,982,000	5.9
Ethnic religionists	228,367,000	3.8
Muslims	1,188,243,000	19.6
Hindus	811,336,000	13.3
Jews	14,434,000	0.2
Sikhs	23,258,000	0.4
Nonreligious	768,159,000	12.7
Atheists	150,090,000	2.5

*World Population Adjusted to Year 2000.

Source: *World Christian Trends: AD 30–AD 2200*, Pasadena, CA: William Carey Library, 2001.

intrinsically good or evil. Although they are powerful, they are not worshipped as gods, and because of their humanlike qualities, they can be manipulated— wheedled, frightened away, or appeased—by using the proper magic rituals. For example, many Native American and South American Indian societies (as well as many other cultures in the world) think sickness is caused by evil spirits. Shamans, or medicine men or women, supposedly can effect cures because of their special relationships with these spirits and their knowledge of magic rituals. If the shamans are good at their jobs, they supposedly can persuade or force the evil spirit to leave the sick person or to discontinue exerting its harmful influence. In our own culture, some people consult mediums, spiritualists, and Ouija boards in an effort to contact the spirits and ghosts of departed loved ones.

Theism

Theism is *the belief in divine beings—gods and goddesses—who shape human affairs.* Gods are seen as powerful beings worthy of being worshipped.

Most theistic societies practice **polytheism,** *the belief in a number of gods.* Each god or goddess usually has particular spheres of influence such as childbirth, rain, or war, and generally one is more powerful than the rest and oversees the others' activities. In the ancient religions of Mexico, Egypt, and Greece, for instance, we find a pantheon, or a host, of gods and goddesses.

Monotheism

Monotheism is *the belief in the existence of a single god.* Only three religions are known to be monotheistic: Judaism and its two offshoots, Christianity and Islam. These three religions have the greatest number of believers worldwide (Table 13–1). However, even these faiths are not purely monotheistic. Christianity, for example, includes belief in such divine or semi-divine beings as angels, the devil, saints, and the Virgin Mary. Nevertheless, because all three religions contain such a strong belief in the supremacy of one all-powerful being, they are considered to be monotheistic.

Abstract Ideals

Some religions are based on abstract ideals rather than a belief in supernatural forces, spirits, or divine beings. **Abstract ideals** *focus on the achievement of personal awareness and a higher state of consciousness through correct ways of thinking and behaving, rather than by manipulating spirits or worshipping gods.* Such religions promote devotion to religious rituals and practices and adherence to moral codes of behavior. Buddhism is an example of a religion based on abstract ideals. The Buddhist's ideal is to become "one with the universe," not through worship or magic, but by meditation and correct behavior.

A Sociological Approach to Religion

When sociologists approach the study of religion, they focus on the relationship between religion and society. The functionalist sociologists have examined the functions that religion plays in social life, whereas conflict theorists have viewed religion as a means for justifying the political status quo.

The Functionalist Perspective

Since at least 60,000 years ago, as indicated by the Neanderthal burials at Shanidar Cave, religion has played a role in all known human societies. The question that interests us here is, what universal functions does religion have? Sociologists have identified four categories of religious function: satisfying individual needs, promoting social cohesion, providing a worldview, and helping adaptation to society.

Satisfying Individual Need   Religion offers individuals ways to reduce anxiety and to promote emotional integration. Although Sigmund Freud (1918, 1928) thought religion to be irrational, he saw it as helpful to the individual in coming to terms with impulses that induce guilt and anxiety. Freud argued that through a belief in lawgiving, powerful deities could help people reduce their anxieties by providing strong, socially reinforced inducements for controlling dangerous or immoral impulses.

Further, in times of stress, individuals can calm themselves by appealing to deities for guidance or even for outright help, or they can calm their fears by trusting in God. In the face of so many things that are beyond human control and yet may drastically affect human fortunes (such as droughts, floods, or other natural disasters), life can be terrifying. It is comforting to "know" the supernatural causes of both good fortune and bad. Some people attempt to control supernatural forces through magic rituals.

Perhaps this is why best-selling author Michael Crichton (a medical doctor and author of *Jurassic Park*) eats the same meal for lunch every day while working on a new novel and why former New York Giants football coach Bill Parcells would stop and buy coffee at two different coffee shops on his way to the stadium before every game.

Social Cohesion   Émile Durkheim, one of the earliest functional theorists, noted the ability of religion to bring about group unity and cohesion. According to Durkheim, all societies have a continuing need to reaffirm and uphold their basic sentiments and values. This is accomplished when people come together and communally proclaim their acceptance of the dominant belief system. In this way people are bound to one another, and as a result, the stability of the society is strengthened.

Not only does religion in itself bring about social cohesion, but often the hostility and prejudice directed at its members by outsiders also helps strengthen bonds between those members. For example, during the 1820s, Joseph Smith, a young farmer from Vermont, claimed that he had received visits from heavenly beings. He said the apparitions enabled him to produce a 600-page history, known as the *Book of Mormon*, of the ancient inhabitants of the Americas. Shortly after the establishment of the Mormon Church, Smith had a revelation that Zion, the place where the Mormons would prepare for the millennium, was to be established in Jackson County, Missouri. Within two years, 1,200 Mormons had bought land and settled in Jackson County. The other residents in the area became concerned about the influx, and in 1833 they published their grievances in a document that became known as the manifesto, or secret constitution.

They charged the Mormons with a variety of transgressions and pledged to remove them from Jackson County. Several episodes of conflict followed, which eventually forced the Mormons to move into an adjoining county. These encounters with a hostile environment produced a sense of collective identity at a time when it was desperately needed. The church was less than two years old and included individuals from diverse religious backgrounds. There was a great deal of internal discord, and without the unity that resulted from the conflict with the townspeople, the group might have disappeared altogether (MacMurray & Cunningham, 1973).

Durkheim's interest in the role of religion in society was aroused by his observation that religion, like the family, seemed to be a universal human institution. This universality meant that religion must serve a vital function in maintaining the social order.

© Richard Lord

Religious rituals fulfill a number of social functions. They bring people together physically, promote social cohesion, and reaffirm a group's beliefs and values.

© Royalty-Free/CORBIS

Émile Durkheim believed that when people recognize or worship supernatural entities, they are really worshipping their own society.

Durkheim felt that he could best understand the social role of religion by studying one of the simplest kinds—the totemism of the aboriginal Australian. A **totem** is *an ordinary object, such as a plant or animal, that has become a sacred symbol to and of a particular group or clan, who not only revere the totem but also identify with it*. Thus, reasoned Durkheim, religious symbols such as totems, as well as religion itself, arose from society itself, not from outside it. When people recognize or worship supernatural entities, they are really worshipping their own society. They do not realize that their religious feelings are actually the result of the intense emotions aroused when people gather together at a clan meeting (a crowd reaction), for example. They look for an outside source of this emotional excitement and may settle on a nearby, familiar object as the symbol of both their religion and their society. Thus, society—the clan—is the origin of the clan members' shared religious beliefs, which in turn help cement together their society.

Durkheim saw religious ritual as an important part of this social cement. Religion, through its rituals, fulfills a number of social functions: It brings people together physically, promoting social cohesion; it reaffirms the group's beliefs and values; it helps maintain norms, mores, and prohibitions so that violation of a secular law—murder or incest, for instance—is also a violation of the religious code and may warrant ritual punishment or purification; it transmits a group's cultural heritage from one generation to the

Figure 13–1 Society, Religion, and the Individual: A Functionalist View

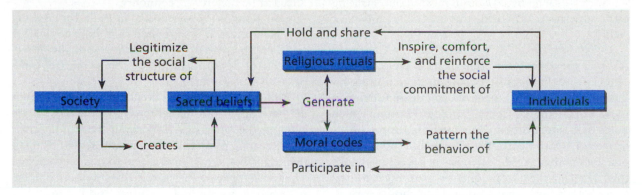

next; and it offers emotional support to individuals during times of stress and at important stages in their lives, such as puberty, marriage, and death (Figure 13–1).

In Durkheim's view, these functions are so important that even a society that lacks the idea of the sacred must substitute some system of shared beliefs and rituals. Indeed, some theorists see communism as such a system. Soviet communism had its texts and prophets (Karl Marx, Friedrich Engels), its shrines (V. I. Lenin's tomb), its rituals (May Day parade), and its unique moral code. Durkheim thought that much of the social upheaval of his day could be attributed to the fact that religion and ritual no longer played an important part in people's lives, and that without a shared belief system, the social order was breaking down.

Although many sociologists today take issue with Durkheim's explanation of the origins of religion based on totemism, they nevertheless recognize the value of his functional approach to understanding the vital role of religion in society.

Secular society depends on external rewards and pressures for results, whereas religion depends on the internal acceptance of a moral value structure. Durkheim believed that because religion is effective in bringing about adherence to social norms, society usually presents those norms as an expression of a divine order. For example, in ancient China, as in France until the late eighteenth century, political authority—the right to rule absolutely—rested securely on the notion that emperors and kings ruled because it was divine will that they do so—the divine right of kings. In Egypt, the political authority of pharaohs was unquestioned because they were more than just kings; they were believed to be gods in human form.

Religion serves to legitimize more than just political authority. Although many forms of institutionalized inequality do not operate to the advantage of the subgroups or individuals affected by them, they help perpetuate the larger social order and often are justified by an appeal to sacred authority. In such situations, although religion serves as a legitimator of social inequality, it does function to sustain societal stability. Thus, the Jews in Europe were kept from owning land and were otherwise persecuted because of the myth "they had killed Christ" (Trachtenberg, 1961); and even slavery has been defended on religious grounds. In 1700, Judge John Saffin of Boston wrote of the

> Order that God hath set in the world, who hath Ordained different degrees and orders of men, some to the High and Honorable, some to be Low and Despicable . . . yea, some to be born slaves, and so to remain during their lives, as hath been proved. (Montagu, 1964a)

Religions do not always legitimize secular authority. In feudal Europe, the church had its own political structure, and there often was tension between church and state. Indeed, just as the church often legitimized monarchs, it also excommunicated those who failed to take its wishes into account. However, the fact remains that religious institutions usually do dovetail neatly with other social institutions legitimizing and helping sustain them. For example, although church and state medieval Europe were separate structures and often conflicted, the church nonetheless played an important role in supporting the entire feudal system.

Establishing Worldviews According to Max Weber in his classic book *The Protestant Ethic and the Spirit of Capitalism*, religion responds to the basic human need to understand the purpose of life. In doing so, religion must give meaning to the social world within which life occurs. This means creating a worldview that can have social, political, and economic consequences. For instance, there is the issue of whether salvation can be achieved through active mastery (hard work, for example) or through passive contemplation (meditation). The first approach can be seen in Calvinism, and the second is evident in several of the

Eastern religions. Another major issue in creating a worldview is whether salvation means concentrating on a supernatural world, this world, or an inner world.

Using these ideas, Weber theorized that Calvinism fostered the Protestant ethic of hard work and asceticism and that Protestantism was an important influence on the development of capitalism. Calvinism is rooted in the concept of predestination, which holds that before they are born, certain people are selected for heaven and others for hell. Nothing anyone does in this world, Calvinists believe, can change this. The Calvinists consequently were eager to find out whether they were among those chosen for salvation. Worldly success—especially the financial success that grew out of strict discipline, hard work, and self-control—was seen as proof that a person was among the select few. Money was accumulated not to be spent but to be displayed as proof of one's position among the chosen. Capitalist virtues became Calvinist virtues. It was Weber's view that even though capitalism existed before Calvinist influence, it blossomed only with the advent of Calvinism.

Weber's analysis has been criticized from many standpoints. Calvinist doctrines were not as uniform as Weber pictured them, nor was the work ethic confined to the Protestant value system. Rather, it seems to have been characteristic of the times, promoted by Catholics and Protestants. Finally, one could just as well argue the reverse: that the social and economic changes leading to the rise of industrialism and capitalism stimulated the emergence of the new Protestantism—a position that Marxist analysts have taken. Today it is generally agreed that although religious beliefs did indeed affect economic behavior, the tenets of Protestantism and capitalism tended to support each other. However, the lasting value of Weber's work is his demonstration of how religion creates and legitimizes worldviews and how important these views are to human social and political life.

Adaptations to Society Religion can also be seen as having adaptive consequences for the society in which it exists. For example, many would view the Hindu belief in the sacred cow, which may not be slaughtered, as a strange and not particularly adaptive belief. The cows are permitted to wander freely and defecate along public paths.

Marvin Harris (1966) has suggested that there may be beneficial economic consequences in India from not slaughtering cattle. The cows and their offspring provide a number of resources that could not be provided easily in other ways. For example, a team of oxen is essential to India's many small farms. Oxen could be produced with fewer cows, but food production would have to be devoted to feeding those cows. With the huge supply of sacred cows,

although they are not well fed, the oxen are produced at no cost to the economy.

Cow dung is also necessary in India for cooking and as fertilizer. It is estimated that dung equivalent to 45 million tons of coal is burned annually. Alternative sources of fuel, such as wood or oil, are scarce or costly.

Although most Hindus do not eat beef, cattle that die naturally or are slaughtered by non-Hindus are eaten by the lower castes. Without the Hindu taboo against eating beef, these other members of the Indian hierarchy would not have access to this food supply. Therefore, because the sacred cows do not compete with people for limited resources and because they give birth to the oxen that are a cheap source of labor, fuel, and fertilizer, the taboo against slaughtering cattle may be quite adaptive.

When societies are under great stress or attack, their members sometimes fall into a state of despair analogous, perhaps, to that of a person who becomes depressed. Institutions lose their meaning for people, and the society is threatened with what Durkheim called anomie, or "normlessness." If this continues, the social structure may break down, and the society may be absorbed by another society, unless the culture can regenerate itself. Under these conditions, revitalization movements sometimes emerge. **Revitalization movements** are *powerful religious movements that stress a return to the traditional religious values of the past.* Many of them can be found in the pages of history and even exist today.

In the 1880s, for instance, the once free Plains Indians lived in misery, crowded onto barren reservations by soldiers of the U.S. government. Cheated out of the pitiful rations that had been promised them, they lived in hunger—and with memories of the past. Then a Paiute by the name of Wovoka had a vision, and he traveled from tribe to tribe to spread the word and demonstrate his Ghost Dance.

> Give up fighting, he told the people. Give up all things of the white man. Give up guns, give up European clothing, give up alcohol, and give up all trade goods. Return to the simple life of the ancestors. Live simply—and dance! Once the Indian people are pure again, the Great Spirit will come, all Indian ancestors will return, and all the game will return. A big flood will come, and after it is gone, only Indians will be left in this good time. (Brown, 1971)

Wovoka's Ghost Dance spread among the defeated tribes. From the Great Plains to California, Native American communities took up the slow, trancelike dance. Some believed that the return of the ancestors would lead to the slaughter of all whites. For others, the dance just rekindled pride in their heritage. For whatever reasons, the Ghost Dance could

GLOBAL SOCIOLOGY

The Worst Offenders of Religious Freedom

Religious freedom is a universal value that almost all of the world's nations give lip service to. In actual fact many countries severely restrict and deny people the ability to practice their religion. Even with widespread religious intolerance throughout the world, the United States has singled out nine of the worst offenders that it calls Countries of Particular Concern or CPCs. These countries stand out in their refusal to allow religious freedom. They include:

China—Officially an atheist country. Religious activities by groups that have not registered are illegal and may be punished. The government is particularly harsh on Tibetan Buddhists, Catholics faithful to the Vatican, underground Protestants Muslims, and Falun Gong. Many religious believers are in prison for their faith, and beatings, torture, and the destruction of places of worship have taken place.

North Korea—Religious freedom does not exist. Religious believers, particularly Christians, often face imprisonment, torture, or even execution. Christians have been imprisoned and tortured for reading the Bible and talking about God. Some reports indicate that Christians endured biological warfare experiments.

Burma—The government severely represses and violates religious freedom. Buddhist monks have been imprisoned. Christian clergy face arrest and the destruction of their churches. The government has destroyed some mosques and Muslims face considerable discrimination and state-sponsored violence.

Iran—Religious minorities such as Sunni Muslims, Baha'is, Mandaeans, Jews, and Christians face imprisonment, harassment, intimidation, and discrimination based on their religious beliefs.

Sudan—The government continues to attempt to impose strict Muslim law on non-Muslims in some parts of the country, and non-Muslims face discrimination and restrictions on the practice of their faith.

Eritrea—Since 2002 only four religious groups have been officially recognized. Activities by other religious groups are banned. Over 200 Protestant Christians and Jehovah's Witnesses have been imprisoned for their faith. Severe torture has been used to pressure believers to renounce their faith. Others have been detained and interrogated.

Saudi Arabia— Freedom of religion does not exist in Saudi Arabia. The government requires all citizens to be Muslim and prohibits all public manifestations of non-Muslim religions. Islamic practice generally is limited to that of a school of the Sunni branch of Islam as interpreted by Muhammad Ibn Abd Al-Wahhab, an eighteenth-century Arab religious reformer, and practices contrary to this interpretation are suppressed. Non-Muslim worshippers risk arrest, imprisonment, or deportation for engaging in religious activities.

Vietnam—Members of the Buddhist, Catholic, Protestant, Hoa Hao, and Cao Dai faiths are in prison for practicing their faith. Protestants have been subjected to physical abuse and pressured to renounce their faith. The government has closed hundreds of churches and places of worship.

Source: "The International Religious Freedom Report for 2004," Washington, D.C.: Department of State, Office of International Religious Freedom, September 15, 2004.

---

not be contained, despite the government's attempts to ban it.

On December 28, 1890, the people of a Sioux village camped under federal guard at Wounded Knee, South Dakota, and began to dance. They ignored orders to stop and continued to dance until someone suddenly fired a shot. The soldiers opened fire, and soon more than 200 of the original 350 men, women, and children were killed. The soldiers' losses were 29 dead and 33 wounded, mostly from their own bullets and shrapnel. This slaughter was the last battle between the Indians of the Great Plains and the soldiers of the dominant society (Brown, 1971). (For a discussion of religious tolerance in other countries, see "Global Sociology: The Worst Offenders of Religious Freedom.")

The Conflict Theory Perspective

Karl Marx asserted that the dominant ideas of each age have always been the ideas of the ruling class (Marx & Engels, 1961/1848), and from this it was a small step to his assertion that the dominant religion of a society is that of the ruling class, an observation that has been borne out by historical evidence. Marxist scholars emphasize religion's role in justifying the political status quo by cloaking political authority with sacred legitimacy and thereby making opposition to it seem immoral.

The concept of alienation is an important part of Marx's thinking, especially in his ideas of the origin and functions of religion. **Alienation** is *the process by which people lose control over the social institutions they themselves invented.* People begin to feel like strangers—aliens—in their own world. Marx further believed that religion is one of the most alienating influences in human society, affecting all other social institutions and contributing to a totally alienated world.

According to Marx, "Man makes religion, religion does not make man" (Marx, 1967/1867a). The function of God thus was invented to serve as the model of an ideal human being. People soon lost sight of this fact, however, and began to worship and fear the ideal they had created as if it were a separate, powerful supernatural entity. Thus, religion, because of the fear people feel for the god they themselves have created, serves to alienate people from the real world.

Marx saw religion as the tool that the upper classes used to maintain control of society and to dominate the lower classes. In fact, he referred to it as the "opiate of the masses," believing that through religion, the masses were kept from actions that might change their relationship with those in power. The lower classes were distracted from taking steps toward social change by the promise of happiness through religion. If they followed the rules established by religion, they expected to receive their reward in heaven, and so they had no reason to try to change or improve their condition in this world. These religious beliefs made it easy for the ruling classes to continue to exploit the lower classes: Religion served to legitimize upper-class power and authority. Although modern political and social thinkers do not accept all of Marx's ideas, they recognize his contribution to the understanding of the social functions of religion.

Although religion performs a number of vital functions in society—helping maintain social cohesion and control while satisfying the individual's need for emotional comfort, reassurance, and a worldview—it also has negative, or dysfunctional, aspects.

Marx would be quick to point out a major dysfunction of religion: Through its ability to make it seem that the existing social order is the only conceivable and acceptable way of life, it obscures the fact that people construct society and therefore can change society. Religion, by imposing the acceptance of supernatural causes of conditions and events, tends to conceal the natural and human causes of social problems. In fact, in this role of justifying or legitimating the status quo, religion may very well hinder much-needed changes in the social structure. By diverting attention from injustices in the existing social order, religion discourages the individual from taking steps to correct these conditions.

An even more basic and subtle dysfunction of religion is its insistence that only one body of knowledge and only one way of thinking are sacred and correct, thereby limiting independent thinking and the search for further knowledge.

Organization of Religious Life

Several ways of organizing religious groups are found in society.

The Universal Church

A **universal church** *includes all the members of a society within one united moral community* (Yinger, 1970). It is fully a part of the social, political, and economic status quo and therefore accepts and supports (more or less) the secular culture. In a preliterate society, in which religion is not really a separate institution but rather part of the entire fabric of social life, a person belongs to the church simply by being a member of the society. In more complex societies, the church cuts across divisions and binds all believers into one moral community. A universal church, however, does not seek to change the conditions of social inequality created by the secular society and culture, and indeed, it may even legitimize them. (An example is the Hindu religion of India, which used to perpetuate a rigid caste system.)

The Ecclesia

An **ecclesia** is *a church that shares the same ethical system as the secular society and has come to represent and promote the interest of the society at large.* Like the universal church, an ecclesia extends itself to all members of a society, but because it has so completely adjusted its ethical system to the political structure of the secular society, it comes to represent and promote the interests of the ruling classes. In this process, the ecclesia loses some adherents among the lower social classes, who increasingly reject it for membership in sects, be they sacred or "civil" (Yinger, 1970). An ecclesia is usually the official or national religion. For most people, membership is by birth, rather than conscious decision. Ecclesias have been common throughout human history. Examples include the Catholic Church in Spain and the Roman Empire; the Anglican Church, which is now the official Church of England; Islam in Saudi Arabia; and Confucianism, which was the state religion in China until early in the twentieth century.

The Denomination

A **denomination** *tends to limit its membership to a particular class, ethnic group, or religious group, or at least to have its leadership positions dominated by members of such a group.* It has no official or unofficial connection with the state, and any political involvement is purely a matter of choice by the denomination's leaders, who may either support or oppose any or all of the state's actions and political positions. Denominations do not withdraw themselves from the secular society. Rather, they participate actively in secular affairs and also tend to cooperate with other religious groups. These two characteristics distinguish them from sects, which are separatist and unlikely to be tolerant of other religious persuasions (Yinger, 1970). (For that matter, universal churches, by their very nature, also typically dismiss other religions.) In America, Lutheranism, Methodism, other Protestant groups, Catholicism, and Judaism embody the characteristics of a denomination.

The Sect

A **sect** is *a small group that adheres strictly to religious doctrine that often includes unconventional beliefs or forms of worship.* Sects generally represent a withdrawal from secular society and an active rejection of secular culture (Yinger, 1970). For example, the Dead Sea Scrolls show clearly that the beliefs of both early Christian and Jewish sects, such as the Essenes, were rooted in a disgust with society's self-indulgent pursuit of worldly pleasures and in a rejection of the corruption perceived in the prevailing religious hierarchy (Wilson, 1969).

Early in their development, sects often are so harsh in their rejection of society that they invite persecution. Some actually thrive on martyrdom, which causes members to intensify their fervent commitment to the faith. Consider, for example, the Christian martyrs in Rome before the conversion to Christianity of the Emperor Constantine. (For a discussion of the continuing evolution of religion, see "Social Change: Religion Is Constantly Changing.")

Millenarian Movements

Millenarian movements *typically prophesy the end of the world, the destruction of all evil people and their works, and the saving of the just.* Millenarian (from the Latin word for "thousand") prophecies often are linked with the symbolic number 1,000 or multiples thereof.

Throughout human history, religious leaders have emerged in times of stress, foretelling the end of the world and asking everyone to stop whatever they are doing to follow the bearers of the message.

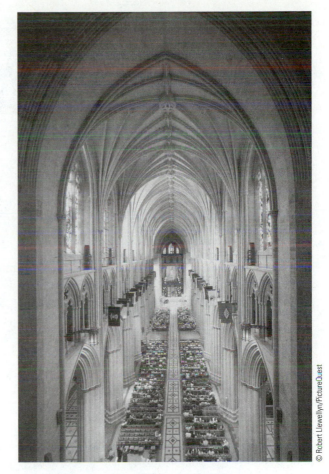

In the United States, Lutheranism and Methodism (and other Protestant groups), Catholicism, and Judaism all embody the characteristics of a denomination.

With the advent of the year A.D. 1000, Christendom in Medieval Europe was thrown into a panic by religious doomsday preachers who predicted that the end of the world was imminent. As a result, people abandoned their homes and crops, and mobs of the devout took refuge in churches or fled on pilgrimages to the Holy Land.

Similar predictions occurred with the advent of the year 2000. Many Christian groups believed the new millennium would see monumental events predicted in the New Testament book of Revelation, which predicts a thousand-year kingdom of peace and plenty and a new heaven and Earth.

Why would anyone want to believe that the end of the world is near? Part of the answer is that the apocalypse usually does not mean pure destruction, but rather the destruction of evil and the victory of good. Believers feel they are the center of the universe and the cosmic drama is coming to its ultimate conclusion in their time. Life is filled with meaning.

Apocalyptic beliefs also appeal to people who feel that life has not been fair to them. One of the great appeals of apocalyptic views is that they help to

SOCIAL CHANGE

Religion Is Constantly Changing

It's tempting to conceive of the religious world as being made up primarily of a few well-delineated and static religious blocs: Christians, Jews, Muslims, Buddhists, Hindus, and so on. But that's dangerously simplistic. It assumes a stability in the religious landscape that is completely at odds with reality. New religions are born all the time. Old ones transform themselves dramatically. Schism, evolution, death, and rebirth are the norm.

And this doesn't apply only to religious groups that one often hears referred to as cults. Today, hundreds of widely divergent forms of Christianity are practiced around the world. Islam is usually talked about in monolithic terms (or, at most, in terms of the Shia-Sunni divide), but one almost never hears about the 50 million or so members of the Naqshabandiya order of Sufi Islam, which is strong in Central Asia and India, or about the more than 20 million members of various schismatic Muslim groups around the world. Think, too, about the strange rise and fall of the Taliban. Buddhism, far from being an all encompassing glow radiating benignly out of the East, is a vast family of religions made up of more than 200 distinct bodies, many of which don't see eye-to-eye at all. Major strands of Hinduism were profoundly reshaped in the nineteenth century, revealing strong Western and Christian influences.

There's no reason to think that the religious movements of today are any less subject to change than were the religious movements of hundreds or even thousands of years ago. History bears this out. Early Christianity was deemed pathetic by the religious establishment: Pliny the Younger wrote to the Roman Emperor Trajan that he could get nothing out of Christian captives but "depraved, excessive superstition." Islam, initially the faith of a band of little-known desert Arabs, astonished the whole world with its rapid spread.

Protestantism started out as a note of protest nailed to a door. In 1871 Ralph Waldo Emerson dismissed Mormonism as nothing more than an "after-clap of Puritanism." Up until the 1940s Pentecostals were often dismissed as "holy rollers," but today the *World Christian Encyclopedia* suggests that by 2050 there may be more than a billion people affiliated with the movement. The implication is clear: What is now dismissed as a fundamentalist sect, a fanatical cult, or a mushy New Age fad could become the next big thing.

Source: Excerpted from "Oh, Gods!" by Toby Lester, *The Atlantic Monthly*, February 2002, pp. 37–45. Used by permission of the author.

---

make sense of the world. All events start to have significance. But they also can lead to conspiracy theories because every detail is linked to a much larger drama in which good and evil are at odds. There is no such thing as chance.

What happens when the chosen date for the end of the world or other millennial event finally arrives —and nothing happens? Some people may just accept it and move on. More often, they will try to find a way to prove that their prophecy did come true, but not exactly in the way predicted. Others may reset their apocalyptic clocks to another date.

We should not be too quick to dismiss millennial movements. Christopher Columbus believed the world would end in 1650. He considered his discovery of the "New World" part of a divine plan to establish a millennial paradise. "God made me the messenger of the new heaven and the new Earth of which he spoke in the Apocalypse of St. John," Columbus wrote in his journal, "and he showed me the spot where to find it" (Sheler, 1997). Many major religious movements received attention because of

apocalyptic predictions. Ironically, whereas the apocalypse that millennialists prophecy may not come true, they often succeed: The world is a different place even if the catastrophic event does not take place. In many ways it was the end of the world as it had been known before.

Aspects of American Religion

The Pilgrims in 1620 sought to build a sanctuary where they would be free from religious persecution, and the Puritans who followed 10 years later intended to build a community embodying all the virtues of pure Protestantism, a community that would serve as a moral guide to others.

Thus, religion pervaded the social and political goals of the early English-speaking settlers in America and played a major role in shaping colonial society. Today, the four main themes that characterize religion in America are religious diversity, widespread belief, secularism, and ecumenism.

Religious Diversity

The United States has always been a land of many religions. The European settlers encountered a wide diversity of native religions. The early immigrants included British Anglicans, Spanish and French Catholics, and Quakers. The Chinese and Japanese who came to work on the West Coast brought with them Buddhist, Taoist, and Confucian traditions. European Jews and Irish and Italian Catholics arrived in large numbers with the nineteenth-century immigration. The past 30 years have seen an even greater expansion in American religious diversity. Buddhists have come from Thailand, Vietnam, and Cambodia; Hindus from India and East Africa; and Muslims from Indonesia, Bangladesh, and the Middle East. Immigrants from Haiti and Cuba have brought religious traditions that blend Catholic and African beliefs. Even though the United States continues to be a predominantly Christian country, it has become the world's most religiously diverse nation (Eck, 2001).

Widespread Belief

Americans generally take religion for granted. Although the religious affiliation and degree of church attendance differ widely, almost all Americans (92%) claim to believe in God. Nine of every 10 Americans have a religious preference and say that they attend services on at least some occasions. Two-thirds of Americans maintain an affiliation with a church or synagogue and 6 in 10 consider religion to be of high importance in their personal lives (Gallup Poll, August 2000). Americans also attend church or synagogue regularly, with 44% claiming to do so each week (CBS News Poll, July 2005).

Evidence as to whether America is experiencing a religious revival, as some have claimed, is contradictory. When asked in 2000 if they think religion's influence on American life is increasing or decreasing, 35% of respondents said it was increasing, and 58% said it was decreasing. This contrasts with a poll taken in 1957 when 69% said religion was increasing in influence and 14% said it was decreasing. Yet nearly half of Americans are devout practitioners of their faith, as noted by attending church or synagogue at least once a week. Two-thirds believe the Bible answers all or most of the basic questions of life (Gallup Poll, March 2000). More than three-fourths of those Americans who believe in heaven think that their chances of getting there are good or excellent (Gallup Poll, June 2001). By and large, and despite dire warnings about the erosion of religiosity because of new ways of thinking and innovative lifestyles, for most Americans religion is a very important part of their lives. (See Table 13–2 for a comparison of how people in various countries see America's religiosity.)

Although Americans differ widely in religious affiliation and degree of church attendance, almost all Americans claim to believe in God.

© Najlah Feanny/CORBIS

Table 13–2

What Others Say about American Religiosity

	People in the United States Are:	
	Too Religious	**Not Religious Enough**
United States	21	58
France	61	26
Netherlands	57	25
Great Britain	39	28
Germany	39	31
Canada	35	38
India	32	57
Spain	31	40
Russia	27	38
Poland	6	56
Jordan	*	95

Source: *Pew Global Attitudes Survey,* June 23, 2005.

NEWS YOU CAN USE

A Nation of Believers

The American people are truly a nation of believers with the vast majority of people believing in God, heaven, miracles, and angels (Table 13–3). In addition, belief in the devil has increased in the last few years from 63% in 1997 to 71% in 2004. Women are more likely to be believers in all the categories than men and younger people are more likely to be believers than older people. There appear to be political differences in beliefs also, with Republicans more likely than Democrats to believe in heaven, hell, and the devil. Democrats are more likely to believe in reincarnation, astrology, and ghosts. The majority of both groups believe religion should play a larger part in people's lives.

Table 13–3

American Believers

Category	Percentage Who Believe In
God	92%
Heaven	85%
Miracles	82%
Angels	78%
Hell	74%
The Devil	71%
Ghosts	34%
Witches	24%

Source: Dana Blanton, "More Believe in God Than Heaven." Opinion Dynamics Corporation/Fox News Poll, June 18, 2004.

More than half of all religiously affiliated individuals belong to a Protestant denomination, clearly reflecting America's colonial history. However, other denominations are also well represented, especially Catholicism and Judaism. There are more than 200 formally chartered religious organizations in America today. Such pluralism is not typical of other societies and has resulted primarily from the waves of European immigrants who began to arrive in the postcolonial era. Americans' traditional tolerance of religious diversity can be seen as a reflection of the constitutional separation of church and state, so that no one religion is recognized officially as better or more acceptable than any other. (For more on American beliefs, see "News You Can Use: A Nation of Believers.")

Secularism

Many scholars have noted that modern society is becoming increasingly **secularized,** that is, *less influenced by religion.* Religious institutions are being confined to ever-narrowing spheres of social influence, while people turn to secular sources for moral guidance in their everyday lives (Berger, 1967). This shift is reflected in the lack of religious knowledge of Americans, who for the most part are notoriously indifferent to, and ignorant of, the basic doctrines of their faiths.

Of course, social and political leaders still rely on religious symbolism to influence secular behavior. The American Pledge of Allegiance tells us that we are "one nation, under God, indivisible," and our currency tells us that "In God We Trust." Since the turn of the century, however, modern society has turned increasingly to science, rather than religion, to point the way. Secular political movements have emerged that attempt to provide most, if not all, of the functions that religion traditionally fulfilled. For example, communism prescribes a belief system and an organization that rivals those of any religion. Like religions, communism offers a general concept of the nature of all things and provides symbols that, for its adherents, establish powerful feelings and attitudes and supply motivation toward action. Thus, some political movements lack only a sacred or supernatural component to qualify as religions. In this increasingly secular modern world, however, sacred legitimacy appears to be unnecessary for establishing meaning and value in life.

Ecumenism

Ecumenism refers to *the trend among many religious communities to draw together and project a sense of unity and common direction.* It is partially a response to secularism and is a tendency evident among many religions in the United States.

Unlike religious groups in Europe, where issues of doctrine have fostered sectlike hard-line separatism among denominations, most religious groups in America have focused on ethics—that is, how to live an ethical and moral life. There is less likelihood of disagreement over ethics than over doctrine. Hence, American Protestant denominations typically have had rather loose boundaries, with members of congregations switching denominations rather easily and churches featuring guest appearances by ministers of other denominations. In this context, ecumenism has flourished in the United States far more than in Europe.

Major Religions in the United States

Nowhere is the diversity of the American people more evident than in their religious denominations. There are hundreds of different religious groups in the United States, and they vary widely in practices, moral views, class structure, family values, and attitudes. A recent survey found surprisingly large and persistent differences among even the major religious groups.

The U.S. census is prohibited from asking about religion, so the U.S. government generally has little to say on the matter. However, since 1972 the National Opinion Research Center has been conducting the General Social Surveys, which do give us a way of examining American religious attitudes and practices. This group has correlated information on a variety of issues with religious affiliation. Some of its findings are summarized here.

It is useful to think of American Protestant religious denominations as ranked on a scale measuring their degree of traditionalism. Conservative Protestant denominations include the fundamentalists (Pentecostals, Jehovah's Witnesses, and so on), Southern Baptists, and other Baptists. The moderates include Lutherans, Methodists, and inter- or nondenominationalists. Liberal Protestants are represented by Unitarian Universalists, Congregationalists, Presbyterians, and Episcopalians. This distinction among Protestants is important because the various branches often differ so markedly in their attitudes, especially toward social issues, that they resemble other religions more than the various denominations of their own. For example, in June 2000, the Southern Baptist Convention—America's largest Protestant denomination, with 15.8 million members—passed a new Baptist statement of faith saying that the Bible is without error and women should not be pastors (Rawls, 2000). This is at a time when 71% of Americans are in favor of having women pastors, ministers, priests, or rabbis (Gallup Poll, August 2000).

With respect to the hereafter, 79% of Americans believe that a day will come when God judges whether one goes to heaven or hell (Gallup Poll, August 2000). Nearly 90% of fundamentalists and Baptists believe in an afterlife. This falls to 80% among the moderate and liberal denominations. Catholics are similar to liberal Protestants in that 75% believe in an afterlife. Among people with no religious affiliation, 46% believe in an afterlife.

A strong belief in sin is typical of fundamentalists and Baptists. This causes them to condemn extramarital and premarital sex, homosexuality, and to favor outlawing pornography. There is greater sexual permissiveness among the moderate denominations and considerably more among the liberal denominations. Catholics tend to resemble the Protestant moderates, and Jews tend to be more liberal than the liberal Protestant denominations in this area. Attitudes toward drugs and alcohol follow the same pattern, with smoking, drinking, or the frequenting of bars least common among fundamentalists and Baptists.

There are substantial class differences among the major denominations. Jews and Episcopalians have the highest median annual household incomes and Baptists have the lowest. The pattern is the same for occupational prestige and education, with Jews and Episcopalians averaging three more years of education than fundamentalists and Baptists.

Given the wide differences in values and attitudes among religious groups, the relative proportion of the population that belongs to each group helps determine the shape of society. Protestants make up about 52% of the adult population. Among the major Protestant groups, the largest is composed of Baptists, who account for 18% of the adult population. Second are the Methodists with 9% (George W. Bush is the first Methodist to be elected U.S. president since William McKinley, who served from 1897 to 1901), next are the Lutherans with 7%, followed by Episcopalians at 3%, and the Church of Christ at 2% (Gallup Poll, March 2000).

Roman Catholics, representing 27% of the population, make up the largest single religious denomination. Jews represent 2% and Mormons 1.5% (72% live in Utah) followed by Muslims and a host of other religions such as Eastern Orthodox, Hindu, Sufi, and Baha'i.

These percentages are in constant flux, however, because demographic factors such as birthrates and migration patterns may influence the numbers of people in any given religion. Religious conversion can also affect these numbers. Fundamentalism, for example, is gaining among the young and winning converts.

Despite trends toward ecumenicalism, it seems that the magnitude of religious differences, the persistence of established faiths, and the continual

TECHNOLOGY AND SOCIETY

Seeking God on the Web

Nearly two-thirds of Internet users in the United States have looked for religiously oriented material online (Table 13–4). These people send e-mails with spiritual content; look for news accounts of religious events; look for information on how to celebrate religious holidays; and seek places to attend religious services.

The typical American who uses the Internet for religious purposes is likely to be white, female, college educated, and with a higher income than the average population. Many are evangelical Christians.

Though the link between religion and technology may seem new, the phenomenon actually goes back to ancient times. People have always found innovative technologies to communicate their religious views. The Sumerians etched their beliefs in stone. The Egyptians wrote on papyrus scrolls. In the first century A.D., Old Testament scrolls were turned into primitive books called codices. Use of these codices gave Christianity, in its beginnings, a technological advantage over Roman paganism. For the next millennium cloistered Christian monks toiled for years transcribing copy after copy of the Bible into leather-bound books. The process was a painfully slow one. A monk would slowly copy a page from a Latin Bible, and after that he and his brothers would ink and gild elaborately illustrated pages, completing only about one each day. These elaborate Bibles, as they circulated around Europe, opened the door for a standard way to distribute information (Ramo, 1996).

Attempting to use these handwritten manuscripts to convert people of other faiths was extraordinarily difficult. These Bibles were fragile, rare, and only in Latin. Proselytizing using the written word didn't become easier until Johannes Gutenberg invented the moveable-type press. And appropriately, the first book he produced was the Bible. His simple invention revolutionized the way ideas were communicated. Before the printing press, ideas could only be shared in person, but they now could be spread easily over all of Europe. There were fewer than 30,000 books in all of Europe when the first Bible was produced by Gutenberg's press. Fifty years later, there were 9 million mostly religiously oriented books (Ramo 1996).

Table 13–4

American Faithful Online

- 64% of the nation's 128 million Internet users have done things online that relate to religious or spiritual matters.

- Those who use the Internet for religious or spiritual purposes are more likely to be women, white, middle aged, college educated, and relatively well-to-do.

- The "online faithful" are devout and they use the Internet for *personal* spiritual matters more than for traditional religious functions or work related to their churches. But their faith activity online seems to augment their already strong commitments to their congregations.

- 26% of the online faithful seek information about the religious faith of others. Most are doing this out of curiosity.

- The majority of online faithful describe themselves as "spiritual and religious."

- Evangelicals are among the most fervent Internet users for religious and spiritual purposes.

Sources: Hoover, Stewart M., Lynn Schofield Clark, and Lee Rainie. *Faith Online: 64% of Wired Americans Have Used the Internet for Spiritual and Religious Purposes.* Washington, DC: Pew Internet & American Life Project, April 7, 2004; Ramo, J. C. "Finding God on the Web," December 16, 1996, *Time*, pp. 60–64, 66–67.

---

development of new faiths will ensure that this pattern of religious diversity will continue (Smith, 1984). (For a discussion of how religions have adapted to the Internet, see "Technology and Society: Seeking God on the Web.")

Protestantism

The United States is a majority Protestant country. That is changing, however, and the percentage of the population that is Protestant has been falling. In 1993, 63% of Americans were Protestant. By 2002, that percentage was down to 52%. There are a number of reasons for why this is happening. Part of it is caused by immigration that has been largely from Asian and Latin American countries in recent decades. A second reason is the number of people raised Protestant and remaining so in adulthood has declined to 83%. In addition, the percentage of people claiming no religion has jumped from 9% to 14% in the last decade (NORC Public Affairs, 2004).

Because American Protestantism is so fragmented, many sociologists simply have classified all non-Catholic Christian denominations in the general category of Protestant. However, differences exist among the various denominations. Membership in the more liberal denominations has been dropping and the more conservative ones have increased their membership. Over the past 10 years membership in the Presbyterian Church has dropped by almost 12%, the United Church of Christ dropped by 15%, and the United Methodist Church, the third largest U.S. Christian denomination, lost 7% of its membership. At the same time, membership in the more conservative denominations has increased dramatically, with The Church of Jesus Christ of Latter-day Saints (Mormons) growing by 19.3%, to 4.2 million. There have also been rapid increases in evangelical groups such as the Churches of Christ, at 18.6%, and Assemblies of God, the largest Pentecostal denomination, at 18.5%. Pentecostal churches emphasize "gifts of the Holy Spirit," which include healing and speaking in tongues. Some of the growth in these churches is caused by the increase in ethnic communities (Lampman, 2002).

Since the late 1980s, fundamentalist and evangelical Christians have become a visible and vocal segment of the Protestant population, and their presence has been felt through the media and through their support of political candidates. Why are the fundamentalist and evangelical churches gaining in popularity? Some of their appeal may lie in the sense of belonging and the comfort they offer through their belief in a well-defined and self-assured religious doctrine—no ambiguities and, hence, few moral choices to be made.

The growth of these churches reflects religion's role as a social institution, changing over time and from place to place, partly in response to concurrent social and cultural changes and partly itself acting as an agent of social change.

Catholicism

Catholics represent the largest single religious denomination in the United States, with 62 million believers in about 22,000 congregations. That is a 16% increase since 1990.

The Catholic population, which has traditionally been based in the Midwest and Northeast, has started to move to the South and Southwest. Rhode Island is the "most Catholic state" with almost 52% of the population being Catholic. The Catholic Church and Evangelical churches have been competing for the allegiance of the growing Hispanic community ("Religious Congregations," 2002).

One of the most striking things about Catholics in the United States is their youth: Twenty-nine percent are younger than 30, 36% are between 30 and 49, and 35% are older than 50. In contrast, 24% of Protestants are younger than 30, and 41% are older than 50. The higher birthrates among Catholics in the baby-boom generation and among Hispanic Catholics

account for a large part of this difference. Another part of the explanation is the difficulty that mainline Protestant denominations have had in retaining young people.

American Catholics have long been an immigrant people, and that tradition is continuing. One in five Catholics is a member of a minority group. Hispanics now make up 16% of American Catholics. Another 3% are black, and an additional 3% describe themselves as nonwhite. Very few Hispanics identify themselves as nonwhite, so the data suggest that the influx of Catholic immigrants from Southeast Asia is starting to show in national surveys. Among Protestants, 14% are black and 2% are Hispanic. This means that although the percentage of blacks among Protestants is five times that among Catholics, a higher percentage of Catholics overall come from minority groups.

Since the mid-1960s, Catholics have equaled Protestants in levels of education and income. The overall figures for Protestants mask significant differences between denominations. When we compare Catholics with individual Protestant denominations, we find them still ranking behind Presbyterians and Episcopalians, about on a par with Lutherans and Methodists, and well ahead of Baptists on scales that measure education and income. This comparison is striking in light of the fact that large numbers of lower-income minorities are included in the overall Catholic figures.

Catholics remain an urban people, with only one in four living in rural areas. A higher percentage of Catholics (39%) than of any other major denomination live in central cities, and 35% live in suburbs. The vast majority of Catholics are concentrated in the Northeast and Midwest.

Catholics have historically favored larger families than have other Americans, but by 1985 the difference in ideal family size between Catholics and Protestants had disappeared, with both groups considering two children the ideal. Despite the Catholic Church's condemnation of artificial means of birth control, American Catholics have favored information about and access to contraceptives in the same proportion as the rest of the population since the 1950s (Gallup & Castelli, 1987).

One of the most important developments in the recent history of Catholicism was the ecumenical council called by Pope John XXIII (Vatican II), which met from 1962 to 1965 and thoroughly reexamined Catholic doctrine. This ecumenical council led to many changes, often referred to as liberalization, including the substitution of common language for Latin in the Mass. One unintended consequence (or latent function) of Vatican II was that the centralized authority structure of the Catholic Church was questioned. Laypeople and priests felt free to dispute the

At the Wailing Wall in Jerusalem, thousands of Jews gather each day to pray and mourn the destruction of the Second Temple by the Romans in 70 A.D.

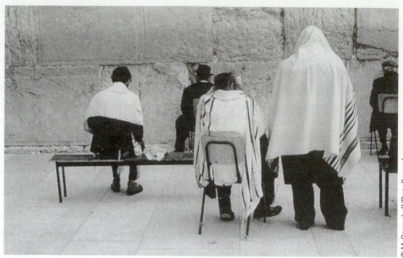

© M. Oppersdorff/Photo Researchers

doctrinal pronouncements of bishops and even of the pope. What began in the 1960s with a seemingly modest effort of reform has ended with every aspect of Catholic tradition under question. America's biggest single denomination now consists of traditional Catholics and "cafeteria Catholics," who pick and choose what to practice.

Surveys show that "people who identify as Catholic are more liberal on sexual morals than Protestants as a whole." But both Protestants and Catholics tend to have similar birthrates and opinions on abortion (Ostling, 1999).

Under the leadership of Pope John Paul II, the Catholic Church took a more conservative turn. It continued to condemn all forms of birth control except the rhythm method and rejected high-technology aids to conception, such as artificial insemination, in vitro fertilization, or surrogate motherhood. Calls for a greater role for women in the church or their ordination to the priesthood were rejected. Women were told to seek meaning in their lives through motherhood and giving love to others.

Judaism

There is a strong identification among Jews on both cultural and religious levels. This sense of connectedness is an important factor in understanding current trends within the religion.

Jews can be divided into three groups on the basis of the manner in which they approach traditional religious precepts. Orthodox Jews observe traditional religious laws very closely. They maintain strict dietary laws and do not work, drive, or engage in other everyday practices on the Sabbath. Reform Jews, by contrast, allow for major reinterpretations of religious practices and customs, often in response to changes in society. Conservative Jews represent a compromise between the two extremes. They are

less traditional than the Orthodox Jews are but not as willing to make major modifications in religious observance as the Reform Jews. In addition, a large secularized segment of the Jewish population still identifies itself as Jewish but refrains from formal synagogue affiliation.

As among Protestants, social-class differences exist among the various Jewish groups. Reform Jews are the best educated and have the highest incomes. For Orthodox Jews, religious rather than secular education is the goal. They have the lowest incomes and the least amount of secular education. As might be expected, Conservative Jews are situated between these two poles.

The state of Israel has played a major role in shaping current Jewish thinking. For many Jews, identification with Israel has come to be a secular replacement for religiosity. Support for the country is tied to many deep psychological and emotional responses. To many, Israel and its continued existence represent a way of guaranteeing that never again will millions of Jews perish in a holocaust. The country is seen as a homeland that can help defend world Jewry from the unwarranted attacks that have occurred throughout history. For many Jews, identification with, and support for, Israel is important to the development of cultural or religious ties.

The Jewish community has had to deal with the issue of ordaining women rabbis. Reform Jews have moved in this direction without a great deal of difficulty, and today women lead Reform congregations around the country. However, the issue of women rabbis initially produced a bitter fight among Conservative Jews. Today, an increasing number of Conservative women rabbis have become the heads of congregations. Orthodox Jews have not had to address the issue, because for them the existence of a female rabbi would represent too radical a departure from tradition to be contemplated.

According to the Bible, God told the Jewish people to be fruitful and multiply. In the United States, it seems the group is not doing so. Recent surveys estimate that 5.2 million Jews live in the United States, a decline from 5.5 million in 1990 (United Jewish Communities, 2002). Demographers predict that the population will continue to decline. The American Jewish community is thus facing a crisis.

The reasons for the lack of growth in the Jewish community are varied. For one thing, Jews in the United States are not bearing enough children to replace themselves: The Jewish birthrate is 1.8 children per woman per lifetime—well below the replacement rate of 2.1. The Jewish population is also quite old, with a median age of 41 and about 19% of American Jews being older than 65. Jewish immigration from the former Soviet Union and Israel has increased the population somewhat but is unpredictable. Finally, substantial numbers of young Jews are choosing to marry outside the faith, although this does not necessarily lead to a loss of Jewish identity.

Whether population erosion is occurring, and what should be done about it, is an ongoing debate in the American Jewish community. Jewish religious groups have attempted to liberalize the definition of who is a Jew, and—in a radical break with tradition—have decided to seek converts.

A common fear is that a further decline in the number of American Jews would lessen their ability to defend their political interests. Others suggest that the United States would lose the contributions of Jewish scientists, artists, and performers. Even though Jews account for about 2% of the U.S. population (Harris Poll, 2000), they make up 20% of America's Nobel laureates.

Although many of the reasons for the shrinking Jewish population are demographic, the issue of how to stem the tide has been the cause of major rifts among the branches of the religion. Recent statistics, for example, put the rate of Jewish intermarriage at 50%. Jewish opinion is not uniform on the subject but shows a sharp division between Orthodox Jews, the most strictly observant branch of Judaism, whose members are a small minority of the American Jewish population, and non-Orthodox. "For example, 64% of Orthodox Jews surveyed said they strongly disapproved of interfaith marriages, as opposed to 15% of Conservative Jews, 3% of Reform Jews, and 2% of those who identified themselves as 'just Jewish'" (Niebuhr, 2000). Interfaith couples have little difficulty today finding rabbis willing to officiate at the marriage ceremony.

A subject that has provoked even more controversy is the Reform movement's break with the tradition of matrilineal, or motherly, descent. In 1983, the Reform movement declared that people could be considered Jewish if either the father or mother was Jewish.

Before that time, Judaism could be passed on to the child only from the mother.

American Jews also differ with respect to converts. The Reform movement's outreach goes against centuries of Jewish tradition in which proselytizing was disdained. The outreach program is quite low-key. There is no advertising or airwave sermonizing, which is more common with Christian evangelical movements. Instead, outreach sessions resemble comparative religion discussion groups, in which introductory classes in Judaism are available to interfaith couples, as well as converts.

Islam

Islam, an Arabic word that means *surrender* or *submission* is the name given to the religion preached by the Prophet Muhammad, who taught from A.D. 612 until his death in 632. A Muslim is someone who has accepted the Islamic declaration of faith, or *shahadah,* that "There is no god but Allah, and Mohammed is His Prophet." Like Jews and Christians, Muslims are monotheists, and all three religions share the prophets of the Old Testament.

From its roots in the Arab world, Islam has spread in virtually every direction. Islam is the world's second largest religion behind Christianity, with more than 1.1 billion followers. Almost one-fourth of the world population may be Muslim in the near future if current high levels of fertility continue. Today, Muslims live in every country in the world. Although Islam began in Arabia, more than half of the world's Muslims live in South and Southeast Asia. The countries with the largest Muslim populations are Indonesia, India, Bangladesh, and Pakistan. About one-fourth of all Muslims live in the Middle East.

The formal acts of worship called the Five Pillars of Islam provide the framework for all aspects of a Muslim's life. The pillars consist of (1) *shahadah,* or faith, (2) prayer, (3) almsgiving, (4) fasting, and (5) pilgrimage.

There are three historic divisions in Islam. The great majority of Muslims belong to the Sunni division. They follow a traditional interpretation of Islam. Most of the conservative Muslims that Westerners call fundamentalists are Sunnis. The next largest division is the Shiah i-Ali, whose members are called Shii Muslims or Shiites. Shiites honor Ali, the cousin and son-in-law of Muhammad, and Ali's descendants, whom they believe should be the leaders of the Muslim community. The Kharijites make up the smallest division of Islam. Their name is based on an Arabic word that means *secessionists.* They were former followers of Ali who broke away in 657. Kharijites are strict Muslims whose beliefs are based on precise adherence to the teachings of the Quran,

Islam is the third largest faith in the United States. Despite estimates of millions of U.S. Muslims, there are only a thousand mosques and community centers to serve them.

or Muslim scripture, and *sunnah* (customs) as their community interprets them (Cornell, 2001).

Among the five most populous nations, the United States has by far the fewest Muslims. It is, however, the third largest faith in the United States, and there are already more Muslims than Presbyterians nationwide. Nonetheless, it is difficult to ascertain how many Muslims live in the United States. The census does not collect reliable data on religious preference. Some data do exist on church attendance, but Islam does not require attendance for formal membership in the way that many Christian groups do. Estimates of the number of Muslims in the United States stand at about 2.8 million. Yet despite the estimates of millions of U.S. Muslims, there are only a thousand mosques and community centers to serve them. Islam faces the same challenge as Judaism faced with the influx from Eastern Europe toward the end of the nineteenth century. New-style congregations had to be invented, new buildings built, and new schools started to train a new type of rabbi. U.S. Islam is just beginning to create the communal organizations that have served Judaism so well. There is no coherent association of mosques to unite immigrants with native-born blacks, and national organizations of other types are new. Muslims are divided the way Protestants have been, by ethnicity, race, and language.

Only in 1996 did U.S. Muslims establish a school for training clergy at the graduate level to parallel the Jewish and Christian seminaries (Ostling, 1999).

About one in four Muslims in the United States is African American. African American Muslims are of two types—"those who follow mainstream Islamic doctrine and those who follow the teachings of the Nation of Islam. This is an important distinction because the Nation of Islam's teachings are very different from those in the Quran" (El-Badry, 1994).

The American public has a more favorable view of Muslim Americans than Islam as a religion. Fifty-one percent of the public has a favorable view of Muslim Americans, but only 39% has a favorable view of Islam.

Muslims tend to be socially conservative, favor a close-knit family, and support religious education. The conservatism is based partly on religious beliefs but also arises from cultural mores. Samia El-Badry (1994) explains that to some Muslims "sexual permissiveness is seen as a reflection on the family, rather than on the individual. The family is seen as the ultimate authority and is therefore responsible for the individual's behavior." Many Muslims also believe that the eldest man is the head of the family.

Muslims living in America often have difficulty reconciling the American way of life with the traditions and ideas about morality that exist within their belief. Immigrant Muslim parents are often at odds with their Americanized offspring about the use of alcohol, which is banned in Islam; dating, which is forbidden; and lack of respect for elders (El-Badry, 1994).

All religions and denominations are affected by the current mood of the country. A heightened social consciousness results in demands for reform, whereas stressful times often produce a movement toward the personalization of religion. In any event, though traditional forms and practices of religion may be changing in the United States, religion itself is likely to continue to function as a basic social institution.

Social Correlates of Religious Affiliation

Religious affiliation seems to be correlated strongly with many other important aspects of people's lives: Direct relationships can be traced between membership in a particular religious group and a person's politics, professional and economic standing, educational level, family life, social mobility, and attitudes toward controversial social issues. For example, Jews are proportionally the best-educated group; they also have higher incomes than Christians in general; and a greater proportion are represented in business and the professions. Despite their high socioeconomic and educational levels, Jews, like Catholics, occupy relatively few of the highest positions of power in the corporate world and politics: These fields generally are dominated by white Anglo-Saxon Protestants.

Attitudes toward social policy also seem to be correlated, to some extent, with religious affiliation. The fundamentalist and evangelical Protestant sects generally are more conservative on key issues than are the major Protestant sects.

Although it is clear that religious associations show definite correlations with people's political, social, and economic lives, we must be careful not to ascribe a cause-and-effect relationship to such data, which at most can be considered an indicator of an individual's attitudes and social standing.

The social and political correlates of religious affiliation have had a significant effect on the directions of the various religious denominations and sects in the United States.

SUMMARY

- Religion is a system of beliefs, practices, and philosophical values shared by a group of people that defines the sacred, helps explain life, and offers salvation from the problems of human existence.
- Religion is one of society's most important institutions.
- Durkheim observed that all religions divide the universe into two mutually exclusive categories: The profane consists of empirically observable things, or things knowable through common, everyday experiences. In contrast, the sacred consists of things that are awe inspiring and knowable only through extraordinary experience.
- Almost anything may be designated as sacred. Sacred traits or objects symbolize important shared values.
- Patterns of behavior or practices that are related to the sacred are known as rituals.
- All religions have formalized social rituals, but many also feature private rituals such as prayer, which is a means for individuals to address or communicate with supernatural beings or forces.
- One of the functions of ritual and prayer is to produce an appropriate emotional state.
- All religions endorse a belief system that usually includes a supernatural order and a set of values to be applied to daily life.
- Sociologists focus on the relationship between religion and society.
- Functionalists examine the utility of religion in social life. Religion, they say, offers individuals ways to reduce anxiety and promote emotional integration.
- Functionalists also believe religion provides for group unity and cohesion not only through its own practices, but also sometimes through the hostility and prejudice directed at members of a religious group by outsiders.
- Durkheim felt that religion serves a vital function in maintaining the social order.
- Durkheim studied totemism. A totem is an ordinary object such as a plant or animal that has become a sacred symbol to and of a particular group or clan, who not only revere the totem but also identify with it.
- Durkheim believed that when people recognize or worship supernatural entities, they are really worshipping their own society.
- Religion frequently legitimizes the structure of the society within which it exists.
- Religion also establishes worldviews that help people to understand the purpose of life. These worldviews can have social, political, and economic consequences.
- Religion can also help a society adapt to its natural environment or to changing social, economic, and political circumstances.
- Conflict theorists emphasize religion's role in justifying the political status quo by cloaking political authority with sacred legitimacy and thereby making opposition to it seem immoral.

- Alienation, the process by which people lose control over the social institutions they themselves have invented, plays an important role in the origin of religion according to this view.
- Conflict theorists believe religion tends to conceal the natural and human causes of social problems in the world and discourages people from taking action to correct these problems.
- The four main themes that characterize religion in America are religious diversity, widespread belief, secularism, and ecumenism.
- Religious affiliation seems to be correlated with other important aspects of people's lives: Direct relationships can be traced between membership in a particular religious group and a person's politics, professional and economic standing, educational level, family life, social mobility, and attitudes toward controversial social issues.

Media Resources

The Companion Website for *Introduction to Sociology*, Ninth Edition

http://sociology.wadsworth.com/tischler9e

Supplement your review of this chapter by going to the companion website to take one of the Tutorial Quizzes, use the flash cards to master key terms, and check out the many other study aids you'll find there. You'll also find special features such as Wadsworth's Sociology Online Resources and Writing Companion, GSS data, and Census 2000 information at your fingertips to help you complete that special project or do some research on your own.

CHAPTER THIRTEEN STUDY GUIDE

KEY CONCEPTS AND THINKERS

Match each concept with its definition, illustration, or explanation presented below.

a. Denomination
b. Magic
c. Polytheism
d. Mana

e. Rituals
f. Supernaturalism
g. Religion
h. Ecumenism

i. Millenarian movements
j. Religious taboo
k. Animism
l. Alienation

_____ 1. A system of beliefs, practices, and philosophical values shared by a group of people that defines the sacred, helps explain life, and offers salvation from the problems of human existence.

_____ 2. Patterns of behavior or practices that are related to the sacred.

_____ 3. An active attempt to coerce spirits or to control supernatural forces.

_____ 4. Belief in the existence of nonpersonalized supernatural forces that can and often do influence human events.

_____ 5. A diffuse, nonpersonalized force that acts through anything that lives or moves.

_____ 6. A sacred prohibition against touching, mentioning, or looking at certain objects, acts, or people.

_____ 7. Belief in inanimate, personalized spirits or ghosts of ancestors that take an interest in and actively work to influence human affairs.

_____ 8. Belief in a number of gods.

_____ 9. The process by which people lose control over the social institutions they themselves invented.

_____ 10. A level of religious organization that tends to limit its membership to a particular class, ethnic, or religious group.

_____ 11. Movements that typically prophesy the end of the world, the destruction of all evil people and their works, and the saving of the just.

_____ 12. The trend among many religious communities to draw together and project a sense of unity and common direction.

Match the thinkers with their main ideas or contributions.

a. Max Weber
b. Karl Marx

c. Bronislaw Malinowski
d. Émile Durkheim

e. Marvin Harris

_____ 1. The first sociologist to distinguish between the sacred and the profane who discussed religion's role in promoting social cohesion.

_____ 2. Discussed, in *The Protestant Ethic and the Spirit of Capitalism,* how the ideology of Calvinism had influenced the development of capitalism.

_____ 3. Showed that the Hindu belief in the sacredness of cows is a positive strategy for adapting to the environment in India and therefore quite rational.

_____ 4. Saw religion as a tool that the upper classes use to maintain control of society and to dominate the lower classes.

_____ 5. An anthropologist who explained the functional differences between religion and magic, with the former uniting a group of believers and the latter helping the individual who used magic.

CENTRAL IDEA COMPLETIONS

Following the instructions, fill in the appropriate concepts and descriptions for each of the questions posed below.

1. Which of the religions discussed in this chapter has the goal of becoming "one with the universe"?

   a. _____

   By what means does that religion expect this "oneness" to occur?

   b. _____

2. Define and provide an example of each of the following basic religious organizations:

   a. Sect: _____

   b. Church: _____

   c. Ecclesia: _____

   d. Denomination: _____

   e. Millenarian movement: _____

3. Define and provide an example of each of the following terms:

   a. Supernaturalism: _____

   b. Animism: _____

   c. Theism: _____

   d. Abstract ideals: _____

4. Discuss the similarities and differences between magic and religion:

   a. Magic: _____

   b. Religion: _____

5. Millenarian movements stress the end of the world. What reasons does your text give as to why one would want to believe the end of the world is near?

   _____

6. What are the characteristics of three major religions in the United States?

   _____

   _____

7. What does the conflict theory perspective have to say about religion? How does it differ from the functionalist perspective?

   a. Conflict theory perspective: _____

   b. Differences with functionalism: _____

8. Rank these religions from 1 to 7 based on the number of adherents worldwide.

   a. Protestants: _____

   b. Roman Catholics: _____

    c. Jews: _____

    d. Muslims: _____

    e. Hindus: _____

    f. Buddhists: _____

    g. Chinese folk religions: _____

9. Define and provide a brief example of the sacred and the profane.

    a. Sacred: _____

    b. Profane: _____

10. Present two differences between Muslims and Catholics.

    a. _____

    b. _____

CRITICAL THOUGHT EXERCISES

1. In a short essay of one page or less, draw on the information presented in the box "Social Change: Religion Is Constantly Changing" to explain how religions are not static.

2. Some people have claimed that a "civil religion" involving good citizenship and moral behavior has replaced traditional religion. What examples of civil religion have you encountered on your campus? If your school is religiously affiliated, discuss how these aspects of civil religion can coexist in a denominational organization.

3. Given that each individual has his or her own beliefs, can religion really be studied? Why or why not? In what ways can the idea of the sociological imagination discussed in Chapter 1 be a useful tool for studying religion?

4. Durkheim argued that rituals and the sacred were the essential elements of religion, and Tischler points out that even nonreligious objects, like Babe Ruth's bat, can take on sacred qualities. Keeping this in mind, compare a secular ritual (a pep rally, a birthday party, a parade, and so on) with religious rituals. How are they the same, and how are they different?

INTERNET ACTIVITIES

1. Visit http://www.hartfordinstitute.org, the home page of the Hartford Institute for Religion Research. Explore the site. What are the topics of research and theory presented? What can sociologists provide for nonsociologists regarding the study of religion?

2. Visit http://www.pluralism.org, the site of the Pluralism Project. Compare the topics listed at this site with those from the site noted in exercise 1. What are the similarities and differences between these two sites devoted to the sociological analysis of religion?

3. http://www.adherents.com is not exactly sociological but it provides information on religious affiliation. (Did you know that Batman was Catholic or that Wolverine was an atheist?)

ANSWERS TO KEY CONCEPTS

1.g 2.e 3.b 4.f 5.d 6.j 7.k 8.c 9.l 10.a 11.i 12.h

ANSWERS TO KEY THINKERS

1.d 2.a 3.e 4.b 5.c

ThomsonNOW™

Reviewing is as easy as ❶ ❷ ❸

1. Before you do your final exam, take the ThomsonNOW diagnostic quiz to help you identify the areas on which you should concentrate. You will find information on ThomsonNOW and instructions on how to access all of its great resources on the foldout at the beginning of the text.

2. As you review, take advantage of ThomsonNOW's study videos and interactive Map the Stats exercises to help you master the chapter topics.

3. When you are finished with your review, take ThomsonNOW's posttest to confirm you are ready to move on to the next chapter.

© Bill Losh/Taxi/Getty Images

14

Education

Learning Objectives

After studying this chapter, you should be able to do the following:

- Describe the manifest and latent functions of education.
- Explain the nature of education from the conflict theory view.
- Explain the causes and effects of racial segregation in the public schools.
- Identify issues related to students who speak English as a second language.
- Discuss the extent to which high school dropouts are a social problem.
- Discuss the issue of standardized testing.
- Evaluate the idea of special programs for gifted students.

It is June 24, graduation day for Seward Park High School in New York City, and teacher Jessica Siegel has a pocketbook full of bobby pins and tissues. The bobby pins are for her students; the tissues are for herself.

The high school sits empty and silent in the late afternoon, a breeze nudging litter down the street outside, for the commencement is being held 80 blocks north in Manhattan, at Hunter College. There the streets unfold with boutiques, diplomatic mansions, and elegant apartment buildings, with awnings and doormen and gilt doors that glint in the summer sun.

With footsteps both urgent and tentative, the students and their families arrive. Six rise from the Lexington Avenue subway, palms shading their eyes. Five pile out of a Dodge Dart piloted by an uncle. Four emerge from the gypsy cab on which they have splurged, with its smoked windows and air freshener. All of them pause to gape, and the cabbie, too, waits an extra moment before driving on, leaning out the window to admire the promised land. Their world is the Lower East Side, its streets crammed with tenements and bodegas and second-story sweatshops, its belly bursting with immigrants, its class and tongues and customs so foreign at Hunter, only 20 minutes from home.

The seniors have traveled to a border, to a frontier. Some will never cross it. They will have their diplomas and shrink back southward, back to the comforting, familiar things, back to the many menaces. However, others will wield their diplomas as passports and will step into the new, the strange, the almost inconceivable. A few will even attend college (Freedman, 1990).

Jessica Siegel was thrilled to see so many of her students graduate from high school, a major accomplishment that only took place because of her dedication and effort. By helping these students get an education, she had accomplished her goals. What are those goals specifically?

It may sound like a simple question, but if you were asked why people should get an education, what would you say? When people across the country are asked just such a question, the most common reason they give is to get a "better job" or "a better paying job" (42%). In fact, 57% believe that schools are not doing enough to develop job skills.

Very few people think the reason to attend school is "to acquire knowledge" (10%), "to learn basic skills" (3%), "to develop an understanding and appreciation for culture" (1%), or "to develop critical thinking skills" (1%) (Gallup Poll, 1994).

Two people responsible for the development of sociology as a field also struggled with trying to decide the purpose of education. Herbert Spencer, the author of the first textbook in sociology and a proponent of social Darwinism, believed education was only important if it had practical value and prepared people for everyday life. Lester Frank Ward, the first president of the American Sociological Society, believed the main purpose of education was to equalize society.

In the 1850s, Herbert Spencer asked, "What knowledge is of most worth?" and concluded that the purpose of education was "to prepare us for complete living." Every study must be judged by whether it had "practical value." The most important knowledge, he believed, was knowledge for gaining a job or trade, being a parent, and carrying out one's civic duties. Spencer believed that the study of science, which was not taught in schools at the time, was the most useful subject that could be studied and should be added to the curriculum. All other subjects had to be judged by how useful they would be in later life.

When Spencer was putting these ideas forth, he enjoyed enormous prestige in the United States because of his views on social Darwinism, which suggested that little could done to help those at the bottom of the social class ladder. His emphasis on practical education was also applauded by a population that was already inclined to doubt the value of book learning.

Lester Frank Ward believed the source of inequality was the unequal distribution of knowledge. The main purpose of education was to equalize society by diffusing knowledge to all. The greatest advances in civilization had been made by people who had the opportunity for an education and the leisure to think. Ward maintained that the potential giants of the intellectual world could be the unskilled workers of today. He wrote, "the number of individuals of exceptional usefulness will be proportionate to the number possessing the opportunity to develop their powers." Ward believed the entire society would benefit if there were more and better educational opportunity. Against both scholarly and popular opinion, he defended "intellectual egalitarianism." The differences between those at the top and the bottom of the social ladder were not due to any difference in intellect, he said, but to differences in knowledge and education. Ward believed the main job of education was to ensure that the heritage of the past was transmitted to all members of society (Ravitch, 2000).

In this chapter, as we examine the role of education in our society, we will contrast the functionalist and conflict theorist approaches to understanding the American educational system.

Functionalists stress the importance of education in socializing the young, transmitting the culture, and developing skills. Conflict theorists, on the other hand, note that education preserves social class distinctions, maintains social control, and promotes inequality. We will also examine the impact of some of the contemporary issues facing education.

Education: A Functionalist View

What social needs does our education system meet? What are its tasks and goals? Education has several *manifest functions* (see Chapter 1)—intended and predetermined goals such as the socialization of the young or the teaching of academic skills.

There are also some latent functions, which are unintended consequences of the educational process. These may include child care, the transmission of ethnocentric values, and respect for the American class structure.

Socialization

In the broadest sense, all societies must have an educational system. That is, they must have a way of teaching the young the tasks that are likely to be expected of them as they develop and mature into adulthood. If we accept this definition of an educational system, then we must believe that there really is no difference between education and socialization. As Margaret Mead (1943) observed, in many preliterate societies no such distinction is made. Children learn most things informally, almost incidentally, simply by being included in adult activities.

Traditionally, the family has been the main arena for socialization. As societies have become more complex, the family has been unable to fulfill all aspects of its socialization function. Thus, there is a need for the formal educational system to extend the socialization process that starts in the family. In modern industrialized societies, a distinction is made between education and socialization. In ordinary speech, we differentiate between socialization and education by talking of bringing up and educating children as two separate tasks. In modern society, these two aspects of socialization are quite compartmentalized: Whereas rearing children is an informal activity, education or schooling is formal. The role prescriptions that determine interactions between students and teachers are clearly defined, and the curriculum to be taught is explicit. Obviously, the educational process goes far beyond just formalized instruction. In addition, children also learn things in their families and among their peers. In school, children's master status (see Chapter 6) is that of student, and their primary task is to learn.

Schools, as differentiated, formal institutions of education, emerged as part of the evolution of civilization. However, until about 200 years ago, education did not help people become more productive in practical ways, and thus it was a luxury that very few could afford. This changed dramatically with the industrialization of Western culture. Workers with specialized skills were required for production jobs, as were professional, well-trained managers.

When the Industrial Revolution moved workers out of their homes and into factories, the labor force consisted not only of adults but also of children. Subsequently, child labor laws were passed to prohibit children from working in factories.

Public schools eventually emerged as agencies dedicated to socializing students, teaching them proper attitudes and behaviors, and encouraging conformity to the norms of social life and the workplace.

Cultural Transmission

The most obvious goal of education is **cultural transmission,** *in which major portions of society's knowledge are passed from one generation to the next.* In relatively small, homogeneous societies, in which almost all members share the culture's norms, values, and perspectives, cultural transmission is a matter of consensus and needs few specialized institutions. In a complex, pluralistic society like ours, with competition among ethnic and other minority groups for economic and political power, the decision about what aspects of the culture will be transmitted is the outgrowth of a complicated process.

If we consider that schools are one of the major means of cultural transmission—both vertically (passing knowledge between generations) and horizontally (disseminating knowledge to adults)—this lack of consensus concerning every important aspect of schooling points to something beyond the educational system itself, to a crisis in the whole of American culture. It is important to understand something of the nature of this cultural crisis before we turn to a narrower discussion of the functions of education in America today. Our pluralistic society contains many cultural differences among its ethnic groups and social classes. Nevertheless, for a society to hold together, there must be certain core values and goals—some common traits of culture—that its constituent social elements share to a greater or lesser degree (see Chapter 3). In America, it seems that this core culture itself is changing rapidly.

A school's curriculum often reflects the ability of organized groups of concerned citizens to impose their views on an educational system, whether local, statewide, or nationwide. Thus, it was a political process that caused African American history to be introduced into elementary, high school, and college curricula during the 1960s. Similarly, it was political activism that caused the creation of women's studies programs in many colleges. Moreover, even though the concept of evolution is a cornerstone of modern scientific knowledge, it is political pressure (from Christian fundamentalists) that causes some textbooks to refer to it as the "theory" of evolution and prevents it from being taught in certain counties.

In recent years, bilingual education has become an educational and political issue. Proponents believe that it is crucial for children whose primary language is not English to be given instruction in their native tongues. They believe that by acknowledging students' native languages, the school system is helping them make the transition into the all-English mainstream—and is helping preserve the diversity of American culture.

Others see a danger in these programs. They believe that many bilingual education programs never provide for the transition into English, leaving many youngsters without the basic skills needed to earn a living and participate in our society.

In the end, the debate centers on how closely our sense of who we are as a nation hinges on the language our children speak in school. For the time being, the only agreement between the two sides is that language is the cornerstone for cultural transmission.

Academic Skills

Another crucial function of the schools is to equip children with the academic skills they need to function as adults—to hold down a job, to balance a checkbook, to evaluate political candidates, to read a newspaper, to analyze the importance of a scientific advance, and so on. Have the schools been successful in this area? Most experts believe they have not.

In 1983, the National Commission on Excellence in Education issued a report titled *A Nation at Risk,* which bitterly attacked the effectiveness of American education. The message of the report was clear and sobering: "The educational foundations of our society are presently being eroded by a rising tide of mediocrity that threatens our very future." As a result of this report, reforms were instituted in all 50 states, which stressed the teaching of the "three Rs" and the elimination of frivolous electives that waste valuable student and teacher time. In addition, high school graduation requirements were raised in 40 states, and in 19 states students were required to pass minimum competency tests before they could receive their high school diplomas. Forty-eight states also required new teachers to prove their competence by passing a standardized test.

Although the back-to-basics movement has grown, its success has been limited. Five years after *A Nation at Risk* appeared, a follow-up report was issued titled *American Education: Making It Work* (1988). According to the report, "the precipitous downward slide of previous decades has been arrested and we have begun the long climb back to reasonable standards." However, despite this progress, especially among minority groups, the report condemned the performance of American schools as unacceptably

An important function of education is to equip children with the academic skills that they need to function in society.

low. "Too many students do not graduate from our high schools, and too many of those who do graduate have been poorly educated. . . . Our students know little, and their command of essential skills is too slight," the report stated.

Particularly troublesome was student performance in math and science. According to a study done by the Nation's Report Card (National Center for Education Statistics, 1988), an assessment group that is part of the Educational Testing Service, the math performance of 17-year-olds is "dismal." The study found that although half the nation's 17-year-olds have no trouble with junior high school math, they flounder when asked to solve multistep, high-school-level problems or those involving algebra or geometry. Fewer than 1 in 15 students was able to answer these problems correctly. At fault might be the very back-to-basics movement that was supposed to rescue our educational system from failure in the early 1980s. Though rote learning has helped improve the scores of the lowest level students, it left others totally unprepared to analyze complex problems.

The Nation's Report Card also found American students' understanding of science "distressingly low." In a condemnation of the performance of America's schools in the area of science, the report found that most 17-year-olds did not have the skills to handle today's technologically based jobs and that only 7% could cope with college-level science courses.

A survey by the National Education Goals Panel echoed these findings:

> American first-graders may already be academically behind their counterparts in Japan and Taiwan. We know that not only is the content studied by our students less challenging than what students in high-achieving countries are exposed to, but American parents expect less of their children academically (1998).

In 1994 Congress passed the Education America Act, intended to set a new direction for our schools. The act set eight educational goals that the nation should achieve quickly:

- All children should enter school ready to learn.
- The U.S. high school graduation rate should reach at least 90%.
- U.S. students should lead the world in mathematics and science performance.
- Every adult should be competent as a citizen and a worker.
- Schools should be disciplined environments free of drugs and violence.
- Students should demonstrate growing academic competencies in specific areas as they progress through school.
- Parents should become more deeply involved in their children's educational welfare.
- Teachers should be helped to expand and perfect their professional skills throughout their careers.

Some of these goals are no different from those presented by then President Bush at a 1989 education summit.

By 2002 not much had changed. Between 1966 and 2002, the federal government had spent $321 billion (in today's dollars) to help educate disadvantaged children. Yet, despite increased spending:

- Less than one-third of U.S. fourth graders could read proficiently.
- Reading performance had not improved in more than 15 years.
- Less than 20% of U.S. twelfth graders were proficient in math.
- And, among the industrialized nations of the world, U.S. twelfth graders ranked near the bottom in science and math.

In response to these facts in 2002, President George W. Bush signed into law the No Child Left Behind Act. This law changed the federal government's role in K–12 education.

The act contains four basic education reform principles: stronger accountability for results, increased flexibility and local control, expanded options for parents, and an emphasis on teaching methods that have been proven to work.

In an effort to hold education systems responsible for children's education, all states must implement statewide accountability systems, which will:

- Set academic standards in each content area for what students should know and be able to do.
- Gather specific, objective data through tests aligned with those standards.
- Use test data to identify strengths and weaknesses in the system.
- Report school academic achievement to parents and communities.
- Empower parents to take action based on school information.
- Recognize schools that make real progress.

The law requires that there be real consequences for districts and schools that fail to make progress. It also makes it possible for parents with a child enrolled in a failing school to transfer their child to a better-performing public school or public charter school.

It will be interesting to see whether this massive effort will produce results. Past attempts at education reform have not been particularly successful.

Innovation

A primary task of educational institutions is to transmit society's knowledge, and part of that knowledge consists of the means by which new knowledge is to be sought. Learning how to think independently and creatively is probably one of the most valuable tools the educational institution can transmit. This is especially true of the scientific fields in institutions of higher education. Until well into this century, scientific research was undertaken more as a hobby than a vocation. This was because science was not seen as a socially useful pursuit. Gregor Mendel (1822–1884), who discovered the principles of genetic inheritance by breeding peas, worked alone in the gardens of the Austrian monastery where he lived. Albert Einstein supported himself between 1905 and 1907 as a patent office employee while making several trailblazing discoveries in physics, the most widely known of which is the theory of relativity.

Today, science obviously is no longer the undertaking of part-timers. Modern scientific research typically is pursued by highly trained professionals, many of whom frequently work as teams; and the technology needed for exploration of this type has become so expensive that most research is possible only with extensive government or corporate funding.

© Jean-Claude Lejeune/Stock, Boston

Learning how to think independently and creatively is probably one of the most valuable tools that our educational system can transmit.

In 2003, the United States spent $284 billion on research and development funding (Bureau of the Census, Statistical Abstract, 2006). In research and development, the areas of national defense, space exploration, and health research receive by far the greatest amount of support. In research alone, the leading three areas are life sciences (biological sciences and agriculture), engineering, and the physical sciences.

The achievements of government and industrial research and development notwithstanding, the importance of the contributions to science by higher academic institutions cannot be overestimated. First, there could be no scientific innovations—no breakthroughs—without the training provided by these schools. In the United States alone, 17,844 doctorates were awarded in 2003 in the physical sciences, mathematics, computer science, and engineering (Bureau of the Census, Statistical Abstract, 2006). Second, the universities of the highest caliber continue to generate some of the most significant research in the biological and the physical sciences. (For a further discussion of innovation in education, see "Technology and Society: College Students and the Internet.")

TECHNOLOGY AND SOCIETY

College Students and the Internet

College students have grown up with the Internet and are heavier users of it than the general public. The current generation of college students was the first to use the Internet for communication, file sharing, and research. In fact, Yahoo, Napster, and other Internet applications were created by college students. The Internet has become such an integrated part of student's lives that it is as ordinary to them as the television and the telephone were to previous generations. The next generation of college students will be even more comfortable with Internet usage. Eighty-seven percent of youth between the ages of 12 and 17 (about 21 million) use the Internet. The vast majority of both the precollege and college students believe the Internet helps them do better in school. Here are some interesting facts about college students and the Internet:

- 20% of today's college students started using the Internet between the ages of 5 and 8.

- 72% check e-mail at least once a day. 66% have at least two e-mail addresses.
- 26% use instant messaging on any given day.
- 85% own their own computers.
- 60% have downloaded music files.
- 46% believe e-mail makes it possible for them to express ideas to a professor they would not express in class.
- 73% say they use the Internet more than the library for research.
- 37% of precollege students said "too many" students use the Internet to cheat.

Sources: Jones, Steve, "The Internet Goes to College," Pew Internet and American Life Project, September 15, 2002. Hitlin, Paul and Lee Rainie, "Teen Use of the Internet at School Has Grown 45% Since 2000" Pew Internet and American Life Project, August 2005. Both available at http://www.pewinternet.org.

It is also worth noting that more women and minorities are earning PhDs than ever before. In 2002, 45% of all PhD recipients were women. In 1988, women received about 33% of the degrees; and in 1967, the total was just 12%. For minorities, the numbers have more than doubled, with the vast majority of those degrees being in education or the social sciences (Noxon, 2003).

In addition to their manifest or intended functions, the schools in America have come to fulfill a number of functions that they were not originally designed to serve.

Child Care

One latent function of many public schools is to provide child care outside the nuclear family. This has become increasingly important since World War II, when women began to enter the labor force in large numbers. As of 2004, 75.6% of married women with school-age children (ages 6 to 17), and 59.3% of those with children under 6 were in the labor force. In addition, women with children worked more hours each week on average in 2004 than they did in 1969 (Statistical Abstract of the United States, 2006).

A related service of schools is to provide children with at least one nutritious meal per day. In 1975, the number of public school pupils in the United States participating in federally funded school lunch programs was 25,289,000, at a cost of $1.28 billion. By 2002, more than 16 million children each month got their lunch through the National School Lunch Program at a cost of $6.06 billion (Statistical Abstract of the United States, 2006).

Postponing Job Hunting

More and more young American adults are choosing to continue their education after graduating from high school. In 2003, 61.2% of male and 66.5% of female recent high school graduates were enrolled in college (Statistical Abstract of the United States, 2006). Even though some of these individuals also work at part-time and even full-time jobs, an important latent function of the American educational system is to slow the entry of young adults into the labor market. This helps keep down unemployment, as well as competition for low-paying unskilled jobs.

Originally, two factors pointed to the possibility that college enrollments would not continue to increase. Because of low birthrates, the number of high school graduates peaked at 3.2 million in 1977 and began a 14-year decline, with only 2.28 million graduating in 1991. That trend has been reversed and 2.75 million graduated in 2004 (Statistical Abstract of the United States, 2006).

Colleges and universities, anticipating enrollment problems, embarked on concerted efforts to ward off disaster. Through hard work and luck, they have succeeded. Total enrollment in two- and four-year

Figure 14-1   Percentage of Adults Age 25–29 Who Have a Bachelor's Degree, 1990 and 2003

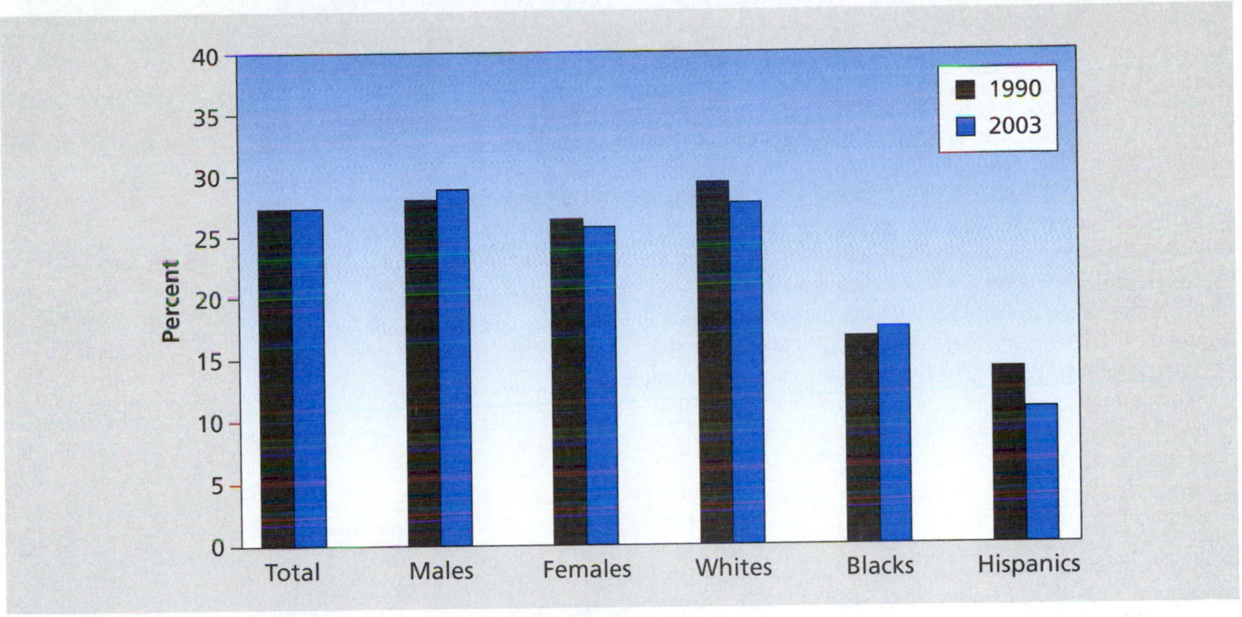

Source: Current Population Reports, *Educational Attainment in the United States: 2003*, June 2004.

colleges rose from 11.5 million in 1977 to more than 16.6 million in 2002 (Statistical Abstract of the United States, 2006). (See Figure 14–1 for a comparison of the percentage of the people with college degrees in 1990 and 2003.)

Colleges have also benefited from the fact that the U.S. economic base has shifted from manufacturing jobs to service jobs. This has caused the demand for professionals and technicians to grow. The salaries for those types of positions are considerably higher than salaries for manufacturing jobs.

Two other trends have also benefited colleges, the first being the increase in the number of women going to college, and the second the increase in older students. Since 1980, the majority of college students have been women. This trend is an outgrowth of changing attitudes about the status of women in our society and the breakdown in gender-role stereotypes (see Chapter 11). The second trend is that people 25 and older represent the most rapidly growing group of college students, accounting for 45% of all undergraduate and graduate students. Women are overrepresented in this group, as are part-time students. Many returning women students are also responding to changing gender-role expectations. (For a discussion of college graduates in other countries, see "Global Sociology: College Graduates: A Worldwide Comparison.")

The Conflict Theory View

To the conflict theorist, society is an arena for conflict, not cooperation. In any society, certain groups come to dominate others, and social institutions become the instruments by which those in power control the less powerful. The conflict theorist thus sees the educational system as a means for maintaining the status quo, carrying out this task in a variety of ways. The educational system socializes students into values dictated by the powerful majority. Schools are seen as systems that stifle individualism and creativity in the name of maintaining order. To the conflict theorist, the function of school "is to produce the kind of people the system needs, to train people for the jobs the corporations require and to instill in them the proper attitudes and values necessary for the proper fulfillment of one's social role" (Szymanski & Goertzel, 1979).

Social Control

In the United States, schools have been assigned the function of developing personal control and social skills in children. Although the explicit, formally defined school curriculum emphasizes basic skills such as reading and writing, much of what is taught is oriented away from practical concerns. Many critics point out that much of the curriculum (other than in special professional training programs) has little direct, practical application to everyday life. This has led conflict theorists and others to conclude that the most important lessons learned in school are not those listed in the formal curriculum but, rather, involve a hidden curriculum. The **hidden curriculum** refers to *the social attitudes and values taught in school that prepare children to accept the requirements of adult life and to fit into the social, political, and economic statuses the society provides.*

To succeed in school, a student must learn both the official (academic) curriculum and the hidden

GLOBAL SOCIOLOGY

College Graduates: A Worldwide Comparison

The percentage of young people with a college degree is higher in the United States than anywhere else in the world (Table 14–1). One reason for this is that there is virtually no difference between the percentage of male and female degree holders. Japan would have more college degree holders if the same pattern held true there. In Japan however, even though substantially more young men have college degrees than men in the United States, Japanese women obtain degrees at only one-third the rate of Japanese men. Traditional gender-role expectations in Japan are limiting the number of women who pursue higher education.

Source: Organization for Economic Co-operation and Development, INES Project, International Indicators Project, *The Condition of Education 1997*, Indicator 23.

Table 14–1

Percentage of People With a College Degree— Ages 25 to 34

Country	Men	Women
Canada	18.0%	18.9%
France	11.9	11.3
Germany	12.7	11.0
Italy	7.7	8.1
Japan	34.2	11.5
United Kingdom	15.7	11.7
United States	23.4	23.5

(social) curriculum. The hidden curriculum is often an outgrowth of the structure within which the student is asked to learn. Within the framework of mass education, it would be impossible to provide instruction on a one-to-one basis or even in very small groups. Consequently, students are usually grouped into relatively larger classes. Because this system obviously demands a great deal of social conformity by the children, those who divert attention and make it difficult for the teacher to proceed are punished. In many respects, the hidden curriculum is a lesson in being docile. For example, an article in *Today's Education*, the journal of the National Education Association, gives an experienced teacher's advice to new teachers:

> During the first week or two of teaching in an inner-city school, I concentrate on establishing simple routines, such as the procedure for walking downstairs. I line up the children, and . . . have them practice walking up and down the stairs. Each time the group is allowed to move only when quiet and orderly.

Social skills are highly valued in American society, and a mastery of them is widely accepted as an indication of a child's maturity. The school is a miniature society, and many individuals fail in school because they are either unable or unwilling to learn or use the values, attitudes, and skills contained in the hidden curriculum. We do a great disservice to these students when we make them feel that they have failed in education, when they have, in fact, only failed to conform to the school's socialization standards.

Screening and Allocation: Tracking

Tracking has existed in American classrooms since the beginning of the twentieth century. First introduced in Britain in the 1920s as "streaming," tracking became widespread in the United States after World War II ("Tracking," 2000).

From its beginning, the American school system, in principle, has been opposed to tracking, or the stratification of students by ability, social class, and various other categories. Educators saw in compulsory education a way to diminish the grip of inherited social stratification by providing the means for individuals to rise as high as their achieved skills would allow. In the words of Horace Mann, an influential American educator of the late nineteenth century, public education was to be "the great equalizer of the conditions of men." Despite the principles on which it is based, the American educational system utilizes tracking. At least two-thirds of U.S. high schools use tracking.

Although tracking is not as formally structured or as completely irreversible in America as in most other industrial societies, it is influenced by many factors, including socioeconomic status, ethnicity, and place of residence. It is also consistently expressed in the differences between public and private schools as well as in the differences among public schools. (In New York City, for example, there are highly competitive math and science-oriented and arts-oriented high schools, neighborhood high schools, and vocational high schools.) Of course, tracking occurs in higher education

SOCIAL CHANGE

Jonathan Kozol on *The Shame of the Nation*

Jonathan Kozol, author of *The Shame of the Nation: The Restoration of Apartheid Schooling*, believes that five decades after the Supreme Court ordered America to integrate its schools, the country has established two separate and unequal school systems divided by class and race, with the gap between them growing larger each year.

Kozol visited 60 schools in 30 different school districts and found that some children were going to schools that mimicked conditions in the third world, while only a short bus ride away were beautiful, well-equipped school campuses. Kozol points out that black and Hispanic students are now concentrated in schools where very few white students exist.

Kozol relies on studies by Harvard professor Gary Orfield and his colleagues to document the scale of educational segregation. "American public schools are now 12 years into the process of continuous resegregation . . . During the 1990's, the proportion of black students in majority white schools has decreased . . . to a level lower than any years since 1968."

There are several reasons why this has taken place. Court-ordered desegregation programs no longer exist in many cities because they do not seem to close the educational gap between white and minority students. Often, in large cities there are not enough white students to form a basis for integration. Immigration patterns have also made integration difficult as the number of minorities to be considered for integration has increased. Finally, some groups oppose integration because they are afraid it will break up their community.

A new dilemma that Kozol is concerned with is the increasing emphasis on testing and the resulting pressure on teachers and children to pass state imposed guidelines. These high stakes tests force teachers to devote more time to test preparation at the expense of subjects such as music, art, social science, and history.

Kozol believes that the real reason for educational inequities is money. He notes that the gaps in school funding are so consistent from one metropolitan area to another—with rich schools in many areas spending more than twice per pupil as poor schools—that it suggests a deliberate pattern. He points out that New York

© Thomas Victor

Jonathan Kozol

City spent $11,627 on each student in 2002–2003 while the wealthier suburb of Manhasset spent $22,311. These inequities Kozol believes lead to a situation like the one in Walton High School in the Bronx where 1,275 ninth-graders entered the school in the fall of 1999. By 2002, there were 400 12th graders and by the spring of 2003 only 188 graduated.

While their children enjoy richer schools, bolstered by higher property tax bases and federal deductions for property taxes and mortgage payments, suburban parents are lulled by politicians who claim that spending more money on inner-city schools will not make any difference.

Kozol points out that "there is a deep-seated reverence for fair play in the United States, but this is not the case in education, health care, or inheritance of wealth. In these elemental areas we want the game to be unfair and we have made it so."

Sources: Jonathan Kozol, *The Shame of the Nation: The Restoration of Apartheid Schooling in America*, New York: Crown Publishers, 2005; Jonathan Kozol, *Savage Inequalities*, New York: Harper Perennial, 1992.

---

in the selection of students by private colleges and universities, state colleges, and junior colleges.

Tracking begins with stratifying students into "fast," "average," and "slow" groups, from first grade through high school. It can be difficult for a student to break out of an assigned category because teachers come to expect a certain level of performance

from an individual. The student, sensing this expectation, will often give the level of performance that is expected. In this way, tracking becomes a self-fulfilling prophecy. (See "Social Change: Jonathan Kozol on *The Shame of the Nation*.")

In one study of this phenomenon, R. Rosenthal and L. Jacobson (1966) gave IQ tests to 650 lower-class elementary school pupils. Their teachers were told that the test would predict which of the students were the "bloomers" or "spurters." In other words, the tests would identify the superior students. This approach was, in fact, not the one employed. Twenty percent of the students were randomly selected to be designated as superior, even though there was no measured difference between them and the other 80% of the school population. The point of the study was to determine whether the teacher's expectations would have any effect on the "superior" students.

At the end of the first year, all the students were tested again. There was a significant difference in the gain in IQ scores between the "superior" group and the control group. This gain was most pronounced among students in the first and second grades. Yet the following year, when these students were promoted to another class and assigned to teachers who had not been told that they were "superior," they no longer made the sort of gains they had evidenced during the previous year.

Nonetheless, the "superior" students in the upper grades continued to gain during their second year, showing that there had been long-term advantages from positive teacher expectations for them. Apparently, the younger students needed continuous input to benefit from the teachers' expectations, whereas the older students needed less.

The most common argument against tracking notes that tracking fosters race and class segregation. Studies have shown that black and Hispanic students are overrepresented in low tracks and underrepresented in high tracks. Critics of tracking argue that student abilities often reflect the effects of placement, rather than vice versa. Critics also claim that tracking can have a negative impact on self-esteem by separating students into "winners" and "losers." Others counter that throwing all students into a mixed ability classroom depresses the achievement of high-ability students and frustrates the lesser ability students ("Tracking," 2000).

The Credentialized Society

Conflict theorists would also argue that we have become a credentialized society (Collins, 1979). A degree or certificate has become necessary to perform a vast variety of jobs. This credential may not necessarily cause the recipient to perform the job better. Even in professions such as medicine, engineering, and law, most knowledge is acquired by performing tasks on the job. However, credentials have become a rite of passage and a sign that a certain process of indoctrination and socialization has taken place. It is recognized that the individual has gone through a process of educational socialization that constitutes adequate preparation to hold the occupational status.

Therefore, colleges and universities act as gatekeepers, allowing those who are willing to play by the rules to succeed, while barring those who may disrupt the social order.

At the same time, advanced degrees are undergoing constant change and becoming less specialized. A law degree from Harvard, Yale, or Columbia is less a measure of the training of a particular candidate than a basis on which leading corporations, major public agencies, and important law firms can recruit those who will maintain the status quo. The degree signifies that the candidate has forged links with the established networks and achieved a grade necessary to obtain a degree.

Colleges and universities are miniature societies more than centers of technical and scientific education. In these environments, students learn to operate within the established order and to accept traditional social hierarchies. In this sense, they provide the power structure with a constantly replenished army of defenders of the established order. According to this view, those who could disrupt the status quo are not permitted to enter positions of power and responsibility.

Issues in American Education

How well have American schools educated the population? The answer depends on the standards one applies. Americans take it for granted that everyone has a basic right to an education and that the state should provide free elementary and high school classes. The United States pioneered this concept long before similar systems were introduced in Europe.

As we have attempted to provide formal education to everyone, we also have had to contend with a wide variety of problems stemming from the diverse population. In this section, we will examine some of the concerns in contemporary American education.

Unequal Access to Education

American minorities have sought equal access to public schools for two centuries. Tracing those efforts over the generations reveals a pattern of dissatisfaction with integrated as well as segregated schools.

African American parents attributed the ineffective instruction at the schools attended by their children to one of two causes. If the schools were all-black, failure was attributed to the racially segregated character of those schools. If whites were attending the schools, African American parents concluded, conditions would be better. This has been the dominant theme in the nineteenth and twentieth centuries.

Discontent also has occurred when African American children have attended predominantly

© Tom and Dee McCarthy/CORBIS

The Coleman Study provided evidence that the home environment, the quality of the neighborhood, and the types of friends one has are more influential in school achievement than is the quality of the school facilities or the skill of the teachers.

white schools. In those instances, the racially integrated character of the school was seen as a problem because white students were thought to be favored by the teachers.

In the 1954 case *Brown v. Board of Education* of Topeka, Kansas, the Supreme Court ruled that school segregation was illegal. The court held that "In the field of public education, the doctrine of separate but equal has no place. Separate educational facilities are inherently unequal." Segregating African American schoolchildren from white schoolchildren was a violation of the equal-protection clause of the Constitution. However, though the court's verdict banned **de jure segregation,** or *laws prohibiting one racial group from attending school with another,* it had little effect on **de facto segregation,** or *segregation resulting from residential patterns.* For example, minority groups often live in areas of a city where there are few, if any, whites. Consequently, when children attend neighborhood schools, they usually are taught in an environment that is racially segregated.

Ten years after the 1954 ruling, the federal government attempted to document the degree to which equality of education had been achieved. It financed a cross-sectional study of 645,000 children in grades 1, 3, 6, 9, and 12 attending some 4,000 different schools nationwide. The results, appearing in James S. Coleman's now-famous report *Equality of Educational Opportunity* (1966), supported unequivocally the conclusion that American education remains largely unequal in most parts of the country, including those where "Negroes form any significant proportion of the population." Coleman noted further that on all tests measuring pupils' skills in areas crucial to job performance and career advancement, not only did Native Americans, Mexican Americans, Puerto Ricans, and African Americans score significantly below whites, but also that the gaps widened in the higher grades. Now for a subtle but extremely important point: Although there are acknowledged wide inequalities of educational opportunity throughout the United States, the discrepancies between the skills of minorities and those of their white counterparts could not be accounted for in terms of how much money was spent on education per pupil, quality of school buildings, number of laboratories or libraries, or even class sizes. Despite good intentions, a school presumably cannot usually outweigh the influence of the family backgrounds of its individual students and of its student population as a whole. The Coleman study thus provided evidence that schools per se do not play as important a role in student achievement as was once thought. It appears that the home environment, the quality of the neighborhood, and the types of friends and associates one has are much more influential in school achievement than is the quality of the school facilities or the skills of teachers.

In effect, then, the areas that schools have least control over—social influence and development—are the most important in determining how well an individual will do in school.

The Coleman report pointed out that lower-class nonwhite students showed better school achievement when they went to school with middle-class whites. Racial segregation, therefore, hindered the educational attainments of nonwhites.

A direct outgrowth of the Coleman report pointing to the harm of de facto segregation was the busing of children from one neighborhood to another to achieve racial integration in the schools. The fundamental assumption underlying school busing was that it would bring about improved academic achievement among minority groups. Nationwide, many parents, black and white, responded negatively to the idea that their school-age children must leave their neighborhoods. For all practical purposes, busing is no longer a major issue. Those schools that needed to desegregate have mostly done so, and other, less disruptive approaches to desegregation are attempted.

One factor that has increased the difficulty of integrating public schools is white flight, the continuing exodus of white Americans by the hundreds of thousands from the cities to the suburbs. White flight has been prompted partly by the migration of African Americans from the South to the inner cities of the North and Midwest since the 1960s, but some authorities strongly maintain that it also is related closely to school desegregation efforts in the large cities.

For example, in a later view of desegregation attempts (1977), Coleman vastly revised the position in his 1966 report, stating that urban desegregation has in some instances had the self-defeating effect of emptying the cities of white pupils. Some authorities (Pettigrew & Green, 1975) took exception to the Coleman thesis, and others believe that what may appear to be flight is related more directly to the characteristic tendency of the American middle class to be upwardly mobile and to constantly seek a better lifestyle. Even though there is some evidence of a countertrend, in which middle-class whites are beginning to regentrify inner cities, there seems to be no abating of this migration. Nor have most established communities relinquished the ideal of self-determination as embodied in the right to maintain neighborhood schools.

Since the Coleman report was issued, the integration of our schools has continued to be a crucial problem, especially in urban areas where the vast majority of minority group members live. According to a study by the National School Boards Association, there has been "no significant progress on the desegregation of black students in urban districts since the mid-1970s," and some areas have shown "severe increases in racial isolation" (Fiske, 1988). Schools in Atlanta and Detroit, for example, are more segregated now than they were a decade ago.

Only about 3% of the nation's white students attend central city schools. As a result, these schools have become almost irrelevant to the nation's white population. There are currently many more minority and far fewer white students in our public schools than in the past. Colleges and universities have had mixed results in increasing minority enrollment. Since the mid-1980s, the number of black undergraduates has not increased appreciably. Many qualified students do not apply because their families cannot afford tuition, even though complete aid packages are available. To overcome this problem, many schools have taken an aggressive recruitment stance, believing that once they find qualified candidates, they can persuade them to attend. At Harvard University, 9% of the freshman class of 2001 was African American —the highest percentage in the school's history. Harvard has the luxury of not turning any student away because of financial need.

Still, thousands of qualified minority students never attend college, and many who do attend fail to graduate. The reasons for low graduation rates include financial problems, poor preparation, and the feeling of being unwelcome. Many cannot afford the loss of income that comes with being a full-time student. For many, family survival depends on the money they contribute. Others are victims of inadequate schools. They simply do not have the skills needed to complete college. Still others drop out because they feel out of place in the predominantly white world of higher education.

Students Who Speak English as a Second Language

The Department of Education reports that 6.3 million children aged 5 to 17, or 14%, speak a language other than English at home. Another 3.2 million elementary and secondary school students are classified as having limited English proficiency. The majority of these students are of Hispanic origin. Virtually all of these students are enrolled in remedial language programs.

The goal of bilingual education is to give immigrant children the opportunity to learn academic subjects in their native languages at the same time as they are learning English. After they become fluent in English, they can join their English-speaking peers at the appropriate grade level. In this way they do not fall behind other students as they learn English. With bilingual education they are not made to feel that their native language and culture are holding them back.

Critics of bilingual education contend the United States has a long history of using the education system to help students assimilate into American society. Immersing students in the language and culture of their new society speeds up the assimilation process. Critics see supporters of bilingual education as people determined to preserve the culture and traditions of another society at the expense of the students. They point to high dropout rates and slow progress in transitioning to English-only classes as evidence that this approach is flawed (Rothstein, 1998).

In general, the younger foreign-born children are when they enter school, the better they will perform. Immigrant children's chances for succeeding in school also improve in direct relation to their parents' educational attainment and income. The obstacles are especially great for children whose parents are illiterate and cannot help with homework.

Bilingual educational research involves many complicated background factors that affect the outcomes of academic studies. Because of the complex nature of educational studies, no single body of research can accurately account for the impact of bilingual education on various aspects of society.

English language proficiency is important today because the work world is more complex than during previous decades. Immigrants now come from a broad range of backgrounds and enter a more complex culture. In the past, less educated people could find well-paid factory jobs. In today's more complex economy, the poorly educated and illiterate are often unemployed.

High School Dropouts

The rate of students dropping out of high school has been declining in recent decades. In 1970, the percentage of 18- to 24-year-olds who dropped out of high school was 17.3%. In 2003, that number fell to 8.4%. Gender differences between high school dropouts have also changed in recent years. In the early 1970s, the trend was for girls to drop out of high school at a higher rate than boys. But in the late 1970s this trend reversed so that now the dropout rate for boys is much higher than for girls. Current U.S. Census Bureau figures show that fewer than 9.9% of women drop out of high school, compared with 12.0% of males in the same age group (AmeriStat, February 2000; Bureau of the Census, Statistical Abstract, 2006).

In 2003, 85% of American adults ages 25 and older had at least completed high school; 27% had a bachelor's degree or higher. In 1975, 63% of adults had a high school diploma, and 14% had obtained a bachelor's degree. Much of the increase in educational attainment levels of the adult population is due to a more educated younger population replacing an older, less educated population.

The percentage of people with at least a high school degree among Hispanics was 89% and among blacks 80%. This represents an increase of 10 percentage points for blacks between 1993 and 2003, and an increase of 5 percentage points for Hispanics (Bergman, 2004).

Dropping out of high school has long been viewed as a serious educational and social problem. By leaving high school before graduation, dropouts risk serious educational deficiencies that severely limit their economic and social well-being. Over the past 95 years, the proportion of people in the adult population who have failed to finish high school has decreased substantially. In 1910, the proportion of the adult population (ages 25 and older) that had completed at least four years of high school was 13.5%. It stood at 24.5% in 1940, at 55.2% in 1970, and at 85.2% in 2004. Among young people (ages 25 to 29) the drop is even more striking, with 87.1% having completed high school in 2004 (Statistical Abstract of the United States, 2006).

Despite these long-term declines in dropout rates, interest in the dropout issue among educators and policy makers has increased substantially in recent years. Legislators and education officials are devoting ever more time and resources to dealing with the issue.

If the long-term trend is that dropout rates are declining, why has the concern for this problem increased lately? First, although the long-term dropout trend has declined, the short-term trend has remained steady and even increased for some groups.

A second reason is that minority populations, who always have had higher dropout rates than whites, are increasing as a proportion of the public high school population. Racial and ethnic minorities represent the majority of students enrolled in most large U.S. cities and more than 90% of all students in such cities as Newark, N.J.; Atlanta; and San Antonio (Bureau of the Census, 2000). Dropout rates are higher for members of racial, ethnic, and language minorities; higher for males than females; and higher for people from the lower socioeconomic classes. Hispanics have the highest dropout rates. In 2003, 33.1% of Hispanics aged 18 to 24 had dropped out of high school, compared with 16.0% for African Americans and 11.8% for whites. Among Hispanics, Puerto Ricans have the highest dropout rates, followed by Mexican Americans and Cuban Americans. Dropout rates are also particularly high among Native Americans (Statistical Abstract of the United States, 2006).

Factors associated with dropping out include low educational and occupational attainment levels of parents, low family income, speaking a language other than English in the home, single-parent families, and poor academic achievement.

The influence of peers is also important, but it has not received much attention in previous research.

Figure 14–2   Median Income by Education Level

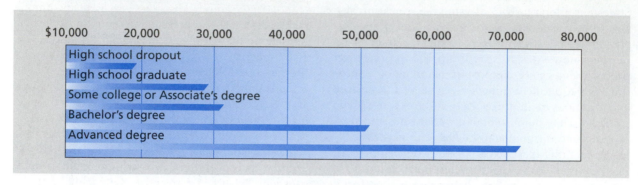

	$10,000	20,000	30,000	40,000	50,000	60,000	70,000	80,000
High school dropout								
High school graduate								
Some college or Associate's degree								
Bachelor's degree								
Advanced degree								

Source: U.S. Census Bureau, *Current Population Survey, Annual Social and Economic Supplement, 2003.*

Nothing undermines the effectiveness of our education system more than unsafe schools.

Many dropouts have friends who are dropouts, but it is not clear to what extent and in what ways a student's friends and peers influence the decision to leave school (Rumberger, 1987).

High school dropouts do more than just damage their employment and earnings potential (Figure 14–2). Dropping out of high school affects not only those who leave school, but also society in general, due to the following reasons:

1. Dropouts pay less in taxes because of their lower earnings.
2. Dropouts increase the demand for social services including welfare, medical assistance, and unemployment compensation.

3. Dropouts are less likely to vote.
4. Dropouts have poorer health.
5. Half of all state prison inmates did not complete high school.

Given these facts, it is small wonder that the U.S. Department of Education has focused an increasing amount of attention on how to improve high school completion rates.

Violence in the Schools

Nothing undermines the effectiveness of our educational system more than unsafe schools. Throughout the country, students bring to school drugs, guns, knives, and other paraphernalia of destruction. Gone are books, pencils, and paper—the tools of learning that must be present for education to take place. Many urban school systems screen students with metal detectors when entering school grounds.

In 2000, students ages 12 through 18 were victims of about 128,000 serious violent crimes at school (that is, rape, sexual assault, robbery, and aggravated assault). There were also 47 school-associated violent deaths in the United States between July 1, 1998 and June 30, 1999, including 38 homicides, 33 of which involved school-aged children.

A disturbing number of students in grades 9 to 12 reported they had carried a weapon at least once during the previous month. Nationwide, 4% of students had missed 1 or more days of school during the 30 days preceding the survey because they had felt unsafe at school or when traveling to or from school (U.S. Centers for Disease Control and Prevention, April 1999).

Gangs have also become a major problem in many schools. Fifteen percent of the students said their school had gangs, and 16% claimed that a student had attacked or threatened a teacher at the school (U.S. Department of Education, 2000). In some schools, where gangs freely sell drugs to students within school buildings, principals vainly chain doors in an effort to keep dealers out and students in. In other schools, students are afraid to use the filthy bathrooms because gang members hang out there.

The good news is that the percentage of students being victimized at school also has declined over the last few years. Between 1995 and 2001, the percentage of students who reported being victims of crime at school decreased from 10 to 6%. However, the prevalence of other problem behavior at school has increased. For example, in 2001, more students reported that they had been bullied at school in the last 6 months, than in 1999. As the rates of criminal victimization in schools have declined or remained constant, students also seem to feel more secure at school now than just a few years ago. The percentage of students who reported carrying a gun decreased from 11.8% in 1993 to 6.1% in 2003 (Statistical Abstract of the United States, 2006).

Home Schooling

Home schooling is emerging as one of the most significant social trends in education. It is an alternative to traditional schooling in which parents assume the primary responsibility for the education of their children. This trend of what is in fact an old practice has occurred for a distinctly modern reason: a desire to wrest control from public education and reestablish the family as central to a child's learning.

Approximately 850,000 to 1 million students are being educated at home, up from 15,000 in the early 1980s.

Home schooling is almost always a matter of choice and not a necessity because of the unavailability of schools. Public schools are there, but home-schooling families choose not to use them. In the past decade there has been an explosive growth in this type of schooling. The numbers are still growing. The number of home schoolers nearly tripled in the 5 years from 1990–1991 to 1995–1996, when there were, according to the best possible estimate, about 700,000 home schoolers (Lines, 2000). The most recent survey released in August of 2003 by the Department of Education estimated that as many as 1.1 million—at least 1.9% of all students nationwide—were being home schooled (Statistical Abstract of the United States, 2006).

The contemporary home-schooling trend began as a liberal, not a conservative, alternative to the public school. Some families in the late 1950s and early 1960s found schools were too rigid and conservative. They instead wanted to pursue a more liberal philosophy of education as advocated by educators such as John Holt, the author of *Why Children Fail*. Holt suggested the best learning took place when children were allowed to pursue their own interests without an established curriculum.

Conservative and religious families joined the home-schooling trend in the 1980s when they decided public schools were undermining their values. Some believed religious duty required them to teach their own children; others sought to integrate religion, learning, and family life. Joining the liberal and conservative wings of the home-schooling movement are families who simply seek the highest quality education for their child, which they believe public and even private schools can no longer provide.

Home schooling is not a new idea or practice. For centuries, children have learned outside formal school settings. Compulsory schooling is relatively new.

> Not until the nineteenth century did state legislatures begin requiring local governments to build schools and parents to enroll their children in them. Even then, compulsory requirements extended to only a few months a year.... Only recently have we begun to treat schooling as a fulltime affair entrusted to professional teachers. Yet in such a short span of time, most of the nation has come to accept classroom schooling as the norm, and so the recent upsurge in home schooling has come to many as a surprise. (Lines, 2000)

The typical home-schooling family is religious, conservative, white, middle income, and better educated than the general population. Home schoolers are more likely to be part of a two-parent family, and the mother typically assumes the largest share of the teaching responsibility, although fathers almost always are involved also.

Most home-schooling children spend time at libraries, museums, or classes offered at a local public school. Normally, parents plan and implement the learning program. The Internet has provided an important resource for these parents and enables them to share information on books and learning opportunities.

How well do home schoolers do? The question is usually related to test scores. But according to Lines (2000), "many home-schooling parents reject this criterion, since their mission is to impart not simply skills but a particular set of values." Approximately 69% of home schoolers go to college compared with 71% of public school graduates (Hammons, 2001). That said, "virtually all of the reported data show that home-schooled children score above average, sometimes well above average." Self-selection may be an important factor here because the parents are often highly motivated in bringing a quality education to their children (Lines, 2000).

The most frequent criticism of home schooling is that it fails to prepare children to deal with the social aspects of life. But there is plenty of evidence to the contrary. Almost all home-schooled children participate in extracurricular activities. Many public schools allow home schoolers to participate in team sports, science labs, or social organizations (Hammons, 2001).

The success or failure of home schooling depends on the success or failure of the family's interpersonal relationships. Home schooling is a complex issue and represents a tremendous commitment on the part of the parents. In most cases the father is the sole earner in the family and the mother spends her time instructing the children.

More research on home schooling is necessary. Up to now research has been limited to case studies of families or self-reports from participants in home schooling. We need a more accurate and thorough assessment of this growing trend in education.

Standardized Testing

In American schools, the standardized test is the most frequently used means of evaluating students' aptitudes and abilities. Every year, more than 100 million standardized tests are administered, ranking the mental talents of students from nursery to graduate school.

Children encounter standardized tests almost from the first day they go to school. Usually, their first experience with testing is an intelligence test. These are given to more than 2 million youngsters each year. Students are also required to take a number of achievement tests, beginning in elementary school. High school and college seniors take college admissions tests that decide whether they will be accepted at universities and graduate schools.

Much criticism has been leveled at standardized tests. The testing services say the tests merely try to chart, scientifically and objectively, different levels of mental achievement and aptitude. The critics assert that the tests are invalid academically and biased against minorities.

The Educational Testing Service's (ETS) Scholastic Aptitude Test (SAT) is the best-known college admissions test and is required by about 1,200 U.S. colleges and universities. Another 2,800 American colleges require or recommend the American College Test (ACT). Students wishing to go to graduate school are required to take other exams, which measure the ability and skills used in the fields that they wish to enter.

The ETS professes to be meticulous in its test construction. It hires college students, teachers, and professors to assist its staff in writing questions. Each of the approximately 3,000 questions that are created each year are reviewed by about 15 people for style, content, or racial bias (see "For Further Thinking: Are College Admissions Tests Fair?").

Figure 14–3 **Average SAT Scores, 1995–2005**

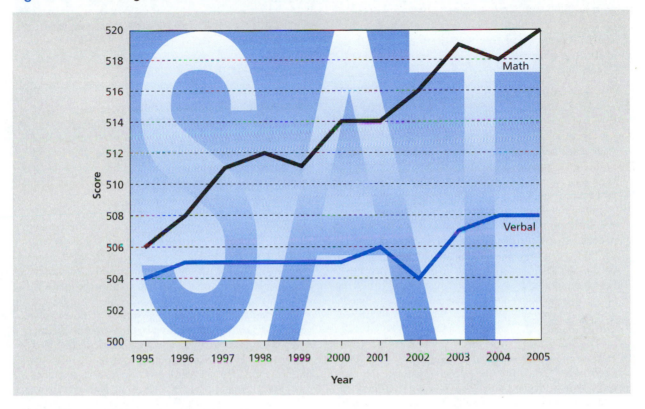

Source: The College Board.

The criticism of standardized tests, however, continues to grow. Many assert that all standardized tests are biased against minorities. The average African American or Hispanic youngster encounters references and vocabulary on a test that are likely to be more familiar to white, middle-class students. Many others oppose the secrecy surrounding the test companies. Groups have pushed for truth in testing, meaning that the test makers must divulge all exam questions and answers shortly after the tests are given. This would enable people to evaluate the tests more closely for cultural bias and possible scoring errors. The testing industry opposes such measures, which would force it to create totally new tests for each administration without the possibility of reusing valid and reliable questions.

No one would contend that standardized tests are perfect measuring instruments. At best, they can provide an objective measure to be used in conjunction with teachers' grades and opinions. At worst, they may discriminate against minorities or not validly measure potential ability. Nonetheless, college admissions officers insist that results from standardized college admissions tests give them a significant tool for evaluating students from a variety of backgrounds and many different parts of the country.

More students are taking the SAT than ever before, a fact that might be expected to result in lower scores.

Instead, 2005 scores are now above the 1980 level, reversing a trend that saw scores drop 90 points between 1963 and 1980. In part, the stabilized scores can be attributed to the more rigorous standards of the back-to-basics movement—a movement that has brought about modest gains among poor performers (Figure 14–3).

Gender Bias in the Classroom

Who is at greater risk of failing in our schools these days—girls or boys? Some researchers (Sadker & Sadker, 1994) believe that sitting in the same classrooms, reading the same textbooks, and listening to the same teachers, boys and girls receive very different educations. A body of research, including most prominently a 1992 study sponsored by the American Association of University Women, *How Schools Shortchange Girls*, has held that girls face much deeper difficulties in school, whereas boys get the lion's share of educational resources and teachers' attention. Over the course of years, the uneven distribution of teacher time, energy, attention, and talent, takes its toll on girls.

Girls are the majority of our nation's schoolchildren. However, each time a girl opens a book and reads a womanless history, she learns she is devalued. Each time the teacher passes over a girl to elicit the

FOR FURTHER THINKING

Are College Admissions Tests Fair?

The Educational Testing Service (ETS), the organization that designs the standardized tests that colleges use for admissions decisions, will administer its Scholastic Assessment Test (SAT) to some 9 million students this year. Admissions testing has never been free from criticism. Today, in fact, its critics are more numerous and more vociferous than ever.

The oldest and most familiar accusation against standardized tests is that they are discriminatory. The prime evidence for this charge is the test results themselves. For many years now, the median score of blacks on the SAT has been 200 points below that of whites. Less dramatic, but no less upsetting to groups like the Center for Women's Policy Studies, has been the persistent 35-point gender gap in scores on the math section of the SAT.

The SAT produces such disparate results, say critics, because the questions favor certain kinds of students over others. Thus, fully comprehending a reading selection might depend on background knowledge naturally available to an upper-middle-class white student (by virtue, say, of foreign travel or exposure to the performing arts) but just as naturally unavailable to a lower-class black student from the ghetto.

At the same time, women are said to be put at a disadvantage by the multiple-choice format itself. Singled out for blame are math questions that emphasize abstract reasoning and verbal exercises based on selecting antonyms, both of which supposedly favor masculine modes of thought. In fact, so biased are the tests, according to their opponents, that they fail to perform even the limited function claimed for them: forecasting future grades. The SAT, consistently "under predicts" the college marks of both women and minorities, which hardly inspires confidence in its ability to measure the skills it purports to identify.

Another line of criticism of the tests grants their accuracy in measuring certain academic skills but challenges the notion that these are the skills most worth having. High test scores, opponents insist, reveal little more than a talent for taking tests. It is also suggested that no mere standardized test can capture the qualities that translate into real-world achievement.

The critics have already won some significant concessions. Faced with both adverse publicity and threats of legal action by activists and the U.S. Department of Education, ETS has tried to remedy differences in group performance. On the Preliminary Scholastic Assessment Test (PSAT), which is used for choosing National Merit Scholars, a new method of scoring was recently introduced in the hope that more women might garner the prestigious award. The old formula, which assigned equal weight to the math and verbal sections of the test, was replaced by an index in which the verbal score, usually the higher one for female test takers, was doubled.

More widely publicized was the massive "recentering" of SAT scores that went into effect with the 1996

---

ideas and opinions of boys, that girl is conditioned to be silent and to defer. As teachers use their expertise to question, praise, probe, clarify, and correct boys, they help those male students sharpen ideas, refine their thinking, gain their voice, and achieve more. When female students are offered the leftovers of teacher time and attention, morsels of amorphous feedback, they achieve less.

Sadker and Sadker point out the following:

- In high school, girls score lower on the SAT and ACT tests, which are crucial for college admission.
- The greatest gender gap is in the crucial areas of science and math.
- Boys are much more likely to be awarded state and national college scholarships.
- Women score lower on all sections of the Graduate Record Exam, which is necessary to enter many postgraduate programs. Women also

trail on most tests needed to enter business school, law school, and medical school.

Christina Hoff Sommers (1994, 2000) believes these examples of bias are exaggerated. She has pointed out that 56% of college students are female. In 1971, women received 43% of bachelor's degrees, 40% of master's degrees, and 14% of doctorates. Today, women receive 53.9% of BAs, 53.6% of MAs, and 41% of doctoral degrees.

In addition, 15 years ago, roughly the same number of boys and girls were taking advanced placement exams in U.S. high schools—a strong indicator of who is enrolling in the most intellectually challenging courses. Boys now have fallen well behind. Boys today are less apt than girls to complete high school, attend college, and stay out of jail. They do less homework and come to school worse prepared. They read and write less well. Save for sports, they are less frequent participants in extracurricular activities of

results. Though the declared aim was to create a better distribution of scores around the test's numerical mid-point, the practical effect was a windfall for students in almost every range.

But because neither "recentering" nor any other such device has succeeded in eliminating disparities in scores, opponents of tests have had to look elsewhere. The law school at the University of California at Berkeley, for instance, has introduced a selection system that will consider a "coefficient of social disadvantage" in ranking applicants.

Some schools go further, hoping simply to do away with standardized tests altogether. There are, they insist, other, less problematic indicators of student merit. High school grades are a starting point, but no less important are essays, interviews, and work portfolios that offer a window into personal traits no standardized test can reveal.

Defenders of the college admissions tests point out that the SAT is hardly the meaningless academic snapshot described by critics. Results from these tests have been shown to correspond with those on a whole range of other measures, including IQ tests. This led the National Academy of Science to conclude that standardized tests display no evidence whatsoever of cultural bias.

Nor do the tests fail to predict how minority students will ultimately perform in the classroom, defenders note. If, indeed, the purported bias in the tests were real, such students would earn better grades in college than what is suggested by their SAT scores; but that is not the case.

Foes of testing are correct when they point out that women end up doing better in college than their scores would indicate. But defenders of the test say the "under prediction" is very slight—a tenth of a grade point on the 4-point scale—and only applies to less demanding schools. For more selective institutions, the SAT predicts the grades of both sexes equally accurately.

As for the claim that test scores depend heavily on income, defenders of the tests claim the facts again tell us otherwise. Though one can always point to exceptions, students who are not of the same race but whose families earn alike tend, on average, to perform very differently.

What about relying less on tests and more on other measuring rods like high school grades? Unfortunately, as everyone knows, high schools across the country vary considerably, not only in their resources but in the demands they make of students. It was precisely to address this problem that a single nationwide test was introduced in the first place.

Source: Excerpted from "The War against Testing," by D.W. Murray, September 1998, *Commentary*. Reprinted by permission from *Commentary*; all rights reserved.

every sort. In those areas—like scores on math and science tests—where boys continue to hold an edge, the gap is fast narrowing. In those areas where girls are out in front, it seems to be widening.

Others, such as Carol Gilligan, have shifted the debate somewhat. She has pointed out that because girls possess a different moral sensibility—being more caring, empathetic, nurturing, sensitive, and community-minded than boys—they are devalued by society and discouraged from entering its most important institutions and honored vocations.

In recent years, even as schools have tried to put in place measures to end the last vestiges of discrimination against students, the debate over whether bias exists in the classroom continues.

The Gifted

The very term *gifted* is emotionally loaded. The word may evoke feelings that range from admiration to resentment and hostility. Throughout history, people have displayed a marked ambivalence toward the gifted. It was not unusual to view giftedness as either divinely or diabolically inspired. Genius was often seen as one aspect of insanity. Aristotle's observation "There was never a great genius without a tincture of madness" continues to be believed as common folklore.

People also tend to believe that intellectualism and practicality are incompatible. That belief is expressed in such sayings as "He (or she) is too smart for his (her) own good" or "It's not smart to be too smart." High intelligence is often assumed to be incompatible with happiness.

There is little agreement on what constitutes giftedness. The most common measure is performance on a standardized test. All those who score above a certain level are defined as gifted, though there are serious problems when this criterion alone is used. Arbitrary approaches to measuring giftedness tend to ignore the likelihood that active intervention could increase the number of candidates among females, minorities, and the disabled, groups that are often underrepresented among the gifted.

Ellen Winner (1996) has proposed that gifted children have three atypical traits. These include (1) *precociousness*—gifted children begin early to master some domain; (2) *nonconformity*, an insistence on doing things according to their own specific rules; and (3) *a rage to master*, or a desire to know everything there is to know about a subject.

Females tend to be underrepresented among the gifted because popular culture holds that high intelligence is incompatible with femininity; thus some girls quickly learn to deny, disguise, or repress their abilities. Minorities are hindered because commonly used assessment tools discriminate against ethnic groups whose members have had different cultural experiences or use English as a second language. The intellectual ability of disabled youngsters is often overlooked. Their physical handicaps may mask or divert attention from their mental potential, particularly when communication is impaired, as this is a key factor in assessment procedures.

Teachers often confuse intelligence with unrelated school behaviors. Children who are neat, clean, and well mannered, have good handwriting, or manifest other desirable but irrelevant classroom traits may often be thought to be very bright.

Teachers often associate giftedness with children who come from prominent families, have traveled widely, and have had extensive cultural advantages. Teachers are likely to discount high intelligence when it might be present in combination with poor grammar, truancy, aggressiveness, or learning disabilities.

The first attempt to deal with the gifted in public education took place in the St. Louis schools in 1868. The program involved a system of flexible promotions enabling high-achieving students not to remain in any grade for a fixed amount of time. By the early 1900s, special schools for the gifted began to appear.

There has never been a consistent, cohesive national policy or consensus on how to educate the gifted. Those special programs that have been instituted have reached only a small fraction of those who conceivably could benefit from them. A serious problem with the education of the gifted arises from philosophical considerations. Many teachers are reluctant to single out the gifted for special treatment, as they feel that the children are already naturally privileged. Sometimes, attention given to gifted children is seen as antidemocratic.

No matter how inadequate it may seem, the effort to provide for the educational needs of learning-disabled children has far exceeded that expended for the gifted. Similarly, the time and money spent on research into educating the slower children far outstrip that set aside for research on materials, methodology for teaching, and so on for the gifted.

When schools do have enrichment programs, they are rarely monitored for effectiveness. Enrichment programs are often provided by teachers totally untrained in dealing with the gifted, for it is assumed that anyone qualified to teach is capable of teaching the gifted. Yet most basic teacher-certification programs do not require even one hour's exposure to information on the theory, identification, or methodology of teaching such children. Most administrators do not have the theoretical background or practical experience necessary to establish and promote successful programs for the gifted.

There is some evidence that the nation's population of gifted children—and possibly, prodigies—is growing. Researchers who test large numbers of children have detected a startling proportion in the 170 to 180 IQ range.

Although psychologists agree that early exceptional ability should be nurtured to thrive, they do not necessarily think that the current movement to produce superbabies by force-feeding a diet of mathematics and vocabulary to infants is a good idea. Pediatricians have begun seeing children with backlash symptoms—headaches, stomachaches, hair-tearing, anxiety, depression—as a result of this pressure to perform.

History has shown that being an authentic child prodigy creates problems enough of its own. The fine line between nurturing genius and trying to force a bright but not brilliant child to be something he or she is not is clearly one that must be walked with care.

Karen Stone McCown (1998) working with a group of Nobel Prize winners found many reported that their social-emotional development was shortchanged. They said that they were so self-motivated to pursue their intellectual passions that almost nothing would have stopped that work—but missing from their lives were the social skills that would help them interact with and connect to family, friends, and the larger world.

It appears that there are more than 2.5 million schoolchildren in the United States who can be described as gifted, or about 3% of the school population. Giftedness is essentially potential. Whether these children will achieve their potential intellectual growth will depend on many factors, not the least of which is the level of educational instruction they receive. We must question why we continue to show such ambivalence toward the gifted and why we are willing to tolerate incompetence and waste in regard to such a valuable resource (Baskin & Harris, 1980).

SUMMARY

- Americans believe the most common reason for getting an education is to get a better job.
- Functionalist sociologists suggest education also consists of activities that are functional for the society as a whole.
- One of the manifest functions of education is socialization.
- In nonindustrial societies, no real distinction is made between education and the socialization that occurs within the family.
- Another function of education is cultural transmission, in which major portions of society's knowledge are passed from one generation to the next.
- A third function of schools is to equip children with the academic skills needed to function as adults.
- Learning how to think independently and creatively is probably one of the most valuable tools the educational institution can transmit.
- In addition to its manifest functions, schooling in America has developed a number of unintended consequences as well.
- One latent function of many public schools is to provide child care outside of the nuclear family. This function has become increasingly important in recent years with the growing number of women in the labor force and the dramatic increase in single-parent families.
- A related service is to provide students with at least one nutritious meal a day.
- Conflict theorists view society as an arena of conflict, in which certain groups dominate others, and social institutions become the instruments by which those in power can control the less powerful.
- Conflict theorists see education as a means for maintaining the status quo by producing the kinds of people the system needs.
- This is accomplished through teaching the hidden curriculum—attitudes and values that

prepare children to accept the requirements of adult life and to fit into the social, political, and economic statuses the society provides.
- Although the overall high school graduation rate has been increasing, the dropout rate for minorities has remained high.
- Factors associated with dropping out include low educational and occupational attainment levels of parents, low family income, speaking a language other than English in the home, single-parent families, and poor academic achievement.
- Violence in schools causes many urban schools to operate under a siege mentality, with doors chained shut and students afraid to be in the wrong place at the wrong time, inside of school or out.
- In American schools, the standardized test is the most frequently used means of evaluating student aptitude, ability, and performance.
- Critics assert that these tests are academically invalid and culturally biased against minorities and the lower class.

 *Media Resources*

The Companion Website for *Introduction to Sociology*, Ninth Edition

http://sociology.wadsworth.com/tischler9e

Supplement your review of this chapter by going to the companion website to take one of the Tutorial Quizzes, use the flash cards to master key terms, and check out the many other study aids you'll find there. You'll also find special features such as Wadsworth's Sociology Online Resources and Writing Companion, GSS data, and Census 2000 information at your fingertips to help you complete that special project or do some research on your own.

CHAPTER FOURTEEN STUDY GUIDE

KEY CONCEPTS AND THINKERS

Match each concept with its definition, illustration, or explanation presented below.

a. De facto segregation   **e.** Cultural transmission   **h.** No Child Left Behind
b. White flight   **f.** The credentialized society   **i.** Tracking
c. "Nation at Risk"   **g.** De jure segregation   **j.** *Brown v. Board Education/Topeka*
d. Hidden curriculum

____   **1.** The process in which major portions of a society's knowledge are passed from one generation to the next.
____   **2.** The social attitudes and values taught in school that prepare children to accept the requirements of adult life and to "fit into" the social, economic, and political statuses the society provides.
____   **3.** The stratification of students by ability, social class, and various other categories.
____   **4.** A form of racial separateness based on laws prohibiting interracial contact.
____   **5.** Supreme Court decision that ended de jure segregation in public schools.
____   **6.** A form of racial separateness resulting from residential housing patterns.
____   **7.** The migration of large numbers of white Americans from the central cities to the suburbs.
____   **8.** The increasing trend in the United States for more and more jobs to require a degree regardless of whether possession of a degree increases job performance.
____   **9.** President George W. Bush's education plan requiring extensive standardized testing.
____   **10.** A 1983 report detailing a "rising tide of mediocrity" in U.S. education.

Match the thinkers with their main ideas or contributions.

a. James Coleman   **c.** R. Rosenthal and L. Jacobson   **e.** Lester Frank Ward
b. Jonathan Kozol   **d.** Herbert Spencer

____   **1.** Conducted a famous study on the effect teachers' expectations had on student performance.
____   **2.** Author of a 1966 survey of 645,000 children that demonstrated substantial class and race differences in educational achievement and opportunity.
____   **3.** Author of several books criticizing public education for shortchanging the poor, especially black and Hispanic children.
____   **4.** Social Darwinist who argued that little could be done to help those at the bottom of the social class ladder and that a school curriculum should be judged on its practical value.
____   **5.** Sociologist who maintained that the purpose of education was to equalize society.

CENTRAL IDEA COMPLETIONS

Following the instructions, fill in the appropriate concepts and descriptions for each of the questions posed in the following section.

1. Remembering your own high school as an example, describe the manifest and latent functions of education.

   a. Manifest: _____

   b. Latent: _____

2. Higher education, looked at from a slightly different perspective, is job postponement via alternative activities. What are the consequences for the individual and the society?

   a. Individual: _____

   b. Society: _____

3. What is standardized testing? What are the purposes that its proponents claim for it? What are some of the problems with standardized testing?

    a. Definition: _____

    b. Purposes: _____

    c. Problems: _____

4. Who are the people that are likely to engage in home schooling? Why has its popularity been increasing in recent years?

    a. Characteristics of home schoolers: _____

    b. Popularity factors: _____

5. What are the major factors contributing to the dropout rate for high school students?

    _____

    _____

    _____

    _____

6. To what extent do American minorities continue to face unequal access to education?

    _____

    _____

    _____

    _____

7. How would a conflict theorist evaluate education in American public high schools?

    _____

    _____

    _____

CRITICAL THOUGHT EXERCISES

1. American schools have been characterized as "being under siege" regarding the levels of violence now common in many places. What factors contribute to violence in the schools? To what extent was violence commonplace in the high school you attended prior to attending this college or university? Go to the documents or educational research section of your library and find a program devoted to reducing high school violence. What are its basic features? How would a sociologist evaluate its effectiveness?

2. Socialization is obviously a manifest function of elementary schools. But in less obvious ways it continues even in higher education. Examine your college or university in terms of its socialization function. What activities are designed to socialize the student population? To what extent are these socialization activities functionally linked to the purposes and goals of your institution? What values, beliefs, norms, and other ideas form the content of this perhaps unstated curriculum? How do these compare with the official aims of the institution as set forth in its mission and goals statement?

3. What implications does the Rosenthal-Jacobson study have for the policy of "tracking"?

INTERNET ACTIVITIES

1. The National Center for Education Statistics (http://nces.ed.gov/surveys/) is a good gateway to a wealth of data on education. One link (http://nces.ed.gov/ccd/bat/) takes you to a page where you can create your own table including only the variables you are particularly interested in.

2. *American Educator* is the magazine of the AFT (the nation's leading teacher's union). The online version (http://www.aft.org/pubsreports/american_educator/index.htm) has articles about primary and secondary education. Their link on "hot topics" (http://www.aft.org/topics/index.htm) gives their view on current controversies in education.

3. For a view the AFT probably would disagree with, the Center for Education Reform (http://www.edreform.com/) provides support for charter schools, No Child Left Behind, and other conservative positions. Their "Issues" link takes you to articles supporting their own view on current controversies in education.

ANSWERS TO KEY CONCEPTS

1.e 2.d 3.i 4.g 5.j 6.a 7.b 8.f 9.h 10.c

ANSWERS TO KEY THINKERS

1.c 2.a 3.b 4.d 5.e

ThomsonNOW™

Reviewing is as easy as ❶ ❷ ❸

1. Before you do your final exam, take the ThomsonNOW diagnostic quiz to help you identify the areas on which you should concentrate. You will find information on ThomsonNOW and instructions on how to access all of its great resources on the foldout at the beginning of the text.
2. As you review, take advantage of ThomsonNOW's study videos and interactive Map the Stats exercises to help you master the chapter topics.
3. When you are finished with your review, take ThomsonNOW's posttest to confirm you are ready to move on to the next chapter.

© Kraft Brooks/CORBIS

15

Political and Economic Systems

Learning Objectives

After studying this chapter, you should be able to do the following:

- Distinguish between authority and coercion.
- Understand the basic functions of the state.
- Know the basic features of capitalism.
- Distinguish between capitalism, socialism, and democratic socialism.

- Describe the basic features of political democracy.
- Contrast the functionalist and conflict theory views of the state.
- Describe the major features of the American political system.

After every presidential election, we require the loser to engage in a public declaration of defeat. Why do we require this invasion of privacy? Why do we not allow the defeated candidate to suffer defeat away from the glare of the media? Could it be that this ritual serves an important social function?

Dwelling on defeat contradicts a basic American commitment to success. After a presidential election, our gaze is on the triumphant winner; but instead of drawing a veil of silence over the crushed hopes of the losing candidate, we require one final ordeal, the concession speech.

The words themselves are less important than the larger purpose they serve. They ritualize the passing of power and the legitimacy of the new authority. Throughout the world, the yielding of power is often a matter of life or death. The concession speech is not merely a report of an election result, it is a reaffirmation of the democratic process (Corcoran, 1994). It affirms the view that the control of the political process must be in the hands of the people and that we are quite fearful of concentrated political power.

The founders of the United States distrusted a strong unified government. The American Constitution, the oldest in the world, established a divided form of government. There was to be a president, two houses of Congress, and a federal high court. These actions represented a deliberate decision to create a weak political system. There were to be varying terms of office. The president was to be chosen every 4 years. Two senators from each state were to be chosen for 6-year terms, with one-third of the seats open every 2 years. The House of Representatives was to be filled every 2 years, with the number allotted to each state roughly proportional to its share of the national population (Lipset, 1996).

Thomas Jefferson thought that even this arrangement had to be reexamined frequently. He suggested that every two decades or so, a new generation should be required collectively to define its own values and redefine those of its forebears. Jefferson believed that a political system had to be taken out periodically, inspected, examined in the harsh daylight, and changed before it was bequeathed to a new generation. If it did not pass muster, then some other political system had to be found that could better guarantee life, liberty, and the pursuit of happiness. A new generation should not be burdened by the traditions of the past or the comfort of well-worn customs (Hart, 1993).

The founders of the United States realized that, in most societies, what laws are passed, or not passed, depend to a large extent on which categories of people have the power. The powerful in a society work hard to pass laws to their liking. They were trying to provide a prescription for how power was to be used and how it was to be passed from one generation to the next. It sounds like a fairly radical and idealistic view of how a country should be governed. However, the United States was born out of radical and idealistic conceptions of politics.

This chapter examines the political institution. **Politics** is *the process by which power is distributed and decisions are made.* This chapter will clarify what is unique about our two-party system and where the American political system fits into the whole spectrum of political institutions. Political systems are often an outgrowth of economic systems. We will also discuss the economy in this chapter.

Politics, Power, and Authority

There are more than 800 candidates for the U.S. House of Representatives every other year. Americans also select senators, governors, and a host of other officials on a regular basis. Candidates ring our doorbells, shake our hands, stuff our mailboxes, and exhort us through our television sets. They make promises they often cannot keep. This is politics, American style. Small wonder that it has been said that politics, like baseball, is the great American pastime. Running for president of the United States is a political activity; so is enacting legislation; so is taxing property owners to subsidize the digging of sewers; so is going to war. The study of the political process, then, is the study of power.

Power

Max Weber (1958) referred to **power** as *the ability to carry out one person's or group's will, even in the presence of resistance or opposition from others.* In this sense, power is the capability of making others comply with one's decisions, often exacting compliance through the threat or actual use of sanctions, penalties, or force.

In some relationships, the division of power is spelled out clearly and defined formally. Employers have specific powers over employees, army officers over enlisted personnel, ship captains over their crews, professors over their students. In other relationships, the question of power is defined less clearly and may even shift back and forth, depending on individual personalities and the particular situation: between wife and husband, among sisters and brothers, or among friends in a social clique.

Power is an element of many types of relationships, and is a complex phenomenon that covers a broad spectrum of interactions. At one pole is **authority**—*power that is regarded as legitimate by those over whom it is exercised, who also accept the authority's legitimacy in imposing sanctions or even in using force if necessary.* For example, in the United States few people are eager to pay income taxes, yet most

do so regularly. Most taxpayers accept the authority of the government not only to demand payment but also to impose penalties for nonpayment.

At the other extreme is **coercion**—*power that is regarded as illegitimate by those over whom it is exerted.* Their compliance is based on fear of reprisals that are not recognized as falling within the range of accepted norms. Power based on authority is quite stable, and obedience to it is accepted as a social norm. Power based on coercion, in contrast, is unstable. People will obey only out of fear, and any opportunity to test this power will be taken. Power based on coercion will fail in the long run. The American Revolution, for example, was preceded by the erosion of the legitimacy of the existing system. The authority of the king of England was questioned, and his power, based increasingly on coercion rather than on acceptance as a social norm, inevitably crumbled.

Political Authority

An individual's authority often will apply only to certain people in certain situations. For example, a professor has the authority to require students to write term papers but no authority to demand the students' votes should he or she run for public office.

In the same sense, Weber pointed out that the most powerful states do not impose their will by physical force alone, but by ensuring that their authority is seen as legitimate. In such a state, people accept the idea that the allocation of power is as it should be and that those who hold power do so legitimately.

Weber (1957) identified three kinds of authority: legal-rational authority, traditional authority, and charismatic authority.

Legal-Rational Authority *is authority derived from the understanding that specific individuals have clearly defined rights and duties to uphold and implement rules and procedures impersonally.* Indeed, that is the key: Power is vested not in individuals but in particular positions or offices. There usually are rules and procedures designed to achieve a broad purpose. Rulers acquire political power through meeting requirements for office, and they hold power only as long as they obey the laws that legitimize their rule.

Traditional Authority *is rooted in the assumption that the customs of the past legitimate the present*—that things are as they always have been and basically should remain that way. Usually, both rulers and ruled recognize and support the tradition that legitimizes such political authority. Typically, traditional authority is hereditary, although this is not always the

The power of a charismatic leader derives from the ruler's force of personality and ability to inspire passion and devotion among followers. President John F. Kennedy was the closet the United States has come to having a charismatic leader.

© AP/Wide World Photos

case. For example, throughout most of English history, the English crown was the property of various families. As long as tradition is followed, the authority is accepted.

Charismatic Authority *derives from a ruler's ability to inspire passion and devotion among followers.* Weber noted that a charismatic leader—who is most likely to emerge during a period of crisis—will emerge when followers (1) perceive a leader as somehow supernatural, (2) blindly believe the leader's statements, (3) unconditionally comply with the leader's directives, and (4) give the leader unqualified emotional commitment. Others (Willner, 1984) have added that charismatic leaders also must perform seemingly extraordinary feats and have outstanding speaking ability.

Sitting Bull and Red Cloud, for example, were charismatic leaders of the Sioux Indians. Their people followed them because they led by example and inspired personal loyalty. However, individuals were free to disagree, to refuse to participate in planned undertakings, and even to leave and look for a group led by people with whom they were more likely to agree (Brown, 1971). This was not true in Russia under V. I. Lenin, in Germany under Adolf Hitler, or in Iran under the Ayatollah Ruholla Khomeini. These men all were charismatic rulers but also had the political authority necessary to enforce obedience or conformity to their demands.

Charismatic authorities and rulers emerge when people lose faith in their social institutions. Lenin led the Russian Revolution in the chaos left in the wake of World War I. Hitler rose to power in a Germany that had been defeated and humiliated in World War I and whose economy was shattered: Inflation was so bad that money was almost worthless. Khomeini rose to power in a country in which rapid modernization had undercut traditional Islamic norms and values, in which great poverty and great wealth existed side by side, and in which fear of the shah's secret police left the populace constantly anxious for its personal safety.

The great challenge facing all charismatic rulers is to sustain their leadership after the crisis subsides and to create political institutions that will survive their death or retirement. Weber pointed out that if the program that the leader has implemented is to be sustained, the leader's charisma must be routinized in some form. For example, after Christ's death—and after it became apparent that his return to earth was not imminent—the apostles began to set up the rudiments of a religious organization with priestly offices.

Government and the State

Governments vary according to the relationship that exists between the rulers and the ruled. In some societies, political power is shared among most or all adults. This is true of the Mbuti pygmies and the !Kung San of Africa, for instance, for whom the group is its own authority and decisions are made by a consensus among adults. Among such societies, the concept of government is meaningless. However, in larger, more complex societies, government does exist.

In complex societies, the **state** is *the institutionalized way of organizing power within territorial limits.* Just as a true government is nonexistent in some societies, so too is the state limited to certain societies. Its presence indicates a high level of social and political development.

Modern thought regarding the nature of government and its forms is derived directly from the ideas of three Greek philosophers: Socrates, his student Plato, and Plato's student Aristotle. Socrates (469–399 B.C.) believed that evil and wrong actions arise from ignorance and the failure to investigate why people act as they do. In the *Republic,* written around 365 B.C., Plato discusses the form of government that would be most just. When Plato (ca. 428–348 B.C.) was writing, Athens had been through a period of political upheaval, so Plato was concerned with maintaining social order. Hence, he rejected democracy or rule by the majority because he believed that this form of government would lead to chaos. He also rejected *autocracy,* or rule by one person, because he thought that no single person could be wise or competent enough to make decisions for a whole society. Rather, he favored what he called **aristocracy** *(a form of oligarchy), or rule by a select few.*

Plato called his proposed ruling class the guardians of society. The guardians, he argued, should be born from the most exemplary parents but then separated from them at birth. They should live in poverty for 30 years while being trained in mind and body, he said, and then they should fill positions of government in which they would execute their responsibilities wisely and without favoritism. It is important not to confound Plato's use of the term *aristocracy* with the modern usage. Plato explicitly rejected the ideal of inherited political power, which he believed inevitably results in power falling to unqualified individuals—leading to an unjust society.

Aristotle (384–322 B.C.) tutored Alexander the Great and was a political scholar. Unlike Plato, he recognized that even in just societies, social-class interests produce class conflicts, which the state must control. Aristotle favored centering political power in the middle class (consisting of merchants, artisans, and farmers), but he insisted on defining the rights and duties of the state in a legal constitution (Laslett & Cummings, 1967).

Functions of the State

Although a preindustrial society can exist without an organized government, no modern industrial society can thrive without those functions that the state performs: establishing laws and norms, providing social control, ensuring economic stability, setting goals, and protecting against outside threats.

Establishing Laws and Norms The state establishes laws that formally specify what the society expects and what it prohibits. The laws often represent a codification of specific norms; for example, one should not steal from or commit violent crimes against others. Establishing laws also means exacting penalties for violating the laws.

Providing Social Control In addition to establishing laws, the state also has the power to enforce them. The police, courts, and various government agencies make sure that violators of those laws are punished. In the United States, the Internal Revenue Service seeks out tax evaders, the courts sentence criminals to prison, and the police attempt to maintain order.

Ensuring Economic Stability In the modern world, no individual can provide entirely for his or her own

One of the tasks a state must perform is to protect itself from outside threats, especially those of a military nature.

needs. Large workforces must be mobilized to build roads, dig canals, and erect dams. Money must be minted, and standards of weights and measures must be set and checked; merchants must be protected from thieves, and consumers from fraud. The state tries to ensure that a stable system of distribution and allocation of resources exists within the society.

Setting Goals The state sets goals and provides a direction for the society. If a society is to curtail its use of oil, for instance, the government must promote this as a goal. It must encourage conservation and the search for alternative energy sources and must discourage (perhaps through taxation or rationing) the use of oil. How is the government able to accomplish these tasks? How can it bring about individual and organizational compliance? Obviously, it would be best if the government could rely on persuasion alone, but this course seldom is enough. In the end, the government usually needs the power to compel compliance.

Protecting against Outside Threats Historic data leave little doubt that the rise of the state was accompanied almost everywhere by the intensification of warfare (Otterbein, 1970, 1973). As early as the fourteenth century, Ibn Khaldun (1958), an Islamic scholar, noted this connection and even attributed the rise of the state to the needs of sedentary farmers to protect themselves from raids by fierce nomads. His views were echoed by Ludwig Gumplowicz (1899):

"States have never arisen except through the subjugation of one stock by another, or by several in alliance." In any event, it is clear that one of the tasks of maintaining a society is to protect it from outside threats, especially those of a military nature. Hence, governments build and maintain armies.

Although there is widespread agreement that the functions just described are tasks that the state should and usually does perform, not all social scientists agree that the state emerged because of the need for those functions.

Types of States

Different types of states exist side by side and must deal with one another constantly in today's shrinking world. To comprehend their interrelationships, it is helpful to understand the structure of each main form of government—autocracy, totalitarianism, and democracy.

Autocracy

In an **autocracy,** *the ultimate authority and rule of the government rest with one person, who is the chief source of laws and the major agent of social control.* For example, the pharaohs of ancient Egypt were autocrats. More recently, the reigns of Ferdinand Marcos of the Philippines, Juan and Eva Peron of Argentina, and Saddam Hussein of Iraq have been autocratic.

In an autocracy, the loyalty and devotion of the people are required. To ensure that this requirement is met, dissent and criticism of the government and the person in power are prohibited. The government controls the media and may use terror to prevent or suppress dissent. For the most part, however, no great attempt is made to control the personal lives of the people. A strict boundary is set up between people's private lives and their public behavior. Individuals have a wide range of freedom in pursuing such private matters as religion, family concerns, and many other traditional elements of life. At the same time, virtually all present-day autocracies have witnessed exploitation of the poor by the rich and powerful—a situation supported by the respective governments.

Totalitarianism

In a **totalitarian government,** *one group has virtually total control of the nation's social institutions.* Any other group is prevented from attaining power. Religious institutions, educational institutions, political institutions, and economic institutions all are managed directly or indirectly by the state. Typically,

under totalitarian rule, several elements interact to concentrate political power.

1. *A single political party* controls the state apparatus. It is the only legal political party in the state. The party organization is itself controlled by one person or by a ruling clique.
2. *The use of terror* is implemented by an elaborate internal security system that intimidates the populace into conformity. It defines dissenters as enemies of the state and often chooses, arbitrarily, whole groups of people against whom it directs especially harsh oppression (for instance, the Jews in Nazi Germany or minority tribal groups in several recently created African states).
3. *The control of the media* (television, radio, newspapers, and journals) is in the hands of the state. Differing opinions are denied a forum. The media communicate only the official line of thinking to the people.
4. *Control over the military apparatus,* both the military personnel and the use of its weapons, is monopolized by those who control the political power of the totalitarian state.
5. *Control of the economy* is wielded by the government, which sets goals for the various industrial and economic sectors and determines both the prices and the supplies of goods.
6. *An elaborate ideology,* in which previous sociopolitical conditions are rejected, legitimizes the current state and provides more or less explicit instructions to citizens on how to conduct their daily lives. This ideology offers explanations for nearly every aspect of life, often in a simplistic and distorted way (Friedrich & Brzezinski, 1965).

Democracy

Democracy has not always been regarded as the best form of government. People have often approved of the aims of democracy, yet have argued that democracy was impossible to attain. Others argued that it was logically unsound. Today, however, there is hardly a government anywhere in the world that does not claim to have some sort of democratic authority. In the United States, we regard our political system as democratic, and the same claim is made by leaders in communist countries. The word *democracy* seems to have so many different meanings today that we face the problem of distinguishing democracy from other political systems.

Democracy comes from the Greek words *demos,* meaning "people," and *kratia,* meaning "authority." By democracy, then, the Greeks were referring to a system in which rule was by the people rather than a few selected individuals. Because of the growth in population, industrialization, and specialization, it has become impossible for citizens to participate in politics today as they did in ancient Athens. Today, **democracy** refers to *a political system operating under the principles of constitutionalism, representative government, majority rule, civilian rule, and minority rights.*

Constitutionalism *means that government power is limited.* It is assumed that there is a higher law, which is superior to all other laws. The various agencies of the government can act only in specified legal ways. Individuals possess rights, such as freedom of speech, press, assembly, and religion, which the government cannot take away.

A basic feature of democracy is that it is rooted in **representative government,** which means that *the authority to govern is achieved through, and legitimized by, popular elections.* Every government officeholder has sought, in one way or another, the support of the **electorate** *(those citizens eligible to vote)* and has persuaded a majority of that group to grant its support *(through voting).* The elected official is entitled to hold office for a specified term and generally will be reelected as long as that body of voters is satisfied that the officeholder is adequately representing its interests.

Representative institutions can operate freely only if certain other conditions prevail. First, there must be what sociologist Edward Shils (1968) calls *civilian rule*—that is, every qualified citizen has the legal right to run for and hold an office of government. Such rights do not belong to any one class (say, of highly trained scholars, as in ancient China), caste, sect, religious group, ethnic group, or race. These rights, with certain exceptions, belong to every citizen. Further, there must be public confidence that such organized agencies as the police and the military will not intervene in, or change the outcome of, elections.

In addition, *majority rule* must be maintained. Because of the complexity of a modern democracy, it is not possible for the people to rule directly. One of the most important ways for people to participate in the political life of the country is to vote. For this to happen people must be free to assemble, to express their views and seek to persuade others, to engage in political organizing, and to vote for whomever they wish.

Democracy also assumes that *minority rights* must be protected. The majority may not always act wisely, and it may be unjust. The minority abides by the laws as determined by the majority, but the minority must be free to try to change these laws.

Democratic societies contrast markedly with totalitarian societies. Ideally, democratic societies are open and culturally diverse, dissent is not viewed as disloyalty, there are two or more political parties, and terror and intimidation are not an overt part of the political scene.

The economic bases of democratic societies can vary considerably. Democracy can be found in a capitalistic country like the United States and in a more socialistic one like Sweden. However, it appears necessary for the country to have reached an advanced level of economic development before democracy can evolve. Such societies are most likely to have the sophisticated population and stability necessary for democracy (Lipset, 1960).

Functionalist and Conflict Theory Views of the State

Functionalists and conflict theorists hold very different ideas about the function of the state. As our discussion of social stratification in Chapter 8 revealed, functionalist theorists view social stratification—and the state that maintains it—as necessary devices that recruit workers to perform the tasks necessary to sustain society. Individual talents must be matched to jobs that need doing, and those with specialized talents must be given sufficiently satisfying rewards. Functionalists therefore maintain that the state emerged because society grew so large and complex that only a specialized, central institution (that is, the state) could manage society's increasingly complicated and intertwined institutions (Davis, 1949; Service, 1975).

Marxists and other conflict theorists take a different view. They argue that technological changes creating surplus production brought about the production of commodities for trade (as opposed to products for immediate use). Meanwhile, certain groups were able to seize control of the means of production and distribution of commodities, thereby establishing themselves as powerful ruling classes that dominated and exploited workers and serfs. Finally, the state emerged to coordinate the use of force, which allows the ruling classes to protect their institutionalized supremacy from the resentful and potentially rebellious lower classes. As Lenin (1949) explained, "The state is a special organization of force: it is an organization of violence for the suppression of some class."

There is evidence to support this view of the state's origins. The earliest legal codes of ancient states featured laws protecting the persons and properties of rulers, nobles, landholders, and wealthy merchants. The Code of Hammurabi of Babylon, dating to about 1750 B.C., prescribed the death penalty for burglars and for anybody who harbored a fugitive slave. The code regulated wages, prices, and fees to be charged for services. It provided that a commoner be fined six times as much for striking a noble or a landholder as for striking another commoner. It also condemned to death women who were proved by their husbands to be uneconomical in managing household resources (Durant, 1954).

Nevertheless, the functionalist view also has value. The state provides crucial organizational functions such as carrying out large-scale projects and undertaking long-range planning, without which complex society probably could not exist. Because it provides a sophisticated organizational structure, the state can—and does—fulfill many other important functions. In most modern societies, the state supports a public school system to provide a basic, uniform education for its members.

The health and well-being of its citizens also have become the concern of the state. In our own country, as in many others, the government provides some level of medical and financial support for its young, old, and disabled, and it sponsors scientific and medical research for the welfare of its people. Regulating industry and trade to some degree also has become a function of the modern state, and it has devised ways (different for each kind of state) of establishing, controlling, and even safeguarding the civil rights and liberties of its citizens. Certainly, one of the most important functions of any state is the protection of its people. Long gone are the days when cave dwellers, lords of the manor, or pioneers defended their territories and other members of their groups from attack or encroachment; the state now provides such protection through specialized agencies: armies, militias, and police forces.

When groups in a society develop sufficient dissatisfaction with their government and achieve the strength to influence the direction of that system, changes can take place. After such changes, the state may perform the same functions, but it may do so in a different way and under different leadership.

The Economy and the State

Philosophers have been writing about the relationship between the economy and the state for thousands of years. The Greek philosopher Plato, quoting Socrates, noted that both wealth and poverty are bad for society: Socrates suggested that a guard be placed at the gates of the city to keep wealth and poverty out. Wealth and poverty were not seen as two evils but as different sides to the same evil, for the wealth of the rich Plato believed was the cause of the poverty of the poor. For Plato this happens because the high consumption of the rich creates shortages for the poor.

For economist Adam Smith (1723–1790) poverty was due to the insufficient production of real wealth.

The obvious solution to this problem was to increase the wealth of the community. "No society can surely be flourishing and happy, of which the far greater part of the members are poor and miserable" (quoted in Clark, 2002). These two views represent the starting points for the economic systems that came to be known as capitalism and socialism.

In its simplest terms, the economy is *the social institution that determines how a society produces, distributes, and consumes goods and services.* Money, goods, and services do not flow of their own accord. People work at particular jobs, manufacture particular products, distribute the fruits of their labor, purchase basic necessities and luxury items, and decide to save or spend their money.

A very simple society may only produce and distribute food, water, and shelter. As a society becomes more complex and productive, the products produced and distributed become increasingly more elaborate. To be useful, all these goods and services must be distributed throughout the society. We depend on impersonal distribution systems to bring us such essential items as food, water, housing, clothing, health care, transportation, and communication, all of which we consume according to our ability to pay for them.

The political problem for any society is to decide how much to be involved in the production and distribution of goods and services. In most economies, markets play the major role in determining what is produced, how, and for whom. Which means, is there a demand for something? Who is willing to produce it? How much are people willing to pay for it?

Capitalism

In its classic form, **capitalism** is *an economic system based on private ownership of the means of production, in which resource allocation depends largely on market forces.* The government plays only a minor role in the marketplace, which works out its own problems through the forces of supply and demand.

There are two basic premises behind capitalism. The first, as Weber noted, is production "for the pursuit of profit and ever renewed profit." Capitalism entitles people to pursue their own self-interests, and this activity is desirable and eventually benefits society through the "invisible hand" of capitalism. For example, pharmaceutical companies may have no other goal in mind than profit when they develop new drugs. The fact that their products eventually benefit society is an indirect benefit brought about by this invisible hand.

The second basic premise behind capitalism is that the free market will determine what is produced and at what price. If people can profit from the production of a product, it will be produced. Adam

TECHNOLOGY AND SOCIETY

Selling Human Life

Even though we live in a capitalist society, there are certain things we believe should not be for sale. These include love, body parts such as hearts and kidneys, and of course children. We believe that it is wrong for people to profit from the sale of these things. The prohibition against the sale of children appears to be particularly strong. After all, how could you put a price on a child?

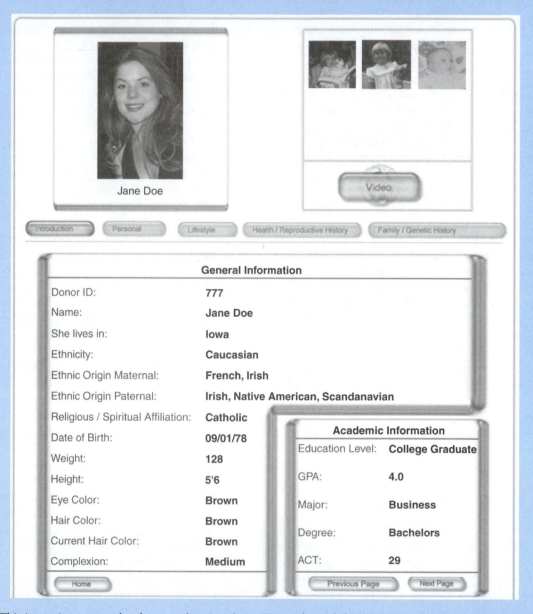

Jane Doe

Video

Introduction   Personal   Lifestyle   Health / Reproductive History   Family / Genetic History

General Information

Donor ID:	**777**
Name:	**Jane Doe**
She lives in:	**Iowa**
Ethnicity:	**Caucasian**
Ethnic Origin Maternal:	**French, Irish**
Ethnic Origin Paternal:	**Irish, Native American, Scandanavian**
Religious / Spiritual Affiliation:	**Catholic**
Date of Birth:	**09/01/78**
Weight:	**128**
Height:	**5'6**
Eye Color:	**Brown**
Hair Color:	**Brown**
Current Hair Color:	**Brown**
Complexion:	**Medium**

Academic Information

Education Level:	**College Graduate**
GPA:	**4.0**
Major:	**Business**
Degree:	**Bachelors**
ACT:	**29**

Home   Previous Page   Next Page

This image is an example of an egg donor website. To see the real thing go to www.eggdonation.com.

Yet, if we were to examine the situation more closely we would see that children and babies are indeed sold. The only difference is that the language we use is different from the language we use for normal business transactions.

Children from Latin America or China are not sold to the adopting parents—they are "matched" with their "forever families." The fee for adopting a healthy child from Guatemala is about $23,000. Eggs are "donated" by women who want to help infertile couples. The compensation they receive can range from $5,000 to $50,000, the latter being paid to a high-intelligence Ivy League graduate. Surrogate mothers will carry an embryo and give birth to a child that will be turned over to a couple. The total cost for this service averages $59,000. Many poor women have found the possibility of being paid to be a surrogate attractive. Currently 30% of surrogate mothers are women of color who are carrying white embryos.

A variety of firms exist that will process sperm to increase the likelihood of producing a male or female. Embryos are manipulated to reduce or produce certain genetic traits. The people involved in these activities believe they are helping individuals in need. Yet the motives or the words cannot hide the fact that these people are engaging in business transactions. Law firms, doctors, individuals acting as go-betweens, egg donors, and various lab technicians are all making money from these activities. We could divide the fertility and child trade business into three parts.

1. Supplier of the basic elements for making a baby: eggs, hormones, sperm, wombs, and embryos.
2. Production sites: fertility specialists, research firms, and clinics that supervise the baby-making process.
3. Intermediaries: adoption agencies, lawyers, and egg donor firms who match couples with the suppliers of the service.

In a typical year 40,000 babies are born through in vitro fertilization, 3,000 from donor eggs, and 1,300 are carried by surrogate mothers. Another 20,000 children are adopted from abroad. All these transactions involve payments and it is clear that a large business exists in the sale of human life.

Source: Debora Spar and Cate Reavis, "The Business of Life," Boston: Harvard Business School Publishing, 2004.

Typical Fees for Surrogate Mothers

Surrogate Compensation	$20,000.00
(Second Time Surrogate)	$23,000.00 and over
Maternity Clothing Allowance	$500.00 ($750.00 if carrying twins)
Transfer Fee	$500.00 for the first 2; $750.00 thereafter
Monthly Allowance	$200.00 and over
Twins Fee	$3,000.00
Triplets Fee	$6,000.00
C-Section Fee	$2,000.00
Lost Wages	100% reimbursement as incurred less disability if available
Selective Reduction/Termination	$750.00
CVS/ Amniocentesis/D&C/Miscarriage Fee	$500.00
Other: Housekeeping/Child Care/Take-Out Food	100% when necessary

Source: Center for Surrogate Parenting, http://www.creatingfamilies.com/surrogacy/smfin.html. Used by permission.

Smith (1723–1790) is regarded as the father of modern capitalism. He set forth his ideas in his book *The Wealth of Nations* (1776), which is still used as a yardstick for analyzing economic systems in the Western world. According to Smith, capitalism has four features: private property, freedom of choice, freedom of competition, and freedom from government interference.

Private Property Smith believed that the ability to own private property acts as an incentive for people to be thrifty and industrious. These motivations, although selfish, will benefit society, because those who own property will respect the property rights of others.

Freedom of Choice Along with the right to own property comes the right to do with it what one pleases as long as it does not harm society. Consequently, people are free to sell, rent, trade, give away, or retain whatever they possess.

Freedom of Competition Smith believed society would benefit most from a free market featuring unregulated competition for profits. Supply and demand would be the main factors determining the course of the economy.

Freedom from Government Interference Smith believed government should promote competition and free trade and keep order in society but should not regulate business or commerce. The best thing the government can do for business, Smith said, is to leave it alone. *This view that government should stay out of business is referred to as* **laissez-faire capitalism.** The French words *laissez-faire* mean "allow to act."

In the United States, the government plays a vital role in the economy. Therefore, the U.S. system cannot be seen as an example of pure capitalism. Rather, many have referred to our system as modified capitalism, also known as a mixed economy.

A **mixed economy** *combines free-enterprise capitalism with governmental regulation of business, industry, and social welfare programs.* Although private property rights are protected, the forces of supply and demand are not allowed to operate with total freedom. Resources are distributed through a combination of market and governmental forces.

Because there are few nationalized industries in this country (the Tennessee Valley Authority and Amtrak are two exceptions), the government uses its regulatory power to guard against private-industry abuse. Our government is also involved in such areas as antitrust violations, the environment, and minority employment. Ironically, this involvement may be even greater than it is in some of the more socialistic European countries.

Most countries have a mixed economy. Some countries, such as the United States and Taiwan, are closer to the free-market end of the continuum, whereas China and Cuba are closer to the command economy or highly planned end. The absolute extremes of a complete command economy or complete capitalism do not exist, whether it be in Cuba or the United States. Command economy countries let consumers choose some of the goods they buy and allow private agricultural markets to some extent. In the United States, in addition to regulating economic activity by setting minimum wage levels, requiring safety standards for the workplace, and enacting antitrust laws and farm price supports, the federal government has also assumed partial or total control of privately owned businesses when their collapse would significantly affect the people.

The Marxist Response to Capitalism

Smith believed that ordinary people would thrive under capitalism; not only would their needs for goods and services be met, but they also would benefit by being part of the marketplace. In contrast, Karl Marx was convinced that capitalism produces a small group of well-to-do individuals, while the masses suffer under the tyranny of those who exploit them for profit.

Marx argued that capitalism causes people to be alienated from their labor and from themselves. Under capitalism, he said, the worker is not paid for part of the value of the goods produced. Instead, he said, this "surplus value" is taken by the capitalist as profit at the expense of the worker.

Workers also are alienated, according to Marx, by doing very small specific jobs, as on an assembly line, and thereby not feeling connected to the final product. The worker feels no relationship or pride in the product and merely works to obtain a paycheck and survive—a far cry from work being the joyous fulfillment of self that Marx believed it should be.

Marx believed that nineteenth-century capitalism contained several contradictions that were the seeds of its own destruction. The main problem with capitalism, he contended, is that profits will decline as production expands. This, in turn, will force the industrialist to exploit the laborers and pay them less in order to continue to make a profit. As the workers are paid less or are fired, Marx said, they are less able to buy the goods being produced. This causes profits to fall even further, leading to bankruptcies, greater unemployment, and even a full-scale depression. After an increasingly severe series of depressions, Marx argued, the workers will rise up and take control of the state. They then will create a socialist form of government in which private property is abolished and turned over to the state. The workers will then control the means of production, and the exploitation of workers will end.

The reality of capitalism has not matched Marxist expectations. As we have seen, earlier capitalist economies have become much more mixed economies than the capitalist model developed by Smith. This has prevented some of Marx's prophesied outcomes. The level of impoverishment that Marx predicted for the workers has not taken place, as labor unions have obtained higher wages and better working conditions for labor. Marx thought these changes could come about only through revolution. Labor-saving machinery has also led to higher profits without the predicted unemployment, and the production of goods to meet consumer demands has increased accordingly.

Marxists have offered a number of explanations for capitalism's continued success. Some have suggested that capitalism has survived because Western societies have been able to sell excess goods to developing countries, and that these sales have enabled the capitalists to maintain high prices and profits. However, Marxists see this as only a temporary solution to the inevitable decline of capitalism. Eventually the whole world will be industrialized, they contend, and the contradictions in capitalism will be revealed. They believe the movement toward socialism has not been avoided, but only postponed.

In addition, critics of capitalism argue that true democracy is an impossible dream in a capitalist society. They claim that although in theory all members of a capitalist society have the same political rights, in fact capitalist society is inherently stratified, and therefore wealth, social esteem, and even political power are unequally distributed (see Chapter 8). Because of this, some critics contend, true democracy can be achieved only under socialism to which we now turn (Schumpeter, 1950).

Socialism

Socialism is *an economic system in which the sources of production—including factories, raw materials, and transportation and communication systems— are collectively owned.* Socialism is an alternative to and a reaction against capitalism. Whereas capitalism views profit as the ultimate goal of economic activity, socialism is based on the belief that economic activity should be guided by public needs rather than private profit. Under an ideal socialist system there would be: public ownership of production and property, government control of the economy without a profit motive, and central planning.

Public Ownership of Production and Property The government owns the factories and apartment buildings. Housing and goods are then made available to everyone at a reasonable price.

Government Control of the Economy without a Profit Motive Production and distribution are oriented toward output rather than profit. This orientation ensures that key industries run smoothly and that the public good is met. Individuals are heavily taxed to support a range of social-welfare programs that benefit every member of the society. Many socialist countries are described as having a "cradle-to-grave" welfare system.

Central Planning Instead of relying on the marketplace to determine prices, under socialism prices for major goods and services are set by government agencies. Socialists believe that major economic, social, and political decisions should be made by elected representatives of the people in conjunction with the broader plans of the state. The aim is to influence the economic system so that wealth and income are distributed as equally as possible. The belief is that everyone should have such essentials as food, housing, medical care, and education before some people can have luxury items, such as cars and jewelry.

The Capitalist View of Socialism

Capitalists view the centrally planned economies of socialist societies as inefficient and concentrating power in the hands of one group whose authority is based on party position. Among the problems they note are: no incentive to increase production, waste of resources, overregulation and inflexibility, and corruption of power.

No Incentive to Increase Production Capitalists claim the main stimulus to investment is competition; the threat that if you do not improve your product or your production process, your rivals will, and they will take your market share. Under capitalism it is necessary to cut costs and raise output in order to compete. Socialist economies must achieve planned targets. There are no rewards for doing better than the target. Capitalists claim that if the producers of goods and services are immune from competition they have few incentives to produce high-quality products.

Waste of Resources Socialist economies need to divert resources into planning, rather than actually producing. In the former Soviet Union, *Gosplan*, the state planning organization, needed to calculate 12 million prices a year, and plan the output of 24 million products.

Overregulation and Inflexibility Capitalists argue that socialist economies cannot be easily adjusted to take account of changing circumstances. They also note that if the state subsidizes essential goods and services and consumers do not pay full cost, nothing will prevent consumers from using more than they are entitled to and taking advantage of the system.

Corruption of Power   Lenin argued that to consolidate power, socialists must use strong repressive measures against the old capitalist governments—in fact, build a dictatorship. As Lenin (1949) noted:

> The proletariat needs state power, a centralized organization of force, and organization of violence, both to crush the resistance of the exploiters and to lead the enormous mass of the population . . . in the work of organizing a socialist economy.

In China, North Korea, Cuba, and more recently in African and Southeast Asian countries, socialist revolutions have resulted in dictatorships. Members of the previous ruling classes have been executed, jailed, "reeducated," or exiled, and their properties have been seized and redistributed. In none of these societies has the dictatorship proved to be temporary, nor has the state gradually withered away, as Marx and Friedrich Engels predicted it would after socialism was firmly established. Many Marxists contend that this will happen in the future, especially once capitalism has been defeated worldwide and socialist states no longer need to protect themselves against counterrevolutionary subversion and even direct military threats by capitalist nations. However, it is fair to observe that even the ancient Greeks knew that power corrupts and that those who have power are unlikely ever to give it up voluntarily.

Democratic Socialism

Democratic socialism is *a convergence of capitalist and socialist economic theory in which the state assumes ownership of strategic industries and services, but allows other enterprises to remain in private hands.* In Western Europe, democratic socialism has evolved as a political and economic system that attempts to preserve individual freedom in the context of social equality and a centrally planned economy.

With the parliamentary system of government present in many European countries, social democratic political parties have been able to win representation in the government. They have been able to enact their economic programs by being elected to office, as opposed to producing a workers' uprising against capitalism. The social democrats have also attempted to appeal to middle-class workers and highly trained technicians, as well as to industrial workers.

Under democratic socialism, the state assumes ownership of only strategic industries and services, such as airlines, railways, banks, television and radio stations, medical services, colleges, and important manufacturing enterprises. Certain enterprises can remain in private hands as long as government policies can ensure that they are responsive to the nation's common welfare. High tax rates prevent excessive profits and the concentration of wealth. In return, the population receives extensive welfare benefits, such as free medical care, free college education, or subsidized housing.

Democratic socialism flourishes to varying degrees in the Scandinavian countries, in Great Britain, and in Israel. These countries all have a strong private (that is, capitalist) sector in their economies, but they also have extensive government programs to ensure the people's well-being. Those programs pertain to such things as national health service, government ownership of key industries, and the systematic tying of workers' pay to increases in the rate of inflation. Many observers believe that the American political economy has been moving in this direction. The social democratic movement is an example of the convergence of the capitalist and socialist economic theories, a trend that has been evident for some time. Capitalist systems have seen an ever-greater introduction of state planning and government programs, and socialist systems have seen the introduction of market forces and the profit motive. The growing economic interdependence of the world's nations will help continue this trend toward convergence.

Political Change

Political change can occur when there is a shift in the distribution of power among groups in a society. It is one facet of the wider process of social change, the topic of the last part of this book. Political change can take place in a variety of ways, depending on the type of political structure the state has and the desire for change among the people.

People may attempt to produce change through established channels within the government, or they may rise up against the political power structure with rebellion and revolution. Here we shall consider briefly three forms of political change: institutionalized change, rebellion, and revolution.

Institutionalized Political Change

In democracies, the institutional provision for the changing of leaders is implemented through elections. Usually, candidates representing different parties and interest groups must compete for a particular office at formally designated periods. There may also be laws that prevent a person from holding the same office for more than a given number of terms. If a plurality or a majority of the electorate is

People may attempt to produce change through established channels within the government, or they may rise up against the political structure.

Coke Witworth/AP/Wide World Photo

dissatisfied with a given officeholder, they can vote the incumbent out of office. Thus, the laws and traditions of a democracy ensure the orderly changeover of politicians and, usually, of parties in office.

In dictatorships and totalitarian societies, if a leader unexpectedly dies, is debilitated, or is deposed, a crisis of authority may occur. In dictatorships, illegal, violent means must often be used by an opposition to overthrow a leader or the government, because no democratic means exist whereby a person or group can be legally voted out of power. Thus, we should not be surprised that revolutions and assassinations are most likely to occur in developing nations that have dictatorships. Established totalitarian societies, such as the former Soviet Union and China, are more likely than dictatorships to offer normatively prescribed means by which a ruling committee decides who should fill a vacated position of leadership.

Rebellions

Rebellions are *attempts—typically through armed force—to achieve rapid political change that is not possible within existing institutions.* Rebellions typically do not call into question the legitimacy of power, but, rather, its uses. For example, consider Shays' Rebellion. Shortly after the American colonies won their independence from Britain, they were hit by an economic depression followed by raging inflation. Soon, in several states, paper money lost almost all its value. As the states began to pay off their war debts (which had been bought up by speculators), they were forced to increase the taxation of farmers, many of whom could not afford to pay those new taxes and consequently lost their land. Farmers began

to band together to prevent courts from hearing debt cases, and state militias were called out to protect court hearings. Desperate farmers in the Connecticut River Valley armed themselves under Daniel Shays, an ex-officer of the Continental Army (Blum et al., 1981). This armed band was defeated by the Massachusetts militia, but its members eventually were pardoned and the debt laws were loosened somewhat (Parkes, 1968). Shays' Rebellion did not intend to overthrow the courts or the legislature; rather, it was aimed at effecting changes in their operation. Hence, it was a typical rebellion. (For a discussion of a current form of conflict designed to bring about change, see "Global Sociology: Does Suicide Terrorism Make Sense?")

Revolutions

Revolution is a powerful word. It evokes vivid images and strong emotions. It contains a mix of hope, excitement, and terror. Small wonder that many great works of art, literature, and film have been inspired by revolution. In contrast to rebellions, **revolutions** are *attempts at rapidly and dramatically changing a society's previously existing structure.* Sociologists further distinguish between political and social revolutions.

Political Revolutions Relatively rapid transformations of state government structures that are not accompanied by changes in social structure or stratification are known as **political revolutions** (Skocpol, 1979). The American war of independence is a good example of a political revolution. The colonists were not seeking to change the structure of society or even necessarily to overthrow the ruling order. Their goal was to put a stop to the abuse of power by the British.

GLOBAL SOCIOLOGY

Does Suicide Terrorism Make Sense?

Why would someone become a suicide terrorist? How can we explain that someone would want to engage in an act that would surely kill themselves as well as dozens or hundreds of innocent people? How can we possibly deter such fanaticism?

Robert Pape examined 188 suicide-terrorist attacks that occurred between 1980 and 2001. They took place throughout the world in places such as Lebanon, Israel, Sri Lanka, Chechnya, India, and Turkey. Pape found that far from being the irrational acts of desperate people, suicide terrorism is guided by identifiable strategic goals. Pape found that 95% of the incidents he studied were part of an organized political campaign. They were not unplanned acts of wanton cruelty. We may not understand the logic of the individual suicide bomber, but those who recruit and train them have a clear plan.

The terrorist organizations behind the attacks are trying to achieve specific political goals. These include coercing the targeted government to change its policy, and mobilizing additional recruits and financial support. The attacks may seem irrational for the individual terrorists, but they are meant to demonstrate that more and greater attacks are still to come. They are instruments of coercion, just as are air power and the dropping of bombs during a war.

Pape believes there are five principles to keep in mind when trying to understand suicide terrorists:

1. Suicide terrorism is strategic. The vast majority of the attacks are not isolated or random acts by individual fanatics. They are part of an organized campaign to achieve specific goals.

2. Suicide terrorism is designed to coerce democracies to make significant concessions. Pape found that every act of suicide terrorism since 1980 has been against a democratic form of government. In general the goal has been to achieve specific territorial goals.

3. Suicide terrorism has increased because the organizers have found it works.

4. The leaders of terrorist organizations have been able to credit suicide attacks with significant progress toward their goals.

5. The gains are moderate and do not lead to complete victory. The power of suicide attacks is the threat of more attacks. The attacks do not cause nations to abandon compelling national interests.

6. The best way to reduce suicide terrorism is to increase security efforts.

7. Pape believes attacks will decrease if the attackers lose confidence in their ability to carry them out.

The main goal of suicide terrorism is to inflict enough pain that the interest in resisting the terrorist demands will weaken and the government will concede to the demands.

Source: Robert A. Pape, "The Strategic Logic of Suicide Terrorism," *American Political Science Review,* Vol. 97, No. 3, August 2003, pp. 1–19.

After the war they created a new form of government, but they did not attempt to change the fact that landowners and wealthy merchants held the reins of political power—just as they had before the shooting started. In the American Revolution, then, a lower class did not rise up against a ruling class. Rather, it was the American ruling class going to war in order to shake loose from inconvenient interference by the British ruling class. The initial result, therefore, was political, not social, change.

Social Revolutions In contrast, **social revolutions** are *rapid and basic transformations of a society's state and class structures.* They are accompanied and in part carried through by class-based revolts (Skocpol, 1979). Hence, they involve two simultaneous and interrelated processes: (1) the transformation of a society's system of social stratification, brought about by upheaval in the lower class(es), and (2) changes in

the form of the state. Both processes must reinforce each other for a revolution to succeed. The French Revolution of the 1790s was a true social revolution. So were the Mexican Revolution of 1910, the Russian Revolution of 1917, the Chinese Revolution of 1949, and the Cuban Revolution of 1959—to name some of the most prominent social revolutions of the twentieth century. In all these revolutions, class struggle provided both the context and the driving force. The old ruling classes were stripped of political power and economic resources, wealth and property were redistributed, and state institutions were thoroughly reconstructed (Wolf, 1969).

Although the American Revolution did not arise from class struggle and did not result immediately in changes in the social structure, it did mark the beginning of a form of government that eventually modified the social stratification of eighteenth-century America.

The American Political System

The United States' political system is unique in a number of ways, growing out of a strong commitment to a democratic political process and the influence of a capitalist economy. It has many distinctive features that are of particular interest to sociologists. In this section, we will examine the role of the electorate and how influence is exerted on the political process.

The Two-Party System

Few democracies have only two main political parties. Besides the United States, there are Australia, New Zealand, and Austria. The other democracies all have more than two major parties, thus providing proportional representation for a wide spectrum of divergent political views and interests. In most European democracies, if a political party receives 12% of the vote in an election, it is allocated 12% of the seats in the national legislature. Such a system ensures that minority parties are represented.

The American two-party political system, however, operates on a winner-take-all basis. Therefore, groups with differing political interests must face not being represented if their candidates lose. Conversely, candidates must attempt to gain the support of a broad spectrum of political interest groups, because a candidate representing a narrow range of voters cannot win. This system forces accommodations between interest groups on the one hand and candidates and parties on the other.

Few, if any, individual interest groups (like the National Organization for Women, the National Rifle Association, or the Conservative Caucus) represent the views of a majority of an electorate, be it local, state, or national. Hence, it is necessary for interest groups to ally with political parties in which other interest groups are involved, hoping thereby to become part of a majority that can succeed in electing one or more candidates. Each interest group then hopes that the candidate(s) it has helped elect will represent its point of view. For most interest groups to achieve their ends more effectively, they must find a common ground with their allies in the party they have chosen to support. In doing so, they often have to compromise some strongly held principles. Hence, party platforms often tend to be composed of mild and noncontroversial issues, and party principles tend to adhere as closely as possible to the center of the American political spectrum.

Some have argued that the two-party system filters out extreme political views. A multiparty allows the extreme, and sometimes destabilizing, elements into the political system. The Nazi party gained credibility after being brought into a German coalition government when none of the major parties could gain a clear electoral victory.

When either party attempts to move away from the center to accommodate a very strong interest group with left- or right-wing views, the result generally is disaster at the polls. This happened to the Republican party in 1964, when the politically conservative Barry Goldwater forces gained control of its organizational structure and led it to a landslide defeat. In that year's presidential election, the Democrats captured 61.1% of the national vote. Eight years later, the Democratic party made the same mistake. It nominated George McGovern, a distinctly liberal candidate who, among other things, advocated a federally subsidized minimum income. The predictable landslide brought the Republicans and Richard Nixon 60.7% of the vote.

The candidates themselves have other problems. To gain support within their parties, they must somehow distinguish themselves from the other candidates. In other words, they must stake out identifiable positions. Yet to win state and national elections, they must appeal to a broad political spectrum. To do this, they must soften the positions that first won them party support. Candidates thus often find themselves justly accused of double-talking and vagueness as they try to finesse their way through this built-in dilemma. This is most true of presidential candidates. It is no accident that once they have their party's nomination, some candidates may express themselves differently and far more cautiously than they did before.

Voting Behavior

In totalitarian societies, strong pressure is put on people to vote. Usually there is no contest between the candidates, as there is no alternative to voting for the party slate. Dissent is not tolerated, and nearly everyone votes. In the United States, there is a constant progression of contests for political office, and in comparison with those of many other countries, the voter turnout is quite low. Since the 1920s, the turnout for presidential elections has ranged from about 50% to nearly 70% of registered voters. A record 122.3 million people, or 60.0% of those eligible, cast a vote for president in 2004 (Bureau of the Census, 2005).

Considering the emphasis that Americans place on a democratic society, it is interesting that participation in national elections is declining steadily, a cause of grave concern to social scientists and political observers alike.

Voting rates vary with the characteristics of the people. For example, those 45 and older and with a college education and a white-collar job have high rates of voter participation. Hispanics, the young, and the unemployed have some of the lowest voter participation rates (see Table 15–1 for voter participation

Table 15–1

Voter Participation by Selected Characteristics, 2004

Characteristic	Percentage Voting
Sex	
Male	56.3
Female	60.1
Race	
White	65.8
Black	56.3
Asian	29.8
Hispanic	28.0
Age	
18–20	41.9
25–34	46.9
35–44	56.9
45–54	64.5
55–64	69.7
65–74	70.8
Education Attained	
Fewer than 9 years	23.6
High-school graduate	52.4
College graduate	72.6
Employment	
Unemployed	46.4
Employed	60.0

Source: U.S. Census Bureau, Current Population Survey, November 2004

by selected characteristics). The age group with the highest proportion of voters is 55- to 74-year-olds, with more than 7 in 10 casting ballots. The lowest voting rates belong to 18- to 24-year-olds, who in 2004 even though they voted at higher rates than usual still had fewer than half of their numbers vote.

The Democratic party has tended to be the means through which the less privileged and the unprivileged have voted for politicians whom they hoped would advance their interests. Since 1932, the Democratic party has tended to receive most of its votes from the lower class, the working class, blacks, those of Southern and Eastern European descent, Hispanics, Catholics, and Jews. Thus, almost all the legislation that has been passed to aid these groups has been promoted by the Democrats and opposed by the Republicans: Democrats have sponsored legislation supporting unions, Social Security, unemployment compensation, disability insurance, antipoverty legislation, Medicare and Medicaid, civil rights, and consumer protection.

The Republican party has tended to receive most of its votes from the upper-middle and lower-middle classes, Protestants, and farmers. Americans younger than 30 tend to vote the Democratic ticket; and those older than 49 tend to vote for the Republican party, even though the Democratic party has been responsible for almost all the legislation to benefit older citizens. These voting patterns are, of course, generalizations and may change during any specific election.

Both parties tend to be most responsive to the needs of the best-organized groups with the largest sources of funds or blocks of votes. Thus, we would expect the Republicans to represent best the interests of large corporations and well-funded professional groups (such as the American Medical Association). The Democrats would logically be more closely aligned with the demands of unions.

Factors other than the social characteristics of voters and the traditional platforms of parties may affect the way people vote (Cummings & Wise, 1981). Indeed, the physical attributes, social characteristics, and personality of a candidate may prompt some people to vote against the candidate of the party they usually support. More important, the issues of the period may cause voters to vote against the party with which they usually identify. When people are frustrated by factors such as war, recession, inflation, and other international or national events, they often blame the incumbent president and the party he represents.

Since the 1960s there have been efforts to increase the number of minority members who register to vote and to improve their voting rate. The greater prominence of minority candidates has helped with this effort, as minority groups are more likely to vote if they feel that the elections are relevant to their lives. As minority group members increase their voting rates, they also become successful in electing members of their groups. Figure 15–1 shows the consistent rise in the number of African American and Hispanic elected officials.

Women also have been successful in increasing their representation in state legislatures. Figure 15–2 shows that the number of women holding such offices nearly tripled between 1975 and 2003.

Despite these advances, the members of Congress are still overwhelmingly white men older than 40. In 2003, the Senate had only 14 women members and no African Americans; the other 86 members were white men. The lack of women and African Americans in the House of Representatives was also striking (see Table 15–2 for a description of selected characteristics of members of Congress).

It would be wrong to conclude from the figures cited that Americans are politically inactive. There is more to political activity than voting. A study of American political behavior identified four modes of participation (Verba, 1972): (1) Some 21% of Americans eligible to vote do so more or less regularly in municipal, state, and national elections but

Figure 15–1 African American and Hispanic Elected Officials, 1985–2001

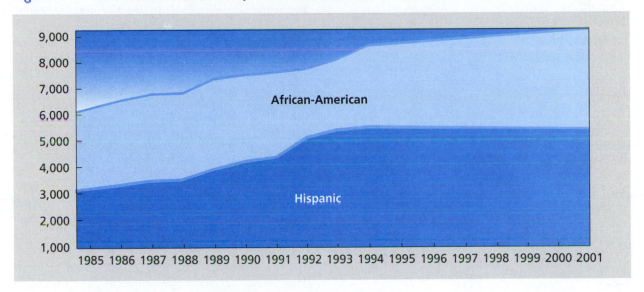

Source: *Statistical Abstract of the United States: 2004-2005* (p. 255), U.S. Bureau of the Census, 2005, Washington, DC: U.S. Government Printing Office.

Figure 15–2 Women Holding Office within State Legislatures, 1975–2004

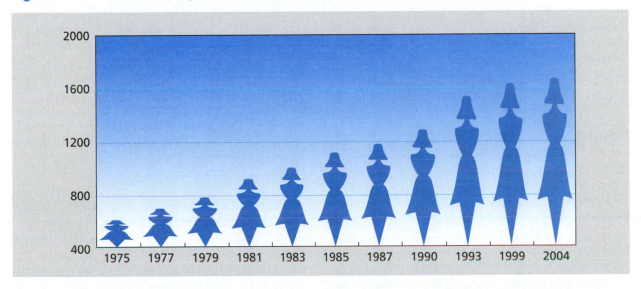

Source: *Statistical Abstract of the United States: 2006* (p. 261), U.S. Bureau of the Census, 2006, Washington, DC: U.S. Government Printing Office; Center for American Women and Politics, Rutgers University.

do not engage in other forms of political behavior. (2) Roughly 4% of the American electorate not only vote but also make the effort to communicate their concerns directly to government officials in an attempt to influence the officials' actions. (3) Some 15% of the electorate vote and periodically take part in political campaigns. However, they do not involve themselves in ongoing local political affairs. (4) About 20% of eligible voters are relatively active in ongoing community politics. They vote but do not get involved in political campaigns.

African Americans as a Political Force

African Americans have increased their presence at the polls from 37% in 1994 to 56.3% in 2004 (Bureau of the Census, 2004). Part of the explanation for this increase may be that African Americans always seem to receive attention before major political campaigns, as each candidate attempts to convince them that he or she takes their interests to heart. Yet African Americans have often been skeptical of those preelection promises, sensing that once

Table 15–2

Selected Characteristics of Members of Congress 2003

	Senators	Representatives
Sex		
Male	86	376
Female	14	59
Race		
White	98	369
Black	0	39
Asian-Pacific Islander	2	5
Hispanic	0	22
Age		
Younger than 40	19	
40–49	12	86
50–59	29	174
60–69	34	121
70 and over	29	32

Source: *Statistical Abstract of the United States: 2004-2005* (p. 250), U.S. Bureau of the Census, 2005, Washington, DC: U.S. Government Printing Office.

the election is won, their concerns will again be given low priority.

Indeed, the statistics on African American progress in economic areas are still grim. Although progress has occurred in a variety of areas in recent years, the economic gap between whites and African Americans remains large.

The picture is somewhat more promising for education, but when one considers such areas as infant health, adult mortality rates, and income, the movement of African Americans into mainstream middle-class America is still a long way off.

Statistics show an improvement in income levels of African American married-couple families, educational attainment and school enrollment, and home ownership among blacks during the previous decade. There have been setbacks also, demonstrated by high African American unemployment, sharply increased divorce and separation rates, and a rise in family households maintained by black women.

A cursory look at income statistics makes blacks appear better off than they are. African American married couples have closed the gap with white incomes, but the proportionate share of married couples in the total black population is smaller than a decade ago (McKinnon & Humes, 2000).

There has been a fragmentation of the African American family. Forty-seven percent of all African American families were married couples, 45% were maintained by women without a husband present, and 8% were maintained by men without a wife present. When we look at married-couple African American families, we find that only about 7% are below the poverty level (McKinnon & Humes 2000).

African Americans also have fallen further behind whites in the accumulation of wealth. Part of this was because they were less likely to own property and therefore were not able to take advantage of rapidly rising real estate values over the previous decade. Regardless of income level, African Americans tend to have less wealth than whites. Even blacks who already own a home or eventually manage to purchase one must contend with the fact that most homes owned by blacks are in central cities, areas in which homes are less likely to appreciate than in the suburbs.

In the political arena, African Americans are not just another interest group. They experience deep-seated economic differences that make them skeptical of political promises. Those differences have made African Americans cautious about supporting white Democrats, no matter how liberal the candidates' voting records are.

Hispanics as a Political Force

There were 27.1 million Hispanics of voting age in November 2004. However, only 34.3% were registered to vote compared with 67.7% of whites and 63.5% of African Americans (Bureau of the Census, 2004).

Hispanic voters are crucial in many presidential elections. The main reason is the geographical distribution of the voters. Nearly half of all Hispanic voters are registered in six major states—New York, New Jersey, Florida, Illinois, Texas, and California—which together can give a presidential candidate most of the electoral votes needed to win the election.

Hispanic voters have traditionally been overwhelmingly Democratic. Hispanics tend to support Democratic candidates because of the support those candidates give to social programs that help the poor. The fact that Hispanics, like African Americans, have a harder time climbing the socioeconomic ladder keeps them in the Democratic camp.

With the Hispanic population expected to grow rapidly over the next few decades, Democrats and Republicans will be paying more attention to the needs of this group. It is also likely the Hispanics will be electing greater numbers of Hispanic candidates to political office.

A Growing Conservatism

Public opinion tends to change as the age composition of the population changes. Older people are typically more conservative than younger people. As the average age of the U.S. population has been increasing, we can expect that conservatism will increase. In addition, if the younger population is also

OUR DIVERSE SOCIETY

Comparing the Political and Moral Values of the 1960s with Today

We often hear about the era of the 1960s and the massive social changes that emerged from that period. It was a time of sweeping changes in civil rights legislation. The youth culture was challenging the moral authority of government in many areas. Social changes, whether in the area of family, sex, religion, or race, were taking place everywhere. Have we become more conservative as a society than we were during that period? In some areas, we clearly have moved away from the liberal views of the 1960s. Surprisingly, in other areas we are actually a much more tolerant society today than we were in the 1960s.

Examples of the more conservative mood include the distrust of the government. During the 1960s, only a distinct minority of Americans was distrustful of government. Today two-thirds of Americans do not trust government to do the right thing (Table 15–3). The proportion of Americans who see big government as the biggest threat to the country's future has grown from 35% in the 1960s to 59% today. In addition, far fewer Americans today want government to do more to help minorities reduce income differences than in the 1960s.

On the other hand, there have been striking changes in the tolerance Americans express for different groups in society. Today, 87% of Americans are against laws that would prohibit marriages between African Americans and whites. In the 1960s, only 36% felt that way. As late as 1977, only 56% of Americans thought homosexuals should have equal rights in terms of job opportunities. Today, 87% support equal job opportunities. Americans have even become more willing to vote for an atheist for president, increasing from 18 to 44%.

Table 15–3

American Values—The 1960s and Today

	The 1960s	Today
Do not trust government in Washington to do the right thing	23%	66%
Would like government to do more to help minorities	50%	37%
Government should reduce differences in income	48%	30%
There should be no laws against blacks and whites being able to marry	36%	87%
Homosexuals should have equal rights in terms of job opportunities	56%	87%
Would vote for a well-qualified atheist for president	18%	44%

Source: "The 60s and the 90s: Americans' Political, Moral, and Religious Values Then and Now," by R. J. Blendon, J. M. Benson, M. Brodie, D. E. Altman, R. Moran, Claudia Deane, & N. Kjellson, Spring 1999, *Brookings Review,* pp. 14–17.

growing more conservative in its own right, as many have claimed, the impact on conservatism doubles.

Data to help us decide whether these assumptions are correct exist in the General Social Survey (GSS) from the National Opinion Research Center (NORC). Thirty years of GSS results leave little doubt that a conservative mood has pervaded the country. All age groups participated in the shift toward conservatism. Overall, self-described liberals dropped from 30% of Americans in 1975 to 25% in 1985, and 19% in 1994, while self-described conservatives increased from 30% to 33%.

The leading edge of the post–World War II baby boom, those 54 to 59 years old in 2005, are already more conservative on capital punishment, the courts, welfare spending, gun control, pornography, and spending on the poor, the cities, and space exploration than they were in 1975 when they were in their 20s. There is only one issue on which they become more liberal as they age: spending for foreign aid. This same group, when asked to categorize themselves as conservative, moderate, or liberal, has grown decidedly more conservative. (See "Our Diverse Society: Comparing the Political and Moral Values of the 1960s with Today.")

The Democratic party has undergone a good deal of soul-searching in recent years. Some have criticized the party for moving to the right in response to perceived shifts in public opinion. If the Democratic party is to continue as a viable political force, it must contend with the changing political climate. Labor unions, long a bastion of Democratic support, are in decline, and the new college-educated service workers are more likely to be Republican. The ranks of those who call themselves Republicans or independents are growing, whereas those who call themselves Democrats are declining. We are witnessing a major turning point in American politics.

The Role of the Media

The media have contributed to a radical transformation of election campaigns in the United States. This transformation involves changes in how political candidates communicate with the voters, and in the information journalists provide about election campaigns.

On one side, we have candidates who are trying to present self-serving and strategically designed images of themselves and their campaigns. On the other side are television and print journalists who believe that they should be detached, objective observers, motivated primarily by a desire to inform the U.S. public accurately.

Although the norm is often violated, there is no doubt that the journalistic perspective is markedly different from that of the candidate. Campaign coverage is far more apt to be critical and unfavorable toward a candidate than favorable. Journalists strive to reveal a candidate's flaws and weaknesses and to uncover tasty tidbits of hushed-up information.

Journalists exercise considerable political power in four important ways. First, the most obvious way involves deciding how much coverage to give a campaign and the candidate involved. A candidate who is ignored by the media has a difficult time becoming known to the public and acquiring important political resources, such as money and volunteers. Such candidates have little chance of winning.

In the beginning stages of presidential nomination campaigns, for example, a candidate's goal is typically to do something that will generate news coverage and stimulate campaign contributions. These contributions can then be used for further campaigning, helping convince the press and the public that the candidate is credible and newsworthy.

Second, the media decide which of many possible interpretations to give to campaign events. Because an election is a complex and ambiguous phenomenon, different conclusions can be drawn about its meaning. Here the journalists help the public form specific impressions of the candidates.

For example, presidential candidates are always concerned with how the results of presidential primaries are interpreted. Candidates want to be seen as winners who are gaining momentum. At the same time, a candidate seen as the front-runner too early is open to being shot down later.

Third, the media exercise discretion in how favorably candidates are presented in the news. Although norms of objectivity and balance prevent most campaign coverage from including biased assertions, a more subtle and pervasive slant or theme to campaign coverage is possible and can be significant.

Finally, newspaper editors and publishers may officially endorse a candidate. This support can be particularly important to the candidate if the newspaper is one of the major national publications.

Politicians need the mass media to get the coverage they need to win. At the same time, they can very easily fall victim to the intense scrutiny that is likely to result.

Although a great deal has been written about the power of the press, little is known about the upper echelons of the newspaper hierarchy—the top decision makers or the boards of directors of newspaper-owning corporations. Much of what is known about these individuals comes from official and unofficial biographies of publishers, histories of particular newspapers, and journalistic accounts. Though these studies suggest that publishers and board members can influence the general tone of a paper as well as specific stories, there has been little systematic research on the characteristics of these people and how (or if) they are connected to other sectors of the U.S. power structure.

Special-Interest Groups

With the government spending so much money and regulating so many industries, special-interest groups constantly attempt to persuade the government to support them financially or through favorable regulatory practices. **Lobbying** refers to *attempts by special-interest groups to influence government policy.* Farmers lobby for agricultural subsidies, labor unions for higher minimum wages and laws favorable to union organizing and strike actions, corporate and big business interests for favorable legislation and less government control of their practices and power, the National Rifle Association to prevent the passage of legislation requiring the registration or licensing of firearms, consumer-protection groups for increased monitoring of corporate practices and product quality, the steel industry for legislation taxing or limiting imported steel, and so on.

Lobbyists Of all the pressures on Congress, none has received more publicity than the role of the Washington-based lobbyists and the groups they represent. The popular image of a lobbyist is an individual with unlimited funds trying to use devious methods to obtain favorable legislation. The role of today's lobbyist is far more complicated.

The federal government has tremendous power in many fields, and changes in federal policy can spell success or failure for special-interest groups. With

Special interest groups often attempt to persuade the government to support them with favorable legislation.

the expansion of federal authority into new areas and the huge increase in federal spending, the corps of Washington lobbyists has grown markedly. The number of registered lobbyists swelled to 37,319 in 2005, twice the number in 1999. The lobbyists spent 2.1 billion dollars trying to get legislation passed (Congressional Budget Office, 2005).

Lobbyists usually are personable and extremely knowledgeable about every aspect of their interest group's concerns. They cultivate personal friendships with officials and representatives in all branches of the government, and they frequently have conversations with these government people, often in a semisocial atmosphere, such as over drinks or dinner.

The pressure brought by lobbyists usually has self-interest aims—that is, to win special privileges or financial benefits for the groups they represent. On some occasions the goal may be somewhat more objective, as when the lobbyist is trying to further an ideological goal or to put forth a group's particular interpretation of what is in the national interest.

There are certain liabilities associated with lobbyists. The key problem is that they may lead Congress to make decisions that benefit the pressure

group but may not serve the interests of the public. A group's influence may be based less on the arguments for their position than on the size of the membership, the amount of their financial resources, or the number of lobbyists and their astuteness.

Lobbyists might focus their attention not only on key members of a committee but also on the committee's professional staff. Such staffs can be extremely influential, particularly when the legislation involves highly technical matters about which the member of Congress may not be knowledgeable. Lobbyists also exert their influence through testimony at congressional hearings. Those hearings may give the lobbyist a propaganda forum and access to key lawmakers who could not have been contacted in any other way. The lobbyists may rehearse their statements before the hearing, ensure a large turnout from their constituency for the hearing, and may even give leading questions to friendly committee members so that certain points can be made at the hearing.

Lobbyists do perform some important and indispensable functions, such as helping inform Congress and the public about problems and issues that normally may not get much attention, thereby stimulating public debate and making known to Congress who would benefit and who would be hurt by specific legislation. Many lobbyists believe that their most important and useful role, both to the groups they represent and to the government, is the research and detailed information they supply. In fact, many members of the government find the data and suggestions they receive from lobbyists to be valuable in studying issues, making decisions, and even in voting on legislation.

Political Action Committees Special-interest groups called **political action committees (PACS)** are *organized for the purpose of raising and spending money to elect and defeat candidates*. Most PACs represent business, labor, or ideological interests. PACs have been around since 1944, when the first one was formed to raise money for the reelection of President Franklin D. Roosevelt. PACs contribute millions of dollars to congressional and senatorial political campaigns.

Several criticisms have been leveled at these special-interest groups. Among the most prominent is that they represent neither the majority of the American people nor all social classes. Most PACs represent groups of affluent and well-educated individuals or large organizations. Only about 10% of the population is in a position to exert this kind of pressure on the government. Disadvantaged groups—those that most need the ear of the government—have no access to this type of political action (Cummings & Wise, 1981).

PACs also tend to favor incumbents. Two-thirds of all PAC contributions in recent elections have gone to incumbents. Challengers therefore end up being much more dependent on small donations from individuals or from the Democratic and Republican National Committees. PACS ultimately may diminish the role of the individual voter.

SUMMARY

- In most societies, what laws are passed or not passed depends largely on which categories of people have power.
- Politics is the process by which power is distributed and decisions are made.
- Power, as Weber defined it, is the ability of a person or group to carry out its will, even in the face of resistance or opposition.
- Power is exercised in a broad spectrum of ways.
- At one pole is authority, or power that is regarded as legitimate by those over whom it is exercised, who also accept the authority's legitimacy in imposing sanctions or even in using force if necessary.
- At the other extreme is coercion, or power that is regarded as illegitimate by those over whom it is exerted.
- Power based on authority is quite stable, and obedience to it is accepted as a social norm.
- Power based on coercion is unstable. It is based on fear; any opportunity to test it will be taken, and in the long run it will fail.
- In modern, complex societies, however, government is necessary, and the state is the institutionalized way of organizing power within territorial limits.
- The state has a variety of functions.
- One is to establish laws that formally specify what is expected and what is prohibited in the society.
- A second function is to enforce those laws and make sure that violations are punished.
- In modern societies, the state must also try to ensure that a stable system of distribution and allocation of resources exists.
- The state also sets goals and provides a direction for society, usually through the power to compel compliance.
- Finally, the state must protect a society from outside threats, especially those of a military nature.

- Politics and economics are intricately linked, and the political form a state takes is tied to the type of economy.
- The economy is the social institution that determines how a society produces, distributes, and consumes goods and services.
- Democracy refers to a political system operating under the principles of constitutionalism, representative government, civilian rule, majority rule, and minority rights.
- Functionalists view social stratification—and the state that maintains it—as necessary for the recruitment of workers to perform the tasks required to sustain society.
- Marxists and other conflict theorists argue that the state emerged as a means of coordinating the use of force, by means of which the ruling classes could protect their institutionalized supremacy from the resentful and potentially rebellious lower classes.
- Growing out of a strong commitment to a democratic political process and the influence of a capitalist economy, the political system in the United States is unique in that it has a winner-take-all two-party system.
- In this system, political candidates are forced to gain the support of a broad spectrum of interest groups to be elected.
- Special-interest groups must find common ground and join together in coalitions or face the prospect of not being represented at all if their candidate loses.

 Media Resources

The Companion Website for *Introduction to Sociology,* Ninth Edition

http://sociology.wadsworth.com/tischler9e

Supplement your review of this chapter by going to the companion website to take one of the Tutorial Quizzes, use the flash cards to master key terms, and check out the many other study aids you'll find there. You'll also find special features such as Wadsworth's Sociology Online Resources and Writing Companion, GSS data, and Census 2000 information at your fingertips to help you complete that special project or do some research on your own.

CHAPTER FIFTEEN STUDY GUIDE

KEY CONCEPTS AND THINKERS

Match each concept with its definition, illustration, or explanation below.

a. Revolution
b. Aristocracy
c. Socialism
d. Traditional authority
e. Democracy

f. Laissez-faire capitalism
g. Rebellion
h. Power
i. Capitalism

j. Democratic socialism
k. Charismatic authority
l. Legal-rational authority
m. Coercion

_____ 1. The ability of people or groups to get their way, even in the face of resistance or opposition.
_____ 2. Power that is regarded as illegitimate by those over whom it is exerted.
_____ 3. A form of authority derived from the understanding that specific individuals have clearly defined rights and duties to uphold and implement rules and procedures impersonally.
_____ 4. A form of authority rooted in the assumption that the customs of the past legitimate the present.
_____ 5. A form of authority derived from a ruler's ability to inspire passion and devotion among followers.
_____ 6. A form of government in which a select few rule.
_____ 7. An economic system based on private ownership of the means of production, and resource allocation through the market.
_____ 8. The view that government should stay out of business affairs.
_____ 9. An economic system in which the government owns the sources of production and sets production and distribution goals.
_____ 10. A political system operating under the principles of constitutionalism, representative government, majority rule, civilian rule, and minority rights.
_____ 11. A convergence of capitalist and socialist economic theory in which the state assumes ownership of strategic industries and services, but allows other enterprises to remain in private hands.
_____ 12. An attempt—typically through armed force—to achieve rapid political change not possible within existing institutions.
_____ 13. An attempt to change a society's previously existing structure rapidly and dramatically.

Match the thinkers with their main idea or contributions.

a. Adam Smith
b. Plato

c. Thomas Jefferson
d. Karl Marx

e. Max Weber

_____ 1. Developed the sociological definitions of power and authority.
_____ 2. Constructed a philosophical argument for aristocracy as the best form of government.
_____ 3. Regarded as the father of modern capitalism; discussed many of the basic premises of this system.
_____ 4. A severe critic of capitalism who argued that it was based on alienation and exploitation.
_____ 5. Argued a political system had to be taken out periodically, inspected, and examined and changed if necessary before passing it on to the next generation.

CENTRAL IDEA COMPLETIONS

Following the instructions, fill in the appropriate concepts and descriptions for each of the questions posed in the following section.

1. Which of these three groups—white, black, Hispanic—has the highest voter participation? _____

   a. Which of the three groups has the lowest voter participation? _____

   b. Explain why: _____

2. How are age and education connected with voter participation in the United States?

_____

_____

3. Are rates of voter participation higher in totalitarian states or in democracies? Why?

_____

_____

4. What are the five functions of the State?

a. _____

b. _____

c. _____

d. _____

e. _____

5. What are the four features of capitalism outlined by Adam Smith?

a. _____

b. _____

c. _____

d. _____

6. What are the six specific elements totalitarian governments use to concentrate political power?

a. _____

b. _____

c. _____

d. _____

e. _____

f. _____

7. Compared with the 1960s, how are political and social ideas in the United States today:

a. More conservative? _____

b. More liberal? _____

CRITICAL THOUGHT EXERCISES

1. Write an essay in which you give examples from U.S. politics and government that display Weber's three types of authority.

2. Look at some of the institutions of your life (school, family, workplace, and so on). In what ways do these perform functions similar to those of the State? How well do different models of government (democracy, aristocracy, and so on) fit with the way these institutions run?

3. The box on Technology and Society says that the distribution of sex, body parts, and children should not occur according to free-market principles that apply to other goods (like automobiles) and services (like plumbing). Why not? Are there other goods and services that are better kept apart from the forces of the market?

INTERNET ACTIVITIES

1. Visit http://www.crab.rutgers.edu/~goertzel/polsoctheories.htm. This site discusses theoretical frameworks in political sociology. Select any one of these frameworks and apply it to the outcome of the most recent national election in the United States. In your discussion, be sure to include consideration of how specifically that theory helps us understand that election's outcome.

2. Visit http://www.politicalresources.net/. This site contains resources on political parties across the globe. Explore the site and, once you become familiar with how it is organized, visit three different global locations and present a comparison of the activities of political parties within the countries you selected. In what ways are political party activities common across the three countries? In which ways is each nation's political parties' behavior in manners unique to that country?

ANSWERS TO KEY CONCEPTS

1.h   2.m   3.l   4.d   5.k   6.b   7.i   8.f   9.c   10.e   11.j   12.g   13.a

ANSWERS TO KEY THINKERS

1.e   2.b   3.a   4.d   5.c

ThomsonNOW™

Reviewing is as easy as ❶ ❷ ❸
1. Before you do your final exam, take the ThomsonNOW diagnostic quiz to help you identify the areas on which you should concentrate. You will find information on ThomsonNOW and instructions on how to access all of its great resources on the foldout at the beginning of the text.
2. As you review, take advantage of ThomsonNOW's study videos and interactive Map the Stats exercises to help you master the chapter topics.
3. When you are finished with your review, take ThomsonNOW's posttest to confirm you are ready to move on to the next chapter.

16

Population and Urban Society

© Lynn Saville/Photonica/Getty Images

Learning Objectives

After studying this chapter, you should be able to do the following:

- Describe the phenomenon of exponential growth.
- Define the three major components of population change.
- Contrast the Malthusian and Marxist theories of population.
- Summarize the demographic transition model and explain why there might be a second demographic transition.

- Discuss the determinants of fertility and family size.
- Discuss the problems of overpopulation and possible solutions.
- Describe the history of urbanization.
- Contrast preindustrial and industrial cities.
- Describe the various theories of urban development.
- Understand the issues surrounding homelessness in American cities.
- Describe trends in urban growth in the United States.

The Chinese words *wan, xi, shao* mean later, longer, fewer: later marriage, longer periods of time between pregnancies, and fewer children.

In 1971, that slogan launched a birth control campaign—unprecedented in scale in all of history—in the People's Republic of China. Posters proclaiming the message appeared virtually everywhere, even in the smallest village outposts.

At first the goal was not specified, but a three-child family was considered acceptable. Then, two. Now the emphasis is on the last word: *shao,* fewer—as few as possible, as fast as possible. *Shao* has been reinterpreted three times from the original "fewer" to "one is best" and finally to "one is enough."

When Mao Ze Dong was in power he said there could never be too many Chinese. He assumed that the sheer number of people would be China's greatest defense in the imminent third world war. As a result, the population grew from 540 million in 1950 to more than 850 million by 1970. In the 1970s, the average Chinese woman of childbearing age gave birth to six children (Hesketh & Zhu, 1997).

As the political situation in China changed, the government decided to take drastic action to ensure future survival, by setting a population goal of no more than one child per couple in most urban areas. To accomplish this goal, in 1979 China instituted a nationwide campaign to persuade couples to follow the government's guidelines. At the time, China had one-fourth of the world's population with only 7% of the world's arable land (Hesketh, Lu, and Xing, 2005).

To promote acceptance of the one-child limit, the Chinese leaders devised a reward-punishment system that makes daily life easier and richer for those who comply and burdensome for those who do not. A nationwide campaign to promote the one-child family included such slogans as "With two children you can afford a 14-inch TV, with one child you can afford a 21-inch TV" (Hesketh & Zhu, 1997).

Chinese couples had to obtain permission to be married as well as to have a child. Couples were also encouraged to obtain the Glorious One Child Certificate, which shows that they have signed an agreement to limit their family to a single child in exchange for extensive benefits. The certificate entitles the couple to a monthly childrearing allowance until the child reaches age 14.

In some areas, particularly cities, the one-child policy is often promoted through incentives, such as extra salary or larger houses for couples who pledge to have just one child. The government generally pays for birth control and abortions (and a woman

419

who has an abortion receives a vacation with pay). Failure to abide by the policy may result in job loss or demotion.

Only certain families may have more than one child: rural families whose first child is a daughter, families where both parents are themselves only children, and families in which the first child is handicapped.

Others who have a second child pay stiff fines. In Shanghai, the fine is three times the combined annual salary of the parents (Rosenthal, 1998). If a couple has a second child without permission or paying a fine, the second child cannot be registered and, therefore, does not legally exist. The child cannot attend school and later will have difficulty obtaining permission to marry, to relocate, and for other life choices requiring the government's permission (Cato Institute, 2000).

Penalties are particularly harsh for those who have a third child. Ten percent of their pay or work points are deducted from the fourth month of pregnancy until the third child is 14 years old. The same penalties are imposed on couples who have a child out of wedlock.

An additional punishment for the three-or-more-child couple is denial of any job promotion and loss of all work bonuses for at least three years. This is to make sure that couples who exceed the state limit for children do so at the price of personal sacrifice. They cannot prosper by doing extra work. In 1996, only 6.6% of Chinese births were third- or higher-order children (Population Reference Bureau, 1997).

There is no question that these policies are working. The one-child policy has reduced China's population by 250 million (UND Women's Center News, 2003). Yet the policy had an unintended side effect. Limiting Chinese couples to one child produced an alarming trend, a shortage of baby girls. It is estimated that as the one-child population matures there may be 29 million to 33 million unmarried males ages 15–34 by 2020 (Wiseman, 2002).

Population Dynamics

Chinese leaders have given the one-child policy top priority, ahead of competing social goals, because population growth overshadows all other issues in that country. As of 2005, China's 1.306 billion people accounted for 21% of the world's inhabitants (CIA Factbook, 2005). The dramatic impact that the one-child policy can have on population growth can be seen if we compare population projections based on a three-child family and a one-child family. If, for the next 100 years, couples were to have an average of three children, China's population would reach 4.2 billion by the year 2080. Even with the one-child policy, China's population is still increasing at the rate of 10 million people a year (Hesketh, Lu, and Xing, 2005).

China is responding to a basic demographic fact: in certain parts of the world population problems become progressively more pressing because of what is referred to as exponential growth. The yearly increase in population is determined by a continuously expanding base. Each successive addition of 1 million people to the population requires less time than the previous addition required, even if the birthrate does not increase. The best way to demonstrate the effects of exponential growth is to use a simple example.

Let us assume that you have a job that requires you to work eight hours a day, every day, for 30 days. At the end of that time, your job is over. Your boss offers you a choice of two different methods of payment. The first choice is to be paid $100 a day for a total of $3,000. The second choice is somewhat different. The employer will pay you 1 cent for the first day, 2 cents for the second day, 4 cents for the third, 8 cents on the fourth, and so on. Each day you will receive double what you received the day before. Which form of payment would you choose? The second form of payment would yield a significantly higher total payment—so high, in fact, that no employer could realistically pay the amount. Through this process of successive doubling, you would be paid $5.12 for your labor on the 10th day. Only on the 15th day would you receive more than the flat $100 a day you could have received from the first day under the alternative payment plan: The amount that day would be $163.84. However, from that day on, the daily pay increase is quite dramatic. On the 20th day you would receive $5,242.88, and on the 25th day your daily pay would be $167,772.16. Finally, on the 30th day, you would be paid $5,368,709.12, bringing your total pay for the month to more than $10 million.

This example demonstrates how the continual doubling in the world's population produces enormous problems. The annual growth rate in the world's population has declined from a peak of 2.04% in the late 1960s to 1.5% in 1997. This difference between global birth and death rates means that the world's population now doubles every 51 years instead of every 35 years (Figure 16–1). Although this is an improvement for the world as a whole, many portions of the world have not seen any improvement in their growth rates; in fact, the reverse may be true. In some countries in Africa, the national fertility rates have actually increased in the past decade, and Africa is now the area of the world with the most rapid population growth. In Ethiopia, for example, the average woman now has 6.7 children. When this fact is combined with the declining infant mortality rate, the country's population could balloon from 67.7 million in 2002 to 256.4 million in 2046. Enormous growth has been projected for Yemen, Niger, and Chad. (See Table 16–1 for projected yearly population growth percentages for selected countries.)

Most of the world's population growth in the next few decades will occur in developing countries.

© A. Ramey/Woodfin Camp and Associates

Figure 16–1 **The Population Explosion**

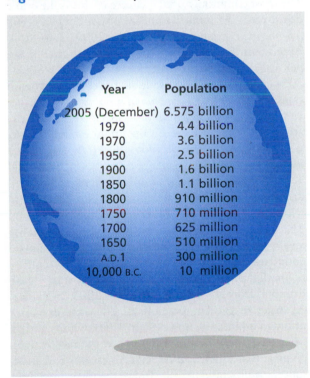

Year	Population
2005 (December)	6.575 billion
1979	4.4 billion
1970	3.6 billion
1950	2.5 billion
1900	1.6 billion
1850	1.1 billion
1800	910 million
1750	710 million
1700	625 million
1650	510 million
A.D.1	300 million
10,000 B.C.	10 million

Sources: Data through 1970, *The World Almanac*, New York: Newspaper Enterprises, 1980, p. 734; data for 1979–2000, U.S. Bureau of the Census, *International Data Base*.

Population growth rates differ greatly by region. The rate is highest in countries in the center of sub-Saharan Africa, where the rate is 3% a year, and lowest in Eastern Europe at −0.5%. Most of these former Soviet republics are experiencing population loss—that is, more deaths than births. If sub-Saharan Africa continued its current rate of growth, it would take it 23 years to double in population. By contrast, at current growth rates, Western Europe's population would need 612 years to double in size (Population Reference Bureau, 2002a).

As you may have deduced, most of the world's growth in population in the next few decades will take place in developing countries. The population of the richer countries will increase by 200 million by the year 2050, while the developing areas will have added about 6 billion (Figure 16–2).

Symbolizing the shift in population from the developed to the less developed world is Sao Paulo, Brazil. In 1950, this city was smaller than Manchester, England, but by the year 2000 Sao Paulo's population had reached about 24 million, making it one of the largest cities in the world. London, which was the second-largest city in the world in 1950, was not in the top 20 largest cities in 2000, due to current growth trends.

Our discussion of the population shifts in various countries falls into what is called demography. **Demography** is *the study of the size and composition of human populations, as well as the causes and consequences of changes in these factors.* Demography is influenced by three major factors: fertility, mortality, and migration.

Fertility

Fecundity is *the physiological ability to have children.* Most women between the ages of 15 and 45 are capable of bearing children. During this time, a woman could potentially have up to 30 children; however, the

Table 16–1

Annual Percentage Increase and Years Needed to Double Current Population

	Yearly Percentage Increase	Years Needed to Double the Population
World		
More developed countries	0.1	809
Less developed countries	1.6	44
Country		
Marshall Islands	3.7	19
Palestinian Territory	3.5	20
Chad	3.3	21
Yemen	3.3	21
Saudi Arabia	2.9	24
Somalia	2.9	24
Oman	2.9	24
Nicaragua	2.8	25
Syria	2.6	27
Iraq	2.5	28
Ethiopia	2.5	28
Libya	2.4	29
Afghanistan	2.4	29
Malawi	2.4	29
El Salvador	2.3	31
Jordan	2.3	31
Philippines	2.2	32
Pakistan	2.1	33
Mozambique	2.0	35
Kenya	2.0	35
Israel	1.5	45
Argentina	1.1	62
Thailand	0.8	84
Singapore	0.8	84
United States	0.6	116
Ireland	0.6	116
France	0.4	178
Canada	0.3	204
Japan	0.2	462
United Kingdom	0.1	546
Italy	0.0	—
Sweden	0.0	—
Russia	−0.7	—
Bulgaria	−0.5	—
Ukraine	−0.8	—

Source: Population Reference Bureau, *2002 World Population Data Sheet.*

realistic maximum number of children a woman can have is about 15. This number is a far cry from real life, though, where health, culture, and other factors limit childbearing. Even in countries with high birthrates, the average woman rarely has more than eight children (McFalls, 1991).

Whereas fecundity refers to the biological potential to bear children, **fertility** refers to *the actual number of births in a given population.* One common way of measuring fertility is by using the **crude birthrate**, *the number of annual live births per 1,000 people in a given population.* The crude birthrate for the United States fell from 24.1 in 1950 to 14.1 in 2003 (Statistical Abstract of the United States, 2006).

Another indicator of reproductive behavior is the **fertility rate**, *the annual number of births per 1,000 women of childbearing age, usually defined as 15 to 44.* In 2003, the fertility rate for the United States was 66.1 (Statistical Abstract of the United States, 2006).

As you will see later in this chapter, the fertility rate is linked to industrialization. Fertility declines with modernization, but not immediately. This lag is a source of tremendous population pressure in developing nations that have benefited from the introduction of modern medical technology, which immediately lowers mortality rates.

Mortality

Mortality is *the frequency of deaths in a population.* The most commonly used measure of this is the **crude death rate**, *the annual number of deaths per 1,000 people in a given population.* In 2003, 2,444,000 Americans died, producing a crude death rate of 8.4 per 1,000 (Statistical Abstract of the United States, 2006). Demographers also look at **age-specific death rates**, which *measure the annual number of deaths per 1,000 people in a population at specific ages.* For example, one measure used is the **infant mortality rate**, which *measures the number of children who die within the first year of life per 1,000 live births.* Of 1,000 babies born this year in Eastern Africa, 97 will die within 1 year. In the world's more developed countries, it will take 60 years for these 97 deaths to occur. The difference reflects a continuing gap in mortality levels between the world's more and less developed countries (Population Reference Bureau, 2002a).

In the United States, the infant mortality rate dropped from 47 in 1940 to 6.9 in 2003 (Statistical Abstract of the United States). This rate does not apply to all infants, however. In 1940, whites had an infant mortality rate of 43.2 (below the national average at that time), and African Americans had an infant mortality rate of 73.8. Today, the rates are 5.8 for whites and 14.4 for blacks (Statistical Abstract of the United States). The lowest infant mortality rates are for infants born to Chinese mothers (2.9) and for Japanese mothers (3.4). Those figures suggest that differing cultural patterns of childrearing may also affect infant mortality as well as the availability of medical care.

Infectious diseases caused most child deaths in the early part of the twentieth century. Millions of children used to die from respiratory diseases, gastrointestinal

Figure 16–2   Past and Projected World Population, A.D. 1–2150

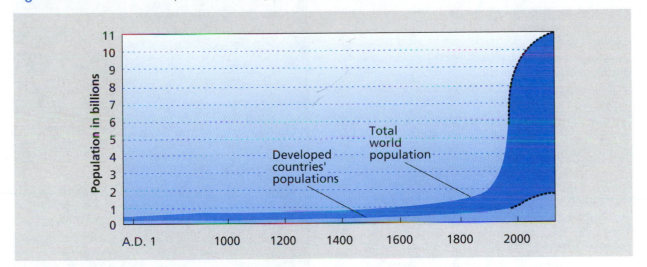

diseases, and tuberculosis. These deaths have been reduced through the introduction of penicillin and other antibiotics and vaccines helped prevent illnesses that killed or crippled large numbers of children. Today accidents, congenital anomalies/deformations/ chromosomal abnormalities, and cancer are the greatest threats to children (AmeriStat, August 2002, "Century").

Although the infant mortality rate in the United States is considerably lower than the rates of developing countries, it is higher than those in some Asian or European countries. The lowest infant mortality rates in the world are found in Singapore (2.2) and Japan (3.2). Mortality is reflected in people's **life expectancy,** *the average number of years a person born in a particular year can expect to live* (see Table 16–2 for life expectancies in selected countries). The average life expectancy at birth in the United States is 77 years. The world's longest life expectancy is in Japan, at 82 years. The shortest life expectancies are in Botswana, Swaziland, and Lesotho (Population Reference Bureau, 2005b).

Maternal mortality is another factor that varies widely throughout the world. There are continuing differences in lifetime risk of maternal mortality between developed and developing countries. An African woman's lifetime risk of dying from pregnancy-related causes is 1 in 16; in Asia, 1 in 65; and in Europe, 1 in 1,400 (United Nations, 2001, *World's Women*).

Life expectancy is usually determined more by infant mortality than by adult mortality. Once an individual survives infancy, life expectancy improves dramatically. In the United States, for example, only when individuals reach their 60s do their chances of dying approximate those of their infancy (Bureau of the Census, 1997a).

Table 16–2

Countries with the Highest and Lowest Life Expectancies

Highest	Years	Lowest	Years
Japan	82	Botswana	35
Iceland	81	Lesotho	35
Sweden	81	Swaziland	35
Australia	80	Zambia	37
Canada	80	Angola	40
France	80	Sierra Leone	40
Italy	80	Zimbabwe	41
Norway	80	Afghanistan	42
Spain	80	Liberia	42
Switzerland	80	Mozambique	42

Source: Population Reference Bureau, *2005 World Population Data Sheet.*

A fact that is often overlooked is that the rapid increase in population growth in third-world countries is caused by sharp improvements in life expectancy, not by rising birthrates. Disease took a dramatic toll on life expectancy in the United States in the not-too-distant past. For example, Abraham Lincoln's mother died when she was 35 and he was 9. Prior to her death she had three children: Abraham Lincoln's brother died in infancy and his sister died in her early 20s. Of the four sons born to Abraham and Mary Todd Lincoln, only one survived to maturity.

In developing countries, the proportion of infant and child deaths is quite high, resulting in a significantly lower life expectancy than that in developed countries. In Bangladesh, infant deaths account for more than one-third of all deaths. In the United States, that figure is about 1%. The high proportion of deaths in developing countries can be attributed to impure drinking water and unsanitary conditions. In addition, the diets of pregnant women and nursing mothers often lack proper nutrients, and babies and children are not fed a healthy diet. Flu, diarrhea, and pneumonia are common, as are typhoid, cholera, malaria, and tuberculosis. Many children are not immunized against common childhood disease (such as polio, measles, diphtheria, and whooping cough), and the parents' income is often so low that when the children do fall ill, they cannot provide medical care.

Life expectancies in developing nations vary greatly. They range from an average regional low of about 52 years in Africa to 75 years in Europe (Population Reference Bureau, 2005b). A major contributor to the overall death rate is infant mortality. A rapid decrease in infant mortality in a developing nation will result in a significant rise in the overall rate of population growth.

Migration

Migration is *the movement of populations from one geographical area to another.* We call it **emigration** *when a population leaves an area* and **immigration** *when a population enters an area.* All migrations, therefore, are both emigrations and immigrations.

Of the three components of population change—fertility, mortality, and migration—migration historically exerts the least impact on population growth or decline.

Most countries do not encourage immigration. When they do permit immigration, it is often viewed as a way to provide needed skilled labor, or to provide unskilled labor for jobs the resident population no longer wishes to do. Exceptions to this trend are the traditional receiver countries, such as the United States, Australia, and Canada. These countries owe much of their growth to immigrant populations.

Where migration is a significant factor, it is necessary to take into account the age and sex of the immigrants and emigrants, as well as the number of migrants. Those characteristics tell us the number of potential workers among the migrants, the number of women of childbearing age, the number of school-age children, the number of elderly, and other factors that will affect society.

Sometimes it is important to distinguish between those movements of populations that cross national boundary lines from those that are entirely within a country. To make this distinction, sociologists use the term **internal migration** for *movement within a nation's boundary lines*—in contrast with immigration, in which boundary lines are crossed.

In only three decades, Korean immigration has changed the face of commercial areas in New York City and Los Angeles.

Since 1970, population growth in the United States has been greatest in the Sunbelt states, reflecting continued migration patterns toward the South and West. California, Texas, and Florida are growing significantly faster than the United States as a whole because they attract many northeastern and midwestern residents. There is some indication, however, that internal migration patterns may be starting to change. Typically, northern and midwestern migrants moved to the three major Sunbelt states and then distributed themselves to the surrounding states.

Theories of Population

The study of population is a relatively new scholarly undertaking; it was not until the eighteenth century that populations as such were examined carefully. The first person to do so, and perhaps the most influential, was Thomas Malthus.

Malthus's Theory of Population Growth

Malthus (1776–1834) was a British clergyman, philosopher, and economist who believed that population growth is linked to certain natural laws. The core of the population problem, according to Malthus, is that populations will always grow faster than the available food supply. With a fixed amount of land, farm animals, fish, and other food resources, agricultural production can be increased only by cultivating new acres, catching more fish, and so on—an additive process that Malthus believed would increase the food supply in an arithmetic progression (1, 2, 3, 4, 5, and so on). Population growth, by contrast, increases at a geometric rate (1, 2, 4, 8, 16, and so on) as couples have 3, 4, 5, and more children. (A stable population requires 2 individuals to produce no more than 2.1 children: 2 to reproduce themselves and 0.1 to make up for those people who remain childless.) Thus, if left unchecked, human populations are destined to outgrow their food supplies and suffer poverty and a never-ending "struggle for existence" (a phrase coined by Malthus that later became a cornerstone of Darwinian and evolutionary thought).

Malthus recognized the presence of certain forces that limit population growth, grouping these into two categories: preventive checks and positive checks. **Preventive checks** are *practices that would limit reproduction.* Preventive checks include celibacy, the delay of marriage, and such practices as contraception within marriage, extramarital sexual relations, and prostitution (if the latter two are linked with abortion and contraception). **Positive checks** are *events that limit reproduction either by causing the deaths of individuals before they reach reproductive age or by causing the deaths of large numbers of people, thereby lowering the overall population.* Positive checks include famines, wars, and epidemics. Malthus's thinking assuredly was influenced by the plague that wiped out so much of Europe's population during the fourteenth and fifteenth centuries.

Malthus refuted the theories of the utopian socialists, who advocated a reorganization of society to eliminate poverty and other social evils. Regardless of planning, Malthus argued, misery and suffering are inevitable for most people. On the one hand, there is the constant threat that population will outstrip the available food supplies; on the other hand, there are the unpleasant and often devastating checks on this growth, which result in death, destruction, and suffering.

History proved Malthus wrong, at least for developed countries. Technological breakthroughs in the nineteenth century enabled Europe to avoid many of Malthus's predictions. The newly invented steam engine used energy more efficiently, and labor production was increased through the factory system. An expanded trade system provided raw materials for growing industries and food for workers. Fertility declined and emigration eased Europe's population pressures. By the end of the nineteenth century, Malthus and his concerns had been all but forgotten.

Marx's Theory of Population Growth

Karl Marx and other socialists rejected Malthus's view that population pressures and their attendant miseries are inevitable. Marxists argue that the sheer number of people in a population is not the problem. Rather, they contend, it is industrialism (and in particular, capitalism) that creates the social and economic problems associated with population growth. Industrialists need large populations to keep the labor force adequate, available, flexible, and inexpensive. In addition, the capitalistic system requires constantly expanding markets, which can be assured only by an ever-increasing population.

As the population grows, large numbers of unemployed and underemployed people compete for the few available jobs, which they are willing to take at lower and lower wages. Therefore, according to Marxists, the norms and values of a society that encourages population growth are rooted in its economic and political systems. Only by reorganizing the political economy of industrial society in the direction of socialism, they contend, is there any hope of eliminating poverty and the miseries of overcrowding and scarce resources for the masses.

Demographic Transition Theory

Sweden has been keeping records of birth and deaths longer than any other country. Throughout the centuries, Swedish birth and death rates fluctuated widely. There were periods of rapid population growth, followed by periods of slow growth, and even population declines during famines. In the late 1800s, Sweden's death rate began a sustained decline, while the birthrate remained high. Eventually, Sweden's birthrate also declined, so that today its births and deaths are virtually in balance.

The shift in Sweden's population can be explained by a theory of population dynamics developed by Warren Thompson. According to the **demographic transition theory,** *societies pass through four stages of population change from high fertility and high mortality to relatively low fertility and low mortality.* During Stage 1, high fertility rates are counterbalanced by a high death rate due to disease, starvation, and natural disaster. The population tends to be very young, and there is little or no population growth. During Stage 2, populations rapidly increase as a result of a continued high fertility that is linked to the increased food supply, development of modern medicine, and public health care. Slowly, however, the traditional institutions and religious beliefs that support a high birthrate are undermined and replaced by values stressing individualism and upward mobility. Family planning is introduced, and the birthrate begins to fall.

According to Thomas Malthus, population will always grow faster than the available food supply.

This is Stage 3, during which population growth begins to decline. Finally, in Stage 4 both fertility and mortality are relatively low, and population growth once again is stabilized (Figure 16–3).

As long as many developing nations remain in Stage 2 of demographic transition (high fertility but falling mortality), they will continue to be burdened by overpopulation, which slows economic development and creates widespread severe hunger.

Overpopulation undermines economic growth by disproportionately raising the **dependency ratio,** *the number of people of nonworking age in a society for every 100 people of working age.* Because populations at Stage 2 have a high proportion of children, as compared with adults, they have fewer able-bodied workers than they need. For example, 51% of Uganda's population is below the age of 15, compared with 21% in the United States (Population Reference Bureau, 2005b). The economic development of countries with high dependency ratios is slowed further by the channeling of capital away from industrialization and technological growth and toward mechanisms for feeding the expanding populations.

Applications to Industrial Society The first wave of declines in the world's death rate came in countries experiencing real economic progress. Those declines gradually gained momentum as the Industrial Revolution proceeded. Advances in agriculture, transportation, and commerce made it possible for people to have a better diet, and advances in manufacturing made adequate clothing and housing more widely available. A rise in people's real income facilitated improved public sanitation, medical science, and public education.

Although the preceding explanation applies well to Western society, it does not explain the population trends in the underdeveloped areas of today's world. Since 1920, those areas have experienced a much faster drop in death rates than Western societies without a comparable rate of increase in economic development. The rapid rate of decline in the death rate in these countries has been due primarily to the application of medical discoveries made in and financed by the industrial nations. For example, the most important death threat being eliminated is infectious disease. Those diseases have been controlled through the introduction of vaccines, antibiotics,

Figure 16-3   The Demographic Transition Theory

The demographic transition theory states that societies pass through four stages of population change. Stage 1 is marked by high birthrates and high death rates. In Stage 2, populations rapidly increase as death rates fall, but birthrates stay high. In Stage 3, birthrates begin to fall. Finally, in Stage 4, both fertility and mortality rates are relatively low.

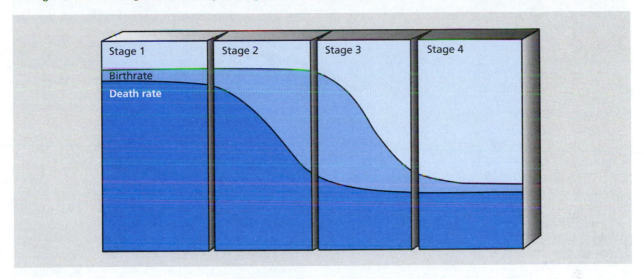

Table 16-3

Countries with the Largest Populations, 2005 and 2050

2005			2050		
Rank	Country	Population (millions)	Rank	Country	Population (millions)
1	China	1,304	1	India	1,628
2	India	1,104	2	China	1,437
3	United States	296	3	United States	420
4	Indonesia	222	4	Indonesia	308
5	Brazil	184	5	Pakistan	295
6	Pakistan	162	6	Brazil	260
7	Bangladesh	144	7	Nigeria	258
8	Russia	143	8	Bangladesh	231
9	Nigeria	132	9	Dem. Rep. of Congo	183
10	Japan	128	10	Ethiopia	170

Sources: Population Reference Bureau, *2005 World Population Data Sheet*, U.S.

and other medical advances developed in the industrial nations. Those who are used to paying high prices for private medical care will find it hard to believe that preventive public health measures in underdeveloped countries can save millions of lives at costs ranging from a few cents to a few dollars per person.

Because the mortality rates in underdeveloped countries have been significantly reduced, the birthrate, which has not fallen as fast or as consistently, has become an increasingly serious problem. Often, this problem persists and worsens despite monumental government efforts to disseminate birth control information and contraceptive devices. In India, for example, despite the government's commitment to controlling population size through birth control and sterilization, the population is expected to increase from 717 million in 1982 to 1.3 billion in 2016. With the current rate of growth, India could surpass China as the most populated country somewhere between 2030 and 2035 (Population Reference Bureau, 1997). (See Table 16–3 for the countries with the largest populations in 2005 and projections for 2050.)

NEWS YOU CAN USE

Do Men without Women Become Violent?

In China and India with their cultural preferences for males, girls have been victims of infanticide or given substandard care for centuries. In the last two decades ultrasound machines have allowed women to abort female fetuses, producing a record number of males. In China, it is estimated that 120 boys are born for every 100 girls. A more typical sex ratio would be 105 boys per 100 girls.

These boys have the potential to become what the Chinese call "bare branches," young men who will never marry or have children. The sheer number of potential "bare branches" is greater than ever because India and China together account for over 38% of the world's population.

Two researchers, Valerie Hudson and Andrea M. den Boer, argue that large numbers of unattached young men have the potential to produce violent social unrest in their own countries and elsewhere. If these unattached young men also have no jobs, Hudson and den Boer argue, the governments become more likely to engage in aggressive military actions to absorb and occupy them.

Hudson and den Boer and point out that countries with few marriageable women often adopt authoritarian political systems and become more violent toward women. Historically, when large groups of men do not marry and hang out together they become marauding bands that wreak havoc on society. For example, during the Nien Rebellion in 1851 and the Boxer Rebellion in 1900, gangs of "bare branches" used violence to get what they wanted.

Hudson and den Boer believe we are at the threshold of large numbers of young men becoming a volatile social force producing social instability. To counter this threat, the governments of India and China may move in a more authoritarian direction. The first group of "bare braches" is moving into their 20s now, and the next decade will decide whether den Boer and Hudson are correct.

Source: Valerie M. Hudson and Andrea M. den Boer. (2004). *Bare Branches: The Security Implications of Asia's Surplus Male Population.* Cambridge, MA: M.I.T. Press.

These failures have shown that the birthrate can be brought down only when attention is paid to the complex interrelationships of biological, social, economic, political, and cultural factors. For example, a study in Pakistan revealed that even among women who already had six children, if all six were daughters, there was a 46% chance that the mother would want more children. If, however, the six children were boys, there was only a 4% chance that she would want additional children (Weeks, 1994).

A Second Demographic Transition

The original demographic transition theory ends with the final stage involving an equal distribution of births and deaths and a stable population. Some people (van de Kaa, 1987) have suggested that Europe has gone beyond the original theory and entered a second demographic transition. The start of this second demographic transition is arbitrarily set at 1965; the principal feature of this transition is the decline in fertility to a level well below replacement.

If fertility stabilizes below replacement, as seems likely in most of Europe, and barring major changes in immigration, the populations of those countries will decline. This second demographic transition is already the case in Estonia, Latvia, Hungary, Bulgaria, the Ukraine, and Italy, as well as most of Eastern Europe (Population Reference Bureau, 2005b).

The United States is not expected to experience a population decline in the near future. United Nations population projections to the year 2025 indicate that the United States will continue to grow modestly, even with continued low fertility, because of immigration levels.

The reasons for the second demographic transition center around a strong desire for individual advancement and improvement. In European societies, as in American society, this advancement is dependent on education and a commitment to develop and use one's talents. This holds for both men and women. Marrying and having children present a number of trade-offs, especially for women. A child may interrupt the parents' career plans, as well as add to financial costs. In European societies, children no longer are expected or required to support the parents in old age or to help with the family finances. Therefore, the emotional satisfaction of parenthood can usually be satisfied by having one or perhaps two children. Multiplied on a large scale, this trend produces a birthrate below replacement level.

Pronatalist Policies In most of this chapter, we are presenting information about population growth rates that appear to be out of control. Yet scores of developed nations face an equally ominous threat: dwindling population because of low fertility. In some 50 countries, the average number of children born to each woman has fallen below 2.1, the number required to maintain a stable population. Nearly all of these countries are in the developed world, where couples have been discouraged from having large families by improved education and health care, widespread female employment, and rising costs of raising and educating children.

The implications are particularly dire for Europe, where the average fertility rate has fallen to 1.4 children per woman. Even if the trend reversed itself and the fertility rate returned to 2.1, the continent would have lost a quarter of its current population before it stabilized around the middle of the century.

With fewer children being born, the ratio of older people to younger people already is growing. These countries face the prospect of shrinking workforces and growing retiree populations, along with slower economic growth and domestic consumption.

Some European governments are quite concerned. Italy, France, and Germany have introduced generous child subsidies, in the form of tax credits for every child born, extended maternal leave with full pay, guaranteed employment upon resumption of work, and free child care.

Although it is most pronounced in Europe, the birth dearth affects a few countries in other parts of the world as well. Japan's population may fall from 126 million today to 55 million over the next century if its 1.35 fertility rate remains unchanged.

Apart from encouraging childbirth, the only other way a government can stop population loss is to open its doors to immigrants. Although immigration has always played a major role in the United States, most European countries are more homogeneous and resistant to immigration. Japan has been particularly inhospitable to immigrants (Population Reference Bureau, 2004). (See "News You Can Use: Do Men without Women Become Violent?")

Pronatalist policies hardly ever lead to spectacular long-term effects on the birthrate. They may, however, contribute toward slowing down the fertility decline and improving the living conditions of the parents and their children.

Population Growth and the Environment

When people discuss population growth they invariably use the term *overpopulation*. The very word *overpopulation* is a problem for people in less

Table 16–4

World Population Clock, 2005

Time	Births	Deaths	Natural Increase
Year	130,013,274	56,130,242	73,883,032
Month	10,834,440	4,677,520	6,156,919
Day	356,201	153,781	202,419
Hour	14,842	6,408	8,434
Minute	247	107	141
Second	4.1	1.8	2.3

Source: U.S Census Bureau, http://www.census.gov/cgi-bin/ipc/pcwe, accessed September 21, 2005.

developed countries. If the world is overpopulated, then who are the unneeded? If people in the richer countries view those in poorer ones as surplus, then a bias is built into the term. (See Table 16–4.)

When most people link population growth and environmental degradation, they are usually referring to *less developed countries,* where most of the world's people live and population growth is high. At first glance, the connection between population growth and environmental problems seems clear. More humans consume more resources and generate more waste. Twelve billion people could do a great deal more environmental damage than the 6 billion in the world of 2000. A closer look at the situation reveals that it is more complex than that. Some people have a far greater environmental impact than others. Most of the environmental degradation takes place in industrialized countries.

The world's richest countries, with 20% of global population, account for 86% of private consumption; the poorest 20% account for just 1.3%. A child born in an industrialized country will add more to consumption and pollution over his or her lifetime than 30 to 50 children born in developing countries (United Nations, 2001c). The United States has 5% of the world's population, yet uses an estimated 33% of the world's resources, and causes an estimated 33% of the world's pollution. The average American uses at least 30 times the amount of resources as a person living in India.

Modern cities are often referred to as "heat centers" and blamed for contributing to the destruction of the world's ozone layer. Although motor vehicles are the primary cause of pollution in cities, increased demand for energy to run air conditioning and electrical appliances is contributing to pollution in many cities. Producing the energy required to run modern urban systems often involves burning fossil fuels, which releases such greenhouse gases as carbon monoxide, carbon dioxide, and nitrogen oxides. These emissions lead to global warming, which can cause climate change, rising sea levels, changes in vegetation, and severe weather events. Oxides of sulfur and nitrogen emitted to the atmosphere from cities have led to acid rain that has killed lakes and forests in North America and Northern Europe.

As the less developed countries industrialize they adopt consumption patterns similar to those of the developed countries. Already, elites in the less developed countries mimic the consumption patterns of rich Americans or Europeans. Consumption has surged in China and India since the 1980s and, with the fall of the former Soviet Union, Eastern Europeans have increased their usage of consumer goods.

The most rapid growth in energy consumption now occurs in less developed countries because of rising affluence, consumption, and population (Population Reference Bureau, 2002, *Human Population*).

In the late 1990s, the United States, home to only about 5% of the world's population, was responsible for roughly one-fourth of global carbon dioxide emissions. Less developed countries, however, are releasing a growing percentage of global carbon dioxide emissions (Nash & De Souza, 2002).

Some people have suggested that instead of being concerned about overpopulation, we should focus on the world's carrying capacity instead (Carty, 1994). How will the world's carrying capacity problems be resolved in the future? Neo-Malthusians paint a gloomy picture of what lies ahead, contending that as we head toward the end of this century, the population inevitably will outpace the supply of food.

In the 1970s, Paul Ehrlich and Anne Ehrlich, outspoken critics of population growth and unlimited consumption by the wealthy countries, predicted global shortages in the future. "It seems certain," they noted, "that energy shortages will be with us for the rest of the century, and that before 1985 mankind will enter a genuine age of scarcity." Crucial materials would be nearly depleted during the 1980s, they predicted. "Starvation among people will be accompanied by starvation of industries for the materials they require" (Ehrlich & Ehrlich, 1974).

Also in the 1970s, The Club of Rome, a group of scientists, businesspeople, and academics, used elaborate computer models to predict the world of the future. The Club concluded that if the then current trends continued, the limits of growth on this planet would be reached within the next hundred years. The result, they predicted, would be a sudden and uncontrollable decline in population and production capacity:

We have tried in every doubtful case to make the most optimistic estimate of unknown qualities, and we have also ignored discontinuous events such as wars or epidemics, which might act to bring an end to growth even sooner than our model would indicate. In other words, the model is biased to allow growth to continue longer than it probably can continue in the real world. We can thus say with some confidence that under the assumption of no major change in the present system, population and industrial growth will certainly stop within the next century, at the latest. (Meadows et al., 1972)

Others believe that we have the technological means to provide all the world's people with food. They speak of a "Green Revolution," in which new breeds of grain and improved fertilizers will raise harvest yields and eliminate the threat of a food shortage. The only thing holding back the revolution is poor international cooperation in planning the production and distribution of food. (See "Global Sociology: What if the Population Problem Is Not Enough People?")

Sources of Optimism

Critics point to a number of logical fallacies in the doomsday predictions and argue that the dire pronouncements ignore the role of the marketplace in helping produce adjustments that bring population, resources, and the environment back in balance. In their optimistic view, population growth is a stimulus, not a deterrent, to economic advance. If imbalances exist, they are because markets are not allowed to operate freely to permit innovators, investors, and entrepreneurs to provide solutions (Simon & Kahn, 1984).

Others have questioned the basic assumptions of the doomsday model, namely exponential growth in population and production and absolute limits on natural resources and technological capabilities. The following are some of these counterarguments.

1. *Wider application of existing technology.* Greater efficiency and wider application of technology on a worldwide basis could continue to supply the world's needs far into the future. For example, between 1961 and 1994, global production of food doubled, and greater efficiency in land cultivation will ensure the ability to feed the world's population for many years. Prices for food have continually decreased since the end of the eighteenth century.
2. *Discovery of new resources.* The supply of natural resources, according to some critics, is not really as fixed as the doomsday predictors claim. These critics maintain that new resources will come into play that were not previously anticipated. Technology will stay ahead of resource use, and the doomsday scenario will never be enacted. For example, raw materials and energy resources are generally more

GLOBAL SOCIOLOGY

What if the Population Problem Is Not Enough People?

The birthrate is declining throughout the world. For a country to maintain its population the average woman must have 2.1 children. This is known as the total fertility rate (TFR). The United States is the only developed country that is close to that number. Every other developed country is below 2.1 and consequently losing population. For example, in Canada the total fertility rate is 1.5, in Germany 1.35, in Japan 1.32, in Italy 1.23, and in Spain it is 1.15.

These low birthrates produce a number of changes in society. First, the median age of the population increase as older people outnumber the young. Second, the countries start to shrink. By 2050, Europe will have 100 million fewer people than in 2005. Japan will lose one-fourth of its population during that time.

As the average age of the population increases there are fewer workers and many more people who need to be supported through pension programs. The potential for economic problems also increases. Europe and Japan have experienced prolonged periods of economic stagnation with only limited prospects of recovery. An older population also increases the demands on the health care system and puts strain on the government and those that have to pay the increased costs.

The national identity and the political influence of the country that is losing population also suffers. Although Europe continues to think of itself as a major force in world politics, that can no longer be maintained with a substantial population decline.

There are a few things countries can do to deal with population decline.

1. Adopt policies that would encourage people to have children. This can be done through tax incentives or child-care benefits. France has a number of policies that encourage families to have children and consequently has the highest fertility rate in Europe.
2. Loosen up immigration policies so that more immigrants enter the country and help the population to grow. To a certain extent this is the case for the United States. European countries have been more wary of immigrants and worry about such a policy because it could threaten their national identities.
3. Raise the retirement age and limit pension and health benefits to the elderly. This would produce a significant outcry from those affected.

The continuing population decline in the developed world will raise a whole new set of population issues that are at odds with those of the developing world.

Sources: Population Reference Bureau, "Transitions in World Population," Vol. 59, No.1, March 2004; and Phillip Longman, *The Empty Cradle: How Falling Birthrates Threaten World Prosperity and What to Do About It*, New York: Basic Books, 2004.

---

abundant and less expensive today than they were 20 years ago. Companies have become more adept at discovering new resources and exploiting old ones. "Exploring for oil used to be a hit or miss proposition, resulting in many dry holes. Today, oil companies can use seismic waves to help them create precise computer images of the earth" (Sagoff, 1997). In effect, the more advanced the technology, the more reserves become known and recoverable. In addition, the more we learn about materials, the more efficiently we use them. Refrigerators sold in 2005 were more efficient than those sold in 2000 and significantly more efficient than those sold in 1980.

3. *Exponential increase in knowledge.* Just as the doomsday predictors claim there will be exponential growth in population and production, others say there also will be exponential growth in knowledge that will enable societies to solve the problems associated with growth. New

technological information and discoveries, furthermore, will alleviate new problems as they arise. For example, computer game consoles today have more computing power than the 1976 Cray supercomputer, which the United States tried to keep away from the Soviet Union for security reasons (Sagoff, 1997).

Population issues will continue to affect the developed and developing areas of the world. The likeliest path to helping populations is through economic development. Poor people who are unable to acquire food and fuel do whatever they have to in order to survive. Sensitive ecological systems such as rain forests often fall victim to these needs. Raising living standards produces lower fertility and lowers environmental deterioration from inefficient resource depletion. There is a strong trend toward an improving situation. In the next decade, we will see whether this trend is developing fast enough to prevent further outbreaks of famine and misery for millions of people.

Urbanization and the Development of Cities

According to archaeologists, people have been on earth for a couple of million years. During the vast majority of these years, human beings lived without cities. Although we accept cities as a fundamental part of human life, cities are a relatively recent addition to the story of human evolution, appearing only within the past 7,000 to 9,000 years.

The city's dominance in social, economic, and cultural affairs is even more recent. Nonetheless, what we label as "civilization" emerged only during the time span that coincides with the city, encompassing the whole history of human triumphs and tragedies. The very terms *civilization* and *civilized* come from the Latin *civis*, which means "a person living in a city."

The cities of the past still were very unusual in an overwhelmingly rural world of small villages. In 1800, 97% of the world lived in rural areas of fewer than 5,000 people. By 1900, 86% of the world still lived in rural areas (Palen, 1992).

England was the first country to undergo urban transformation. One hundred years ago it was the only predominantly urban country. Not until 1920 was the United States that urbanized. Today we are on the threshold of living in a world that will for the first time be more urban than rural. The most rapid change is occurring in the developing world. Within the next decade, more than half of the world's population, an estimated 3.3 billion, will be living in urban areas. As recently as 1975, just over one-third of the world's people lived in urban areas.

Not all parts of the world are urbanizing at the same pace, though. In the more industrialized areas—North America, Europe, and the republics of the former Soviet Union—urban growth has stopped or slowed considerably. For example, in 1970, 73.5% of the U.S. population lived in urban areas, whereas in 1990, the figure had only increased to 75.2% (Bureau of the Census, 1997).

The greatest urban growth is now in the nonindustrial world, such as Africa, Latin America, the Middle East, and Asia. "The United Nations projects that world population will increase from 6.1 billion in 2000 to 7.8 billion in 2025. Ninety percent of this growth will occur in urban areas of less developed countries. By 2020, a majority of the population of less developed countries will live in urban areas" (Brockerhoff, 2000).

The population of the less developed countries will become increasingly concentrated in large cities of 1 million or more people. Already there are hundreds of cities of more than 1 million people in the less developed countries. Most of us have never heard the names of more than a few of those cities. In 2000, there were an estimated 292 "million-plus" cities in less developed countries (Brockerhoff, 2000) (Figure 16–4).

Some cities are very large and are what the United Nations calls *megacities* (over 8 million people). The number of megacities is also growing rapidly. There were just 8 megacities in 1985, but the number more than doubled to 19 by 2000. The UN projects an additional 15 new megacities by 2015—all in less developed countries. Just 6 megacities (Los Angeles, Moscow, New York, Osaka [Japan], Paris, and Tokyo) will be in the more developed world in 2015—the same number as in 1985. In 1970, just 4 of the world's 10 largest cities were in less developed countries. By 2015, 8 of the 10 largest cities will be in less developed counties (see Table 16–5 for a list of the world's largest megacities, 1995 and 2015).

In some respects, the growth is similar to what took place a century ago in Europe and North America. Many of these cities sustained growth, which was as fast as that now underway in the developing world. What is different, however, is the number of countries undergoing rapid urbanization and the sheer number of people involved. As a measure of comparison, consider that in 1950 only two cities in the world had populations that exceeded 8 million, New York and London. By 2000, there were 19 such cities. By 2015, there will be 34 such megacities. Only six of these megacities will be in the developed world, the same as in 1985 (United Nations, 2000).

Cities have certain advantages. Historically cities have produced innovations in science and technology. Dense urban centers also make the distribution of goods and services cheaper. Urban centers also have lower fertility rates than rural areas (Brockerhoff, 2000).

The rapid transformation from a basically rural to a heavily urbanized world and the urban lifestyles that accompany this shift are having a dramatic impact on the world's peoples. In this section, we will examine the historical development of cities and urbanization trends.

The Earliest Cities

Two requirements had to be met for cities to emerge. The first was that there had to be a surplus of food and other necessities. Farmers had to produce more food than their immediate families needed to survive. This surplus made it possible for some people to live in places where they could not produce their own food and had to depend on others to supply their needs. Those settlements could become relatively large, densely populated, and permanent.

The second requirement was that there had to be some form of social organization that went beyond the family. Even though there might be a surplus of

Figure 16–4   **Number of Cities with 1 Million or More Residents, 1975, 1995, and 2015**

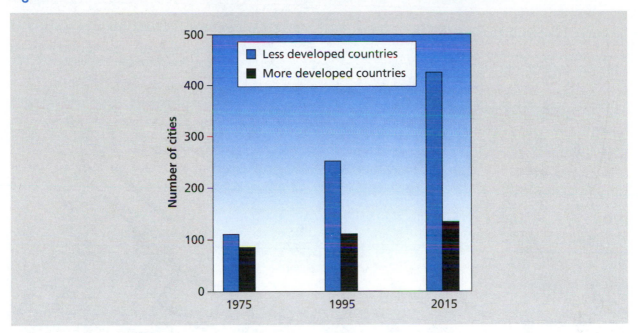

Source: United Nations, *World Urbanization Prospects: The 1999 Revision* (2000).

Table 16–5

World's Largest Megacities, 1995 and 2015

1995		2015	
City	**Population (millions)**	**City**	**Population (millions)**
Tokyo-Yokohama, Japan	28.4	Bombay, India	28.2
Mexico City, Mexico	23.9	Tokyo-Yokohama, Japan	26.4
Sao Paulo, Brazil	21.5	Lagos, Nigeria	23.2
Seoul, South Korea	19.1	Dhaka, Bangladesh	23.0
New York City, United States	14.6	Sao Paulo, Brazil	20.4
Osaka-Kobe-Kyoto, Japan	14.1	Karachi, Pakistan	19.8
Bombay, India	13.5	Mexico City, Mexico	19.2
Calcutta, India	12.9	Delhi, India	17.8
Rio de Janeiro, Brazil	12.8	New York City, United States	17.4
Buenos Aires, Argentina	12.2	Jakarta, Indonesia	17.3

Source: Bureau of the Census, *Statistical Abstract of the United States: 1997*, U.S. Bureau of the Census, 1997, Washington, DC: U.S. Government Printing Office; United Nations, *World Urbanization Prospects: The 1999 Revision*, 2000.

food, there was no guarantee that it would be distributed to those in need of it. Consequently, a form of social organization adapted to those kinds of living environments had to emerge.

The world's first fully developed cities arose in the Middle Eastern area, mostly in what is now Iraq, which was the site of the Sumerian civilization. The land is watered by the giant Tigris and Euphrates rivers, and it yielded an abundant food surplus for the people who farmed there. In addition, this area (called Mesopotamia) lay at the crossroads of the trade networks that already, 6,000 years ago, tied together East and West. Not only material goods, but also the knowledge of technological and social innovations, traveled along these routes.

Sumerian cities were clustered around temple compounds that were raised high up on brick-sheathed mounds called ziggurats. The cities and

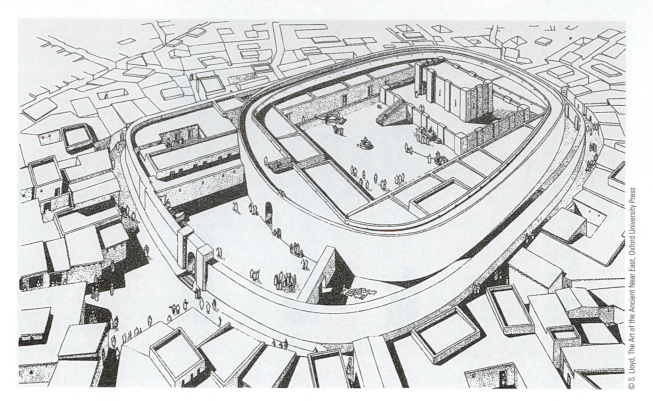

© S. Lloyd, The Art of the Ancient Near East, Oxford University Press

The early cities had walls that were closed off at night for protection.

their surrounding farmlands were believed to belong to the city god, who lived inside the temple and ruled through a class of priests who organized trade caravans and controlled all aspects of the economy. In fact, these priests invented the world's first system of writing as well as numerical notation late in the fourth millennium B.C. to keep track of their commercial transactions. Because warfare both among cities and against marauders from the deserts was chronic, many of these early cities were walled and fortified, and they maintained standing armies. In time, the generals who were elected to lead these armies were kept permanently in place, and their positions evolved into hereditary kingships (Frankfort, 1956).

The early Sumerian cities had populations that ranged between 7,000 and 20,000. However, one Sumerian city, Uruk, extended over 1,100 acres and contained as many as 50,000 people (Gist & Fava, 1974). By today's standards, the populations of those early cities seem rather small. They do, however, present a marked contrast with the small nomadic and seminomadic bands of individuals that existed prior to the emergence of these cities.

Within the next 1,500 years, cities arose all across the ancient world. Memphis was built around 3200 B.C. as the capital of Egypt, and between 2500 and 2000 B.C. major cities were built in what is now Pakistan. The two largest, Harappa and Mohenjo-Daro, were the most advanced cities of their day. They were carefully planned in a grid pattern with central grain warehouses and elaborate water systems, including wells and underground drainage. The houses of the wealthy were large and multistoried. Built of fired brick, they were in neighborhoods separated from the humble dried-mud dwellings of the common laborers. Like the Sumerian cities, these cities were supported by a surplus-producing agricultural peasantry and were organized around central temple complexes.

By 2400 B.C., cities were established in Europe; by 1850 B.C., in China. No fully developed cities were erected in the Americas until some 1,500 years later during the so-called Late Preclassic times (300 B.C. to A.D. 300). In Africa, cities of prosperous traders appeared around A.D. 1000 in what are now Ghana and Zimbabwe.

Preindustrial Cities

Preindustrial cities—*cities established prior to the Industrial Revolution—often were walled for protection and densely packed with residents whose occupations, religion, and social class were clearly evident from symbols of dress, heraldic imagery, and manners.* Power typically was shared between the feudal lords and religious leaders. Preindustrial cities housed only 5 to 10% of a country's population. Most had populations of fewer than 10,000.

These cities often served as the seats of political power and as commercial, religious, and educational centers. Their populations were usually stratified into a broad-based pyramid of social classes: A small ruling elite sat at the top; a small middle class of entrepreneurs rested just beneath; and a very large, impoverished class of manual laborers (artisans and peasants) was at the bottom. Religious institutions were strong, well established, and usually tightly interconnected with political institutions, the rule of which they supported and justified in theological terms. Art and education flowered (at least among the upper classes), but these activities were strongly oriented toward expressing or exploring religious ideologies.

Gideon Sjoberg (1956) has noted that three things were necessary for the rise of preindustrial cities. First, there had to be a favorable physical environment. Second, advanced technology in either agricultural or nonagricultural areas had to have developed to provide a means of shaping the physical environment—if only to produce the enormous food surplus necessary to feed city dwellers. Finally, a well-developed system of social structures had to emerge so that the more complex needs of society could be met: an economic system, a system of social control, and a political system.

Industrial Cities

Industrial cities are *cities established during or after the Industrial Revolution are characterized by large populations that work primarily in industrial and service-related jobs.*

We use the term *Industrial Revolution* to refer to the application of scientific methods to production and distribution, wherein machines came to perform work that had formerly been done by humans or farm animals. Food, clothing, and other necessities could be produced and distributed quickly and efficiently, freeing some people—the social elites—to engage in other activities.

The Industrial Revolution of the nineteenth century forever changed the face of the world. It created new forms of work, new institutions, and new social classes, and multiplied many times over the speed with which humans could exploit the resources of their environment. In England, where the Industrial Revolution began in about 1750, the introduction of the steam engine was a major stimulus for such changes. This engine required large amounts of coal, which England had, and made it possible for cities to be established in areas other than ports and trade centers through its use for transportation vehicles. Work could take place wherever there were coal deposits, industries grew, and workers streamed in to fill the resulting jobs. Thus, industrial cities arose, cities with populations that were much larger than those of preindustrial cities.

Nineteenth-century urban industrialization produced industrial slums, which were seen as some of the worst results of capitalism. Friedrich Engels, a close associate of Karl Marx, described the horrors of one of these areas.

> The view from this bridge—mercifully concealed from smaller mortals by a parapet as high as a man—is quite characteristic of the entire district. At the bottom the Irk flows, or rather stagnates. It is a narrow, coal-black stinking river full of filth and garbage, which it deposits on the lower-lying bank. In dry weather, an extended series of the most revolting blackish green pools of slime remain standing on this bank, out of whose depths bubbles of miasmatic gases constantly rise and give forth a stench that is unbearable even on the bridge forty or fifty feet above the level of the water. (Engels, 1845)

Modern industrial cities are large and expansive, often with no clear physical boundary separating them from surrounding towns and suburbs. Like the preindustrial cities before them, industrial cities are divided into neighborhoods that reflect differences in social class and ethnicity. (See Table 16–6 for a comparison of the preindustrial and industrial city.)

The industrial cities of today have become centers for banking and manufacturing. Their streets are designed for autos and trucks as well as for pedestrians, and they feature mass transportation systems. They are stratified, but class lines often become blurred. The elite is large and consists of business and financial leaders as well as some professionals and scientists. There is a large middle class consisting of white-collar salaried workers and professionals such as sales personnel, technicians, teachers, and social workers.

Formal political bureaucracies with elected officeholders at the top govern the industrial city. Religious institutions no longer are tightly intertwined with the political system, and the arts and education are secular with a strong technological orientation. Mass media disseminate news and pattern the consumption of material goods as well as most aesthetic experiences. Subcultures proliferate, and ethnic diversity often is great.

The Structure of Cities

The community and the city have been two of the primary areas of study since the beginning of American sociology. In the 1920s, classical *human ecology* blossomed under the leadership of Robert E. Park and Ernest W. Burgess at the University of Chicago. The early human ecologists were attempting to systematically apply the basic theoretical scheme of plant and animal ecology to human communities.

Theories of human communities were developed that were analogous to theories explaining plant and animal development. For example, if you were to

Table 16–6

A Comparison of the Preindustrial City and the Industrial City

	Preindustrial City	Industrial City
Physical Characteristics	A small, walled, fortified, densely populated settlement, containing only a small part of the population in the society	A large, expansive settlement with no clear physical boundaries, containing a large proportion of the population in the society
Transportation	Narrow streets, made for travel by foot or horseback	Wide streets, designed for motorized vehicles
Functions	Seat of political power; commercial, religious, and educational center	Manufacturing and business center of an industrial society
Political Structure	Governed by a small, ruling elite, determined by heredity	Governed by a larger elite made up of business and financial leaders and some professionals
Social Structure	A rigid class structure	Less rigidly stratified but still containing clear class distinctions
Religious Institutions	Strong, well established, tightly connected with political and economic institutions	Weaker, with fewer formal ties to other social institutions
Communication	Primarily oral, with little emphasis on record keeping beyond mercantile data; all records handwritten	Primarily written, with extensive record keeping; use of mechanical print media
Education	Religious and secular education for upper-class men	Secular education for all classes but with differences related to social class

drive from the mountains to the desert, you would find that different soil, water, and climate conditions produce entirely different types of vegetation. By analogy, driving from a city's business district to its suburbs, you also will notice different types of communities based on a competition for specific types of land uses.

In fact, the human ecologists told us, human communities could be understood from a Darwinian perspective. Communities and cities have evolved and changed as a consequence of competition for prime space, invasion, succession, and segregation of new groups.

Park and Burgess and other members of the Chicago school of sociologists studied the internal structure of cities as revealed by what they called the ecological patterning (or spatial distribution) of urban groups. In investigating the ways in which cities are patterned by their social and economic systems and by the availability of land, these sociologists proposed a theory based on concentric circles of development.

Concentric Zone Model The concentric zone model, sometimes irreverently called the bull's-eye model, is illustrated in Figure 16–5 (Park, Burgess, & McKenzie, 1925). The **concentric zone model** is *a theory of city development in which the central city is made up of (1) a business district, and radiating from this district is (2) a zone of transition with low-income, crowded*

and unstable residential housing with high crime rates, prostitution, gambling, and other vices; (3) a working-class residential zone; (4) a middle-class residential zone; and (5) an upper-class residential zone in what we would now think of as the suburbs. These zones reflect the fact that urban groups are in competition for limited space and that not all space is equally desirable in terms of its location and resources.

The concentric zone model initially was quite influential in that it did reflect the structure of certain cities, especially those like Chicago that developed quickly early in the Industrial Revolution, before the development of mass transportation and the automobile introduced the complicating factor of increased mobility. It did not, however, describe many other cities satisfactorily, and other models were needed.

Sector Model In the 1930s Homer Hoyt (1943) developed a modified version of the concentric zone model that attempted to take into account the influence of urban transportation systems. He agreed with the notion that a business center lies at the heart of a city but abandoned the tight geometrical symmetry of the concentric zones. Hoyt suggested that the structure of the city could be better represented by a **sector model**, *in which urban groups establish themselves along major transportation arteries (railroad*

The legal boundaries of a city seldom encompass all the people and businesses that depend on the city or have an impact on it. The U.S. Bureau of the Census has realized that it is necessary to consider the entire population in and around the city.

Figure 16–5 **Concentric Zone Model**

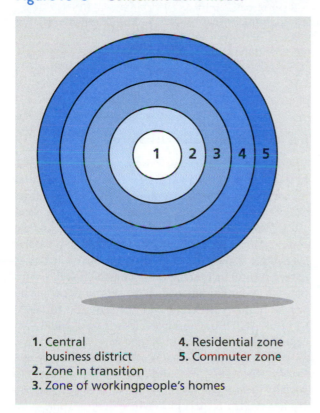

1. Central business district
2. Zone in transition
3. Zone of workingpeople's homes
4. Residential zone
5. Commuter zone

lines, waterways, and highways). Then, as the city becomes more crowded and desirable land is even farther from its heart, each sector remains associated with an identifiable group but extends its boundaries toward the city's edge (Figure 16–6).

Multiple Nuclei Model A third ecological model, developed at roughly the same time as the sector model, stresses the impact of land costs, interest rate schedules, and land-use patterns in determining the structure of cities. This multinuclei model (Harris & Ullman, 1945) emphasizes the fact that different industries have different land-use and financial requirements, which determine where they establish themselves (Figure 16–7). Thus, *the* **multiple nuclei** **model** *holds that as similar industries are established near one other, the immediate neighborhood is shaped by the nature of its typical industry, becoming one of a number of separate nuclei that together constitute the city.* For example, some industries, such as scrap metal yards, need to be near railroad lines. Others, such as plants manufacturing airplanes or automobiles, need a great deal of space. Still others, such as dressmaking factories, can be squeezed into several floors of central business district buildings. In this model, a city's growth is marked by an increase in the number and kinds of nuclei that compose it.

The limitation of the ecological approach to studying urban structure is that it downplays variables that often strongly influence urban residential and land-use patterns. For instance, the ethnic composition of a city may be a powerful influence on its structure: A city with but one or two resident ethnic groups will look very different from a city with many groups. Another important variable is the local culture—the history and traditions that attach certain meanings to specific parts of the city. For example, the north end of Boston has become the Italian section of the city. People of Italian descent who normally might move to the suburbs have remained in the neighborhood because of strong ties to the traditions associated with that area. Indeed, cultural factors are important contributors to the continuing trend of urbanization.

The early ecologists could not predict some of the trends that have taken place since World War II. Since that time, our eastern and midwestern cities have declined in population while the sprawling Sunbelt cities of the South and West have gained

Figure 16–6 Sector Model

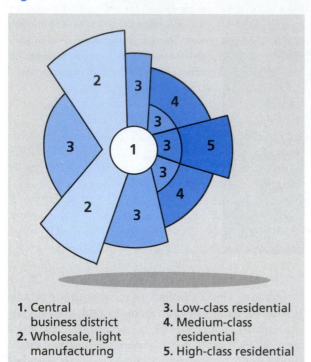

1. Central business district
2. Wholesale, light manufacturing
3. Low-class residential
4. Medium-class residential
5. High-class residential

Figure 16–7 Multiple Nuclei Model

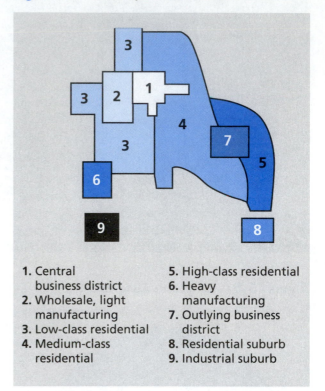

1. Central business district
2. Wholesale, light manufacturing
3. Low-class residential
4. Medium-class residential
5. High-class residential
6. Heavy manufacturing
7. Outlying business district
8. Residential suburb
9. Industrial suburb

population and business. In addition, cities everywhere are more decentralized because of the automobile. The central business districts of cities have become less important over time. As the cities spread out, urban areas become linked to one another in a manner more complex than the early urban ecologists could have imagined.

Contemporary urban ecologists (Berry & Kasarda, 1977; Hawley, 1981; Exline, Peters, & Larkin, 1982) have developed more advanced theories that take into account some contemporary developments. Computers and modern statistical techniques are used now to analyze the variables that influence the growth and development of a city.

The Nature of Urban Life

Ever since sociologists began writing about communities, they have been concerned with differences between rural and urban societies and with changes that take place as society moves away from small, homogeneous settlements to modern-day urban centers. These changes have been accompanied by a shift in the way people interact and cooperate.

The Chicago school of sociology, as it was called, produced a large number of studies dealing with human interaction in city communities. Those sociologists were interested in discovering how the sociological, psychological, and moral experiences of city life reflected the physical environment.

Social Interaction in Urban Areas

The anonymity of social relations and the cultural heterogeneity of urban areas give the individual a far greater range of personal choices and opportunities than typically are found in rural communities. People are less likely to inherit their occupations and social positions. Rather, they can pick and choose and even improve their social positions through education, career choice, or marriage. Urbanism creates a complicated and multidimensional society, with people involved in many different types of jobs and roles.

Louis Wirth proposed what is now a widely accepted definition of city in his classic essay "Urbanism as a Way of Life" (1938). Wirth defined the city as a "relatively large, dense, and permanent settlement of socially heterogeneous individuals." For years urban studies tended to accept Wirth's view of the city as an alienating place where, because of population density, people hurry by one another without personal contact. However, in *The Urban Villagers* (1962), Herbert Gans helped refocus the way sociologists see urban life. Gans showed that urbanites can and do participate in strong and vital community cultures, and a number of subsequent studies have supported this view. For example, researchers in Britain found that people who live in cities actually have a greater number of social relationships than do rural folk (Kasarda & Janowitz, 1974). Other investigators have discovered that the high population density typical of city neighborhoods need not be a deterrent to the formation of friendships; under certain circumstances, city crowding may even enhance the likelihood that such relationships will occur. Gerald Suttles (1968) showed that in one of the oldest slum areas of Chicago, ethnic communities flourish with their own cultures—with norms and values that are well adapted to the poverty in which these people live.

Of course, increased population size can lead to increased superficiality and impersonality in social relations. People interact with one another because they have practical rather than social goals. For example, adults patronize neighborhood shops primarily to purchase specific items rather than to chat and share information. As a result, urbanites rarely know a significant number of their neighbors. As Georg Simmel (1955) noted, in rural society people's social relationships are rich because they interact with one another in terms of several role relationships at once (a neighbor may be a fellow farmer, the local baker, and a member of the town council). In urban areas, by contrast, people's relationships tend to be confined to one role set at a time (see Chapter 6).

With increasing numbers of people, it also becomes possible for segments or subgroups of the population to establish themselves—each with their own norms, values, and lifestyles—as separate from the rest of the community. Consequently, the city becomes culturally heterogeneous and increasingly complex. As people in an urban environment come into contact with so many different types of people, typically they also become more tolerant of diversity than do rural people.

Although urban areas may be described as alienating places in which lonely people live in crowded, interdependent, social isolation, there is another side to the coin, one that points to the existence of vital community life in the harshest urban landscapes. Further, urban areas still provide the most fertile soil for the arts in modern society.

The close association of large numbers of people, wealth, communications media, and cultural heterogeneity provide an ideal context for aesthetic exploration, production, and consumption.

Urban Neighborhoods

People sometimes talk of city neighborhoods as if they were all single, united communities, such as Spanish Harlem or Little Italy in New York City, or Chinatown in San Francisco. Those communities do display a strong sense of identity, but to some extent this notion is rooted in a romantic wish for the good old days when most people still lived in small towns and villages that were in fact communities and that gave their residents a sense of belonging. Yet although the sense of community that does develop in urban neighborhoods is not exactly like that in small, closely knit rural communities, it is very much present in many sections throughout a city. Urban dwellers have a mental map of what different parts of their city are like and who lives in them.

Gerald Suttles (1972) found that people living in the city draw arbitrary (in terms of physical location) but socially meaningful boundary lines between local neighborhoods, even though these lines do not always reflect ethnic group composition, socioeconomic status, or other demographic variables. In Suttle's view, urban neighborhoods attain such symbolic importance in the local culture because they provide a structure according to which city residents organize their expectations and their behavior. For example, in New York City the neighborhood of Harlem (once among the most fashionable places to live) begins east of Central Park on the north side of Ninety-sixth Street and is a place that has symbolic significance for all New Yorkers. Whites often think of it as a place where they may not be welcome and that is inhabited by African American and Spanish-speaking people. For many African Americans and Hispanics, however, Harlem represents the "real" New York City and is where most of their daily encounters take place.

Even those urban neighborhoods that are well known, that have boundaries clearly drawn by very distinctive landmarks, and that have local and even national meaning are not necessarily homogeneous communities. For example, Boston's Beacon Hill neighborhood is divided into four (or possibly five) subdistricts (Lynch, 1960), and New York's Greenwich Village consists of several communities defined in terms of ethnicity, lifestyle (artists), and subculture (especially homosexual).

On the whole, Jane Jacobs's observations in *The Death and Life of American Cities* (1961) generally seem to hold true. She has argued that the social control of public behavior and the patterning of social interactions in terms of what might be called community life are to be found on the level of local blocks rather than entire neighborhoods. Once city dwellers venture beyond their own block, they tend to lose their feelings of identification. In fact, one of the typical features of urban life is the degree to which people move through many neighborhoods in their daily comings and goings rushing here and there without much attention or attachment to their surroundings. Occasionally a city as a whole may have meaning to all or most of its residents and may, for this reason, assume some community-like qualities. Consider, for example, the community spirit expressed in spontaneous celebrations for homecoming World Series or Super Bowl winners.

Although urban blocks and neighborhoods may offer a rich context for community living, there are some inescapably unpleasant facts about urban America that make many people decide to live elsewhere. Cities and urban areas in general can be crowded, noisy, and polluted; they can be dangerous;

OUR DIVERSE SOCIETY

Disorderly Behavior and Community Decay

What contributes to the decay of neighborhoods? Crime, violence, drugs, or minor offenses? At a time when people suggest we should do more to solve serious crime problems, George Kelling and Catherine Coles have suggested that the minor offenses may be more serious than we may think.

Starting in the 1960s, disorderly behavior began to be seen as a sign of cultural pluralism. In 1967, President Johnson's Commission on Crime pointed out that even though we have to do something about disorder, it is not the primary business of the police. The role of the police, according to the commission, was to arrest people for serious crimes and put them into the criminal justice system. Of course, it is nice to think that the police are not going to be too intrusive. But the result was that city streets were abandoned by police, and therefore citizens as well.

What is disorder? In its broadest sense, disorder is incivility—boorish and threatening behavior that disturbs life, especially urban life. Urban life is characterized by the presence of many strangers, and in such circumstances citizens need minimum levels of order. . . . Most citizens have little difficulty balancing civility, which implies self-imposed restraint and obligation with freedom. Yet a few are either unable or unwilling to accept any limitations on their own behavior.

Kelling and Coles describe extreme disorder as "predatory criminals who murder, assault, rape, rob, and steal." Less extreme is disorderly behavior such as panhandling, drunkenness, and sleeping on the street. Most of the latter are either ignored or punishable by fines or community service.

How should we respond to a panhandler or a drunk person walking down the street? Should we be sympathetic, helpful, or fearful? Some would see these individuals as reminders of the shortcomings of society, who are making a statement about how society has failed them.

Others may not think of these less severe offenses in political terms, but as minor inconveniences with little in the way of serious consequences.

George Kelling and Catherine Coles have suggested that we use the image of broken windows to explain how neighborhoods might decay into disorder and even crime if no one attends faithfully to their maintenance. If a window in a factory is broken and not fixed, passersby will conclude that no one cares and no one is in charge of taking care of the property.

Over time, youths in the neighborhood throw rocks at the building and break some more windows. Eventually all the windows are broken. Now it seems that not only is no one in charge of the building, but no one is in charge of the street that it faces. The message is that only the foolhardy or criminal have any business on this unprotected street, and so more and more citizens abandon the street to those they assume prowl it.

In this way small disorders lead to larger ones and eventually to crime. However, if a neighborhood appears orderly, people feel safer, and wrongdoers are less likely to commit crimes.

Source: Based on *Fixing Broken Windows: Restoring Order and Reducing Crime in Our Communities*, by G. L. Kelling & C. M. Coles, 1996, New York: Martin Kessler Books (The Free Press).

and they may have poorer schools than those in the suburbs. Consequently, many families, especially those with children, choose the suburbs as an alternative to urban life. Other city dwellers, such as the elderly living on fixed incomes, may be forced to remain despite their wish to move.

Urban Decline

A grim circle of problems threatens to strangle urban areas. Since World War II there has been a migration of both white and African American middle-class families out of the cities and into the suburbs. The number of African American middle-class families moving to the suburbs increased sharply in the aftermath of the civil rights movement of the 1960s. This migration pattern of the middle class has led to

a greater concentration of poor people in the central cities, which is reflected in the loss of revenues for many large cities. As the more affluent families leave urban areas, so do their tax dollars and the money they spend in local businesses. In fact, many businesses have followed the middle class to the suburbs, taking with them both their tax revenues and the jobs that are crucial to the survival of urban neighborhoods. Urban poverty in the United States is a characteristic of central cities. A person living in the central city is twice as likely to be poor as a person residing in a suburb. Job growth has occurred faster in suburbs than in central cities. This has meant a shrinking of the central cities' tax base, while at the same time creating conditions (such as the loss of jobs) that force people to rely on government assistance. (See "Our Diverse Society: Disorderly Behavior and Community Decay.")

Some cities have experienced a revival since the early 1990s. Many middle-class young adults have begun to find urban life attractive again. Most of these people are single or are married with no children or older couples whose children have left home.

This trend has produced *an upgrading of previously marginal urban areas and the replacement of some poor residents with middle-class ones, a process known as* **gentrification.** Critics contend that gentrification depletes the housing supply for the poor. Others counter that it improves neighborhoods and increases a city's tax base. So far, however, this trend has been confined to a limited number of cities. If the trend continues to grow, however, it clearly will have a major impact on the future of urban life and could serve as a convincing argument against doomsday predictions about the city.

Homelessness

Three decades ago every city had its "skid row" with "bums," "derelicts," and "vagrants." Aside from the occasional story about the executive who became an alcoholic and ended up sleeping in "flophouses," little interest or sympathy was expressed for the denizens of these marginal areas of the city.

Today, not only have the words that we use to describe these people changed, but so have our thinking and attitudes about them. The "bum" or "hobo" of the past has become today's "homeless person." The sense of personal responsibility for the fate attributed to them before has been replaced with a view that the homeless are the victims of a selfish, even ruthless society.

What has really changed? Has society become more heartless and created more victims, or have we become more compassionate and become more aware of the problem? Have the numbers of homeless gone up so much that we are forced to recognize the issue?

The movement to the suburbs of post–World War II America emphasized the suburban ideal of a single-family home. The city was where people worked during the day; once nightfall came, they left for the safety of the suburban community.

If the movement to the suburbs required abandoning the downtown streets at nightfall, there were many people who did not leave the central city. There are, of course, the working-class neighborhoods of the older cities where family life goes on in close proximity to the central business district, under somewhat less private and more crowded conditions than in the suburbs. There are also the marginal people for whom downtown provided alternatives not available elsewhere. Commercial and industrial areas, as well as fringe areas in decaying working-class districts, have tended to provide the housing vital to poor people not living in conventional

Advocates for the homeless often claim that these people are on the street because of a lost job, a low minimum wage, or a lack of affordable housing. Others point to problems that cannot be corrected by economic measures.

families. Single-room-occupancy hotels, rooming houses, and even skid-row flophouses have provided low-cost single accommodations for those who might not be able to come by them elsewhere.

Downtowns in many older cities traditionally have contained the cities' skid rows and red-light districts, which provided shelter and a degree of tolerance for deviant individuals and activities. Being close to transportation and requiring little initial outlay (often renting by the week), single-room housing traditionally has been utilized by the elderly poor, the seasonally employed, the addicted, and the mentally handicapped.

As the old skid rows decrease in size or disappear, the traditional skid-row population of single older men is being supplemented with large numbers of people of both sexes, many of whom are mentally ill or drug addicted. The deinstitutionalization of the mentally ill caused many of those who previously would have been committed to institutions to become homeless street people. They are not necessarily physically dangerous, but rather are disturbed or marginally competent individuals without supportive families.

In recent years, the downtown sections of many American cities have undergone extensive renovation and revitalization. This gentrification movement has been both hailed as an urban renaissance and condemned for disrupting urban neighborhoods and displacing inner-city residents. As city land becomes more desirable, space declines in what is usually considered to be the nation's least desirable housing stock, namely single-room-occupancy hotels, rooming houses, and shelters. Although these places have long

CONTROVERSIES IN SOCIOLOGY

What Produces Homelessness?

Are the homeless like you and me? Is the lack of affordable housing the main reason for the problem? Many advocates for the homeless state that a little bit of bad luck could cause any one of us to suffer the same plight that befalls many of those we see in our urban centers. Christopher Jencks, the John D. MacArthur Professor of Sociology at Northwestern University, has done a thorough study of homelessness. He notes that mental illness is one of the main reasons that homelessness has increased over the past two decades.

As soon as Americans noticed more panhandlers and bag ladies on the streets, they began trying to explain the change. Since the most noticeable of these people behaved in quite bizarre ways, and since everyone knew that state mental hospitals had been sending their chronic patients "back to the community," many sidewalk sociologists initially assumed that the new homeless were mostly former hospital inmates.

Although deinstitutionalization mostly meant that patients were released from mental hospitals after a few weeks instead of remaining there for months, years, or even a lifetime, it also meant that some people who would once have been sent to a mental hospital were now sent to the psychiatric service of a general hospital or were treated as outpatients.

It follows that considerably more than a quarter of today's homeless might have spent time in a mental hospital if we still ran the system the way we ran it in the 1950s.

Before blaming homelessness on deinstitutionalization, however, we must explain one awkward fact: hospitalization rates for mental illness began to fall in the late 1950s, not in the late 1970s or early 1980s. Since deinstitutionalization caused very little homelessness from 1955 to 1975, how could it have suddenly begun to cause a lot of homelessness after that? The answer is that deinstitutionalization was not a single policy but a series of different policies, all of which sought to reduce the number of patients in state mental hospitals but each of which did so by moving these patients to a different place. The policies introduced before 1975 worked quite well. Those introduced after 1975 worked very badly.

By 1975 most state hospitals had discharged almost everyone they thought they could house elsewhere. Their 200,000-odd remaining inmates were of two kinds: long-term residents who were so disturbed nobody else would take them, and short-term patients who were admitted, medicated, observed for a couple of weeks, and discharged. Some of the short-term patients were readmitted fairly regularly, often because they stopped taking their medication, but they spent the bulk of their time outside hospitals.

Although the number of patients in state mental hospitals fell from 468 per 100,000 adults in 1950 to

been seen as the very symbols of urban decay, they serve the vital needs of people with few resources or alternatives. Gentrification has placed these powerless people in direct competition for inner-city space. These trends offer a partial explanation for the growing ranks of the homeless on the streets of many cities.

Advocates for the homeless often assert that these people are on the streets because of a lost job, a low minimum wage, or a lack of affordable housing—all things outside the control of the homeless. Certainly there are many situations where these conditions are the cause, particularly among those homeless for a short spell. However, for the broader category of the homeless, we will find very few auto workers laid off from well-paid jobs. Most homeless people have histories of chronic unemployment, poverty, family disorganization, illiteracy, crime, mental illness, and welfare dependency—problems that cannot be corrected by quick or simple economic measures.

The largest single category of homeless people is made up of the mentally ill. Many of today's homeless

were dumped onto the streets as a result of the deinstitutionalization process. Others who have reached adulthood since then have never been institutionalized, but would have been during previous decades.

Homelessness was not the intended result of this process. The original idea was to free the patients from the wretched and abusive conditions in mental hospitals and to allow them, instead, to be treated in the community. However, funding for community treatment never materialized and many people ended up on the streets, uncared for and unable to care for themselves. (For more on the cause of homelessness, see "Controversies in Sociology: What Produces Homelessness?")

The National Coalition for the Homeless believes that on a typical night there are about 700,000 homeless people sleeping on the streets or in shelters. That number is a broad estimate, based on street interviews and counts at soup kitchens, shelters, and other services for the homeless. Other groups estimate the number of homeless people to be closer to 300,000.

119 in 1975, advocates of deinstitutionalization were far from satisfied. Rather than simply continuing their campaign to alter physicians' clinical judgments about who should be hospitalized, reformers increasingly turned to the courts, challenging physicians' right to commit anyone at all. These challenges began to influence medical practice in some states during the early 1970s, but their main impact came in the late 1970s, when they precipitated a fourth round of deinstitutionalization.

Once America restricted involuntary commitment, many seriously disturbed patients began leaving state hospitals even when they had nowhere else to live. When their mental condition deteriorated, as it periodically did, these patients were also free to break off contact with the mental health system. In many cases they also broke with the friends and relatives who had helped them deal with public agencies. The mentally ill are seldom adept at dealing with such agencies on their own, so once they lost touch with the people who had acted as their advocates, they often lost (or never got) the disability benefits to which they were theoretically entitled. In due course some ended up not only friendless but penniless and homeless.

If the courts had not limited involuntary commitment and if state hospitals had not started discharging patients with nowhere to go, the proportion of the adult population living in state hospitals would probably be about the same today as in 1975. Were that the case, state hospitals would have sheltered 234,000 mental patients on an average night rather than 92,000. It follows that 142,000 people who would have been sleeping in a state hospital under the 1975 rules are sleeping somewhere else today. On any given night, some of these people were in the psychiatric wards of general hospitals, and a few were in private psychiatric hospitals, but many were in shelters or on the streets.

Almost everyone agrees that what happened to the mentally ill after 1975 was a disaster. Both liberals and conservatives blame this disaster on their opponents, and both are half right. It was the insidious combination of liberal policies aimed at increasing personal liberty with conservative polices aimed at reducing government spending that led to catastrophe.

Had politicians been committed to keeping the mentally ill off the streets, they could have used the money that hospitals once spent on these patients to provide SRO [single-room occupancy] rooms and out-patient services. Some states did try this. In most states, however, political leaders mouthed clichés and looked the other way.

Source: Reprinted by permission of the publisher from *The Homeless* by Christopher Jencks, pp. 21–40, Cambridge, Mass.: Harvard University Press, Copyright © 1994 by Christopher Jencks.

The homeless have a variety of problems. Homeless people are more likely than the general population to be mentally ill, suffer from chronic illness, and be addicted to drugs or alcohol. A 1999 survey by the U.S. government's Housing and Urban Development office found that:

- Nearly two-thirds of homeless people suffer from chronic or infectious diseases. Another 9% are suffering from AIDS.
- Thirty-nine percent are mentally ill.
- One-quarter were abused as children and 21% were homeless as children.

In recent years, many cities have imposed restrictions on the homeless as "compassion fatigue" has set in. People have grown disillusioned with programs for the homeless because they appear to have little effect. Instead, policy makers and the public have used more aggressive ways of dealing with the problem. By 1999, all 50 of the largest cities in the United States had approved regulations limiting activities associated with the homeless, such as loitering and sleeping on sidewalks.

The deinstitutionalized, the ex-offender, the addicted, the poor, the sick, and the elderly all bring to the central city a lifestyle incompatible with that of the new urban middle class. Yet these people will not go away simply because their housing is eliminated. They remain on our streets and tax the strained resources of the remaining shelters. Unlike the suburb, the newly gentrified inner city cannot close its gates to marginal members of society. It therefore becomes imperative that new alternatives be provided.

Future Urban Growth in the United States

What will metropolitan areas look like in the future? Which cities will grow the most? Which cities will lose the most population? One way to answer such questions would be to extend into the future the trends of the 1970s, 1980s, and 1990s. Yet doing so would miss some important changes that have been taking place in the United States.

Two important trends have had an important impact on cities. The first trend is the sharp rise in immigration to the United States. Each year about 1 million people, predominantly Latin American and Asian in origin, arrive in the United States, most settling in urban areas (Frey, 2000).

The impact of immigration is apparent by looking at urban areas experiencing the greatest population gains between 1990 and 2000. The population gains in Los Angeles, Houston, San Diego, Miami, and Dallas came entirely from international migration and natural increase. Were it not for immigration, the population of these areas would have been far smaller or would have outright declined.

The second trend involves the aging of the baby-boomer generation, the 76 million people born between 1946 and 1964. The immigrants are settling in the cities while the baby boomers are aging in the suburbs.

These two trends complicate urban/suburban relationships as well as race and ethnic dynamics. The city of Los Angeles provides an example of how the two trends interact. In Los Angeles, the out-migration of whites to the suburbs has coincided with waves of immigration of new ethnic minorities. Los Angeles County's elderly population is still mostly white, its working-age population is only about one-third white, and its child population is predominantly Hispanic but includes children from other racial and ethnic groups. Reflecting their age, the growing racial and ethnic groups in Los Angeles are concerned about affordable housing, good schools, and neighborhoods conducive to the raising children. The older white population is more concerned with health and social support services for an aging, dependent population.

"Whatever else the new urban population profiles show, the old models of dealing with cities and suburbs will need to be revised to adapt to new demographic forces in America today" (Frey, 2000).

Metropolitan trends vary widely according to region. Some metropolitan areas are expected to maintain high growth rates over the next two decades. Those areas are primarily in the South and West. However, constraints to this expansion are appearing. The rate of growth for Houston and other southern and western metropolitan areas should slow somewhat, and the same pattern will occur in most other Sunbelt areas.

Suburban Living

Suburbs are *incorporated or unincorporated spatial communities that lie outside the central city but within the metropolitan area.* According to this definition, 60 million people lived in the suburbs in 1960, 74 million in 1970, and more than 143 million in 2000. In fact, most residents of metropolitan areas live in suburbs rather than in the central cities (Palen, 1995). (See Figure 16–8.)

Originally, farmers mocked the small-scale agriculture of suburban gardeners. City newspaper editors derided the lack of cultural facilities in the suburbs. Suburbanites inhabited a territory that did not fit any traditional definition of city or country.

In many respects, at least until recently, suburbs have served as a dramatic contrast to city life. Suburbs generally are cleaner than cities, less crowded, less noisy, and less crime ridden. Often, their school systems are newer and better. Many characteristics of urban life, however, have followed people to the suburbs.

Today we accept without question the sprawling landscape of single-family houses on small lots and assume that suburbs have always been with us. We give little thought to the suburb's origin. Suburbs, as we know them, developed relatively recently and largely without any planning. They were a direct response to changes that made commuting easier. Well-surfaced roads and hundreds of new bridges began to appear in the early nineteenth century. The steamship also changed matters dramatically in the New York City region. Individuals discovered a new lifestyle that included steaming down the Hudson River in the early morning and cruising slowly upstream in early evening.

Railroads were also instrumental in the development of the suburbs. Trains made it possible to travel in all kinds of weather, prompting thousands of upper-class Americans to use them. Originally, the railroad companies discouraged short-haul commuting, but eventually demand created the commuter train. Suburbs began to blossom all along the railroad routes that led to the city. Thousands of middle-income families settled in the suburbs, which offered the advantages of both rural and urban living.

The typical suburban house was part farmhouse, part urban residence, and it reflected a desire for open space, sanitation, and security. It usually had many large closets, a large cellar, an attic, a pantry, and a back hall—even guest rooms.

Suburbanites also started to copy the front lawn that was typical of estates and English country homes. The lawn mower began to appear in the 1880s, and magazines explained the use of the new machines. By the 1920s the idea of a smooth, green lawn became widely accepted. Lawn-mower manufacturers determined that grass height should be about 1.5 inches, and suburbanites were quick to follow the recommendation.

Suburban growth slowed during the depression years and the subsequent war years when gasoline rationing took place. Once the war ended, however, suburban growth resumed. By the mid-1950s, the automobile made it possible for suburbs to exist far

Figure 16–8 Urban and Rural Populations, 1950–2030

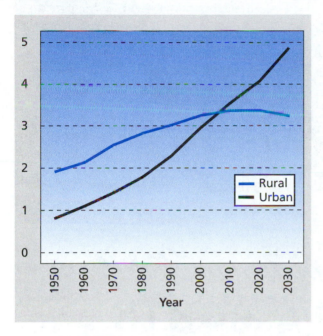

Source: UN, World Urbanization Prospects: The 2003 Revision (2004).

beyond the range of railroads and trolleys. Popular television shows of that era such as *Father Knows Best* and *Leave It to Beaver* responded to the public sentiment, which was turning against the city, and shifted their locales to the suburbs.

The way in which homes were purchased also influenced the growth of the suburbs. Before the 1920s, a home buyer borrowed 30 to 40% of the cost of the property. Mortgage interest payments of 5 or 6% were made semiannually. At the end of anywhere from three to eight years, the principal was repaid in one lump sum or the mortgage was renegotiated. After the 1920s, savings and loan associations replaced the informal, individual lender. Following World War II, the government guaranteed mortgages to veterans. Ex-GIs needed only the smallest of down payments to purchase a home, and a building boom emerged.

For the children of those families, the suburbs represented the world, a world ruled by women more than men. Men went to work in the city and returned every evening to the suburban landscape. As commuting time increased, the fathers looked for jobs in the suburbs and began deserting the city altogether. When white-collar workers started to appear in the suburbs in large numbers, suburbs entered their present stage.

The growth of the nation's suburbs continued throughout the century. The share of the U.S. population that lived in the suburbs doubled from 1900 to 1950 and doubled again from 1950 to 2000.

The flight of more affluent city residents to suburbs has been attributed to numerous social, political, and economic factors. These factors include racial tension in cities, superior educational and recreational facilities in suburbs, low government investment in central-city services and industry, and the construction of highways (which facilitated movement between suburban homes and central-city jobs).

Since the 1980s, however, suburbanization has slowed, or even reversed, in some areas. Suburbs continue to grow more rapidly than central cities in most U.S. metropolitan areas, but their share of metropolitan growth has declined.

Meanwhile, many "inner-ring" suburbs that were developed in the 1950s and 1960s, are experiencing more severe problems, including crime, job loss, and disinvestments. Such older suburbs are losing population in many cities.

People used to believe the stereotype that suburbs were predominantly white. This is no longer the case and ignores the heterogeneous growth patterns within the suburbs. In the nation's largest metropolitan areas (those with populations exceeding 500,000), minorities accounted for 27% of the suburban populations in 2000, compared with 19% in 1990 (Frey, 2001). During the 1980s and 1990s, some areas saw explosive growth in the number of African American suburbanites. Today, there are more than 40 metropolitan areas with at least 50,000 African American suburbanites. The largest suburban black population is found in Washington, D.C. A second large African American suburb is located outside of Atlanta, where suburban African Americans live in such neighborhoods as Brook Glen, Panola Mill, and Wyndham Park. Los Angeles also has large numbers of suburban African Americans. New York, Los Angeles, Chicago, San Francisco, and Miami suburbs have seen growth driven by increases in the Asian and Hispanic populations.

One characteristic that still separates suburban from city neighborhoods is that suburbs tend to be homogeneous with regard to the stage of development of families in their natural life cycle. For example, in the suburbs, retired couples rarely live among young couples that are just starting to have children. Aside from the rather dulling sameness of many suburban tracts that results, the situation also creates problems in the planning of public works. For example, a young suburb might well invest money in school buildings, which, some 20 years later, are likely to stand empty when the children have left home.

A problem of the late twentieth century has been suburban sprawl. Suburban sprawl involves the expansion of new residential subdivisions and commercial areas, increased use of cars, and segregated use of land according to activities.

The increased dependence on cars produced by sprawl has changed how we live. Many Americans now think of the second car as a necessity, not a luxury. For suburbanites, sprawl increases commuting time, raises the cost of living, and reduces the standard of living. An average suburban household drives 3,300 more miles per year than their central-city counterparts. Suburban residents of Denver used 12 times as much gasoline as those in Manhattan.

Sprawled suburbs have road costs up to 33% higher and utility costs 18 to 25% higher than sprawl-free communities. Sprawled development consumes 25 to 67% more open land than nonsprawl development, and produces about one-third more water pollution (Burchell et al., 1999).

Sprawl also worsens conditions for poor central-city residents because the movement of businesses to suburbs creates a mismatch between where this untapped supply of potential workers lives and where jobs are located. Public transport is not bridging this gap.

Now in the twenty-first century, most Americans live in suburbs. But they have not exactly left the cities behind.

Cities and suburbs alike face daunting challenges. Our suburbs—from wealthy gated communities, to gritty blue-collar bungalow and industrial communities, to new housing tract developments on the urban fringe—are inexorably linked to the fate of nearby cities. In fact, many of the places we call suburbs are really small cities. They, too, confront serious problems: fiscal challenges, poverty, environmental concerns, crime, housing shortages, traffic congestion, ethnic tensions, and struggling public schools (Villaraigosa, 2000).

Exurbs

Problems that were once associated with central cities—traffic congestion, overcrowded schools, and the loss of open space—have emerged in many suburbs. This situation has motivated some people to move to the more rural areas also known as exurbs.

Exurbs are *middle- and upper-middle-class communities that can be found in outlying semirural suburbia.* These areas are located in a newer, second ring beyond the old suburbs. These communities sometimes form around old villages or small towns. The inhabitants, as a rule, are affluent, well-educated professionals (Palen, 1995).

Most of the nation's rapid-growth counties are exurbs in expanding metropolitan areas. Exurbs differ sharply from the traditional suburb. For one thing, development in the exurbs is much less dense, emerging near farms and rural land in the remotest fringes of metropolitan areas. For another, suburbs traditionally are dependent on the city for jobs and services. Not so in the exurb, which creates its own economic base in its shopping malls, office complexes, and decentralized manufacturing plants. Some examples of exurbs include Douglas County (Denver); Loudoun County, Virginia (Washington, D.C.); and Rockwall, Collin, and Williamson counties in Texas (the former two in metro Dallas, the latter outside Austin).

Why are so many people moving still farther out from the central city? Recent research suggests a variety of reasons, but on the whole, reasons that are not very surprising when we consider the suburban exodus of the early postwar era. Basically, exurbanites are similar to people a century ago who were seeking the better life in romantic suburbs and rustic settings. Today these people expect the services, schools, and shopping areas to follow them out of the city.

In short, the exurb, like the suburb before it, is seemingly another step in the quest for the American dream.

Urbanism has become a way of life. Although the shape and form of metropolitan areas continue to change, their influence continues to dominate the manner in which we interact with our environment.

SUMMARY

- The size of a population tends to become a progressively greater problem because of exponential growth, in which a continuously expanding base rapidly doubles.
- The annual growth rate in world population has recently declined so that the world's population now doubles every 51 years instead of every 35 years. In most developing countries, however, the rate of growth is much faster.
- Much of the world's growth in population in the next few decades will take place in developing countries.
- Demography is the study of the size and composition of human populations as well as the causes.
- Demography is influenced by three major factors: fertility, mortality, and migration.
- Life expectancy is usually determined more by infant than adult mortality.
- Rapid population growth in the third world is largely due to an increase in life expectancy rather than a rise in the birthrate.
- Migration is the movement of populations from one geographical area to another.
- Historically, migration has had less impact on population change in an area than either fertility or mortality.
- The core of the population problem, according to Thomas Malthus, is that populations will always grow faster than the available food supply, because resources increase arithmetically while population increases geometrically.
- According to demographic transition theory, societies pass through four stages of population change as they move from high fertility and mortality to relatively low fertility and mortality.
- Demographic transition theory accurately describes the population changes that have occurred in Western society with the advance of industrialism.
- Demographic transition theory does not, however, explain population trends in the underdeveloped world today.
- Developing countries have experienced a much faster drop in death rates than Western societies did, without a comparable rate of increase in economic development.
- In developing countries the birthrate, which has not fallen as fast or as consistently as it did in Western countries, has become an increasingly serious problem.

- Some people believe that the population crisis will be avoided through wider application of existing technologies, the discovery of new resources, or an exponential growth in knowledge.
- Population growth is leading to increasing urbanization and the pace is accelerating in the nonindustrial world.
- For vast majority of time that humans have lived on earth, they have lived without cities. Cities have appeared only within the past 7,000 to 9,000 years, coinciding with the rise of what we know as civilization.
- The industrialized areas of the world are already heavily urban, and the rate of urbanization is slowing.
- Two requirements had to be met for cities to emerge:
  - There had to be a surplus of food and other necessities, so that some people could afford to live in settlements where they did not produce their own food but could depend on others to meet their needs.
  - Some form of social organization beyond the family that was capable of distributing the surplus to those who needed it.
- The world's first fully developed cities arose 6,000 years ago in the Middle East in what is now Iraq, which was the site of the Sumerian civilization.
- Preindustrial cities—cities established before the Industrial Revolution—often were walled for protection and densely packed with residents whose occupations, religion, and social class were clearly evident from symbols of dress and manners.
- Industrial cities are cities established during or after the Industrial Revolution and are characterized by large populations that work primarily in industrial and service-related jobs.
- The Industrial Revolution of the nineteenth century created new forms of work, as well as new institutions and social classes, and multiplied many times over the speed with which humans could exploit the resources of their environment.

- Modern industrial cities are thus large and expansive, often with no clear physical boundary separating them from surrounding towns and suburbs. Modern industrial cities have become centers for banking and manufacturing.
- The vast majority of people in the United States live in urban areas, though not in cities.
- A city is a unit that typically has been incorporated according to the laws of the state within which it is located.
- The legal boundaries of a city seldom encompass all the people and businesses that may be affected by the social and economic aspects of an urban environment.
- Although some urban blocks and neighborhoods offer a rich community life, there are also many serious problems in the cities—crime, pollution, noise, poor educational and social services, and so on—that make many people want to live elsewhere.
- Levittown, Pennsylvania, became the symbol of the post–World War II suburban building boom.
- With little or no money down and low monthly payments attracting former GIs and their families, the U.S. urban landscape and society were changed.

 ## Media Media Resources

The Companion Website for *Introduction to Sociology,* Ninth Edition

http://sociology.wadsworth.com/tischler9e

Supplement your review of this chapter by going to the companion website to take one of the Tutorial Quizzes, use the flash cards to master key terms, and check out the many other study aids you'll find there. You'll also find special features such as Wadsworth's Sociology Online Resources and Writing Companion, GSS data, and Census 2000 information at your fingertips to help you complete that special project or do some research on your own.

CHAPTER SIXTEEN STUDY GUIDE

KEY CONCEPTS AND THINKERS

Match each concept with its definition, illustration, or explanation presented below.

a. Fertility
b. Crude death rate
c. Age-specific death rate
d. Fecundity
e. Crude birthrate

f. Age-specific birthrate
g. Dependency ratio
h. Positive checks
i. Second demographic transition

j. Concentric zone model
k. Multiple nuclei model
l. Sector model
m. Gentrification

_____ **1.** The annual number of births relative to total population.

_____ **2.** The annual number of births relative to the number of women of childbearing age.

_____ **3.** The decline of fertility below replacement.

_____ **4.** The physiological ability to have children.

_____ **5.** The actual number of births.

_____ **6.** The number of nonworking-age people relative to the number of people of working age.

_____ **7.** A model of urban development in which areas with distinct characteristics evolve based on their distance from core of the city.

_____ **8.** A model of urban development in which groups establish themselves along transportation arteries.

_____ **9.** The number of deaths relative to total population.

_____ **10.** The number of deaths relative to the numbers in each age category.

_____ **11.** Middle-class people moving into and upgrading a once poorer neighborhood.

_____ **12.** Events like war, famines, and epidemics that Malthus believed limited population growth by causing large number of deaths.

_____ **13.** A model of urban development stressing the location of many different industries as centers of different kinds of activity.

Match the thinkers with their main ideas or contributions.

a. Paul Ehrlich
b. Thomas Malthus
c. Karl Marx

d. Herbert Gans
e. Louis Wirth

f. Jane Jacobs
g. Robert Park and Ernest Burgess

_____ **1.** In *Urbanism as a Way of Life,* he argued that because of its population density the city was alienating and made for impersonal relationships.

_____ **2.** A pioneer in the study of population, he believed natural laws governed their growth and that population would grow faster than food supply.

_____ **3.** Argued that industrial capitalism was the real cause of overpopulation.

_____ **4.** Founded the Chicago school of sociology; developed the sector model of urban growth.

_____ **5.** Challenged the prevailing view of cities as alienating and, through his research, demonstrated that urban residents do participate in strong, vital villagelike community cultures.

_____ **6.** In *The Death and Life of Great American Cities,* argued that the block, rather than the neighborhood, was the basic unit of city life and social control.

_____ **7.** A twentieth-century writer who warned about the catastrophic consequences of unchecked population growth.

CENTRAL IDEA COMPLETIONS

Following the instructions, fill in the appropriate concepts and descriptions for each of the questions posed in the following section.

1. Identify the four stages of the population according to demographic transition theory.

   a. _____

   b. _____

   c. _____

   d. _____

2. Define and present an example of "the second democratic transition."

   _____

   _____

3. Identify three factors that allowed Europe to avoid the demographic fate predicted by Malthus.

   a. _____

   b. _____

   c. _____

4. With fewer children being born in certain countries, the ratio of older people to younger people is growing. What are three consequences of this changing ratio?

   a. _____

   b. _____

   c. _____

5. Identify three reasons that the dire ecological predictions made in the 1970s were wrong.

   a. _____

   b. _____

   c. _____

6. Today, where would one expect to find the greatest urban growth across the world?

   _____

7. What are the two effects of migration of rural populations into urban centers?

   a. _____

   b. _____

8. In the debate over homelessness, some people blame the personal characteristics of the homeless, others blame public policy. Describe two personal and public factors that contribute to the existence of homelessness.

   a. Personal factors: _____

   b. Policy factors: _____

CRITICAL THOUGHT EXERCISES

1. Some people, including some sociologists, have theorized that city life requires that most relationships be impersonal and narrow. What aspects of cities might lead to such a conclusion? Other people, including some sociologists, argue that urban life is personally as rich and robust as life in rural areas, perhaps more so. What aspects of cities might contribute to this interpersonal robustness?

2. Imagine two conflicting scenarios for the future of the earth and its people, one more pessimistic, the other more optimistic. Both should be based on demographics and the dynamics of population growth. What kinds of policies would each scenario suggest—internal policies for advanced countries, internal policies for underdeveloped countries, and international policies?

INTERNET ACTIVITIES

1. For more on world population, http://www.geography.learnontheinternet.co.uk/topics/popn.html has easily understood graphics and good links. http://www.worldometers.info/ provides running tabulations of demographic statistics.

2. The National Aeronautics and Space Administration (NASA) has a picture of the earth showing where the lights are: http://antwrp.gsfc.nasa.gov/apod/image/0011/earthlights_dmsp_big.jpg. The distribution of lights is a good proxy of energy consumption but also of population. For a similar picture showing sunset over Europe and Africa, see http://www.freemaninstitute.com/sunsetearth.htm.

3. For a graphic representation of the models of urban growth in this chapter, see http://www.cs.ucr.edu/rgl/projects/interactiveCities/interactiveCities.html. It has a PowerPoint presentation of some of these ideas as well as 15-second animations simulating the growth of two types of region. Urban sprawl is discussed at http://www.sprawlcity.org/. At the other end of the spectrum, http://www.carfree.com/ has photos and maps of cities and part of cities with no cars.

4. SimCity is an urban simulation game sold commercially. A free simulation called LinCity (developed originally on a Linux platform but can run on Windows) promises, "You are required to build and maintain a city. You must feed, house, provide jobs and goods for your residents. You can build a sustainable economy with the help of renewable energy and recycling, or you can go for broke and build rockets to escape from a pollution ridden and resource starved planet." The game is available here: http://lincity.sourceforge.net/.

ANSWERS TO KEY CONCEPTS

1.e 2.f 3.i 4.d 5.a 6.g 7.j 8.l 9.b 10.c 11.m 12.h 13.k

ANSWERS TO KEY THINKERS

1.e 2.b 3.c 4.g 5.d 6.f 7.a

ThomsonNOW™

Reviewing is as easy as ❶ ❷ ❸

1. Before you do your final exam, take the ThomsonNOW diagnostic quiz to help you identify the areas on which you should concentrate. You will find information on ThomsonNOW and instructions on how to access all of its great resources on the foldout at the beginning of the text.

2. As you review, take advantage of ThomsonNOW's study videos and interactive Map the Stats exercises to help you master the chapter topics.

3. When you are finished with your review, take ThomsonNOW's posttest to confirm you are ready to move on to the next chapter.

17

Health and Aging

© Archive Holding Inc./The Image Bank/Getty Images

Learning Objectives

After studying this chapter, you should be able to do the following:

- Know what sociologists mean by the sick role.
- Describe the basic characteristics of the U.S. health care system.
- Understand the link between demographic factors and health.
- Describe the three major models of illness prevention.
- Describe the basic demographic features of the older population in the United States.

Tierney looks radiant in a new black-and-white striped dress, smiling and chatting about the gentle bulge in her normally flat stomach. She lies on a table in the obstetrics room of St. Francis Hospital. Her husband Greg sits in a chair beside her, watching as a technician begins the routine ultrasound test that is their last task before a weeklong vacation on Martha's Vineyard.

The technician, Maryann, strokes Tierney's belly with a sonogram wand, using sound waves to create a picture of the life inside. Greg studies the video screen, fascinated by the details emerging from what looks like a half-developed Polaroid: an arm here, a leg there, a tiny face in profile; then the internal organs: brain, liver, kidneys.

"I'm having a hard time seeing the heart. Maybe the baby's turned," Maryann says calmly.

Then it appears, pumping in a confident rhythm. She stops the moving image, capturing a vivid cross-section, and Greg remembers his high-school biology.

"All mammals have four chambers in the heart," Greg thinks to himself. "There are only three chambers there."

"You know," Maryann says tactfully, "I'm not as good at this as some other people. Maybe somebody else should take a look." She tries to mask her alarm as she leaves the room.

The inescapable truth is staring at Tierney and Greg from the silent screen: There is no fourth chamber. There is a hole in the heart. And not just any hole,

they will soon learn. It is a telltale sign of Down syndrome, a genetic stew of physical defects and mental retardation.

A hole in the heart—in their baby's and, suddenly, in their own.

Tierney and Greg Fairchild have just entered a world of technological wizardry and emotional uncertainty called "prenatal screening." It is a confusing place where even the name is misleading; abortion screening is more accurate.

Most disorders tested for today—including Down syndrome, muscular dystrophy, and cystic fibrosis—cannot be corrected. That means the most common question prompted by distressing prenatal test results is not, "How can we fix it?" It is: "Should this pregnancy continue?"

In the weeks ahead, Tierney and Greg will make a journey through uncharted terrain, filled with fears and tears. They will be tested and torn, changing their minds repeatedly as they confront a new reality amid the ache of lost dreams.

Researchers say they have deciphered the blueprint of human development—the genetic code that acts as the operating instructions for creating life. That achievement is expected to drive prenatal screening into the realm of science fiction. Then what? Does a woman carry to term a baby susceptible to mental illness? Cancer? Obesity? Infertility?

Already, hard science has far outpaced the emotional side of the equation. People can learn a great deal about their unborn children, but no one tells them how to handle that knowledge, or what the future might hold (excerpted from Zuckoff, 1999).

Medicine and health care issues are intertwined with our social and cultural life. In this chapter, we will examine these interactions as we explore health and illness.

The Experience of Illness

Illness not only involves the body, but it also affects the individual's social relationships, self-image, and behavior. Being defined as "sick" has consequences that are independent of any physiological effects. Talcott Parsons (1951) has suggested that to prevent the potentially disruptive consequences of illness on a group or society, there exists a sick role. The **sick role** is *a shared set of cultural norms that legitimates deviant behavior caused by the illness and channels the individual into the health care system.* According to Parsons, the sick role has four components. First, the sick person is excused from normal social responsibilities, except to the extent that he or she is supposed to do whatever is necessary to get well. Second, the sick person is not held responsible for his or her condition and is not expected to recover by an act of will. Third, the sick person must recognize that being ill is undesirable and must want to recover. Finally, the sick person is obligated to seek medical care and cooperate with the advice of the designated experts, notably the physicians. In this sense, sick people are not blamed for their illnesses, but they must work toward regaining their health.

The sick role concept is based on the perspective that all roads lead to medical care. It tends to create a doctor-centered picture, with the illness being viewed from outside the individual. Some (Strauss & Glaser, 1975; Schneider & Conrad, 1983) have suggested that the actual subjective experience of being sick should be examined more closely. These researchers suggest that we should focus more on individuals' perceptions of illness, their interactions with others, and the effects of the illness on the person's identity.

At the same time, society tends to define what should be considered a medical issue. Over the years, many issues have become medical issues that were not considered such before. **Medicalization** is *the process by which nonmedical problems become defined and treated as medical problems, usually in terms of illness or disorder.* Of late, such things as alcoholism, drug abuse, battering, gender confusion, obesity, anorexia and bulimia, and a host of reproductive issues from infertility to menopause have undergone medicalization (Conrad, 1992).

Health Care in the United States

The United States has the most advanced health care resources in the world. We can scan a brain for tumors, reconnect nerve tissues and reattach severed limbs through microsurgery, and eliminate diseases like poliomyelitis, which crippled a president. We can do all this and much more; yet many consider our health care system wholly inadequate to meet the needs of all Americans. Critics maintain that the U.S. health care system is one that pays off only when the patient can pay. (See Table 17–1 for changes in life expectancy during the last century.)

The American health care system has been described as "acute, curative, [and] hospital based" (Knowles, 1977). This statement implies that our approach to medicine is organized around the cure or control of serious diseases and repairing physical injuries, rather than caring for the sick or preventing disease. The American medical care system is highly technological, specialized, and increasingly centralized.

Medical care workers include some of the highest paid employees (physicians) in our nation and some of the lowest paid. About three-quarters of all medical workers are women, although the majority of doctors are men. Many of the workers are members of minority groups, and most come from lower-middle-class backgrounds. The majority of the physicians are white and upper middle class.

The medical care workforce can be pictured as a broad-based triangle, with a small number of highly paid physicians and administrators at the top. Those people control the administration of medical care services. As one moves toward the bottom of the triangle, there are increasing numbers of much lower-paid workers with little or no authority in the health care organization. This triangle is further layered with more than 300 licensed occupational categories of medical workers. There is practically no movement of workers from one category to another, because each requires its own specialized training and qualifications (Conrad & Kern, 1986).

Table 17–1

Life Expectancy at Birth and at Age 65 (Years)

	At Birth			At Age 65		
	Total	Male	Female	Total	Male	Female
1900	47.3	46.3	48.3	11.9	11.5	12.2
1950	68.2	65.6	71.1	13.9	12.8	15.0
2003	77.6	74.8	80.1	18.5	16.8	19.8

Source: Hoyert, D. L., Kung, H. C., Smith, B. L. Deaths: Preliminary Data for 2003. *National Vital Statistics Reports,* Vol 53, No. 15. Hyattsville, MD: National Center for Health Statistics, 2005.

Figure 17–1 **Death Rates for Selected Illnesses for People 25–34 by Gender**

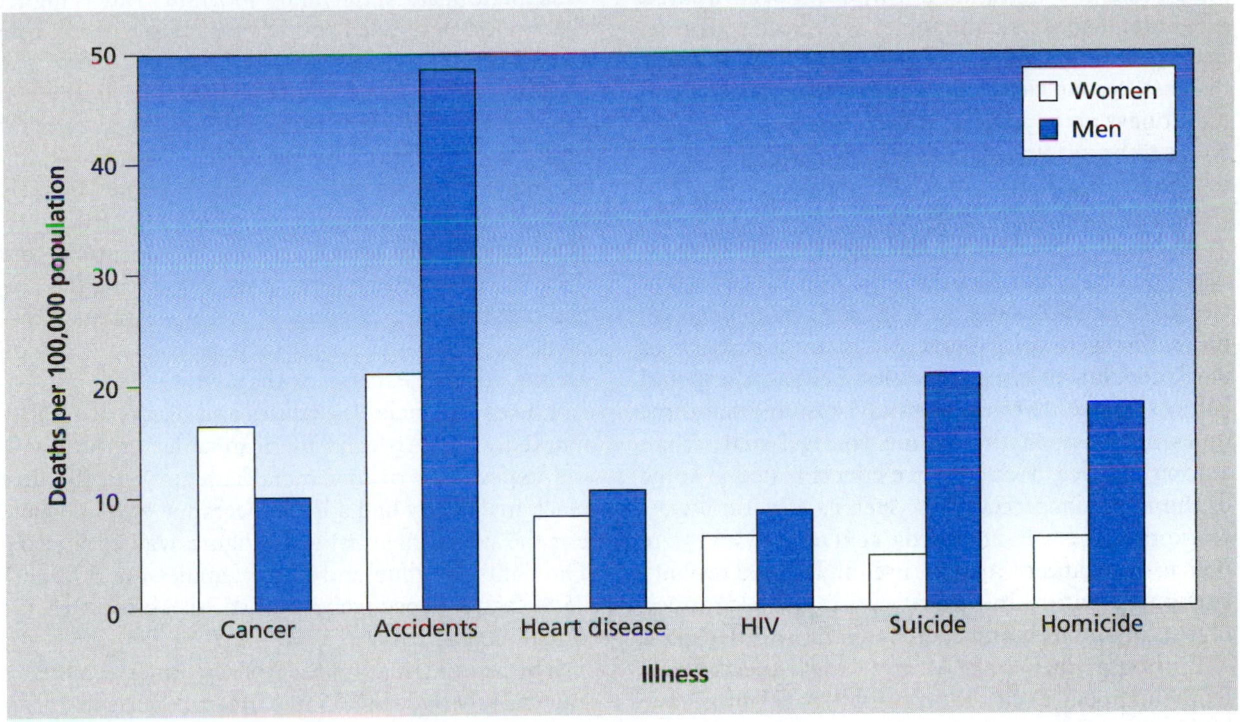

Source: Anderson, R. N. (2002). Deaths: Leading causes for 2000. National Vital Statistics Reports. Vol. 50 No. 16. Hyattsville, MD: National Center for Health Statistics.

Gender and Health

You probably are aware of the striking differences in life expectancy between men and women. The life expectancy of women in the United States has been continuously higher than that for men since records began being kept. Even though the life expectancy for both men and women has increased, the increase has been greater for women. Between 1950 and 2003, life expectancy at birth has increased by 9.3 years for women and 9.4 years for men (National Center for Health Statistics, 2005).

In 1900, women could expect to live an average of two years longer than men. By the 1970s, life expectancy for women was seven years greater than men. That number has since decreased to 5.6 years in 2003 when males lived an average of 74.8 years and women an average of 80.1 (National Center for Health Statistics, 2005).

Women have historically had high rates of death from complications related to pregnancy and childbirth. But improvements in prenatal and obstetric care have dramatically decreased the risk of pregnancy-related deaths.

Today, women of all ages experience lower mortality rates. This difference is due in part to the fact that men are three times more likely to die from unintentional injuries, homicide, or suicide than women. And programs to address these causes of death have achieved very slow progress during the last 50 years

compared with programs for other causes of death. Studies also show that men are less likely to seek medical attention for health-related problems than are women, and when care is sought they are less likely to comply with the given medical instructions (AmeriStat, December 2002). (See Figure 17–1.)

The leading cause of death among both women and men is heart disease, although among women, deaths from heart disease have fallen in recent years. Women's higher levels of estrogen seem to provide protection against the incidence and severity of heart disease. In recent decades, the growth of estrogen replacement therapy for older women has also helped reduce deaths from heart disease.

Women suffer from illness and disability more frequently than men, but their health problems are usually not as life threatening as those encountered by men. Women, of course, do suffer from most of the same diseases as men. The difference is at what point in life they encounter those diseases. For example, coronary heart disease is a leading cause of death for women older than 66, but for men it is the leading killer after age 39.

Men appear to have lower life expectancies than women because of biological and sociological reasons. Males are at a biological disadvantage to females, as seen by the higher mortality rates from the prenatal and neonatal (newborn) stages onward.

Although the percentages may vary from year to year, the chances of dying during the prenatal stage are approximately 12% greater among males than fe-males, and 130% greater during the newborn stage. Neonatal disorders common in male rather than fe-male babies include respiratory diseases, digestive diseases, certain circulatory disorders of the aorta and pulmonary artery, and bacterial infections. The male seems to be more vulnerable than the female even be-fore being exposed to the different social roles and stress situations of later life.

There are a number of sociological factors that also play an important role in the different life expectan-cies of men and women. Men are more likely to place themselves in dangerous situations both at work and during leisure activities. Therefore it should be no surprise that accidents cause more than three times as many deaths among younger males than among females. Men also are concentrated in some of the most dangerous jobs, such as structural steel workers, loggers, bank guards, coal miners, and state police. The rates of alcohol use, high-speed driving, and participation in violent sports are also much higher among men and contribute to the differences.

Although men have shorter life expectancies, women appear to be sick more often. Women have high rates of acute illnesses, such as infectious and parasitic diseases, digestive problems, and respiratory conditions, as well as chronic illnesses, such as hy-pertension, arthritis, diabetes, and colitis. Some have suggested that women may not be sick more often, but may just be more sensitive to bodily discomforts and more willing to report them to a doctor. (See "Global Sociology: Women Live Longer Than Men throughout the World.")

Men are just as vulnerable to psychiatric problems as women. A key difference however is that men, when emotionally disturbed, are likely to act out through drugs, liquor, and antisocial acts, whereas women display behaviors that show an internaliza-tion of their problems, such as depression or phobias (Hoyert et al., 2001)

The gap between male and female life expectancy during most of the twentieth century was attributed primarily to the fact that men smoked more than women. But in recent decades, the prevalence of smok-ing among women has increased while the preva-lence among men has declined (Hoyert et al., 2001).

Race and Health

There are significant differences in the health of the various racial groups in the United States. Asian Americans have the best health profile, followed by whites. African Americans and Native Americans display the worst health data.

There are glaring disparities in childhood mor-tality of the white and black populations. The infant mortality rate for African Americans is more than twice that for white infants. In some cities, such as Washington, D.C., the infant mortality rate is higher than in less developed countries such as Cuba, Costa Rica, and Chile (Bureau of the Census, 2000).

Many of the same problems that face mothers in less developed countries are factors in the high infant mortality rate among African Americans. For exam-ple, black women giving birth are 2.5 times as likely as white mothers to be younger than 18, nearly one-third have fewer than 12 years of education, and 40% have not received prenatal care during the cru-cial first trimester of pregnancy. Consequently, low-birth-weight babies are more than twice as common among African Americans than whites.

Life expectancies for whites and blacks also differ markedly. The African American male has the lowest life expectancy of any racial category. In 2002, a black male baby had a life expectancy of 68.8 years. For the white male baby, the figure was 75.1 years. The data for white and black females are 80.3 and 75.6 years, respectively (Statistical Abstract of the United States, 2006).

The health situation for African American men is particularly bad. In 1999, African American adults were more likely to die from accidents, homicide, and HIV compared with either whites or Hispanics. Among people ages 25 to 34, the non-Hispanic black homicide rate in 1999 was 43 deaths per 100,000, compared with 15 deaths per 100,000 among Hispanics and only 4 deaths per 100,000 among non-Hispanic whites. The death rate for HIV was 30 per 100,000 among non-Hispanic blacks, 8 per 100,000 among Hispanics, and 3 per 100,000 among non-Hispanic whites. The suicide rate was the only cause that was lower among blacks (10) than it was for whites (15). Higher death rates among minorities, particularly among those infected with HIV, stem in part from their limited access to health insurance, and consequently, to medical care (Anderson, 2001).

The health situation for Hispanic Americans is also worse than that for Anglo Americans. Studies show that Hispanic Americans have a higher infant mortality rate, a shorter life expectancy, and higher rates of death from influenza, pneumonia, diabetes, and tuberculosis. Hispanic death rates for heart dis-ease and cancer are lower than those for whites.

Native Americans have shown an improvement in overall health since 1950. Native Americans have the lowest cancer rates in the United States, and their mortality rates from heart disease are lower than those of the general population as well. There are other areas in which Native Americans fare much worse than other groups, however. Their mortality rates from diabetes are 2.3 times that of the general population. The complications from diabetes take a further toll by increasing the probability of kidney

GLOBAL SOCIOLOGY

Women Live Longer than Men throughout the World

The widening of the sex differential in life expectancy has been a central feature of mortality trends in developed countries in the twentieth century. In 1900, in Europe and North America, women typically outlived men by 2 or 3 years.

Today, the average gap between the sexes is roughly 7 years (Figure 17–2), but exceeds 12 years in parts of the former Soviet Union as a result of the unusually high levels of male mortality. This differential reflects the fact that in most nations females have lower mortality than males in every age group and for most causes of death. Female life expectancy now exceeds 80 years in over 30 countries and is approaching this level in many other nations. The gender differential usually is smaller in developing countries, commonly in the 3–6 year range, and even is reversed in some South Asian and Middle East societies where cultural factors (such as low female social status and preference for male rather than female offspring) are thought to contribute to higher male than female life expectancy at birth.

Figure 17–2 **Female Advantage in Life Expectancy at Birth: 2000**

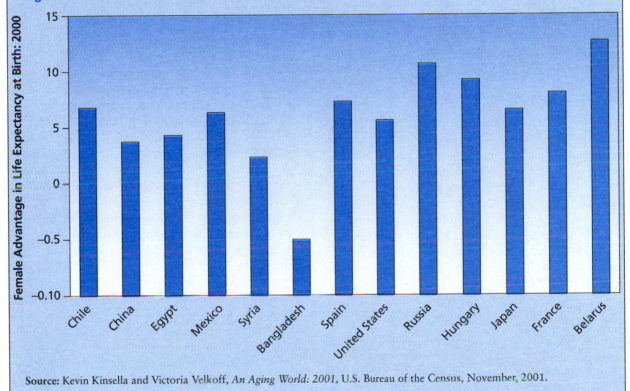

Source: Kevin Kinsella and Victoria Velkoff, *An Aging World: 2001*, U.S. Bureau of the Census, November, 2001.

disease and blindness. Native Americans also suffer from high rates of venereal disease, hepatitis, tuberculosis, and alcoholism and alcohol-related diseases such as cirrhosis of the liver, gastrointestinal bleeding, and dietary deficiency.

The suicide rate for Native Americans is 20% higher than for the general population. Native American suicide victims are generally younger than those of other groups, with their suicide rate peaking between ages 15 and 39. For the general population, suicide peaks after age 40.

Mortality rates for Asian and Pacific Islanders are lower than expected given their social and economic status in the United States. The leading causes of death for this group are heart disease and cancers, rather than injuries, suicide, or homicide as in other ethnic groups. Studies show that Asians experience an "immigrant advantage," which may contribute to their good health. International migrants are usually healthy and optimistic individuals who tend to eat better and take care of their overall health more conscientiously than nonimmigrants. Although the

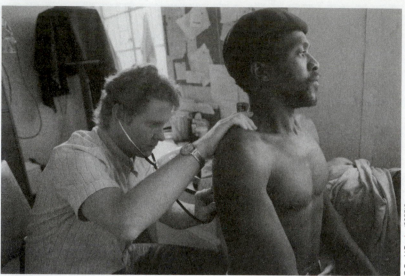

© David Turnley/CORBIS

Asian population in the United States includes a high proportion of recent immigrants, the "immigrant advantage" alone does not fully account for the good health of Asian Americans (National Center for Health Statistics, 2001, "Health").

Social Class and Health

It should surprise no one that poverty and the difficult life circumstances that accompany it, such as inadequate housing, malnutrition, stress, and violence, increase your chance of getting sick. Recognition of the link between poverty and disease has existed for more than a century. One difference, however, is that researchers in the past saw high disease rates among the poor as a moral weakness.

In 1891, John Shaw Billings, the surgeon general of the United States at that time, after examining records from different hospitals, came to the surprising conclusion for that time that lower-income patients had higher death rates than those with more money. Billings decided that this reflected the moral failing of

> [A] distinct class of people who are structurally and almost necessarily idle, ignorant, intemperate and more or less vicious, who are failures or the descendants of failures. (Billings, 1891)

Part of the reason why the poor have higher death rates is because they have less access to high-quality medical care, good nutrition, and are less likely to feel they have control over their life circumstances. They are also more likely to smoke, be overweight, and have low physical activity levels. Studies that have tried to account for these risk factors have shown that this is only part of the explanation, however. This suggests that the relationship between social class and health is more complex (National Center for Health Statistics, 2000).

Poverty contributes to disease and a shortened life span, both directly and indirectly. An estimated 25 million Americans do not have enough money to feed themselves adequately and as a result suffer from serious nutritional deficiencies that can lead to illness and death.

Poverty also produces living conditions that encourage illness. Pneumonia, influenza, alcoholism, drug addiction, tuberculosis, whooping cough, and even rat bites are much more common in poor minority populations than among middle-class ones. Inadequate housing, heating, and sanitation all contribute to those acute medical problems, as does the U.S. fee-for-service system that links medical care to the ability to pay.

An example of how social class can account for race differences with respect to health issues can be seen by examining the health data of Asian Americans. Asian Americans have the highest levels of income, education, and employment of any racial or ethnic minority in the United States, often exceeding those of the general white population. At the same time, the lowest age-adjusted mortality rates in the United States are among Asian Americans. Even though heart disease is the leading cause of death for Asian Americans, their mortality from this disease is less than that for whites and for other minorities. Deaths from homicide and suicide are particularly low for Asian Americans. The infant mortality rates for Asian Americans range between 5 and 6 per 1,000, which are lower than those for whites. Although infant mortality rates are only one indicator of health within a group, they are nevertheless an important measure of the quality of life experienced by that population.

Studies of life expectancy show that on every measure, social class influences longevity. At age 65, white men in the highest income families can expect to live 3.1 years longer than white men in the lowest income families. (See "Our Diverse Society: Why Isn't Life Expectancy in the United States Higher?")

OUR DIVERSE SOCIETY

Why Isn't Life Expectancy in the United States Higher?

The United States has one of the most advanced health care systems in the world. There are thousands of highly qualified doctors, and medical research produces life-saving breakthroughs on a regular basis. Yet, according to the World Health Organization, U.S. life expectancy is twenty-fourth in the world. The Japanese have the longest life expectancy, and people in Australia, France, Spain, and Italy also live longer than they do in the United States.

One of the ways to begin to understand why the U.S. life expectancy is not higher is to realize that the United States includes many groups who have vastly different life experiences. Some groups have life expectancies significantly above the average, others significantly below the average. Some of the reasons why the average life expectancy in the United States is not higher include the following:

- In the United States, some groups, such as Native Americans, rural African Americans,

and the inner-city poor, have extremely poor health, more characteristic of a poor developing country rather than a rich industrialized one.
- The United States is one of the leading countries for cancers related to tobacco, especially lung cancer. Tobacco use also causes chronic lung disease.
- There is a high coronary heart disease rate, which has dropped in recent years but remains high.
- The HIV epidemic causes a higher proportion of death and disability to U.S. young and middle-aged than in most other advanced countries.
- There are fairly high levels of violence, especially homicides, compared with other industrial countries.

Source: *World Health Organization Press Release,* June 4, 2000.

Age and Health

As advances in medical science lengthen the life span of most Americans, the problem of medical care for the aged becomes more acute. In 1900, there were only 3.1 million Americans 65 or older, a group that constituted a mere 4% of the total population. Today, however, there are more than 35 million Americans aged 65 or older, a full 13% of the total population. This change in the age structure of the American population has had important consequences for health and health care (Smith & Tillipman, 2000).

At the turn of the twentieth century, more Americans were killed by pneumonia, influenza, tuberculosis, infections of the digestive tract, and other microorganism diseases than by any other cause. By comparison, only 8% of the population died of heart disease and 4% of cancer. Today, this situation is completely reversed. Heart disease, cancer, and stroke and its related disorders are now the three most common causes of death. Those diseases are tied to the bodily deterioration that is a natural part of the aging process.

The result of these changes in health patterns is increased hospitalization for those older than 65. Only 1% of all Americans are institutionalized in medical facilities. However, of those 65 and older, 5% are institutionalized in convalescent homes, homes for the aged, hospitals, and mental hospitals. The elderly are 30 times as likely to be in nursing homes as people under 65.

The percentage of people 65 and over living in nursing homes is down from 5.1% in 1990, according to the 2000 Census. The decline over the 10-year period was particularly sharp among those age 85 and over: 18.2% resided in nursing homes in 2000; 24.5% did so in 1990 (Bureau of the Census, 2002). (For a discussion of the health problems of another age group, see "News You Can Use: Binge Drinking as a Health Problem.")

Education and Health

We usually assume that the benefits of a college education include higher incomes and a richer intellectual life. We should also mention better health should be added to the list. Death rates for those with a college education are considerably lower than those with less education and particularly those who have not completed high school. Those with a college education do not just live longer, but those extra years are actually healthier and more active years. This is partly because that college-educated people are less likely to smoke or take part in risky behavior (AmeriStat, August 2002).

Cigarette smoking by adults is strongly associated with educational attainment. Adults with less than a high school education were almost three times as likely to smoke as those with a bachelor's degree or more education in 2000. The percentage of high school students who smoke cigarettes increased in the

NEWS YOU CAN USE

Binge Drinking as a Health Problem

Every semester when we read about the deaths of college students from acute alcohol overdoses, we are reminded that despite progress in reducing underage drinking in the United States, alcohol consumption by college students remains a persistent problem.

A 2001 study found that approximately two in five (44.4%) college students reported binge drinking during the previous two weeks. Binge drinking consisted of five or more drinks at a single occasion for men and four or more for women.

These results were similar to those found in 1993. One striking difference however, was that in 1993, students attending all-women's colleges had much lower rates of binge drinking, and attendance at these schools seemed to protect women from a heavy-drinking lifestyle. Since that time, students at these schools have reported significant increases in frequent binge drinking, and they are now narrowing the gap in drinking behavior between all-women's colleges and coeducational schools.

The amount of change in binge drinking habits since 1993 is very small despite significant efforts to combat this problem. The 2001 study found that more than half of all students reported that their school provided them with information about college rules governing alcohol use, the penalties for breaking those rules, and where to get help for alcohol-related problems.

The students also reported a variety of problems resulting from their drinking.

For example, 35% reported they did something they regretted, 29.5% missed a class, and 21.6% fell behind in their schoolwork. Students also reported they suffered from hangovers, forgot where they were or what they did, argued with friends, had been hurt or injured, had unplanned sexual relations, had not used protection when having sex, damaged property, and had been in trouble with the police.

Even the non-binge-drinking students were affected by the drinkers. At schools with high binge levels, the non-drinking students were more likely to experience assaults, property damage, interrupted sleep, unwanted sexual advances, serious quarrels, and having to take care of a drunk student than those at low-binge-level schools.

College-age binge drinking also leads to traffic fatalities. Car crashes are the leading cause of death in the United States for people under age 25. Nearly half of all motor vehicle deaths of people between 15 and 24 are alcohol related. National research comparing the blood alcohol level of drivers in single-vehicle fatal crashes has found that each 0.02% increase in blood alcohol level nearly doubles the risk of a fatal crash. For drivers under age 21, the risk of a fatal crash increases even more rapidly than it does for older drivers. Such drivers have had less road experience and as a group more often take risks such as speeding or failing to wear seat belts.

How should college campuses respond to underage drinking and driving after drinking? This question has been a contentious topic of debate on many campuses. Most of the students who engage in binge drinking do not feel they have a problem and the large majority consider themselves moderate drinkers instead of binge drinkers. They believe five drinks do not constitute a binge and instead believe out-of-control drinking is what defines binge drinking.

Some argue that tough campus alcohol restriction drives alcohol consumption off campus and into the surrounding communities, where it could produce even greater dangers. Some even suggest that the drinking age should be lowered so that teenagers can learn to drink safely before they leave home for college.

To address the problem, students must themselves become involved in the solution. If only city and college officials deal with the issue, the initiatives may appear paternalistic and engender resistance among the students. Student leaders need to be involved in educating their peers about the risks posed by alcohol, not only to the frequent binge drinkers, but to the college community in general. Emphasis should be placed on protecting the rights of those negatively affected by binge drinkers, very much like we emphasize the rights of innocent drunk driving victims.

Sources: Henry Wechsler, Jae Eun Lee, Meichun Kuo, Mark Seibring, Toben F. Nelson, Hang Lee, "Trends in College Binge Drinking During a Period of Increased Prevention Efforts: Findings From Four Harvard School of Public Health College Alcohol Study Surveys: 1993–2001," *Journal Of American College Health,* Vol. 50, No. 5, 2002.

early 1990s. Since 1997 the percentage of students who smoke has declined. In 2001, 29% of high school students reported smoking during the past month (Pastor et al., 2002).

Women in Medicine

Women have always played a major role in U.S. health care, and by the late nineteenth century the United States was a leader in the training of female doctors. In the twentieth century, the number of women in medical schools dropped markedly when a review of medical education closed many medical schools that had served them. There were fewer female physicians in Boston in 1950 than there had been in 1890.

The rise of the women's movement and affirmative action created an atmosphere more conducive to women becoming physicians. In 1960, only 5.8% of incoming medical students were female, but the proportion increased to 13.7% in 1971 following the passage of the Equal Opportunity Act. The number of female physicians in the United States increased 310% between 1970 and 1990, from 25,400 to 104,200. In 1996, 26.4% of all U.S. physicians were women, up from 13% in 1980. A good predictor of this trend continuing is that in 2005, nearly 48.5% of new entrants into medical school were women (American Association of Medical Colleges, 2005).

Even with these advances, some problems still remain. The Association of American Medical Colleges questioned the 2002 medical school graduates and found that 18.2% of the students had been denied opportunities for training or rewards because of their gender; 14.8% had been subjected to unwanted sexual advances by school personnel; 29.2% had been subjected to offensive sexist remarks; and 21.2% believed they received lower evaluations or grades solely because of their gender rather than performance (American Association of Medical Colleges, 2002).

Female physicians earn less than male doctors. According to the American Medical Association, salaries of female family practitioners average 86% of that earned by men. The earnings ratio is higher for women who have been practicing medicine for 20 years or more, but these experienced doctors still earn less than their male peers (Medical Group Management Association, 1999).

Structural discrimination against women accounts for some of this earnings discrepancy, but there are other factors. The average woman in medical practice is younger and less experienced than the average man. She also earns less than a man partly because she works fewer hours, 55 a week compared with 61 for men. Doctors usually do not get paid by the hour, but those in private practice get paid according to how many patients they see; fewer hours translate into fewer patients and less income.

Women doctors may work fewer hours than men for the same child-care-related reasons other women do. Two-thirds of practicing female physicians have children. Although they may have broken some traditional barriers at work, they still remain the primary family caretakers, being in charge of three times as much child care as men and more than twice the other household duties.

Studies also show that female doctors spend more time with each patient. Their per-visit fees are not necessarily smaller, but they see fewer patients per hour than men. Male physicians saw 117 patients per week compared with 97 for women.

This may explain why studies show that patients consider women physicians "more sensitive, more altruistic, and less egoistic" than men. Patients who see a female physician report a significantly higher total satisfaction level than those who see a male physician. These patients may be picking up on the fact that when men were asked why they entered medicine, they responded, "prestige and salary," whereas women said, "helping people." The interesting side effect of this differing approach to practicing medicine is that women physicians are far less likely to be sued for malpractice than are male physicians.

The type of practice a woman is in also determines her earnings. Female physicians are twice as likely as male physicians to be employed by a hospital, health maintenance organization (HMO), or group practice. Men are more likely to be in independent practices. Independent practitioners generally earn more money and tend to work longer hours.

Women also tend to be concentrated in lower-paying specialties such as internal medicine, pediatrics, and family practice. Specialists such as radiologists, surgeons, and cardiologists typically earn more than primary-care practitioners.

As the number of women entering the field continues to increase, we should see changes in the education of physicians and the manner in which medical care is carried out.

Contemporary Health Care Issues

The American health care system is among the best in the world. The United States invests a large amount of social and economic resources into medical care. It has some of the world's finest physicians, hospitals, and medical schools. It is no longer plagued by infectious diseases and is in the forefront in developing medical and technological advances for the treatment of disease and illness.

At the same time, there are many issues that the American health care system must address. In this section we will discuss a few of them.

Acquired Immunodeficiency Syndrome (AIDS)

The Centers for Disease Control (CDC) defines AIDS as a specific group of diseases or conditions that are indicative of severe immunosuppression related to infection with the human immunodeficiency virus (HIV). AIDS prevalence has increased steadily over time: At the end of December 2004, approximately 462,792 persons in the United States were living with AIDS. Another 529,113 had died of the disease.

Of persons living with AIDS, 43% were black, 35% were white, 20% were Hispanic, 1% were Asian/Pacific Islander, and less than 1% were American Indian/Alaska Native. Newly developed antiretroviral drugs have caused the deaths from AIDS to drop over the past five years (Department of Health and Human Services, 2005).

AIDS, a disease now known virtually everywhere in the world, was only identified in 1981, and the retrovirus that causes it was only discovered in 1983. Yet this disease could transform the global future in ways no one imagined even a decade ago. For many developing countries, the impact of this disease could be as great as that of a major war, unless a vaccine or cure can be developed soon.

AIDS is caused by HIV, which is a member of the retrovirus family. HIV gradually incapacitates the immune system by infecting at least two types of white blood cells. The depletion of white blood cells leaves the infected person vulnerable to a multitude of other infections and certain types of cancers. Those infections and cancers rarely occur, or produce only mild illness, in individuals with normally functioning immune systems. HIV also causes disease directly by damaging the central nervous system.

HIV is transmitted through sexual contact, piercing the skin with HIV-contaminated instruments, transfusion of contaminated blood products, and transplantation of contaminated tissue. An infected mother can transmit the virus to her child before, during, or shortly after giving birth. There is no evidence that HIV is transmitted by casual contact. According to the CDC, the majority of AIDS cases have occurred among homosexual or bisexual males (46%) or heterosexual intravenous drug users (25%) (Figure 17–3) (National Center for Health Statistics, 2001, "Health"). The vast majority of those infected are young adults who do not realize they are infected. Within 6 to 10 years, half of those infected will have developed AIDS. Unless treated with the newly developed antiretroviral drugs, death often follows within a few years after the disease emerges.

AIDS is transmitted sexually and through intravenous drug use, but not as easily as is commonly assumed, perhaps because carriers are only intermittently infectious. Women are much more likely to contract the disease from infected men than men from infected women. It takes as long as two years before half of the spouses or regular sex partners of infected people become infected.

The CDC estimates that 800,000 to 900,000 people are infected with HIV in the United States, and most are expected to develop AIDS within 10 to 15 years. This amounts to 1 in 30 men aged 30 to 50 carrying the infection. On a national scale, the spread of AIDS is concentrated in a number of metropolitan areas. Miami (57.8 cases per 100,000 population) and New York City (56.7) have infection rates that are about 10 times higher than those of cities such as Youngstown (6.8) or Pittsburgh (5.7) (Department of Health and Human Services, 2005).

Initially, the majority of U.S. AIDS victims were male homosexuals. AIDS, however, is not a gay disease; it can be transmitted between two people of any sex by the exchange of infected blood or semen. Racial differences are appearing in the transmission of AIDS. Sixty-two percent of white men and 53% of Asian men were exposed to the disease through homosexual contact and fewer than 9% from intravenous drug use. About 40% of Hispanic men and 31% of black men were infected through homosexual contact and about 19% were exposed to the disease through intravenous drug use (U.S. Centers for Disease Control and Prevention, December 2000).

In the United States, the impact of HIV and AIDS in the African American community has been devastating. Representing only an estimated 13% of the total U.S. population, African Americans made up 47% of all AIDS cases reported in the United States in 2000. Blacks are 10 times more likely than whites to be diagnosed with AIDS, and 10 times more likely to die from it.

The CDC now believes that 1 in every 50 black American men is infected with HIV. AIDS is the leading cause of death for African Americans between the ages of 25 and 44.

Hispanic groups are also disproportionately represented in the AIDS population. In 2004, Hispanics represented 13% of the U.S. population, but accounted for 17% of the total number of new U.S. AIDS cases reported that year (Department of Health and Human Services, 2005).

In 2002, the number of AIDS deaths dropped by one-third from 1995 among white men. The majority of AIDS deaths in that year were to black and Hispanic men. We will see if American society will continue to be as concerned about AIDS as the disease shifts from one that affects mainly white gay men to one whose victims are minority-group drug users. Race, class, and gender issues have an impact on how an illness is perceived and treated, and discrimination factors are not insignificant.

Figure 17–3 **AIDS Cases in the United States by Exposure Category**

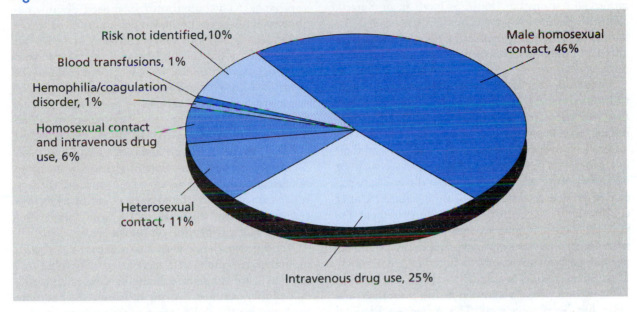

Risk not identified, 10%

Blood transfusions, 1%

Hemophilia/coagulation disorder, 1%

Homosexual contact and intravenous drug use, 6%

Heterosexual contact, 11%

Male homosexual contact, 46%

Intravenous drug use, 25%

Source: *HIV/AIDS Surveillance Reports*, U.S. Centers for Disease Control and Prevention, December 2001, 13(2).

Health Insurance

Most people pay for their health services through some form of insurance. Poor people, however, cannot afford premiums or the out-of-pocket expenses required before insurance coverage begins. They receive coverage through the government-sponsored Medicare and Medicaid programs.

Most of the money spent on medical care in the United States comes in the form of third-party payments as differentiated from direct or out-of-pocket payments. Third-party payments are those made through public or private insurance or charitable organizations. Essentially, insurance is a form of mass financing that ensures that medical care providers will be paid and people will be able to obtain the medical care they need. Insurance involves collecting small amounts of money from a large number of people. That money is put into a pool, and when any of the insured people get sick, that pool pays for the medical services.

The United States has private and public insurance programs. Public insurance programs include Medicare and Medicaid, which are funded with tax money collected by federal, state, and local governments. The country also has two types of private insurance organizations: nonprofit tax-exempt Blue Cross and Blue Shield plans, and for-profit commercial insurance companies. Blue Cross and Blue Shield emerged out of the Great Depression of the 1930s as a mechanism to pay medical bills to hospitals and physicians. People made regular monthly payments to the plan, and if they became sick their hospital bills

were paid directly by the insurance plan. Blue Cross and Blue Shield originally set the cost of insurance premiums by what was called community rating, giving everybody within a community the chance to purchase insurance at the same price. Commercial insurance companies appeared after World War II and based their prices on experience rating, which bases premiums on the statistical likelihood of the individual needing medical care. People more likely to need medical care are charged more than those less likely to need it. Eventually, in order to compete, Blue Cross and Blue Shield had to follow the path of the commercial insurers.

One unfortunate result of the use of experience ratings was that those who most needed insurance coverage, "the old and the sick," were least able to afford or obtain it. Congress created Medicare and Medicaid in 1965 to help with this problem.

Medicare pays for medical care for people older than 65, and Medicaid pays for the care of those too poor to pay their own medical costs. In 2000, adults 18–24 years of age were most likely to lack coverage and those 55–64 years of age were least likely. Persons with incomes below or near the poverty level were almost four times as likely to have no health insurance coverage as those with incomes twice the poverty level or higher. Hispanic persons and non-Hispanic black persons were more likely to lack health insurance than non-Hispanic white persons (Institute of Medicine, 2001).

Even with these forms of government insurance, many people believe that the way health care is delivered in the United States produces problems.

Critics note that the United States is the only leading industrial nation that does not have an organized, centrally planned health care delivery system. Attempts to resolve this problem were met with resistance from the medical establishment and the general public.

In addition, the care that the poor receive is inferior to that received by the more affluent. Many doctors will not accept Medicare or Medicaid assignments, or ask patients to reimburse them for the difference between the insurance coverage and their bill. Moreover, most doctors will not practice in poor neighborhoods, with the result that the poor are relegated to overcrowded, demeaning clinics. Under conditions like these, it is not a surprise that the poor generally wait longer before seeking medical care than do more affluent patients, and many poor people seek medical advice only when they are seriously ill and intervention is already too late.

At various times, a national health insurance program has been suggested. The American Medical Association is a leading opponent of such a plan. Doctors prefer a fee-for-service system of remuneration.

The fee-for-service approach has produced health care delivery problems such as the uneven geographic distribution of doctors and the overabundance of specialists. For example, only 8.3% of the nation's physicians are general practitioners, even though many more are needed to treat the total population, especially the poor and elderly. Part of the reason is that salaries for male general practitioners are about half that of specialists such as radiologists or neurosurgeons (Medical Group Management Association, 2001).

HMOs and other types of managed care have dramatically changed the face of health care in the United States in the past decade. HMOs try to control the cost of medical care by strictly regulating patients' access to doctors and treatments. This shift has meant lower insurance premiums for many individuals and employers. Many people, however, say those savings have come at too great a price. The HMOs have been accused of being more concerned with profits than patients.

Critics of HMOs and managed care include the public, doctors' groups such as the American Medical Association, and many lawmakers. The public and lawmakers have supported efforts to regulate HMOs. These patients' rights laws are designed to give patients more access to care and hold the HMOs accountable for their decisions.

Preventing Illness

Our cultural values cause us to approach health and medicine from a particular vantage point. We are conditioned to distrust nature and assume that aggressive medical procedures work better than other approaches. This situation has caused about one-quarter of all births in the United States to be by cesarean section. The rate of hysterectomy in the United States is twice that in England and three times that in France. Sixty percent of these hysterectomies are performed on women younger than 44. Our rate of coronary bypass operations is five times that of England. We tend to be a can-do society in which doctors emphasize the risk of doing nothing and minimize the risk of doing something. This approach is coupled with the fact that Americans want to be in perfect health. The result is that far more surgery is performed in the United States than in any other country. We tend to think that if something is removed, then our health will return (Payer, 1988).

Yet our experience with heart disease has shown us that significant health benefits can be obtained from such adjustments as changing diets or engaging in healthier practices. At the moment, the best way to deal with the AIDS crisis is through prevention techniques that limit the spread of the illness.

If the health care system is to reorient its perspective from one that seeks cures to one that focuses on prevention, it must place greater emphasis on sociological issues. Illness and disease are socially as well as biophysiologically produced. During the past century, the medical system has devoted a great deal of effort to combating germs and viruses that cause specific illnesses. We are starting to see that there are limitations to this viewpoint. We must now investigate environments, lifestyles, and social structures for the causes of diseases with the same commitment we have shown to investigating germ theory. This is not to say that we should ignore established biomedical knowledge; rather, we should focus greater attention on the interaction of social environments and human physiology.

At first glance, most of the factors that come to mind when thinking about preventing disease are little more than healthful habits. For example, if people adopt better diets, with more whole grains and less red meat, sugar, and salt; and if people stop smoking, exercise regularly, and keep their weight down, they will surely prevent illness.

Being overweight and obesity are risk factors for a variety of chronic health conditions, including hypertension and diabetes mellitus. Despite public health efforts to encourage Americans to attain and maintain a healthy weight, it appears that much work remains to be done.

Over one-half of all American adults (54.7%) are overweight and 1 in 5 (19.5%) are obese. Women (49.5%) are more likely than men (36.3%) to be of healthy weight although men and women are equally likely to be obese. Obesity is most prevalent among middle-aged adults, among black non-Hispanic adults

Figure 17–4   Percentage of People Overweight by Gender and Education

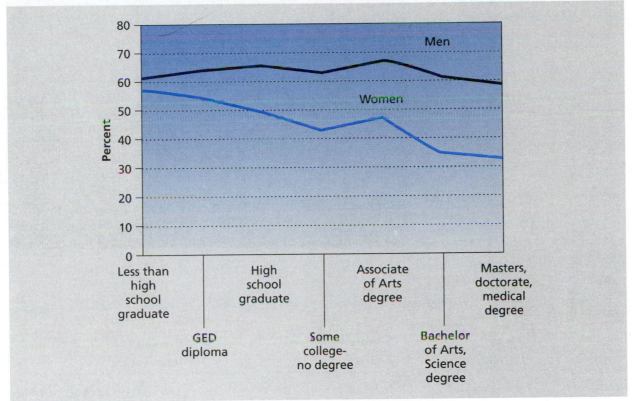

Source: Schoenborn, Charlotte A., Patricia Adams and Patricia Barnes. (2002). "Body Weight Status of Adults: United States, 1997–1998. Advance Data from Vital Health and Statistics, September, 2002. U.S. Centers for Disease Control and Prevention.

and Hispanic adults, and among adults with less education and lower income.

Among women, African American women have the highest percentage of overweight individuals (64.5%), followed by Hispanic women (56.8%), white women (43%), and Asian/Pacific Islander women (25.2%). In addition, black women are nearly twice as likely as white women and more than five times as likely as Asian/Pacific Islander women to be obese. There is little difference among white, Hispanic, or African American men, with more than 60% of each group being overweight. Only Asian/ Pacific Islander men had significantly lower rates of being overweight (35.2%) than each of the other three groups.

Being overweight or obese is correlated with educational and social-class levels. The prevalence of being overweight ranged from approximately 60% among women who had not finished high school, down to about 29% for women who had earned postgraduate college degrees. In contrast, among men, the prevalence of being overweight remained high (and even increased) at all levels of education, dropping noticeably only among men who had attained college degrees. Even then, well over 50% of the men were overweight.

Among men and women, rates of obesity were particularly high for those who had not graduated from high school (27.4%) and those who had earned a GED (26.1%). Rates declined steadily with increasing education and were markedly lower among women who had earned a bachelor's degree (12.5%) or an advanced academic degree (10.5%). Similarly among men, rates of obesity were highest among those with less education and lowest among those with college degrees, although the association was not as strong as for women.

Income level is also related to obesity. This association is stronger among women. Obesity among women ranges from 28.7% for those living below poverty to 13.7% for those in the highest income group. By comparison, rates of obesity among men range from 21.7% for those living below poverty to 17.7% of those in the highest income category (Schoenborn, Adams, & Barnes, 2002). (See Figure 17–4.)

We must also think of illness prevention as involving at least three levels: medical, behavioral, and structural (Table 17–2). Medical prevention is directed at the individual's body, behavioral prevention is directed at changing people's behavior, and structural prevention is directed at changing the society or

For the first time in our history, there are more people ages 65 and older in the population than there are teenagers.

© Walter Hodges/CORBIS

Table 17–2

Types of Illness Prevention

Level of Prevention	Type of Intervention	Site of Intervention	Examples of Intervention
Medical	Biophysiological	Individual's body	Vaccinations; medical procedures
Behavioral	Social psychological	Individual's behavior or lifestyle	Change of habits or behaviors
Structural	Sociological	Social structure and social systems	Legislative controls; environmental changes

Source: *The Sociology of Health and Illness,* 2d ed., by P. Conrad & R. Kern, eds., 1986, New York: St. Martin's Press. Used by permission.

the environments within which people work and live.

We hear a great deal about trying to prevent disease on a behavioral level. Although this is an important level of prevention, we have little knowledge about how to change people's unhealthful habits. Education is not sufficient. Most people are aware of the health risks of smoking or not wearing seat belts, yet roughly 30% of Americans smoke, and 80% do not use seat belts regularly. Sometimes individual habits are responses to complex social situations, such as coping mechanisms to stressful and alienating work environments. Behavioral approaches to prevention focus on the individual and place the burden of change on the individual.

The structural factors related to health care in the United States do not get as much attention as the medical and behavioral factors. We discussed the issues of gender, race, and class earlier in this chapter. Increasingly, people are realizing that health care and the prevention of illness take place on a number of different levels.

The Aging Population

The United States is experiencing a major population shift. For the first time in our history, there are more people aged 65 and over in the population than there are teenagers. When the first U.S. Census was taken in 1790, there were about 50,000 Americans over age 65, representing 2% of the population of 2.5 million. One hundred years later, the over-65 population had grown to 2.4 million, or just under 4% of the population.

The life expectancy of a newborn white child was only 50 years then; it was less than 35 for African Americans. In just three decades, the average life expectancy shot up to 60 years for whites and 50 years for African Americans. The number of elderly people more than doubled, to 6.7 million, about 5% of the total population.

By 1960, the elderly population more than doubled again. There were 16.7 million people over age 65, about 9% of the population. By 2000, the number had doubled again to 35 million, representing 12% of the population.

Figure 17–5   United States Population Age 65 and Over, 1900–2050

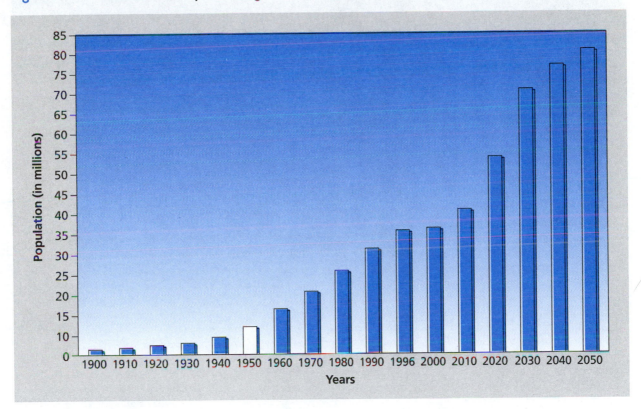

Sources: *Historical Statistics of the United States: Colonial Times to 1970*, Bureau of the Census, 1976; and Current Population Reports, pp. 25–1104, Table 2.

By 2030, demographers estimate that one in five Americans will be age 65 or older, which is nearly four times the proportion of elderly 100 years earlier, in 1930. The effects of this change will reverberate throughout the American economy and society.

During the 1990s, the most rapid growth in the older population occurred among the oldest age groups: The population 85 years and over increased by 38%, from 3.1 million in 1990 to 4.2 million in 2000. (See Figure 17–5.)

There are demographic and medical reasons for the growth in the older population. Many of the elderly were born when the birthrate was high or immigrated to the United States before World War II. The growth in the elderly population has also been influenced by the improvements in medical technology that have created a dramatic increase in life expectancy.

In the next 20 years the growth in the elderly population is going to be influenced by the aging of "baby-boom population," those people born between 1946 and 1964. This population bulge of 76 million people will place substantial demands on society as it tries to address their needs, whether they be for health care, housing, or leisure activities.

Until the last 50 years, most gains in life expectancy came as the result of improved child mortality. The survival of larger proportions of infants and children to adulthood radically increased average life expectancy in the United States and many other countries over the past century. Now, gains are coming at the end of life as greater proportions of 65-year-olds are living until age 85, and more 85-year-olds are living into their 90s. Increasing life expectancy, especially accompanied by low fertility, changes the structure of families. Families have fewer members in each generation, but more generations alive at any one time. Historically, families have played a prominent role in the lives of elderly people. Is this likely to change? These changes raise a multitude of questions: How will these years of added life be spent? Will increased longevity lead to a greater role for the elderly in our society? What are the limits of life expectancy? (See "Social Change: The Discovery of a Disease.")

Composition of the Older Population

At the dawn of the twentieth century, three demographic trends—high fertility, declining infant and child mortality, and high rates of international immigration—were acting in concert in the United States and were keeping the population young. All

SOCIAL CHANGE

The Discovery of a Disease

Something had to be done with the woman. Frau August D. was fifty-one years old, socially prominent, and she was becoming an embarrassment to her family. She raged at her husband, accusing him of infidelity. She raged at her doctors, accusing them of rape. She wandered the city. She screamed in the streets. So, one day in 1906, in the city of Frankfort in Germany, her family brought her to see a doctor, a promising forty-two-year-old Bavarian neuropathologist. Stout and balding, peering out at the world through thick spectacles, the doctor almost was a parody of the owlish academic. His name was Alois Alzheimer.

Alzheimer's first thought was that the woman was suffering from "presenile dementia" a condition only recently defined, but one known to medical science from antiquity. The most recent studies available to Alzheimer suggested that the condition had something to do with a mysterious process by which the brain seemed to atrophy.

Alzheimer studied the woman for nearly a year. At the end, she was screaming at him in a harsh alien voice. She knew this doctor. She knew what he wanted. He wanted what they all wanted. He wanted to cut her up.

And, after she died, he did.

He took tissue samples from her brain. He found them riven with sticky plaques and tangles containing a mysterious substance that would later be identified as a protein called beta-amyloid. The neurons around these areas were utterly destroyed, as if each one had been burned away.

In 1907, Alzheimer published his paper. He argued that the woman had not been suffering from an unspecified condition known as premature senility. He maintained that she had been the victim of a specific disease. It sparked a great debate. Medical conferences coveted him as a speaker. In 1915, he died of heart failure, the same age as Frau D. was when her family had her brought to his clinic because she was screaming in the streets.

Source: Excerpted from Charles Pierce, *Hard to Forget: An Alzheimer's Story*, New York: Random House, 2000, pp. xvii–xix.

---

this has changed and today the United States has seen its elderly population grow more than tenfold during the twentieth century.

People tend to think of the older population as a homogeneous group with common concerns and problems. This is not the case. We can divide the older population into three groups: the young-old, ages 65 to 74 (18.7 million); the middle-old, ages 75 to 84, (11.4 million); and the old-old, ages 85 and over (3.8 million). The older population as a whole is itself getting older. Whereas in 1960 about one-third of the older population was over 75, by the year 2000, nearly 50% of the older people were over 75. In that same period, the old-old will have nearly tripled in numbers.

Elderly Americans are among the wealthiest and among the poorest in our nation. They come from a variety of racial and ethnic backgrounds. Some are employed full time, whereas others require full-time care. Although general health has improved, many elderly suffer from poor health.

The older population in the twenty-first century will come to later life with different experiences than did older Americans in the last century—more women will have been divorced, more will have worked in the labor force, more will be childless.

Aging and the Sex Ratio

Women outnumber men at every age category among the elderly. Women at any age are less likely to die than men. There are approximately 105 male babies born for every 100 female babies, but higher male death rates cause the sex ratio to decline as age increases, and around age 35, females outnumber males in the United States. At ages 85 and older, the ratio is 41 men per 100 women. This has helped produce a substantial sex ratio imbalance in the over-65 group, which continues to become more apparent as we move further up the age scale. In 2000, there were three women for every two men age 65 or older. Among those over age 85, there is one man for every three women. Most elderly women will outlive their spouses and are less likely to remarry than men. They are more likely than older men to be poor, live alone, or need to enter a nursing home or be dependent on the care of others (Himes, 2001).

Aging and Race

About 82.4% of the people over 65 in the United States are white, 8.2% are African American; 5.7% Hispanic; 2.8% Asian or Pacific Islander, and 1%

Figure 17–6 Elderly Americans by Race and Ethnicity, 2003 and 2050

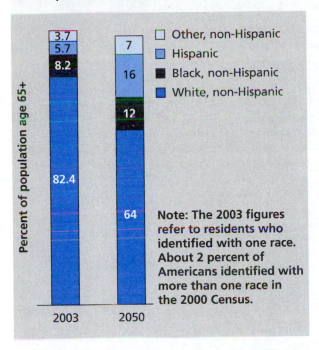

Note: The 2003 figures refer to residents who identified with one race. About 2 percent of Americans identified with more than one race in the 2000 Census.

Sources: U.S. Department of Health and Human Services, Administration on Aging, Racial and Ethnic Composition, http://www.aoa.gov/prof/Statistics/profile/2004/7.asp accessed September 30, 2005; U.S. Census Bureau, *Census 2000 Demographic Profile* (2001); and U.S. Census Bureau, "Projections of the Resident Population by Age, Sex, Race and Hispanic Origin, 1999–2100." http://www.census.gov/population/www/projections/natdel-D1A.html, accessed September 19, 2001.

Native American. "By 2050, the proportion of elderly who are non-Hispanic white is projected to drop to 64 percent as the growing minority populations move into old age" (U.S. Dept. of Health and Human Services, 2004). (See Figure 17–6.)

The African American percentage of the over-65 group is lower than might be expected. That is, if African Americans make up 13% of the total population, then we would expect them to make up 13% of the older population. This difference is caused by the higher recent birthrate among African Americans, as well as the fact that life expectancy is lower among blacks than it is among whites (Bureau of the Census, 2001). In 2000 life expectancy for white women was five years longer than for black women. For white men it was seven years longer than for black men (National Center for Health Statistics, 2001, "Health").

These differences narrow at older ages, however, so that the black-white difference in life expectancy at age 65 is about 1.7 years, and it falls to zero by age 85. The death rates of elderly blacks fall below those of elderly whites at advanced ages. At ages 90 and older, black men and women have slightly more years

of additional life expected than do their white counterparts (National Center for Health Statistics, 2000, "Health").

Why does the black-white life span gap disappear and even reverse as the two races get older? One explanation for the crossover in mortality for black Americans is that blacks who are still alive at older ages are the hearty survivors of extraordinary mortality risks at younger ages. Because older whites were not exposed to the same mortality risks, it is believed they are more frail than blacks of the same age (Manton & Stallard, 1984). There is no real proof for this view however.

Aging and Marital Status

More than 95% of older Americans have been married. Marriage is important for older Americans for several reasons. The presence of a spouse provides a variety of resources in the household. Married elderly are less likely to be poor, to enter a nursing home, or to be in poor health. Spouses are the primary caregivers to their partners.

Women in the United States are more likely than men to outlive their spouse because they live longer. Seventy percent of women over age 75 are widows, whereas less than 20% of men that age are widowers. In addition to the longer life expectancy of women, this disparity also is caused by the fact that men tend to marry women younger than themselves. This situation also is explained by the fact that a widow will find few older men available to marry whereas a widower has many older women available to remarry. Consequently, it is more likely that those older people living alone or in institutions are women.

In 2001, about 75% of men age 65 or older were currently married, compared with 44% of women. At the oldest ages, both men and women are less likely to have a surviving spouse, but the gender gap is even wider: Thirteen percent of women, compared with 53% of men, age 85 or older were married in 2001. Although the likelihood of having experienced a divorce has increased over time, only a small percentage of Americans age 65 or older are divorced: 7% in 2001 (Fields & Casper, 2001).

Aging and Wealth

Special-interest groups lobby Washington politicians claiming that millions of older adults could be thrown into poverty if there are cuts in Social Security. Meanwhile, cruise ship companies, automobile manufacturers, and land developers present images of energetic, attractive, silver-haired couples with substantial amounts of money.

Which of these two images is correct? Both of these presentations are accurate depictions of some

OUR DIVERSE SOCIETY

Stereotypes about the Elderly

The elderly are the most heterogeneous of all age groups because we tend to age differently biologically, sociologically, and psychologically. Yet people tend to have fairly specific stereotypes of the elderly and what it must be like to be old. Surprisingly, the elderly hold the same negative stereotypes about their age group that the general population does.

In one study, people over 65 were asked if any of the following was a serious problem for them: fear of crime, not having enough money to live on, poor health, loneliness, being needed, and keeping busy. Next, they were asked if they thought these issues were serious problems for other elderly people. Table 17–3 presents the results.

There were vast differences between how serious a problem these issues were for the elderly person being questioned and how much of a problem the person thought it was for other elderly people. For example, 57% of the elderly people questioned thought poor health was a serious problem for other elderly people, but only 15% said it was a serious problem for them. Fifty-five percent thought not having enough money was a serious problem for other elderly people, but only 12% said it was a serious problem for them.

We could understand that younger people would have distorted stereotypes about the elderly, but why would the elderly themselves also be so one-sided in their views? Part of the answer is that the ageism of the larger society, the belief that the elderly represent the least capable, least healthy, and least alert members of society, has seeped into the thinking of the elderly as

well. In addition, there is an ironic twist here in that when elderly people believe that other older people are worse off than they are, it produces a higher level of life satisfaction. In a sense, holding the ageist stereotypes makes some elderly people feel better if they can say the stereotype does not apply to them. It is time for us all, no matter what our age, to rethink the distorted views we hold of the elderly.

Table 17–3

Percentage of People Over 65 Who Believe the Problem Is "Very Serious"

Problem	Serious for Others	Serious for Self	Difference
Fear of crime	69%	37%	32%
Not enough money	55	12	33
Loneliness	46	6	40
Poor health	57	15	42
Being needed	41	8	33
Keeping busy	26	4	22

Source: "Images of Aging in America," American Association of Retired People, 1994.

older Americans. The lobbyists are talking about the poorest 20% of households containing the elderly. These households have an average net worth of about $3,400. Luxury cruise lines are interested in the richest 20%, who are worth almost 90 times as much.

Some of the confusion over the economic condition of older Americans stems from the ways in which they use and save money. The median income of the average American household peaks when its primary working members are between the ages of 45 and 54. The incomes of households headed by people age 65 and older are two-fifths as high, on average. This would seem to indicate that older Americans are significantly less well off than younger Americans.

Yet lower incomes do not necessarily mean less spending power. Although the median income of households headed by people aged 65 and older is only about 40% that of households headed by 45- to 54-year-olds, the older group has a much smaller average

household size, so its per capita discretionary income is actually higher than it is among the younger group.

To get a more accurate picture, we should not focus on income, but instead on the financial condition of older Americans—that is, net worth, or the market value of all assets minus all debts. This gives a more realistic picture of wealth than does income alone. For example, older adults can finance major purchases by cashing in on stocks, real estate, or other assets.

A Census Bureau survey found that median net worth rises with the age of the household, moving from $7,428 for households younger than 35, to $61,248 for those 45 to 54, to $106,408 for those 65 to 69. From this point on it starts to drop, but even those households headed by people over age 75 have a median net worth of $81,600 (Davern & Fisher, 2001). Looking at net worth shows us that the oldest households control more wealth than do households in other age groups. Moreover, the share of wealth

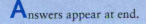

GLOBAL SOCIOLOGY

Global Aging Quiz

Answers appear at end.

1. *True or false?* In the year 2000, children under the age of 15 still outnumbered elderly people (aged 65 and over) in almost all nations of the world.
2. China has the world's largest total population (more than 1.2 billion people). Which country has the world's largest elderly (65+) population?
   a. Japan **b.** Germany **c.** China **d.** Nigeria
3. *True or false?* More than half of the world's elderly today live in the industrialized nations of Europe, North America, and Japan.
4. Of the world's developed countries, which had the highest percentage of elderly people in the year 2000?
   a. Sweden **b.** Turkey **c.** Italy **d.** France
5. *True or false?* Current demographic projections suggest that 35% of all people in the United States will be at least 65 years of age by the year 2050.
6. In which country are elderly people least likely to live alone?
   a. The Philippines **b.** Hungary **c.** Canada **d.** Denmark
7. *True or false?* In developing countries, older men are more likely than older women to be illiterate.

Answers

1. True. Although the world's population is aging, children still outnumber the elderly in all major nations except six: Bulgaria, Germany, Greece, Italy, Japan, and Spain.

2. c. China also has the largest elderly population, numbering nearly 88 million in 2000.
3. False. Although industrialized nations have higher percentages of elderly people than do most developing countries, 59% of the world's elderly now live in the developing countries of Africa, Asia, Latin America, the Caribbean, and Oceania.
4. c. Italy, with 19.1% of all people aged 65 or over. Monaco, a small principality of about 32,000 people located on the Mediterranean, has more than 22% of its residents aged 65 and over.
5. False. Although the United States will age rapidly when the baby boomers (people born between 1946 and 1964) begin to reach age 65 after the year 2010, the percentage of population aged 65 and over in the year 2050 is projected to be slightly above 20% (compared with about 13% today).
6. a. The Philippines. The percentage of elderly people living alone in developing countries is usually much lower than that in developed countries; levels in the latter may exceed 40%.
7. False. Older women are less likely to be literate. In China in 1990, for example, only 11% of women aged 60 and over could read and write, compared with half of men aged 60 and over.

Source: Kevin Kinsella and Victoria Velkoff, *An Aging World: 2001,* U.S. Bureau of the Census, November, 2001.

controlled by working-age Americans is eroding, whereas the share controlled by the elderly is increasing.

Three factors have caused the elderly to control a substantial and increasing portion of the nation's wealth. First, the share of households headed by the elderly has been increasing, thereby increasing the aggregate wealth of older Americans. Second, the stock market growth has benefited the affluent elderly who control a large portion of individual stock holdings. Finally, the escalation in home values in many states has boosted the net worth of the elderly because most older Americans own their own homes.

These facts paint a picture that is different than what is normally thought of as being the case. As with so much in sociology, the conclusion one draws

appears to depend on which factors one considers most important in understanding the issue (Longino & Crown, 1991). (See "Our Diverse Society: Stereotypes about the Elderly.")

In general, the elderly have more assets than the nonelderly, and their poverty rate is also below that of the general population. There is, however, considerable racial variation.

Global Aging

The numerical and proportional growth of older populations around the world is indicative of major achievements—decreased fertility rates, reductions in infant and maternal mortality, reductions in infectious and parasitic diseases, and improvements in nutrition and education—that have occurred, although unevenly, on a global scale.

The rapidly expanding numbers of older people represent a social phenomenon without historical precedent. The world's elderly population numbers half a billion people today and is expected to exceed 1 billion by the year 2020. In most countries, the elderly population is growing faster than the population as a whole. Almost half of the world's elderly live in China, India, the United States, and the countries of the former Soviet Union. China alone is home to more than 20% of the global total.

The oldest-old (85 plus) are the fastest growing segment of the population in many countries worldwide. Unlike the elderly as a whole, the oldest-old today are more likely to reside in developed than developing countries, although this trend too is changing.

The percentage of the elderly population living alone varies widely among nations. In developed countries, percentages generally are high, ranging from 9% in Japan to 40% in Sweden. In developing countries, few elderly people live alone. In China 3% of the elderly live alone, in South Korea 2%, and in Pakistan 1%. Living alone in those countries is often the result of having a spouse, siblings, and even children who have died.

To date, population aging has been a major issue mainly in the industrialized nations of Europe, Asia, and North America. In at least 30 of these countries, 15% or more of the entire population is 60 and older. Those nations have experienced intense public debate over elder-related issues such as social security costs and health care provisions (Figure 17–7).

Future Trends

Population aging is already having major consequences and implications in all areas of day-to-day human life, and it will continue to do so. In the economic arena, population aging will affect economic growth, savings, investment and consumption, labor markets, pensions, taxation and the transfers of wealth, property, and care from one generation to another. Population aging will continue to affect health and health care, family composition and living arrangements, housing and migration. In the political arena, population aging has already produced a powerful voice in developed countries, as it can influence voting patterns and representation. Older voters usually read, watch the news, and educate themselves about the issues, and they vote in much higher percentages than any other age group.

The radical shift in the U.S. population age structure over the last 100 years provides only one part of the story of the U.S. elderly population. Another remarkable aspect is the rapid growth in the number of

Figure 17–7   **People 65 and Older in Selected Countries**

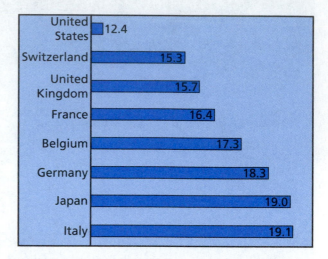

Source: U.S. Census Bureau, International Data Base. http://www.Census.gov/ipc/www/idbnew.html, accessed September 30, 2005.

elderly, and the increasing numbers of Americans at the oldest ages, above ages 85 or 90.

The most rapid growth in the 65-or-older age group occurred between the 1920s and the 1950s. During each of these decades, the older population increased by at least 34%, reaching 16.6 million in 1960. The percentage increase slowed after 1960, and between 1990 and 2000, the population age 65 or older increased by just 12%, because the growth of the older population largely reflects past patterns of fertility, and U.S. fertility rates plummeted in the 1930s.

As the baby-boom population ages, its large numbers will cause cycles of relative growth and decline at each stage of life. The aging of the baby-boom generation will push the median age, now at 34, to more than 38 by 2050.

In 2010, the oldest members of the baby boom will be nearing 65, and the youngest will be around 50. As the baby-boom cohorts begin to reach age 65 starting in 2011, the number of elderly people will rise dramatically.

In about 2030, the final phase of the elderly explosion caused by the baby boom will begin. At that point, the population aged 85 and older will be the only older age group still growing. It will increase from 2.7 million now to 8.6 million in 2030, to more than 16 million by 2050 (Bureau of the Census, 1999, "Resident Population"). To put it another way, today 1 in 100 people is 85 years or older; in 2050, 1 in 20 people could be so old. People 85 and older could constitute close to one-quarter of the older population by then.

Concerns associated with problems of the older population are exacerbated by the large excess of

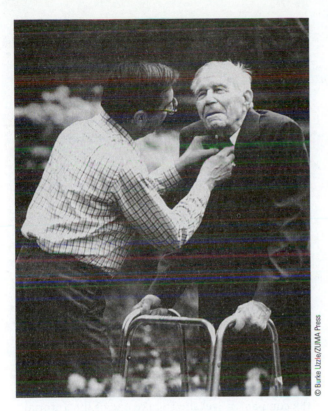

Increased longevity and the aging of the baby-boom generation
will mean that many people will find themselves caring for very
old persons after they themselves have reached retirement age.

© Burke Uzzle/ZUMA Press

women over men in the older ages. Among the aged,
women outnumber men 3 to 2. That imbalance increases to more than 2 to 1 for people 85 and over.

Increased longevity and the aging of the baby-
boom generation will mean that many people will find
themselves caring for very old people after they themselves have reached retirement age. Assuming that
generations are separated by about 25 years, people
85, 90, or 95 would have children who are anywhere from 60 to 70 years old. It is estimated that
every third person 60 to 64 years old will have a living elderly parent by 2010.

The rising cost of caring for the elderly will become
a bigger problem. Without serious health care reform, the United States could spend 20% of its gross
national product on health care.

Now, there are more than twice as many children
as there are elderly people. By 2030, the proportion
of children will have shrunk and the proportion of elderly will have grown until these two groups are approximately equal, at just over one-fifth of the population each (Bureau of the Census, 1999, "Resident
Population").

There are few certainties about the future, but
the demographic outlook seems relatively clear, at
least for people already born. Careful consideration
of the impact of our aging population can be an important tool in planning for the future.

SUMMARY

- Medicine and health care issues are intertwined with social and cultural customs and reflect the society of which they are a part.
- Illness not only involves the body, but it also affects an individual's social relationships, self-image, and behavior.
- Talcott Parsons has suggested that to prevent the potentially disruptive consequences of illness for a group or society, there exists a sick role, which is a shared set of cultural norms that legitimates deviant behavior caused by illness and channels the individual into the health care system.
- The United States has the most advanced health care resources in the world. Critics maintain that the system pays off, though, only when the patient can pay.
- The system has been described as acute, curative, and hospital based, in that the focus is on curing or controlling serious diseases rather than on maintaining health.
- Male death rates exceed female death rates at all ages and for the leading causes of death such as heart disease, cancer, cerebrovascular diseases, accidents, and pneumonia. Women seem to suffer from illness and disability more than men, but their health problems are usually not as life threatening.
- Some have suggested that women may not be sick more often, but may in fact be more sensitive to bodily discomforts and more willing to report them to a doctor.
- Male infants are biologically more vulnerable than female infants, in both the prenatal and neonatal stages.
- Sociologically, men are more likely to have dangerous jobs and more likely than women to place themselves in dangerous situations during both work and leisure.
- Men and women are equally vulnerable to psychiatric problems, but emotionally disturbed men are likely to act out through drugs, liquor, and antisocial acts, whereas women display behaviors such as depression or phobias that indicate an internalization of their problems.
- Asian Americans have the best health profile, followed by whites.
- African Americans and Native Americans have the worst health profiles. On average, life expectancy for African American infants is about six years less than that of white infants, for both males and females.
- Poverty contributes to disease and a shortened life span both directly and indirectly.
- An estimated 25 million Americans do not have enough money to feed themselves adequately,

and therefore suffer from serious nutritional deficiencies that lead to illness and death.

- Diseases such as tuberculosis, influenza, pneumonia, whooping cough, and alcoholism are more common among the poor than the middle class.
- Inadequate housing, heating, and sanitation all contribute to those medical problems, as does the U.S. fee-for-service system that links medical care to the ability to pay.
- Most Americans pay for their health services through some form of health insurance or third-party payments. Individuals—or their employers—pay premiums into a pool, which is used to finance the medical care of those covered by the insurance.
- Public insurance programs include Medicare, for those older than 65, and Medicaid, for those who are below or near the poverty level.
- The United States is the only leading industrial nation that does not have an organized, centrally planned health care delivery system.
- The care that the poor receive is inferior to that received by the more affluent.
- Recently, there has been a shift among health care providers from the cure orientation to a prevention orientation.
- Prevention has three levels: Medical prevention is directed at the individual's body; behavioral prevention is directed at changing the habits and behavior of individuals; structural prevention involves changing the social environments where people work and live.
- For the first time in our history, there are more people aged 65 and older in the population than there are teenagers.

- The older population can be divided into the young-old (age 65 to 74), the middle-old (age 75 to 84), and the old-old (age 85 and older). By 2000, 50% of the older population was older than 75.
- Because mortality rates are higher for men, there is a substantial sex ratio imbalance in the over-65 group that becomes more apparent as one goes up the age scale.
- Due to higher mortality rates among African Americans and other minorities, the older population is disproportionately white.

 Media Resources

The Companion Website for *Introduction to Sociology,* Ninth Edition

http://sociology.wadsworth.com/tischler9e

Supplement your review of this chapter by going to the companion website to take one of the Tutorial Quizzes, use the flash cards to master key terms, and check out the many other study aids you'll find there. You'll also find special features such as Wadsworth's Sociology Online Resources and Writing Companion, GSS data, and Census 2000 information at your fingertips to help you complete that special project or do some research on your own.

CHAPTER SEVENTEEN STUDY GUIDE

KEY CONCEPTS

Match each concept with its definition, illustration, or explanation presented below.

a. Sick role
b. Medicalization
c. Talcott Parsons
d. Third-party payments

e. Experience rankings
f. WHO (World Health Organization)
g. Medicare
h. Medicaid

i. CDC (Centers for Disease Control)
j. Managed care
k. HMO (Health Maintenance Organization)

_____ **1.** A system in which the costs of an individual's health care are paid for by some form of public or private insurance or charitable organization.

_____ **2.** A medical organization run by an insurance company, with doctors paid a salary and costs tightly controlled.

_____ **3.** The process by which problems become defined and treated as medical problems, usually in terms of illness or disorder.

_____ **4.** The U.S. government program to pay for medical costs of the poor.

_____ **5.** A process used by health companies to distribute insurance coverage across different groups based on their likelihood of illness.

_____ **6.** American sociologist who developed the concept of the sick role.

_____ **7.** A system where insurance companies set maximum fees they will pay to doctors.

_____ **8.** A shared set of cultural norms that legitimates deviant behavior caused by illness and channels the individual into the health care system.

_____ **9.** A part of the U.S. Public Health Service, this agency is charged with monitoring communicable ailments.

_____ **10.** An international organization that monitors health issues.

_____ **11.** The U.S. government program to pay for medical costs of the elderly.

CENTRAL IDEA COMPLETIONS

Following the instructions, fill in the appropriate concepts and descriptions for each of the questions posed in the following section.

1. What factors could account for the fact that Asian and Pacific Islanders have much lower mortality rates than would be expected?

_____

_____

2. Identify at least three factors that cause African Americans to have much higher mortality rates than those of other groups.

   a. _____

   b. _____

   c. _____

3. Most of us think of health in terms of becoming ill. Identify three indicators government agencies and other official organizations use to assess the health of populations.

   a. _____

   b. _____

   c. _____

4. What changes in AIDS since its discovery in the early 1980s might result in changes in how Americans view the seriousness of this disease?

_____

_____

_____

5. What biological and sociological factors account for the differences in life expectancy between men and women?

   a. Biological: _____

   _____

   _____

   b. Sociological: _____

   _____

   _____

6. Identify at least two factors that result in U.S. life expectancy ranking in the lower half of the global scale of healthiest societies.

   a. _____

   b. _____

7. Identify at least three differences between male and female doctors.

   a. _____

   b. _____

   c. _____

8. Why is marriage important for older Americans?

   _____

   _____

   _____

CRITICAL THOUGHT EXERCISES

1. Consider the three levels of prevention—medical, behavioral, and structural. Discuss how each type might be related to public policy. How much does each type of intervention contribute to the differences in health among countries with similar levels of economic development?

2. Discuss binge drinking as a national health concern across colleges and universities in the United States.

   a. How common is binge drinking at your home institution?

   b. What are the major factors associated with binge drinking?

   c. If you were in charge of developing a program to reduce binge drinking, what would you include in your plan?

   d. How might the type of institution, for example, a small college or a state university, be a factor in the type of program you have developed?

3. Present a detailed discussion explaining the connection between education and health in the United States. How does the number of years of schooling affect health, and why? Consider this from the perspective of all three levels of illness prevention.

4. Discuss the changes in patterns of death among Americans since the turn of the twentieth century to the present. How have these changes affected social class and racial groups in America?

5. Regarding elderly Americans:

    a. What special concerns does this population face?

    b. Are the present demographic trends likely to continue in the next couple of decades?

    c. What are the likely outcomes of such demographic trends?

INTERNET ACTIVITIES

1. Visit http://www.who.or.jp/index.html. This is the WHO Kobe Centre. It conducts investigations on topics related to global health and development. The issues presented at the center are relevant not only to your current chapter on Health and Aging, but also to those chapters covering research methods and culture, to name two.

2. Paying for health care is a controversial topic, especially in the United States today. Most other advanced countries have some form of national health care. For comparisons on health and aging (and a variety of other topics), go to the OECD website (http://www.oecd.org/maintopic/0,2626,en_2649 _201185_1_1_1_1_1,00.html) and click on the topic that interests you. More pointed comparisons between the United States and other countries can be found here: http://www.huppi.com/kangaroo/ L-healthcare.htm. PBS has a series of shows on "The Health Care Crisis," with accompanying interviews and other materials here: http://www.pbs.org/healthcarecrisis/index.htm. For more personal experiences of health care systems in the United States, Canada, and France, see the discussion between two writers, Adam Gopnick and Malcolm Gladwell, in the *Washington Monthly* (http://www.washingtonmonthly.com/features/2000/0003.gladwellgopnik.html).

3. Visit http://www.nia.nih.gov/. This is the site of the National Institute on Aging. It contains almost everything imaginable on the topic and probably some topics you may have never thought about—for example, growth hormones and sex, dietary supplements, employment opportunities, and cognitive abilities in the elderly.

4. The CDC has a wide variety of information on their website, http://www.cdc.gov/nchs/Default.htm.

ANSWERS TO KEY CONCEPTS

1.d 2.k 3.b 4.h 5.e 6.c 7.j 8.a 9.i 10.f 11.g

ThomsonNOW™

Reviewing is as easy as ① ② ③

1. Before you do your final exam, take the ThomsonNOW diagnostic quiz to help you identify the areas on which you should concentrate. You will find information on ThomsonNOW and instructions on how to access all of its great resources on the foldout at the beginning of the text.

2. As you review, take advantage of ThomsonNOW's study videos and interactive Map the Stats exercises to help you master the chapter topics.

3. When you are finished with your review, take ThomsonNOW's posttest to confirm you are ready to move on to the next chapter.

18

Collective Behavior and Social Change

© Reuters NewMedia Inc./CORBIS

Learning Objectives

After studying this chapter, you should be able to do the following:

- Describe the source of social change in society.
- Describe the attributes and types of crowds.
- Know what the various dispersed forms of collective behavior are.
- Describe the major types of social movements.
- Understand the life cycle of social movements.

- Contrast differing ideologies and show how they influence social change.
- Explain the processes of cultural diffusion and forced acculturation.
- Understand the impact of technological innovation.
- Summarize the changes that will occur in the U.S. labor force in the future.

Even though social change is a common fact of life and many people embrace it, equal numbers fear it. As common as social change may be, rebelling against it is also common. In 1811, the Luddites, English textile weavers, who had been displaced by the invention of the automatic loom, smashed those machines in a revolt against the Industrial Revolution. Many written works that we hold in high esteem were really expressions of distress over the way science was going to change the world for the worse. Henry David Thoreau's book *Walden* is essentially about one man's rebellion against the complicated world coming about because of social change. Mary Shelley's novel *Frankenstein*, published more than 150 years ago, is not really just about a monster created in the laboratory, but about a world where scientists can use technology to create havoc. Stanley Kubrick's classic film *2001* warns us of the same danger when we see HAL, the robot computer, going out of control. Ted Kaczynski, the person known as the Unabomber, who sent letter bombs to those he thought were responsible for some of the technological changes in society, was also rebelling in a vicious way against a changing world. All around us we have academics, writers, and journalists who refuse to use computers to write on and insist on using their manual typewriters or even legal pads to produce their works. All these people are saying that social change, though inevitable, should not be embraced wholeheartedly.

Twenty-five centuries ago, the Greek philosopher Heraclitus lectured on the inevitability of change. "You cannot step into the same river twice," he said, "for . . . waters are continually flowing on." Indeed, he observed, "everything gives way and nothing stays fixed" (quoted in Wheelright, 1959). Social and technological change is an important topic for sociological study. Sociologists must be aware of the impressive number of technological changes and the social changes growing out of them. In this chapter, we will try to examine how, why, and in what specific direction societies are likely to change.

Society and Social Change

What, exactly, is social change? The best way to analyze how sociologists define social change is through example. The invention of the steam locomotive was not in itself a social change, but the acceptance of the invention and the spread of railroad transportation were. Martin Luther's indictments of the Catholic Church, which he nailed to the door of Wittenberg Cathedral in 1517, were not in themselves social change, but they helped give rise to one of the major social changes of all time, the Protestant Reformation.

© Brown Brothers

The invention of the airplane was not in itself a social change, but the acceptance of the invention and the spread of air travel were.

Adam Smith's great work *An Inquiry Into the Nature and Causes of the Wealth of Nations* (first published in 1776) was not in itself social change, but it helped initiate a social change that altered the world—the Industrial Revolution.

Thus, individual discoveries, actions, or works do not themselves constitute social change, but they may lead to alterations in shared values or patterns of social behavior or even to the reorganization of social relationships and institutions. When this happens, sociologists speak of social change. Hans Gerth and C. Wright Mills (1953) defined social change as "whatever happens in the course of time to the roles, the institutions, or the orders comprising a social structure, their emergence, growth and decline." To put it simply, using terms we defined in Chapter 6, **social change** consists of *any modification in the social organization of a society in any of its social institutions or social roles.*

Social change is often the result of **collective behavior,** *the relatively spontaneous social actions that occur when people respond to unstructured and ambiguous situations.* Collective behavior has the potential for causing the unpredictable, and even the improbable, to happen. Collective actions are capable of unleashing powerful social forces that catch us by surprise and change our lives, at times temporarily but at other times even permanently. It is the more dramatic forms of collective behavior that we tend to remember: riots, mass hysterias, lynchings, or panics. Fads, fashion, and rumor, however, also are forms of collective behavior.

Some collective behavior is violent and dramatic, like the French Revolution of 1789 or the 1917 Russian Revolution. However, not all cases of violent social or collective behavior are instances of social change. Thus, the U.S. race riots of the late 1960s and early 1970s were not in themselves examples of social change. For that matter, not all social change need be violent. For example, the transformation of the family into its modern forms over the past 200 years represents an enormous social change that has profoundly affected both the general nature of society and each person's childhood and adult experiences. Similarly, the rise of computer technology and developments in telecommunications over the past few decades have resulted in the emergence of what sociologist Edward Shils (1971) described as "mass society," in which vast numbers of individuals share collectively in the community and in a common language.

Sources of Social Change

What causes social change? Sociologists have cited several factors. Those factors, which we shall consider here, are categorized as internal and external sources of change.

Internal Sources of Social Change

Internal sources of social change include *those factors that originate within a specific society and that singly or in combination produce significant alterations in its social organization and structure.* The most important internal sources of social change are technological innovation, ideology, cultural conflicts, and institutionalized structural inequality.

Technological Innovation Technological change in industrial society is advancing at a dizzying pace, carrying social organizations and institutions along with it. Computer- and communications-based electronic information technology has already transformed American life as we know it in the home, family, workplace, and school.

The Internet has rapidly changed our lives. By 2001, 104 million adults (56% of the adult population) and 30 million children (45% of the under-18 population) in the United States were using the Internet. On a typical day, 58 million Americans were logging on, 9 million more than the previous year. Women, minorities, and people earning $30,000 to $50,000 were among the usage segments that were growing the most rapidly.

Even so, there are significant disparities in online access by income and age. Eighty-two percent of those living in households earning more than $75,000 have Internet access, compared with 38% of those in households earning less than $30,000. Meanwhile, 75% of 18- to 29-year-olds have Internet access, compared with 15% of those 65 and over (Stellin, February 2001).

Homes have been transformed into technological workstations in which people both work and live. By receiving and sending information through computer terminals workers can do their work anywhere in the world.

Already, signs of these trends are plentiful. We can use our home computers to transfer funds and pay bills without ever handling money. The computer has even changed crime patterns. Clever thieves have diverted funds using computer codes in banks. In fact, so much about society is changing so quickly as a result of computer technology that scholars are beginning to wonder whether humans have the psychological resilience to adapt to the social changes that must follow.

Technology increases the complexity in our lives. When taken individually, the new devices that surround us seem reasonable and helpful: microwave ovens, voicemail systems, and the Internet just help us get more done, save time, and communicate with friends and colleagues more easily. But when taken together, and when used not just by us but also by everybody else in our society, these devices sometimes make our lives more complex.

These effects of new technology are not obvious at first. With each new device the manner in which we communicate with others increases. It used to be that the main way to reach someone was to use the land-line telephone. Now we can also communicate by way of cell phones, e-mail, instant messages, video conferences, and through a number of other devices. People are accessible to us and we are more accessible to them. The same technologies that save us time may actually increase our obligations. (Homer-Dixon, 2000). (Some technological innovations change the entire structure of work. See "News You Can Use: The McDonaldization of Society.")

Ideology The term **ideology** most often refers to *a set of interrelated religious or secular beliefs, values, and norms that justify the pursuit of a given set of goals through a given set of means.* Throughout history, ideologies have played a major role in shaping the direction of social change.

Conservative (or traditional) **ideologies** try to *preserve things as they are.* Indeed, conservative ideologies may slow down social changes that technological advances are promoting.

For a brief period in the 13th Century Korea led the world in printing technology, introducing

the use of metal for making printing blocks. This distinguished position was short-lived because Korean scholars refused to accept a 25-character phonetic alphabet that King Sejong developed to replace the thousands of Chinese ideographic characters then in use. . . . Korea's printers were soon left behind by developments elsewhere. (Jacobs et al., 1997)

Liberal ideologies seek *limited reforms that do not involve fundamental changes in the social structure of society.* Affirmative action programs, for example, are intended to redress historical patterns of discrimination that have kept women and minority groups from competing on an equal footing with white males for jobs. Although far-reaching, these liberal ideological programs do not attempt to change the economic system that more radical critics believe is at the heart of job discrimination.

Radical (or revolutionary) ideologies reject liberal reforms as mere tinkerings that simply make the structural inequities of the system more bearable and therefore more likely to be maintained. Like the socialist political movement described in Chapter 15, **radical ideologies** seek *major structural changes in society.* Interestingly, radicals sometimes share the objectives of conservatives in their opposition to liberal reforms that would lessen the severity of a problem, thereby making major structural changes less likely. For example, conservative as well as many radical groups bitterly attacked President Franklin D. Roosevelt's New Deal policies in which federal funds were used to create jobs and bring the country out of the Great Depression. Conservatives attacked the New Deal as "creeping socialism," and radicals saw it as a desperate (and successful) attempt to save the faltering capitalist system and stave off a socialist revolution.

Reactions to Institutionalized Inequality When groups perceive themselves to be the victims of unjust and unequal societal patterns or laws, they are likely to demand social, economic, political, and cultural reforms. Such pressure for social change exists in U.S. society in a variety of forms. African Americans, Hispanics, and other minorities, for example, have often been the victims of institutionalized inequality. As these groups have asserted their rights, society has been forced to change. For example, federal, state, and local laws have been passed to make it illegal to discriminate against minorities in voting; in access to schools, the labor force, housing; and in other sectors of American life. The labor and civil rights movements, for example, arose because of institutionalized inequalities in American society.

NEWS YOU CAN USE

The McDonaldization of Society

Sociologist George Ritzer believes McDonald's, the fast-food restaurant chain, is undoubtedly one of the most influential creations of twentieth-century America. This is true not just as a fast-food restaurant, but also as a model for many other institutions, and for its principles that have been widely adopted throughout American society and much of the rest of the world.

George Ritzer uses the term *McDonaldization* to describe the process by which the principles of the fast-food restaurant are coming to dominate more and more sectors of American society, as well as the rest of the world. McDonaldization is not only affecting the restaurant business, but also education, work, travel, leisure-time activities, dieting, politics, religion, the family, and virtually every other sector of society. Ritzer believes McDonaldization has shown every sign of being an irresistible process as it sweeps through institutions and parts of the world that one would have thought impervious to it.

McDonald's has achieved its exalted position in the world because virtually all Americans, and many people from other countries, have passed through its golden arches. Furthermore, Ritzer has noted, we have been bombarded by commercials extolling McDonald's virtues. Thus, in a survey of schoolchildren, 96% were able to identify Ronald McDonald, making him second only to Santa Claus in name recognition.

Ritzer believes the McDonald's model has succeeded because it offers the consumer efficiency and predictability, and because it seems to offer the diner a good value. It also has flourished because it has been able to exert control, through nonhuman technologies, over both employees and customers, getting them to behave the way the organization wishes them to. Thus, there are good, solid reasons why the process of McDonaldization continues unabated.

Ritzer notes that the McDonaldization process as it permeates many areas of modern society also has a downside. We can think of efficiency, predictability, quantification, and control through nonhuman technology as the basic components of a rational system. However, rational systems often spawn irrationalities—that is, rational systems serve to deny human reason.

The fast-food restaurant, Ritzer points out, is often a dehumanizing setting in which to eat and work. People lining up for a burger or waiting in the drive-through line often feel as if they are dining on an assembly line. Assembly lines are hardly human settings in which to eat, and they have been shown to be inhuman settings in which to work.

Ritzer believes we should question the headlong rush toward McDonaldization in our society and throughout the world. True, there are great gains to be made from McDonaldization. But there are great costs and enormous risks also, as Ritzer has pointed out. Ultimately, we must ask whether, in creating rationalized systems, we are not also creating an even greater number of irrationalities. At the minimum, Ritzer would like us to be aware of the costs associated with McDonaldization.

Source: Adapted from *The McDonaldization of Society*, by G. Ritzer, 1992, Newbury Park, CA: Pine Forge Press. Used with permission.

External Sources of Social Change

As we noted in Chapter 3, diffusion, the process of transmitting traits from one culture to another, occurs when groups with different cultures come into contact and exchange items and ideas with one another. Diffusion is thus an example of an **external source of social change,** *changes within a society produced by events emanating from outside that society.*

It does not take the diffusion of many cultural traits to produce profound social changes, as the anthropologist Lauriston Sharp (1952) demonstrated with regard to the introduction of steel axes to the Yir Yoront, a Stone Age tribe inhabiting southeastern Australia. Before European missionaries introduced steel axes to the Yir Yoront, the natives made axes by chipping and grinding stone—a long, laborious process. Axes were very valuable, had religious importance, and also were the status symbol of tribal leaders. Women and young men had to ask permission from a leader to use an ax, which reinforced the patriarchal authority structure. However, anybody could earn a steel ax from the missionaries simply by impressing them as being "deserving." With women and young men thus having direct access to superior tools, the symbols representing status relations between male and female as well as young and old were devalued, and the norms governing those traditional relationships were upset. In addition, introducing into the tribe valuable tools that did not have religious sanctions governing their use led to a drastic rise in the incidence of theft. In fact, the entire

Cultural diffusion inevitably results when people from one group or society come into contact with another.

moral order of the Yir Yoront was undermined because their myths explained the origins of all important things in the world, but did not account for the arrival of steel axes. This, as Sharp observed, caused conditions fertile for the introduction of a new religion, a happy circumstance for the missionaries.

Diffusion occurs wherever and whenever different cultures come into contact with one another, though contact is not essential for traits to diffuse from one culture to another. For example, Native American groups below the Arctic smoked tobacco long before the arrival of the Europeans. However, in Alaska the Inuit (Eskimos) knew nothing of its pleasures. European settlers brought tobacco back to Europe, where it immediately became popular and diffused eastward across Central Europe and Eurasia, up into Siberia, and eventually across the Bering Strait to the Inuit.

Today, of course, when so many of the world's peoples increasingly are in contact with one another through all forms of mass communication, cultural traits spread easily from one society to another. The direction of diffusion, however, rarely is random or balanced among societies. In general, traits diffuse from more powerful to weaker peoples, from the more technologically advanced to the less so. *A social change that is imposed by might or conquest on weaker people* is called **forced acculturation.**

Crowd Behavior and Social Change

Some social change involves crowds. A **crowd** is *a temporary concentration of people who focus on some thing or event, but who also are attuned to one another's behavior.* There is a magnetic quality to a crowd: It attracts passersby, who often interrupt whatever they are doing to join. Think, for example, of the crowds that gather "out of nowhere" at fires or accidents. Crowds also fascinate social scientists, because crowds always have within them the potential for unpredictable behavior and group action that erupts quickly and often seems to lack structure or direction—either from leaders or from institutionalized norms of behavior.

Attributes of Crowds

In his study *Crowds and Power* (1978), Elias Canetti attributed to crowds the following traits:

1. *Crowds are self-generating.* Crowds have no natural boundaries. When boundaries are imposed artificially—for example, by police barricades intended to isolate a street demonstration—there is an ever-present danger that the crowd

will erupt and spill over the boundaries, thereby creating chaos. So, in effect, crowds always contain threats of chaos, serious disorder, and uncontrollable force.

2. *Crowds are characterized by equality.* Social distinctions lose their importance within crowds. Indeed, Canetti believes that people join crowds specifically to achieve the condition of equality with one another, a condition that carries with it a charged and exciting atmosphere.

3. *Crowds love density.* The circles of private space that usually surround each person in the normal course of events shrink to nothing in crowds. People pack together shoulder to shoulder, front to back, touching one another in ways normally reserved for intimates. Everyone included within the crowd must relinquish a bit of his or her personal identity to experience the crowd's fervor. With a "we're all in this together" attitude, the crowd discourages isolated factions and detached onlookers.

4. *Crowds need direction.* Many crowds are in motion. They may move physically as they do in a marching demonstration or emotionally as at a rock concert. The direction of movement is set by the crowd's goals, which become so important to crowd members that individual and social differences lessen or disappear. This constant need for direction contains the seeds of danger. Having achieved or abandoned one goal, the crowd may easily seize on another, perhaps destructive, one. The direction that a crowd will take depends on the type of crowd involved.

Types of Crowds

In his essay on collective behavior, Herbert Blumer (1946) classified crowds into four types: acting, expressive, conventional, and casual.

Acting Crowd An **acting crowd** is *a group of people whose passions and tempers have been aroused by some focal event, who come to share a purpose, and who feed off one another's arousal, often erupting into spontaneous acts of violence.*

In 2000, Fat Tuesday celebrations in Seattle illustrate typical acting crowd behavior:

Hundreds of Fat Tuesday revelers taunted and threw beer bottles at police in riot gear early Wednesday as Mardi Gras festivities in Pioneer Square turned ugly. . . . Under the glare of a helicopter spotlight, police gradually pushed crowds away from the epicenter of the disorder where as many as 500 had gathered to celebrate the final day of Mardi Gras. . . . The disturbance began around midnight when about 300 to 500 people gathered at First Avenue and Yesler Street. During this scene, a woman standing on a newspaper vending box fell and hit her head. Officers who tried to come to her aid were pelted with rocks and bottles. (Associated Press, 2000)

Acting crowds can become violent and destructive, as 400 million worldwide television viewers discovered in the summer of 1985. Sixty thousand soccer fans had assembled in Brussels to watch the European Cup Finals between Italy and Great Britain. Verbal taunts quickly turned into rocks and bottles being thrown. Suddenly, British fans stormed the fence and surged toward the Italian fans, trampling hundreds of helpless spectators. Before the horror could be stopped, 38 people were dead and another 400 injured (Lacayo, 1985).

A **threatened crowd** is *an acting crowd that is in a state of alarm, believing that some kind of danger is present.* Such a crowd is in a state of panic, as when a crowded nightclub catches fire and everybody tries to get out, jamming exits and trampling one another in their rush to escape. A threatened crowd created havoc when a busboy accidentally ignited an artificial palm at the Coconut Grove Night Club in Boston on November 28, 1942, spreading fire instantaneously throughout the club. The fire lasted only 20 minutes, but 488 people died. Most died needlessly when panic gripped the crowd. Fire investigators found that the club's main entrance—a revolving door—was jammed by hundreds of terrified patrons. With their escape route blocked, those people died of burns and smoke inhalation only feet away from possible safety (Veltfort & Lee, 1943).

In February of 2003 a nightspot known as the Station was consumed by fire minutes after the band Great White lit off an unlicensed pyrotechnic display at the beginning of its show. The pyrotechnics ignited highly flammable polyurethane packing foam that had been placed around the stage walls and ceiling as sound insulation. The fire engulfed the building within minutes and produced thick, toxic smoke that quickly overcame patrons trying to flee. Of the 350 people in the building at the time, 99 were killed and 184 were injured (Rowland, 2003). In this as well as other threatened crowds, there is a lack of communication regarding escape routes.

Expressive Crowd An **expressive crowd** is *drawn together by the promise of personal gratification through active participation in activities and events.* For example, many rock concert audiences are not content simply to listen to the music and watch the show. In a very real sense, they want to be part of the show. Many dress in clothing calculated to draw attention to themselves, take drugs during the performance, body surf or slam dance in packed masses up against the stage, and delight in giving problems to security personnel.

Conventional Crowd A **conventional crowd** is *a gathering in which people's behavior conforms to some well-established set of cultural norms, and gratification results from a passive appreciation of an event.* Such crowds include the audiences attending lectures, the theater, and classical music concerts, where everybody is expected to follow traditional norms of etiquette.

Casual Crowd A casual crowd is the inevitable outgrowth of modern society, in which large numbers of people live, work, and travel closely together. A **casual crowd** is *any collection of people who happen, in the course of their private activities, to be in one place at the same time and focus attention on a common object or event.* On Fifth Avenue in New York City at noon, many casual crowds gather to watch an accident, a purse snatcher, the construction of a new building, or a theatrical performer. A casual crowd has the potential of becoming an acting crowd or an expressive crowd; the nature of a crowd can change if events change.

The Changeable Nature of Crowds

Although the typology presented is useful for distinguishing kinds of crowds, it is important to recognize that any crowd can shift from one type to another. For example, if a sidewalk musician starts playing a violin on Fifth Avenue, part of the aggregate walking by will quickly consolidate into a casual crowd of onlookers. Or an expressive crowd at a rock concert will become a threatened crowd if a fire breaks out.

Changing times may also affect the nature of crowds. For example, until the 1970s, British soccer matches generally attracted conventional crowds who occasionally turned into expressive crowds chanting team songs. Since the late 1970s, however, British soccer fans have become active crowds: Fighting in the stands is epidemic, charging onto the field to assault players and officials has become common, and rioting has taken place. In 2000, Danish police and British soccer fans clashed in Copenhagen prior to a match between the Turkish and British soccer teams. Police used tear gas, dogs, and batons on about 100 British soccer fans.

Minutes before that clash, British fans fought with Turkish fans near Copenhagen's town square. As a precaution against further rioting, authorities erected iron fencing outside the 39,000-seat stadium to separate British and Turkish fans. Another fence was put up inside around the playing field (CNN News, 2000).

Because they are relatively concentrated in place and time, crowds present rich materials for sociological study (even if much of the data must be tracked down after the dust has settled). However, when collective behavior is widely dispersed among large numbers of people whose connection with one another is minimal or even elusive, the sociologist must then deal with phenomena that are extremely difficult to study, including fads and fashions, rumors, public opinion, panics, and mass hysteria.

Dispersed Collective Behavior

In this age of mass media, with television, the Internet, and other systems of communication spreading information instantaneously throughout the entire population, collective behavior shared by large numbers of people who have no direct knowledge of one another has become commonplace. Sociologists use the term **mass** to describe *a collection of people who, although physically dispersed, participate in some event either physically or with a common concern or interest.*

A nationwide television audience watching a presidential address or a Super Bowl game is a mass. So are those individuals who rush out to buy the latest best-selling CD and the fashion-conscious whose hemlines, lapel widths, and clothes always reflect the "in" look. In other words, dispersed forms of collective behavior seem to be universal. (See "Technology and Society: Dispersed Collective Behavior on the Internet.")

Fads and Fashions

Fads and fashion are transitory social changes (Vago, 1980), patterns of behavior that are widely dispersed among a mass but that do not last long enough to become fixed or institutionalized. Yet it would be foolish to dismiss fads and fashions as unimportant just because they fade relatively quickly. In modern society, fortunes are won and lost trying to predict fashions and fads—in clothing, in entertainment preferences, in eating habits, or choices of investments.

Probably the easiest way to distinguish between fads and fashions is to look at their typical patterns of diffusion through society. Fads are social changes with a very short life span marked by a rapid spread and an abrupt drop in popularity. This was the fate of the Hula Hoop in the 1950s and the dance known as the "twist" in the 1960s. The roller-skating fad that emerged in 1979 rolled off into the pages of history

TECHNOLOGY AND SOCIETY

Dispersed Collective Behavior on the Internet

More than half of all Americans use the Internet to send e-mail or to gather information. With the growth of the Internet, collective behavior shared by large numbers of people who have no direct knowledge of one another has become commonplace. Sociologists call this dispersed collective behavior. The early users of the Internet were often well-educated white males. Today this stereotype no longer holds and women now make up half of all Internet users. Hispanic and African Americans are now using the Internet at accelerating rates also. All this has produced a revolutionary change in the way people interact with each other and learn about the world (Horrigan, 2002). (See Table 18–1.)

Source: John B. Horrigan, "New Internet Users: What They Do Online, What They Don't, and Implications for the 'Net's Future," Pew Internet and American Life Project.

Table 18–1

Internet Activities

Activity	Percent of Those With Internet Access
Send email	93%
Use an online search engine to find information	85
Research a product or service before buying it	75
Surf the web for fun	65
Buy a product	61
Buy or make a reservation for travel	50
Send an instant message	46
Check sports scores	44
Play a game	37
Download music files to the computer	32
Chat in a chat room or in an online discussion	25
Visit an adult website	13
Take a class online for college credit	7
Gamble	5

Source: Pew Internet & American Life Project Tracking surveys (March 2000—present).

---

sometime in the 1980s, as did the Rubik's cube, and Coleco, the company that made Cabbage Patch dolls, which went bankrupt in 1988. The Tickle Me Elmo stuffed toys were quickly forgotten after the 1996 Christmas season. Whereas you or your parents may remember such past fads as yo-yos and Mr. Potato Head, recent fads include the introduction of Furby, Teletubbies, Pokemon, and Dragonball Z.

Some fads may seem particularly absurd. During the Great Depression of the 1930s, when many Americans were having trouble putting food on the table, college students started engaging in the practice of swallowing goldfish. The fad was started by a Harvard freshman who swallowed a single, live fish as fellow students looked on in disgust. Three weeks later a student at Franklin and Marshall College swallowed three fish. The practice quickly escalated and new records were set daily, with 89 being swallowed in one sitting at Clark University. Eventually a pathologist at the U.S. Public Health Service cautioned that goldfish may contain tapeworms that

could cause intestinal problems and anemia. The fad disappeared shortly thereafter (Levin, 1993).

A fad that is especially short-lived may be called a **craze.** The Mohawk hairstyle among both young males and females was a relatively short-lived craze, as was streaking, or running naked down a street or through a public gathering, in 1974. One streaker even ran on stage during the Academy Awards presentation.

At the peak of their popularity, fads and crazes may become competitive activities. For example, when streaking was a craze, individual streaking was followed by group streaking, streaking on horseback, and parachuting naked from a plane.

On other occasions, what appears to be a fad actually signals a trend and a change in social values. In 1922, newspapers reported the shocking news that smoking was common among female college students. The University of Wisconsin's dean of women said the smoking fad, most popular among women of the "idle, blase, disappointed class," was

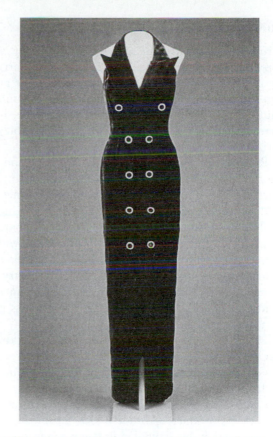

Fashions relate to standards of dress and manners during a particular time. Here we see how the dress of royalty has changed over time. At the left is a court dress worn between 1765 and 1775. At the right is a dinner dress worn by Diana, Princess of Whales, in 1991.

already passing. She pointed out that an intelligent woman "cannot see herself rocking a baby or making a pie with a cigarette in her mouth, flicking ashes in the baby's face or dropping them in the pie crust" (Schwarz, 1997).

Fashions relate to *the standards of dress or manners in a given society at a certain time.* They spread more slowly and last longer than fads. In his study of fashions in European clothing from the eighteenth to the present century, Alfred A. Kroeber (1963) showed that though minor decorative features come and go rapidly (that is, are faddish), basic silhouettes move through surprisingly predictable cycles that he correlates with degrees of social and political stability. In times of great stress, fashions change erratically; but in peaceful times, they seem to oscillate in cycles lasting about 100 years.

Georg Simmel (1957) believed that changes in fashion (such as dress or manners) are introduced or adopted by the upper classes, who seek in this way to keep themselves visibly distinct from the lower classes. Of course, those immediately below them observe these fashions and also adopt them in an attempt to identify themselves as upper class. This process repeats itself again and again, with the fashion slowly moving down the class ladder, rung by rung. When the upper classes see that their fashions have become commonplace, they take up new ones, and the process starts all over again.

Blue jeans have shown that this pattern is no longer true today. Jeans started out as sturdy work pants worn by those engaged in physical labor. Young people then started to wear them for play and everyday activities. College students wore them to class. Eventually, fashion designers started to make fancier, higher-priced versions, known as designer jeans, worn by the middle and upper classes. In this way the introduction of blue jeans into the fashion scene represents movement in the opposite direction from what Simmel noted.

Of course, the power of the fashion business to shape consumer taste cannot be ignored. Fashion designers, manufacturers, wholesalers, and retailers earn money only when people tire of their old clothes and purchase new ones. Thus, they promote certain colors and widen and narrow lapels to create new looks, which consumers purchase.

Indeed, the study of fads and fashions provides sociologists with recurrent social events through which to study the processes of change. Because they so often involve concrete and quantifiable objects, such as consumer goods, fads and fashions are much easier to study and count than are rumors, another common form of dispersed collective behavior.

Rumors

A **rumor** is *information that is shared informally and spreads quickly through a mass or a crowd.* It arises in situations that, for whatever reasons, create ambiguity with regard to their truth or their meaning. Rumors may be true, false, or partially true, but characteristically they are difficult or impossible to verify.

Rumors are generally passed from one person to another through face-to-face contact, but they can be started through television, radio, and the Internet as well. However, when the rumor source is the mass media, the rumor still needs people-to-people contact to enable it to escalate to the point of causing widespread concern (or even panic). Sociologists see rumors as one means through which collectivities try to bring definition and order to situations of uncertainty and confusion. In other words, rumors are "improvised news" (Shibutani, 1966).

Hard-to-believe rumors usually disappear first, but this is not always the case. For 103 years, Procter & Gamble, the maker of familiar household products such as Mr. Clean and Tide laundry detergent, used the symbol of the moon and 13 stars as a company logo on its products. Around 1979, a rumor began circulating that this symbol indicated a connection between the giant corporation and satanic religion. There was no evidence to substantiate this rumor, but unable to dispel it, the company finally decided to remove the logo from its products in 1985 (Koenig, 1985). In 1997 Procter & Gamble was still plagued by the rumor and filed the latest in a series of lawsuits, this time against Amway Corporation and some of its distributors for allegedly spreading rumors that Procter & Gamble is affiliated with the Church of Satan. Since the early 1980s, the company said it has filed 15 lawsuits to fight the rumors.

Public Opinion

The term *public* refers to a dispersed collectivity of individuals concerned with or engaged in a common problem, interest, focus, or activity. An opinion is a strongly held belief. Thus, **public opinion** refers to *the beliefs held by a dispersed collectivity of individuals about a common problem, interest, focus, or activity.* It is important to recognize that a public that forms around a common concern is not necessarily united in its opinions regarding this concern. For example, Americans concerned about abortion are sharply divided into pro and con camps.

Whenever a public forms, it is a potential source for, or opposition to, whatever its focus is. Hence, it is extremely important for politicians, market analysts, public relations experts, and others who depend on public support to know the range of public opinion on many different topics. Those individuals often are not willing to leave opinions to chance, however. They seek to mold or influence public opinion, usually through the mass media. Advertisements are attempts to mold public opinion, primarily in the area of consumption. They may create a need where there was none, as they did with fabric softeners, or they may try to convince consumers that one product is better than another when there is actually no difference. *Advertisements of a political nature, seeking to mobilize public support behind one specific party, candidate, or point of view* technically are called **propaganda** (but usually by only those people in disagreement). For example, radio broadcasts from the former Soviet Union were habitually called "propaganda blasts" by the American press, but similar Voice of America programs were called "news" or "informational broadcasts" by the same American press.

Opinion leaders are *socially acknowledged experts whom the public turns to for advice.* The more conflicting sources of information there are on an issue of public concern, the more powerful the position of opinion leaders becomes. The leaders weigh various news sources and then provide an interpretation in what has been called the two-step flow of communication. Those opinion leaders can have a great influence on collective behavior, including voting (Lazarsfeld et al., 1968), patterns of consumption, and the acceptance of new ideas and inventions. Typically, each social stratum has its own opinion leaders (Katz, 1957). Jesse Jackson, for example, is an opinion leader in the African American community. The mass media have turned news anchors into accepted opinion leaders for a broad portion of the American public. Rush Limbaugh has emerged as one of the more influential opinion leaders, as the fortunes of political candidates are determined by his loyal listeners.

When rumor and public opinion grip the public imagination so strongly that facts no longer seem to matter, terrifying forces may be unleashed. Mass hysteria may reign, and panic may set in.

Mass Hysteria and Panic

At a summer program in Florida, 150 children would gather every day in a dining hall where they were served prepackaged lunches. One girl said the sandwich tasted funny and that she felt sick. Other children soon said they felt sick also. The lunch room aid told the rest of the children to stop eating because the food might be poisoned. Within the hour over 60 more children said they were sick. Ambulances rushed the children to three different hospitals. All tests revealed there was nothing wrong with the food and the children quickly recovered (Feldman

According to Herbert Blumer, a crowd is a collectivity of people more or less waiting for something to happen. Eventually something stirs them, and people react without the kind of caution and critical judgments they would normally use.

& Feldman, 1998). It appears the children were all victims of a case of mass hysteria (Feldman & Feldman, 1998).

Mass hysteria occurs when *large numbers of people are overwhelmed with emotion and frenzied activity or become convinced that they have experienced something for which investigators can find no discernible evidence.* A **panic** is *an uncoordinated group flight from a perceived danger,* as in the public reaction to Orson Welles's 1938 radio broadcast of H. G. Wells's *War of the Worlds.*

According to Irving Janis and his colleagues (1964), people generally do not panic unless four conditions are met. First, people must feel that they are trapped in a life-threatening situation. Second, they must perceive that the threat to their safety is so large that they can do little else but try to escape. Third, they must realize that their escape routes are limited or inaccessible. Fourth, there must be a breakdown in communication between the front and rear of the crowd. Driven into a frenzy by fear, people at the rear of the crowd make desperate attempts to reach the exit doors, and their actions often completely close off the possibility of escape.

The perception of danger that causes a panic may come from rational as well as irrational sources. A fire in a crowded theater, for example, can cause people to lose control and trample one another in their attempt to escape. This happened when fire broke out at the Beverly Hills Supper Club in Southgate, Kentucky, on May 28, 1977. When employees discovered an out-of-control fire, they warned the 2,500 patrons and tried to usher them out of the building. A panic resulted as people attempted to escape the overcrowded, smoke-filled building. People trampled each other trying to reach the exits, and 165 people died in the process.

Such extreme events are not very common, but they do occur often enough to present a challenge to social scientists, some of whom believe there is a rational core behind what at first glance appears to be wholly irrational behavior (Rosen, 1968). For example, sociologist Kai T. Erikson (1966) looked for the rational core behind the wave of witchcraft trials and hangings that raged through the Massachusetts Bay Colony beginning in 1692. Erikson joins most other scholars in viewing this troublesome episode in American history as an instance of mass hysteria (Brown, 1954). He accounts for it as one of a series of symptoms, suggesting that the colony was in the grip of a serious identity crisis and needed to create real and present evil figures who stood for what the colony was

Leadership is an important ingredient in the emergence of a social movement.

© New York Daily News Picture Collection

not—thus enabling the colony to define its identity in contrast and build a viable self-image.

Mass hysterias account for some of the more unpleasant episodes in history. Of all social phenomena, they are among the least understood—a serious gap in our knowledge of human behavior.

Social Movements

A **social movement** is *a form of collective behavior in which large numbers of people are organized or alerted to support and bring about, or to resist, social change.* By their very nature, social movements are an expression of dissatisfaction with the way things are or with changes that are about to take place.

Participation in a social movement is, for most people, only informal and indirect. Usually, large numbers of sympathizers identify with and support the movement and its program without joining any formal organizations associated with the movement. For people to join a social movement, they must think that their own values, needs, goals, or beliefs are being stifled or challenged by the social structure or specific individuals. The people feel that this situation is undesirable and that something must be done to set things right. Some catalyst, however, is needed to mobilize the discontent that people feel. Two major theories, relative deprivation theory and resource mobilization theory, attempt to explain how social movements emerge.

Relative Deprivation Theory

Relative deprivation is a term that was first used by Samuel A. Stouffer (1950). It refers to the situation in which deprivation or disadvantage is measured not by objective standards, but by comparison with the condition of others with whom one identifies or thinks of as a reference group.

Thus, **relative deprivation theory** assumes *social movements are the outgrowth of the feeling of relative deprivation among the large numbers of people who believe they lack certain things they are entitled to*—such as better living conditions, working conditions, political rights, or social dignity.

From the standpoint of relative deprivation theory, the actual degree of deprivation people suffer is not automatically related to whether people feel deprived and therefore join a social movement to correct the situation. Rather, deprivation is considered unjust when others with whom the people identify do not suffer the deprivation (Gurr, 1970).

Karl Marx expressed this view when he noted,

A house may be large or small; as long as the surrounding houses are equally small it satisfies all social demands for a dwelling. But let a palace arise beside the little house, and it shrinks from a little house to a hut. . . .Our desires and pleasures spring from society; we measure them therefore, by society and not by the objects which serve for their satisfaction. Because they are of a social nature, they are of a relative nature. (Marx, 1968)

There is a flaw in the theory of relative deprivation, in that often the people who protest a situation or condition may not be deprived. Sometimes people protest because a situation violates their learned standards of justice. The white civil rights marchers in the 1960s were not personally the victims of antiblack discrimination. We could argue, however, that they were experiencing deprivation in the sense that the reality was not judged to be what it ought to be. They were experiencing a moral as opposed to a material or personal social deprivation (Rose, 1982).

Resource Mobilization Theory

The **resource mobilization theory** assumes that *social movements arise at certain times and not at others because some people know how to mobilize and channel the popular discontent.* Although discontent exists virtually everywhere, a social movement will not emerge until specific individuals actually mobilize resources available to a group, by persuading people to contribute time, money, information, or anything else that might be valuable to the movement. An organizational format must also be developed for allocating these resources. Leadership, therefore, becomes a crucial ingredient for the emergence of a social movement.

The leadership tries to formulate the resources and ideology in such an attractive fashion that many others will join the movement. It is not enough to make speeches and distribute fliers; the messages have to strike a respondent chord in others (Ferree & Hess, 1985).

Types of Social Movements

In the politics chapter, we discussed rebellions and revolutions, which certainly qualify as social movements; but there are other kinds of social movements as well. Scholars differ as to how they classify social movements, but some general characteristics are well recognized. We shall discuss these characteristics according to William Bruce Cameron's (1966) four social-movement classifications: reactionary, conservative, revisionary, and revolutionary. In addition, we shall examine the concept of expressive social movements first developed by Herbert Blumer (1946).

Although this classification is useful to sociologists in their studies of social movements, in practice it is sometimes difficult to place a social movement in only one category. This is because any social movement may possess a complex set of ideological positions in regard to the many different features of the society, its institutions, the class structure, and the different categories of people within that society.

Reactionary Social Movement **Reactionary social movements** *embrace the aims of the past and seek to return general society to yesterday's values.* Using

For people to join a social movement, they must think that their own values, needs, goals, or beliefs are being stifled or challenged by the social structure or by specific individuals.

slogans like the "good old days" and our "grand and glorious heritage," reactionaries abhor the changes that have transformed society and are committed to re-creating a set of valued social conditions that they believe existed at an earlier time. Reactionary groups such as the neo-Nazis and the Ku Klux Klan hold racial, ethnic, and religious values that are more characteristic of a previous historic period.

The Klan has sought to uphold white dominance and what it sees as traditional morality. To do this it has threatened, flogged, mutilated, and, on occasion, murdered. The main purpose of the Klansmen has been to defend and restore what they conceived as traditional cultural values. Their values legitimize prejudice and discrimination based on race, ethnicity, and religion, patterns that are now neither culturally legitimate nor legal (Chalmers, 1980).

Recently, this kind of reactionary fanaticism has been displayed by organized groups of teenagers who are part of a neo-Nazi political movement. The teens, known for violence, are white supremacists who commit a variety of hate crimes against blacks, immigrants, and Jews across the country.

Conservative Social Movement **Conservative social movements** *seek to maintain society's current values by reacting to change or threats of change that they believe will undermine the status quo.* Many of the evangelical religious groups hold conservative views. For example, they often oppose the forces that promulgate equal rights for women. To preserve what they consider traditional values of the family and religion, these groups have threatened to boycott advertisers that sponsor television programs containing sex and violence and have mounted successful campaigns to defeat political candidates who oppose their views.

Conservative movements are most likely to arise when traditional-minded people perceive a threat of change that might alter the status quo. Reacting to what might happen if another movement achieves its goals, members of conservative movements mobilize an "anti" movement crusade. Thus, antigun control groups have waged political war against any group seeking to restrict access to guns, whether they be handguns or assault weapons. Although reactive in nature, conservative movements are far different from true reactionary movements, which attempt to restore values that have already changed.

Revisionary Social Movement **Revisionary social movements** *seek partial or slight changes within the existing order but do not threaten the order itself.* The women's movement, for example, seeks to change the institutions and practices that have imposed prejudice and discrimination on women. The civil rights movements, the antinuclear movement, and the ecology movement are all examples of revisionary social movements.

Revolutionary Social Movement **Revolutionary social movements** *seek to overthrow all or nearly all of the existing social order and replace it with an order they consider more suitable.* For example, the black guerrilla movement in Zimbabwe (formerly Rhodesia) was a revolutionary movement. Through the use of arms and political agitation, the guerrillas were successful in forcing the white minority to turn over political power to the black majority and in creating a new form of government that guaranteed 80 of the 100 seats in the country's legislative body would be held by blacks.

Although revolutionary and revisionary social movements both seek change in society, they differ in the degree of change they seek. The American Revolution, for example, which sought to overthrow British colonial rule and led to the formation of our own government, differed significantly from the women's movement, which seeks change within the existing judicial and legislative structures.

Expressive Social Movement Though other types of movements tend to focus on changing the social structure in some way, **expressive social movements** *stress personal feelings of satisfaction or well-being and typically arise to fill some void or to distract people from some great dissatisfaction in their lives.* The Unification Church and the Hare Krishnas are religious movements of this type.

Revisionary social movements try to make changes within society, but do not threaten the social order.

The Life Cycle of Social Movements

Social movements, by their nature, do not last forever. They rise, consolidate, and eventually succeed, fail, or change. Armand L. Mauss (1975) suggested that social movements typically pass through a series of five **life cycle stages:** *(1) incipiency, (2) coalescence, (3) institutionalization, (4) fragmentation, and (5) demise.* However, these stages are by no means common to all social movements.

Incipiency *The first stage of a social movement,* called **incipiency,** *begins when large numbers of people become frustrated about a problem and do not perceive any solution to it through existing institutions.* Incipiency occurred in the nineteenth century when American workers, desperate over their worsening working conditions, formed the U.S. labor movement. This stage is a time of disorder, when people feel the need for something to give their lives direction and meaning or to channel their behavior toward achieving necessary change. Disruption and violence may mark a social movement's incipiency. In 1886 and 1887, as the labor movement grew, workers battled private Pinkerton agents and state militia and called nationwide strikes. Although physically beaten, the workers continued to organize.

Incipiency is also a time when leaders emerge. Various individuals offer competing solutions to the perceived societal problem, and some people are more persuasive than others. According to Max Weber (see Chapter 15), many of the more successful leaders have charismatic qualities derived from exceptional personal characteristics. Samuel Gompers, who launched the American Federation of Labor (AFL) in 1881, was such a leader, as was the Reverend Martin Luther King, Jr.

Coalescence During *the second stage,* known as **coalescence,** *groups form around leaders to promote policies and to promulgate programs.* Some groups join forces; others are defeated in the competition for new members. Gradually, a dominant group or coalition of groups emerges that establishes itself in a position of leadership. Its goals become the goals of many, its actions command wide participation, and its policies gain influence. The labor movement coalesced in 1905, when William D. Haywood organized the Industrial Workers of the World (IWW), which led its increasingly dissatisfied members in a number of violent strikes. Labor coalescence continued in 1935, when such militant industrial union leaders as John L. Lewis of the United Mine Workers and David Dubinsky of the International Ladies Garment Workers founded the Committee for Industrial Organization (CIO), which rapidly organized the steel, automobile, and other basic industries. Thus, through coalescence, the labor movement gradually created several large, increasingly powerful organizations.

Institutionalization During *the third stage,* known as **institutionalization,** *social movements reach the peak of their strength and influence and become firmly established.* Their leadership no longer depends on the elusive quality of charisma to motivate followers. Rather, it has become firmly established in formal, rational organizations (see Chapter 6) that have the power to effect lasting changes in the social order. At this point, the organizations themselves become part of the normal pattern of everyday life.

When the institutionalization of the U.S. labor movement became formalized with the legalization of unions in the 1930s, union leaders no longer used the revolutionary rhetoric that was necessary when unions were neither legitimate nor legal. Instead, they talked in pragmatic terms, worked within the political power structure, and sought reforms within the structure of the existing democratic, capitalistic system.

Not all social movements become institutionalized. In fact, social movements fail and disappear more often than they reach this stage. Institutionalization depends to a great degree on how the members feel about the movement—whether it reflects their goals and has been successful in achieving them—and on the extent to which the movement is accepted or rejected by the larger society.

Ironically, the acceptance of a social movement may also mark its end. Many members drop out or lose interest once a movement's goals have been reached. It can be argued that a certain amount of opposition from those in power reminds the members that they still must work to accomplish their goals. Movement leaders often hope for a confrontation that will clarify the identity of the opposition and show the members against what and whom they must fight. Movements that evoke an apathetic or disinterested response from the institutions controlling the power structure have few resources with which to unite their membership.

Fragmentation **Fragmentation** *is the fourth stage of a social movement, when the movement gradually begins to fall apart.* Organizational structures no longer seem necessary because the changes they sought to bring about have been institutionalized or the changes they sought to block have been prevented. Disputes over doctrine may drive out dissident members, as when the United Auto Workers (UAW) and the Teamsters left the AFL-CIO. Also, demographic changes may transform a once-strong social movement into a far less powerful force. Economic changes have been largely responsible for the fragmentation of the American labor movement. Unions now represent the smallest share of the labor force since World War II, even though the workforce continues to expand. Their lost power is due, in part, to a sharp decrease in the percentage of more easily unionized blue-collar workers in the labor force and a dramatic increase in the percentage of white-collar employees, who are largely resistant to unionization.

Demise **Demise,** *the last stage, refers to the end of a social movement.* The organizations that the movement created and the institutions they introduced may well survive—indeed, their goals may become official state policy—but they are no longer set apart from the mainstream of society. Transformed from social movements into institutions, they leave behind well-entrenched organizations that guarantee their members the goals they sought. This pattern of social-movement demise has occurred in parts of the American labor movement. The United Auto Workers, for example, is no longer a social movement fighting for the rights of its members from the outskirts of the power structure. Rather, it is now an institutionalized part of society. All unions have not followed this course. Labor is still very much a social movement, for it is trying to organize such previously unorganized groups as farm workers, nonunionized clerical professional workers, and all workers in the traditionally nonunion South. The American Federation of State, County, and Municipal Employees is a recent example of the labor movement's continued organizational efforts.

Globalization and Social Change

Globalization is *the worldwide flow of goods, services, money, people, information, and culture. It leads to a greater interdependence and mutual awareness among the people of the world* (Guillen, 2001).

Globalization, however, is also an ideology with multiple meanings. It is often identified with technological solutions to economic development and reform. The term has also been adopted by cross-border advocacy organizations defending human rights, the environment, women's rights, or world peace (Keck & Sikkink, 1998). The environmental movement, in particular, has raised the banner of globalism in its struggle for a clean planet, as in its "Think Global, Act Local" slogan. Thus, globalization is often thought of as an impersonal and inevitable force. Its policies or behaviors are often criticized, however praiseworthy some of them might be (Held et al., 1999).

It is unclear when globalization began. One could argue that globalization began with the dawn of history. Most research on the topic, however, has tended to view globalization as beginning at the end of World War II, with the coming of the nuclear age, and the expansion of worldwide trade and investment (Guillen, 2001).

People also believe that globalization leads to the rise of a global culture. This idea goes back to Marshall McLuhan's (1964) concept of the "global village." The global culture produces a lifestyle that has spread throughout the world and results in the standardization of tastes and values. For example, former secretary of the treasury Robert Reich (1991) stated that "national economies" are disappearing and companies no longer have a nationality; only people do. There are, however, some skeptics, who argue that the economy is becoming more international but the world is as fragmented as ever. Anthropologist Clifford Geertz (1998) observes that the world is "growing both more global and more divided, at the same time."

Globalization is neither a monolithic nor an inevitable phenomenon. Its impact varies across countries, societies, and time. Therefore, it is difficult to know exactly what the unexpected and unintended consequences will be. (See Table 18–2 for a ranking of countries experiencing high levels of globalization based on international travel, telephone calls, transfer of money, and Internet use)

Table 18–2

Countries Experiencing High Levels of Globalization

Singapore

Netherlands

Sweden

Switzerland

Finland

Ireland

Austria

United Kingdom

Norway

Canada

*Based on international travel and tourism, international telephone traffic, cross-border transfers of money and Internet use.

Source: A. T. Kearney, Inc., "Measuring Globalization," *Foreign Policy Magazine*, January/February 2001.

Social Change in the United States

There is virtually no area of life in the United States that has not changed in some respect since the relatively simple days of the 1950s. In addition, the pace of change will quicken even more as the turn of the century has passed. The following are a few of the major forces that are shaping future life in the United States.

Technological Change

In the past two decades, the personal computer and related technology has transformed the workplace and our lives. Biotechnology has helped us discover genetically engineered vaccines and a host of other benefits to society. Researchers increasingly tap the body itself as a new source of medications that genetic engineers can copy and improve on. New insights into human diseases should essentially make it possible to prevent such autoimmune conditions as rheumatoid arthritis, multiple sclerosis, and insulin-dependent diabetes, in which the body mistakenly attacks its own tissue. What could possibly top what has already happened and the changes that have been produced? The immediate future will produce not just more technology, but also some startling discoveries.

Even the individual scientist can now conduct research that was impossible just a short time ago. With a personal computer, compact computer disks that hold enormous amounts of information, and the Internet, huge databases can be tapped, making hard-to-find information available in minutes.

Computers themselves have already advanced to the point where they enable scientists to "see" objects on a smaller scale than microscopes can. This should make a vital contribution to chemistry, chemical engineering, molecular biology, and other fields. Computers also will be able to describe complex events, such as the chemistry involved in photosynthesis, in greater detail than is possible with today's instruments. Visual computing in effect reproduces the world around us mathematically within a computer. The objects can then be both seen and manipulated in all sorts of ways. A researcher can then describe the nature and behavior of an object or phenomenon with equations and present it visually. The objects simulated can include anything from the wing of an airplane to the interior of the sun. This kind of computing will create new areas of scientific inquiry and the development of new consumer products (Bylinsky, 1988).

The introduction of new technologies into our lives takes place at an ever more rapid pace. It took 12 years for 10% of the public to have a color television set; 10 years for 10% to have a videocassette recorder; 8 for a cell phone; and 4 for a personal computer (Horrigan, 2005).

We must be careful to realize that technology *influences* rather than determines social change. Technological innovation always occurs in the context of other forces—political, economic, or historical—which themselves help shape the technology and its uses. Technological innovation takes hold only when there is some need for it and social acceptance for it. Technology itself is neutral; people decide whether and how to use it (Rybczynski, 1983).

The Workforce of the Future

Employment opportunities are affected by population trends, and changes in the size and composition of the U.S. population between 2000 and 2010 will influence the demand for goods and services. For example, the 16- to 24-age group will grow more rapidly than the overall population—a turnaround that began in the mid-1990s. The 55- to 64-age group will increase by 11 million persons—more than any other group (Fullerton, 2001).

The projected labor force growth will be affected by the aging of the baby-boom generation, persons born between 1946 and 1964. In 2010, the baby-boom cohort will be ages 46 to 64, and this age group will show significant growth over the 2000–2010 period. The median age of the labor force will continue to rise, even though the youth labor force (aged 16 to 24) is expected to grow more rapidly than the overall labor force for the first time in 25 years. The baby-boom generation is expected to stay in the labor force longer than the previous generation of workers, because they are more highly educated and identify more strongly with their jobs.

The proportion of minorities and immigrants in the workforce will increase. Many more workers will be of African American, Hispanic, or Asian heritage, reflecting higher birthrates among African Americans, Hispanics, and substantial numbers of immigrants. Non-Hispanic whites have historically been the largest sector of the workforce, but their share has been dropping and is expected to fall from 73% in 2000 to 69% in 2010.

The Hispanic male labor force exceeded that of black men in 2000. Also, as the Hispanic population continues to grow, its labor force will show substantial growth over that of blacks. The Asian population also is growing rapidly.

Women will continue to join the labor force in growing numbers between 2000 and 2010. The growth of women in the labor force will be more rapid than that for men. Women's share of the labor force will increase from 47% in 2000 to 48% in 2010 (Fullerton, 2001).

The workforce of the next decade will be influenced by a new generation of sophisticated information and communication technologies currently being introduced into a wide variety of work situations.

Nowhere is the effect of the computer revolution and reengineering of the workplace more pronounced than in the manufacturing sector. In the 1950s, 33% of all U.S. workers were employed in manufacturing. By 2010, only 11.4% of the workforce will be engaged in blue-collar manufacturing work. Contrary to popular opinion, the loss of manufacturing jobs is not just due to cheap foreign labor, but more so to automation and technologically based efficiency. At the same time as the number of manufacturing jobs has been declining, manufacturing productivity has been soaring (see Figures 18–1 and 18–2 for the fastest growing and fastest declining occupations).

Service-producing industries will account for virtually all of the job growth between 2000 and 2010. Certain occupational groups are projected to grow faster than average between now and the year 2010. Computer and mathematical occupations are projected to add 2 million jobs. The demand for computer-related occupations will continue to increase as a result of the rapid advances in computer technology and the continuing demand for new computer applications, including those for the Internet and intranets. In addition, demand is expected to increase for home health care workers, medical assistants, and corrections officers.

Among the slowest growing groups are office and administrative support occupations; production occupations; and farming, fishing, and forestry occupations (Fullerton, 2001).

Occupations that require the highest levels of education and skill make up an increasing proportion of jobs. Of the 30 fastest growing occupations, 21 require

Figure 18–1   Where the Jobs Are

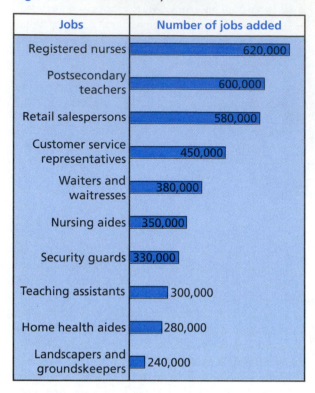

Jobs	Number of jobs added
Registered nurses	620,000
Postsecondary teachers	600,000
Retail salespersons	580,000
Customer service representatives	450,000
Waiters and waitresses	380,000
Nursing aides	350,000
Security guards	330,000
Teaching assistants	300,000
Home health aides	280,000
Landscapers and groundskeepers	240,000

Source: U.S. Department of Labor, Bureau of Labor Statistics, *Occupational Outlook Handbook, 2004–2005.* http://www.bls .gov/oco/images/ocotjc08.gif.

Figure 18–2   Where the Jobs Are Not

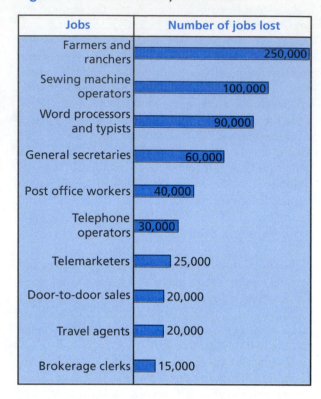

Jobs	Number of jobs lost
Farmers and ranchers	250,000
Sewing machine operators	100,000
Word processors and typists	90,000
General secretaries	60,000
Post office workers	40,000
Telephone operators	30,000
Telemarketers	25,000
Door-to-door sales	20,000
Travel agents	20,000
Brokerage clerks	15,000

Source: U.S. Department of Labor, Bureau of Labor Statistics, *Occupational Outlook Handbook, 2004–2005.* http://www.bls .gov/oco/images/ocotjc09.gif.

a college degree. There will still be many jobs available for those with only a high school diploma. However, the opportunities for those who do not finish high school will be increasingly limited, and people who cannot read may not be considered for most jobs. Those who have not completed high school are likely to have low-paying jobs with little advancement potential.

SUMMARY

- Social change consists of any modification in the social organization of a society in any of its social institutions or social roles.
- It can take place gradually or relatively quickly and it can be accomplished through violent or peaceful means.
- Some social change involves collective behavior.
- Collective behavior refers to relatively spontaneous social actions that occur when people respond to unstructured and ambiguous situations.
- A crowd is a temporary concentration of people who focus on some thing or event but who also are attuned to one another's behavior.
- Crowds have the potential for unpredictable behavior and group action that erupts quickly and often seems to lack structure or direction.

- Because the mass media today spread information quickly among millions of people, collective behavior is often shared by large numbers of people who have no direct knowledge of or contact with one another.
- A social movement is a form of collective behavior in which large numbers of people are organized or alerted to support and bring about or to resist social change.
- For people to join a social movement, they must feel that their own values, needs, goals, or beliefs are being stifled or challenged by the social structure or specific individuals and that things must be set right.
- Some catalyst, however, is needed to mobilize the discontent people feel.
- In theory, social movements can be classified according to type; in practice, a given movement may possess a complex ideology that places parts of it in several different classifications.
- Relative deprivation theory assumes social movements are the outgrowth of the feeling of relative deprivation among the large numbers of people who believe they lack certain things they are entitled to—such as better living conditions, working conditions, political rights, or social dignity.
- Resource mobilization theory assumes that social movements arise at certain times and not

at others because some people know how to mobilize and channel the popular discontent.

- Social change can occur as a result of internal or external forces.
- Technological innovation is an important source of internal change that has transformed the way we live and work.
- Ideologies are an important internal source of change.
- Ideologies are interrelated religious or secular beliefs, values, and norms that justify the pursuit of a given set of goals through a given set of means.
- External sources of social change originate outside of a society.
- Diffusion, the process of transmitting the traits of one culture to another, is an external source of social change.
- There is virtually no area of life in the United States that has not changed in some respect since the 1950s. Moreover, the pace of change is likely to quicken.

 *Media Resources*

The Companion Website for *Introduction to Sociology*, Ninth Edition

http://sociology.wadsworth.com/tischler9e

Supplement your review of this chapter by going to the companion website to take one of the Tutorial Quizzes, use the flash cards to master key terms, and check out the many other study aids you'll find there. You'll also find special features such as Wadsworth's Sociology Online Resources and Writing Companion, GSS data, and Census 2000 information at your fingertips to help you complete that special project or do some research on your own.

CHAPTER EIGHTEEN STUDY GUIDE

KEY CONCEPTS AND THINKERS

Match each concept with its definition, illustration, or explanation presented below.

a. Expressive social movement
b. Craze
c. Reactionary social movement
d. Crowd
e. Revolutionary social movement
f. Acting crowd
g. Fad
h. Incipiency

i. Social movement
j. Resource mobilization theory
k. Revisionary social movement
l. Institutionalized inequality
m. Cultural diffusion
n. Public opinion
o. Expressive crowd

p. Mass
q. Mass hysteria
r. Relative deprivation theory
s. Opinion leaders
t. Coalescence
u. McDonaldization
v. Globalization

_____ 1. The explanation of social movements that emphasizes the actions and abilities of leaders.
_____ 2. A group of people passionately aroused by some focal event, who often erupt into unplanned violence.
_____ 3. The increasing interdependence of people, businesses, and institutions in different countries.
_____ 4. Any collection of people who just happen, in the course of their private activities, to be in one place at the same time and focus their attention on a common object or event.
_____ 5. A social change with a very short life span marked by a rapid spread and an abrupt drop in popularity.
_____ 6. A fad that is especially short-lived.
_____ 7. Beliefs held by a dispersed collectivity of individuals about a common problem, interest, focus, or activity.
_____ 8. A large group of people who, although not physically gathered in one place, share participation in some event or share a common interest.
_____ 9. A sustained, organized effort by large numbers of people to support and bring about, or to resist, social change.
_____ 10. The transformation of traditional institutions by organizing them with an emphasis on efficiency, calculability, and predictability.
_____ 11. The first stage of a social movement, when many people feel frustrated by the lack of currently available solutions to a problem.
_____ 12. Recognized experts who influence what the public thinks.
_____ 13. A movement that embraces the aims of the past and seeks to return the general society to yesterday's values.
_____ 14. A movement that seeks small changes but does not challenge the legitimacy of existing social arrangements.
_____ 15. The theory that says social movements arise when people are dissatisfied with their position in society compared with others.
_____ 16. A movement that seeks to overthrow all or nearly all of the existing social order and replace it with an order its members consider more suitable.
_____ 17. A movement that stresses personal feelings, especially those of satisfaction or well-being.
_____ 18. The stage of a social movement when groups form around leaders and begin to devise proposals and strategies.
_____ 19. The spread of an idea or technology from one society to another.
_____ 20. Differences between groups caused by arrangements in the social system rather than by individual acts of discrimination.

Match the thinkers with their main ideas or contributions.

a. Elias Canetti
b. Irving Janis

c. Armand L. Mauss
d. George Ritzer

e. Kai Erikson
f. Samuel Stouffer

_____ 1. Described the conditions under which people collectively panic.
_____ 2. Coined the term *McDonaldization* and used it as a concept to understand changes in contemporary life.
_____ 3. Developed the concept of relative deprivation.

_____ 4. Explained the seventeenth-century Salem witchcraft trials as an episode of mass hysteria created by the Massachusetts Bay Colony's identity crisis.

_____ 5. Proposed a model of the "life cycle" of social movements from creation to demise.

_____ 6. Described the important traits of crowds.

CENTRAL IDEA COMPLETIONS

Following the instructions, fill in the appropriate concepts and descriptions for each of the questions posed in the following section.

1. What does your text mean when it says, "There is a flaw in the theory of relative deprivation"?

   _____

   _____

2. Differentiate between fads and fashions, providing two specific examples of each, one in the area of clothing and one in some area of behavior.

   a. Fad: _____

   Clothing: _____

   Behavior: _____

   b. Fashion: _____

   Clothing: _____

   Behavior: _____

3. What are the stages in the life cycle of a social movement?

   a. _____

   _____

   b. _____

   _____

   c. _____

   _____

   d. _____

   _____

   e. _____

   _____

4. State and provide an example of each of the four attributes of crowds.

   a. Example: _____

   _____

   b. Example: _____

   _____

   c. Example: _____

   _____

   d. Example: _____

   _____

CRITICAL THOUGHT EXERCISES

1. Consider the growing use of "chat rooms" as social interaction. Develop an argument, based specifically on the theoretical material presented in this chapter, in which you support or dispute the idea that chat rooms can be an integral part of a social movement.

2. How do you think the structure of relations between unmarried young men and women (for example, dating, courtship) has changed over the last few generations? What are the sources for this social change, and how do these fit into the categories of social change outlined in this chapter?

3. Movements surround some of the controversial issues on the current political landscape—abortion, homosexuality, evolution, war, environmentalism, guns, and others. Choose one of these and analyze it as a social movement using the concepts in this chapter. Show how concepts like relative deprivation, life cycle, and resource mobilization help you understand the movement and allow you to compare it with others.

INTERNET ACTIVITIES

1. Social change is easier to predict after it happens. Predicting the future is more difficult, but staying ahead of social trends can be very important, especially for business enterprises and the media, who need to know where the market and technology will be going. Look at http://www.hfac.uh.edu/ MediaFutures/home.html for predictions about the media—predictions from the past (such as Edison's 1922 prediction that movies would soon replace textbooks) and predictions about the future. If you have any predictions about the media, you can submit your own forecast.

2. The Internet is itself an important piece of technology. Not surprisingly, several websites follow technology and its relation to social trends. Check out http://www.tecsoc.org/culture/culture.htm (or its parent site http://www.tecsoc.org/).

3. For information about specific social movements, http://www2.fmg.uva.nl/sociosite/topics/activism .html#ACTIVISM has links to a variety of groups.

ANSWERS TO KEY CONCEPTS

1.j 2.f 3.v 4.d 5.g 6.b 7.n 8.p 9.i 10.u 11.h 12.s 13.c 14.k 15.r 16.e 17.a 18.t
19.m 20.l

ANSWERS TO KEY THINKERS

1.b 2.d 3.f 4.e 5.c 6.a

ThomsonNOW™

Reviewing is as easy as ❶ ❷ ❸
1. Before you do your final exam, take the ThomsonNOW diagnostic quiz to help you identify the areas on which you should concentrate. You will find information on ThomsonNOW and instructions on how to access all of its great resources on the foldout at the beginning of the text.
2. As you review, take advantage of ThomsonNOW's study videos and interactive Map the Stats exercises to help you master the chapter topics.
3. When you are finished with your review, take ThomsonNOW's posttest to confirm you are ready to move on to the next chapter.

Glossary

Abstract ideals Aspects of a religion that focus on correct ways of thinking and behaving, rather than on a belief in supernatural forces, spirits, or beings.

Achieved statuses Statuses obtained as a result of individual efforts.

Acting crowd A group of people whose passions and tempers have been aroused by some focal event, who come to share a purpose, who feed off one another's arousal, and who often erupt into spontaneous acts of violence.

Activity theory The view that satisfaction in later life is related to the level of activity the person engages in.

Adaptation The process by which human beings adjust to the changes in their environment.

Adult socialization The process by which adults learn new statuses and roles. Adult socialization continues throughout the adult years.

Affiliation Meaningful interaction with others.

Age-specific death rates The annual number of deaths per 1,000 people at specific ages.

Agricultural societies Societies that use the plow in food production.

Alienation The process by which people lose control over the social institutions they themselves invented.

Analysis The process through which scientific data are organized so that comparisons can be made and conclusions drawn.

Anglo conformity A form of assimilation that involves the renunciation of the ancestral culture in favor of Anglo-American behavior and values.

Animism The belief in animate, personalized spirits or ghosts of ancestors that take an interest in, and actively work to influence, human affairs.

Annihilation The deliberate practice of trying to exterminate a racial, religious, or ethnic group; also known as genocide.

Anomie The feeling of some individuals that their culture no longer provides adequate guidelines for behavior; a condition of "normlessness" in which values and norms have little impact.

Aristocracy The rule by a select few; a form of oligarchy.

Ascribed statuses Statuses conferred on an individual at birth or on other occasions by circumstances beyond the individual's control.

Assimilation The process whereby groups with different cultures come to have a common culture.

Associations Purposefully created special interest groups that have clearly defined goals and official ways of doing things.

Attachment disorder Children who are unable to trust people and form relationships with others.

Authoritarian leader A type of instrumental leader who makes decisions and gives orders.

Authority Power regarded as legitimate by those over whom it is exercised, who also accept the authority's legitimacy in imposing sanctions or even in using force, if necessary.

Autocracy A political system in which the ultimate authority rests with a single person.

Bilateral descent system A descent system that traces kinship through both female and male family members.

Bilocal residence Marital residence rules allowing a newly married couple to live with either the husband's or wife's family of origin.

Blind investigator A researcher who does not know whether a specific subject belongs to the group of actual cases being investigated or to a comparison group. This is done to eliminate researcher bias.

Bourgeoisie The term used by Karl Marx to describe the owners of the means of production and distribution in capitalist societies.

Bureaucracy A formal, rationally organized social structure with clearly defined patterns of activity in which, ideally, every series of actions is fundamentally related to the organization's purpose.

Capitalism An economic system based on private ownership of the means of production, in which resource allocation depends largely on market forces.

Caste system A rigid form of social stratification based on ascribed characteristics that determines its members' prestige, occupation, residence, and social relationships.

Casual crowd A crowd made up of a collection of people who, in the course of their private activities, happen to be in the same place at the same time.

Charismatic authority The power that derives from a ruler's force of personality. It is the ability to inspire passion and devotion among followers.

City A unit typically incorporated according to the laws of the state within which it is located.

Class system A system of social stratification that contains several social classes and in which greater social mobility is permitted than in a caste or estate system.

Closed society A society in which the various aspects of people's lives are determined at birth and remain fixed.

Coalescence The second stage in the life cycle of a social movement, when groups begin to form around leaders, promote policies, and promulgate programs.

Coercion A form of conflict in which one of the parties in a conflict is much stronger than the others and imposes its will because of that strength.

Cognitive culture The thinking component of culture, consisting of shared beliefs and knowledge of what the world is like—what is real and what is not, what is important and what is trivial; one of the two categories of nonmaterial culture.

Cohabitation Unmarried couples living together out of wedlock.

Collective behavior Relatively spontaneous social actions that occur when people respond to unstructured and ambiguous situations.

Collective conscience A society's system of fundamental beliefs and values.

Command economy An economy in which the government makes all the decisions about production and consumption.

Communism The name commonly given to totalitarian socialist forms of government.

Companionate marriage Marriage based on romantic love.

Competition A form of conflict in which individuals or groups confine their conflict within agreed-upon rules.

Concentric zone model A theory of city development in which the central city is made up of a business district and, radiating from this district, zones of low-income, working-class, middle-class, and upper-class residential units.

Conditioning The molding of behavior through repeated experiences that link a desired reaction with a particular object or event.

Conflict The opposite of cooperation. People in conflict struggle against one another for some commonly prized object or value.

Conflict approach The view that the elite use their power to enact and enforce laws that support their own economic interests and go against the interests of the lower classes.

Conflict theory This label applies to any of a number of theories that assume society is in a constant state of social conflict, with only temporarily stable periods, and that social phenomena are the result of this conflict.

Consensus approach An approach to law that assumes laws are merely a formal version of the norms and values of the people.

Conservative ideologies Ideologies that try to preserve things as they are.

Conservative social movement A social movement that seeks to maintain society's current values.

Constitutionalism Government power that is limited by law.

Contagion theory The theory that states members of a crowd acquire a crowd mentality, lose their characteristic inhibitions, and become highly receptive to group sentiments.

Context The conditions under which an action takes place, including the physical setting or place, the social environment, and the other activities surrounding the action.

Contractual cooperation Cooperation in which each person's specific obligations are clearly spelled out.

Conventional crowd A crowd in which people's behavior conforms to some well-established set of cultural norms and in which people's gratification results from passive appreciation of an event.

Convergence theory Views collective behavior as the outcome of situations in which people with similar characteristics, attitudes, and needs are drawn together.

Cooperative interaction A form of social interaction in which people act together to promote common interests or achieve shared goals.

Craze A fad that is especially short-lived.

Crime Behavior that violates a society's legal code.

Criminal justice system Personnel and procedures for arrest, trial, and punishment to deal with violations of the law.

Cross-sectional study An examination of a population at a given point in time.

Crowd A temporary concentration of people who focus on some thing or event but who also are attuned to one another's behavior.

Crude birthrate The number of annual births per 1,000 population.

Crude death rate The annual number of deaths per 1,000 population.

Cultural lag A situation that develops when new patterns of behavior conflict with traditional values. Cultural lag may occur when technological change (material culture) is more rapid than are changes in norms and values (nonmaterial culture).

Cultural relativism The position that social scientists doing cross-cultural research should view and analyze behaviors and customs within the cultural context in which they occur.

Cultural traits Items of a culture such as tools, materials used, beliefs, values, and typical ways of doing things.

Cultural transmission The transmission of major portions of a society's knowledge, norms, values, and perspectives from one generation to the next. Cultural transmission is an intended function of education.

Cultural universals Forms or patterns for resolving the common, basic, human problems that are found in all cultures. Cultural universals include the division of labor, the incest taboo, marriage, the family, rites of passage, and ideology.

Culture All that human beings learn to do, to use, to produce, to know, and to believe as they grow to maturity and live out their lives in the social groups to which they belong.

Culture shock The reaction people may have when encountering cultural traditions different from their own.

De facto segregation Segregation of community or neighborhood schools that results from residential patterns in which minority groups often live in areas of a city where there are few whites or none at all.

De jure segregation Segregation that is an outgrowth of local laws that prohibit one racial group from attending school with another.

Demise The last stage in the life cycle of a social movement, when the movement comes to an end.

Democracy A political system operating under the principles of constitutionalism, representative government, majority rule, civilian rule, and minority rights.

Democratic leader A type of instrumental leader who attempts to encourage group members to reach a consensus.

Democratic socialism A political system that exhibits the dominant features of a democracy, but in which the control of the economy is vested in the government to a greater extent than under capitalism.

Demographic transition theory A theory that explains population dynamics in terms of four distinct stages from high fertility and high mortality to relatively low fertility and low mortality.

Demography The study of the dynamics of human populations.

Denomination A religious group that tends to draw its membership from a particular socially acceptable class or ethnic group, or at least to have its leadership positions dominated by members of such a group.

Dependency ratio The number of people of nonworking age in a society for every 100 people of working age.

Dependency theory A theory that proposes that the economic positions of rich and poor nations are linked and cannot be understood in isolation from each other. Global inequality is due to the exploitation of poor societies by the rich ones.

Dependent variable A variable that changes in response to changes in the independent variable.

Deviant behavior Behavior that fails to conform to the rules or norms of the group in which it occurs.

Dictatorship A totalitarian government in which all power rests ultimately in one person, who generally heads the only recognized political party.

Diffusion One of the two mechanisms responsible for cultural evolution. Diffusion is the movement of cultural traits from one culture to another.

Directed cooperation Cooperation characterized by a joint effort under the control of people in authority.

Discrimination Differential treatment, usually unequal and injurious, directed at individuals who are assumed to belong to a particular category or group.

Diversion Steering youthful offenders away from the juvenile justice system to nonofficial social agencies.

Double-blind investigator A researcher who does not know either the kind of subject being investigated or the hypothesis being tested.

Dramaturgy The study of the roles people play to create a particular impression in others.

Dyad A small group that contains only two members.

Ecclesia A church that shares the same ethical system as the secular society and that has come to represent and promote the interests of the society at large.

Economy An institution whose primary function is to determine the manner in which society produces, distributes, and consumes goods and services.

Ecumenism The trend among many religions to draw together and project a sense of unity and common direction.

Ego In Freudian theory, one of the three separately functioning parts of the self. The ego tries to mediate in the conflict between the id and the superego and to find socially acceptable ways for the id's drives to be expressed. This part of the self constantly evaluates social realities and looks for ways to adjust to them.

Electorate Those citizens eligible to vote.

Emergent norm theory A theory that notes that even though crowd members may have different motives for participating in collective behavior, they acquire common standards by observing and listening to one another.

Emigration The movement of a population from an area.

Empirical question A question that can be answered by observation and analysis of the world as it is known.

Empiricism The view that generalizations are valid only if they rely on evidence that can be observed directly or verified through our senses.

Endogamy Societal norms that limit the social categories from within which one can choose a marriage partner.

Environmental determinism The belief that the environment dictates cultural patterns.

Estate A segment of a society that has legally established rights and duties.

Estate system A closed system of stratification in which social position is defined by law and membership is based primarily on inheritance. A very limited possibility of upward mobility exists.

Ethnic group A group that has a distinct cultural tradition with which its members identify and which may or may not be recognized by others.

Ethnocentrism The tendency to judge other cultures in terms of one's own customs and values.

Ethnomethodology The study of the sets of rules or guidelines people use in their everyday living practices. This approach provides information about a society's unwritten rules for social behavior.

Ethology The scientific study of animal behavior.

Evolution The continuous change from a simpler condition to a more complex state.

Exchange interaction An interaction involving one person doing something for another with the express purpose of receiving a reward or return.

Exogamy Societal norms that require an individual to marry someone outside his or her culturally defined group.

Experiment An investigation in which the variables being studied are controlled and the researcher obtains the results through precise observation and measurement.

Expressive crowd A crowd that is drawn together by the promise of personal gratification for its members through active participation in activities and events.

Expressive leadership A form of leadership in which a leader works to keep relations among group members harmonious and morale high.

Expressive social movement A social movement that stresses personal feelings of satisfaction or well-being and that typically arises to fill some void or to distract people from some great dissatisfaction in their lives.

Expulsion The process of forcing a group to leave the territory in which it resides.

Extended families Families that include, in addition to nuclear family members, other relatives such as the parents' parents, the parents' siblings, and in-laws.

External means of social control The ways in which others respond to a person's behavior that channel his or her behavior along culturally approved lines.

External source of social change Changes within a society produced by events external to that society.

Exurbs The fast-growing area located in a newer, second ring beyond the old suburbs.

Fad A transitory social change that has a very short life span marked by a rapid spread and an abrupt drop from popularity.

Family of orientation The nuclear family in which one is born and raised; also known as family of origin.

Family of procreation The family that is created by marriage.

Fascism A political-economic system characterized by totalitarian capitalism.

Fashion A transitory change in the standards of dress or manners in a given society.

Fecundity The physiological ability to have children.

Felonies Offenses punishable by a year or more in a state prison.

Fertility The actual number of births in the population.

Fertility rate The number of annual births per 1,000 women of childbearing age in a population.

Folkways Norms that permit a rather wide degree of individual interpretation as long as certain limits are not overstepped. Folkways change with time and vary from culture to culture.

Forced acculturation The situation that occurs when social change is imposed on weaker peoples by might or conquest.

Forced migration The expulsion of a group of people through direct action.

Formal negative sanctions Actions that express institutionalized disapproval of a person's behavior, such as expulsion, dismissal, or imprisonment. They are usually applied within the context of a society's formal organizations, including schools, corporations, and the legal system.

Formal positive sanctions Actions that express social approval of a person's behavior, such as public gatherings, rituals, or ceremonies.

Formal sanctions Sanctions that are applied in a public ritual, usually under the direct or indirect leadership of social authorities; examples: the award of a prize or the announcement of an expulsion.

Fragmentation The fourth stage in the life cycle of a social movement, when the movement gradually begins to fall apart.

Functionalism (structural functionalism) One of the major sociological perspectives, which assumes that society is a system of highly interrelated parts that operate (function) together rather harmoniously.

Funnel effect The situation in our criminal justice system whereby many crimes are committed, but few criminals seem to be punished.

Game stage According to George Herbert Mead, the stage in the development of the self when the child learns that there are rules that specify the proper and correct relationship among the players.

Gemeinschaft A community in which relationships are intimate, cooperative, and personal.

Gender The social, psychological, and cultural attributes of masculinity and femininity that are based on biological distinctions.

Gender identity The view of ourselves resulting from our sex.

Gender-role socialization The lifelong process whereby people learn the values, attitudes, and behavior considered appropriate to each sex by their culture.

Generalized others The viewpoints, attitudes, and expectations of society as a whole or of a general community of people that we are aware of and who are important to us.

Genes The set of inherited units of biological material with which each individual is born.

Gentrification A trend that involves wealthier, middle-class people moving to marginal urban areas, upgrading the neighborhood, and displacing some of the poor residents who become priced out of the available housing.

Gesellschaft A society in which relationships are impersonal and independent.

Ghetto A term originally used to refer to the segregated quarter of a city where the Jews in Europe were often forced to live. Today it is used to refer to any kind of segregated living environment.

Globalization The worldwide flow of goods, services, money, people, information, and culture. It leads to a greater interdependence and mutual awareness among the people of the world.

Group A collection of specific, identifiable people.

Hidden curriculum The social attitudes and values learned in school that prepare children to accept the requirements of adult life and to fit into the social, political, and economic statuses of adult life.

Homogamy The tendency to choose a spouse with a similar racial, religious, ethnic, educational, age, or socioeconomic background.

Horizontal mobility Movement that involves a change in status with no corresponding change in social class.

Horticulture societies Societies in which muscle power and handheld tools are used to cultivate gardens and fields.

Hunting and food-gathering societies Societies that survive by foraging for vegetable foods and small game, fishing, collecting shellfish, and hunting larger animals.

Hypothesis A testable statement about the relationship between two or more empirical variables.

I The portion of the self that wishes to have free expression, to be active, and to be spontaneous.

Id In Freudian theory, one of the three separately functioning parts of the self. The id consists of the unconscious drives or instincts that Freud believed every human being inherits.

Ideal norms Expectations of what people should do under perfect conditions. The norm that marriage will last "until death do us part" is an ideal norm in American society.

Ideal type A simplified, exaggerated model of reality used to illustrate a concept.

Ideational culture A term developed by Pitirim A. Sorokin to describe a culture in which spiritual concerns have the greatest value.

Ideologies Strongly held beliefs and values to which group members are firmly committed and which cement the social structure.

Ideology A set of interrelated religious or secular beliefs, values, and norms justifying the pursuit of a given set of goals through a given set of means.

Immigration The movement of a population into an area.

Incest The term used to describe sexual relations within families. Most cultures have strict taboos against incest, which is often associated with strong feelings of horror and revulsion.

Incest taboo A societal prohibition that forbids sexual intercourse among closely related individuals.

Incipiency The first stage in the life cycle of a social movement, when large numbers of people perceive a problem without an existing solution.

Independent variable A variable that changes for reasons that have nothing to do with the dependent variable.

Industrial cities Cities established during or after the Industrial Revolution, characterized by large populations that work primarily in industrial or service-related jobs.

Industrial societies Societies that use mechanical means of production instead of human or animal muscle power.

Industrialism Consists of the use of mechanical means (machines and chemical processes) for the production of goods.

Infant mortality rate The number of children who die within the first year of life per 1,000 live births.

Informal negative sanctions Spontaneous displays of disapproval of a person's behavior. Disapproving treatment is directed toward the violator of a group norm.

Informal positive sanctions Spontaneous actions such as smiles, pats on the back, handshakes, congratulations, and hugs, through which individuals express their approval of another's behavior.

Informal sanctions Responses by others to an individual's behavior that arise spontaneously with little or no formal leadership.

Innovation Any new practice or tool that becomes widely accepted in a society.

Innovators Individuals who accept the culturally validated goal of success but find deviant ways of reaching it.

Instincts Biologically inherited patterns of complex behavior.

Institutionalization The third stage in the life cycle of a social movement, when the movement reaches its peak of strength and influence and becomes firmly established.

Institutionalized prejudice and discrimination Complex societal arrangements that restrict the life chances and choices of a specifically defined group.

Instrumental leadership A form of leadership in which a leader actively proposes tasks and plans to guide the group toward achieving its goals.

Interactionist perspective An orientation that focuses on how individuals make sense of or interpret the social world in which they participate.

Intergenerational mobility Changes in the social level of a family through two or more generations.

Internal means of social control A group's moral code which becomes internalized and becomes part of each individual's personal code of conduct. It operates even in the absence of reactions by others.

Internal migration The movement of a population within a nation's boundary lines.

Internal sources of social change Those factors that originate within a specific society and that singly or in combination produce significant alterations in its social organization and structure.

Interview A conversation between an investigator and a subject for the purpose of gathering information.

Intragenerational mobility Social changes during the lifetime of one individual.

Juvenile crime Refers to the breaking of criminal laws by individuals under the age of eighteen.

Labeling theory A theory of deviance that assumes the social process by which an individual comes to be labeled a deviant contributes to causing more of the deviant behavior.

Laissez-faire capitalism The view of capitalism that believes government should stay out of business.

Laissez-faire leader A type of instrumental leader who is a leader in name or title only and does little actively to influence group affairs.

Latent function One of two types of social functions identified by Robert Merton, referring to the unintended or not readily recognized consequences of a social process.

Laws Formal rules adopted by a society's political authority.

Leader Someone who occupies a central role or position of dominance and influence in a group.

Legal code The formal body of rules adopted by a society's political authority.

Legal-rational authority Authority that derives from the fact that specific individuals have clearly defined rights and duties to uphold and who implement rules and procedures.

Liberal ideologies Ideologies that seek limited reforms that do not involve fundamental changes in the structure of society.

Life expectancy The average number of years that a person born in a particular year can expect to live.

Lobbying Attempts by special-interest groups to influence government policy.

Longitudinal research A research approach in which a population is studied at several intervals over a relatively long period of time.

Looking-glass self A theory developed by Charles Horton Cooley to explain how individuals develop a sense of self through interaction with others. The theory has three stages: (1) We imagine how our actions appear to others, (2) we imagine how other people judge these actions, and (3) we make some sort of self-judgment based on the presumed judgments of others.

Magic Interaction with the supernatural. Magic does not involve the worship of a god or gods, but rather, it is an attempt to coerce spirits or control supernatural forces.

Majority rule The right of people to assemble to express their views and seek to persuade others, to engage in political organizing, and to vote for whomever they wish.

Mana A Melanesian/Polynesian concept of the supernatural that refers to a diffuse, nonpersonalized force that acts through anything that lives or moves.

Manifest function One of two types of social functions identified by Robert Merton, referring to an intended and recognized consequence of a social process.

Marital residence rules Rules that govern where a newly married couple settles down and lives.

Marriage The socially recognized, legitimized, and supported union of individuals of opposite sexes.

Mass A collection of people who, although physically dispersed, participate in some event either physically or with a common concern or interest.

Mass hysteria A condition in which large numbers of people are overwhelmed with emotion and frenzied activity or become convinced that they have experienced something for which investigators can find no discernible evidence.

Mass media Methods of communication, including television, radio, magazines, films, and newspapers, that have become some of society's most important agents of socialization.

Master status One of the multiple statuses a person occupies that dominates the others in patterning that person's life.

Material culture All the things human beings make and use, from small hand-held tools to skyscrapers.

Matriarchal family A family in which most family affairs are dominated by women.

Matrilineal system A descent system that traces kinship through the females of the family.

Matrilocal residence Marital residence rules that require a newly married couple to settle down near or within the wife's mother's household.

Me The portion of the self that is made up of those things learned through the socialization process from the family, peers, school, and so on.

Mechanically integrated society A type of society in which members have common goals and values and a deep and personal involvement with the community.

Mechanisms of social control Processes used by all societies and social groups to influence or mold members' behavior to conform to group values and norms.

Megalopolis Another term for Consolidated Metropolitan Statistical Area.

Metropolitan area An area that has a large population nucleus, together with the adjacent communities that are economically and socially integrated into that nucleus.

Middle-range theories Theories concerned with explaining specific issues or aspects of society instead of trying to explain how all of society operates.

Migration The movement of populations from one geographical area to another.

Millenarian movements Religious movements that prophesy the end of the world, the destruction of all evil people and their works, and the saving of the just.

Minority A group of people who, because of physical or cultural characteristics, are singled out from others in the society in which they live for different and unequal treatment and who therefore regard themselves as objects of collective discrimination.

Minority rights The minority has the right to try to change the laws of the majority.

Mixed economy An economy that combines free-enterprise capitalism with government regulation of business, industry, and social-welfare programs.

Modernization The complex set of changes that take place as a traditional society becomes an industrial society.

Modernization theory Assumes a direct relationship between the extent of modernization in a society and the status and condition of the elderly.

Monogamous marriage The form of marriage in which each person is allowed only one spouse at a time.

Monotheism The belief in the existence of only one god.

Moral code The symbolic system, made up of a culture's norms and values, in terms of which behavior takes on the quality of being good or bad, right or wrong.

Moral order A society's shared view of right and wrong.

Mores Strongly held norms that usually have a moral connotation and are based on the central values of the culture.

Mortality The frequency of deaths in a population.

Multiple marriage A form of marriage in which an individual may have more than one spouse (polygamy).

Multiple nuclei model A theory of city development that emphasizes that different industries have different land-use and financial requirements, which determine where they establish themselves. As similar industries are established close to one another, the immediate neighborhood is strongly shaped by the nature of its typical industry, becoming one of a number of separate nuclei that together constitute the city.

Negative sanctions Responses by others that discourage the individual from continuing or repeating the behavior.

Neolocal residence Marital residence standards that allow a newly married couple to live virtually anywhere, even thousands of miles from their families of origin.

Nonmaterial culture The totality of knowledge, beliefs, values, and rules for appropriate behavior that specifies how a people should interact and how they may solve their problems.

Normal behavior Behavior that conforms to the rules or norms of the group in which it occurs.

Norms Specific rules of behavior that are agreed upon and shared within a culture to prescribe limits of acceptable behavior.

Nuclear family The most basic family form, made up of parents and their children, biological or adopted.

Nurture The entire socialization experience.

Oligarchy Rule by a few individuals who occupy the highest positions in an organization.

Open-ended interview (See semistructured interview.)

Open society A society that provides equal opportunity to everyone to compete for the role and status desired, regardless of race, religion, gender, or family history.

Operational definition A definition of an abstract concept in terms of the observable features that describe the things being investigated.

Opinion leaders Socially acknowledged experts to whom the public turns for advice.

Organically integrated society A type of society in which social solidarity depends on the cooperation of individuals in many different positions who perform specialized tasks.

Organized crime Structured associations of individuals or groups who come together for the purpose of obtaining gain, mostly from illegal activities.

Panic An uncoordinated group flight from a perceived danger.

Pantheon The hierarchy of deities in a religious belief system.

Paradigm A basic model for explaining events that provides a framework for the questions that generate and guide research.

Participant observation A research technique in which the investigator enters into a group's activities while, at the same time, studying the group's behavior.

Pastoral societies Societies that rely on herding and the domestication and breeding of animals for food and clothing to satisfy most of the group's needs.

Patriarchal family A family in which most family affairs are dominated by men.

Patriarchal ideology The belief that men are superior to women and should control all important aspects of society.

Patrilineal system A family system that traces kinship through the males of the family.

Patrilocal residence Marital residence rules that require a newly married couple to settle down near or within the husband's father's household.

Peers Individuals who are social equals.

Personality The patterns of behavior and ways of thinking and feeling that are distinctive for each individual.

Play stage According to George Herbert Mead, the stage in the development of the self when the child has acquired language and begins not only to imitate behavior, but also to formulate role expectation.

Pluralism The development and coexistence of separate racial and ethnic group identities in a society in which no single subgroup dominates.

Political action committees (PACs) Special interest groups concerned with very specific issues who usually represent corporate, trade, or labor interests.

Political revolutions Relatively rapid transformations of state or government structures that are not accompanied by changes in social structure or stratification.

Politics The process by which power is distributed and decisions are made.

Polyandrous family A polygamous family unit in which the central figure is female and the multiple spouses are male.

Polygamous families Nuclear families linked together by multiple marriage bonds, with one central individual married to several spouses.

Polygynous family A polygamous family unit in which the central person is male and the multiple spouses are female.

Polytheism The belief in a number of gods.

Positive checks Events, described by Thomas Robert Malthus, that limit reproduction, either by causing the deaths of individuals before they reach reproductive age or by causing the deaths of large numbers of people, thereby lowering the overall population; examples: famines, wars, and epidemics.

Positive sanctions Responses by others that encourage the individual to continue acting in a certain way.

Postindustrial societies Societies that depend on specialized knowledge to bring about continuing progress in technology.

Power The ability of an individual or group to attain goals, control events, and maintain influence over others—even in the face of opposition.

Power elite The group of people who control policy making and the setting of priorities.

Prayer A religious ritual that enables individuals to communicate with supernatural beings or forces.

Preindustrial cities Cities established before the Industrial Revolution. Those cities were usually walled for protection, and power was typically shared between feudal lords and religious leaders.

Prejudice An irrationally based negative, or occasionally positive, attitude toward certain groups and their members.

Preparatory stage According to George Herbert Mead, the stage in the development of the self characterized by the child's imitating the behavior of others, which prepares the child for learning social-role expectations.

Prestige The approval and respect an individual or group receives from other members of society.

Preventive checks Practices, described by Thomas Robert Malthus, that limit reproduction; examples: contraception, prostitution, and other vices.

Primary deviance A term used in labeling theory to refer to the original behavior that leads to the individual being labeled as deviant.

Primary group A group characterized by intimate, face-to-face association and cooperation. Primary groups involve interaction among members who have an emotional investment in one another and who interact as total individuals rather than through specialized roles.

Primary socialization The process by which children master the basic information and skills required of members of society.

Profane All empirically observable things that are knowable through ordinary everyday experiences.

Proletariat The label used by Karl Marx to describe the mass of people in society who have no resources to sell other than their labor.

Propaganda Advertisements of a political nature seeking to mobilize public support behind one specific party, candidate, or point of view.

Property crime An unlawful act that is committed with the intent of gaining property, but does not involve the use or threat of force against an individual.

Psychoanalysis The form of therapy developed by Sigmund Freud for treating mental illness.

Psychoanalytic theory A body of thought developed by Sigmund Freud that rests on two basic hypotheses: (1) Every human act has a psychological cause or basis, and (2) every person has an unconscious mind.

Public opinion The beliefs held by a dispersed collectivity of individuals about a common concern, interest, focus, or activity.

Race A category of people who are defined as similar because of a number of physical characteristics.

Radical ideologies Ideologies that seek major structural changes in society.

Random sample A sample selected purely on the basis of chance.

Reactionary social movement A social movement that embraces the aims of the past and seeks to return the general society to yesterday's values.

Real norms Norms that allow for differences in individual behavior. Real norms specify how people actually behave, not how they should behave under ideal circumstances.

Rebellions Attempts to achieve rapid political change that is not possible within existing institutions.

Rebels Individuals who reject both the goals of what to them is an unfair social order and the institutionalized means of achieving them. They propose alternative societal goals and institutions.

Recidivism Repeated criminal behavior after punishment.

Reference group A group or social category that an individual uses to help define beliefs, attitudes, and values, and to guide behavior.

Reformulation The process in which traits passed from one culture to another are modified to fit better in their new context.

Relative deprivation theory A theory that assumes social movements are the outgrowth of the feeling of relative deprivation among large numbers of people who believe they lack certain things they believe they are entitled to.

Reliability The ability to repeat the findings of a research study.

Religion A system of beliefs, practices, and philosophical values shared by a group of people that defines the sacred, helps explain life, and offers salvation from the problems of human existence.

Religious taboo A sacred prohibition against touching, mentioning, or looking at certain objects, acts, or people.

Replication Repetition of the same research procedure or experiment for the purpose of determining whether earlier results can be duplicated.

Representative government The authority to govern is achieved through, and legitimized by, popular elections.

Representative sample A sample that has the same distribution of characteristics as the larger population from which it is drawn.

Researcher bias The tendency for researchers to select data that support their hypothesis and to ignore data that appear to contradict it.

Research process A sequence of steps in the design and implementation of a research study, including defining the problem, reviewing previous research, determining the research design, defining the sample and collecting data, analyzing and interpreting the data, and preparing the final research report.

Resocialization An important aspect of adult socialization that involves being exposed to ideas or values that conflict with what was learned in childhood.

Resource-mobilization theory A theory that assumes social movements arise at certain times and not at others because some people know how to mobilize and channel popular discontent.

Retreatists Individuals—such as drug addicts, alcoholics, drifters, and panhandlers—who have pulled back from society altogether and who do not pursue culturally legitimate goals.

Revisionary social movement A social movement that seeks partial or slight changes within the existing order but does not threaten the order itself.

Revitalization movements Powerful religious movements that stress a return to the religious values of the past. Those movements spring up when a society is under great stress or attack.

Revolutionary social movement A social movement that seeks to overthrow all or nearly all of the existing social order and replace it with an order it considers to be more suitable.

Revolutions Relatively rapid transformations that produce change in a society's power structure.

Rites of passage Standardized rituals that mark the transition from one stage of life to another.

Ritualists Individuals who deemphasize or reject the importance of success once they realize they will never achieve it and instead concentrate on following and enforcing rules more precisely than ever was intended.

Rituals Patterns of behavior or practices related to the sacred.

Role conflict The situation in which an individual who is occupying more than one status at the same time is unable to enact the roles of one status without violating those of another.

Roles Culturally defined rules for proper behavior associated with every status.

Role sets The roles attached to a single status.

Role strain The stress that results from conflicting demands within a single role.

Rumor Information that is shared informally and spreads quickly through a mass or a crowd.

Sacred Things that are awe inspiring and knowable only through extraordinary experience. Sacred traits or objects symbolize important values.

Sample The particular subset of a larger population that has been selected for study.

Sampling A research technique in which a manageable number of subjects (a sample) is selected for study from a larger population.

Sampling error The failure to select a representative sample.

Sanctions Rewards and penalties used to regulate an individual's behavior. All external means of control use sanctions.

Sapir-Whorf hypothesis A hypothesis that argues that the language a person uses determines his or her perception of reality.

Science A body of systematically arranged knowledge that shows the operation of general laws. The term also refers to the logical, systematic methods by which that knowledge is obtained.

Scientific method The approach to research that involves observation, experimentation, generalization, and verification.

Secondary analysis The process of making use of data that has been collected by others.

Secondary deviance A term used in labeling theory to refer to the deviant behavior that emerges as a result of a person being labeled as deviant.

Secondary group A group that is characterized by an impersonal, formal organization with specific goals. Secondary groups are larger and much less intimate than are primary groups, and the relationships among members are patterned mostly by statuses and roles rather than by personality characteristics.

Sect A small religious group that adheres strictly to religious doctrine involving unconventional beliefs or forms of worship.

Sector model A modified version of the concentric zone model, in which urban groups establish themselves along major transportation arteries around the central business district.

Secularization The process by which religious institutions are confined to ever-narrowing spheres of social influence, while people turn to secular sources for moral guidance in their everyday lives.

Segregation A form of subjugation that refers to the act, process, or state of being set apart.

Selectivity A process that defines some aspects of the world as important and others as unimportant. Selectivity is reflected in the vocabulary and grammar of language.

Self The personal identity of each individual that is separate from his or her social identity.

Semistructured (open-ended) interview An interview in which the investigator asks a list of questions but is free to vary them or make up new ones that become important during the interview.

Sensate culture A term developed by Pitirim A. Sorokin to describe a culture in which people are dedicated to self-expression and the gratification of their immediate physical needs.

Sex The physical and biological differences between men and women.

Sick role The sick role legitimates the deviant behavior caused by the illness and channels the individual into the health-care system.

Significant others Those people who are most important in our development, such as parents, friends, and teachers.

Signs Objects or things that can represent other things because they share some important quality with them. A clenched fist, for example, can be a sign of anger because fists are used in physical arguments.

Small group A relative term that refers to the many kinds of social groups that actually meet together and contain few enough members so that all members know one another.

Social action Anything people are conscious of doing because of other people.

Social aggregate People who happen to be in the same place but share little else.

Social attachments The emotional bonds that infants form with others that are necessary for normal development. Social attachments are a basic need of human beings and all primates.

Social change Any modification in the social organization of a society in any of its social institutions or social roles.

Social class A category of people within a stratification system who share similar economic positions, similar lifestyles, and similar attitudes and behavior.

Social Darwinism The application of Charles Darwin's notion of "survival of the fittest" to society. Darwin believed those species of animals best adapted to the environment survived and prospered, while those poorly adapted died out.

Social evaluation The process of making qualitative judgments on the basis of individual characteristics or behaviors.

Social function A social process that contributes to the ongoing operation or maintenance of society.

Social group A number of people who have a common identity, some feeling of unity, and certain common goals and shared norms.

Social identity The statuses that define an individual. Social identity is determined by how others see us.

Social inequality The uneven distribution of privileges, material rewards, opportunities, power, prestige, and influence among individuals or groups.

Social institutions The ordered social relationships that grow out of the values, norms, statuses, and roles that organize those activities that fulfill society's fundamental needs.

Social interaction The interplay between the actions of one individual and those of one or more other people.

Socialism An economic system under which the government owns and controls the major means of production and distribution. Centralized planning is used to set production and distribution goals.

Socialization The long and complicated processes of social interactions through which a child learns the intellectual, physical, and social skills needed to function as a member of society.

Social mobility The movement of an individual or a group from one social status to another.

Social movement A form of collective behavior in which large numbers of people are organized or alerted to support and bring about, or to resist, social change.

Social organization The web of actual interactions among individuals and groups in society that defines their mutual rights and responsibilities and differs from society to society.

Social revolutions Rapid and basic transformations of a society's state and class structures that are accompanied, and in part carried through, by class-based revolts.

Social sciences All those disciplines that apply scientific methods to the study of human behavior. The social sciences include sociology, cultural anthropology, psychology, economics, history, and political science.

Social solidarity People's commitment and conformity to a society's collective conscience.

Social stratification The division of society into levels, steps, or positions that is perpetuated by the major institutions of society such as the economy, the family, religion, and education.

Social structure The stable, patterned relationships that exist among social institutions within a society.

Society A grouping of people who share the same territory and participate in a common culture.

Sociobiology An approach that uses biological and evolutionary principles to explain the behavior of social beings.

Sociological imagination The relationship between individual experiences and forces in the larger society that shape our actions.

Sociology The scientific study of human society and social interactions.

Specialization One of the two forms of adaptation. Specialization is developing ways of doing things that work extremely well in a particular environment or set of circumstances.

Spontaneous cooperation The oldest and most common form of cooperation, which arises from the needs of a particular situation.

State The institutionalized way of organizing power within territorial limits.

Statement of association A proposition that changes in one thing are related to changes in another, but that one does not necessarily cause the other.

Statement of causality A proposition that one thing brings about, influences, or changes something else.

Statistical significance A mathematical statement about the probability that some event or relationship is not due to chance alone.

Statuses The culturally and socially defined positions occupied by individuals throughout their lifetimes.

Status inconsistency Situations in which people rank differently (higher or lower) on certain stratification characteristics than on others.

Status offenses Behavior that is criminal only because the person involved is a minor.

Stratified random sample A technique to make sure that all significant variables are represented in a sample in proportion to their numbers in the larger population.

Structural conduciveness One of sociologist Neil Smelser's six conditions that shape the outcome of collective behavior. Structural conduciveness refers to the conditions within society that may promote or encourage collective behavior.

Structural strain One of Smelser's six conditions that shape the outcome of collective behavior. Structural strain refers to the tension that develops when a group's ideals conflict with its everyday realities.

Structured interview An interview with a predetermined set of questions that are followed precisely with each subject.

Subculture The distinctive lifestyles, values, norms, and beliefs of certain segments of the population within a society.

Subgroups Splinter groups within the larger group.

Subjugation The subordination of one group and the assumption of a position of authority, power, and domination by the other.

Suburbs Those territories that are part of a metropolitan statistical area but are outside the central city.

Superego In Freudian theory, one of the three separately functioning parts of the self. The superego consists of society's norms and values, learned in the course of a person's socialization, that often conflict with the impulses of the id. The superego is the internal censor.

Supernaturalism A belief system that postulates the existence of impersonal forces that can influence human events.

Survey A research method in which a population or a sample is studied in order to reveal specific facts about it.

Symbolic interactionism A theoretical approach that stresses the meanings people place on their own and one another's behavior.

Symbols Objects that represent other things. Unlike signs, symbols need not share any of the qualities of whatever they represent.

Taboo A sacred prohibition against touching, mentioning, or looking at certain objects, acts, or people.

Techniques of neutralization A process that makes it possible to justify illegal or deviant behavior.

Technological determinism The view that technological change has an important effect on a society and has an impact on its culture, social structure, and even its history.

Theism A belief in divine beings—gods and goddesses—who shape human affairs.

Theory of differential association A theory of juvenile delinquency based on the position that criminal behavior is learned in the context of intimate groups. People become criminals as a result of associating with others who engage in criminal activities.

Threatened crowd A crowd that is in a state of alarm, believing itself to be in danger.

Total institutions Environments, such as prisons or mental hospitals, in which the participants are physically and socially isolated from the outside world.

Totalitarian capitalism A political-economic system under which the government retains control of the social institutions but allows the means of production and distribution to be owned and managed by private groups and individuals.

Totalitarian government A government in which one group has virtually total control of the nation's social institutions.

Totalitarian socialism In addition to almost total regulation of all social institutions, the government controls and owns all major means of production and distribution.

Totem An ordinary object, such as a plant or animal, which has become a sacred symbol to a particular group that not only reveres the totem but identifies with it.

Tracking The stratification of students by ability, social class, and various other categories.

Tracks The academic and social levels typically assigned to and followed by the children of different social classes.

Traditional authority Power that is rooted in the assumption that the customs of the past legitimize the present.

Traditional cooperation Cooperation that is tied to custom and is passed on from one generation to the next.

Traditional ideology An ideology that tries to preserve things as they are.

Triad A small group containing three members.

Universal church A church that includes all the members of a society within one united moral community.

Urbanization A process whereby a population becomes concentrated in a specific area because of migration patterns.

Urban population The inhabitants of an urbanized area and the inhabitants of incorporated or unincorporated areas with a population of 2,500 or more.

Validity The ability of a research study to test what it was designed to test.

Value-added theory A theory that attempts to explain whether collective behavior will occur and what direction it will take.

Values A culture's general orientations toward life—its notion of what is good and bad, what is desirable and undesirable.

Variable Anything that can change (vary).

Vertical mobility Movement up or down in the social hierarchy that results in a change in social class.

Victimless crimes Acts that violate those laws meant to enforce the moral code.

Violent crime Crime that involves the use of force or the threat of force against the individual.

White-collar crime Crime committed by individuals who, while occupying positions of social responsibility or high prestige, break the law in the course of their work for illegal, personal, or organizational gain.

References

Ahlberg, Dennis A., and Carol J. De Vita. (1992, August). "New Realities of the American Family." *Population Bulletin* 47(2). Population Reference Bureau.

Albright, Joseph, and Marcia Kunstel. (1999, February 2). "Grim Odds for Chinese Girl Babies." *Atlanta Journal-Constitution,* p. 8A.

Amato, Paul, and Jacob Cheadle. (2005, February). "The Long Reach of Divorce: Divorce and Child Well-Being across Three Generations." *Journal of Marriage and Family* 67, pp. 191–206.

American Association of Medical Colleges. (2005). "Applicants, Accepted Applicants, and Matriculants by Sex, 1994–2005." Available at http://www.aamc.org/data/facts/2005/2005summary.htm, accessed December 6, 2005.

American Association of Retired People. (1994). *Images of Aging in America.* Washington, DC.

AmeriStat. (2002, December). "The Gender Gap in U.S. Mortality." Population Reference Bureau.

———. (2002, August). "Higher Education Means Lower Mortality Rates." Population Reference Bureau.

———. (2002, August). "A Century of Progress in U.S. Infant and Child Survival." Population Reference Bureau.

———. (2001). "Solitary Living on the Rise in the United States." Population Reference Bureau.

———. (2001, November). "Regional Variations in the Traditional American Household Population." Population Reference Bureau.

———. (2001, June). "Who Marked More Than One Race in the 2000 U.S. Census?" Population Reference Bureau.

———. (2000, February). "U.S. High School Dropouts: The Gender Gap." Population Reference Bureau.

Anderson, Craig A. (2003). "Violent Video Games: Myths, Facts, and Unanswered Questions." Available at http://www.apa.org/science/psa/sb-andersonprt.html, accessed October 18, 2005.

Anderson, Craig A., and Karen E. Dill. (2000). "Video Games and Aggressive Thoughts, Feelings, and Behavior in the Laboratory and in Life." *Journal of Personality and Social Psychology* 78(4), pp. 772–790.

Anderson, Robert N. (2001). "Deaths: Leading Causes for 2000." *National Vital Statistics Reports* 50(16).

Anderson, Robert N., and Betty L. Smith. (2005, March 7). "Deaths: Leading Causes for 2002." *National Vital Statistics Reports* 53(17).

Annan, Kofi A. (2001). *We the Children: Meeting the Promises of the World Summit for Children.* Available at http://www.unicef.org, accessed February 15, 2002.

Ansolabehere, Stephen, and Shanto Iyengar. (1995). *Going Negative: How Political Advertisements Shrink and Polarize the Electorate.* New York: Free Press.

Antonovsky, Aaron. (1972). "Social Class, Life Expectancy and Overall Mortality." In E. Gatly Jaco (Ed.), *Patients, Physicians and Illness,* 2nd ed., pp. 5–30. New York: Free Press.

Arenson, Karen. (2005, August 31). "SAT Math Scores at Record High, but Those on the Verbal Exam Are Stagnant." *New York Times,* p. 16A.

Aristotle. (1908). *The Politics and Economics of Aristotle.* In Edward English Walford and John Gillies (Trans.). London: G. Bell & Sons.

Armstrong, Karen. (1994). *A History of God.* New York: Ballantine Books.

Asch, Solomon. (1955). "Opinions and Social Press." *Scientific American* 193, pp. 31–35.

Associated Press. (2000, March 9). "Eleven Arrested in Fat Tuesday Celebration."

Association of American Medical Colleges. (2002). *Medical School Graduation Questionnaire.*

A. T. Kearney, Inc. (2001, January/February). "Measuring Globalization." *Foreign Policy.*

Austen, Jane. (1813). *Pride and Prejudice,* reprint ed., 1998. Oxford: Oxford UP.

Austen-Smith, David, and Roland G. Fryer, Jr. (2005, May). "An Economic Analysis of 'Acting White.'" *Quarterly Journal of Economics* 120(2), pp. 551–583.

Axell, Albert. (2002). *Kamikaze: Japan's Suicide Gods.* Boston: Longman.

Axtell, Roger E. (1998). *Gestures: The Do's and Taboos of Body Language around the World.* New York: John Wiley.

Baker, Paul J., Louis E. Anderson, and Dean S. Dorn. (1993). *Social Problems: A Critical Thinking Approach,* 2nd ed. Belmont, CA: Wadsworth.

Baldus, David C., George Woodworth, and Charles A. Pulaski, Jr. (1990). *Equal Justice and the Death Penalty: A Legal and Empirical Study.* Boston: Northeastern University Press.

Bales, R. F. (1958). "Task Roles and Social Roles in Problem-Solving Groups." In E. E. Maccoby, T. M. Newcomb, and E. L. Hartley (Eds.), *Readings in Social Psychology,* 3rd ed. New York: Holt, Rinehart and Winston.

Bandura, Albert. (1969). *Principles of Behavior Modification*. New York: Holt, Rinehart and Winston.

Barnes, Jessica S., and Claudette E. Bennett. (2002, February). *The Asian Population: 2000*. U.S. Bureau of the Census.

Barnett, Cynthia. (2002). "The Measurement of White-Collar Crime Using Uniform Crime Reporting (UCR) Data." Federal Bureau of Investigation, Criminal Justice Information Services (CJIS) Division.

Barnett, Rosalind C., and Grace K. Baruch. (1987). "Social Roles, Gender, and Psychological Distress." In Rosalind C. Barnett, Lois Biener, and Grace K. Baruch (Eds.), *Gender and Stress*. New York: Free Press.

Barnett, R. C., and N. L. Marshall. (1991). "The Relationship between Women's Work and Family Roles and Their Subjective Well-Being and Psychological Distress." In M. Frankenhaeuser, V. Lundberg, and M. Chesney (Eds.), *Women, Work, and Health: Stress and Opportunities*. New York: Plenum.

Baron, Salo W. (1976). "European Jewry before and after Hitler." In Yisrael Gutman and Livia Rochkirchen (Eds.), *The Catastrophe of European Jewry*. Jerusalem: Yad Veshem.

Bartholet, Elizabeth. (1993). *Family Bonds: The Politics of Adoptive Parenting*. Boston: Houghton Mifflin.

———. (1993, Fall). "Blood Knots: Adoption, Reproduction, and the Politics of the Family." *American Prospect* 15.

———. (1991, May). "Where Do Black Children Belong? The Politics of Race Matching in Adoption." *University of Pennsylvania Law Review*.

Bartholomew, Robert E., and Erich Goode. (2000, May/June). "Mass Delusions and Hysterias: Highlights from the Past Millennium." *Skeptical Inquirer Magazine* 24(3), p. 20.

Baskin, Barbara H., and Karen Harris. (1980). *Books for the Gifted Child*. New York: R. R. Bowker.

Bateson, Mary Catherine. (1994). *Peripheral Visions*. New York: HarperCollins.

Becker, Howard. (1963). *Outsiders: Studies in the Sociology of Deviance*. New York: Free Press.

Beckwith, Burnham P. (1986, July/August). "Religion: A Growing or Dying Institution?" *The Futurist* 20(4), pp. 24–25.

Bedau, Hugo Adam. (Ed.). (1997). *The Death Penalty in America: Current Controversies*. New York: Oxford University Press.

Begley, Sharon. (2005, February 4). "People Believe a 'Fact' That Fits Their Views Even If It's Clearly False." *Wall Street Journal*, p. B1.

Bell, Daniel. (1973). *The Coming of the Postindustrial Society: A Venture in Social Forecasting*. New York: Basic Books.

Benedict, Jeffrey. (1997). *Public Heroes, Private Felons: Athletes and Crimes against Women*. Boston: Northeastern University Press.

Benedict, Jeffrey, and Alan Klein. (1997). "Arrest and Conviction Rates for Athletes Accused of Sexual Assault." *Sociology of Sport Journal* 14, pp. 86–94.

Benedict, Ruth. (1961/1934). *Patterns of Culture*. Boston: Houghton Mifflin.

———. (1938). "Continuities and Discontinuities in Cultural Conditioning." *Psychiatry* 1, pp. 161–167.

Berger, Peter. (1967). *The Sacred Canopy*. New York: Doubleday.

———. (1963). *Invitation to Sociology: A Humanistic Perspective*. New York: Doubleday.

Berger, Suzanne E. (1996). *Horizontal Woman*. Boston: Houghton Mifflin.

Bergman, Mike. (2004, June 29). "High School Graduation Rates Reach All-Time High; Non-Hispanic White and Black Graduates at Record Levels." U.S. Bureau of the Census.

Bernard, L. L. (1924). *Instinct*. New York: Holt, Rinehart and Winston.

Berry, Brewton, and Henry L. Tischler. (1978). *Race and Ethnic Relations*, 4th ed. Boston: Houghton Mifflin.

Berry, Brian J., and John D. Kasarda. (1977). *Contemporary Urban Ecology*. New York: Macmillan.

Bertalanffy, Ludwig von. (1968). *General System Theory*. New York: George Braziller.

Besharov, Douglas J. (1999, July 14). "Asking More from Matrimony." *New York Times*.

Best, Joel. (2002). "Monster Hype." *Education Next*, Washington, DC, Hoover Institution. Available at http://www.educationnext.org/20022/50.html#fig2, accessed January 9, 2006.

———. (2002, Summer). "Monster Hype: How a Few Isolated Tragedies—and Their Supposed Causes—Were Turned into a National 'Epidemic.'" *Education Next* 2, pp. 50–55.

———. (2001). *Damned Lies and Statistics*. Berkeley: University of California Press.

Bianchi, Suzanne. (1991, March). "Family Disruption and Economic Hardship." U.S. Bureau of the Census, Series P-70, No. 23.

Bianchi, Suzanne, and Lynne M. Casper. (2000). "American Families." *Population Bulletin* 55(4). Population Reference Bureau.

Bierstadt, Robert. (1974). *The Social Order*, 4th ed. New York: McGraw-Hill.

Billings, John S. (1891, February). "Public Health and Municipal Government." *Annals of the American Academy of Political and Social Science* (Supplement).

Black, Dan, Gary Oates, Seth Sanders, and Lowell Taylor. (2000, May). "Demographics of the Gay and Lesbian Population of the United States: Evidence from Available Systematic Data Sources." *Demography* 37, pp. 139–154.

Blank, Jonah. (1992). *Arrow of the Blue-Skinned God*. Boston: Houghton Mifflin.

Blankenhorn, David. (1995). *Fatherless America*. New York: Basic Books.

Blau, Peter M. (1964). *Exchange and Power in Social Life*. New York: John Wiley.

Blendon, Robert J., John M. Benson, Mollyann Brodie, Drew E. Altman, Richard Moran, Claudia Deane, and Nina Kjellson. (1999, Spring). "The 60s and the 90s: Americans' Political, Moral, and Religious Values Then and Now." *Brookings Review*, pp. 14–17.

Blum, John M., Edmund S. Morgan, Willie Lee Rose, Arthur Schlesinger, Jr., Kenneth M. Stamp, and C. Van Woodard. (1981). *The National Experience: A History of the United States*, 5th ed. New York: Harcourt Brace Jovanovich.

Blumer, Herbert. (1946). "Collective Behavior." In Alfred McClung Lee (Ed.), *Principles of Sociology*. New York: Barnes & Noble.

Blumstein, Phillip, and Pepper Schwartz. (1983). *American Couples*. New York: William Morrow.

Bonczar, Thomas P. (2003, August). Bureau of Justice Statistics, Special Report, "Prevalence of Imprisonment in the U.S. Population, 1974–2001." NCJ 197976.

Borjas, George J. (1999). *Heaven's Door*. Princeton, NJ: Princeton University Press.

Bornstein, Kate. (1994). *Gender Outlaw*. New York: Routledge.

Bowles, Samuel, and Herbert Gintis. (1976). *Schooling in Capitalist America: Educational Reform and the Contradictions of Economic Life*. New York: Basic Books.

Bradsher, Keith. (1995, August 14). "Low Ranking of Poor American Children." *New York Times*, p. A9.

Bramlett, M. D., and W. D. Mosher. (2002). "Cohabitation, Marriage, Divorce, and Remarriage in the United States." *Vital Health Statistics* 23(22). National Center for Health Statistics.

Braus, Patricia. (1994, November). "How Women Will Change Medicine." *American Demographics*, pp. 40–47.

Brehm, Sharon. (1992). *Intimate Relationships*, 2nd ed. New York: Random House.

Brockerhoff, Martin P. (2000). *An Urbanizing World*. Population Reference Bureau.

Brotz, Howard. (1966). *Negro Social and Political Thought 1850–1920*. New York: Basic Books.

Brown, Dee. (1971). *Bury My Heart at Wounded Knee*. New York: Holt, Rinehart and Winston.

Brown, Lester R. (1974). *In the Human Interest*. New York: Norton.

Brown, Roger W. (1954). "Mass Phenomena." In Gardner Lindzey (Ed.), *Handbook of Social Psychology*. Cambridge, MA: Addison-Wesley.

Bullough, Vern L. (1973). *The Subordinate Sex*. Chicago: University of Chicago Press.

Bumpass, Larry L. (1991). "What's Happening to the Family? Interactions between Demographic and Institutional Change." *Demography* 27(4), p. 485E.

Bumpass, Larry, and Lu Hsien-Hen. (2000). "Trends in Cohabitation and Implications for Children's Family Contexts in the U.S." *Population Studies* 54, pp. 29–41.

Bumpass, Larry, R. K. Raley, and J. A. Sweet. (1995). "The Changing Character of Stepfamilies: Implications of Cohabitation and Nonmarital Childbearing." *Demography* 32, pp. 425–436.

Bumpass, Larry L., James A. Sweet, and Andrew Cherlin. (1991, November). "The Role of Cohabitation in Declining Rates of Marriage." *Journal of Marriage and the Family* 53(4), pp. 913–927.

Burchell, Robert W., et al. (1999). *The Costs of Sprawl—Revisited*. Transit Cooperative Research Program (TCRP) Report 39, National Academy of Sciences, Washington DC.

Bureau of the Census. (2005). Current Population Survey, 1960 to 2005 Annual Social and Economic Supplements.

———. (2005, October 11). Public-Use Microdata Sample (PUMS). Available at http://www.census.gov/main/www/pums.html, accessed November 8, 2005.

———. (2005, August 30). Historical Poverty Tables, Table 3. "Poverty Status of People, by Age, Race, and Hispanic Origin: 1959 to 2004." Housing and Household Economic Statistics Division. Available at http://www.census.gov/hhes/www/poverty/histpov/hstpov3.html, accessed August 31, 2005.

———. (2005, August 30). "Income Stable, Poverty Rate Increases, Percentage of Americans without Health Insurance Unchanged." Available at http://www.census.gov/Press-Release/www/releases/archives/income_wealth/005647.html, accessed January 9, 2006.

———. (2005, June 29). "Estimated Median Age at First Marriage, by Sex: 1890 to the Present." Available at http://www.census.gov/population/socdemo/hh-fam/ms2.pdf, accessed October 18, 2005.

———. (2005, June 9). "Hispanic Population Passes 40 Million, Census Bureau Reports." Available at http://www.census.gov/Press-Release/www/releases/archives/population/005164.html, accessed January 9, 2006.

———. (2005, January). Current Population Survey.

———. (2004, November). Current Population Survey.

———. (2004, March). International Data Base. Available at http://www.census.gov/ipc/www/idbnew.html, accessed January 9, 2006.

———. (2003). Current Population Survey, Annual Social and Economic Supplement.

———. (2002). *Statistical Abstract of the United States: 2001,* 121st ed. Washington, DC: U.S. Government Printing Office.

———. (2002, August). "Interracial Tables." Available at http://www.census.gov, accessed January 9, 2006.

———. (2002, June 4). "Number of Foreign-Born Up 57 Percent since 1990, According to Census 2000."

———. (2001). *Statistical Abstract of the United States: 2000,* 120th ed. Washington, DC: U.S. Government Printing Office.

———. (2001, December). "Profile of the Foreign Born Population in the United States: 2000." Available at http://www.census.gov, accessed January 9, 2006.

———. (2001, October). Population Estimates Program. Washington, DC: U.S. Government Printing Office.

———. (2001, September). "Nation's Household Income Stable in 2000, Poverty Rate Virtually Equals Record Low, Census Bureau Reports." Press release, available at http://www.census.gov, accessed January 9, 2006.

———. (2001, August). *The Black Population: 2000.* Census 2000 brief. Available at http://www.census.gov, accessed January 9, 2006.

———. (2001, May). *The Hispanic Population: 2000.* Census 2000 brief. Available at http://www.census.gov, accessed January 9, 2006.

———. (2000). *Statistical Abstract of the United States: 1999,* 119th ed. Washington, DC: U.S. Government Printing Office.

———. (2000, March). "The Foreign Born Population in the United States, March 2000." Available at http://www.census.gov, accessed January 9, 2006.

———. (1999). "Resident Population of the United States: Middle Series Projections. 2015–2030 by Age and Sex." Available at http://www.census.gov/population/projections/nation/nas/npas1530.txt, accessed February 12, 2001.

———. (1999, October). "Region of Birth a Key Indicator of Well-Being for America's Foreign-Born Population, Census Bureau Reports." Press release, available at http://www.census.gov, accessed November 3, 2000.

———. (1999, March). "Educational Attainment in the United States," pp. 205–228.

———. (1998, March). *World Population Profile: 1998.* Washington, DC: U.S. Government Printing Office.

———. (1997a). *Statistical Abstract of the United States: 1997,* 117th ed. Washington, DC: U.S. Government Printing Office.

———. (1997b). *World Population Profile: 1996.* Washington, DC: U.S. Government Printing Office.

———. (1996). *Statistical Abstract of the United States: 1996,* 116th ed. Washington, DC: U.S. Government Printing Office.

———. (1995, April). "Housing of American Indians on Reservations." Available at http://www.census.gov, accessed March 10, 1998.

———. (1991). *Statistical Abstract of the United States: 1990,* 110th ed. Washington, DC: U.S. Government Printing Office.

———. (1981). *Statistical Abstract of the United States: 1980.* Washington, DC: U.S. Government Printing Office.

———. (1976). *Historical Statistics of the United States: Colonial Times to 1970.* Washington, DC: U.S. Government Printing Office.

Bureau of Justice Statistics. (2005). "Nation's Prison and Jail Population Grew by 932 Inmates per Week, Number of Female Inmates Reached More Than 100,000." Available at http://www.ojp.usdoj.gov/bjs, accessed July 21, 2005.

———. (2005). "National Crime Victimization Survey, 2004." Available at http://www.ojp.usdoj.gov/bjs/pub/pdf/cv04.pdf, accessed October 28, 2005.

———. (2002). *Sourcebook of Criminal Justice Statistics 2001,* 29th ed. U.S. Department of Justice. Available at http://www.albany.edu/sourcebook/, accessed March 8, 2004.

———. (2002). "Capital Punishment Statistics 2001, summary findings." U.S. Department of Justice.

———. (2002, September). "The Nation's Two Crime Measures." U.S. Department of Justice press release.

———. (2002, June). "Two-thirds of Former State Prisoners Rearrested for Serious New Crimes." U.S. Department of Justice press release.

———. (1997). *Criminal Victimizations in the United States, 1996.* U.S. Department of Justice.

Bureau of Labor Statistics. (2005). U.S. Department of Labor, *Occupational Outlook Handbook, 2004–05 Edition,* Childcare Workers. Available at http://www.bls.gov/oco/ocos170.htm, accessed August 09, 2005.

———. (2002). "Occupational Outlook Quarterly." *Monthly Labor Review,* Winter 2001–02, U.S. Department of Labor.

———. (2001). Occupational Employment Projections to 2010." *Monthly Labor Review,* November 2001, U.S. Department of Labor.

———. (2000, May). "Highlights of Women's Earnings in 1999." Report 943.

Buss, David M. (1994). *The Evolution of Desire.* New York: Basic Books.

Buttner, E. H., and M. McEnally. (1996). "The Interactive Effect of Influence Tactic. Applicant Gender and Type of Job on Hiring Recommendations." *Sex Roles* 34, pp. 581–591.

Bylinsky, Gene. (1988, July 19). "Technology in the Year 2000." *Fortune,* pp. 92–98.

Califano, Joseph A., Jr. (1994). *Radical Surgery.* New York: Random House.

Cameron, William Bruce. (1966). *Modern Social Movements: A Sociological Outline.* New York: Random House.

Canetti, Elias. (1978/1960). *Crowds and Power.* New York: Seabury Press.

Capizzano, Jeffrey, and Regan Main. (2005, March 31). "Many Young Children Spend Long Hours in Child Care." Washington, DC: Urban Institute.

Caplow, Theodore, Louis Hicks, and Ben J. Wattenburg. (2001). *The First Measured Century: An Illustrated Guide to Trends in America. 1900–2000.* Washington, DC: American Enterprise Institute.

Cardozo Law Innocence Project. Available at http://www.cardozo.yu.edu/innocence_project/, accessed January 9, 2006.

Carli, L. L. (2001). "Gender and Social Influence." *Journal of Social Issues* 57(4), pp. 725–737.

Carli, L. L., S. J. LaFleur, and C. C. Loeber. (1995). "Nonverbal Behavior, Gender, and Influence." *Journal of Personality and Social Psychology* 68(6), pp. 1030–1041.

Carlson, Darren K. (2001, February 14). "Over Half of Americans Believe in Love at First Sight." Princeton, NJ: Gallup News Service.

Carr, P., A. Ash, R. Friedman, et al. (2000). "Faculty Perceptions of Gender Discrimination and Sexual Harassment in Academic Medicine." *Annals of Internal Medicine* 132, pp. 889–896.

Carty, Win. (1994, October). "Population Lingo Can Push 'Hot Buttons.'" *Population Today* 3.

Casper, Lynne M., and Philip Cohen. (2000, May). "How Does POSSLQ Measure Up? Historical Estimates of Cohabitation." *Demography* 37(2), pp. 237–245.

Cassidy, Tina. (1999, September 28). "Job Complaint Recalls 'Racial Charade.'" *Boston Globe,* pp. B1, B5.

"Caste." (2002). *The Columbia Encyclopedia,* 7th ed.

Cato Institute. (2000, December 12). "China's One-Child Policy." *Cato Daily Dispatch.*

CBS News. (2005, April 20). "Faithful See Image of Virgin Mary." Available at http://www.cbsnews.com/stories/2005/04/20/national/main689630.shtml, accessed January 9, 2006.

CBS News Poll. (2005, July 13–14). "Poll: More Concerned about Terror." Available at http://www.cbsnews.com/stories/2005/07/15/opinion/polls/main709488_page2.shtml?CMP=ILC-Search Stories, accessed January 9, 2006.

Centers for Disease Control and Prevention. (2005). "Life Expectancy Hits Record High, Gender Gap Narrows." Available at http://www.cdc.gov/nchs/pressroom/05facts/lifeexpectancy.htm, accessed January 9, 2006.

Center for Surrogate Parenting. (2003). "Financial Considerations." Available at http://www.creatingfamilies.com/surrogacy/smfin.html, accessed February 11, 2005.

Centerwall, B. (1992). "Television and Violence: The Scale of the Problem and Where to Go from Here." *Journal of the American Medical Association* 267, pp. 3059–3061.

Chaiken, Jan M., and Marcia R. Chaiken. (1982). *Varieties of Criminal Behavior.* Santa Monica, CA: RAND Corporation.

Chalmers, David. (1980, August). "The Rise and Fall of the Invisible Empire of the Ku Klux Klan." *Contemporary Review* 237, pp. 57–64.

Chambliss, William J. (1973). "Elites and the Creation of Criminal Law." In William J. Chambliss (Ed.), *Sociological Readings in the Conflict Perspective.* Reading, MA: Addison-Wesley.

Chandler, David L. (1992, November 2). "Polling: The Methods behind the Madness." *Boston Globe,* pp. 35, 37.

Chaudhry, L. (2000, May 23). "Hate Sites Bad Recruiting Tools." *Wired News.*

Cherlin, Andrew J. (1992). *Marriage, Divorce, and Remarriage.* Cambridge, MA: Harvard University Press.

Cherlin, Andrew J., and Frank F. Furstenberg, Jr. (1994, January 1). "Stepfamilies in the United States: A Reconsideration." *Annual Review of Sociology* 20.

Chomsky, Noam. (1975). *Language and Mind.* New York: Harcourt Brace Jovanovich.

Chorover, Stephan L. (1979). *From Genesis to Genocide: The Meaning of Human Nature and the Power of Behavior Control.* Cambridge, MA: MIT Press.

CIA World Factbook. (2005, July). "Rank Order—Population." Available at http://www.cia.gov/cia/publications/factbook/rankorder/2119rank.html, accessed January 9, 2006.

Clark, Charles M. A. (2002. June). "Wealth and Poverty: On the Social Creation of Scarcity." *Journal of Economic Issues* 36(2).

Clark, Robert D., and Elizabeth Hatfield. (1989). "Gender Differences in Receptivity to Sexual Offers." *Journal of Psychology and Human Sexuality* 2, pp. 39–55.

Clarke-Stewart, Alison, and Virginia D. Allhusan. (2005). *What We Know about Childcare*. Cambridge, MA: Harvard University Press.

Cleland, John G., and Jerome K. Van Ginneken. (1988). "Maternal Education and Child Survival in Developing Countries: The Search for Pathways of Influence." *Social Science and Medicine* 27(12), pp. 357–368.

Cloninger, Dale. O., and Roberto Marchesini. (2001). "Execution and Deterrence: A Quasi-Controlled Group Experiment." *Applied Economics* 35(5), pp. 569–576.

CNN News. (2000, May 17). "British Soccer Fans, Danish Police Clash."

Cohen, Patricia. (2000, April 8). "Oops, Sorry: Seems That My Pie Chart Is Half-Baked." Available at http://facstaff.uww.edu/mohanp/junkscience.html, accessed February 3, 2003.

Cohen, Yehudi A. (1981, January/February). "Shrinking Households." *Society* 18(2), p. 51.

Cohn, D'Vera. (2001, August 22). "Count of Gay Couples up 300%; 2000 Census Ranks DC, Arlington, Alexandria among Top Locales." *Washington Post*, p. A3.

Colapinto, John. (2000). *As Nature Made Him*. New York: HarperCollins.

Coleman, James S. (1977). *Parents, Teachers, and Children*. San Francisco: San Francisco Institute for Contemporary Studies.

———. (1966). *Equality of Educational Opportunity*. Washington, DC: U.S. Government Printing Office.

Collins, Randall. (1979). *The Credential Society: An Historical Sociology of Education and Stratification*. New York: Academic Press.

———. (1975). *Conflict Sociology: Toward an Explanatory Science*. New York: Academic Press.

Comte, Auguste. (1968/1851). *System of Positive Policy*, Vol. 1 (John Henry Bridges, Trans.). New York: Burt Franklin.

Congressional Budget Office. (2005). "PoliticalMoneyLine.com Senate Office of Public Records." Available at http://www.politicalmoneyline.com/cgi-win/lb_directory.exe?DoFn, accessed December 12, 2005.

Conrad, Peter. (1992). "Medicalization and Social Control." *Annual Review of Sociology* 18, pp. 209–232.

Conrad, Peter, and Rochelle Kern. (Eds.). (1986). *The Sociology of Health and Illness*, 2nd ed. New York: St. Martin's Press.

Cook, Phillip, and Jens Ludwig. (1997). "Weighing the 'Burden of Acting White': Are There Race Differences in Attitudes towards Education?" *Journal of Public Policy and Analysis* 16, pp. 256–278.

Cooley, C. H. (1909). *Social Organization*. New York: Scribner's.

Cooper, Edith Fairman. (2005). "Missing and Exploited Children: Overview and Policy Concerns." CRS Report for Congress, April 29, 2005, Washington, DC: Congressional Research Service, Library of Congress.

Corcoran, Paul E. (1994). "Presidential Concession Speeches: The Rhetoric of Defeat." *Political Communication* 11, pp. 109–131.

Cornell, Vincent J. (2001). "Islam." *World Book Online Americas Edition*. Available at http://www.aolsvc.worldbook.aol.com/wbol/wbPage/na/ar/co/282380, accessed May 23, 2003.

"Corruption in the Higher Education System of Kazahkistan." (June 2002). KIMEP Times No. 7. Kazahkistant Institute of Management Economics and Strategic Research.

Cortada, J. W. (2002). *Making the Information Society*. Upper Saddle River, NJ: Prentice Hall.

Cose, Ellis. (1997). "Census and the Complex Issue of Race." *Commentary* 34(6), pp. 9–13.

Coser, L. A. (1977). *Masters of Sociological Thought*, 2nd ed. New York: Harcourt Brace Jovanovich.

———. (1967). *Continuities in the Study of Social Conflict*. New York: Free Press.

———. (1956). *The Functions of Social Conflict*. Glencoe, IL: Free Press.

"Covenant Marriage." (1999, May 7). *Issues and Controversies on File*, pp. 187–189.

Crano, William D., and Joel Aronoff. (1978, August). "A Cross-Cultural Study of Expressive and Instrumental Role Complementarity in the Family." *American Sociological Review* 43, pp. 463–471.

Crichton, Judy. (1998). *America 1900*. New York: Henry Holt.

Crosby, F. J. (1991). *Juggling: The Unexpected Advantages of Balancing Career and Home for Women and Their Families*. New York: Free Press.

Crossen, Cynthia. (2003, March 5). "Even Women Didn't Want to Give Women the Vote." *Wall Street Journal*, p. B1.

———. (1994). *Tainted Truth: The Manipulation of Fact in America*. New York: Simon & Schuster.

Cummings, Milton C., and David Wise. (1981). *Democracy under Pressure: An Introduction to the American Political System*, 4th ed. New York: Harcourt Brace Jovanovich.

Current Population Reports. (2004, June). "High School Graduation Rates Reach All-Time High; Non-Hispanic White and Black Graduates at Record Levels." *Educational Attainment in the United States: 2003*. Washington, DC: U.S. Census Bureau.

Current Population Survey. (2002). *Annual Demographic Survey,* March Supplement, 1960–2002. Washington, DC: U.S. Census Bureau.

———. (2001). *Annual Demographic Survey,* March Supplement, U.S. Bureau of the Census.

———. (2000, March). "Poverty Status of People in Families by Type of Family, Age of Householder, and Number of Children: 1999." *Annual Demographic Survey,* U.S. Bureau of the Census.

———. (1998, March). "Marital Status and Living Arrangements." U.S. Bureau of the Census.

Curtiss, Susan. (1977). *Genie: A Psycholinguistic Study of a Modern-Day Wild Child.* New York: Academic Press.

Cuzzort, R. P., and E. W. King. (1980). *Twentieth Century Social Thought,* 3rd ed. New York: Holt, Rinehart and Winston.

Dahrendorf, R. (1959). *Class and Conflict in Industrial Society.* Stanford, CA: Stanford University Press.

———. (1958, September). "Out of Utopia: Toward a Reorientation of Sociological Analysis." *American Journal of Sociology* 64, pp. 158–164.

Dalton, Trumbo. (1971). *Johnny Got His Gun.* New York: Bantam.

Daly, M., and M. Wilson. (1988). *Homicide.* Hawthorne, NY: Aldine de Gruyter.

D'Andrade, Roy G. (1966). "Sex Differences and Cultural Institutions." In Eleanor Emmons Maccoby (Ed.), *The Development of Sex Differences.* Stanford, CA: Stanford University Press.

Darwin, Charles. (1964/1859). *On the Origin of Species.* Cambridge, MA: Harvard University Press.

Davern, Michael E., and Patricia Fisher. (2001, February). *Household Net Worth and Ownership.* U.S. Bureau of the Census.

Davis, F. James. (1979). *Understanding Minority-Dominant Relations.* Arlington Heights, IL: AHM Publishing.

Davis, Kingsley. (1949). *Human Society.* New York: Macmillan.

———. (1940). "Extreme Social Isolation of a Child." *American Journal of Sociology* 45, pp. 554–565.

Davis, Kingsley, and W. E. Moore. (1945). "Some Principles of Stratification." *American Sociological Review* 10, pp. 242–249.

Deacon, Terrence W. (1997). *The Symbolic Species.* New York: Norton.

de Beauvoir, Simone. (1972). *The Second Sex* (H. M. Parshley, Trans.). New York: Penguin.

DeNavas-Walt, Carmen, and Robert Cleveland. (2002, September). *Money Income in the United States, 2001.* U.S. Bureau of the Census.

DeNavas-Walt, Carmen, Bernadette D. Proctor, and Cheryl Hill Lee. (2005). *Income, Poverty, and Health Insurance Coverage in the United States: 2004.* Current Population Reports, P60–229. Washington, DC: U.S. Census Bureau.

Department of Health and Human Services. (2005). "Cases of HIV Infection and AIDS: 2004." *Centers for Disease Control and Prevention HIV Surveillance Report* 16.

Dershowitz, Alan M. (1996). *Reasonable Doubts: The O.J. Simpson Case and the Criminal Justice System.* New York: Simon & Schuster.

Dezhbakhsh, Hashem, Paul H. Rubin, and Joanna M. Shepherd. (2002, January). "Does Capital Punishment Have a Deterrent Effect? New Evidence from Post-moratorium Panel Data." Department of Economics, Emory University.

Diamond, Milton, and H. Keith Sigmundson. (1997, March). "Sex Reassignment at Birth: A Long Term Review and Clinical Implications." *Archives of Pediatric and Adolescent Medicine* 151, pp. 298–304.

Dinan, T. G. (1996). "Serotonin: Current Understanding and the Way Forward." *International Clinical Psychopharmacology* 11(1) Supplement, pp. 19–21.

Ditton, Paula M., and Doris Wilson. (1999). *Truth in Sentencing in State Prisons.* U.S. Department of Justice, Bureau of Justice Statistics.

Domhoff, G. William. (1983). *Who Rules America Now?* Englewood Cliffs, NJ: Prentice Hall.

Doolittle, Teri. (1995). "The Long Term Effects of Institutionalization on the Behavior of Children from Eastern Europe and the Former Soviet Union." Available at http://www.mariaschildren.org/english/babyhouse/effects.html, accessed January 9, 2006.

Dowd, Maureen. (2002, April 10). "The Baby Bust." *New York Times,* p. 27A.

Dreifus, Claudia. (2005, May 31). "Declaring with Clarity, When Gender Is Ambiguous." *New York Times,* p. 2F.

Droege, Kristen. (1995, Winter). "Child Care: An Educational Perspective." *MIJCF, Jobs and Capital* 4, pp. 1–8.

DuBois, W.E.B. (1968). *Autobiography: A Soliloquy on Viewing My Life from the Last Decade of Its First Century.* New York: International Publishers.

Dugger, Celia W. (2001, May 6). "Modern Asia's Anomaly: The Girls Who Don't Get Born." *New York Times,* p. 4.

Duhart, Detis T. (2000). *Urban, Suburban, and Rural Victimization, 1993–98.* U.S. Department of Justice, Bureau of Justice Statistics.

Duncan, Greg J., and Saul D. Hoffman. (1988, November). "What Are the Economic Consequences of Divorce?" *Demography* 25(4), p. 641.

Durant, Will. (1954). "Our Oriental Heritage." *The Story of Civilization* (Vol. 1). New York: Simon & Schuster.

Durkheim, Émile. (1961/1915). *The Elementary Forms of Religious Life.* New York: Collier Books.

———. (1960/1893). *The Division of Labor in Society* (G. Simpson, Trans.). New York: Free Press.

———. (1958/1895). *The Rules of Sociological Method.* Glencoe, IL: Free Press.

Dwork, Deborah, and Robert Jan van Pelt. (2002). *Holocaust: A History.* New York: Norton.

Eberstadt, Nicholas. (1992, January 20). "America's Infant Mortality Problem: Parents." *Wall Street Journal,* p. A14.

Eck, Diana L. (2001). *A New Religious America: How a "Christian Country" Has Become the World's Most Religiously Diverse Nation.* New York: HarperCollins.

Edmondson, Brad. (1996, October). "How to Spot a Bogus Poll." *American Demographics* 18(10), pp. 10, 12–15.

Edwards, D. H., and E. A. Kravitz. (1997). "Serotonin. Social Status and Aggression." *Current Opinions in Neurobiology* 7(6), pp. 812–819. Review.

Eggdonation.com. The Egg Donor Program. Available at http://www.eggdonation.com/SampleDonor/Donor1.asp?DonorID=1026, accessed January 9, 2006.

Ehrlich, Isaac. (1975). "The Deterrent Effect of Capital Punishment: A Question of Life and Death." *American Economic Review* 65(3), pp. 397–417.

———. (1974). *The End of Affluence.* New York: Simon & Schuster.

Ehrlich, Paul, and Ann H. Ehrlich. (1974). *Population, Resource, Environment.* San Francisco: W. H. Freeman.

Ekman, Paul, William V. Friesen, and John Bear. (1984, May). "The International Language of Gestures." *Psychology Today,* pp. 64–69.

El-Badry, Samia. (1994, January). "Understanding Islam in America." *American Demographics,* pp. 10–11.

Elkind, David. (1981). *The Hurried Child.* Reading, MA: Addison-Wesley.

Ellen, Elizabeth Fried. (2002, August). "Identifying and Treating Suicidal College Students." *Psychiatric Times* 19(8).

Ellis, Joseph J. (1997). *American Sphinx: The Character of Thomas Jefferson.* New York: Knopf.

Elshtain, Jean Bethke. (2001). *Jane Addams and the Dream of American Democracy.* New York: Basic Books.

Ember, Carol R., and Melvin Ember. (1981). *Anthropology,* 3rd ed. Englewood Cliffs, NJ: Prentice Hall.

Embree, Edwin R. (1967/1939). *Indians of the Americas.* Boston: Houghton Mifflin.

Enard, Wolfgang, et al. (2002, April 12). "Intra- and Interspecific Variation in Primate Gene Expression Patterns." *Science* 296, pp. 340–344.

Engels, Friedrich. (1973/1845). *The Condition of the Working Class in England in 1844.* Moscow: Progress.

———. (1942/1884). *The Origin of the Family, Private Property and the State.* New York: International Publishers.

Erikson, Erik H. (1968). *Identity, Youth and Crisis.* New York: Norton.

———. (1964). *Childhood and Society.* New York: Norton.

Erikson, Kai T. (1966). *Wayward Puritans: A Study in the Sociology of Deviance.* New York: John Wiley.

Erkel, R. Todd. (1994, July/August). "The Mighty Wedge of Class." *Family Therapy Networker.*

Exline, Christopher H., Gary L. Peters, and Robert P. Larkin. (1982). *The City: Patterns and Processes in the Urban Ecosystem.* Boulder, CO: Westview Press.

Federal Bureau of Investigation. (2005). *Uniform Crime Reports: Supplementary Homicide Reports, 1976–2002.*

———. (2001). *Uniform Crime Reports: Crime in the United States 2001.* Available at http://www.fbi.gov/ucr/01cius.htm, accessed March 18, 2003.

Federal Communications Commission. (2005). "The V-Chip: Putting Restrictions on What Your Children Watch." Available at http://www.fcc.gov/cgb/consumerfacts/vchip.html, accessed October 18, 2005.

Federal Interagency Forum on Aging-Related Statistics. (2000). *Older Americans 2000: Key Indicators of Well-Being.* Washington, DC: U.S. Government Printing Office.

Feldman, M. D., and J. M. Feldman. (1998). *Stranger Than Fiction: When Our Minds Betray Us.* Washington, DC: American Psychiatric Press.

Fenster, M. (1993). "Genre and Form: The Development of the Country Music Video." In Simon Firth, Andrew Goodwin, and Lawrence Grossberg (Eds.), *Sound and Vision: The Music Video Reader.* New York: Routledge.

Ferree, Myra Marx, and Beth B. Hess. (1985). *Controversy and Coalition: The New Feminist Movement.* Boston: G. K. Hall.

Festinger, Leon, Henry W. Rieken, and Stanley Schacter. (1956). *When Prophesy Fails.* New York: Harper Torchbooks.

Field, T. M., F. Scafidi, R. Pickens, M. Prodromidis, M. Pelaez-Nogueras, J. Torquati, H. Wilcox,

J. Malphurs, J. Schanberg, and C. Kuhn. (1998). "Polydrug-using Adolescent Mothers and Their Infants Receiving Intervention." *Adolescence* 33(129), pp. 117–143.

Fields, Jason. (2004). "America's Families and Living Arrangements: 2003." *Current Population Reports*, pp. 20–553.

———. (2001, May). "Living Arrangements of Children: Fall, 1996." *Current Population Reports*, pp. 70–74. Washington, DC: U.S. Census Bureau.

Fields, Jason, and Lynne M. Casper. (2001, June). "America's Families and Living Arrangements." *Current Population Reports*, P20–537. Washington, DC: U.S. Census Bureau.

Fine, Mark A. (1994, May). "An Examination and Evaluation of Recent Changes in Divorce Laws in Five Western Countries: The Crucial Role of Values." *Journal of Marriage and the Family* 56, pp. 249–263.

Firebaugh, G., and D. Sandu. (1998). "Who Supports Marketization and Democratization in Post-Communist Romania?" *Sociological Forum* 13(3), pp. 521–541.

Fischler, Stan. (1980, March 2). "Garden Security and the Return of the Bruins." *New York Times*, p. 25.

Fishbein, Diana. (2001). *Biological Perspectives in Criminology*. Belmont, CA: Wadsworth/Thomson Learning.

Fisher, Helen. (1994, October 16). "'Wilson,' They Said, 'You're All Wet!'" *New York Times Book Review*, pp. 15–17.

———. (1992). *Anatomy of Love*. New York: Ballantine Books.

Fisher, Robert. (1984). *Let the People Decide: Neighborhood Organizing in America*. Boston: G. K. Hall.

Fiske, E. B. (1988). "The Undergraduate Hispanic Experience: A Case of Juggling Two Cultures." *Change* 20(3), pp. 29–33.

Flannery, Kent V. (1968). "Archaeological Systems Theory and Early Mesopotamia." In Betty J. Meggars (Ed.), *Anthropological Archeology in the Americas*. Washington, DC: Anthropological Society of Washington.

———. (1965). "The Ecology of Early Food Production in Mesopotamia." *Science* 147, pp. 1247–1256.

Ford, Clellan S. (1970). "Some Primitive Societies." In Georgene H. Seward and Robert C. Williamson (Eds.), *Sex Roles in Changing Society*. New York: Random House.

Fortes, M., R. W. Steel, and P. Ady. (1947). "Ashanti Survey, 1945–46: An Experiment in Social Research." *Geographical Journal* 110, pp. 149–179.

Fouts, Roger. (1997). *Next of Kin: What Chimpanzees Have Taught Me about Who We Are*. New York: William Morrow.

Frankfort, H. (1956). *The Birth of Civilization in the Near East*. Garden City, NY: Doubleday/Anchor.

Fredrickson, George M. (1971). *The Black Image in the White Mind*. New York: Harper & Row.

Freedman, Samuel G. (1990). *Small Victories*. New York: Harper & Row.

Freeman, James M. (1974, January). "Trial by Fire." *Natural History*, pp. 54–63.

Freud, Sigmund. (1930). "Civilization and Its Discontents." *Standard Edition of the Complete Psychological Works of Sigmund Freud* (Vol. 29). London: Hogarth Press.

———. (1928). *The Future of an Illusion*. New York: Horace Liveright and the Institute of Psycho-Analysis.

———. (1923). "The Ego and the Id." *Standard Edition of the Complete Psychological Works of Sigmund Freud* (Vol. 19). London: Hogarth Press.

———. (1920). "Beyond the Pleasure Principle." *Standard Edition of the Complete Psychological Works of Sigmund Freud* (Vol. 14). London: Hogarth Press.

———. (1918). *Totem and Taboo*. New York: Moffat, Yard.

Frey, Darcy. (1994). *The Last Shot*. Boston: Houghton Mifflin.

Frey, William H. (2001). "Melting Pot Suburbs: A Census 2000 Study of Suburban Diversity." Washington, DC: Brookings Institution, Census 2000 Series.

———. (2000, Summer). "The New Urban Demographics: Race Space and Boomer Aging." *Brookings Review* 18(3), pp. 18–21.

Fried, Morton. (1967). *The Evolution of Political Society*. New York: Random House.

Friedrich, Carl J., and Zbigniew Brezinski. (1965). *Totalitarian Dictatorship and Autocracy* (Vol. 2). Cambridge, MA: Harvard University Press.

Fryer, Roland G., Jr., and Paul Torelli. (2005, May 1). "An Empirical Analysis of 'Acting White.'" Harvard University Society of Fellows and NBER. Available at http://post.economics.harvard.edu/faculty/fryer/papers/fryer_torelli.pdf, accessed January 9, 2006.

Fullerton, Howard N., Jr., and Mitra Toossi. (2001, November). "Labor Force Projections to 2010: Steady Growth and Changing Composition." *Monthly Labor Review*. Bureau of Labor Statistics, U.S. Department of Labor.

Galinsky, Ellen, Caroline Howe, Susan Koutos, and Marybeth Shinn. (1994). *The Study of Children in Family Child Care and Relative Care: Highlights and Findings*. New York: Families and Work Institute.

Gallup, George, Jr., and Jim Castelli. (1987). *The American Catholic People: Their Beliefs, Practices, and Values.* Garden City, NY: Doubleday.

Gallup Poll. (2002, June 18). "Fewer Blacks Say Anti-White Sentiment Is Widespread in Black Community."

———. (2001, September 15). "It's Still a Man's World." *Gallup Management Journal,* Fall 2001.

———. (2001, June 15). "As Much as Ever, in God We Trust." *Gallup Management Journal,* Summer 2001.

———. (2001, May 4). "Few Say It's Ideal for Both Parents to Work Full Time Outside of Home."

———. (2001, February 14). "Over Half of Americans Believe in Love at First Sight."

———. (2000, December 26). "When It Comes to Having Children, Americans Still Prefer Boys."

———. (2000, August 24–27). "Religion."

———. (1997). "Public Attitudes toward Public Schools."

———. (1997, November). "Family Values Differ Sharply around the World."

———. (1997, October). "Many Women Cite Spousal Abuse: Job Performance Affected."

———. (1997, March). "Religious Belief Widespread but Many Skip Church."

———. (1994, July). "The Purpose of an Education."

———. (1972). *The Gallup Poll: Public Opinion 1934–1971.* New York: Random House.

Gandhi, M. K. (1962). *All Religions Are True.* Ed. and published by Anand T. Hingorani. Bombay, India: Bharatiya Vidya Bhavan.

Gans, Herbert J. (1979, May). "Deception and Disclosure in the Field." *The Nation* 17, pp. 507–512.

———. (1977, February 12). "Why Exurbanites Won't Reurbanize Themselves." *New York Times,* p. 21.

———. (1962). *The Urban Villagers.* New York: Free Press.

Gardner, Howard. (1978). *Developmental Psychology.* Boston: Little, Brown.

———. (1972). "Studies of the Routine Grounds of Everyday Activities." In David Snow (Ed.), *Studies in Social Interaction.* New York: Free Press.

Gargan, Edward A. (1988, November 2). "Beijing Admits Easing of Birth Limits." *New York Times,* p. A3.

Garfinkel, Harold. (1972). "Conditions of Successful Degradation Ceremonies." In J. Manis and B. Meltzer (Eds.), *Symbolic Interactionism,* pp. 201–208. New York: Allyn & Bacon.

———. (1967). Studies in Ethnomethodology. Englewood Cliffs, NJ: Prentice Hall.

Geertz, Clifford. (1998). "The World in Pieces: Culture and Politics at the End of the Century." *Focaal: Tijdschrift voor Antropolgie* 32, pp. 91–117.

———. (1973). *The Interpretation of Cultures.* New York: Basic Books.

Gehrke, Robert. (2001, January 8). "Utah Case Tests American Indian Law." *Associated Press.*

Gelles, Richard J. (1996). *The Book of David.* New York: Basic Books.

Gelles, Richard J., and J. R. Conte. (1990, November). "Domestic Violence and Sexual Abuse of Children: A Review of Research in the Eighties." *Journal of Marriage and the Family* 52, pp. 1045–1058.

Gelles, Richard J., and Murray A. Straus. (1988). *Intimate Violence.* New York: Simon & Schuster.

Gerth, Hans, and C. Wright Mills. (1953). *Character and Social Structure.* New York: Harcourt Brace.

Gewertz, Deborah. (1981). "A Historical Reconsideration of Female Dominance among the Chambri of Papua New Guinea." *American Ethnologist* 8(1), pp. 94–106.

Gill, Richard T. (1991, Fall). "Day Care or Parental Care?" *Public Interest.*

Gilligan, Carol. (1982). *In a Different Voice.* Cambridge, MA: Harvard University Press.

Ginsberg, Morris. (1958). "Social Change." *British Journal of Sociology* 9(3), pp. 205–229.

Gist, Noel P., and Sylvia Fleis Fava. (1974). *Urban Society,* 6th ed. New York: Crowell.

Glazer, Nathan, and Daniel P. Moynihan. (Eds.). (1975). *Ethnicity: Theory and Experience.* Cambridge, MA: Harvard University Press.

Gnessous, Mohammed. (1967). "A General Critique of Equilibrium Theory." In Wilburt E. Moore and Robert M. Cooke (Eds.), *Readings on Social Change.* Englewood Cliffs, NJ: Prentice Hall.

Goffman, E. (1971). *Relations in Public.* New York: Basic Books.

———. (1963). *Behavior in Public Places.* New York: Free Press.

———. (1961). *Asylums: Essays on the Social Situation of Mental Patients and Other Inmates.* Chicago: Aldine.

———. (1959). *The Presentation of Self in Everyday Life.* Garden City, NY: Doubleday.

Goldhagen, Daniel. (2002). *A Moral Reckoning: The Role of the Catholic Church in the Holocaust and Its Unfulfilled Duty of Repair.* New York: Knopf.

———. (1996). *Hitler's Willing Executioners.* New York: Knopf.

Goldman, Ari L. (1991). *The Search for God at Harvard.* New York: Times Books.

Goliber, Thomas J. (1997, December). "Population and Reproductive Health in Sub-Saharan Africa." *Population Bulletin* 52(4). Population Reference Bureau.

Good, Kenneth. (1991). *Into the Heart.* New York: Simon & Schuster.

"Good News! A Record Number of Doctoral Degrees Awarded to African Americans." (2005). *Journal of Blacks in Higher Education.* Available at http://www.jbhe.com/news_views/46_blacks_doctoraldegrees.html, accessed September 13, 2005.

Goode, W. J. (1963). *World Revolution and Family Patterns.* New York: Free Press.

———. (1960, August 25). "A Theory of Role Strain." *American Sociological Review,* pp. 902–914.

Gordon, Milton M. (1975/1961). "Assimilation in America: Theory and Reality." In Norman R. Yetman and C. Hoy Steele (Eds.), *Majority and Minority: The Dynamics of Racial and Ethnic Relations.* Boston: Allyn & Bacon.

———. (1964). *Assimilation in American Life.* New York: Oxford University Press.

Gough, Kathleen. (1952). "Changing Kinship Usages in the Setting of Political and Economic Change among the Nayars of Malabor." *Journal of Royal Anthropological Institute of Great Britain and Ireland* 82, pp. 71–87.

Gould, H. (1971). *Caste and Class: A Comparative View.* Reading, MA: Addison-Wesley, pp. 1–24.

Gould, Stephen Jay. (1976, May). "The View of Life: Biological Potential versus Biological Determinism." *Natural History Magazine* 85, pp. 34–41.

Gouldner, Alvin W. (1970). *The Coming Crisis of Western Sociology.* New York: Avon.

Gowdy, V. B., et al. (1998). *Women in Criminal Justice: A Twenty Year Update.* U.S. Department of Justice, Coordination Group for Women.

Graeff, F. G., F. S. Guimaraes, T. G. De Andrade, and J. F. Deakin. (1996). "Role of 5HT in Stress, Anxiety, and Depression." *Pharmocology and Biochemistry of Behavior* 54(1), pp. 129–141.

Granovetter, Mark. (1983). "The Strength of Weak Ties: A Network Theory Revisited." *Sociological Theory* 1, pp. 201–233.

Greenberg, David. (1999, June 15). "White Weddings: The Incredible Staying Power of the Laws against Interracial Marriage." *Slate.* Available at http://slate.msn.com/id/30352/, accessed December 9, 1999.

Greenberg, J. (1980). "Ape Talk: More Than Pigeon English?" *Science News* 117(19), pp. 298–300.

Grieco, Elizabeth M., and Rachel C. Cassidy. (2001). "Overview of Race and Hispanic Origin: 2000." *Census 2000 Brief.* U.S. Census Bureau.

Guillen, Mauro F. (2001). "Is Globalization Civilizing, Destructive or Feeble? A Critique of Five Key Debates in the Social Science Literature." *Annual Review of Sociology* 27, pp. 235–260.

Gumplowicz, Ludwig. (1899). *The Outlines of Sociology.* Philadelphia: American Academy of Political and Social Sciences.

Gurr, Ted Robert. (1970). *Why Men Rebel.* Princeton, NJ: Princeton University Press.

Hall, Edward T. (1981). *The Silent Language.* New York: Doubleday.

———. (1974). *Handbook for Proxemic Analysis.* Washington, DC: Society for the Anthropology of Visual Communication.

———. (1969). *The Hidden Dimension.* Garden City, NY: Doubleday.

Hall, Edward T., and Mildred Reed Hall. (1990). *Hidden Differences.* New York: Anchor Books.

Hammer, Heather, David Finkelhor, and Andrea J. Sedlak. (2002, October). "Runaway/Thrownaway Children: National Estimates and Characteristics." Washington, DC: U.S. Department of Justice, Office of Justice Programs.

Hammons, Christopher W. (2001). "School @ Home." *Education Next,* pp. 1–10.

Hampden-Turner, Charles, and Fons Trompenaars. (2000). *Building Cross-Cultural Competence.* New Haven, CT: Yale University Press.

Hare, Paul A. (1976). *Handbook of Small Group Research,* 2nd ed. New York: Free Press.

Harlow, Harry F. (1959, June). "Love in Infant Monkeys." *Scientific American,* pp. 68–74.

Harlow, Harry F., and M. Harlow. (1962). "The Heterosexual Affectional System in Monkeys." *American Psychologist* 17, pp. 1–9.

Harris, C. D., and E. L. Ullman. (1945). "The Nature of Cities." *Annals of the American Academy of Political and Social Science* 242, p. 12.

Harris, Marvin. (1975). *Culture, People, and Nature: An Introduction to General Anthropology,* 2nd ed. New York: Crowell.

———. (1966). "The Cultural Ecology of India's Sacred Cattle." *Current Anthropology* 7, pp. 51–63.

Harris Poll. (2000). "Religious Affiliation Year 2000." Available at http://www.adherents.com/misc/poll_Harris2000.html, accessed January 9, 2006.

———. (1998, May 15–19). "Prestige of Occupations." Available at http://www.harrisinteractive.com/harris_poll/index.asp?PID=111, acccessed January 9, 2006.

Harrison, Paige M., and Allan Beck. (2005). "Prisoners in 2004." Available at http://www.ojp.usdoj.gov/bjs/pub/pdf/p04.pdf, accessed October 28, 2005.

———. (2002). *Prisoners in 2001.* Washington, DC: U.S. Department of Justice, Bureau of Labor Statistics.

Hart, Gary. (1993). *The Good Fight.* New York: Random House.

Harvard Law Review. (1993, June). "Notes." 106, pp. 1905–1925.

Hassan, S. (1988). *Combating Cult Mind Control.* Rochester, VT: Park Street Press.

"Hate Speech on the Internet." (1999, June 4). *Issues and Controversies on File.*

Hatfield, E. (1988). "Passionate and Companionate Love." In R. J. Steinberg and M. L. Barnes (Eds.), *The Psychology of Love.* New Haven, CT: Yale University Press.

Haub, Carl. (1994, December). "Population Change in the Former Soviet Republics." *Population Bulletin* 49(4). Population Reference Bureau.

Havemann, Ernest. (1967, October). "Computers: Their Scope Today." *Playboy Magazine.*

Hawes, Alex. (1995). "Machiavellian Monkeys and Shakespearean Apes: The Question of Primate Language." *Zoogoer* 24(6), pp. 1–10.

Hawkins, Sanford A., and Robert Hastie. (1990). "Hindsight: Biased Judgments of Past Events after the Outcomes Are Known." *Psychological Bulletin* 107(3), pp. 311–327.

Hawley, Amos. (1981). *Urban Society,* 2nd ed. New York: John Wiley.

Heer, David M. (1980). "Intermarriage." *Harvard Encyclopedia of American Ethnic Groups.* Cambridge, MA: Harvard University Press, pp. 513–521.

Held, D., A. McGrew, D. Goldblatt, and J. Perraton. (1999). *Global Transformations.* Stanford, CA: Stanford University Press.

Henig, Robin Marantz. (1997). *The People's Health: A Memoir of Public Health and Its Evolution at Harvard.* Washington, DC: Joseph Henry Press.

Henshaw, S. (1991). "Induced Abortion: A World Review, 1990." *Family Planning Perspectives* 22(2), pp. 76–89.

Herbers, John. (1986). *The New Heartland.* New York: Times Books.

———. (1981, May 31). "Census Finds More Blacks Living in Suburbs of Nation's Large Cities." *New York Times* 1, p. 48.

Hernandez, Peggy. (1989, July 26). "Firemen Who Claimed to Be Black Lose Appeal." *Boston Globe,* p. 13.

———. (1988, November 7). "Many Chances to Dispute Malones Firefighters' Minority Status Unchallenged for 10 Years." *Boston Globe,* p. 1.

Herring, Susan. (2000, Winter). "Gender Differences in CMC: Findings and Implications." *CPSR Newsletter* 18(1).

———. (1994). "Gender Differences in Computer-Mediated Communication: Bringing Familiar Baggage to the New Frontier." Keynote talk at panel entitled "Making the NetWork: Is There a Z39.50 in Gender Communication?" American Library Association annual convention, Miami, June 27, 1994.

Hesketh, Therese, Li Lu, and Zhu Wei Xing. (2005, September). "The Effect of China's One-Child Family Policy After 25 Years." *The New England Journal of Medicine* 353(11), pp. 1171–1176.

Hesketh, Therese, and Zhu, Wei Xing. (1997). "The One Child Policy: The Good, the Bad, and the Ugly." *British Medical Journal* 314(7095).

Hetherington, E. Mavis. (2002, April 8). "Marriage and Divorce American Style." *American Prospect.*

Hetherington, E. Mavis, and John Kelly. (2002). *For Better or for Worse: Divorce Reconsidered.* New York: Norton.

Higher Education Research Institute. (2004). "The American Freshman: National Norms for Fall 2004." Los Angeles: University of California.

———. (2001). "College Freshmen More Politically Liberal Than in the Past, UCLA Survey Reveals." Available at http://www.gseis.ucla.edu, accessed July 11, 2002.

Himes, Christine L. (2001, December). "The Elderly American." *Population Bulletin* 56(4). Population Reference Bureau.

Hingson, Ralph W. (1998, January/February). "College-Age Drinking Problems." *Public Health Reports,* p. 113.

Hirschi, Travis. (1969). *Causes of Delinquency.* Berkeley: University of California Press.

Hirschi, Travis, and Michael Gottfredson. (1993). "Commentary: Testing the General Theory of Crime." *Journal of Research in Crime and Delinquency* 30, pp. 47–54.

Hirst, P., and G. Thompson. (1996). *Globalization in Question.* London: Polity.

Hitlin, Paul, and Lee Rainie. (2005, August). "Teen Use of the Internet at School Has Grown 45% since 2000." Washington, DC: Pew Internet and American Life Project. Available at http://www.pewinternet.org, accessed January 9, 2006.

Hochschild, Arlie Russell. (1997). *Time Bind: When Work Becomes Home and Home Becomes Work.* New York: Henry Holt.

———. (1997, April 20). "There's No Place Like Work." *New York Times Magazine,* pp. 51–55, 81, 84.

Hoebel, E. Adamson. (1960). *The Cheyennes: Indians of the Great Plains.* New York: Holt, Rinehart and Winston.

Hoecker-Drysdale, Susan. (1992). *Harriet Martineau: First Woman Sociologist.* Oxford, England: Berg.

Holden, Constance. (1997, March). "Changing Sex Is Hard to Do." *Science* 275(5307), p. 1745.

Holmes, Steven A. (2000, March 11). "New Policy on Census Says Those Listed as White and Minority Will Be Counted as Minority." *New York Times.*

Holt, Jim. (2004, November 7). "The Other National Conversation." *New York Times Magazine,* section 6, p. 17.

Homans, G. C. (1950). *The Human Group.* New York: Harcourt Brace.

"Homelessness." (2000, January 21). *Issues and Controversies on File.*

Homer-Dixon, Thomas. (2000). *The Ingenuity Gap.* New York: Knopf.

Hooton, E. A. (1939a). *Crime and the Man.* Cambridge: Harvard University Press.

———. (1939b). *Twilight of Man.* New York: G. P. Putnam's Sons.

———. (1939c). *The American Criminal.* Cambridge: Harvard University Press.

Hoover, Stewart M., Lynn Schofield Clark, and Lee Rainie. (2004, April 7). "Faith Online: 64% of Wired Americans Have Used the Internet for Spiritual and Religious Purposes." Washington, DC: Pew Internet and American Life Project. Available at http://www.pewinternet.org, accessed September 20, 2004.

Horn, Wade F. (1998). *Father Facts,* 3rd ed. Gaithersburg, MD: National Fatherhood Initiative.

Horrigan, John B. (2005, September 24). "Broadband Adoption at Home in the United States: Growing but Slowing." Telecommunications Policy Research Conference, Pew Internet and American Life Project.

———. (2001). "New Internet Users: What They Do Online, What They Don't, and Implications for the 'Net's Future." Washington, DC: Pew Internet and American Life Project. Tracking surveys, March 2000–present.

Howe, Irving. (1976). *World of Our Fathers.* New York: Harcourt Brace.

Howes, C., and C. Hamilton. (1991). "Child Care for Young Children." In B. Spodek (Ed.), *Handbook of Research in Early Childhood Education.* New York: Macmillan.

Hoyert, Donna L, Elizabeth Arias, et al. (2001). "Deaths: Final Data for 1999." *National Vital Statistics Reports* 49(8).

Hoyt, H. (1943). "The Structure of American Cities in the Post-War Era." *American Journal of Sociology* 48, pp. 475–492.

Hudson, Valerie M., and Andrea M. den Boer. (2004). *Bare Branches: The Security Implications of Asia's Surplus Male Population.* Cambridge, MA: MIT Press.

Huntington, Samuel P. (1996). *The Clash of Civilizations and the Remaking of World Order.* New York: Simon & Schuster.

———. (1968). *Political Order in Changing Societies.* New Haven, CT: Yale University Press.

Ibarra, Herminia. (2002). *Working Identity: Unconventional Strategies for Reinventing Your Career.* Cambridge, MA: Harvard Business School Press.

Iceland, John. (2005, August). "Why Concentrated Poverty Fell in the United States in the 1990s." Population Reference Bureau. Available at http://www.prb.org/Template.cfm?Section=PRB&template=/ContentManagement/ContentDisplay.cfm&ContentID=12769, accessed September 12, 2005.

Immerwahr, John, and Jean Johnson. (1996). "Incomplete Assignment. America's Views on Standards: An Assessment by Public Agenda." *Progress of Education Reform* 8.

"Income Gap." (1999, December 17). *Issues and Controversies on File* 4(3), pp. 489–496.

Inkeles, Alex, and David H. Smith. (1974). *Becoming Modern: Individual Change in Six Developing Countries.* Cambridge, MA: Harvard University Press.

Institute of Medicine Committee on the Consequences of Uninsurance. (2001). *Coverage Matters: Insurance and Health Care.* Washington, DC: National Academies Press.

International Conference on Child Labor. (1998). Oslo. October 27–30, 1997. Geneva: International Labour Office.

Irvine, Martha. (1999, April). "Facts about Violence among Youth and Violence in Schools." Washington, DC: U.S. Centers for Disease Control and Prevention.

Itard, J. (1932). *The Wild Boy of Aveyron* (G. Humphrey and M. Humphrey, Trans.). New York: Appleton-Century-Crofts.

Jablon, Robert. (2000, May 19). "Study: Women's Response to Stress May Lead to Healthier Lives." *Associated Press.*

Jacob, H. (1988). *Silent Revolution: The Transformation of Divorce Law in the United States.* Chicago: University of Chicago Press.

Jacobs, Garry, Robert Macfarlane, and N. Asokan. (1997, November 15). "Comprehensive Theory of Social Development." Napa, CA: International Center for Peace and Development.

Jacobs, J. (1961). *The Death and Life of American Cities.* New York: Vintage.

Jacoby, Tamar. (2000, July-August). "In Asian America." *Commentary* 110(1), pp. 58–62.

Jaffe, S., and J. S. Hyde. (2000). "Gender Differences in Moral Orientation." *Psychological Bulletin* 126, pp. 703–726.

Jamison, Kay Redfield. (1995). *An Unquiet Mind.* New York: Knopf.

Janis, I., Dwight W. Chapman, John P. Gillin, and John P. Spiegel. (1964). "The Problem of Panic." In Duane P. Schultz (Ed.), *Panic Behavior.* New York: Random House.

Jarvik, L. F., V. Klodin, and S. Matsuyama. (1973). "Human Aggression and the Extra Y Chromosome: Fact or Fantasy?" *American Psychologist* 28, pp. 674–682.

Jencks, Christopher. (1994). *The Homeless.* Cambridge, MA: Harvard University Press.

Jensen, Malene. (1999). "Want to Marry Money? Take a Class to Meet the Right Class." *Associated Press*.

Johnson, George. (1995, June 6). "Chimp Talk Debate: Is It Really Language?" *New York Times*, p. C1.

Johnson, J. G., et al. (2002). "Television Viewing and Aggressive Behavior during Adolescence and Adulthood." *Science* 295, pp. 2468–2471.

Johnson, Julie. (1988, August 11). "President Signs Law to Redress Wartime Wrong." *New York Times*, p. A16.

Johnson, Tallese and Jane Dye. (2005). "Indicators of Marriage and Fertility in the United States from the American Community Survey: 2000 to 2003." Available at http://www.census.gov/population/www/socdemo/fertility/mar-fert-slides.html, accessed January 9, 2006.

Jones, Jeffrey M. (2002, August 16). "Public Divided on Benefits of Living Together before Marriage." *Gallup News Service*.

Jones, Steve. (2002, September 15). "The Internet Goes to College." Washington, DC: Pew Internet and American Life Project. Available at http://www.pewinternet.org, accessed December 20, 2003.

Josephy, Alvin M., Jr. (1994). *500 Nations: An Illustrated History of North American Indians*. New York: Knopf.

Kallen, Horace M. (1956). *Cultural Pluralism and the American Idea: An Essay in Social Philosophy*. Philadelphia, PA: University of Pennsylvania Press.

Kasarda, John D., and M. Janowitz. (1974). "Community Attachment in Mass Society." *American Journal of Sociology* 48, pp. 328–339.

Katz, Elihu. (1957). "The Two-Step Flow of Communication: An Up-to-Date Report on an Hypothesis." *Public Opinion Quarterly* 21, pp. 61–78.

Keck, Margaret E., and Kathryn Sikkink. (1998). *Activists beyond Borders: Advocacy Networks in International Politics*. Ithaca, NY: Cornell University Press.

Kelling, George L., and Catherine M. Coles. (1996). *Fixing Broken Windows: Restoring Order and Reducing Crime in Our Communities*. New York: Martin Kessler Books (Free Press).

Kennedy, John F. (1961). "Introduction." In William Brandon (Ed.), *The American Heritage Book of Indians*. New York: Dell.

Kennedy, Randall. (2003). *Interracial Intimacies: Sex, Marriage, Identity and Adoption*. New York: Pantheon Books.

———. (1997). *Race, Crime and the Law*. New York: Pantheon Books.

———. (1994, Spring). "Orphans of Separatism: The Painful Politics of Transracial Adoption." *American Prospect* 17, pp. 38–45.

Kent, Mary M., and Mark Mather. (2002, December). "What Drives U.S. Population Growth?" *Population Bulletin* 57(4). Population Reference Bureau.

Khaldun, Ibn. (1958). *The Mugaddimah*. Bolligen Series 43. Princeton, NJ: Princeton University Press.

Kielburger, Craig, and Kevin Major. (1998). *Free the Children*. Toronto: McClelland & Stewart.

Kiely, John L. (1995, December 15). "Poverty and Infant Mortality: United States, 1988." *Morbidity and Mortality Weekly Report* 44(49), pp. 922–927.

Kilker, Ernest Evans. (1993). "Black and White in America: The Culture and Politics of Racial Classification." *International Journal of Politics, Culture and Society* 7(2), pp. 229–257.

Kincheloe, Joe L. (2002). *The sign of the Burger: McDonald's and the Culture of Power*. Philadelphia: Temple University Press.

King, Lisa J. (2000 Winter). "Gender Issues in Online Communities." *CPSR Newsletter* 18(1).

Kinsella, Kevin, and Victoria Velkoff. (2001, November). *An Aging World: 2001*. U.S. Bureau of the Census.

Klaus, Patsy, and Callie M. Rennison. (2002). *Age Patterns in Violent Victimization, 1976–2000*. Washington, DC: U.S. Department of Justice, Bureau of Justice Statistics.

Klick, Jonathan, and Alexander Tabarrok. (2005, April). "Using Terror Alert Levels to Estimate the Effect of Police on Crime." *Journal of Law and Economics* 48.

Knowles, John H. (1977). *Doing Better and Feeling Worse: Health in the United States*. New York: Norton.

Knox, Richard A. (1996, October 24). "Women Doctors' Research 'Ceiling.'" *Boston Globe*, p. A3.

Koenig, Frederick. (1985). *Rumor in the Marketplace: The Social Psychology of Commercial Heresy*. Dover, MA: Auburn House.

Kohlberg, Lawrence. (1969). "Stage and Sequence: The Cognitive-Developmental Approach to Socialization." In David A. Goslin (Ed.), *Handbook of Socialization Theory and Research*. Chicago: Rand McNally.

———. (1967). "Moral and Religious Education in the Public Schools: A Developmental View." In T. Sizer (Ed.), *Religion and Public Education*. Boston: Houghton Mifflin.

Kohn, Hans. (1956). *Nationalism and Liberty: The Swiss Example*. London: Allen and Unwin.

Kohn, Melvin L., and Carmi Schooler. (1983). *Work and Personality: An Inquiry into the Impact of Social Stratification*. New York: Ablex Press.

Kopel, David B. (1995). "Massaging the Medium: Analyzing and Responding to Media Violence without Harming the First Amendment." *Kansas Journal of Law and Public Policy* 4, p. 17.

Kosmin, Barry A., and Seymour P. Lachman. (1993). *One Nation under God*. New York: Harmony.

Kotkin, Joel, and Yoriko Kishimoto. (1988). *The Third Century: America's Resurgence in the Asian Era*. New York: Crown.

Kozol, Jonathan. (1992). *Savage Inequalities: Children in America's Schools*. New York: Harper Perennial.

———. (1989). *Death at an Early Age*. New York: Plume Books.

Kraft, Susan. (1994, July). "Hide-and-Seek with Illegal Aliens." *American Demographics*, pp. 10–11.

Krause, Michael. (1966). *Immigration: The American Mosaic*. New York: Van Nostrand-Reinhold.

Kreider, Rose M. (2005). *Number, Timing, and Duration of Marriages and Divorces: 2001*. Current Population Reports, P70–97. Washington, DC: U.S. Census Bureau.

Kristof, Nicholas, and Sheryl Wudunn. (1994). *China Wakes*. New York: Random House.

Kroeber, Alfred A. (1963/1923). *Anthropology: Culture Patterns and Processes*. New York: Harcourt, Brace & World.

Kronholz, June. (1998, January 9). "Bilingual Schooling Faces Big Test in California as Voters Consider Measure to Abolish Program." *Wall Street Journal*, p. A18.

Kutner, Lawrence. (1988, June 30). "For Children and Stepparents War Isn't Inevitable." *New York Times*, p. C8.

Lacayo, Richard. (1985, June. 10). "Blood in the Stands." *Time*, p. 35.

Lakshmanan, Indira A. R. (1997, August 2). "In Taiwan's Legislature Democracy Packs Punch." *Boston Globe*, pp. A1, A12.

———. (1997, August 26). "Love No Match for Custom: India's Young Favor Arranged Unions." *Boston Globe*, pp. A1, A10.

Lampman, Jane. (2002, October 10). "Charting America's Religious landscape." *Christian Science Monitor*. Available at http://www.csmonitor.com/2002/1010/p12s01-lire.html, accessed December 18, 2002.

Landes, Richard. (1998). *While God Tarried: Disappointed Millennialism and the Making of the Modern West*. New York: Houghton Mifflin.

———. (1997, April 6). "Countdown 2000." *Boston Globe*.

Lang, Kurt, and Gladys Lang. (1961). *Collective Dynamics*. New York: Crowell.

Lantz, Herman R. (1982, Spring). "Romantic Love in the Pre-Modern Period: A Sociological Commentary." *Journal of Social History*, pp. 349–370.

Larsen, Luke J. (2004). "The Foreign-Born Population in the United States: 2003." *Current Population Reports*, pp. 20–551.

Larson, Jan. (1992, July). "Understanding Stepfamilies." *American Demographics*, pp. 36–40.

Lasch, Christopher. (1977). *Haven in a Heartless World: The Family Besieged*. New York: Basic Books.

Laslett, P. (1965). *The World We Have Lost: England before the Industrial Age*. New York: Scribner's.

Laslett, P., and Phillip W. Cummings. (1967). "History of Political Philosophy." In Paul Edwards (Ed.), *The Encyclopedia of Philosophy* (Vols. 5, 6). New York: Macmillan.

Lasswell, Thomas. (1965). *Class and Stratum*. Boston: Houghton Mifflin.

Lattimore, Pamela K., and Cynthia A. Nahabedian. (2000). "The Nature of Homicide: Trends and Changes." *Sourcebook of Criminal Justice Statistics: 1999*. U.S. Department of Justice. Washington, DC: U.S. Government Printing Office.

———. (1997). *The Nature of Homicide: Trends and Changes*. U.S. Department of Justice. Washington, DC: U.S. Government Printing Office.

Laverty, Kassin, Alice Burton, Marcy Whitebook, and Dan Bellm. (2001). *Current Data on Child Care Salaries and Benefits in the United States*. Washington, DC: Department of Health and Human Services Center for Child Care Workforce.

Lazarsfeld, Paul F., Bernard Berelson, and Hazel Gaudet. (1968). *The People's Choice*, 3rd ed. New York: Columbia University Press.

Leaf, Clifton. (2002, March 18). "Enough Is Enough: They Lie, They Cheat, They Steal, and They've Been Getting Away with It for Too Long." *Fortune*.

Leakey, Richard E., and Roger Lewin. (1977). *Origins*. New York: Dutton.

Le Bon, Gustave. (1960/1895). *The Crowd: A Study of the Popular Mind*. New York: Viking.

Lee, Alfred McClung. (1978). *Sociology for Whom?* New York: Oxford University Press.

Lee, Richard Borshay. (1980). *The !KungSan*. Berkeley and Los Angeles: University of California Press.

Lee, Sharon M., and Barry Edmonston. (2005). "New Marriages, New Families: U.S. Racial and Hispanic Intermarriage." *Population Bulletin* 60(2). Population Reference Bureau.

Leland, Elizabeth. (2002, June 30). "Born White, Raised Black." *Charlotte Observer*, p. 7.

Lemert, Edwin. (1972). *Human Deviance, Social Problems and Social Control*, 2nd ed. Englewood Cliffs, NJ: Prentice Hall.

Lenin, Vladimir I. (1949/1917). *The State and Revolution*. Moscow: Progress.

Lenski, Gerhard. (1966). *Power and Privilege: A Theory of Social Stratification.* New York: McGraw-Hill.

Lenski, Gerhard, and Jean Lenski. (1982). *Human Societies,* 4th ed. New York: McGraw-Hill.

Leonard, Ira M., and R. D. Parmet. (1972). *American Nativism: 1830–1860.* New York: Van Nostrand-Reinhold.

Leslie, Gerald R. (1979). *The Family in Social Context,* 4th ed. New York: Oxford University Press.

Lester, Toby. (2002, February). "Oh. Gods!" *Atlantic Monthly,* pp. 37–45.

Levin, Jack. (1993). *Sociological Snapshots.* Newbury Park, CA: Pine Forge Press.

Levine, Robert. (1993, February). "Is Love a Luxury?" *American Demographics,* pp. 27, 29.

Levinson, Daniel J. (with Judy Levinson). (1996). *The Seasons of a Woman's Life.* New York: Knopf.

Levinson, Daniel J. (with Charlotte N. Darrow, Edward B. Klein, Maria H. Levinson, and Braxton McKee). (1978). *The Seasons of a Man's Life.* New York: Ballantine.

Lewandowsky, Stephan Werner G. K. Stritzke, Klaus Oberauer, and Michael Morales. (2005, March). "Memory for Fact, Fiction, and Misinformation: The Iraq War 2003." *Psychological Science* 16(3).

Lewin, Kurt. (1948). *Resolving Social Conflicts.* New York: Harper.

Lewis, Bernard. (2002). *What Went Wrong: Western Impact and Middle Eastern Response.* New York: Oxford University Press.

Lewis, David Levering. (2000). *W. E. B. DuBois: The Fight for Equality and the American Century, 1919–1963.* New York: Henry Holt.

———. (1993). *W. E. B. DuBois: Biography of a Race/1868–1919.* New York: Henry Holt.

Lichter, Daniel T., and Martha L. Crowley. (2002). "Poverty in America: Beyond Welfare Reform." *Population Bulletin* 57(2). Population Reference Bureau.

Lindesmith, Alfred R., and Anselm L. Strauss. (1956). *Social Psychology.* New York: Holt, Rinehart and Winston.

Lines, Patricia M. (2000, Summer). *Public Interest* 140, 74–85.

Linton, R. (1936). *The Study of Man.* New York: Appleton-Century-Crofts.

Lipset, Seymour Martin. (1996). *American Exceptionalism: A Double-Edged Sword.* New York: Norton.

———. (1960). *Political Man.* Garden City, NY: Doubleday.

Lipset, Seymour Martin, and Gabriel Salman Lenz. (2000). "Corruption, Culture and Markets." In Lawrence E. Harrison and Samuel P. Huntington (Eds.), *Culture Matters.* New York: Basic Books, pp. 112–124.

Lipsey, R. G., and P. D. Steiner. (1972). *Economics.* New York: Harper & Row.

Liska, Allen E. (1991). *Perspectives on Deviance,* 3rd ed. Englewood Cliffs, NJ: Prentice Hall.

Lombroso-Ferrero, Gina. (1972). *Criminal Man,* reprint ed. Montclair, NJ: Patterson Smith.

Longino, Charles F., Jr. (1994, August). "Myths of an Aging America." *American Demographics,* pp. 36–43.

Longino, Charles F., Jr., and William H. Crown. (1991, August). "Older Americans: Rich or Poor?" *American Demographics,* pp. 48–53.

Longman, Phillip. (2004). *The Empty Cradle: How Falling Birthrates Threaten World Prosperity and What to Do about It.* New York: Basic Books.

Lovejoy, Anna. (2001, May 3). "The NICHD Study of Early Child Care."

Lown, Bernard. (1996). *The Lost Art of Healing.* Boston: Houghton Mifflin.

Ludwig, Jack. (2000, February 28). "Perceptions of Black and White Americans Continue to Diverge Widely on Issues of Race Relations in the U.S." Gallup Organization.

Lutz, Wolfgang. (1994). *The Future Population of the World: What Can We Assume Today?* London: Earthscan.

Lynch, K. (1960). *The Image of the City.* Cambridge, MA: MIT Press.

Maccoby, Eleanor Emmons, C. M. Buchanan, R. H. Mnookin, and S. M. Dornbusch. (1993). "Post-divorce Roles of Mothers and Fathers in the Lives of Children." *Journal of Family Psychology* 7, pp. 24–38.

Maccoby, Eleanor Emmons, and Carol Nagy Jacklin. (1975). *The Psychology of Sex Differences.* Stanford, CA: Stanford University Press.

Machel, Grac'a. (1996, August). *Impact of Armed Conflict on Children.* New York: United Nations.

Mackenzie, Doris Layton. (2000). "Sentencing and Corrections in the 21st Century: Setting the Stage for the Future." Washington, DC: U.S. Department of Justice.

MacMurray, V. D., and P. H. Cunningham. (1973). "Mormons and Gentiles." In Donald E. Gelfand and Russell D. Lee (Eds.), *Ethnic Conflicts and Power: A Cross-National Perspective.* New York: John Wiley.

Maddux, Cleborne D. (1997). "The World Wide Web and School Culture: Are They Incompatible?" *Computers and the Schools* 13(1/2), pp. 7–10.

Madsen, William. (1973). *The Mexican-Americans of South Texas,* 2nd ed. New York: Holt, Rinehart and Winston.

Maguire, Kathleen, and Ann L. Pastore (Eds.). (2001). *Sourcebook of Criminal Justice Statistics: 2000.* U.S. Department of Justice, Bureau of Justice Statistics. Washington, DC: U.S. Government Printing Office.

———. (2001). Uniform Crime Reports; FBI, Crime in the United States, 1976 through 2000. U.S. Department of Justice, Bureau of Justice Statistics. Washington, DC: U.S. Government Printing Office.

———. (2000). *Sourcebook of Criminal Justice Statistics: 1999.* U.S. Department of Justice. Bureau of Justice Statistics. Washington, DC: U.S. Government Printing Office.

Maines, David R., and Monica J. Hardesty. (1987, September). "Temporality and Gender: Young Adults' Career and Family Plans." *Social Forces* 66(1), pp. 102–120.

Males, Mike. (1993, September 20). "Public Enemy Number One." *In These Times.*

Malinowski, Bronislaw. (1954). *Magic, Science and Religion.* New York: Free Press.

———. (1922). *Argonauts of the Western Pacific.* New York: Dutton.

Manning, Wendy D., and Daniel T. Lichter. (1996). "Parental Cohabitation and Children's Economic Well-Being." *Journal of Marriage and the Family* 58, pp. 998–1010.

Manton, K. G., and E. Stallard. (1984). *Recent Trends in Mortality Analysis.* Orlando, FL: Academic Press.

Martin, Philip, and Elizabeth Midgley. (1994, September). "Immigration to the United States: Journey to an Uncertain Destination." *Population Bulletin* 49(2). Population Reference Bureau.

Marx, Karl. (1968). "Wage, Labour and Capital." In Karl Marx and Friedrich Engels, *Selected Works in One Volume.* New York: International Publishers.

———. (1967/1867a). *Capital: A Critique of Political Economy.* Friedrich Engels (Ed.). New York: New World.

———. (1967/1867b). *Das Kapital* (Vols. 1–3). Friedrich Engels (Ed.). New York: International Publishers.

———. (1959/1847). *Class and Class Conflict in Industrial Society.* Stanford, CA: Stanford University Press.

———. (1906). *The Process of Capitalist Production.* The Modern Library. New York.

Marx, Karl, and Friedrich Engels. (1961/1848). "The Communist Manifesto." In Arthur P. Mendel (Ed.), *Essential Works of Marxism.* New York: Bantam Books.

Matschiner, M., and S. K. Murnen. (1999). "Hyperfemininity and Influence." *Psychology of Women Quarterly* 23, pp. 631–642.

Mauss, Armand I. (1975). *Social Problems of Social Movements.* Philadelphia: Lippincott.

Mayberry, Maralee. (1991). "Conflict and Social Determinism: The Reprivatization of Education." Paper presented at the American Educational Research Association Meeting, Chicago.

Maybury, Kelly. (2002, January 22). "I Do? Marriage in Uncertain Times." *Gallup Poll.*

McCown, Karen Stone, Joshua M. Freedman, and Marsha C. Rideout. (1998). *The Emotional Intelligence Curriculum.* San Mateo, CA: Six Seconds Press.

McDonald's Worldwide. (2004). *Corporate Responsibility Report 2004.* Oak Brook, IL: McDonald's Corporation.

McFalls, Joseph A. (1991 October). "Population: A Lively Introduction." *Population Bulletin* 46(2). Population Reference Bureau.

McKinnon, Jesse, and Karen Humes. (2000, September). *The Black Population in the United States.* Washington, DC: U.S. Bureau of the Census.

McLuhan, M. (1964). *Understanding Media.* London: Routledge.

McNeill, William H. (1976). *Plagues and People.* New York: Anchor/Doubleday.

Mead, George H. (1934). *Mind, Self, and Society.* C. W. Morris (Ed.). Chicago: University of Chicago Press.

Mead, Margaret. (1943). "Our Educational Emphases in Primitive Perspectives." *American Journal of Sociology* 48, 633–639.

———. (1935). *Sex and Temperament in Three Primitive Societies.* New York: William Morrow.

Meadows, Donelle H., Dennis L. Meadows, Jorgan Randers, and William Behrens III. (1972). *The Limits of Growth: A Report of the Club of Rome's Project on the Predicament of Mankind.* New York: Universe Books.

Mediascope. (1996). *National Television Violence Study.* Findings available at http://www.mediascope .org, accessed September 14, 1998.

Medical Group Management Association. (2001). *Academic Practice Compensation and Production Survey: 2002 Report Based on 2001 Data.* Press release, Denver, Colorado.

———. (1999). *Academic Practice Compensation and Production Survey: 2000 Report Based on 1999 Data.* Press release, Denver, Colorado.

Mednick, Sarnoff A. (1977). "A Biosocial Theory of the Learning of Law-Abiding Behavior." In Sarnoff A. Mednick and Karl O. Christiansen (Eds.), *Biosocial Bases of Criminal Behavior.* New York: Gardner Press.

Mednick, Sarnoff A., Terrie E. Moffitt, and Susan A. Stacks. (Eds.). (1987). *The Causes of Crime: New Biological Approaches.* Cambridge, England: Cambridge University Press.

Merton, Robert K. (1969/1949). *Social Theory and Social Structure*. New York: Free Press.

———. (1938). "Social Structure and Anomie." *American Sociological Review* 3, pp. 672–682.

Michels, R. (1966/1911). *Political Parties* (Eden Paul and Adar Paul, Trans.). New York: Free Press.

Migration News. (2000, February). "Integration, Census." 7(2).

Miller, Thomas A.W., and Geoffrey D. Feinberg. (2002, March/April). "Culture Clash: Personal Values Are Shaping Our Times." *Public Perspective*, pp. 6–9.

Mills, C. Wright. (1959). *The Sociological Imagination*. New York: Oxford University Press.

Minino, Arialdi M., and Betty L. Smith. (2001). "Deaths: Preliminary Data for 2000." *National Vital Statistics Reports* 49(12).

Mishel, Lawrence, and Gared Bernstein. (1995). *The State of Working America, 1994–95*. Washington, DC: Economic Policy Institute Service, M. E. Sharpe.

Mocan, Naci, and Kaj Gittings. (2001). "Pardons, Executions and Homicide." Working Paper 8639, National Bureau of Economic Research.

Monaghan, Peter. (1993, November 11). "Free After 6 Months: Sociologist Who Refused to Testify Is Released." *Chronicle of Higher Education*.

Money, John, and Paul Tucker. (1975). *Sexual Signatures: On Being a Man or Woman*. Boston: Little, Brown.

Montagu, Ashley (Ed.). (1973). *Man and Aggression*, 2nd ed. London: Oxford University Press.

———. (1964a). *The Concept of Race*. New York: Collier Books.

———. (1964b). *Man's Most Dangerous Myth: The Fallacy of Race*. New York: Meridian.

Moore, Stephen. (1999, October 13). "Defusing the Population Bomb." *Washington Times*.

Moore, Wilbert E. (1965). *The Impact of Industry*. Englewood Cliffs, NJ: Prentice Hall.

Morris, Desmond. (1970). *The Human Zoo*. New York: McGraw-Hill.

Moskos, Charles C., and John Sibley Butler. (1996). *All That We Can Be: Black Leadership and Racial Integration in the Army*. New York: Basic Books.

Mthethwa, Thulani. (2004, September 6). "Swazi Schools Closed for Wedding." BBC News. Available at http://news.bbc.co.uk/2/hi/africa/3420775.stm, accessed July 20, 2005.

Murdock, George P. (1949). *Social Structure*. New York: Macmillan.

———. (1937). "Comparative Data on the Division of Labor by Sex." *Social Forces* 15(4), pp. 551–553.

Murray, Charles. (1994, December). "What to Do about Welfare." *Commentary* 98(6), pp. 26–34.

Murray, Christopher J. L., and Alan D. Lopez (Eds.). (1996). *The Global Burden of Disease*. Cambridge, MA: Harvard School of Public Health.

Murray, David W. (1998, September). "The War against Testing." *Commentary*.

Murray, David W., Joel Schwartz, and S. Robert Lichter. (2001). *It Ain't Necessarily So: How Media Make and Unmake the Scientific Picture of Reality*. New York: Rowman and Littlefield.

Muson, H. (1979, February). "Moral Thinking—Can It Be Taught?" *Psychology Today*, pp. 26–29.

Myrdal, Gunnar. (1969). *Objectivity in Social Research*. New York: Pantheon Books.

Nakao, Keoko, and Judith Treas. (1993). *General Social Surveys, 1972–1991: Cumulative Codebook*. Chicago: National Opinion Research Center, pp. 827–835.

———. (1990). "Occupational Prestige in the United States Revisited: Twenty-five Years of Stability and Change." Paper presented at the annual meeting of the American Sociological Association, Washington, DC.

Nash, Jonathan G., and Roger-Mark De Souza. (2002). *Making the Link: Population, Health, Environment*. Population Reference Bureau.

National Association of Black Social Workers. (1994). "Preserving African-American Families." Position Paper.

National Center for Education Statistics. (1990). "The Nation's Report Card: 1988."

National Center for Health Statistics. (2002). "The Condition of Education 2001." U.S. Department of Education.

———. (2002). "Health, United States. 2001."

———. (2002, June 26). "Births, Marriages, Divorces, and Deaths: Provisional Data for October 2001. *National Vital Statistics Reports* 50(11).

———. (2001). "Urban and Rural Health Chartbook."

———. (2001). "Health, United States. 2000, with Adolescent Chartbook."

———. (2000). "Indicators of School Crime and Safety, 2000." U.S. Department of Education.

———. (1999). "Health, United States, 1998, with Socioeconomic Status and Health Chartbook."

National Commission on Excellence in Education. (1998). *The NICHD Study of Early Child Care*.

———. (1988). *American Education: Making it Work*. Washington DC: U.S. Government Printing Office.

———. (1983). *A Nation at Risk*. Washington, DC: U.S. Government Printing Office.

National Education Goals Panel. (1998). "1998 Key Findings." Available at http://govinfo.library.unt.edu/negp/page7-3.htm, accessed January 16, 2000.

National Institute of Child Health and Human Development (NICHD). (2001)."Results of NICHD Study of Early Child Care."

National Institutes of Health (2001). "Women's Health." Health Resources and Services Administration.

———. (1997, April 3). "Results of NICHD Study of Early Child Care Reported at Society for Research in Child Development Meeting." *NIH News Release.*

National Public Radio, Kaiser Family Foundation, and Harvard University Kennedy School of Government. "Poverty in America." Available at http://www.npr.org/programs/specials/poll/poverty/summary.html, accessed June 27, 2002.

Neugarten, Bernice, L. (1979). "Timing, Age, and the Life Cycle." *American Journal of Psychiatry* 136, pp. 887–894.

Newburger, Eric C. (2001, September). "Home Computer and Internet Use in the United States: August 2000." *Current Population Reports,* pp. 23–27. Washington, DC: U.S. Census Bureau.

Niebuhr, Gustav. (2000, October 31). "Marriage Issue Splits Jews, Poll Finds." *New York Times.*

Nisbett, Richard E., and Dov Cohen. (1996). *Culture of Honor: The Psychology of Violence in the South.* Boulder, CO: Westview Press.

Nock, Albert Jay. (1996). *Jefferson.* New York: John Day Company.

Nolan, Patrick, and Gerhard E. Lenski. (1999). *Human Societies: An Introduction to Macrosociology.* New York: McGraw-Hill.

NORC. GSS 1972–2002 Cumulative Datafile. Available at http://sda.berkeley.edu:7507/quicktables/quickoptions.do, accessed January 9, 2006.

NORC Public Affairs. (2004). "America's Protestant Majority is Fading, University of Chicago Research Shows." Available at http://www-news.uchicago.edu/releases/04/040720.protestant.shtml, accessed July 20, 2004.

Novit-Evans, Bette, and Ashton Wesley Welch. (1983). "Racial and Ethnic Definition as Reflections of Public Policy." *Journal of American Studies* 17(3), pp. 417–435.

Noxon, Bill. (2003, December 3). "National Survey Reveals Continuing Decline in Science and Engineering Doctoral Degrees." Chicago: University of Chicago News Office.

Ogburn, William F. (1964). *On Culture and Social Change.* Chicago: University of Chicago Press.

Ogunwole, Stella U. (2002 February). *The American Indian and Alaska Native Population: 2000.* Washington, DC: U.S. Bureau of the Census.

O'Hare, William P. (1996, September). "A New Look at Poverty in America." *Population Bulletin* 5(2). Population Reference Bureau.

O'Hare, William P., and William H. Frey. (1992, September). "Booming, Suburban, and Black America." *American Demographics,* pp. 30–38.

Oliver, Melvin L., and Thomas M. Shapiro. (1997). *Black Wealth/White Wealth.* New York: Routledge.

Organization for Economic Co-operation and Development. (1997). INES Project, International Indicators Project. *The Condition of Education 1997* (Indicator 23).

Ortner, Sherry. (1974). "Is Female to Male as Nature Is to Culture?" In Michelle Zimbalist Rosaldo and Louise Lampheres (Eds.), *Woman, Culture and Society.* Stanford, CA: Stanford University Press.

Ostling, Richard N. (1999, Spring). "America's Ever Changing Religious Landscape." *Brookings Review,* pp. 10–13.

Otterbein, Keith. (1973). "The Anthropology of War." In John J. Honigmann (Ed.), *Handbook of Social and Cultural Anthropology.* Chicago: Rand McNally.

———. (1970). *The Evolution of War.* New Haven, CT: Human Relations Area Files.

Oxana, Malaya. (2005). "The Ukrainian Dog Girl." Available at http://www.feralchildren.com/en/showchild.php?ch=oxana, accessed October 18, 2005.

Palen, John J. (1995). *The Suburbs.* New York: McGraw-Hill.

———. (1992). *The Urban World,* 2nd ed. New York: McGraw-Hill.

Pankhurst, Alula. (1999). "'Caste' In Africa: The Evidence from South-Western Ethiopia Reconsidered." Edinburgh, Scotland: Edinburgh University Press.

Pape, Robert A. (2003, August). "The Strategic Logic of Suicide Terrorism." *American Political Science Review* 97(3), pp. 1–19.

Papert, S., and G. Caperton. (1999). "Vision for Education: The Caperton-Papert Platform." Paper presented to the 91st annual National Governors' Association meeting, August 1999.

Papert, Seymour. (1993). *The Children's Machine.* New York: Basic Books.

Park, R., E. Burgess, and R. McKenzie. (Eds.). (1925). *The City.* Chicago: University of Chicago Press.

Parker, R. N. (1995). *Alcohol and Homicide: A Deadly Combination of Two American Traditions.* Albany, NY: State University Press.

Parkes, Henry Bamford. (1968). *The United States of America: A History,* 3rd ed. New York: Knopf.

Parrillo, Vincent N. (1997). *Strangers to These Shores,* 5th ed. Boston: Allyn & Bacon.

Parsons, Talcott. (1971). *The System of Modern Societies.* Englewood Cliffs, NJ: Prentice Hall.

———. (1966). *Societies: Evolutionary and Comparative Perspectives.* Englewood Cliffs, NJ: Prentice Hall.

———. (1954). *Essays in Sociological Theory,* rev. ed. New York: Free Press.

———. (1951). *The Social System.* New York: Free Press.

———. (1937). *The Structure of Social Action: A Study in Social Theory with Special Reference to a Group of Recent European Writers.* New York: McGraw Hill.

Parsons, Talcott, and Robert F. Bales. (1955). *Family Socialization and Interaction Process.* New York: Free Press.

Pastor, P. N., D. M. Makuc, C. Reuben, and H. Xia. (2002). "Chartbook on Trends in the Health of Americans." *Health, United States, 2002.* Hyattsville, MD: National Center for Health Statistics.

Patterson, Orlando. (1997). *The Ordeal of Integration.* Washington, DC: Civitas Counterpoint.

Patterson, T. E. (2002). *The Vanishing Voter.* New York: Knopf.

———. (2002, August 25). "Disappearing Act." *Boston Globe,* pp. D1, D2.

Pavlov, I. P. (1927). *Conditioned Reflexes* (G. V. Anrep, Trans.). New York: Oxford University Press.

Payer, Lynn. (1988). *Medicine and Culture.* New York: Henry Holt.

Perry, T. (1993–1994). "The Transracial Adoption Controversy: An Analysis of Discourse and Subordination." *New York University Review of Law and Social Change* 21, pp. 30–34.

Petersen, William. (1975). "On the Subnations of Western Europe." In Nathan Glazer and Daniel P. Moynihan (Eds.), *Ethnicity: Theory and Experience.* Cambridge, MA: Harvard University Press.

Peterson, Richard R. (1996, June). "A Re-Evaluation of the Economic Consequences of Divorce." *American Sociological Review* 61, pp. 528–536.

———. "Reply to Weitzman." (1996, June). *American Sociological Review* 61, pp. 539–540.

Pettigrew, Thomas F., and Robert C. Green. (1975). "School Desegregation in Large Cities: A Critique of the Coleman White Flight Thesis." *Harvard Educational Review* 46(1), pp. 1–53.

Pew Global Attitudes Project. (2005, June 23). "American Character Gets MixedReviews." Available at http://pewglobal.org/reports/display .php?ReportID=247, accessed January 9, 2006.

Pew Research Center of People and the Press. (2005, July 26). "Views of Muslim-Americans Hold Steady after London Bombings." Available at http:// people-press.org/reports/display.php3?ReportID= 252, accessed January 9, 2006.

Piaget, J., and B. Inhelder. (1969). *The Psychology of the Child.* New York: Basic Books.

Pierce, Charles. (2000). *Hard to Forget: An Alzheimer's Story.* New York: Random House.

Pine, D. S., J. D. Coplan, G. A. Wasserman, L. S. Miller, J. E. Fried, M. Davies, T. B. Cooper, L. Greenhill, D. Shaffer, and B. Parsons. (1997). "Neuroendocrine Response to Fenfluramine Challenge in Boys. Associations with Aggressive Behavior and Adverse Rearing." *Archives of General Psychiatry* 54(9), pp. 839–846.

Pinker, Steven. (2002). *The Blank Slate.* New York: Viking.

———. (1994). *The Language Instinct.* New York: Harper Perennial.

Plucker, J. A. (Ed.). (2003). "Human Intelligence: Historical Influences, Current Controversies, Teaching Resources." Available at http://www .indiana.edu/~intell, accessed July 12, 2005.

Pollan, M. (1997, December 14). "Town-Building Is No Mickey Mouse Operation." *New York Times Magazine,* pp. 56–63, 76, 78, 80–81, 88.

Popenoe, David. (2005). "The Top Ten Myths of Marriage." National Marriage Project. Available at http://marriage.rutgers.edu/Publications/Print/ Print%20Myths%20of%20Marriage.htm, accessed October 18, 2005.

———. (2005). "The Top Ten Myths of Divorce." National Marriage Project. Available at http:// marriage.rutgers.edu/Publications/Print/Print%20 Myths%20of%20Divorce.htm, accessed October 18, 2005.

———. (1994). "The Evolution of Marriage and the Problem of Stepfamilies." In A. Booth and J. Dunn (Eds.), *Stepfamilies: Who Benefits? Who Does Not?* Hillsdale, NJ: Lawrence Erlbaum, pp. 3–27.

Popenoe, David, and Barbara Dafoe Whitehead. (2004, June). "The State of Our Unions 2004." The National Marriage Project. Available at http://marriage.rutgers.edu/Publications/SOOU/ SOOU2004.pdf, accessed October 14, 2005.

"Population and the Environment." (1998, July 17). *CQ Researcher* 8(26).

Population Reference Bureau. (2005a). *Human Development Report 2005.*

———. (2005b). *2005 World Population Data Sheet.*

———. (2004, March). *Transitions in World Population.* 59(1).

———. (2002a). *2002 World Population Data Sheet.*

———. (2002b). *Human Population: Fundamentals of Growth Environmental Relationships.*

———. (2000). *2000 World Population Data Sheet.*

———. (1997). *1997 World Population Data Sheet.*

———. (1995, April). "Women, Children, and AIDS." *Population Today* 23(4).

Power, Carla, and Sudip Mazumdar. (2000, July 3). "Caste Struggle." *Newsweek International*, p. 30.

Prejean, Helen. (1993). *Dead Man Walking*. New York: Random House.

Proctor, Bernadette D., and Joseph Dalaker. (2001, September). *Poverty in the United States: 2001*. Washington, DC: U.S. Bureau of the Census.

Propp, K. M. (1995). "An Experimental Examination of Biological Sex as a Status Cue in Decision-Making Groups and Its Influence on Information Use." *Small Group Research* 26, pp. 451–474.

Provence, Sally. (1972). "Psychoanalysis and the Treatment of Psychological Disorders of Infancy." In S. Wolman (Ed.), *A Handbook of Child Psychoanalysis: Research, Theory, and Practice*. New York: Van Nostrand-Reinhold.

Provine, Robert A. (2000). *Laughter: A Scientific Investigation*. New York: Penguin Books.

Putka, Gary. (1984, April 13). "As Jewish Population Falls in U.S., Leaders Seek to Reverse Trend." *Wall Street Journal*, pp. 1, 10.

Putnam, Robert D. (2000). *Bowling Alone*. New York: Simon & Schuster.

———. (1996, Winter). "The Strange Disappearance of Civic America." *American Prospect* 24.

Pyeritz, R., C. Madansky, H. Schreier, L. Miller, and J. Beckwith. (1977). "The XYY Male: The Making of a Myth." In Ann Arbor Science for the People (Ed.), *Biology as a Social Weapon*.

Quain, Anthony. (Ed.). (2002). *The Political Reference Almanac 2001–2002*. Arlington. VA: Keystone.

Quinney, Richard. (1974). *Critique of Legal Order*. Boston: Little, Brown.

Radelet, Michael L., Hugo Adam Bedau, and Constance E. Putnam. (1992). *In Spite of Innocence*. Boston: Northeastern University Press.

Raines, Howell. (1988, June 17). "British Government Devising Plan to Curb Violence by Soccer Fans." *New York Times*, p. A1.

Rainwater, Lee, and T. M. Smeeding. (1995). "Doing Poorly: The Real Income of Children in Comparative Perspective." Working Paper No. 127. Luxembourg Income Study. Maxwell School of Citizenship and Public Affairs. New York: Syracuse University.

Ramo, Joshua Cooper. (1996, December 16). "Finding God on the Web." *Time*, pp. 60–64, 66–67.

Rash, Wayne, Jr. (1997). *Politics on the Net*. New York: W. H. Freeman, pp. 95–97.

Ravitch, Diane. (2000). *Left Back: A Century of Failed School Reforms*. New York: Simon & Schuster.

Rawls, Phillip. (2000, November 15). "Baptists Leaders Affirm New Creed." *Associated Press*.

Reich, Robert. (1991). *The Work of Nations: Preparing Ourselves for 21st Century Capitalism*. New York: Knopf.

Reid, Sue Titus. (1991). *Crime and Criminology*, 6th ed. Fort Worth, TX: Harcourt Brace Jovanovich.

Reiman, Jeffrey H. (1990). *The Rich Get Richer and the Poor Get Prison: Ideology, Class, and Criminal Justice*, 3rd. ed. New York: John Wiley.

Reinharz, Shulamit. (1993, September). *A Contextualized Chronology of Women's Sociological Work*, 2nd ed. Waltham, MA: Brandeis University Women's Studies Program, Working Papers Series.

Riesman, D. (1950). *The Lonely Crowd: A Study of the Changing American Character*. New Haven, CT: Yale University Press (in collaboration with Nathan Glazer and Reuel Denney).

"Religious Congregations and Membership in the United States: 2000." (2002). Cincinnati, OH: Glenmary Research Center.

"Religious Freedom Abroad." (2000, January 21). *Issues and Controversies on File*, pp. 17–24.

Rennison, Callie M. (2002). "Criminal Victimization 2001: Changes 2000–2001 with Trends 1993–2001." Washington, DC: U.S. Department of Justice, Bureau of Justice Statistics.

———. (2001). "Criminal Victimization 2000: Changes 1999–2000 with Trends 1993–2000." Washington, DC: U.S. Department of Justice, Bureau of Justice Statistics.

Reuters News Service. (1997, July 11). "Television Industry Submits a Voluntary System of Parental Guidelines for Rating Television Programming to the FCC for Review."

Rheingold, Howard. (1999, January). "Look Who's Talking." *Wired*, pp. 35–39.

Rhymer, Russ. (1993). *Genie: An Abused Child's Flight from Silence*. New York: Basic Books.

Riche, Martha Farnsworth. (1987, November). "Behind the Boom in Mental Health Care." *American Demographics*, pp. 34–37, 60–61.

Richerson, Peter J. and Robert Boyd. (2004). *Not by Genes Alone: How Culture Transformed Human Evolution*. Chicago, IL: University of Chicago Press.

Ricks, Thomas E. (1997). *Making the Corps*. New York: Scribner's.

Rifkin, Jeremy. (1995). *The End of Work: The Decline of the Global Labor Force and the Dawn of the Post-Market Era*. New York: Putnam.

Riley, Glenda. (1991). *Divorce: An American Tradition*. New York: Oxford University Press.

Riley, Nancy E. (1997, May). "Gender, Power, and Population Change." *Population Bulletin* 52(1). Population Reference Bureau.

———. (1996, February). "China's 'Missing Girls': Prospects and Policy." *Population Today* 24(2), pp. 4–5.

Ritzer, George. (1992). *The McDonaldization of Society*. Newbury Park, CA: Pine Forge Press.

Rob, A. K. Ubaidur (1988). "Community Characteristics, Leaders, Fertility and Contraception in Bangladesh." *Asia-Pacific Population Journal* 3(2), pp. 55–72.

Robinson, Jacob. (1976). "The Holocaust." In Yisrael Gutman and Livia Rothkirchen (Eds.), *The Catastrophe of European Jewry*. Jerusalem: Yad Veshem.

Roos, Patricia. (1995, August). "Occupational Feminization, Occupational Decline?" Paper presented at the annual meeting of the American Sociological Association.

Rosaldo, Michelle Zimbalist. (1974). "Woman, Culture and Society: A Theoretical Overview." In Michelle Zimbalist Rosaldo and Louise Lamphere (Eds.), *Woman, Culture and Society*. Stanford, CA: Stanford University Press.

Rose, Elaina. (2004, March). Education and Hypergamy in Marriage Markets Department of Economics. Paper #353330. University of Washington, Seattle.

Rose, Jerry D. (1982). *Outbreaks: The Sociology of Collective Behavior*. New York: Free Press.

Rosen, George. (1968). *Madness in Society*. New York: Harper Torchbooks.

Rosenthal, Elizabeth. (1998, November 1) "For One Child Policy: China Rethinks Iron Hand." *New York Times*, pp. 1, 20.

Rosenthal, R., and L. Jacobson. (1966). "Teachers' Expectancies: Determinants of Pupils' I.Q. Gain." *Psychological Reports* 18, pp. 115–118.

Rosenzweig, Jane. (1999, July/August). "Can TV Improve Us?" *American Prospect* 45.

Rossides, Daniel W. (1990). *Social Stratification*. Englewood Cliffs, NJ: Prentice Hall.

Rostow, W. W. (1960). *The Stages of Economic Growth*. Cambridge, England: Cambridge University Press.

Roth, Philip. (2000). *The Human Stain*. New York: Houghton Mifflin.

———. (1998). *American Pastoral*. New York: Vintage Books.

Rothstein, Richard. (1998, May). "Bilingual Education: The Controversy." *Phi Delta Kappan*.

Rowland, Christopher. (2003, March 10)." R.I. Club's Exits at Issue in Fire Probe." *Boston Globe*, p. A1.

Rubin, Zick. (1973). *Liking and Loving*. New York: Holt, Rinehart and Winston.

———. (1970). "Measurement of Romantic Love." *Journal of Personality and Social Psychology* 16(2), pp. 265–273.

Rubinstein, Moshe. (1975). *Patterns of Problem Solving*. Englewood Cliffs, NJ: Prentice Hall.

Rumberger, Russell W. (1987, Summer). "High School Drop-outs: A Review of Issues and Evidence." *Review of Educational Research* 57(2), pp. 101–121.

Rybczynski, Witold. (1983). *Taming the Tiger: The Struggle to Control Technology*. New York: Viking Press.

Sabol, William J. (1999, May). *Crime Control and Common Sense Assumptions Underlying the Expansion of the Prison Population*. Washington, DC: Urban Institute.

Sadker, Myra, and David Sadker. (1994). *Failing at Fairness*. New York: Scribner's.

Sagoff, Mark. (1997, June). "Do We Consume Too Much?" *Atlantic Monthly*.

Sahlins, Marshall D., and Elman R. Service. (Eds.). (1960). *Evolution and Culture*. Ann Arbor: University of Michigan Press.

Sailer, Steve. (1997, July 14). "Is Love Colorblind?" *National Review*.

"Same-Sex Partnerships." (2000, February 18). *Issues and Controversies on File*, pp. 49–56.

Samovar, Larry A., Richard Porter, and Nemi C. Jain. (1981). *Understanding Intercultural Communication*. Belmont, CA: Wadsworth.

Sapir, Edward. (1961). *Culture, Language and Personality*. Berkeley and Los Angeles: University of California Press.

Sappenfield, Mark. (2002, March 29). "Mounting Evidence Links TV Viewing to Violence." *Christian Science Monitor*.

Sax, L. J., et al. (2001). *The American Freshman: National Norms for Fall 2001*. Los Angeles: Higher Education Research Institute.

Scarce, Rik. (1994, July). "(No) Trial (But) Tribulations." *Journal of Contemporary Ethnography* 23(2), pp. 123–149.

Scheck, Barry, Peter Neufeld, and Jim Dwyer. (2000). *Actual Innocence*. New York: Random House.

Schmidley, A. Diane. (2003). "The Foreign-Born Population in the United States: March 2002." *Current Population Reports*, P20–539. Washington, DC: U.S. Census Bureau.

———. (2001, December). *Profile of the Foreign-Born Population in the United States: 2000*. Washington, DC: U.S. Census Bureau.

Schneider, Joseph W., and Peter Conrad. (1983). *Having Epilepsy: The Experience and Control of Illness*. Philadelphia: Temple University Press.

Schoenborn, Charlotte A., Patricia Adams, and Patricia Barnes. (2002, September). "Body Weight Status of Adults: United States, 1997–1998. Advance Data from Vital Health and Statistics." Washington, DC: U.S. Centers for Disease Control and Prevention.

Schumpeter, Joseph A. (1950). *Capitalism, Socialism and Democracy*, 3rd ed. New York: Harper Torchbooks.

Schur, Edwin M., and Hugo A. Bedau. (1974). *Victimless Crimes: Two Sides of a Controversy*. Englewood Cliffs, NJ: Prentice Hall.

Schwarz, Frederic. (1997, February/March). "Women Who Smoke and the Men Who Arrest Them." *American Heritage*, pp. 108–111.

Scott, Elizabeth. (1990). "Rational Decision Making about Marriage and Divorce." *Virginia Law Review* 76, pp. 9–94.

Severson, Margaret and Christine Wilson Duclos. (2005). "American Indian Suicides in Jail: Can Risk Screening Be Culturally Sensitive." Available at http://www.ncjrs.org/pdffiles1/nij/207326.pdf, accessed October 18, 2005.

Service, Elman R. (1975). *Origins of the State and Civilization*. New York: Norton.

Shapiro, David A. (1987). "Learning by the Book." *Changes* 5, pp. 314–315.

Sharp, Lauriston. (1952). "Steel Axes for Stone-Age Australians." *Human Organization* 11, pp. 17–22.

Shattuck, R. (1980). *The Forbidden Experiment*. New York: Farrar, Straus & Giroux.

Shaw, Clifford R., and Henry D. McKay. (1942). *Juvenile Delinquency and Urban Areas*. Chicago: University of Chicago Press.

———. (1931). "Social Factors in Juvenile Delinquency." In *National Committee on Law Observance and Law Enforcement, Report on the Causes of Crime* (Vol. 2). Washington, DC: U.S. Government Printing Office.

Sheaffer, Robert. (1998, September/October). "Massive Uncritical Publicity for Supposed 'Independent UFO Investigation' Demonstrates Media Gullibility Once Again." *Skeptical Inquirer*.

Sheldon, W. H., E. M. Hartl, and E. McDermott. (1949). *The Varieties of Delinquent Youth*. New York: Harper.

Sheldon, W. H., and S. S. Stevens. (1942). *The Varieties of Temperament*. New York: Harper.

Sheldon, W. H., and W. B. Tucker. (1940). *The Varieties of Human Physique*. New York: Harper.

Sheler, Jeffrey L. (1997, December 15). "Dark Prophecies." *U.S. News and World Report*, pp. 62–63, 64, 68–71.

Shelton, Deborah L. (1999, March 2). "Scientists Study Why Women Get Certain Diseases More Readily Than Men Do." *Los Angeles Times*.

Shepardson, Mary. (1963). *Navajo Ways in Government*. Manasha, WI: American Anthropological Association.

Shermer, Michael. (2000). *How We Believe: The Search for God in an Age of Science*. San Francisco: W. H. Freeman.

Shibutani, Tamotsu. (1966). *Improvised News: A Sociological Study of Rumor*. Indianapolis, IN: Bobbs-Merrill.

Shils, Edward. (1971/1960). "Mass Society and Its Culture." In Bernard Rosenberg and David Manning (Eds.), *Mass Culture Revisited*. New York: Van Nostrand-Reinhold.

———. (1968). *Political Development in the New States*. The Hague: Mouton.

Shorto, Russell. (1997, December 7). "Muslims in the United States." *New York Times Magazine*, pp. 60–61.

Silverman, Morton M., Meyer, P. M., Sloane, F., et al. (1997). "The Big Ten Student Suicide Study: A 10-year Study of Suicides on Midwestern University Campuses." *Suicide Life Threat Behavior* 27(3), pp. 285–303.

Simmel, Georg. (1957). "Fashion." *American Journal of Sociology* 62, pp. 541–588.

———. (1955). *Conflict: The Web of Group Affiliations*. Glencoe, Illinois: The Free Press.

———. (1950). *The Sociology of Georg Simmel*. Kurt Wolff (Ed.). New York: Free Press.

Simon, Julian L., and Herman Kahn (Eds.). (1984). *The Resourceful Earth: A Response to Global 2000*. New York: Basil Blackwell.

Simon, Rita J., and Howard Alstein. (1987). *Transracial Adoptees and Their Families: A Study of Identity and Commitment*. New York: Praeger.

Simpson, George E., and Milton Yinger. (1972). *Racial and Cultural Minorities: An Analysis of Prejudice and Discrimination*, 4th ed. New York: Harper & Row.

Sjoberg, Gideon. (1956). *Preindustrial City: Past and Present*. New York: Free Press.

Skocpol, Theda. (1979). *States and Social Revolutions*. New York: Cambridge University Press.

Slater, Philip. (1966). *Microcosm: Structural, Psychological, and Religious Evolution in Groups*. New York: John Wiley.

Smeeding. T., L. Rainwater, and G. Burtless. (2001, May). "United States Poverty in Cross-National Context." Prepared for the IRP Conference Volume: "Understanding Poverty in America: Progress and Problems."

Smelser, Neil J. (1971). "Mechanisms of Change and Adjustment to Change." In George Dalton (Ed.), *Economic Development and Social Change*. Garden City, NY: Natural History Press.

———. (1962). *Theory of Collective Behavior*. New York: Free Press.

Smith, Adam. (1776/1976). *An Inquiry into the Nature and Causes of the Wealth of Nations*. Oxford: Oxford University Press.

Smith, Denise, and Hava Tillipman. (2000, September). "The Older Population in the United States:

1999." *Current Population Reports.* Washington, DC: U.S. Census Bureau.

Smith, Kristin. (2002). "Who's Minding the Kids? Child Care Arrangements: Spring 1997." *Current Population Reports,* pp. 70–86. Washington, DC: U.S. Census Bureau.

Smith, Ted J., and Melanie Scarborough. (1992, May 19). "A Startling Number of American Children in Danger of Starving: A Case History of Advocacy Research." Paper presented at the meetings of the American Association for Public Opinion Research.

Smith, Tom. (1984, June). "America's Religious Mosaic." *American Demographics,* pp. 19–23.

Smock, Pamela J. (2000). "Cohabitation in the United States: An Appraisal of Research Themes: Findings and Implications." *Annual Review of Sociology* 26.

Sniffen, Michael J. (1999, December 6). "Most Female Crime Is Simple Assault." *Associated Press.*

Social Security Administration. (2005, May 6). Most Popular Baby Names, 1880–2004. Available at http://www.ssa.gov/OACT/babynames/http://www.ssa.gov/OACT/babynames/, accessed January 9, 2006.

Society Magazine. (March/April 1997).

Solecki, Ralph. (1971). *Shanidar: The First Flower People.* New York: Knopf.

Solomon, Amy L., Vera Kachnowski, and Avi Bhati. (2005, March 31). "Does Parole Work? Analyzing the Impact of Postprison Supervision on Rearrest Outcomes." Washington, DC: Urban Institute.

Solomon, Robert C. (2002, March). "Reasons for Love." *Journal for the Theory of Social Behaviour* 32(1), pp. 1–28.

Sommers, Christina Hoff. (2000). *The War against Boys: How Misguided Feminism Is Harming Our Young Men.* New York: Simon & Schuster.

———. (1994). *Who Stole Feminism?* New York: Simon & Schuster.

Sosin, Michael. (1986). *Homelessness in Chicago.* Chicago: School of Social Service Administration, University of Chicago.

Sourcebook of Criminal Justice Statistics 2003, Table 4.19. Available at http://www.albany.edu/sourcebook/pdf/t419.pdf, accessed January 9, 2006.

Spar, Debora, and Cate Reavis. (2004). *The Business of Life.* Boston: Harvard Business School Publishing.

Spengler, Oswald. (1932). *The Decline of the West.* New York: Knopf.

Spicer, Edward H. (1962). *Cycles of Conquest.* Tucson: University of Arizona Press.

Spitz, Rene A. (1945). "Hospitalism: An Inquiry into the Genesis of Psychiatric Conditions in Early Childhood." In Anna Freud et al. (Eds.), *The Psychoanalytic Study of the Child.* New York: International University Press.

Sprecher, S., and S. Metts. (1989). "Development of the 'Romantic Beliefs Scale' and Examination of the Effects of Gender and Gender-Role Orientation." *Journal of Personal and Social Relationships* 6, pp. 387–411.

Squire, Peverill. (1988, Spring). "Why the 1936 Literary Digest Poll Failed." *Public Opinion Quarterly* 52, pp. 125–133.

Stark, Rodney, and William Sims Bainbridge. (1985). *The Future of Religion: Secularization, Revival, and Cult Formation.* Berkeley: University of California Press.

Stark, Rodney, and Charles Y. Glock. (1968). *American Piety: The Nature of Religious Commitments.* Berkeley and Los Angeles: University of California Press.

Starr, Paul. (1992). "Social Categories and Claims in the Liberal State." *Social Research* 59(2), pp. 263–296.

———. (1982). *The Social Transformation of American Medicine.* New York: Basic Books.

Statistical Abstract of the United States: 2004–2005, 124th ed. (2005). U.S. Bureau of the Census, 2005, Washington, DC: U.S. Government Printing Office.

Stats. (February 4, 2000). "Flaws in Survey Undermine Expose on AIDS among Priests." (Originally appeared in *Tampa Tribune.*)

———. (February 1, 2000). "Bad News on Priests and AIDS." Available at http://www.stats.org, accessed June 23, 2002.

———. (March 1, 2000). "AIDS/Priests: A Rebuttal to the Kansas City Star."

Stellin, Susan. (2001, October 29). "Most Schools Are Wired." *New York Times,* p. C7.

———. (2001, February 19). "Number of New Internet Users Is Growing." *New York Times,* p. C3.

Steward, Julian H. (1955). *The Theory of Culture Change: The Methodology of Multilineal Evolution.* Urbana: University of Illinois Press.

Stinchcombe, Arthur L. (1969). "Some Empirical Consequences of the Davis-Moore Theory of Stratification." In Jack L. Roach, Llewellyn Gross, and Orville R. Gursslin (Eds.), *Social Stratification in the United States.* Englewood Cliffs, NJ: Prentice Hall.

Stoll, Clifford. (1999). *High Tech Heretic.* New York: Doubleday.

———. (1995). *Silicon Snake Oil.* New York: Doubleday.

Stouffer, Samuel A. (Ed.). (1950). *The American Soldier.* Princeton, NJ: Princeton University Press.

Stout, David. (1999, September 10). "U.S. Report Details World's Religious Persecution." *New York Times*, p. A14.

Strauss, Anselm, and Barney Glaser. (1975). *Chronic Illness and the Quality of Life*. St. Louis: Mosby.

Stulgoe, John R. (1984, February/March). "The Suburbs." *American Heritage* 35(2), pp. 21–36.

Sullivan, Andrew. (1995). *Virtually Normal: An Argument about Homosexuality*. New York: Knopf.

Sulloway, F. J. (1996). *Born to Rebel: Birth Order, Family Dynamics and Creative Lives*. New York: Pantheon Books.

Susskind, Ron. (1998). *A Hope in the Unseen*. New York: Broadway Books.

Sutherland, Edwin H. (1961). *White Collar Crime*. New York: Holt, Rinehart and Winston.

———. (1940). "White Collar Criminality." *American Sociological Review* 40, pp. 1–12.

———. (1924). *Criminology*. New York: Lippincott.

Sutherland, Edwin H., and D. R. Cressey. (1978). *Principles of Criminology*, 10th ed. Chicago: Lippincott.

Suttles, G. (1972). *The Social Construction of Communities*. Chicago: University of Chicago Press.

———. (1968). *The Social Order of the Slum*. Chicago: University of Chicago Press.

Sutton, P. D., and M. L. Munson. (2005). "Births, Marriages, Divorces, and Deaths: Provisional Data for February 2005." *National Vital Statistics Reports* 54(3). Hyattsville, MD: National Center for Health Statistics.

Suzuki, David, and Peter Knudtson. (1989). *Genetics: The Clash between the New Genetics and Human Values*. Cambridge, MA: Harvard University Press.

Swarthmore College Peace Collection. "Introduction to an Exhibit of Photographs of Jane Addams, Her Family, and Hull-House." Available at http://www.swarthmore.edu, accessed March 7, 2002.

"Swazi King Marries Eleventh Wife." (2005, May 30). BBC News. Available at http://news.bbc.co.uk/2/hi/africa/4592961.stm, accessed July 24, 2005.

Sykes, Gresham, and David Matza. (1957). "Techniques of Neutralization: A Theory of Delinquency." *American Sociological Review* 22(6), pp. 664–670.

Szymanski, Albert T., and Ted George Goertzel. (1979). *Sociology: Class, Consciousness, and Contradictions*. New York: Van Nostrand-Reinhold.

Talbot, Margaret. (1997, December 22). "Married in a Mob: Washington Scene." *New Republic* 217(25), pp. 14–15.

Tannen, Deborah. (1998). *The Argument Culture: Moving from Debate to Dialogue*. New York: Random House.

———. (1994). *Talking from 9 to 5*. New York: William Morrow.

———. (1990). *You Just Don't Understand: Women and Men in Conversation*. New York: William Morrow.

Taylor, I., P. Walton, and J. Young. (1973). *The New Criminology*. London: Routledge & Kegan Paul.

Taylor, John Wesley. (1987). *Self-Concept in Home Schooling Children*. Doctoral dissertation, Andrews University.

Taylor, O. L. (1990). *Cross-Cultural Communication: An Essential Dimension of Effective Education*, rev. ed. Washington, DC: Mid-Atlantic Equity Center.

Teigen, K. H. (1986). "Old Truths or Fresh Insight? A Study of Students' Evaluation of Proverbs." *British Journal of Social Psychology* 25, pp. 43–50.

Terrace, Herbert S., L. A. Petitto, R. J. Sanders, and T. G. Bever. (1979). "Can an Ape Create a Sentence?" *Science* 206, pp. 891–902.

Thernstrom, Stephen, and Abigail Thernstrom. (1997). *America in Black and White: One Nation Indivisible*. New York: Simon & Schuster.

Thomas, Judy L. (2000). "Catholic Priests Are Dying of AIDS, Often in Silence." Available at http://www.kcstar.com/projects/priests/priest.htm, accessed January 29, 2000.

Thomas, Karen. (1994, April 6). "Learning at Home: Education outside School Gains Respect." *USA Today*, p. 5D.

Thomas, W. I. (1928). *The Child in America*. New York: Knopf.

Tierney, John J., Jr. (2000, August). "The World of Child Labor." *The World and I* 15, p. 54.

Tiger, Lionel, and Robin Fox. (1971). *The Imperial Animal*. New York: Holt, Rinehart and Winston.

Tilly, Charles. (1985). "Does Modernization Breed Revolution?" In Jack A. Goldstone (Ed.), *Revolutions: Theoretical, Comparative, and Historical Studies*. New York: Harcourt Brace Jovanovich.

———. (1981). *As Sociology Meets History*. New York: Academic Press.

———. (1978). *From Mobilization to Revolution*. Reading, MA: Addison-Wesley.

Times Mirror Center for the People and the Press. (1991). *The Pulse of Europe: A Survey of Political and Social Values and Attitudes* (sec. VIII). Washington, DC: Times Mirror Center.

Tischler, Henry L. (1994, March). "The Message behind the Image: A Comparison of Rock Music Videos with Country Music Videos." Paper presented at the Eastern Sociological Society Meetings.

Tönnies, Ferdinand. (1963). *Community and Society*. New York: Harper & Row. (Originally published in German as *Gemeinschaft und Gesellschaft* in 1887.)

Torrey, E. Fuller. (1988, March). "Homelessness and Mental Illness." *USA Today Magazine,* pp. 26–30.

Townsend, Bickley. (1987, March). "Back to the Future." *American Demographics,* p. 10.

Toynbee, Arnold. (1946). *A Study of History.* New York and London: Oxford University Press.

Trachtenberg, Joshua. (1961). *The Devil and the Jews.* New York: Meridian Books.

"Tracking." (2000, September 15). *Issues and Controversies on File.* Facts on File.

Transparency International. (2000, September 13). "TI Press Release: Transparency International Releases Year 2000 Corruption Perception Index." Berlin.

Turnbull, Colin. (1972). *The Mountain People.* New York: Simon & Schuster.

Turner, Ralph H. (1964). "Collective Behavior." In R.E.L. Faris (Ed.), *Handbook of Modern Sociology.* Chicago: Rand McNally.

Tylor, E. (1958/1871). *Primitive Culture: Researches into the Development of Mythology, Philosophy, Religion, Art and Custom* (Vol. 1). London: John Murray.

Ulc, Otto. (1975/1969). "Communist National Minority Policy: The Case of the Gypsies in Czechoslovakia." In Norman R. Yetman and C. Hoy Steele (Eds.), *Majority and Minority: The Dynamics of Racial and Ethnic Relations.* Boston: Allyn & Bacon.

UNAIDS. (2005). *AIDS Epidemic Update 2004.* Available at http://www.unaids.org/wad2004/report_pdf, accessed September 20, 2005.

———. (2002a). "New UNAIDS Report Warns AIDS Epidemic Still in Early Phase and Not Leveling Off in Worst Affected Countries." New York: United Nations. Available at http://www.unaids.org, accessed April 3, 2003.

———. (2002b, December). *AIDS Epidemic Update.* New York: United Nations.

———. (2000). "One Million More Living with HIV/AIDS in Sub-Saharan Africa, New Report Reveals." New York: United Nations.

———. (2000, November). "AIDS Epidemic Explodes in Eastern Europe." New York: United Nations.

———. (1997). "New UN World AIDS Day Report Warns That HIV Epidemic Is Far Worse than Previously Thought." New York: United Nations.

———. (1993). *The Progress of Nations: 1993.* New York: United Nations Statistical Office and Population Division.

UND Women's Center News. (2003, January). Available at http://www.und.edu/dept/womenctr/newsletters/January2004.pdf, accessed February 12, 2004.

Unicef.org. (2005). "Factsheet: Early Marriage." Available at http://www.unicef.org/protection/files/earlymarriage.pdf, accessed January 9, 2006.

UNICEF. (2005). "Child Poverty in Rich Countries, 2005." Innocenti Report Card No. 6. Florence, Italy: UNICEF Innocenti Research Centre.

Uniform Crime Reports; FBI Supplement. (2005). "Homicide Reports 1976–2002."

Uniform Crime Reports. (2005). "Untangling the Statistics: Numbers Don't Lie—But They Can Deceive." Available at http://www.whitehousedrugpolicy.gov/publications/whos_in_prison_for_marij/untangling_the_stats.pdf, accessed August 31, 2005.

United Jewish Communities. (2002). "U.S. Jewish Population Fairly Stable over Decade, According to Results of National Jewish Population Survey 2000–01." Press release, October 8, 2002. Available at http://www.ujc.org, accessed June 12, 2003.

United Nations. (2005, June). "World Population Prospects: The 2004 Revision." *Population Newsletter* 79.

———. (2004). *World Urbanization Prospects: The 2003 Revision.* New York: United Nations Publications.

———. (2002). *UN Human Development Report: Deepening Democracy in a Fragmented World.* New York: United Nations Publications.

———. (2002, April). "Population Aging: Facts and Figures." Second World Assembly on Ageing. New York: United Nations Publications.

———. (2002, March-May). *UN Chronicle* 39(1). New York: United Nations Publications.

———. (2002). *World Urbanization Prospects: The 2001 Revision.* New York: United Nations Publications.

———. (2001a). *The State of World Population, 2001.* New York: United Nations Publications.

———. (2001b). "*The World's Women: Trends and Statistics.*" New York: United Nations Publications.

———. (2001c). *World Population Prospects: The 2001 Revision.* New York: United Nations Publications.

———. (2000). *World Population Prospects: The 1999 Revision.* New York: United Nations Publications.

———. (1997). *Women's Indicators and Statistics Database.* New York: United Nations Publications.

———. (1996). "*World Population Prospects: The 1996 Revision (Annex I and II).*" New York: United Nations.

———. (1948). *Convention on the Prevention and Punishment of the Crime of Genocide.* General Assembly resolution, December 9, 1948.

U.S. Bureau of the Census. (2005). "Unmarried and Single Americans Week September 18–24, 2005." Available at http://www.census.gov/Press-Release/www/releases/archives/facts_for_features_special_editions/005384.html, accessed September 12, 2005.

———. (2005). "2004 American Community Survey." Available at http://www.census.gov/acs/www/, accessed January 9, 2006.

U.S. Census Bureau. (2005). "African-American History Month: February 2005." Available at http://www.census.gov/Press-Release/www/releases/archives/facts_for_features_special_editions/003721.html, accessed January 9, 2006.

———. (2005). "Current Population Survey, 2004 and 2005." *Annual Social and Economic Supplements*, accessed January 9, 2006.

———. (2004). "50th Anniversary of 'Wonderful World of Color' TV." Available at http://www.census.gov/Press-Release/www/releases/archives/facts_for_features/001702.html, accessed October 18, 2005.

U.S. Census Bureau News. (2005). "Income Stable, Poverty Rate Increases, Percentage of Americans Without Health Insurance Unchanged." Available at http://www.census.gov/Press-Release/www/releases/archives/income_wealth/005647.html, accessed August 30, 2005.

U.S. Centers for Disease Control and Prevention. (2002, October). *HIV Surveillance Supplement Report. AIDS Cases by State and Metropolitan Area of Residence.* 8(2).

———. (2002, July). *Injury Fact Book 2001–2002: Working to Prevent and Control Injury in the United States.*

———. (2001). *HIV/AIDS Surveillance Report.* 7(1).

———. (2000, December). *HIV/AIDS Surveillance Report.* 12(1).

———. (1999, April 21). "Facts about Violence among Youths and Violence in Schools."

U.S. Department of Agriculture. (2004). Economic Research Service, Rural Development Research Report 100. Washington, DC.

———. (2000). "National School Lunch Program Fact Sheet." Available at http://www.fns.usda.gov/cnd/Lunch/, accessed March 30, 2001.

U.S. Department of Education. (2003). *Digest of Education Statistics, 2002.* Washington, D.C.: NCES.

———. (2000). *Indicators of School Crime and Safety, 2000.*

U.S. Department of Health and Human Services, Administration on Aging. (2004). "A Profile of Older Americans: 2004. Racial and Ethnic Composition." Available at http://www.aoa.gov/prof/Statistics/profile/2004/7.asp, accessed September 30, 2005.

U.S. Department of Justice. (2004, November). Bulletin NCJ 206627. Capital Punishment, 2003, p. 9, Table 9.

U.S. Department of Justice, Federal Bureau of Investigation. (2005). "Crime in the United States, 2004." Available at http://www.fbi.gov/ucr/cius_04/offenses_cleared/index.html, accessed October 28, 2005.

———. (2005). "Crime in the United States, 2004, Table 38." Available at http://www.fbi.gov/ucr/cius_04/documents/04tbl38a.xls, accessed January 9, 2006.

U.S. Department of Justice, Office of Justice Programs. (1996). *The Nature of Homicide: Trends and Changes.* Washington, DC: U.S. Government Printing Office.

U.S. Department of Labor. (2005). "Women in the Labor Force in 2004." Available at http://www.dol.gov/wb/factsheets/Qf-laborforce-04.htm, accessed November 18, 2005.

U.S. Department of Labor, Bureau of Labor Statistics. (2005). *Occupational Outlook Handbook, 2004–2005.*

U.S. Department of State. (2002, October 7). "The International Religious Freedom Report for 2002." Washington, DC: Office of International Religious Freedom.

U.S. Office of Refugee Resettlement. (1995). "Annual Report to Congress."

Usdansky, Margaret L. (1993, April 12). "Gay Couples by the Numbers." *USA Today*, p. 8A.

Vago, Steven. (1988). *Law and Society*, 2nd ed. Englewood Cliffs, NJ: Prentice Hall.

———. (1980). *Social Change.* New York: Holt, Rinehart and Winston.

Van de Kaa, Dirk J. (1987, March). "Europe's Second Demographic Transition." *Population Bulletin*. Population Reference Bureau.

van den Haag, Ernest. (1991). *Punishing Criminals: Concerning a Very Old and Painful Question.* New York: University Press of America.

———. (1986). "The Ultimate Punishment: A Defense." *Harvard Law Review*, pp. 124–128.

———. (1981). "Punishment as a Device for Controlling the Crime Rate." *Rutgers Law Review*, pp. 706, 719.

Vanfossen, Beth E. (1979). *The Structure of Social Inequality.* Boston: Little, Brown.

Van Lawick-Goodall, Jane. (1971). *In the Shadow of Man.* Boston: Houghton Mifflin.

Veltfort, Helene, and George E. Lee. (1943, April). "The Coconut Grove Fire: A Study in Scapegoating." *Journal of Abnormal and Social Psychology, Clinical Supplement* 38, pp. 138–154.

Ventura, Stephanie J., and Christine A. Bachrach. (2000, October 18). "Nonmarital Childbearing in the United States: 1940–1999." *National Vital Statistics Reports* 48(16).

Verba, Sidney. (1972). *Small Groups and Political Behavior: A Study of Leadership.* Princeton, NJ: Princeton University Press.

Villaraigosa, Antonio R. (2000, Summer). "America's Urban Agenda: A View from California," *Brookings Review* 18(3), pp. 46–49.

VitalSTATS. (2002). "Phony Numbers on Child Abduction." Available at http://www.stats.org/record.jsp?type=news&ID=46, accessed August 1, 2002.

Walker, T. B., and L. D. Elrod. (1993). "Family Law in the Fifty States: An Overview." *Family Law Quarterly* 26, pp. 319–421.

Wallerstein, Immanuel. (1991). *Geopolitics and Geoculture: Essays on the Changing World.* Cambridge, England: Cambridge University Press.

———. (1980). *Mercantilism and the Consolidation of the European World-Economy, 1600–1750.* New York: Academic Press.

———. (1979). *The Capitalist World Economy.* New York: Cambridge University Press.

———. (1974). *The Modern World System: Capitalist Agriculture and the Origins of European World Economy in the Sixteenth Century.* New York: Academic Press.

Wallerstein, Judith S., Julia M. Lewis, and Sandra Blakeslee. (2000). *The Unexpected Legacy of Divorce.* New York: Hyperion.

Watkins, Paul. (1993). *Stand before Your God: A Boarding School Memoir.* New York: Random House.

Watson, James L. (Ed.). (1997). *Golden Arches East: McDonald's in East Asia.* Berkeley: University of California Press.

Watson, J. B. (1925). *Behavior.* New York: Norton.

Weber, Max. (1958/1921). *The City.* New York: Collier.

———. (1957). *The Theory of Social and Economic Organization.* New York: Free Press.

———. (1956). "Some Consequences of Bureaucratization." In J. P. Mayer (Trans.), *Max Weber and German Politics,* 2nd ed. New York: Free Press.

———. (1930/1920). *The Protestant Ethic and the Spirit of Capitalism* (Talcott Parsons, Trans.). New York: Scribner's.

———. (1968/1922). *Economy and Society* (Ephraim Fischoff, Trans.). New York: Bedminster Press.

Wechsler, Henry. (2000). "1999 College Alcohol Studies." Harvard School of Public Health.

Wechsler, Henry, Jae Eun Lee, Meichun Kuo, Mark Seibring, Toben F. Nelson, and Hang Lee. (2002). "Trends in College Binge Drinking during a Period of Increased Prevention Efforts: Findings from 4 Harvard School of Public Health College Alcohol Study Surveys: 1993–2001." *Journal of American College Health* 50(5).

Weeks, John R. (1994). *Population,* 5th ed. Belmont, CA: Wadsworth.

Weiss, Robert S. (1979). *Going It Alone: The Family Life and Social Situation of the Single Parent.* New York: Basic Books.

Weitzman, Lenore J. (1985). *The Divorce Revolution: The Unexpected Social and Economic Consequences for Women and Children in America.* New York: Free Press.

Westoff, Charles F., and Elise F. Jones. (1977, September/October). "The Secularization of Catholic Birth Control Practice." *Family Planning Perspectives* 9, pp. 96–101.

Wheelwright, Philip. (1959). *Heraclitus.* Princeton, NJ: Princeton University Press.

Whitebook, M., D. Phillips, and C. Howe. (1993). *National Child Care Staffing Study Revisited: Four Years in the Life of Center-Based Care.* Oakland, CA: Child Care Employee Project.

Whitebook, Marcy, Laura Sakai, Emily Gerber, and Carollee Howes. (2001). *Then and Now: Changes in Child Care Staffing 1994–2000.* Washington, DC: Center for Child Care Workforce.

Whitehead, Barbara Dafoe, and David Popenoe. (2002). "Why Men Won't Commit: Exploring Young Men's Attitudes about Sex, Dating, and Marriage." *The State of Our Unions: The Social Health of Marriage in America.* Piscataway, NJ: Rutgers.

Whorf, B. (1956). *Language, Thought, and Reality.* Cambridge, MA: MIT Press.

Whyte, W. H. (1988). *City: Rediscovering the Center.* New York: Doubleday.

———. (1956). *The Organization Man.* New York: Simon & Schuster.

Whyte, William Foote. (1943). *Street Corner Society.* Chicago: University of Chicago Press.

Wiarda, Howard J. (1987). *Latin America at the Crossroads: Debt, Development, and the Future.* American Enterprise Institute for Public Policy Research.

Williams, Lena, and Charlene Hunter-Gault. (2002). *It's the Little Things: Everyday Interactions That Anger, Annoy, and Divide the Races.* New York: Harcourt.

———. (1997, December 14). "It's the Little Things." *New York Times.*

Willner, Ruth Ann. (1984). *The Spellbinders: Charismatic Political Leadership.* New Haven, CT: Yale University Press.

Wilson, Barbara Foley, and S. C. Clarke. (1992, June). "Remarriages: A Demographic Profile." *Journal of Family Issues* 13, pp. 123–141.

Wilson, Doris, James and Paula M. Ditton. (1999, January). "Truth in Sentencing in State Prisons." U.S. Department of Justice, Office of Justice Programs, Bureau of Justice Statistics. Special Report. NCJ 1700–32.

Wilson, Edmund. (1969). *The Dead Sea Scrolls 1947–1969,* rev. ed. London: W. H. Allen.

Wilson, Edward O. (1994). *Naturalist.* Washington, DC: Island Press.

———. (1979). *Sociobiology,* 2nd ed. Cambridge, MA: Belknap.

———. (1978). *On Human Nature.* Cambridge, MA: Harvard University Press.

———. (1975). *Sociobiology: The New Synthesis.* Cambridge, MA: Harvard University Press.

Wilson, James Q. (1996, March). "Against Homosexual Marriage." *Commentary.*

Wilson, James Q., and Richard J. Herrnstein. (1985). *Crime and Human Nature.* New York: Simon & Schuster.

Wilson, G. Willow. (2005, May 29). "The Comfort of Strangers." *New York Times,* p. 22.

Wilson, Woodrow. (1915, May 10). "Americanism." Speech given at Convention Hall, Philadelphia.

Winkler, Anne E. (1998, April). "Earnings of Husbands and Wives in Dual-Earner Families." *Monthly Labor Review.*

Winner, Ellen. (1996). *Gifted Children: Myths and Realities.* New York: Basic Books.

Wirth, Louis. (1944, March). "Race and Public Policy." *Scientific Monthly* 58(4), p. 303.

———. (1938). "Urbanism as a Way of Life." *American Journal of Sociology* 64, pp. 1–24.

Wiseman, Paul. (2002, November 19). "China Thrown off Balance as Boys Outnumber Girls." *USA Today.*

Woessman, Ludger. (2001, Summer). "Why Students in Some Countries Do Better." *Education Next.*

Wolf, Eric. (1969). *Peasant Wars of the Twentieth Century.* New York: Harper & Row.

World Bank. (2001). "Carbon Dioxide Information Analysis Center, U.S. Department of Energy." *World Development Report 2001.* New York: Oxford University Press.

———. (1984). "Measuring the Value of Children." In *World Development Report 1984.* New York: Oxford University Press.

Yearbook of Immigration Statistics: 2004. (2005). U.S. Department of Homeland Security.

Yinger, J. Milton. (1970). *The Scientific Study of Religion.* New York: Macmillan.

Yu, Ping. (1995, April). "Chinese Youth Favor One-Child Families." *Population Today.* Population Reference Bureau.

Zaslow, Martha J., and Kathryn Tout. (2002, April 8). "Child-Care Quality Matters." *American Prospect* 13(7).

Zeitlin, I. M. (1981). *Social Condition of Humanity.* New York: Oxford University Press.

Zimring, Franklin E., and Gordon Hawkins. (1997). *Crime Is Not the Problem: Lethal Violence in America.* New York: Oxford University Press.

Zoll, Rachel. (2002, October 9). "U.S. Jewish Population Declining." Information about the 2000–2001 National Jewish Population Survey. Available at http://www.ujc.org/content-display.html?ArticleID 60346, accessed February 21, 2003.

Zuckoff, Mitchell. (1999, December 12). "A Hole in the Heart." In the series "Choosing Naia: A Family's Journey." *Boston Globe.*

Credits

Chapter 1. 2: AP/Wide World Photos; **4:** © Picture Contact/Alamy; **9:** © Frans Lemmens/Lineair/Peter Arnold, Inc.; **12:** © Bettmann/CORBIS; **13 (both):** Boston Filmworks; **14:** © Bettmann/CORBIS; **15:** © Bettmann/CORBIS; **17:** © Brown Brothers; **20:** The Art Archive/Culver Pictures

Chapter 2. 30: © Taxi/Getty Images; **34:** © Steve Chenn/CORBIS; **35:** © James Carroll/Stock Boston; **36:** © Gary Conner/PhotoEdit; **37:** © Antonio Mari; **40:** © Janine Wiedel Photolibrary/Alamy

Chapter 3. 54: © Paul Sauders/Stone/Getty Images; **56:** © Tony Freeman/PhotoEdit; **58:** © Gianluigi Guercia/AFP/Getty Images; **59:** © Lynsey Addario/CORBIS; **60 (left to right):** K. W. Westermann/CORBIS; © George Holton/Photo Researchers; **61:** Foto Du Monde; **66 (left to right):** © Wartenberg/Picture Press/CORBIS; © Jonathan Nourok/PhotoEdit; **68:** © James Holland/Stock, Boston; **72:** © Warwick Kent/Photolibrary/PictureQuest; **74:** © Kenneth Good

Chapter 4. 80: © SuperStock/PictureQuest; **82:** © Big CheesePhoto LLC/Alamy; **84:** The Granger Collection; **87:** © Hoby Finn/PhotoDisc/Getty Images, Inc.; **88:** Boston Filmworks; **90:** © Department of Special Collections, University of Chicago Library; **92:** © David Young-Wolff/Alamy; **93:** © Robert Frerck/Odyssey/Chicago; **95:** © Owen Franken/CORBIS; **97:** © Tony Anderson/Taxi/Getty Images; **99:** © Chad Johnston/Masterfile; **103:** © Flash! Light/Stock, Boston

Chapter 5. 110: © Alexander Walter/Taxi/Getty Images; **112:** © Paul Damien; **113:** © Bob Thomas/Stone/Getty Images; **116:** © Frank Sitman/Stock, Boston; **117:** AP/Wide World Photos; **121:** © Rick Gomez/CORBIS; **124 (left to right):** © Jeff Greenberg/PictureQuest; © Arnold Gold/New Haven Register/The Image Works

Chapter 6. 134: © Ryan McVay/Stone/Getty Images; **141:** © Bob Daemmrich/The Image Works; **144:** © Martin Barraud/Taxi/Getty Images; **146:** © Mark Richards/PhotoEdit

Chapter 7. 152: Linda Hayes/Boston Filmworks; **154:** AP/Wide World Photos; **155 (top to bottom):** © Stockbyte/PictureQuest; © Andre Jenny/PictureQuest; **156:** © Robert Capa/Magnum; **166:** © Jonathan Kaplan/Panos Pictures; **168:** Boston Filmworks; **172:** © Bettmann/CORBIS; **179:** AP/Wide World Photos

Chapter 8. 190: © Britt Erlanson/The Image Bank/Getty Images; **192:** © Christopher Morris/Black Star Publishing/PictureQuest; **193:** © Topham/The Image Works; **197 (top to bottom):** © Wally McNamee/CORBIS; © John Lei/Stock, Boston; **206:** © Michael Dyer/Stock, Boston; **207 (left to right):** © Gloria Karlson; © Neal Preston/CORBIS

Chapter 9. 218: Sue Cunningham/Alamy; **220:** © Andrew Holbrooke/Black Star; **223:** © Peter Johnson/CORBIS; **225:** © David Turnley/CORBIS; **234:** © Banana Stock/SuperStock

Chapter 10. 240: © Patrik Giardino/CORBIS; **242:** © Brownie Harris/CORBIS; **247:** © Ragu Rai/Magnum; **255 (both):** © National Anthropological Archives, Smithsonian Institution, Washington D.C.; **256:** © Brown Brothers; **260:** © Bettmann/CORBIS; **263:** © Monika Graff/The Image Works; **267:** © Brown Brothers; **269:** © Pete Saloutos/CORBIS

Chapter 11. 278: Jack Hayes/Boston Filmworks; **282 (top to bottom):** © Dana White/PhotoEdit; © The Hutchinson Library; **285:** © Robert Brenner/PhotoEdit; **289:** Boston Filmworks; **292:** © Mike Adaskaveg/Boston Herald

Chapter 12. 302: Boston Filmworks; **308:** © Raghu Rai/Magnum; **311:** Boston Filmworks; **313:** © CORBIS; **323:** © Michelle D. Birdwell/PhotoEdit; **327:** © Amy Etra/PhotoEdit

Chapter 13. 336: © Eric van den Brulle; **338:** © Kim Kyung-Hoon/Reuters/CORBIS; **342:** © Tim Hall/PhotoDisc/Getty Images; **344 (top to bottom):** © Richard Lord; © Royalty Free/CORBIS; **349:** © Robert Llewellyn/PictureQuest; **351:** © Najlah Feanny/CORBIS; **356:** © M. Oppersdorff/Photo Researchers; **358:** © AFP/Getty Images

Chapter 14. 366: © Bill Losh/Taxi/Getty Images; **370:** © Ariel Skelley/Blend Images/Getty Images; **371:** © Jean-Claude Lejeune/Stock, Boston; **375:** © Thomas Victor; **377:** © Tom and Dee McCarthy/CORBIS; **381:** © Mario Villafuerte/Getty Images

Chapter 15. 392: © Kraft Brooks/CORBIS; **395:** © AP/Wide World Photos; **396:** © CORBIS; **405:** Coke Witworth/AP/Wide World Photo; **413:** © Todd A. Gipstein/CORBIS

Chapter 16. 418: © Lynn Saville/Photonica/Getty Images; **421:** © A. Ramey/Woodfin Camp and Associates; **424:** © Don Smetzer/Stone/Getty Images; **426:** © AP/Wide World Photos; **434:** S. Lloyd, The Art of the Ancient Near East, Oxford University Press; **437:** © Walter Meayers/National Geographic/Getty Images; **441:** AP/Wide World Photos; **446:** © Levitt Homes

Chapter 17. 452: © Archive Holding Inc./The Image Bank/Getty Images; **458:** © David Turnley/CORBIS; **466:** © Walter Hodges/CORBIS; **473:** © Burke Uzzle/ZUMA Press

Chapter 18. 478: © Reuters NewMedia Inc./CORBIS; **480:** © Brown Brothers; **483:** © Louise Gubb/The Image Works; **487 (left):** © The Elizabeth Day McCormick Collection, Museum of Fine Arts, Boston; **489:** © Chung Sung-Jun/Getty Images; **490:** © New York Daily News Picture Collection; **491:** © Michael Siluk; **492:** © Nicholas Kamm/AFP/Getty Images

Index

Page numbers with "t" denote tables; those with "f" denote figures; and those with "b" denote boxes.

Practice Tests

CHAPTER ONE   THE SOCIOLOGICAL PERSPECTIVE

SOCIOLOGY AS A POINT OF VIEW

1. **T** F According to Tischler, the social science most closely related to sociology is cultural anthropology.

2. **T** F A sociologist studying high school shootings in the United States would first want to know whether such behaviors were actually increasing or decreasing before formulating an explanation of why they were occurring.

3. T **F** The term "the sociological imagination" refers to the use of sociology to solve complex social issues.

4. T **F** The social sciences focus on phenomena which are far less complex than those studied by the physical sciences.

5. **T** F To say that an argument is not "empirical" is to say that is not based on observable evidence.

6. The trouble with common sense propositions is that
   a. they are usually wrong.
   b. they are usually irrelevant to real life.
   c. there are usually other common sense propositions which contradict them.
   d. they usually apply to minor matters.

7. Although sociology and psychology have similarities, they differ in that psychology usually emphasizes
   a. human society.
   b. individual motivation and behavior.
   c. the differences among different cultures.
   d. the operations of government.

8. One difference between sociologists and journalists is that
   a. sociologists use data.
   b. sociologists address important topics.
   c. sociologists don't worry about their reports being popular.
   d. sociologists have a narrower time frame.

9. Who developed the concept of the sociological imagination?
   a. Auguste Comte
   b. Émile Durkheim
   c. C. Wright Mills
   d. W.E.B. DuBois
   e. Karl Marx

10. Systematically arranged knowledge demonstrating the operation of general laws is
    a. general laws production.
    b. science.
    c. a solution to a problem.
    d. a gremlin.

11. Which of the following is not a component of the scientific method?
    a. experimentation
    b. generalization
    c. verification
    d. personalization
    e. observation

12. Which of the following questions about domestic assault would a sociologist be most likely to ask?
    a. What are the personality characteristics of the worst offenders?
    b. What kind of treatment is most effective with offenders?
    c. Is violence more common in single-earner marriages than in families where the woman works outside the home?
    d. Should offenders be punished as harshly as drug offenders?

13. Sociology is a social science because of all the following EXCEPT
     a. its use of systematic methods.
     b. its selection of socially relevant problems to investigate.
     c. its refusal to use unverified findings.
     d. its use of the scientific method.

14. What is the main difference between sociology and social work?
     a. Sociology uses theory and social work does not.
     b. Social work overlaps with psychology while sociology does not.
     c. Social workers help people solve their problems while sociologists try to understand why the problems exist.
     d. There really is no difference between sociology and social work.

THE DEVELOPMENT OF SOCIOLOGY

15. T **F** The term *sociology* dates back to the philosophers of ancient Greece.

16. T **F** Sociology emerged as a separate field of study during the Renaissance.

17. **T** F Auguste Comte invented the term "sociology."

18. **T** F Comte thought that sociology would move society toward perfection.

19. T **F** The doctrine of Social Darwinism calls on the government to intervene actively to help the less fortunate in society.

20. T **F** Jane Addams' Hull House was largely based on the ideas of Social Darwinism.

21. T **F** Karl Marx thought that the most important fact about social class was the lifestyle a person chose.

22. T **F** Max Weber and Karl Marx were in close agreement in their explanations of the role of religion in society.

23. T **F** Durkheim's research on suicide relied on several in-depth studies of individuals who had killed themselves.

24. **T** F Max Weber maintained that a system of beliefs he called "the Protestant Ethic" enabled the development of capitalism where the "ethic" existed.

25. **T** F Whereas Marx believed that socialism and communism would ultimately bring an end to capitalistic exploitation, Weber believed bureaucracy would characterize both socialist and capitalist societies.

26. T **F** Weber predicted that religion would lead to a gradual turning away from science as the main mode of accounting for what was occurring in the real world.

27. **T** F Suicides resulting from a sense of feeling disconnected from society's values are termed *anomic suicides.*

28. **T** F Latent functions are the unintended or not readily recognized consequences of social processes.

29. **T** F W.E.B. DuBois said that it was his experience with racial problems that led him to question the entire social order.

30. **T** F Functionalists view society as a system of highly interrelated parts that generally operate together harmoniously.

31. The emergence and initial growth of sociology was a response to which of the following?
     a. the Reformation
     b. the effects of the Industrial Revolution
     c. widespread social distress that came with the worldwide depression in the 1930s
     d. the decline of religion in the years following World War II
     e. cultural changes of the 1960s

32. According to Durkheim, social integration refers to how
    a. individuals are bonded to social groups.
    b. educational institutions are racially integrated.
    c. the component parts of the personality are aligned.
    d. the social order is without conflict.

33. Karl Marx's primary focus was on
    a. the reasons behind human social interaction.
    b. the relationship of people to the means of production.
    c. human social groups.
    d. human social life.
    e. helping persons solve personal problems.

34. Which of the following best describes the ideas of Auguste Comte?
    a. He preferred "armchair philosophy" to empirical data.
    b. He believed all societies move through fixed stages of development.
    c. He thought suicide was an aspect of the most advanced stage of society.
    d. He separated the analysis of society into latent and manifest functions.
    e. He predicted that the proletariat would rise to overthrow the capitalists.

35. Which of the following is *not* true of the work of Harriet Martineau?
    a. She worked with Jane Adams to establish Hull House in Chicago during the 19th Century.
    b. She translated into English Comte's major work.
    c. Her major research involved observing day-to-day life in the United States.
    d. She compared social stratification systems in Europe with those in America.
    e. These are all true of the work of Harriet Martineau.

36. Durkheim identified three types of suicide based on
    a. the personality of the individual who might commit suicide.
    b. the particular method (gun, pills, etc.) of committing suicide.
    c. the involvement or lack thereof with social groups.
    d. the age of the individual.

37. Kamikaze pilots and suicide bombers exemplify which type of suicide?
    a. Anomic
    b. Egoistic
    c. Altruistic

38. Durkheim would say that Kamikaze pilots and suicide bombers were products of
    a. a very high level of social solidarity.
    b. a very low level of social solidarity.
    c. a lack of social solidarity.
    d. an ideology of evil and hate.

39. Many people who commit suicide in the U.S. today are depressed and confused. They represent which type of suicide?
    a. Anomic
    b. Egoistic
    c. Altruistic

40. Which aspect of U.S. society would Durkheim see as contributing to egoistic suicide?
    a. the high level of patriotism
    b. the high standard of living
    c. the emphasis on individualism and self-reliance
    d. the gap between rich and poor

41. Which of the following represents Marx's view of the law?
    a. The law is impartial and serves the interests of the entire society.
    b. The law is created by the wealthy, and it protects their interests from the rest of society.
    c. The law is the main tool the poor can use to protect themselves from exploitation by the wealthy.
    d. The law represents the ideas of the most intelligent and most fair-minded people in the society.

42. According to Marx, all history is the history of
    a. class struggle.
    b. technological progress.
    c. intellectual progress.
    d. religious morality.

THEORETICAL PERSPECTIVES

43. **T** F A "paradigm" is an overall framework that shapes the questions a sociologist is likely to ask about a social issue.

44. T **F** Symbolic interactionists usually base their studies on data from national surveys.

45. **T** F Sociologists who are influenced by Marx's ideas are most likely to emphasize the conflict theory paradigm.

46. A sociologist studied higher education in the U.S. to find out how well universities trained students to do important and necessary work that benefitted the society. This sociologist was working mainly from
    a. the functionalist paradigm.
    b. the conflict theory paradigm.
    c. the social interactionist paradigm.

47. A sociologist studied higher education to see how students tried in subtle ways to influence teachers into giving less work and better grades. This sociologist was working mainly from
    a. the functionalist paradigm.
    b. the conflict theory paradigm.
    c. the social interactionist paradigm.

48. An auto executive once proclaimed, "what's good for the country is good for General Motors and vice versa." This notion that GM and the entire U.S. have identical interests is most likely to be opposed by sociologists using
    a. the functionalist paradigm.
    b. the conflict theory paradigm.
    c. the symbolic interactionist paradigm.

49. Egoistic suicide comes from
    a. overinvolvement with others.
    b. a general uncertainty from norm confusion.
    c. overall feelings of depression resulting from economic setbacks.
    d. low group solidarity and underinvolvement with others.

ESSAY QUESTIONS

50. How might each of the main sociological paradigms look at the sociology class you are now in? What aspects of the class would each paradigm be most interested in, and how would each go about exploring those questions?

51. Compare the way a psychologist might look at the issue of suicide with the sociological approach. What are the differences in the kinds of question each would ask and the kinds of data they would use in answering these questions?

CHAPTER TWO   DOING SOCIOLOGY: RESEARCH METHODS

TRUE/FALSE

1. T **F** Independent and dependent variables are found in statements of causality but not in statements of association.

2. **T** F A testable statement about the relationship between two or more empirical variables is known as a hypothesis.

3. T **F** Most research problems of interest to sociologists are most easily and accurately studied by means of controlled experiments.

4. **T** F An empirical question is an issue posed in such a way that it can be studied through observation.

5. T **F** The mean is the number that occurs most often in a data set.

6. T **F** The phenomenon studied by the researcher is called the independent variable.

7. **T** F A self-fulfilling prophecy is produced when a researcher who is strongly inclined toward a particular point of view communicates that attitude to the research subjects so that their responses end up consistent with the initial point of view.

8. **T** F A structured interview is a form of research conversation in which the questionnaire is followed rigidly.

9. **T** F Validity is the extent to which a study tests what it was intended to test.

10. **T** F One reason forecasters inaccurately predicted the winner of the 1936 Landon-Roosevelt presidential election was that their sample was too small.

11. **F** F Secondary analysis is useful for collecting historical and longitudinal data.

12. **T** F Factors affecting the accuracy of the *Kansas City Star* Catholic priest study was the low response rate and large percentage of non-respondents in the study.

13. **T** F The major factor affecting the response rate in the *Kansas City Star* study was the voluntary nature of priest participation in the survey.

14. **T** F Sociologists try to provide answers to two general questions: "Why did it happen?" and "Under what circumstances is it likely to happen again?"

15. T **F** Normally, the research method one selects for an investigation is designed specifically for that particular research problem.

16. T **F** The first step in the research process is to develop one or more hypotheses.

17. T **F** An hypothesis is a specific statement about an abstract concept in terms of the observable features that describe the thing being studied.

18. T **F** An operational definition is not required if the term being studied is widely understood.

19. **T** F Complete objectivity in research may be impossible to achieve but it is still a reasonable goal to strive for.

20. T **F** In setting up tables, it is often all right to omit the headings for rows and columns.

21. T **F** The American Sociological Association's Code of Ethics states that since sociologists are not free from arrest, confidentiality can be broken if there is a risk of being sent to jail for contempt of court.

22. T **F** Social scientists regard deception of research participants as a necessary evil in most social research.

23. T **F** One advantage of social research is that it frequently benefits the research subjects directly and immediately.

24. **T** F Most subjects of sociological research belong to groups with little or no power.

25. **T** F An empirical question can be answered by observing the world as it is known.

26. T **F** A statement of association says that one thing causes another.

27. T **F** A variable that changes another variable is a dependent variable.

28. **T** F Survey research is not as effective as interviewing or participant observation for allowing the researcher to understand the feelings and attitudes of the respondents.

29. **T** F Participant observation studies are usually less objective and less replicable than are studies that use survey research.

MULTIPLE CHOICE

30. Which of the following questions would one ask to determine the quality of any opinion survey or poll?
    a. Did you ask the right people?
    b. What is the margin of error?
    c. What were the questions that were asked?
    d. all of the above
    e. none of the above

31. A subset of the population that exhibits, in correct proportion, the significant characteristics of the population as a whole is known as a
    a. dependent variable.
    b. nonrandom sample.
    c. representative sample.
    d. cross-sectional sample.

32. The figure that falls in the exact middle of a ranked series of scores is the
    a. median.
    b. mean.
    c. mode.
    d. average.

33. A longitudinal study is research
    a. aimed at predicting the future.
    b. investigating a population over a period of time.
    c. examining a population at a given point in time.
    d. conducted without the participants' knowledge.

34. Which of the following steps in the research process must come **last**?
    a. developing hypotheses
    b. reviewing previous research
    c. defining the problem
    d. determining research design
    e. analyzing the data and drawing conclusion

35. Which of the following is a measure of central tendency that is commonly referred to as the "average"?
    a. median
    b. mean
    c. mode
    d. meridian

36. Which of the following research methods is the most subjective?
    a. surveys
    b. controlled experiments
    c. participant observation
    d. secondary analysis

37. Which of the following is a statement of causality?
    a. Rural areas have fewer services than urban.
    b. This sociology course is difficult.
    c. Poverty produces low self-esteem.
    d. Mean income in New York is higher than in Florida.

38. Researchers ride in unmarked police cars to collect data on drug dealers. They are using the _____ method of research.
    a. longitudinal survey
    b. participant observation
    c. laboratory experiment
    d. semi-structured interview

39. Consultants for political candidates often use polls to judge how their candidate's positions are being perceived by the public. This method of research is closest to
    a. survey research
    b. interviews
    c. experiments
    d. participant observation

40. Consultants for political candidates sometimes use "focus groups" to judge how their candidate's positions are being perceived by the public. In a focus group, a small number of people (a dozen or so) are questioned and encouraged to talk freely about specific issues. This method of research is closest to
    a. survey research
    b. interviews
    c. experiments
    d. participant observation

41. What element of a statistical table should tell you the subject of the data?
    a. footnotes
    b. column headings
    c. headnotes
    d. title
    e. source

42. Some critics of IQ tests claim that the lower-class people do poorly not because they are less intelligent but because they have not been exposed to the kinds of cultural things that the tests ask about. The critics are claiming that as a test of intelligence, IQ tests are not
    a. valid.
    b. reliable.
    c. blind.
    d. double-blind.

43. Using the following quiz scores, **2, 4, 6, 7, 7, 10** calculate the mode.
    a. 4
    b. 7
    c. 5
    d. 6
    e. 8

44. Bill Gates walks into your classroom. What statistic about the income of the population in the classroom has been most affected?
    a. the mean
    b. the median
    c. the mode
    d. the reliability

45. A researcher publishes findings that several other researchers doing similar research cannot replicate. The first researcher's data therefore were not
    a. valid.
    b. reliable.
    c. scientific.
    d. objective.

46. People are selected from a group in such a way that every person has the same chance of being selected. The people who are selected make up what type of sample?
    a. representative
    b. cross-sectional
    c. random
    d. unintentional
    e. stratified

ESSAY QUESTIONS

47. Drawing on your reading, discuss the techniques you would use to determine whether an opinion poll is "bogus."

48. Discuss three important criteria that a person could use to determine whether the results of a research study were accurate.

49. Generate a hypothesis about a current social issue. Identify the dependent and independent variables and indicate whether the hypothesis is a statement of causality or association.

CHAPTER THREE   CULTURE

THE CONCEPT OF CULTURE

1. T F Human beings, like most other species, pass a wide variety of behavioral patterns from one generation to the next through their genes.

2. T F Social scientists have identified some societies that, for all practical purposes, do *not* have a culture.

3. T F Genetically determined behavior is innate; culturally determined behavior is learned or acquired.

4. T F In 2001, the first human gene involved specifically in language was discovered.

5. T F The culture of a group is generally shared by all group members rather than being a matter of individual preference.

6. If we based our study of another society on the principle of cultural relativism, we would
    a. pay close attention to how people treated their relatives.
    b. try to see whether that society was superior to others we had studied.
    c. try to understand that society on its own terms and withhold moral judgment.
    d. look at the art, music, and other high culture to see if it was as good as ours.

7. The behavior of most nonhuman animal species is largely determined by
    a. norms.
    b. culture.
    c. instinct.
    d. mores.

8. Anthropologist Kenneth Good seeing a Yanomoma woman about to be raped—a commonplace event—said, "I stood there, my heart pounding, uncertain what to do." This response is an example of
    a. cultural lag.
    b. culture shock.
    c. instinct.
    d. material culture.

9. The process of making judgments about other cultures based on the customs and values of one's own culture is called
    a. ethnocentrism.
    b. cultural innovation.
    c. cultural relativism.
    d. reformulation.

10. An example of nonmaterial American culture is
    a. patriotism.
    b. Big Macs.
    c. iMacs.
    d. Mack Trucks.

COMPONENTS OF CULTURE

11. T F Ideal norms are expectations of what people should do under perfect conditions.

12. T F The Sapir-Whorf hypothesis suggests language affects how humans perceive the world around them.

13. T F Real norms are strongly held rules of behavior that have a moral connotation.

14. T F Culture is transmitted to children by DNA and other genetic material.

15. T F According to Lipset, the same value in a culture can produce both positive and negative behaviors.

16. The view that "time is money" is an example of
    a. a material aspect of culture.
    b. a universal shared by all cultures.
    c. cultural relativism.
    d. a nonmaterial aspect of culture.

17. The gift of a clock is a sign of respect in the U.S.; in China it is considered bad luck. This shows that the two countries differ in their
    a. nonmaterial culture.
    b. material culture.
    c. ethnocentrism.
    d. cultural lag.

18. It is considered polite to send regrets when you are invited to a party but cannot attend. Failure to do so, however, is not considered a moral lapse. This rule about RSVPs is the type of norm called a
    a. value.
    b. taboo.
    c. folkway.
    d. subculture.

19. In American culture, such things as freedom, individualism, and equal opportunity are deemed to be highly desirable. In sociological terms these concepts are
    a. values.
    b. folkways.
    c. beliefs.
    d. norms.

20. The Hopi use the same word for everything that flies except birds (planes, flying insects, etc.). According to the Sapir-Whorf hypothesis, this means that
    a. the Hopi language is generally inferior to English.
    b. Hopi are less likely to perceive differences between planes and helicopters than are most Americans.
    c. Hopi are afraid of flying.
    d. the Hopi have much to teach us about flying.

21. The study of U.S. and Saudi values showed that these two cultures differed in the importance they place on all the following EXCEPT
    a. religion
    b. modesty
    c. freedom
    d. individualism
    e. tradition

22. Samuel Huntington, in the book cited in the chapter, sees the conflict between some Arab countries and the U.S. as largely a matter of
    a. oil revenues.
    b. possession of weapons of mass destruction.
    c. education.
    d. culture.

THE SYMBOLIC NATURE OF CULTURE

23. **T F** A gesture like a "thumbs up" is a symbol.

24. **T F** Within a culture, there is a wide variation among people as to what most symbols mean.

25. **T F** Some simple gestures, like a nod of the head to mean yes, are universal and always have the same meaning in all cultures and societies.

26. **T F** Tischler's overall conclusion regarding culture is that all aspects of culture—nonmaterial and material—are symbolic.

CULTURE AND ADAPTATION

27. **T F** Cultural adaptation is a response to changes in the environment.

28. **T F** Cultures that are isolated are unlikely to experience cultural diffusion.

29. **T F** The fewer cultural items in a society's inventory, the less likely it is to see cultural innovation.

30. **T F** Just as chimpanzees cannot speak, they also cannot learn simple sign language.

31. **T F** Although chimpanzees can recognize icons on a computer, they cannot make new combinations of these icons to express their desires.

32. Humans are superior to other animals in the area of
    a. surviving in natural environments.
    b. using symbols.
    c. physical strength.
    d. physical agility.
    e. self-preservation.

33. In America, you can now get "Hawaiian pizza"—pizza with pineapple on it. This is an instance of
    a. ethnocentrism.
    b. cultural relativism.
    c. cultural reformulation.
    d. cultural lag.

34. Although cell phones have been around for years now, the norms regarding how, when, and where they should be used are still in flux. The failure of society to formulate norms in response to technological change is an example of
    a. innovation.
    b. diffusion.
    c. cultural lag.
    d. ethnocentrism.

SUBCULTURES

35. **T F** Subcultures are the ways of life of smaller groups within the larger society.

36. **T F** An immigrant group that retains many of the ways of the old country constitutes a "deviant subculture."

37. **T F** Subcultures always represent a threat to the dominant culture's major values.

38. Novelist F. Scott Fitzgerald once commented that "the rich really are different from you and me." He was implying that the rich were
   a. ethnocentric.
   b. a deviant subculture.
   c. a product of innovation.
   d. a social class subculture.

39. According to Nisbett and Cohen's study cited in this chapter, the South has higher murder rates than other regions because of its
   a. higher rates of poverty.
   b. higher rates of gun ownership.
   c. cultural beliefs regarding honor.
   d. history of slavery.

40. According to Nisbett and Cohen, the South's early settlers were livestock herders from Scotland and Ireland, who saw violence as necessary to protect livestock from thieves. The carrying of these views across the ocean and over many generations in America is an example of
   a. cultural diffusion.
   b. cultural innovation.
   c. ethnocentrism.
   d. cultural relativism.

UNIVERSALS OF CULTURE

41. T F About 20% of cultures throughout the world do not have an incest taboo.

42. T F Although all cultures have marriage as an institution, the specific form of marriage and the rules of marriage vary greatly.

43. T F Chimpanzees and other primates are generally more cooperative and less self-sufficient than humans.

44. According to Tischler, the most important function of the incest taboo is that
   a. it prevents immoral behavior.
   b. it protects against genetic mutation and disease caused by inbreeding.
   c. it expands social bonds outside the family, creating larger, stronger groups.
   d. it prevents conflict and jealousy within the nuclear family.

45. Confirmations, weddings, and funerals are examples of
   a. division of labor.
   b. rites of passage.
   c. cultural diffusion.
   d. cultural innovation.

46. The belief that America is the land of opportunity for all is part of the American
   a. ideology.
   b. division of labor.
   c. social class subculture.
   d. material culture.

47. "The conflict between being a researcher and a human being" is basically the conflict between
   a. cultural innovation and cultural lag.
   b. cultural relativism and ethnocentrism.
   c. norms and values.
   d. mores and folkways.

ESSAY QUESTIONS

48. Think of some subculture of which you are or have been a member—a high school clique, a student subculture at your college, a group at work. How do the norms, beliefs, and values of this subculture differ from those of the dominant institution in which the subculture exists? How does the group try to reinforce its cultural ideas when these are in conflict with those of the institution?

49. Is there a culture of violence in the U.S.? The U.S. has higher rates of violent crime than do other economically advanced countries. It also spends more on its military than does the rest of the world combined. Is violence an American value? What beliefs about violence do Americans share? How might American values and beliefs account for its high rate of violence?

CHAPTER FOUR   SOCIALIZATION AND DEVELOPMENT

1. **T F** Most of our bodily processes, such as how tall we will be, are completely controlled by our genes and uninfluenced by factors in the environment.

2. **T F** Sociobiologists such as Edward O. Wilson believe that human behavior can be understood as continuing attempts to ensure the transmission of one's genes to a new generation.

3. **T F** The concept of instinct is more important for biological explanations of behavior rather than cultural explanations.

4. **T F** Behavioral psychologists like John B. Watson believed that almost all human behavior was controlled by principles of conditioning.

5. **T F** According to Piaget, the stage when the infant relies on touch and the manipulation of objects for information about the world is called the operational stage.

6. **T F** Moral order is a society's shared view of right and wrong.

7. **T F** According to Kohlberg, concepts such as good and bad, right and wrong, once established, carry the same meaning for us throughout our lives.

8. **T F** Piaget was more interested in the development of logical thought rather than the development of moral reasoning.

9. **T F** In Kohlberg's theory, the highest stage of moral development is an orientation toward the law exactly as it is written.

10. **T F** According to Kohlberg's research, once a person has achieved a higher level of moral development, he or she never regresses to a lower level of moral reasoning.

11. **T F** In Freudian theory, the part of the psyche that we are most conscious of is the superego.

12. **T F** In Freudian theory, the ego is the repository of the thoughts and feelings that we are not aware of.

13. **T F** According to Mead, the portion of the self that is made up of everything learned through the socialization process is called the "me."

14. **T F** Resocialization often involves exposure to people with ideas and values different from those that a person was raised with.

15. **T F** Overall, the fields of academic sociology and anthropology have not been quick to accept the positions advanced by theorists in sociobiology.

16. **T F** Young people who feel ignored by their parents seem to be more vulnerable to peer pressure.

17. **T F** In Mead's terms, the general society as an agent of socialization is called a "significant other."

18. **T F** Marriage is an important event in adult socialization.

19. **T F** The majority of American families fit the traditional model, with the husband as wage-earner outside the home, and the wife as homemaker.

20. **T F** The terms "personal identity" and "social identity" mean essentially the same thing.

21. **T F** Attachment disorder results in children being unable to form relationships.

22. **T F** Total institutions are environments in which people are physically and socially isolated from the outside world.

23. **T F** Studies of mating strategies show that men and women are looking for the same qualities in their romantic and sexual partners.

24. **T F** According to research cited by Tischler, peer groups of black high school students encourage their members to try to get better grades than whites.

25. **T F** In most communities in the U.S., schools are havens isolated from the conflicting values of the wider society.

26. Tischler writes that "as the authority of the family diminishes under the pressures of social change, peer groups move into the vacuum and substitute their own morality for that of the parents." The youth who changes his or her morality in this way is undergoing a process of
    a. attachment disorder.
    b. cognitive development.
    c. resocialization.
    d. preparatory stage.

27. The debate about "nature vs. nurture" is basically a debate about the relative importance of
    a. Darwin and E. O. Wilson.
    b. biology and culture.
    c. personality and nourishment.
    d. peer group and school.

28. Gould's critique of sociobiology maintained that human behavior was
    a. almost entirely controlled by genes.
    b. evolved through natural selection in the same way that physical characteristics evolved.
    c. completely unaffected by genes.
    d. limited by but not absolutely controlled by genetics.

29. Harlow studied monkeys raised in isolation, with no contact with other monkeys. When these monkeys were placed among other monkeys
    a. they eventually learned to interact like the others.
    b. they sought out a single other monkey to play with.
    c. they could interact only in order to mate with monkeys of the opposite sex.
    d. they never learned to interact normally with other monkeys.

30. In Mead's terms, a "significant other" is
    a. a romantic partner.
    b. a representative of society in general.
    c. anyone important to a person's development.
    d. the "looking-glass self."

31. In Mead's theory, the play of children was
    a. an important source of the formation of the self.
    b. a temporary release from the demands of forming a self.
    c. a time when children often acquired the wrong kinds of self.
    d. irrelevant to the development of self.

32. Pre-school age children of parents who work are most frequently cared for by
    a. day-care centers.
    b. relatives.
    c. babysitters, nannies, or other non-relatives.
    d. nobody; they stay home alone.

33. Compared with children cared for at home by their parents, children cared for in day-care centers are less
    a. intelligent.
    b. verbal.
    c. aggressive.
    d. respectful.

34. Which of the following is **not** one of the components of the *looking-glass self,* according to Cooley?
    a. our imagination of what we must really be like
    b. our imagination of how our actions appear to others
    c. our imagination of how others judge our actions
    d. a self-judgment in reaction to the imagined judgments of others

35. Each individual's *social* identity consists of
    a. all the behaviors they have learned to imitate.
    b. their changing yet enduring view of themselves.
    c. the sum total of all the statuses they occupy.
    d. all the ways that other people view them.

36. The process of social interaction that teaches a child the intellectual, physical, and social skills needed to function as a member of society is called
    a. identification.
    b. social adjustment.
    c. socialization.
    d. social conditioning.

37. Studies cited by Tischler on the effects of television support the idea that
    a. violent TV shows are responsible for much real violence.
    b. watching non-violent shows like "Sesame Street" made children less violent.
    c. regardless of the content of the shows they watch, children who watch more television are more likely to be violent than are children who watch less TV.
    d. there is no relationship between television and real-life acts of violence.

38. Children raised with little contact with others often suffer a psychological condition known as
    a. attachment disorder.
    b. identity crisis.
    c. id.
    d. operant conditioning.

39. Stephen J. Gould was a leading critic of
    a. moral development.
    b. Freudian psychology.
    c. socialization.
    d. sociobiology.

40. The ideas of sociobiology are inspired mostly by theories of
    a. Mead.
    b. Cooley.
    c. Darwin.
    d. Freud.

41. Social identity, unlike personal identity, does not involve a person's
    a. ideas.
    b. occupation.
    c. sex.
    d. statuses.

42. Students were asked why it was wrong to plagiarize a paper. Which response represents the *lowest* level on Kohlberg's scale of moral reasoning?
    a. because I'd be looked down on by my family
    b. because I'd fail the course if I got caught
    c. because it wouldn't be fair to others who wrote their own papers
    d. because I want my grade to represent my own abilities

43. Which of the following theorists coined the term "identity crisis"?
    a. Pavlov
    b. Freud
    c. Erikson
    d. Mead

44. Erikson thought that the "identity crisis" occurred most frequently among
    a. children.
    b. adolescents.
    c. middle-aged people.
    d. the elderly.

45. The structure of "total institutions" often makes them very effective places for
    a. cognitive development.
    b. evolutionary psychology.
    c. attachment disorder.
    d. resocialization.

46. An older man was required to retire; then his wife died; his children were grown and no longer need his parenting. These facts were most important for his
    a. social identity.
    b. moral reasoning.
    c. cognitive development.
    d. superego.

CHAPTER FIVE   SOCIETY AND SOCIAL INTERACTION

UNDERSTANDING SOCIAL INTERACTION

1. **T F** Human behavior is not random. It is patterned and, for the most part, predictable.

2. **T F** Goffman's studies on dramaturgy reveal that Americans enjoy theater more than persons of other nationalities.

3. **T F** Until the norms of social distance are broken, most people are unaware that these rules exist even though they abide by them.

4. **T F** When students acted like boarders in their own homes, family members knew them quite well, so the family members were rarely upset by the students' behavior.

5. To an observer, two people engaging in a fistfight on the street would mean something entirely different from the same two people fighting in a boxing ring. This illustrates how the meaning of social interaction is dependent on
   a. personalities.
   b. competition.
   c. status.
   d. context.

6. A group of autoworkers meets to discuss various labor problems. During the meeting two of the workers remain silent. In sociological terms these two would be described as
   a. having no influence on the interaction.
   b. inactive members of the group.
   c. conveying a message through their silence.
   d. being uncooperative.

7. In a large lecture class, it is generally expected that students will raise their hands and be called upon before speaking. This illustrates the operation of
   a. bureaucracy.
   b. norms.
   c. roles.
   d. statuses.

8. The process of two or more people taking each other's actions into account is called
   a. social accounting.
   b. social organization.
   c. social interaction.
   d. social action.

9. If a white teacher is talking to an African American child who is looking toward the wall, we are seeing an example of
   a. a child daydreaming.
   b. a child with oppositional deviance disorder.
   c. a difference of social norms.
   d. poor teaching skills.

TYPES OF SOCIAL INTERACTION

10. **T F** When people do something for each other with the express purpose of receiving something in return, they are engaged in exchange.

11. **T F** Communication by means of gesture is pretty much the same the world over.

12. **T F** Coercion is a form of conflict in which one of the parties is much stronger than the other and can impose its will on the weaker party.

13. **T F** Competition is a form of conflict within agreed-upon rules.

14. **T F** A successful society is one in which conflict has been eliminated.

15. **T F** Relations within a small group are based on either cooperation or conflict, never both.

16. We tend to maintain the most eye contact when we are speaking to someone of _____ status than us.
   a. higher
   b. approximately the same
   c. lower

17. Staring, smiling, nodding one's head, and using one's hands while talking are all examples of
   a. nonverbal communication.
   b. instinctive behavior.
   c. cooperation.
   d. exchange.

18. "I'm certainly not going to watch Marge's kids this afternoon. She didn't even bring them to Jason's birthday party last week." The person speaking is thinking of her relation to Marge in terms of
   a. competition.
   b. conflict.
   c. cooperation.
   d. exchange.

19. The main difference between cooperation and exchange is that
   a. cooperation is based upon shared goals.
   b. relationships of exchange are voluntarily.
   c. cooperation has no material awards.
   d. exchange provides benefits for individuals only.

20. Using power in a way that is not considered legitimate by those on whom it is used is called
   a. conflict.
   b. competition.
   c. exploitation.
   d. coercion.

21. A sociologist was interested in impostors and how they managed to create the impression that they were doctors or lawyers or pilots or other things they were not. This sociologist's perspective is that of
   a. dramaturgy.
   b. conflict.
   c. exchange.
   d. competition.

ELEMENTS OF SOCIAL INTERACTION

22. T F Socially defined positions such as teacher, student, athlete, and daughter are called statuses.

23. T F For many students, keeping up good grades, holding down a job, and spending enough time with significant others involves role conflict.

24. T F A master status refers to one of the multiple statuses a person occupies that seems to dominate the others in patterning his or her life.

25. T F The status of *mayor* is basically defined by any person who occupies that position.

26. T F When people feel comfortable in a situation, a sociologist would say that they are not playing a role.

27. College professors are usually expected to engage in at least three types of activities: teaching effectively, conducting research and publishing the results, and devoting time to institutional governance and civic activities. Taken together, these activities constitute a
   a. status set.
   b. master status.
   c. role set.
   d. secondary group.

28. A college professor finds herself taking time away from class preparation in order to keep producing published research, or skipping meetings in order to grade papers for class. She is experiencing
   a. role strain.
   b. role conflict.
   c. status anxiety.
   d. status inconsistency.

29. Which of these multiple statuses would **most likely** function as a master status?
   a. President of the United States
   b. husband
   c. father
   d. none, as all are equal statuses

30. A professor said of a class that did not learn much, "I taught good, but boy did they learn lousy." The professor is ignoring his
   a. role set.
   b. role conflict.
   c. role strain.
   d. status set.

31. Which of the following is an ascribed status?
   a. male
   b. employee
   c. student
   d. shortstop

32. The relationship between roles and statuses is that
   a. a status may include a number of roles.
   b. a role may include many statuses.
   c. not all statuses have roles attached.
   d. statuses are dynamic while roles are not.

33. Marissa, a police officer, begins to suspect that her teenage son is involved in selling drugs. She is torn between responding as a parent and responding as a police officer. This is an example of
    a. role conflict.
    b. status conflict.
    c. role strain.
    d. role playing.

34. Some colleges have a policy of "legacy admissions," where the children of alumni are admitted even though their grades and exam scores may be lower than that of other applicants who were rejected. The policy is based on
    a. conflict.
    b. exchange.
    c. achieved status.
    d. ascribed status.

35. A primary school made sure that a child whose father was a teacher in the school was not assigned to be in his father's class. The policy was designed to avoid
    a. role conflict.
    b. role strain.
    c. achieved status.
    d. exchange.

TYPES OF SOCIETY

36. **T F** The type of society that produces the greatest degree of selfishness is hunter-gatherer.

37. **T F** Human societies have existed for no more than 10,000 years.

38. **T F** Post-industrial society has less class division than any other type of society in the history of humankind.

39. **T F** The main difference between horticultural and agricultural societies is that agricultural societies use a plow to till the soil, thereby making it more productive.

40. **T F** Most of the Native American Indian tribes of North America were pastoral societies.

41. **T F** Patterned relationships that fulfill the basic needs of a society are called "institutions."

42. After an African American church was destroyed by arson, the reverend said, "They didn't burn down the church. They burned down the building." His remark illustrates the concept of the church as
    a. an institution.
    b. an aggregate.
    c. a Gesellschaft.
    d. a group.

43. According to Tischler, the large increase in the percentage of women working outside the home led to a change in
    a. Gemeinschaft.
    b. social aggregates.
    c. social organization.
    d. oligarchy.

44. For most of its existence, the human species has lived in what type of society?
    a. hunter-gatherer
    b. agricultural
    c. pastoral
    d. industrial
    e. post-industrial

45. The type of society in which the greatest amount of equality among all people of the society is
    a. hunter-gatherer
    b. agricultural
    c. pastoral
    d. industrial
    e. post-industrial

46. The rise of cities occurred in which type of society?
    a. hunter-gatherer
    b. agricultural
    c. pastoral
    d. industrial
    e. post-industrial

47. According to Tischler, the rise of horticulture was a response to
    a. technological improvement.
    b. climate change.
    c. overcrowding.
    d. changes in family structure.

48. One difference between industrial and post-industrial society is
    a. the degree of social inequality.
    b. the structure of the family.
    c. the proportion of service jobs.
    d. the amount of political freedom.

ESSAY QUESTIONS

49. Discuss the influence of television on American social interaction.

50. Discuss the ways in which the context of the college classroom structures at least three different types of social interaction.

51. Create an essay in which you describe the ways blacks and whites in American society offend each other without realizing.

52. Tischler says that a role is "a collection of rights and obligations." Choose some role that you play and detail the rights and obligations it includes.

53. Robert Putnam argues that television detracts from social interaction and the general level of citizen involvement. But among young people today, some television time has given way to time spent online. Is sitting in front of a computer screen different from sitting in front of a TV screen? How does the Internet affect social interaction?

CHAPTER SIX   SOCIAL GROUPS AND ORGANIZATIONS

THE NATURE OF GROUPS

1. T F Because they share a particular characteristic common only to them, left-handed people constitute what sociologists would classify as a *social group.*

2. T F Social groups and social aggregates both cease to exist when members are apart from one another.

3. T F Primary groups are usually more willing than secondary groups to tolerate members who deviate from group expectations.

4. T F An aggregate is a group of people with close emotional ties and long-term relationships.

5. T F The difference between primary and secondary groups lies in the kinds of relationships the members have with one another.

6. T F Cooley called primary groups the nursery of human nature because they are effective during early childhood.

7. T F A secondary group may have specific goals, but offers less intimacy among the members than primary groups.

8. T F Primary groups are more likely than are secondary groups to rely on formal procedures and sanctions as a means of social control.

9. T F Secondary groups are more likely than are primary groups to deal with deviance by expelling the rule violator.

10. Which of the following would most likely be a secondary group?
    a. a family
    b. a juvenile gang
    c. a college club
    d. a high school clique

11. Which of the following is **not** a fundamental characteristic of social groups?
    a. Members have a common identity.
    b. Members enjoy each other's company.
    c. There is some feeling of unity.
    d. There are common goals and shared norms.

12. What is the sociological name for strangers waiting in line to buy movie tickets?
    a. group
    b. social aggregate
    c. clique
    d. social category

13. The most important characteristic of primary groups that is missing in secondary groups is
    a. intimacy.
    b. interaction.
    c. small size.
    d. shared expectations.

FUNCTIONS OF GROUPS

14. **T F** A reference group is a purposefully created special-interest group that has clearly defined goals and official ways of doing things.

15. **T F** According to Tischler, a group can usually get along quite well without expressive leadership.

16. **T F** According to Tischler, personality tests can now determine who will be a good leader in a group.

17. Tischler says that by doing tasks, members may also increase their commitment to one another and to the group. This function of increasing commitment would be classified as
    a. instrumental.　　　　　　　　　　c. primary.
    b. expressive.　　　　　　　　　　　d. goal-oriented.

18. Some adults fear that youths may choose "gangsta rap" artists as role models, dressing and talking like them, displaying the attitudes promoted in rap music, and even asking themselves, "What would Tupac do?" as a guide to behavior. Such youths would be using the rappers as
    a. a primary group.　　　　　　　　c. a reference group.
    b. a secondary group.　　　　　　　d. an aggregate.

19. "Why do I have to go to Uncle Mike's party? I barely know him, and I don't really like him," says a youngster. "Because you're a member of this family," answers his parent. What essential function of groups is the parent invoking in her reasoning?
    a. defining boundaries　　　　　　　c. setting goals
    b. choosing leaders　　　　　　　　d. assigning tasks

20. During jury deliberations, two members get into an angry argument about a piece of evidence. The jury foreman says, "I think tempers are a bit short; let's take a ten minute break, cool off, and discuss this calmly. The foreman is exhibiting
    a. task leadership.　　　　　　　　　c. expressive leadership.
    b. instrumental leadership.　　　　　d. legal leadership.

SMALL AND LARGE GROUPS

21. **T F** The smallest group is a dyad.

22. **T F** New members are more threatening to small groups than to large groups.

23. **T F** Large groups are more likely to rely on formal procedures than are small groups.

24. **T F** People in small groups are more likely to have specific and clearly spelled-out tasks than are people in large groups.

25. **T F** In Tönnies' view, we can expect to find more individualism and less of a cooperative spirit in Gemeinschaft rather than Gesellschaft.

26. **T F** Gesellschaft is characterized by a far greater number of secondary groups than occurs in Gemeinschaft.

27. **T F** Mechanical solidarity is more characteristic of small, preliterate societies than is organic solidarity.

28. **T F** Durkheim felt that the trouble with modern society was that it had no collective conscience.

29. Three students always study together. This group is a
    a. trio.　　　　　　　　　　　　　　c. dyad.
    b. triage.　　　　　　　　　　　　　d. triad.

30. In *Gemeinschaft and Gesellschaft*, Tönnies was most concerned with the change from
    a. capitalism to socialism.
    b. democracy to dictatorship.
    c. dyads to triads.
    d. rural society to urban society.

31. A good example of a Gesellschaft-like society in the U.S. is
    a. the Goth subculture.
    b. the Republican party.
    c. the March of Dimes.
    d. the Amish.

32. The core beliefs and values shared by the members of a society, in Durkheim's terms, is the
    a. Gesellschaft.
    b. collective unconscious.
    c. collective conscience.
    d. guilty conscience.

33. The coordination of many different and specialized positions in a society, in Durkheim's terms, is
    a. mechanical integration.
    b. organic integration.
    c. racial integration.
    d. social disintegration.

34. Durkheim thought that the collective conscience did NOT appear in
    a. simple, preliterate societies.
    b. agricultural societies.
    c. industrial societies.
    d. post-industrial societies.
    e. none of these; he thought it appeared in all societies.

BUREAUCRACY

35. T F Many people tend to think that only government agencies are bureaucracies, but most large private corporations are bureaucracies as well.

36. T F All bureaucracies have networks of people who help one another by bending rules and taking procedural shortcuts.

37. T F According to Weber's model of bureaucracy, bureaucrats (people who work in a bureaucracy) are likely to have special training and competency for the office they hold.

38. T F In bureaucracies that resemble Weber's ideal type, workers are especially likely to give special favorable treatment to their friends and family members but not to strangers.

39. T F In the view of Robert Michels, democracy is nearly impossible to maintain in a large organization.

40. T F The more like an oligarchy an organization becomes, the closer the leaders are to the ordinary members of the organization.

41. T F Michels felt that the "iron law of oligarchy" could be avoided by using personality tests to get the right kinds of leaders.

42. T F A change in social organization usually brings about a change in ideas.

43. Which of the following is **not** an essential aspect of bureaucracy as outlined in Weber's ideal type?
    a. extensive division of labor
    b. personalized service
    c. impartiality toward everyone seeking service
    d. hierarchy of authority

44. One advantage of the division of labor in a bureaucracy is that it allows for
    a. a high level of specialization and expertise.
    b. the ability to get around official regulations.
    c. a friendly atmosphere.
    d. personalized service.

45. "Boy, did I get the runaround in the administration building. They kept sending me from one office to the next till I finally found someone who knew how to solve my problem." The problem referred to here is caused by what aspect of bureaucracy?
    a. division of labor
    b. hierarchy of authority
    c. impartiality
    d. efficiency

46. After an African American church was destroyed by arson, the reverend said, "They didn't burn down the church. They burned down the building." His remark illustrates the concept of the church as
   a. an institution.
   b. an aggregate.
   c. a Gesellschaft.
   d. an oligarchy.

47. According to Tischler, the large increase in the percentage of women working outside the home led to a change in
   a. Gemeinschaft.
   b. social aggregates.
   c. social organization.
   d. oligarchy.

ESSAY QUESTIONS

48. Distinguish between primary and secondary groups, and weigh the relative advantage of each. That is, describe at least two circumstances or contexts in which primary groups are more necessary, advantageous, or effective than secondary groups, or vice versa.

49. Responding to an ad in the newspaper, a group of strangers gets together to form a software designers club. Discuss from a sociological point of view the six tasks these people must successfully accomplish to create a viable group.

50. Describe how a relatively informal group such as a family might look if you were to impose bureaucratic structures on it. Pay special attention how the group fulfills the functions mentioned in the text.

51. What groups serve as reference groups for you? Describe the ways in which these are reference groups and the nature of your relationship with them.

CHAPTER SEVEN   DEVIANT BEHAVIOR AND SOCIAL CONTROL

DEFINING NORMAL AND DEVIANT BEHAVIOR

1. **T F** Sociologically speaking, behavior is classified as normal or deviant only with reference to the group in which it occurs.

2. **T F** Durkheim maintained that a society without any deviant behavior is both desirable and possible.

3. **T F** In a large society, subgroups have differences of opinion as to which acts are deviant and which are not.

4. Which of the following is a dysfunction of deviance?
   a. It makes social life difficult and unpredictable.
   b. It diverts valuable resources that could be used elsewhere.
   c. It causes confusion about the norms and values of society.
   d. It undermines trust.
   e. all of the above

MECHANISMS OF SOCIAL CONTROL

5. **T F** Internal mechanisms of control are more effective and cost-efficient than are external mechanisms of control.

6. **T F** The term *sanctions* refers only to negative punishments, not positive rewards.

7. Which of the following is an internal means of social control?
   a. ridicule
   b. guilt
   c. imprisonment
   d. exclusion from the group

8. At a mostly black high school the principal instituted a program where students who got all A's were given $100 at an awards ceremony (see Chapter 4 "Our Diverse Society"). The sanctions the program uses are
a. formal positive.
b. formal negative.
c. informal positive.
d. informal negative.

9. At the awards assemblies for A students, other students often jeered and called the recipients names like "nerd" and "Whitey." The sanctions used by the students were
a. formal positive.
b. formal negative.
c. informal positive.
d. informal negative.

THEORIES OF CRIME AND DEVIANCE

10. T F Lombroso's theory of atavism is based on the idea that criminal behavior is learned and that people become criminals because others teach them the wrong kinds of behavior.

11. T F Shaw and McKay's study of Chicago neighborhoods showed that as the ethnicity of an area changed, its crime rate changed dramatically, and an area high in crime in one period might be low in crime a decade later.

12. T F Neutralization theory emphasizes the idea that the law should be neutral on matters of race and social class.

13. "Sure I lifted his wallet. Anybody who gets drunk and flashes a lot of money around is just asking to get ripped off." This is an example of which technique of neutralization?
a. denial of responsibility
b. appeal to a higher principle
c. denial of injury
d. denial of the victim

14. Which of the following theories is less concerned with the causes of norm violations than with the way others react to the deviance?
a. psychoanalytic theory
b. anomie theory
c. cultural transmission theory
d. labeling theory
e. control theory

15. Freudian theory sees a source of crime in the failure to develop a proper superego. The superego is
a. an internal mechanism of control.
b. an external mechanism of control.
c. a mechanism of control that is sometimes internal, sometimes external.
d. not a mechanism of control at all.

16. Which of the following emphasizes the social bond between the individual and the society as the most important factor in determining whether someone will become criminal?
a. psychoanalytic theory
b. anomie theory
c. cultural transmission theory
d. labeling theory
e. control theory

17. A youth who wants to achieve the sort of affluent lifestyle he sees on television but, unable to find a well-paying job, turns to drug dealing to pay for it would be classified as a(n) _____ in Merton's typology.
a. retreatist
b. rebel
c. innovator
d. conformist
e. danger to society

18. Sykes and Matza's neutralization theory emphasizes which of the following as a factor in crime?
a. thought processes
b. biological and genetic make-up
c. punishment and deterrence
d. attachment to society

19. Durkheim saw anomie as a condition of
  a. weak law enforcement.
  b. overemphasis on the welfare of the group.
  c. normlessness.
  d. dependency.

20. Lombroso and Sheldon, each in his own way, attempted to explain deviant behavior on the basis of
  a. psychological orientation.
  b. early childhood experiences.
  c. anatomical characteristics.
  d. differential association.

21. "People commit crimes because they see the rewards as outweighing the risks, and because it is more profitable than anything someone with their abilities might do." This statement is most consistent with which type of theory?
  a. behavioral
  b. psychoanalytic
  c. rational choice
  d. biological

22. Fourteen-year-old Janet is arrested for shoplifting. Even though the charges are later dropped, Janet's teachers designate her a troublemaker and someone not to be trusted. Because the teachers make school an unwelcoming place for her, Janet begins to skip school frequently and to get into fights when she is there. Lemert and others would call these latter behaviors
  a. primary deviance.
  b. secondary deviance.
  c. recidivism.
  d. anomie.

23. The research of Shaw and McKay, which linked crime to certain types of urban neighborhoods, provided the foundation for _____ theories of deviance.
  a. control
  b. labeling
  c. genetic transmission
  d. cultural transmission

THE IMPORTANCE OF LAW

24. Conflict theory maintains that laws have the effect of
  a. protecting law-abiding people from harm.
  b. protecting poor people against exploitation by the wealthy and powerful.
  c. protecting the wealthy and powerful from losing their position of dominance.
  d. protecting the entire society from harmful conflicts.

25. "The law is merely a formal and enforceable statement of widely accepted norms and values." The idea expressed here is most consistent with
  a. control theory.
  b. labeling theory.
  c. conflict theory.
  d. consensus theory.

CRIME IN THE UNITED STATES

26. **T F** Because robbery involves the taking of property, it is classified as a "property crime."

27. **T F** The FBI's Uniform Crime Reports is based on a nationwide, door-to-door survey of crime victims.

28. **T F** The major reason that the Uniform Crime Reports gives an inaccurate count of the number of crimes in the U.S. is that victims of crime frequently do not call the police.

29. **T F** The United States violent crime rate in 2005 reached the ***highest level*** since the Bureau of Justice Statistics started measuring it in 1973.

30. **T F** Violent crimes are committed far more frequently than property crimes and far outnumber property crimes in both the UCR and the NCVS.

31. **T F** The UCR's measure of serious crime does not include white-collar crimes like bribery and fraud.

32. **T F** The amount lost to white-collar crime far exceeds the amount lost to street crimes like burglary and robbery.

33. **T F** The United States has much higher rates of murder than do other countries with similar economic and political systems (Western European countries, Canada, etc.)

34. **T F** The most frequent reason given by victims for not reporting crime to the authorities is the belief that the crime was not important enough.

35. **T F** Levin concluded from his research on *serial murderers* that the majority of such murderers suffer from insanity.

36. **T F** A major difference between adult and juvenile crime is that juveniles are much more likely to commit offenses in groups.

37. **T F** Females are much more likely to be victims of serious crimes than are males.

38. Which of the following crimes is most likely to be reported to the police?
    a. robbery
    b. rape
    c. motor vehicle theft
    d. burglary

39. Relatively minor crimes that are usually punishable by a fine or less than a year's confinement are called
    a. civil offenses.
    b. misdemeanors.
    c. larcenies.
    d. felonies.

40. A "status offense" is an act that is
    a. against the law if committed by a juvenile but not if committed by an adult.
    b. committed by a criminal in order to achieve higher status.
    c. committed by a person of high status.
    d. committed by a person of low status.

41. Violation of laws meant to enforce the moral code, such as public drunkenness, prostitution, gambling, and possession of illegal drugs, are called _____ crimes.
    a. moral
    b. victimless
    c. organized
    d. status

CRIMINAL JUSTICE IN THE UNITED STATES

42. **T F** Unlike policing in countries with a national police force, policing in the U.S. is controlled largely at the local level.

43. **T F** Truth-in-sentencing laws are an attempt to reduce prison overcrowding.

44. **T F** Women are more likely to commit property as opposed to violent crimes.

45. **T F** About half of all inmates on death row come from middle-class backgrounds.

46. The increased police presence during terror alerts had the greatest impact on which crime?
    a. murder
    b. rape
    c. motor vehicle theft
    d. fraud

47. Steering offenders away from the justice system and into social agencies is known as:
    a. diversion.
    b. labeling.
    c. recidivism.
    d. rehabilitation.

48. We can get an idea of whether prisons deter criminals who are sent there by looking at rates of
    a. diversion.
    b. recidivism.
    c. incarceration.
    d. capital punishment.

49. Which of the following is a goal of imprisonment?
    a. punishment of criminal behavior
    b. separation of criminals from society
    c. deterrence of criminal behavior
    d. rehabilitation of criminals
    e. these are all goals of imprisonment.

50. Which of the following countries executes more people than does the U.S.?
   a. Japan
   b. France
   c. Mexico
   d. India
   e. none of these

51. The process in which a large number of crimes committed results in only a small number of criminals going to prison is called the _____ effect.
   a. diversion
   b. deterrence
   c. funnel
   d. recidivism

52. At which stage in the "funnel effect" do the greatest number of crimes disappear?
   a. victims not reporting crimes to the police
   b. police not arresting the criminal
   c. arrested criminals not being convicted
   d. judges giving convicted criminals sentences other than prison

ESSAY QUESTIONS

53. Since the 1960s, America has seen a succession of youth cultures, each of which has been self-consciously deviant from mainstream attitudes and values. Evaluate the functional and dysfunctional aspects of deviant youth culture.

54. Using the rules and regulations on your own campus as examples, compare and contrast the *conflict* with the *consensus* theories of law.

55. Describe Merton's five modes of individual adaptation to the discrepancy between cultural goals and institutionalized means.

CHAPTER EIGHT   SOCIAL CLASS IN THE UNITED STATES

AMERICAN CLASS STRUCTURE

1. T F Social class is a fact of all industrialized societies, even in socialist and communist countries.

2. T F In recent years the income gap between the rich and the poor in the U.S. has increased.

3. T F Members of a social class usually have roughly similar lifestyles.

4. T F In his essay on changing social class, Todd Erkel claims that education has made it easy for him to move from one social class to another.

5. T F Historically, the upper class in the United States have been predominantly Jewish.

6. T F In the U.S., members of the upper class, despite their great wealth, often live, work, and associate socially with members of the upper-middle and lower-middle class.

7. T F According to Todd Erkel, the struggles working-class children experience develop high self-esteem and tend to generate high expectations.

8. T F Lower-class people typically do not share the American desire for advancement and achievement found in the other social classes.

9. T F Because most Americans work for a living, well over 50% of the population is usually considered working class.

10. T F Great Britain, with its long history of nobility and aristocracy, has a much greater gap between the rich and the rest of society than does the U.S.

11. Nearly all sociologists agree that there are _____ social classes in the U.S.
    a. 3
    b. 4
    c. 5
    d. 6
    e. There is no number of social classes that sociologists agree on.

12. About what proportion of the total amount of income in the U.S. goes to the richest one-fifth of the population?
    a. 20%   **c.** 50%
    b. 35%   **d.** 75%

13. The upper class of the United States is approximately _____ of the population.
    a. 1 to 3%   **c.** 13 to 15%
    b. 7 to 9%   **d.** 21 to 23%

14. The term for stocks, real estate, and other owned property not including the money a person earns at work is called
    a. income.   **c.** prosperity.
    b. wealth.   **d.** mobility.

15. Managers and other people who work in the professional and technical fields are members of the _____ class.
    a. upper   **c.** lower-middle
    b. upper-middle   **d.** working

16. If you wanted to show a larger degree of inequality between the rich and the rest of the population which data would you look at?
    a. distribution of income
    b. distribution of wealth
    c. neither, they show the same degree of inequality

17. In which of these countries is income inequality—the difference between rich and poor—greatest?
    a. Sweden
    b. Germany
    c. France
    d. Italy
    e. the United States

18. Bob is a master carpenter. He holds a high school diploma, owns a modest home, but struggles to give his family the things they need and want. He considers himself politically conservative and moderately religious. According to your text, what social class does he belong to?
    a. upper-middle   **c.** lower
    b. lower-middle   **d.** working

POVERTY

19. **T F** Basically, poverty refers to a condition in which people maintain a standard of living that includes the basic necessities, but none of the extras and luxuries.

20. **T F** Most government agencies define a family as poor if its income is less than half of the average American family income.

21. **T F** More than one American in ten was below the poverty line in 2004.

22. **T F** The poverty index includes only cash income and does not take account of noncash benefits like Medicaid and food stamps.

23. **T F** Mothers who are divorced are usually economically worse off than mothers who were never married.

24. **T F** Poor people who work are often not eligible for the assistance that jobless poor people are eligible for.

25. T F Since the 1970s, the percentage of the elderly living in poverty has decreased while the percentage of children living in poverty has increased.

26. T F The minimum wage in the U.S. is set so that a single parent of two children who works full time at a minimum-wage job will have an income above the poverty index.

27. Since the passage of the Welfare Reform Act in 1996, the number of families receiving aid has
   a. increased greatly.
   b. decreased greatly.
   c. remained about the same.

28. Because of the way the government computes the official poverty index—the income level below which a family is defined as poor—this index can be affected by a change in the cost of
   a. food.
   b. clothing.
   c. shelter.
   d. fuel.
   e. all of these

29. Of the following nations, the highest rate of poverty among children is found in
   a. the United States.
   b. Canada.
   c. Germany.
   d. the United Kingdom.
   e. Sweden.

30. According to the Bureau of Census, in 2002 a family of four was living in poverty if their income was below
   a. $18,100.
   b. $24,300.
   c. $28,200.
   d. $32,000.

31. Poverty is more frequent
   a. among men than women.
   b. among whites than minorities.
   c. among the elderly than middle-aged people.
   d. in rural than in urban areas.
   e. in the Northeast than in the South.

32. The feminization of poverty refers to
   a. the trend in which females represent an increasing proportion of the poor.
   b. a federal program eliminating aid to unwed mothers.
   c. an increase of feminist politicians involved in aid programs.
   d. the vast overrepresentation of women among social workers who work with the poor.

GOVERNMENT ASSISTANCE PROGRAMS

33. T F The amount of government benefits going to the middle class (Social Security, Medicare, etc.) is greater than the amount going to the poor in welfare programs.

34. T F The amount of government money going to female-headed families in poverty is less than one-tenth of the amount going to Social Security payments to the elderly.

35. T F Welfare programs are generous enough that they provide a comfortable lifestyle for most of the poor.

36. If a government program is "means tested,"
   a. it is a beta version, still being tested.
   b. it is available to all people regardless of income.
   c. it is available only to people who can prove that they earn below a certain income.
   d. it is available to people with an average income.

37. The largest amount of the federal government assistance goes to
    a. social security retirement.
    b. aid to families with dependent children.
    c. Medicaid.
    d. food stamps.

38. United States government programs to combat poverty have been most successful among
    a. children.
    b. working-age adults.
    c. the elderly.
    d. women.

39. When people are asked to assess the causes of poverty, the poor are more likely than other groups to mention
    a. education.
    b. "the system."
    c. drugs.
    d. loose morals.

CONSEQUENCES OF SOCIAL STRATIFICATION

40. **T F** Even when a middle-class person and a poor person commit similar crimes, the poor person is more likely to be arrested.

41. **T F** Because of the right to a free lawyer, a poor person charged with a crime is no more likely to be convicted than is a middle-class person.

42. **T F** Rich-poor differences in health do not begin until late childhood. Babies of the poor are no more likely to die in the first year than are babies of the middle or upper classes.

43. **T F** Prison systems are heavily populated by the poor.

44. **T F** The crimes committed by the poor do much more financial harm to society than the crimes of those in the middle and upper classes.

45. In which social class are parents most likely to treat sons and daughters similarly?
    a. lower class
    b. working class
    c. middle class
    d. none of these; there are no real differences among social classes when it comes to child rearing

THEORIES

46. **T F** The functionalist explanation maintains that inequality exists and persists because it benefits the society as a whole.

47. **T F** Kingsley Davis and Wilbert Moore contend that social stratification exists in most industrialized nations except the United States.

48. **T F** According to functionalist theory, a society must pay some jobs more in order to attract people to important positions.

49. **T F** According to functionalist theory, low-paying jobs are low-paying because they are not as necessary to the society as are better-paying jobs.

50. **T F** Conflict theorists argue that inequality persists because those with wealth and power use that wealth and power to protect and further their own interests, even at the expense of the interests of others in the society.

51. **T F** In Marx's terms, the "bourgeoisie" is made up of the owners of the means of production.

52. **T F** Max Weber, unlike Marx, argued that status and power could be separated from economic position.

53. Conflict theories of inequality are often based on the ideas of
    a. George W. Bush and Ronald Reagan.
    b. Kingsley Davis and Wilbert Moore.
    c. Karl Marx.
    d. Émile Durkheim.

54. According to Marx, what was the relation between the owners of the means of production (factories, land, etc.) and those who worked for them?
    a. Owners, if they were wise, made sure that workers were satisfied with their jobs.
    b. Owners usually took workers' interests into consideration.
    c. Owners and workers cooperated since the success of a business meant increased income for everyone in it.
    d. Owners exploited workers and paid them as little as possible.

CHAPTER NINE   GLOBAL STRATIFICATION

STRATIFICATION SYSTEMS

1. **T F** More than half the people in the world live on less than $2 a day.

2. **T F** In most countries, the gap between rich and poor has gotten narrower over the last half century.

3. **T F** In all societies, the basis of social stratification is income.

4. **T F** Social class systems are more open than social caste systems.

5. **T F** The caste system in India is just as rigid today as it was 100 years ago.

6. In India, the *jatis* (subcastes) are in principle based on
    a. age.
    b. sex.
    c. religion.
    d. occupation.
    e. income.

7. In India, as in most societies with a caste system, the justification of the caste system is based mostly on
    a. legal principles.
    b. religious ideas.
    c. economic necessity.
    d. the teachings of a charismatic leader.

8. **T F** The principle of "untouchability" and the untouchable caste (*Dalitis*) is currently illegal in India.

9. Despite being made illegal nearly 60 years ago, untouchability in India still persists, especially in
    a. business.
    b. poor neighborhoods of large cities.
    c. rural areas.
    d. government.
    e. none of these; untouchability has been nearly totally eliminated everywhere.

10. **T F** Estate systems are usually more rigid than the caste systems.

11. The system of stratification prevalent in Europe during the Middle Ages was
    a. a caste system.
    b. a class system.
    c. an estate system.
    d. a religious hierarchy.
    e. a classless society.

12. A person with great ambition would fit best in
    a. a caste system.
    b. a class system.
    c. an estate system.
    d. a religious hierarchy.
    e. a classless society.

13. **T F** In medieval Europe, the clergy was not considered an estate.

14. **T F** During the Middle Ages, merchants were not actually a formally recognized estate.

15. **T F** The estate system of the Middle Ages provided some limited social mobility.

THEORIES OF GLOBAL STRATIFICATION

16. **T F** Modernization theory plays down the importance of a society's cultural environment when describing the modernization process.

17. **T F** According to modernization theory, developed countries are actually very important in helping less developed countries modernize.

18. **T F** Critics of modernization theory see it as a defense of capitalism.

19. Which of the following would modernization theorists see as contributing most to the economic development of a poor country?
    a. strong religion
    b. strong cultural traditions
    c. education
    d. high birth rates producing lots of potential workers
    e. all of these

20. **T F** According to modernization theory, changes in government policy have little or no impact on their country's economic development.

21. **T F** Dependency theory is an outgrowth of a Marxist view of the interconnection among world systems.

22. Which theory locates the causes of world poverty in the actions and policies of wealthy developed countries?
    a. modernization theory
    b. dependency theory
    c. neither; they both emphasize factors in the underdeveloped countries themselves

23. **T F** Dependency theory views capitalism as having a global impact rather than in one specific country only.

24. Dependency theory gets its name because it emphasizes the idea that
    a. rich countries are dependent on poor countries for raw materials.
    b. rich countries are dependent on poor countries for cheap labor.
    c. rich countries are dependent on poor countries as buyers of finished goods.
    d. poor countries are dependent on rich countries for capital.

25. Tischler says that dependency theory is "an outgrowth of a Marxist view." In what way does dependency theory resemble other Marxist ideas?
    a. It emphasizes the conflict between the interests of the rich and the interests of the poor.
    b. It predicts the fall of global capitalism.
    c. It emphasizes the role of religion as a factor in development.
    d. It says that the most important factor in poor countries is the attitude of people there.

26. **T F** Critics of dependency theory argue that many countries that are poor have little contact with rich countries, so rich countries could not be a cause of their poverty.

27. Even with the diversity throughout the world, what factors are common to a significant percentage of the population in most developing nations?
    a. poor sanitation
    b. no access to modern health services
    c. inadequate housing
    d. many children not attending school
    e. all of the above

28. **T F** In the last century, the worldwide average life expectancy has at least doubled.

29. T F Because of great advances in medicine and communication, the majority of people in the world now have access to professional heath care.

30. Approximately what percent of the population in China today may be expected to reach age 70?
   a. less than 10%
   b. about 25%
   c. 60%
   d. 90%

31. T F Because of the existence of antibiotics, approximately the same relative percentage of persons is dying in developing nations as in the developed countries.

32. In developing nations, the most frequent cause of death for children under age five is
   a. child abuse.
   b. malnutrition.
   c. predatory animals.
   d. infectious diseases.
   e. accidents.

33. In developed countries (Europe, the U.S., etc.), Infectious diseases like measles and diarrhea cause about what percent of all childhood death?
   a. 1%
   b. 25%
   c. 40%
   d. more than 50%

34. T F Most childhood deaths in developing countries could be prevented by making sanitary drinking water and inexpensive medicines available.

35. T F In rich countries and poor, an increase in the mother's education is associated with decrease in child mortality.

36. In which area of the world are HIV/AIDS rates highest (i.e., the largest percentage of the population is infected)?
   a. East Asia
   b. North America
   c. Russia and Eastern Europe
   d. sub-Saharan Africa
   e. the Caribbean

37. T F Despite the increase of HIV/AIDS worldwide, in wealthier countries the number of AIDS deaths has declined since the 1990s.

38. In countries outside the Europe and North America, the most common source of HIV/AIDS transmission is
   a. intravenous drug use.
   b. homosexual sex.
   c. heterosexual sex.
   d. blood transfusion.

39. Roughly how many people are there currently on planet Earth?
   a. 200 million
   b. 1–2 billion
   c. 6–7 billion
   d. 10–12 billion

40. The population of the Earth did not begin to grow until about _____ years ago.
   a. 1 million
   b. 50,000
   c. 10,000
   d. 1,000
   e. 400

41. T F The increase in population on Earth is accounted for mostly by countries in the developing world.

42. T F There is no connection between the age at which a woman marries and the number of children she has. Fertility is determined by economic and health factors.

43. In countries with little modern contraception (Pakistan, Bangladesh, sub-Saharan Africa) the greatest factor in lowering fertility is
   a. abortion.
   b. abstinence.
   c. breast feeding.
   d. herbal contraceptive potions.

44. Worldwide, the most widely used form of birth control is
   a. condoms.
   b. the pill.
   c. abstinence.
   d. abortion.
   e. sterilization.

45. If you wanted to reduce fertility in a country, you would do best to increase the years of education of
   a. women.
   b. men.
   c. there is no difference

46. T F In the developing world, birth rates are generally higher in cities than in rural areas.

47. T F By the year 2050, the number of elderly people will exceed the number of younger persons worldwide.

48. T F In the developing world, a factor promoting the desirability of boys over girls is the expectation that sons will be more likely to assume the care of elderly parents.

49. T F The patriarchal family structure of China and India is important in the preferences for boy children over girl children.

50. T F Throughout all countries, people with the greatest income usually have more children than those with lower income, largely because of wealthier people have the ability to support additional family members.

51. In which of these countries are elderly people most likely to be living alone?
   a. Japan
   b. Sweden
   c. Pakistan
   d. China

52. Barring deliberate human intervention that affects sex ratios, which sex has a better survival rate from birth on?
   a. males
   b. females
   c. males in the first six years, then females
   d. none of these; there is no difference

CHAPTER TEN   RACIAL AND ETHNIC MINORITIES

THE CONCEPT OF RACE

1. T F Racial classifications are not simple or cut-and-dried; a person considered "white" in one society might be categorized as "black" in another.

2. T F Differences in traits such as skin color, hair texture, and nose type have proved useful in making biological classifications of human beings.

3. T F Even back in the 1700s, some physiologists argued that racial categories did not reflect the actual divisions among human groups.

4. T F In recent years we have seen the rise in hate sites on the World Wide Web.

5. T F Prior to 2000, the U.S. census allowed a person to check only one category for race; nobody could be classified as multi-racial.

6. T F Nearly half of the U.S. population now classifies themselves as multi-racial.

7. T F The number of interracial marriages in the U.S. has increased greatly in the last three decades.

8. T F Despite changes in race relations and laws over the last 50 years, a majority of Americans still say they disapprove of interracial marriage.

9. The term *race* refers to a category of people who are defined as similar because they
   a. have a unique and distinctive genetic makeup.
   b. share a number of physical characteristics.
   c. exhibit similar behaviors.
   d. express comparable attitudes.

10. The U.S. census counts a person as African American if he or she
   a. has at least one African American parent (1/2 black).
   b. has at least one African American grandparent (1/4 black).
   c. has at least one African American great-grandparent (1/16 black).
   d. defines himself or herself as African American regardless of ancestry.

11. When were laws prohibiting interracial marriage declared unconstitutional by the Supreme Court?
   a. shortly after the founding of the U.S.
   b. shortly after the Civil War
   c. shortly after World War II
   d. in the 1960s
   e. none of these; the Court has never declared anti-intermarriage laws unconstitutional

12. About what percentage of married couples in the U.S. is interracial?
   a. 5%   **c.** 50%
   b. 25%   **d.** 75%

13. The text's example of Philip and Paul Malone was used to illustrate
   a. the failure of multiculturalism.
   b. a humorous look at the problems of birth records.
   c. the difficulties in accurately assigning race.
   d. that racial appearance and racial definition generally match closely.

THE CONCEPT OF ETHNIC GROUP

14. T F To qualify as an ethnic group, members must be unified politically.

15. According to Tischler, the distinguishing features of an ethnic group are usually
   a. cultural   **c.** economic
   b. physical   **d.** political

THE CONCEPT OF MINORITIES

16. Women can be regarded as a minority group in American society because
   a. There are fewer women than there are men.
   b. They belong to a variety of different ethnic and racial groups.
   c. They are discriminated against because of their sex.
   d. None of these; they cannot be regarded as a minority by any definition.

PROBLEMS IN RACE AND ETHNIC RELATIONS

17. T F Prejudice is a negative attitude that is nevertheless rational.

18. T F Prejudice serves some positive functions for the people who hold that prejudice.

19. When psychologists say that prejudice allows for "projection," they mean that the prejudiced person
   a. sees what are really his own faults in some other group.
   b. speaks loudly about his opinions of other groups.
   c. prefers movies about the groups he dislikes.
   d. feels guilty about his own hatreds.

20. T F According to Tischler, people with prejudiced attitudes frequently do not engage in discriminatory behavior.

21. Social arrangements that restrict a group's life chances even though there is no apparent hatred, prejudice, or stereotyping is called
   a. racism.   **c.** institutionalized prejudice.
   b. socialism.   **d.** projection.

22. A major difference between prejudice and discrimination is that
    a. prejudice involves ethnic groups; discrimination involves race.
    b. prejudice involves thoughts; discrimination involves actions.
    c. prejudice involves economics; discrimination involves culture.
    d. prejudice can be reversed; discrimination is permanent.

23. **T F** Prejudicial feelings are more likely to develop between groups that are competing against each other for scarce resources.

24. Many white shopkeepers in U.S. Southern towns during the 1950s and 1960s depended upon African American customers for a large part of their business but considered them social inferiors. These merchants are examples of
    a. unprejudiced nondiscriminators.
    b. unprejudiced discriminators.
    c. prejudiced nondiscriminators.
    d. prejudiced discriminators.

25. J.T. is a member of a minority group. She is unable to obtain a well-paying, secure job not because of outright racism, but rather because, like many others of her minority group, she attended a less-than-adequate school and lacks "connections." J.T. is a victim of
    a. subtle and unrecognized personal prejudice.
    b. unfortunate accidental discrimination.
    c. institutionalized prejudice and discrimination.
    d. bad luck that has nothing to do with her being a minority.

26. An individual who feels very uncomfortable when friends tell a racist joke and yet does not speak out for fear of being ridiculed would be classified by Merton as a(n)
    a. unprejudiced nondiscriminator.
    b. unprejudiced discriminator.
    c. prejudiced nondiscriminator.
    d. prejudiced discriminator.

PATTERNS OF RACIAL AND ETHNIC RELATIONS

27. Canada's maintaining two official languages, English and French, is an example
    a. pluralism.
    b. assimilation.
    c. Anglo conformity.
    d. subjugation.

28. During the nineteenth century, Native Americans were pushed off of land desired by white settlers and onto small and distant reservations. This exemplifies all of the following EXCEPT
    a. forced migration.
    b. segregation.
    c. assimilation.
    d. expulsion.

29. A 1946 government report on Native Americans called for "their complete integration into the mass of the population." The report was advocating
    a. assimilation.
    b. pluralism.
    c. forced migration.
    d. segregation.

30. "Not the elimination of differences but the perfection and conservation of differences" is a statement favoring
    a. pluralism.
    b. assimilation.
    c. Anglo conformity.
    d. subjugation.

31. "Genocide," a practice banned by international law, is another term for
    a. forced migration.
    b. segregation.
    c. subjugation.
    d. annihilation.

32. Most of the nineteenth-century immigrants to America had distinctive subcultures with their own unique language, style of dress, norms, and values. The children of these immigrants, however, rapidly learned to speak English and to adopt mainstream American cultural styles. The children's behavior is an example of
    a. segregation.
    b. pluralism.
    c. assimilation.
    d. subjugation.

33. The development of ethnic neighborhoods like Chinatowns and Little Italys is an example of voluntary
    a. segregation.
    b. Anglo conformity.
    c. subjugation.
    d. assimilation.

34. The policy of the Nazis toward the Jews during the Holocaust of the 1930s and 1940s was one of
    a. pluralism.
    b. Anglo conformity.
    c. annihilation.
    d. subjugation.

35. In order to advance his career in radio, Jaime Fernandez learns to speak English without a trace of a Spanish accent and changes his name to Jim Fox. This is an example of
    a. subjugation.
    b. Anglo conformity.
    c. Chicano conformity.
    d. annihilation.

RACIAL AND ETHNIC IMMIGRATION TO THE UNITED STATES

36. T F Even during its more restrictive periods, the United States has had one of the more open immigration policies in the world.

37. T F Each year, the U.S. takes in more immigrants than does the rest of the world combined.

38. T F Hispanics are now the largest ethnic group in the U.S.

39. The largest proportion of Hispanics in the U.S. came from
    a. Cuba.
    b. The Dominican Republic.
    c. Mexico.
    d. Argentina.

40. According to Tischler, compared with other Hispanics, Cuban Americans
    a. are less educated.
    b. have a lower average income.
    c. are more likely to resist assimilation.
    d. are more likely to vote Democratic.

41. The earliest large wave of immigrants from China worked mostly in
    a. restaurants.
    b. laundries.
    c. opium dens.
    d. railroad construction.

42. Which ethnic or regional group has the highest average levels of education?
    a. African Americans
    b. Italian Americans
    c. Asian Americans
    d. Hispanics

43. Italian Americans today are most likely to trace their ancestors back to
    a. the old migration.
    b. the new migration.
    c. the modern migration.
    d. the postmodern migration.

44. About what proportion of U.S. residents were born in foreign countries?
    a. 2%
    b. 10%
    c. 25%
    d. 44%

45. In recent years, some states have proposed laws making English the official state language and eliminating or restricting bilingual education. The aim of these measures is
    a. annihilation.
    b. Anglo conformity.
    c. exploitation.
    d. pluralism.
    e. segregation.

46. The largest proportion of immigrants to the U.S. today is accounted for by people from
    a. Europe.
    b. Africa.
    c. Asia.
    d. Latin America.
    e. Antarctica.

47. T F More than half of all Native Americans live on or near reservations administered by the U.S. government.

48. *Old migration* to the United States consisted of people from
    a. the ancient civilizations of the Mediterranean.
    b. northern Europe who came prior to 1880.
    c. eastern Europe who came after 1880.
    d. age 50 upward.

49. The immigration act of 1924 placed quotas specifically designed to discriminate against potential immigrants from
    a. Ireland.
    b. Africa.
    c. Southern and Eastern Europe.
    d. South America.
    e. China.

50. The largest number of African Americans is found in what part of the U.S.?
    a. Northeast
    b. Midwest
    c. South
    d. Mountain states
    e. West Coast

51. According to Tischler, a major reason for the increasing income gap between African Americans and whites is
    a. racism.
    b. the decline of unionized jobs.
    c. the influx of African immigrants.
    d. the increase in female-headed families among African Americans.

52. According to Tischler, the most successfully racially integrated institution in the U.S. is
    a. the church.
    b. education.
    c. the corporate world.
    d. the military.

53. The majority of all illegal immigrants to the United States come from
    a. Mexico.
    b. Central America.
    c. Southeast Asia.
    d. the Caribbean area.

54. In the year 2000, African Americans made up approximately _____ % of the total population of the United States.
    a. 2.5
    b. 12.3
    c. 24.6
    d. 43.3

ESSAY QUESTION

55. Choose a specific group, category, or subculture and explain how it does or does not qualify as a race, ethnic group, or minority.

CHAPTER ELEVEN   GENDER STRATIFICATION

ARE THE SEXES SEPARATE AND UNEQUAL?

1. T F Traditional Chinese society had much greater equality between the sexes than did traditional European society.

2. T F Critics of sociobiology are likely to place much more emphasis on learned behavior and socialization.

3. T F The biblical story of creation has been used as a theological justification for patriarchal ideology.

4. T F Pioneering sociologist Auguste Comte differed from most other thinkers of his time in that he saw women as intellectually equal to men.

5. T F Although societies may treat the sexes unequally, diseases like diabetes and multiple sclerosis affect men and women with equal frequency.

6. T F In her book *Sex and Temperament,* Margaret Mead tried to provide support for biological and genetic theories of male-female differences.

7. T F Unlike modern Western society, many simpler, preliterate societies have no division of labor by sex. Men and women do pretty much the same things.

8. The study of animal behavior is called
a. ethology.
b. ethnology.
c. ethnography.
d. animology.
e. sociobiology.

9. The branch of science that tries to establish the genetic bases of human behavior is called
a. biogenetics.
b. ethology.
c. ethnography.
d. sociobiology.

10. Ethology is the study of
a. women named Ethel.
b. ethical behavior.
c. animal behavior.
d. ethnic groups.

11. In Clark and Hatfield's study of pick-ups, which group was most likely to agree to have sex with an attractive stranger of the opposite sex?
a. men
b. women
c. there was no difference

12. In Clark and Hatfield's study of pick-ups, which group was most likely to agree to a date with an attractive stranger of the opposite sex?
a. men
b. women
c. there was no difference

13. A patriarchal ideology is
a. the study of powerful males in past societies.
b. the belief that there are differences in the social behavior of men and women.
c. an attempt to find a genetic basis for human behavior.
d. the belief that men are superior to women and should control all aspects of society.

14. On a form, the category that asks whether you are male or female should properly be labeled
a. sex
b. gender
c. either
d. neither

15. Dr. John Money asserted that "the gender identity gate is open at birth." This statement, along with Money's research and practice, suggests that
a. a child must remain in the gender indicated by its genitals at the time of birth.
b. all children are born with ambiguous sexual characteristics.
c. sex-role identity is not something a child is born with but something that is learned early in life.
d. anatomy is destiny.

16. T F The biographies of Dr. John Money's patients and others who received gender-changing operations in early life have generally supported Money's idea that gender roles can by changed.

17. Many academics protested when a university president implied that women were "naturally" not as good at math as were men. The critics accusing him of placing too much emphasis on biology thought he had
a. used the concept sex when he should have used the concept of gender.
b. used achieved status when he should have used ascribed status.
c. emphasized learned behavior when he should have emphasized physiology.
d. used genetic ideas when he should have used religious ideas.

18. Gender is best understood as a(n) _____ status.
a. achieved
b. ascribed
c. ideal
d. peripheral

19. In stressful situations
    a. males and females react with similar intensity.
    b. males and females react with similar behaviors, though women are more intense.
    c. men react more slowly.
    d. women react more slowly.

20. Critics of sociobiology assert that
    a. even among animals and insects, much gender-related behavior is learned, not innate.
    b. sex differences have no biological basis.
    c. gender differences do not exist among nonhuman primates.
    d. it is not valid to generalize from animal to human behavior.

21. **T F** Differences in hormones in men's and women's brains might explain why aging women suffer cognitive decline (e.g., Alzheimer's disease) at much higher incidence than men.

22. Research suggests that "befriending" behavior among females may be caused by the way their bodies process oxytocin. This research supports the _____ explanation of male-female difference.
    a. cultural
    b. social
    c. socialization
    d. biological

WHAT PRODUCES GENDER INEQUALITY?

23. Functionalists argue that the family functions best when
    a. the father focuses on things outside the home while the mother focuses on relationships within the family.
    b. gender roles are more equal.
    c. men share more equally in the internal life of the family.
    d. the father deals with sons' emotions while mother deals with daughter's emotions.

24. The "instrumental" role in the family usually involves
    a. accompanying family singing.
    b. nurturing children.
    c. earning money.
    d. listening carefully to one's spouse's emotional needs.

25. The "expressive" role in the family usually involves
    a. earning money.
    b. dealing with interpersonal conflicts.
    c. paying the bills.
    d. cleaning and fixing up the house.

26. Conflict theorists argue that gender inequality is _____ based.
    a. biologically
    b. functionally
    c. economically
    d. psychologically

27. Conflict theorists see the traditional male-female relationship as one of
    a. cooperation.
    b. exploitation.
    c. mutuality.
    d. passive-aggressive behavior.

28. Karl Marx's sometime collaborator Friederich Engels argued that the basis for inequality between the sexes was
    a. biology.
    b. capitalism.
    c. socialism.
    d. communism.

29. **T F** Anthropologist Michelle Rosaldo argues that the status of females will be lowest in societies that have a "firm differentiation between domestic and public spheres."

SOCIALIZATION

30. **T F** Recent research now indicates that a majority of women and men now favor female bosses over male bosses in the workplace.

31. Women today, says Tischler, are encouraged to pursue careers before, during, and after marriage. The increased presence of women in the labor force, according to this theory, arises from changes in
    a. the economy.
    b. socialization.
    c. education.
    d. hormones.

32. Maines and Hardesty say that men live in a "linear temporal world," while women live in a "contingent temporal world." By this they mean that men, more so than women,
    a. focus on career goals.
    b. align their career with family demands.
    c. complete things on time.
    d. keep their tempers in line.

33. Carol Gilligan argues that women's ethical systems are characterized by a(n)
    a. inability to make clear moral choices.
    b. emphasis on the interconnectedness of actions and relationships.
    c. lower level of moral reasoning than do men.
    d. application of absolute principles.

34. Social psychologist Erik Erikson argued that in Western society it is more difficult for girls than for boys to
    a. achieve a positive identity.
    b. learn to moderate their innate aggression.
    c. learn to be nurturing.
    d. develop behaviors to attract a suitable mate.

GENDER INEQUALITY AND WORK

35. **T F** Even when women and men have similar job titles and do equivalent work, women on average receive lower pay.

36. **T F** Women are more likely than men to work in low-paying jobs.

37. **T F** By the year 2000, women were just as likely as men with college degrees to win career-enhancing promotions.

38. On the average, women with two-year college degrees earn
    a. more than men with college two-year degrees.
    b. about the same as men with college degrees.
    c. only slightly more than men with high school diplomas.
    d. less than men with high school diplomas.

39. In terms of college, which gender earns the greatest share of undergraduate degrees?
    a. females
    b. males
    c. there is no difference as they are exactly equal

40. According to Deborah Tannen, men, more so than women, use language to
    a. make jokes with their buddies.
    b. come off as better than others.
    c. blow off steam.
    d. convey information.

41. According to Deborah Tannen, women, more so than men, use language to
    a. convey information.
    b. connect with other people.
    c. solve problems.
    d. complain.

42. According to Deborah Tannen, when someone brings up a personal problem in conversation, men are more likely to offer _____ , while women are more likely to offer _____ .
    a. money; ideas
    b. cynicism; sincerity
    c. bafflement; knowledge
    d. solutions; sympathy

43. **T F** Because Internet users need not disclose their true identities, men and women have similar styles when they contribute to discussions on the Internet.

ESSAY QUESTION

44. Present at least two arguments in support of the view that American society is biased against assertive women.

45. Discuss at least three problems posed by traditional gender-role socialization for women and men.

46. Describe the major patterns of job discrimination in the U.S. that are related to gender.

47. Gilligan studied moral reasoning, and Tannen studied conversation. Write an essay, based on these studies and any other information in this chapter, about general differences in the ways that men and women think and interact.

CHAPTER TWELVE   MARRIAGE AND ALTERNATIVE FAMILY ARRANGEMENTS

THE NATURE OF FAMILY LIFE

1. T F Murdock's study of 250 societies found that in a small number of these societies—about six out of the 250—the family does not exist.

2. T F Every known human society has an incest taboo.

3. T F The relationships covered by the incest taboo vary from one society to another.

4. T F The nuclear family is found in all societies regardless of degree of industrialization.

5. T F Polyandry is a very rare form of family structure, existing in only a few societies.

6. T F Most of the world's societies have bilateral systems of descent.

7. T F Patriarchal family systems are far more common than matriarchal systems.

8. Patriarchy and matriarchy refer to
  a. whether a child takes its name from its mother or father.
  b. whether a couple lives near the husband's family or the wife's.
  c. the distribution of power between men and women in the family.
  d. whether the father or mother has primary responsibility for socializing children.

9. A(n) _____ family structure consists of a married couple and their children.
  a. patriarchal
  b. nuclear
  c. matrilineal
  d. extended

10. A society that traces descent through the mother's side of the family would be characterized as a _____ system.
  a. matriarchal
  b. matrilineal
  c. matrilocal
  d. polygynous

11. Dwayne and Katrina are married. They have two children and live with Katrina's brother's family. This is an example of a(n) _____ family.
  a. extended
  b. polygamous
  c. nuclear
  d. blended

MARRIAGE

12. T F Many of the world's cultures allow individuals, mostly males, to have more than one spouse.

13. T F Almost all societies allow for divorce.

14. T F Marriage for love has been the most common type of marriage throughout the history of the human species.

15. **T F** Many European nations have had laws restricting marriage between people of different religions; the U.S. has not.

16. **T F** Societies that see marriage as an economic and political institution are less likely to emphasize romantic love.

17. According to anthropologist Marvin Harris, the major factor restraining the amount of polygamy in most cultures of the world is
    a. social disapproval.
    b. the likely increase in marital discord with more than one spouse.
    c. the high cost of maintaining more than one spouse.
    d. strict laws against more than one marriage at a time.

18. One negative consequence of marriage based on romantic love is that it
    a. is less likely to produce children.
    b. is less likely to make for a strong husband-wife relationship.
    c. weakens ties to families of origin.
    d. is less suited to socializing children.

19. Homogamy refers to
    a. gay marriage.
    b. marriage in which family functions are shared equally between spouses.
    c. marriage based on love.
    d. marriage between people of similar social backgrounds.

20. In the United States today, the most common form of residence for newly married couples is
    a. patrilocal.
    b. matrilocal.
    c. bilocal.
    d. neolocal.
    e. uptown local.

21. The incest taboo ensures a certain degree of
    a. homogamy.
    b. exogamy.
    c. endogamy.
    d. monogamy.

22. Which of the following is a feature of marriage in all societies?
    a. marriage is a public, socially approved relationship
    b. the husband and wife are expected to love one another
    c. the husband and wife must support one another emotionally
    d. the husband and wife must socialize the children

23. The most common type of interracial marriage in the U.S. involves a marriage between a
    a. white man and an African American woman.
    b. white man and a Hispanic.
    c. white man and a woman of a race other than African American.
    d. white woman and an African American man.
    e. white woman and an Asian man.

24. When did the Supreme Court rule that state laws banning interracial marriage were unconstitutional?
    a. shortly after the founding of the country
    b. shortly after the Civil War
    c. during the immigration waves of the late 1890s
    d. during the 1960s
    e. none of these; such laws have not been declared unconstitutional

25. "Stick to your own kind," sings a character in *West Side Story*. She is advocating
    a. exogamy.   c. matriarchy.
    b. polyandry.   d. homogamy.

26. Since the 1950s, the age at which Americans first marry
    a. has been getting younger.
    b. has been getting older.
    c. has remained relatively unchanged.
    d. has fluctuated randomly with no discernible pattern.

27. Marriage in the U.S. has become less homogamous on all the following variables EXCEPT
    a. race
    b. religion
    c. age
    d. none of these; all exhibit a decreasing degree of homogamy

28. The family in which a person is raised is his or her
    a. nuclear family.
    b. extended family.
    c. family of orientation.
    d. family of procreation.
    e. family of aggravation.

29. According to Tischler, one of the latent functions of the educational system in the U.S. is to maintain
    a. racial exogamy.
    b. social class homogamy.
    c. monogamy.
    d. romantic love.

THE TRANSFORMATION OF THE FAMILY

30. **T F** The United States has the highest divorce rate in the world.

31. **T F** Since 1975, the absolute number of marriages in the U.S. has increased.

32. **T F** Since 1975, the proportion of marriageable people who actually marry has decreased.

33. **T F** Less than 1/4 of all households in the U.S. consist of a "traditional family" (married couples with children).

34. **T F** Slightly more than half of all households in the U.S. consist of a "traditional family" (married couples with children).

35. **T F** Nearly all studies of cohabitation show that couples who live together before marriage are less likely to get divorced than are couples who do not try living together before marriage.

36. **T F** Households in 19th century America were much more likely than those today to include people not related to the family.

37. **T F** In divorce cases in the United States today, legal custody is given to the father about as often as it is given to the mother.

38. **T F** A large majority of divorced persons remarry.

39. **T F** Joint custody laws require children to spend equal time living with each of the divorced parents.

40. **T F** The modern period has witnessed the transfer of functions from the family to outside institutions.

41. **T F** Children who are victims of abuse are more likely to be abusive as adults than children who have not already experienced family violence.

42. **T F** The presence of children lowers the probability of remarriage for women, but not for men.

43. **T F** Step-parents are more likely to abuse children than are biological parents.

44. **T F** The majority of gay and lesbian households include children.

45. According to William J. Goode, the shift from agricultural to industrial society created pressures for the decline of
    a. the nuclear family.
    b. the extended family.
    c. companionate marriage.
    d. exogamy.
    e. all of these

46. A term for marriage based on romantic love is
    a. homogamy
    b. alternate marriage
    c. companionate marriage
    d. compassionate marriage
    e. compatible marriage

47. According to Tischler, World War II affected marriage in the U.S. because
    a. it led to the baby boom.
    b. it led to an economic boom.
    c. it moved women into the paid labor force.
    d. it militarized the country.

48. Which of the following is a feature of the transformation of the American family in the six decades following the end of World War II?
    a. average family size increased
    b. the family reverted to a more extended family structure
    c. the family decreased in importance in socializing children
    d. family roles became less equal

49. Which of the following variables is NOT related to family violence
    a. social class
    b. religion
    c. number of children
    d. unemployment

50. No-fault divorce laws
    a. have made it more difficult to get a divorce.
    b. have made it easier to get a divorce.
    c. have had no affect on divorce.
    d. have made it easier for men, but more difficult for women, to get a divorce.

51. According to Judith Wallerstein's research, when are the negative effects of divorce on children most likely to occur?
    a. in the first year or so after the divorce
    b. during early adolescence (ages 12–14)
    c. in adulthood
    d. none of these; Wallerstein found no negative effects of divorce

52. T F Cohabitation refers to a situation in which a newly married couple settles down in a home that is near neither the bride's family nor the groom's family.

53. T F Surveys that ask people how happy they are in their marriage find that on average people are happier in their marriages today than people were twenty years ago.

54. T F Children who grow up experiencing violence are more likely to become violent adults than children from non-violent households.

CHAPTER THIRTEEN   RELIGION

THE NATURE OF RELIGION

1. T F According to Durkheim, the profane consists of objects that people are prohibited from touching, looking at, or even mentioning.

2. T F The concept "God" has had a fixed, unchanging meaning through human history.

3. Émile Durkheim observed that all religions, regardless of their particular doctrines, divide the universe into two mutually exclusive categories:
    a. the natural and the unnatural.
    b. the good and the bad.
    c. the ugly and the beautiful.
    d. the sacred and the profane.

4. A group of teenagers met each month at the full moon to worship Satan, curse God, and place voodoo hexes on their teachers. Durkheim would classify this as an example of
    a. the sacred.
    b. the profane.
    c. evil.
    d. rationality.

5. Compared with sacred things, things that are profane are more likely to be
   a. solid.
   b. large.
   c. useful.
   d. expensive.

6. Standardized behaviors or practices such as receiving holy communion, the singing of hymns, praying while bowing toward Mecca, and the Bar Mitzvah ceremony are examples of
   a. totems.
   b. magic.
   c. rituals.
   d. shamanism.

7. Which of the following, according to Durkheim, is an element in all religions?
   a. priests
   b. a holy book
   c. ritual
   d. a house of worship

8. All religions
   a. rely on the teachings of a sacred book.
   b. utilize magic in their rituals.
   c. promote social equality.
   d. demand some public, shared participation.
   e. all of these

MAGIC

9. One difference between magic and religion is that
   a. magic relies on the supernatural, religion is more rational.
   b. magic is usually used to benefit the individual, religion benefits the group.
   c. magic uses special objects and words, religion does not.
   d. magic is rational, religion is supernatural.

10. **T F** According to Stark and Bainbridge, Christianity has always been opposed to magic.

11. According to Stark and Bainbridge, the church's saints and shrines for specialized miracles, such as healing at Lourdes, are really a form of
    a. magic.
    b. the profane.
    c. totemism.
    d. Protestantism.
    e. rationalism.

12. **T F** Reliance on magic is more common among peoples who have less actual control over their circumstances.

MAJOR TYPES OF RELIGIONS

13. **T F** Only three world religions are known to be monotheistic: Judaism and its two offshoots, Christianity and Islam.

14. An ordinary object that has become a sacred symbol for a group or clan is said to be its
    a. mana.
    b. mama.
    c. mojo.
    d. totem.

15. **T F** The earliest archaeological evidence of religious practice has been found in northern Europe.

16. In "The Exorcist," a priest heals a girl by forcing a demon to leave her body. This idea behind exorcism is closest to
    a. monotheism.
    b. polytheism.
    c. animism.
    d. totemism.

17. The largest of the world's religions, in terms of the size of its membership, is
    a. Christianity.
    b. Islam.
    c. Hinduism.
    d. Confucianism.

18. "The carcasses of every beast which divideth the hoof, and is not cloven-footed, nor cheweth the cud, are unclean unto you: every one that toucheth them shall be unclean." (Leviticus 11:26) In other words, a person who touches an "unclean" animal such as a pig becomes himself unclean. This statement combines the concepts of
   a. ritual and prayer.
   b. cleanliness and godliness.
   c. taboo and mana.
   d. animism and theism.

19. Animism involves a belief in the power of
   a. animals.
   b. vegetables.
   c. minerals.
   d. spirits.

20. Most theistic societies practice
   a. monotheism.
   b. totemism.
   c. polytheism.
   d. ecumenism.

A SOCIOLOGICAL APPROACH TO RELIGION

21. What did Karl Marx refer to as "the opiate of the masses"?
   a. capitalism
   b. communism
   c. religion
   d. drugs

22. T F Durkheim thought that even atheistic societies would develop rituals that served the same functions as religious rituals.

23. T F According to Durkheim, when people worship supernatural entities such as God, they are really worshiping their own society.

24. T F Alienation refers to the process by which people lose control over the social institutions they themselves invented.

25. Sigmund Freud emphasized which function of religion?
   a. helping individuals to deal with guilt and anxiety
   b. bringing about group cohesion
   c. bringing spiritual enlightenment to individuals
   d. getting individuals to accept inequality in the society at large

26. For Durkheim, the most important function of religion was to
   a. provide comfort for the individual.
   b. foster social cohesion.
   c. suppress social revolt.
   d. prevent suicide.

27. According to Max Weber, which of the following belief systems fostered a world view that promoted the development of capitalism?
   a. Judaism
   b. Catholicism
   c. Confucianism
   d. Calvinism

28. According to Weber, which of the following provided the background in which European capitalism could flourish?
   a. Jewish values that emphasized education and learning.
   b. Catholic values that emphasized glorifying God with expensive churches, clothes, art, etc.
   c. Protestant values that emphasized hard work and self-denial.
   d. Pagan values that emphasized innovation and problem-solving.

29. Which of the following is **not** a function common to most religions?
   a. emotional integration and the reduction of personal anxiety
   b. providing charity for the poor
   c. legitimizing arrangements in the secular society
   d. establishing a world view that helps to explain the purpose of life

30. According to Marvin Harris, the Hindu taboo against eating beef is most important because
    a. it promotes public health since many cattle in India carry disease.
    b. it promotes public health by eliminating a source of cholesterol.
    c. it allows for more economic sources of food and fuel.
    d. it promotes social cohesion.
    e. it gives support to the animal-rights movement.

31. The Ghost Dance of the Plains Indians is an example of a
    a. millenarian movement.
    b. revitalization movement.
    c. religious sect.
    d. universal church.

32. The best-selling *Left Behind* books and movie suggest that we are near the "end of time" when the faithful will be taken to Heaven and the rest left behind to a dismal fate. The ideas behind *Left Behind* are an example of
    a. millenerianism.
    b. revitalization.
    c. denominationalism.
    d. totemism.

ORGANIZATION OF RELIGIOUS LIFE

33. The animistic beliefs and rituals of a Native American tribe, in which all members of the tribe participate, are an example of a(n)
    a. ecclesia.
    b. universal church.
    c. sect.
    d. denomination.

34. The Church of England, or Anglican Church, is the official church of that country, and its titular head is the king or queen of England. This would make the Anglican Church a(n)
    a. ecclesia.
    b. universal church.
    c. sect.
    d. denomination.

35. Which of the following types of religious organization is most likely to reject ideas of the dominant society?
    a. ecclesia
    b. universal church
    c. sect
    d. denomination

36. The Puritans who left England for America because they felt they were being persecuted are an example of a(n)
    a. ecclesia.
    b. universal church.
    c. sect.
    d. denomination.

37. Once established in Massachusetts, the Puritans became dominant, and their religious ideas became the basis for politics and society. The Puritans of Massachusetts thus constituted a(n)
    a. ecclesia.
    b. universal church.
    c. sect.
    d. denomination.

ASPECTS OF AMERICAN RELIGION

38. T F Although most Americans say that religion is losing its influence in society, only about 10% of U.S. citizens say they have no religion.

39. Since 1997, the percentage of Americans believing in the existence of the devil has
    a. decreased greatly (more than 25%).
    b. decreased slightly (0–25%).
    c. remained unchanged.
    d. increased.

40. T F Although most Americans believe in God, only a minority, less than half, still believe in angels.

41. In their degree of religiosity, Americans are most similar to citizens of
    a. Canada.
    b. Great Britain.
    c. Spain.
    d. China.
    e. Iraq.

42. T F The U.S. has a much broader diversity of denominations than do European countries.

43. During the 1960s, many religious organizations and many different churches came together to work for civil rights. This cooperation is an example of
  a. secularism.
  b. ecumenism.
  c. denominationalism.
  d. an ecclesia.

44. T F Unlike in Europe, ecumenism has flourished in the United States because the boundaries between denominations here are less rigid and more fluid.

MAJOR RELIGIONS IN THE UNITED STATES

45. Which of these religions has the most followers in the U.S. today?
  a. Catholicism
  b. Protestantism
  c. Judaism
  d. Buddhism
  e. Atheism

46. T F Although the Catholic Church officially condemns artificial means of birth control, most American Catholics do not support this ban.

47. Which type of Protestant denomination has been gaining members more rapidly?
  a. conservative (Mormons, evangelicals)
  b. liberal (Episcopalians, Presbyterians)
  c. neither; they have both lost membership in recent years

48. Which denomination has the highest proportion of foreign-born members?
  a. conservative Protestant
  b. liberal Protestant
  c. Catholic
  d. Jewish

49. Which denomination is decreasing in numbers?
  a. conservative Protestant
  b. liberal Protestant
  c. Catholic
  d. Jewish
  e. none; they are all growing, though at different rates

50. Worldwide, which of the following religions has the lowest membership?
  a. Jewish
  b. Hindus
  c. Sikhs
  d. Anglican
  e. Buddhists

51. T F Because of the great diversity of religions in the U.S., religious affiliation is a very poor predictor of political attitudes.

52. T F Because of their high levels of education, Jews have come to dominate the top positions in the corporate world and in politics.

ESSAY QUESTIONS

53. List and define the basic elements of religion. Which element do you see as most important?

54. Choose a religion with which you are familiar and show how it fulfills the four major functions of religion.

CHAPTER FOURTEEN   EDUCATION

EDUCATION: A FUNCTIONALIST VIEW

1. **T F** The functionalist perspective stresses the role of schools in perpetuating class differences from generation to generation.

2. **T F** Preliterate societies which have no schools do not fulfill the basic functions of education.

3. **T F** Many preliterate societies do not distinguish between education and socialization.

4. **T F** The American educational system helps to slow the entry of young adults into the labor market.

5. **T F** In medieval times, education was more of a trade school, emphasizing practical, productivity-oriented learning.

6. **T F** Women and people older than 25 are the fastest-growing groups of college students.

7. **T F** Lester Frank Ward believed the main purpose of education was to equalize society.

8. To say that schools are places of socialization means that
   a. schools provide a place for children to interact with friends.
   b. schools are anti-capitalist.
   c. schools inculcate the ways of the society in children.
   d. schools should focus on basic academic skills.

9. Which of the following is a latent function of education?
   a. reducing unemployment rates by keeping youths out of the labor market
   b. teaching basic academic skills
   c. transmitting cultural knowledge
   d. generating innovation

10. To say that child care is a "latent" function of schools means that child care
    a. does not come until later in life.
    b. is necessary for children whose parents work late.
    c. is not an officially stated goal of schools.
    d. is unimportant.

11. Opponents of bilingual education argue that it
    a. is too expensive.
    b. hurts immigrant children by not giving them necessary language skills.
    c. favors non-native children at the expense of native-born children.
    d. confuses children as to which language is the best.

12. What, according to Lester Frank Ward is the main source of inequality in society?
    a. the intellectual abilities of those at the top and bottom of society
    b. the differences in the way rich and poor families socialize their children
    c. the unequal distribution of knowledge
    d. the poor academic skills of teachers in the inner cities
    e. all of the above

13. What was the basic message of the report titled *A Nation At Risk*?
    a. U.S. schools were the best in the world but would soon face competition from Asian countries.
    b. U.S. schools had held steady in their performance.
    c. U.S. schools were improving, but not rapidly enough.
    d. U.S. schools were bad and getting worse.

14. Major assessments of U.S. students have found them most lacking in ability in
    a. math.
    b. English composition.
    c. English grammar.
    d. geography.
    e. athletics.

15. The single most important element in the phenomenon of continuing innovation in American society is the
    a. work done by garage and basement hobbyists.
    b. continuous effort to recruit foreign geniuses.
    c. performance of high-caliber academic and research universities.
    d. increased attention to standardized testing in science education.

16. The Nation's Report Card asked high school students to solve high-school math problems involving multi-step logic, algebra, or geometry. About what percentage of U.S. high school students could solve such problems?
    a. less than 10%   c. 50%
    b. 25%   d. 75%

17. According to the Nation's Report Card estimate, what percentage of U.S. high school students are adequately prepared for college science courses?
    a. less than 10%   c. 50%
    b. 25%   d. 77%

18. **T F** For the last quarter-century, the majority of college students in the U.S. have been female.

THE CONFLICT THEORY VIEW

19. **T F** To the conflict theorist, the function of school is to produce the kind of people the system needs.

20. **T F** According to conflict theorists, what is important about obtaining a degree from Harvard or Yale or some other elite college is that it is a guarantee that a person has received quality training.

21. **T F** According to conflict theorists, the hidden curriculum of schooling subtly promotes creativity and imagination and downplays rote learning.

22. In the view of conflict theory, the most important function of schools is to
    a. provide childcare.
    b. provide skills so that children will become successful.
    c. provide employment for teachers.
    d. preserve the existing class system.

23. The "hidden curriculum," according to conflict theory includes
    a. advanced placement classes for the lucky few.
    b. after-school programs.
    c. getting students to accept society as it is.
    d. secret clubs like the Dead Poets' Society.

24. "Tracking" in schools, according to conflict theory, is a way of
    a. keeping track of problem students.
    b. increasing racial and class inequality.
    c. adjusting teaching to students' abilities.
    d. finding talented but underprivileged students.
    e. winning track meets.

25. Rosenthal and Jacobson found that student performance is substantially affected by the
    a. location of their school.
    b. occupational status of their parents.
    c. level of their innate intelligence.
    d. expectations of their teachers.

26. Which of the following is an aspect of "The Credentialized Society"?
    a. Credentials are a sign of competence.
    b. For employers, socialization matters more than does competence.
    c. The U.S. is moving away from an emphasis on credentials.
    d. Credentials are a way for less privileged people to get ahead.

27. Ted Brown has returned to school for certification in computer programming despite the fact that he has ten years of work experience in this field. The schooling will do very little for his performance, but it will get him a raise. This is an example of
  a. functional illiteracy.
  b. the hidden curriculum.
  c. de facto segregation.
  d. the credentialized society.

ISSUES IN AMERICAN EDUCATION

28. T F Cross-district busing of schoolchildren was a direct outgrowth of the *Coleman Report* of 1966.

29. T F The United States adopted the concept of mass public education only after it had been accepted in Europe.

30. T F The *Coleman Report* of 1966 found that minority students perform better when they go to school with others like them in predominantly minority schools.

31. T F The high school dropout rate has been increasing gradually since 1980.

32. T F The *Coleman Report* and much subsequent research have shown a strong positive correlation between the amount of money a school district spends and the achievement of its students.

33. T F Teachers often associate giftedness with children who come from prominent families, who have traveled widely, and who have extensive cultural advantages.

34. T F Standardized tests can accurately measure intelligence and abilities, especially among younger children.

35. T F The level of violence in schools has been decreasing since the mid-1990s.

36. T F In the last twenty years, the difference between males and females on the math part of the SAT has all but disappeared.

37. The Supreme Court decision declaring school segregation policies unconstitutional was handed down
  a. in the early days of the republic.
  b. shortly after the Civil War.
  c. shortly after World War II.
  d. in the 1950s.
  e. None of these; the Court has never declared segregation unconstitutional.

38. Prior to the 1950s, state policy in the South required that blacks and whites attend separate schools. This system was known as
  a. de facto segregation.
  b. de jure segregation.
  c. de minimis segregation.
  d. de gustibus segregation.

39. The major cause of de facto segregation is
  a. the hidden curriculum.
  b. No Child Left Behind.
  c. de jure segregation.
  d. residential segregation.
  e. tracking.

40. What is the relationship between education and median income?
  a. Each higher level of education attained brings higher median income.
  b. Level of education attained has little effect on median income.
  c. While obtaining a high school diploma increases median income, going to college results in little additional earnings.
  d. While obtaining a four-year college degree increases median income, post-graduate degrees add little in median incomes.

41. Carol Gilligan argues that girls are devalued by society because
  a. they posses a different moral sensibility.
  b. they take jobs and positions males will not take.
  c. more males graduate with bachelor's degrees than females.
  d. young girls are not as serious about school grades as are young boys.
  e. all of the above

42. Which of the following is **not** a social consequence of dropping out of high school?
   a. increased crime
   b. decreased tax revenues
   c. increased intergenerational mobility
   d. reduced political participation

43. T F Nearly 20% of all students in grades 9–12 reported they had carried a weapon at least once during the previous month.

44. The *Coleman Report* of 1966 concluded that
   a. the quality of school experiences for black and white students had become approximately equal.
   b. the only way to improve the quality of school experiences for blacks was to spend more money on their schools.
   c. schools play a less important role in student academic achievement than once thought.
   d. academic success is most powerfully influenced by individual merit rather than social factors.

45. Dropping out of high school affects not only those who leave school, but also society in general because dropouts
   a. pay less in taxes, because of their lower earnings.
   b. are less likely to vote.
   c. have poorer health.
   d. increase the demand for social services.
   e. all of the above

46. In its famous *Brown v. Board of Education* decision in 1954, the United States Supreme Court banned
   a. de jure segregation.
   b. busing.
   c. standardized testing.
   d. tracking.

47. Which of the following is associated with dropping out of high school?
   a. speaking a language other than English
   b. low family income
   c. low educational attainment of parents
   d. poor academic achievement
   e. these are all associated with dropping out of high school

48. About _____ % of the nation's white students attend central city schools.
   a. 3
   b. 20
   c. 35
   d. 45
   e. 58

49. Over the last 75 years the percentage of Americans completing high school has
   a. declined slightly.
   b. risen dramatically.
   c. risen slightly.
   d. remained about the same.

50. The country with the highest percentage of college graduates among its youth is
   a. Germany.
   b. Japan.
   c. the United States.
   d. Great Britain.
   e. Sweden.

51. The use of standardized tests like the SAT was intended to
   a. ensure that only a small number of minorities would get into elite colleges.
   b. help colleges compare students from high schools whose standards for grades varied widely.
   c. give high school students an incentive to study harder, especially in math and science.
   d. ensure that equal numbers of males and females were admitted to colleges.
   e. provide jobs for test makers.

ESSAY QUESTIONS

52. Compare and contrast the functionalist and conflict theory views of education as socialization.

53. Discuss the issue of gender bias in the classroom. Based upon your reading of this chapter, what is your position on this issue?

54. Discuss the pro and con issues connected with the phenomenon of home schooling.

CHAPTER FIFTEEN   POLITICAL AND ECONOMIC SYSTEMS

POLITICS, POWER, AND AUTHORITY

1. T F Power that is regarded as illegitimate by those over whom it is exercised is called coercion.

2. T F Power based on fear is the most stable and enduring form of power.

3. T F Legitimacy refers to the condition in which people accept the idea that the allocation of power is as it should be.

4. T F Thomas Jefferson believed that, once established, a political system should only be changed in unusually dire circumstances.

5. T F Even small tribes like the Mbuti pygmies, who make decisions by group consensus, constitute a "state."

6. T F In most societies, what laws are passed, or not passed, depends to a large extent on which categories of people have power.

7. Regardless of his personal abilities or popularity, Charles, Prince of Wales, will become King of England when his mother, Elizabeth II, abdicates the throne or dies. This is an example of _____ authority.
a. rational-legal
b. appointive
c. traditional
d. charismatic

8. Establishing durable institutions of government is a problem most likely to present a problem for rule based on
a. charismatic authority.
b. traditional authority.
c. rational-legal authority.

9. Although Ronald Reagan was considered by many to have a magnetic personality and to be a symbol of the conservative movement in America, as president he was nevertheless limited by the Constitution and the system of checks and balances established there. Thus, in Weber's terms, he is best thought of as a _____ authority.
a. legal-rational
b. representative
c. traditional charismatic

10. _____ is the ability to carry out one's will, even in the face of opposition.
a. Power
b. Politics
c. Force
d. Authority

GOVERNMENT AND THE STATE

11. Plato's ideal society would be one ruled by
a. the people as a whole.
b. a trained aristocracy.
c. a military elite.
d. the wisest person among all the people.

12. Plato thought that the most important problem facing government was
    a. providing for the common defense.
    b. promoting the general welfare.
    c. ensuring domestic tranquility.
    d. securing the blessings of liberty to all citizens.

13. Which of the following is **not** one of the basic functions of the state?
    a. establishing laws and norms
    b. ensuring economic stability
    c. protecting against outside threats
    d. socializing the young

14. The institutionalized way of organizing power within territorial limits is known as
    a. politics.
    b. authority.
    c. the state.
    d. the market.

THE ECONOMY AND THE STATE

15. **T F** In theory, "the invisible hand of the market" means that a multiplicity of self-interested acts by individuals produces a social benefit.

16. **T F** In socialist societies, consumer goods tend to be expensive, while necessities are kept affordable.

17. According to Adam Smith, everyone in a society benefits most from
    a. competition among producers.
    b. centralized government planning.
    c. local government control of economic processes.
    d. democratic decision making in the workplace.

18. Which of the following is characteristic of mixed economies?
    a. all economic decisions are made by central planners
    b. private property is virtually abolished
    c. the government intervenes to prevent industry abuses
    d. government planners decide on the best mix of public good like education and private consumer goods like televisions.

19. Marx predicted that, as production expands in capitalist economies,
    a. profits will decline and wages will fall, leading to revolution.
    b. wages will increase and profits will decline, leading to bankruptcy.
    c. profits and wages will both increase, leading to inflation.
    d. profits and wages will become irrelevant, as a decent standard of living for all is attained.

20. In many localities in the U.S., primary and secondary schools are public; that is, they are provided by and ultimately run by the government. This form of education is an example of
    a. capitalism.
    b. laissez-faire.
    c. socialism.
    d. democracy.

21. Which of the following is one of the basic premises behind capitalism?
    a. Producers attempt to serve the best interests of society as a whole.
    b. Consumers attempt to serve the best interests of society as a whole.
    c. Democratically elected leaders decide what should be produced to meet the needs of the society.
    d. Free markets decide what is produced, and for what price.

22. Marx thought that there was an inherent conflict between the interests of capitalists and the interests of
    a. owners.
    b. consumers.
    c. professionals.
    d. workers.

23. _____ is a mechanism for determining the supply, demand, and price of goods and services through consumer choice.
    a. Command economy
    b. Legal-rational authority
    c. Market
    d. Representative government

24. The Federal Communications Commission is supposed to regulate broadcasting in the public interest. One FCC commissioner said, "The public interest is what interests the public." His statement is most in keeping with
    a. a mixed economy.
    b. laissez-faire capitalism.
    c. socialism.
    d. state capitalism.

25. In politics, people who speak most favorably about the virtues of "the market" are most likely to support
    a. grocery shopping.
    b. socialism.
    c. laissez-faire capitalism.
    d. a command economy.

26. The U.S. economic system is best described as
    a. laissez-faire capitalism.
    b. democratic socialism.
    c. a mixed economy.
    d. a command economy.

27. Which of the following is a basic principle of socialism?
    a. Everyone should have the essentials before some can have luxury items.
    b. Major decisions should be made by professional economists appointed by the government leaders.
    c. The government should give the greatest economic rewards to those who contribute the most to the society.
    d. Individual freedom cannot exist without economic freedom.

TYPES OF STATES

28. **T F** Democracy has always been regarded as the best form of government.

29. **T F** There are both capitalist and socialist states that are totalitarian in nature.

30. **T F** Representative government is a form of government in which a select few rule.

31. **T F** According to Edward Shils, civilian rule means that no member of the military may hold public office.

32. **T F** Democracy can exist only in capitalist societies.

33. Under democratic socialism,
    a. private ownership of means of production is abolished.
    b. taxes are kept low.
    c. the state assumes ownership of strategic industries.
    d. little effort is made to expand social welfare programs or redistribute income.

34. _____ refers to the principle of limited government.
    a. The invisible hand
    b. Autocracy
    c. Constitutionalism
    d. Democracy

35. A major difference between an autocratic government and a totalitarian government is that under an autocratic government
    a. there is less repression of political dissent.
    b. people have more freedom in nonpolitical areas.
    c. there is greater economic planning.
    d. there is more influence by the military.

36. Minimum-wage laws require that employers pay workers at least a certain amount even if workers are willing to work for less. The government's setting wages violates the principle of
    a. the market.
    b. democracy.
    c. totalitarianism.
    d. socialism.

37. Totalitarian capitalism is often referred to as
    a. fascism.
    b. a mixed economy.
    c. laissez-faire capitalism.
    d. a command economy.

38. Which of the following is **not** a characteristic of democratic political systems?
    a. majority rule
    b. civilian rule
    c. direct government by the people
    d. constitutionalism

39. Socialists argue that
    a. true democracy is impossible in capitalist society.
    b. with appropriate modifications capitalism can be democratic.
    c. democracy is just an illusion in any form of society.
    d. socialism and democracy are in theory incompatible.

FUNCTIONALIST AND CONFLICT THEORY VIEWS OF THE STATE

40. T F Functionalists maintain that the state emerged to manage and stabilize an increasingly complex society.

41. T F Historical evidence indicates that the earliest legal codes were enacted to protect the property of the wealthy.

42. Conflict theorists see which of the following as the key to the origin of the state?
    a. the need to coordinate increasingly large and complex societies
    b. the desire of populations to control their own destiny
    c. the nature of human nature, in which some will always dominate others
    d. the need for a way to protect elite control over surplus production

43. One main difference between the functionalist theory of the origins of the state and the conflict theory is that conflict theory emphasizes the idea that
    a. the state arose in order to deal fairly with conflicts.
    b. the state arose in order to make life better for all citizens.
    c. the state arose to protect the interests of the powerful.
    d. the state arose to reduce inequalities that could lead to conflict.

POLITICAL CHANGE

44. When the Chinese Communists under Mao Zedong took power in China in 1949, they sought to institute an entirely new way of life for their people, transforming politics, economics, and culture. This was an example of a
    a. rebellion.
    b. political revolution.
    c. political disorder.
    d. social revolution.

45. One thing that democracies have that dictatorships tend to lack is a mechanism for
    a. dealing with dissent.
    b. protection against foreign powers.
    c. implementing laws and official policies.
    d. institutionalized political change.

46. The American Revolution is an example of a
    a. rebellion that produced social change.
    b. rebellion that produced political change.
    c. revolution that produced social change.
    d. revolution that produced political change.

47. Rebellions differ from revolutions in that they
    a. do not make use of force.
    b. question the uses of power, but not its legitimacy.
    c. are rarely successful.
    d. attack social, but not political issues.

THE AMERICAN POLITICAL SYSTEM

48. T F Lobbyists are people paid by special-interest groups to attempt to influence government policy.

49. T F The percentage of women voting is substantially lower than the percentage of men, thus accounting for the small percentage of women elected to high public office.

50. T F Journalists and the media report the news on politics but have no real political influence on their own.

51. T F Men in the U.S. are more involved in politics so they have higher rates of voting than do women.

52. Which of the following groups has the highest rate of voting?
   a. eighteen-to-twenty-year-olds
   b. the unemployed
   c. college graduates
   d. those with high school diplomas but no college education

53. Which of the following statements concerning the political influence of Hispanics is correct?
   a. They have virtually no influence because of their very low numbers in the electorate.
   b. They have rather little influence because most of them don't speak English.
   c. They have disproportionate influence because of their concentration in states with many electoral votes.
   d. They have disproportionate influence because of their large numbers and their relatively high rate of voter registration.

54. Data on members of Congress indicate that they are disproportionately
   a. Hispanic.
   b. young (under 50).
   c. male.
   d. intelligent.

ESSAY QUESTIONS

55. Winston Churchill said, "Democracy is the worst form of government except for all those others that have been tried." What's wrong with democracy? What's wrong with those other forms of government?

56. Airlines, railways, banks, television and radio stations, medical services, colleges, and important manufacturing enterprises—these are some of the industries run by the state in various democratic socialist governments. What are the advantages and disadvantages of having the state run each of these?

CHAPTER SIXTEEN   POPULATION AND URBAN SOCIETY

POPULATION DYNAMICS

1. T F Contemporary Chinese fertility-reduction policies have not succeeded in lowering the birth rate very much.

2. T F Up through the 1970s, China, the world's most populated society, actually had a policy of encouraging population growth.

3. T F The United States has achieved the lowest infant mortality rate in the world.

4. T F The crude death rate is defined as the annual number of deaths per 1,000 people in a given population.

5. T F In many of the wealthier, industrialized countries, population is actually decreasing rather than increasing.

6. T F If two countries have the same crude death rate, their age-specific death rates will also be the same.

7. T F Immigration and emigration contribute much more to differences in population growth than do birth rates and death rates.

8. Differences in life expectancy among countries are most affected by the rate of
   a. infant mortality.
   b. adult disease.
   c. abortion.
   d. heart attacks.
   e. AIDS.

9. The Chinese are attempting to lower their national birthrate to an average of
   a. one child for every three married couples.
   b. one child per married couple.
   c. 2.1 children per married couple.
   d. one child per adult in the family.

10. Which of the following exerts the **least** impact on a country's population growth or decline?
    a. migration
    b. fertility
    c. fecundity
    d. mortality

11. The movement of a population within a nation's boundary lines is known as
    a. internal migration.
    b. immigration.
    c. migration.
    d. emigration.

12. A well-known example of an "age-specific death rate" is
    a. war.
    b. famine.
    c. the crude mortality rate.
    d. the infant mortality rate.

13. Fecundity refers to the
    a. physiological ability to bear children.
    b. the number of women of child-bearing age.
    c. actual number of births in a given population.
    d. total number of children a woman in a given society can be expected to bear.

14. The "crude birth rate" is the annual number of births
    a. per 1,000 people.
    b. to the population over age 21.
    c. to the number of women of child-bearing age.
    d. to the number of abortions.
    e. to the number of pregnancies.

15. The "fertility rate" is the number of births relative to
    a. the number of instances of sexual intercourse.
    b. the total population.
    c. the number of abortions.
    d. the number of women of child-bearing age.
    e. the number of people over age 65.

16. Which is higher, the crude birth rate or the fertility rate?
    a. the crude birth rate
    b. the fertility rate
    c. neither, they are always the same
    d. it depends on the country and the population

17. The rapid rates of population growth in the Third World in recent decades are largely the result of
    a. sharp rises in birthrates.
    b. sharp improvements in life expectancy.
    c. immigration.
    d. internal migration.

18. The infant mortality rate refers to
    a. babies who are born dead.
    b. miscarriages before the ninth month of pregnancy.
    c. babies who die within the first year of life.
    d. babies who die before reaching their tenth birthday.

19. Which continent has the highest rate of population growth?
    a. South America
    b. Europe
    c. Africa
    d. Asia
    e. North America

20. Which area of the world is currently experiencing the *lowest* population growth?
    a. Africa
    b. Southeast Asia
    c. Eastern Europe
    d. South America

THEORIES OF POPULATION

21. T F Malthus theorized that in most societies population grows to be as large as possible without being too large for the supply of available food.

22. T F The second demographic transition has occurred primarily in South America.

23. In Malthus's terms, which of the following would NOT be a "positive check" on population growth?
   a. war
   b. earthquakes and other natural disasters
   c. famine
   d. birth control
   e. none of these; they are all positive checks

24. Malthus lived in England, where he died in 1834. Since then, trends in England's population
   a. have supported Malthus's ideas.
   b. have contradicted Malthus's ideas.
   c. have been irrelevant to Malthus's ideas.

25. Karl Marx believed that the source of the population problem was
   a. the sheer number of people in the world.
   b. insufficient birthrates to reproduce the working class.
   c. the rise of totalitarian socialist regimes.
   d. industrial capitalist exploitation.

26. "Demographic transition theory" is based on changes in
   a. fertility but not mortality.
   b. mortality but not fertility.
   c. both fertility and mortality.
   d. neither fertility nor mortality.

27. Countries who think their "dependency ratio" is too high would prefer to have
   a. higher birth rates.
   b. lower death rates.
   c. fewer immigrants.
   d. political independence.

28. During the first stage of the demographic transition,
   a. fertility rates are low, mortality rates are low, and population size is stable.
   b. fertility rates are low, mortality rates are high, and population size is declining.
   c. fertility rates are high, mortality rates are low, and population size is increasing rapidly.
   d. fertility rates are high, mortality rates are high, and population size is stable.

29. In demographic transition theory, the first change away from stability is
   a. a decline in population caused by new diseases.
   b. a decline in population caused by family planning.
   c. an increase in population caused by higher birth rates.
   d. an increase in population caused by lower death rates.

30. The demographic transition refers to the
   a. simultaneous decline in birth and death rates as a country industrializes.
   b. decline in birth rates followed later by decline in death rates as a country industrializes.
   c. decline in death rates followed later by a decline in birth rates as a country industrializes.
   d. rise in death rates followed later by a decline in birth rates as a country industrializes.

31. Population increases most rapidly in stage _____ of the demographic transition.
   a. 1
   b. 2
   c. 3
   d. 4

32. In many underdeveloped countries, according to Tischler, medical advances have cut mortality rates, but birth rates have remained high. These countries are in which stage of demographic transition?
   a. first
   b. second
   c. third
   d. fourth

33. The "second demographic transition" is characterized by
   a. birth rates below replacement.
   b. increasing birth rates.
   c. increasing death rates.
   d. high birth rates and high mortality rates.

34. In countries where a second demographic transition is occurring, governments have begun to institute
    a. pronatalist policies.
    b. antinatalist policies.
    c. preventive checks.
    d. immigration restrictions.

35. In the 1970s, environmental predictions by the Club of Rome, Paul Ehrlich, and others were pessimistic about the environment and natural resources. According to Tischler, these predictions have, for the most part, proved to be
    a. accurate.
    b. inaccurate.
    c. irrelevant.
    d. none of these; it's too soon to tell

36. Optimism about the future of the environment and natural resources looks chiefly to the effects of
    a. conservation and decreased use.
    b. technology.
    c. religion.
    d. birth control.

37. According to Tischler, most harm to the environment comes from
    a. pollution and consumption in the wealthy, industrialized countries like the U.S.
    b. pollution in the newly industrializing countries like China.
    c. pollutions in the underdeveloped countries.
    d. deforestation in the underdeveloped countries.

URBANIZATION AND URBAN LIFE

38. T F Cities have existed in one form or another for as long as humans have lived on this planet.

39. T F Worldwide, the trend toward increasingly large cities has leveled off, and the growth of cities is nowhere near as great as it was from 1900 to 1950.

40. T F Deinstitutionalization of the mentally ill began in the 1950s, two decades before homelessness became a recognized problem.

41. In preindustrial cities, the largest social class consisted of
    a. the ruling elite.
    b. a middle class of shopkeepers.
    c. manual laborers.
    d. slaves.

42. Which of the following is a requirement that had to be met before cities could appear on the social landscape?
    a. the development of a factory system of production
    b. the technology to create paved streets
    c. the capacity to produce a surplus of food
    d. social tolerance for people of differing religious views

43. The world's first fully developed cities arose in
    a. the Middle East, mostly in what is now Iraq.
    b. East Africa, mostly in what is now Kenya.
    c. coastal China.
    d. England near what is now London.

44. T F The concentric zone model is a model of urban development in which distinct, class-identified zones radiate out from a central business district.

45. T F The multiple-nuclei model of urban development holds that industries locate near one another and shape the characteristics of the immediate neighborhood.

46. T F Jane Jacobs argued that social control of public behavior and community life in the city take place on the level of the block, not the neighborhood.

47. Which category of persons makes up the largest number of the urban homeless?
    a. the mentally ill
    b. laid off factory workers
    c. displaced minorities
    d. young rural migrants unable to find employment

48. _____ refers to the trend in which middle-class young adults move back into poorer central city areas, improving the area but displacing the poor.
    a. Urbanization
    b. Gentrification
    c. Mechanical integration
    d. Exurbanization

49. In his book *Urban Villagers,* Herbert Gans showed that
    a. city dwellers often participate in strong community cultures.
    b. city life tends to be alienating and lacking in close personal contacts between people.
    c. the most successful city neighborhoods provided open space where residents could grow their own vegetables.
    d. urban community life is possible only in well-to-do neighborhoods.

50. George Kelling's "Broken Windows" model emphasizes _____ as a source of the decline of urban neighborhoods.
    a. poor education
    b. disorder
    c. poverty
    d. single-parent families

51. Which of the following is NOT a consequence of suburban sprawl?
    a. political disenfranchisement
    b. increased dependence on automobiles
    c. less efficient use of land and natural resources
    d. disruption of labor markets for low-income workers

52. According to Tischler, "deinstitutionalization" was an important cause of homelessness. He is referring to the deinstitutionalization of
    a. prisoners.
    b. the mentally ill.
    c. the physically ill.
    d. students.
    e. all of the above

CHAPTER SEVENTEEN   HEALTH AND AGING

THE EXPERIENCE OF ILLNESS

1. **T F** The sick role is a shared set of cultural norms that legitimates deviant behavior caused by the illness.

2. The concept of the "sick role" suggests that how we behave when we are ill is determined by
    a. medical requirements.
    b. individual preferences.
    c. cultural norms.
    d. scientific procedures.

3. According to Tischler, thinking about illness in terms of the "sick role" assumes the importance of
    a. individual perceptions.
    b. medical care.
    c. good health habits.
    d. public health projects.

4. Which of the following is (are) a component of the sick role, according to Parsons?
    a. The sick person is not held responsible for his or her condition.
    b. The sick person must cooperate with the advice of designated experts.
    c. The sick person is excused from normal responsibilities.
    d. The sick person must want to get better.
    e. These are all components of the sick role, according to Parsons.

5. In the 19th-century America, drunks were seen as morally weak. Today, Alcoholics Anonymous and many other people define them as being "in the grip of a progressive illness." This change is an example of
    a. progress.
    b. ambiguity.
    c. secularization.
    d. medicalization.

HEALTH CARE IN THE UNITED STATES

6. **T F** Due to medical advances, hypertension is no longer a major problem for African Americans.

7. **T F** Studies of life expectancy show that on every measure social class influences longevity.

8. **T F** In the United States, females have higher death rates from accidents than do males.

9. **T F** The United States has the most advanced health-care resources in the world.

10. **T F** Women are more likely than men to have psychiatric disorders.

11. **T F** The infant mortality rate among African Americans is currently double the rate among whites.

12. **T F** The suicide rate is lower for blacks than it is for white.

13. **T F** Because men tend to need more caretaking, more than two-thirds of the nursing home population is male.

14. **T F** Life expectancy in the U.S. is longer than in any other country.

15. **T F** Medical care system in the U.S. features a great amount of mobility; workers gain skills and move up the career ladder.

16. To say that U.S. medical system is "acute, curative, and hospital based" implies that it pays too little attention to
    a. science.
    b. serious illness.
    c. prevention.
    d. cancer.
    e. technology.

17. Over the course of the 20th century, with its great improvements in medicine, the life-expectancy gap between women and men
    a. decreased.
    b. increased.
    c. remained the same.

18. One reason that women have, on average, longer lives than men is that
    a. parents take better care of girls than of boys.
    b. they are less likely to die young from nonmedical causes like accident and homicide.
    c. they spend more of their income on preventive medical care.
    d. women get diseases at an earlier age and thus build up their immunity.

19. Women are more likely than men to have psychological problems emerge in the form of
    a. depression.   c. drug abuse.
    b. obsessive-compulsive disorder.   d. schizophrenia.

20. Which of the following countries has the shortest life expectancy?
    a. the United States   d. Spain
    b. Australia   e. Japan
    c. Italy

21. Which of the following is NOT a factor in black-white differences in infant mortality and low-weight births?
    a. help from families   c. age of mother
    b. access to prenatal care   d. education

22. Compared with white males, African American males have higher death rates from which of the following?
    a. childhood infectious diseases
    b. HIV
    c. homicide
    d. diabetes
    e. African American rates were higher for all of them.

23. Which of the following groups has the best health profile?
    a. white Americans
    b. Native Americans
    c. African Americans
    d. Asian Americans

24. The "immigrant advantage" in health refers to the fact that
    a. U.S. government policy forbids the admission of sick people.
    b. immigrants receive several months of free health care.
    c. the kind of people who immigrate are healthy, more robust, and more likely to take care of themselves.
    d. immigrants tend to locate in healthier areas of the country.

25. One hundred years ago, the poor health of low-income people was attributed to
    a. moral weakness.
    b. lack of medical care.
    c. pollution and other environmental factors in poor neighborhoods.
    d. hazardous work conditions.

26. T F Thanks to medical advances and government programs, in the U.S. today, the health of poor people today is, on average, about the same as the health of middle-class people.

27. T F The farther you have gone in school, the less likely you are to smoke cigarettes.

28. T F In the last thirty years, the number of female doctors in the U.S. has increased faster than the number of male doctors.

29. Which of the following is NOT a reason that female physicians earn less than male doctors?
    a. Female physicians are not as competent as male doctors and therefore must serve a less affluent clientele.
    b. The average female physician works fewer hours.
    c. The average female physician is younger and less experienced.
    d. Female physicians are less likely to be in private practice.
    e. Female physicians are concentrated in lower-paying specialties.

30. In studies comparing patients of male doctors with patients who saw female doctors, who reported greater satisfaction with treatment?
    a. patients of male doctors
    b. patients of female doctors
    c. there was no difference; satisfaction depended entirely on the competence of the doctor

CONTEMPORARY HEALTH CARE ISSUES

31. T F The American Medical Association has been a leading advocate of national health insurance.

32. T F More surgery is performed in the United States than in any other country.

33. T F Asian and Pacific Islanders have much higher mortality rates in the U.S. due to their social and economic status.

34. T F Medicare is a program legislated by Congress to pay the medical bills of people over age 65.

35. T F In recent decades the prevalence of smoking among women has increased while the prevalence among men has declined.

36. In coming years, the greatest pressure on the medical resources of the U.S. will be caused by
    a. the aging of the baby boomers.
    b. the return of wounded Iraq war veterans.
    c. the closing of some medical schools.
    d. unemployment leading to the loss of health insurance.

37. Which group accounts for the fewest AIDS cases in the U.S.?
    a. whites
    b. African Americans
    c. Hispanics
    d. Asian Americans

38. According to current estimates, about how many people in the U.S. are infected with HIV?
    a. one million   **c.** fifty million
    b. ten million   **d.** one hundred million

39. Which of the following groups, in the United States, has the highest death rate for HIV?
    a. Hispanics
    b. whites
    c. blacks
    d. Asians
    e. There are no differences across the above four groups in deaths from HIV.

40. The human immunodeficiency virus causes AIDS by
    a. directly attacking the major organs of the body.
    b. seeding the growth of particularly virulent forms of cancer.
    c. incapacitating the body's immune system by destroying white blood cells.
    d. altering the genetic code in white blood cells so that they attack and eventually destroy the body of origin.

41. Blue Cross and Blue Shield were originally developed to ensure that
    a. everyone had access to affordable health care.
    b. only competent health care professionals provided health services.
    c. physicians and hospitals got paid.
    d. socialized medicine would one day be possible.

42. Third-party payments are
    a. payments made by insurance or health care organizations to doctors.
    b. payments to doctors made by patients with a too-active social life.
    c. payments made to doctors by three-person families.
    d. money that patients pay to insurance companies.

43. Which of the following countries does not have a national health care system?
    a. the United States
    b. Japan
    c. France
    d. Great Britain
    e. Italy

44. Making workplaces safer by enforcing rules on toxic chemicals is an example of
    a. medical prevention.
    b. behavioral prevention.
    c. structural prevention.
    d. all of these

45. Use of the polio vaccine is an example of
    a. medical prevention.   **c.** structural prevention.
    b. behavioral prevention.   **d.** all of these

THE AGING POPULATION

46. T F For the first time in U.S. history, there are more persons over 65 than teenagers.

47. T F The two most common causes of death in America today are AIDS and homicide.

48. T F Almost one in five Americans will be sixty-five or older by 2030.

49. T F The largest proportion of Americans over 65 consists of those aged 85 or older.

50. T F Households of people over 65 have a higher net worth than do households of people in the peak earning ages of 45 to 54.

51. The leading edge of the "baby boom" consists of people born in
    a. 1936.   **c.** 1956.
    b. 1946.   **d.** 1966.

52. African Americans make up about 13% of the total U.S. population. They make up about what percentage of the over-65 population?

a. 8%   **c.** 22%

b. 13%   **d.** 41%

53. Among Americans age 65 and up, who is more likely to be married?

a. men

b. women

c. neither; rates are about the same

54. Surveys that ask college students if they have participated in binge drinking in the past two weeks find about what percentage saying yes?

a. 5%   **c.** 30%

b. 20%   **d.** 45%

55. The biggest change in binge drinking among college students since 1993 is

a. the general decrease in binge drinking.

b. the general increase in binge drinking.

c. the increase in binge drinking at all-women schools.

d. the shift from regular beer to "lite" beer.

CHAPTER EIGHTEEN COLLECTIVE BEHAVIOR AND SOCIAL CHANGE

1. T F The U.S. family has changed considerably since 1776—in its numbers, types of relationships, functions, etc.—but because no person or group explicitly worked for these changes, they do not constitute real social change.

2. T F Because so much of U.S. electronic equipment is made in Asia, technology in the U.S. is classified as an "external" cause of social change.

3. A set of ideas that justifies a set of goals and means to those goals is called

a. norms.   **c.** ideology.

b. religion.   **d.** mobilization.

4. T F According to Tischler's definition of social change, the invention of the automobile was not in itself an example of social change.

5. Conservative ideologies, according to Tischler, represent an attempt to

a. change the way the courts work.

b. change society so as to give the power of the middle class.

c. change the laws so as to give people more individual freedom.

d. prevent things from changing.

6. The difference between "liberal" and "radical" ideologies is that

a. liberal ideology does not seek change in the basic structures of society.

b. liberal ideology is always based on technological change.

c. liberal ideology is always less popular.

d. liberal ideology usually advocates violence as a means of change.

7. Missionaries giving steel axes to a Stone Age tribe caused social change because

a. it undermined the system of status, which had been based on possession of axes.

b. it enabled the tribe to cut down more trees.

c. it provided many lethal weapons for angry tribespeople.

d. it decreased the power of women and younger men by giving the chiefs newer and better tools.

8. A news photo in late 2005 showed an African tribesman standing on the Savannah in traditional tribal clothes but talking on a cell phone. The photo illustrates

a. liberal ideology.   **c.** a fad.

b. diffusion.   **d.** incipiency.

9. After World War II, the U.S. forces occupying Japan required that Japan rewrite its constitution so as to bring about democratic reforms. This change is an example of
   a. conservative ideology.
   b. forced acculturation.
   c. resource mobilization.
   d. McDonaldization.

10. According to Tischler, cultural forms and ideas usually travel
    a. from more powerful societies to weaker ones.
    b. from weaker societies to more powerful ones.
    c. from less advanced societies up to more advanced societies.
    d. from southern regions to northern regions.
    e. none of these; there is no pattern of cultural diffusion.

CROWD BEHAVIOR AND SOCIAL CHANGE

11. A sizable number of passers-by stop to gawk at an auto accident. In sociological terms, this is a(n) _____ crowd.
    a. expressive
    b. casual
    c. conventional
    d. acting

12. Students gathering on the lawn of a campus building between classes would be an example of a(n) _____ crowd.
    a. expressive
    b. conventional
    c. casual
    d. acting

13. **T F** According to Canetti, crowds usually magnify social class and economic differences that already exist among their members.

14. Immediately after a city's team won a basketball championship, large numbers of people spontaneously moved into the streets. What started as a celebration came to include the looting of stores, the overturning of cars, and attacks on police who tried to quiet things down. This is an example of a change from
    a. an acting crowd to a nasty crowd.
    b. an expressive crowd to an acting crowd.
    c. a threatened crowd to an expressive crowd.
    d. a lonely crowd to an angry crowd.

15. Which of the following pairs exemplifies the principle difference between a "crowd" and a "mass"?
    a. The people who cheer at football game and those who merely attend.
    b. The people who attend a football game and those who watch it on television.
    c. The people who play in a football game and those in the stadium.
    d. The people who watch a football game on television and those who do not.

16. A principle difference between "mass hysteria" and "panic" is that
    a. panic is more likely to result from an actual threat.
    b. panic usually lasts longer.
    c. panic is usually directed by a small number of people.
    d. panic requires better communication among all people involved.

17. Which of the following is an external source of social change?
    a. technology
    b. diffusion
    c. ideology
    d. social inequality

DISPERSED COLLECTIVE BEHAVIOR

18. **T F** A craze is a fad that is especially short-lived.

19. **T F** When large numbers of people in a particular part of the country claim to have seen Elvis, we have an example of a fad.

20. **T F** Public opinion seeks to mobilize public support behind one specific party, candidate, or point of view.

21. **T F** Fads and fashions are transitory and have little social impact.

22. **T F** Because books rely on technology that has been around for five hundred years, they cannot be considered to be one of the mass media.

23. Rush Limbaugh, a popular and influential radio political commentator, is an example of a(n)
    a. opinion leader.
    b. mass hysteria.
    c. reactionary social movement.
    d. fad.

24. Sociologist Georg Simmel argued that changes in clothing fashions occur because
    a. people have an insatiable desire for novelty.
    b. changes in the physical environment make new types of clothing necessary.
    c. the young feel a constant need to be different from their elders, who also try to look young.
    d. the upper classes attempt to distinguish themselves from the lower classes, who then try to imitate them.

25. Tischler notes that the wearing of blue jeans started among laborers, then grew to include college students, and finally was picked up by fashion designers who made expensive versions for wealthier customers. This growing of the blue jeans market is an example of
    a. McDonaldization.
    b. mass hysteria.
    c. a fad.
    d. diffusion.

26. Someone yells "Fire!" in a crowded movie theater and people immediately begin a feverish and chaotic run for the exits. This is an example of
    a. mass hysteria.
    b. a panic.
    c. a rumor.
    d. mobilization.

27. The difference between crowds and masses of people is that masses
    a. are inherently unstable.
    b. work to resist social change.
    c. do not require close proximity.
    d. consist of people with lower levels of education.

28. Propaganda is information presented to the public to
    a. prevent the spread of rumors.
    b. evaluate public opinion.
    c. clarify political issues.
    d. deliberately influence opinion.

SOCIAL MOVEMENTS

29. **T F** When a social movement opens a lobbying office in Washington, D.C. it has reached the institutionalization phase.

30. **T F** The idea behind relative deprivation theory is that people feel distressed in comparison with significant reference groups.

31. **T F** One flaw in the theory of relative deprivation is that often the people who protest a situation or condition may not be deprived.

32. In some states in the U.S. that have experienced an influx in Spanish-speaking residents, movements have emerged to make English the "official state language." These movements are best categorized as
    a. reactionary.
    b. conservative.
    c. revolutionary.
    d. expressive.

33. A study found that given a choice between having $100,000 in a world where most people had $50,000 and having $125,000 where most people had $200,000, people usually chose the lower income ($100,000) even though it meant that they wouldn't be able to buy as much stuff. Which of the following best explains their choice?
    a. resource mobilization theory
    b. utilitarianism
    c. resource maximization theory
    d. relative deprivation theory

34. When people in the Watts section of Los Angeles rioted in 1964, some people were surprised because people in Watts had the houses, cars, televisions, etc. and a standard of living far better than that of middle class people in many countries of the world. Those who were surprised were overlooking the concept of
    a. resource mobilization.
    b. relative deprivation.
    c. mass hysteria.
    d. globalization.

35. According to resource mobilization theory, the crucial factor in the creation and success of social movements is
    a. resources.
    b. automobiles.
    c. dissatisfaction and anger.
    d. leadership.

36. A social movement seeks to banish cars from cities so that people will walk and use bicycles. In Tischler's categories, this movement to return to pre-automobile cities would be classified as
    a. reactionary.
    b. conservative.
    c. revolutionary.
    d. expressive.

37. Social movements like the civil rights movement, which accept most of society's values but seek partial change in the existing social order, are called _____ movements.
    a. expressive
    b. revisionary
    c. reactionary
    d. revolutionary

38. The first stage in the life cycle of social movements, in which the need for change is felt but no means for achieving it is readily available, is called
    a. frustration.
    b. fragmentation.
    c. incipiency.
    d. institutionalization.

39. Anti-gun-control groups who protect existing opportunities to buy and carry guns are an example of a(n) _____ movement.
    a. reactionary
    b. conservative
    c. expressive
    d. revisionary

40. In the 1960s, Timothy Leary encouraged his followers to "turn on, tune in, and drop out"—i.e., to take drugs, pay attention to their inner reality, and withdraw their energy from the existing society. This was an example of a(n) _____ movement.
    a. reactionary
    b. expressive
    c. revisionary
    d. revolutionary

41. The charisma of a leader is especially important during a social movement's
    a. institutionalization.
    b. coalescence.
    c. incipiency.
    d. fragmentation.

42. At which stage in the life cycle of a social movement does the charisma of its leader or leaders lose a great deal of importance?
    a. incipiency
    b. coalescence
    c. adolescence
    d. institutionalization

43. After the environmental movement had become nationally recognized and respected, some groups that wanted to adopt violent methods were harshly criticized and rejected by those that wanted to use conventional means. This stage of a social movement is called
    a. coalescence.
    b. de-coalescence.
    c. fragmentation.
    d. fermentation.

44. A social movement that seeks to overthrow all or nearly all of the existing social order and to replace it with an order considered to be more suitable is known as a(n) _____ social movement.
    a. reactionary
    b. expressive
    c. revisionary
    d. revolutionary

45. The period in the life cycle of a social movement when groups begin to form around leaders, promote policies, and promulgate programs is known as
    a. fragmentation.
    b. coalescence.
    c. incipiency.
    d. institutionalization.

46. Some economists maintain that national economies are disappearing and that business is dominated by transnational corporations not firmly rooted in any one country. This trend is also known as

    a. incipiency.

    b. coalescence.

    c. McDonaldization.

    d. globalization.

47. Columnist Thomas Friedman once observed that "no two countries that both had a McDonald's had fought a war against each other, since each got its McDonald's." His observation about the benefits of open trade among nations lends support to the idea of

    a. globalization.

    b. McDonaldization.

    c. coalescence.

    d. technology.

48. The composition of the U.S. labor force in the next decade will be most influenced by

    a. technology.

    b. education.

    c. demography.

    d. psychology.

ESSAY QUESTIONS

49. Define and give an example of each of the following: fad, craze, fashion, rumor, propaganda, mass hysteria, and panic. Briefly discuss why and under what circumstances each occurs.

50. What conditions on a college campus might lead to the creation of a social movement? Describe the possible life cycle of such a movement. What aspects of a university and its population make it more difficult for social movements to take root there? What aspects make it easier for social movements to arise on campus?

Practice Tests Answers

CHAPTER 1

1.T	2.T	3.F	4.F	5.T	6.c	7.b	8.c	9.c	10.b	11.d	12.c
13.a	14.c	15.F	16.F	17.T	18.T	19.F	20.F	21.F	22.F	23.F	24.T
25.T	26.F	27.T	28.T	29.T	30.T	31.b	32.a	33.b	34.a	35.a	36.c
37.c	38.a	39.b	40.c	41.b	42.a	43.T	44.F	45.T	46.a	47.c	48.b
49.d											

CHAPTER 2

1.F	2.T	3.F	4.T	5.F	6.F	7.T	8.T	9.T	10.T	11.T	12.T
13.T	14.T	15.F	16.F	17.F	18.F	19.T	20.F	21.F	22.F	23.F	24.T
25.T	26.F	27.F	28.T	29.T	30.d	31.c	32.a	33.b	34.e	35.b	36.c
37.c	38.b	39.a	40.b	41.d	42.a	43.b	44.a	45.b	46.c		

CHAPTER 3

1.F	2.F	3.T	4.T	5.T	6.c	7.c	8.b	9.a	10.a	11.T	12.T
13.F	14.F	15.T	16.a	17.a	18.c	19.a	20.b	21.a	22.d	23.T	24.F
25.F	26.T	27.T	28.T	29.T	30.F	31.F	32.b	33.c	34.c	35.T	36.F
37.F	38.d	39.c	40.a	41.F	42.T	43.F	44.c	45.b	46.a	47.b	

CHAPTER 4

1.F	2.T	3.T	4.T	5.F	6.T	7.F	8.T	9.F	10.F	11.F	12.F
13.T	14.T	15.T	16.T	17.F	18.T	19.F	20.F	21.T	22.T	23.F	24.F
25.F	26.c	27.b	28.d	29.d	30.c	31.a	32.b	33.d	34.a	35.c	36.c
37.c	38.a	39.d	40.c	41.a	42.b	43.c	44.b	45.d	46.a		

CHAPTER 5

1.T	2.F	3.T	4.F	5.d	6.c	7.b	8.c	9.c	10.T	11.F	12.T
13.T	14.F	15.F	16.a	17.a	18.d	19.a	20.d	21.a	22.T	23.F	24.T
25.F	26.F	27.c	28.a	29.a	30.a	31.a	32.a	33.a	34.d	35.a	36.F
37.F	38.F	39.T	40.F	41.T	42.a	43.c	44.a	45.a	46.b	47.b	48.c

CHAPTER 6

1.F	2.F	3.F	4.F	5.T	6.T	7.T	8.F	9.T	10.c	11.b	12.b
13.a	14.F	15.F	16.F	17.b	18.c	19.a	20.b	21.T	22.T	23.T	24.F
25.F	26.T	27.T	28.F	29.d	30.d	31.c	32.c	33.b	34.e	35.T	36.T
37.T	38.F	39.T	40.F	41.F	42.T	43.b	44.a	45.a	46.a	47.c	

CHAPTER 7

1.T	2.F	3.T	4.e	5.T	6.F	7.b	8.a	9.d	10.F	11.F	12.F
13.d	14.d	15.a	16.e	17.c	18.a	19.c	20.c	21.c	22.b	23.d	24.c
25.d	26.F	27.F	28.T	29.F	30.F	31.T	32.T	33.T	34.T	35.F	36.T
37.F	38.c	39.b	40.a	41.b	42.T	43.F	44.T	45.F	46.c	47.a	48.b
49.e	50.e	51.c	52.a								

CHAPTER 8

1.T	2.T	3.T	4.F	5.F	6.F	7.F	8.F	9.F	10.F	11.e	12.c
13.a	14.b	15.b	16.b	17.e	18.b	19.F	20.F	21.T	22.T	23.F	24.T
25.T	26.F	27.b	28.a	29.a	30.a	31.d	32.a	33.T	34.T	35.F	36.c
37.a	38.c	39.c	40.T	41.F	42.F	43.T	44.F	45.c	46.T	47.F	48.T
49.T	50.T	51.T	52.T	53.c	54.d						

CHAPTER 9

1.T	2.F	3.F	4.T	5.F	6.d	7.b	8.T	9.c	10.F	11.c	12.b
13.F	14.T	15.T	16.F	17.F	18.T	19.c	20.F	21.T	22.b	23.T	24.d
25.a	26.T	27.e	28.T	29.F	30.c	31.F	32.d	33.a	34.T	35.T	36.d
37.T	38.c	39.c	40.c	41.T	42.F	43.c	44.d	45.a	46.F	47.T	48.T
49.T	50.F	51.b	52.b								

CHAPTER 10

1.T	2.F	3.T	4.T	5.T	6.F	7.T	8.F	9.b	10.d	11.d	12.a
13.c	14.F	15.a	16.c	17.F	18.T	19.a	20.T	21.c	22.b	23.T	24.c
25.c	26.b	27.a	28.c	29.a	30.a	31.d	32.c	33.a	34.c	35.b	36.T
37.T	38.T	39.c	40.c	41.d	42.c	43.b	44.b	45.b	46.d	47.T	48.b
49.c	50.c	51.d	52.d	53.a	54.b						

CHAPTER 11

1.F	2.T	3.T	4.F	5.F	6.F	7.F	8.a	9.d	10.c	11.a	12.c
13.d	14.a	15.c	16.F	17.a	18.a	19.d	20.d	21.T	22.d	23.a	24.c
25.b	26.c	27.b	28.b	29.T	30.F	31.b	32.a	33.b	34.a	35.T	36.T
37.F	38.d	39.a	40.d	41.b	42.d	43.F					

CHAPTER 12

1.F	2.T	3.T	4.T	5.T	6.F	7.T	8.c	9.b	10.b	11.a	12.T
13.T	14.F	15.T	16.T	17.c	18.c	19.d	20.d	21.b	22.a	23.c	24.d
25.d	26.b	27.c	28.c	29.b	30.T	31.T	32.T	33.T	34.F	35.F	36.T
37.F	38.T	39.F	40.T	41.T	42.T	43.T	44.F	45.b	46.c	47.c	48.c
49.b	50.b	51.c	52.F	53.F	54.T						

CHAPTER 13

1.F	2.F	3.d	4.a	5.c	6.c	7.c	8.d	9.b	10.F	11.a	12.T
13.T	14.d	15.T	16.c	17.d	18.c	19.d	20.c	21.c	22.T	23.T	24.T
25.a	26.b	27.d	28.c	29.b	30.c	31.b	32.a	33.b	34.a	35.c	36.c
37.a	38.T	39.c	40.F	41.e	42.T	43.b	44.T	45.b	46.T	47.a	48.c
49.d	50.a	51.F	52.F								

CHAPTER 14

1.F	2.F	3.T	4.T	5.F	6.T	7.T	8.c	9.a	10.c	11.b	12.c
13.d	14.a	15.c	16.a	17.a	18.T	19.F	20.F	21.F	22.d	23.c	24.b
25.d	26.b	27.d	28.T	29.F	30.F	31.F	32.F	33.T	34.F	35.T	36.F
37.d	38.b	39.d	40.a	41.a	42.c	43.T	44.c	45.e	46.a	47.e	48.a
49.b	50.c	51.b									

CHAPTER 15

1.T	2.F	3.T	4.F	5.F	6.T	7.c	8.a	9.a	10.a	11.b	12.c
13.d	14.c	15.T	16.T	17.a	18.c	19.a	20.c	21.d	22.d	23.c	24.b
25.c	26.c	27.a	28.F	29.F	30.T	31.F	32.F	33.c	34.c	35.b	36.a
37.a	38.c	39.a	40.T	41.T	42.d	43.c	44.d	45.d	46.d	47.b	48.T
49.F	50.F	51.F	52.c	53.c	54.c						

CHAPTER 16

1.F	2.T	3.F	4.T	5.T	6.F	7.F	8.a	9.b	10.c	11.a	12.d
13.a	14.a	15.d	16.b	17.b	18.c	19.c	20.c	21.F	22.F	23.d	24.b
25.d	26.c	27.a	28.d	29.d	30.c	31.b	32.b	33.a	34.a	35.b	36.b
37.a	38.F	39.F	40.T	41.c	42.c	43.a	44.T	45.T	46.T	47.a	48.b
49.a	50.b	51.a	52.b								

CHAPTER 17

1.T	2.c	3.b	4.e	5.d	6.F	7.T	8.F	9.T	10.T	11.T	12.T
13.F	14.F	15.F	16.c	17.c	18.b	19.a	20.a	21.a	22.e	23.d	24.c
25.a	26.F	27.T	28.T	29.a	30.b	31.F	32.T	33.F	34.T	35.T	36.a
37.d	38.a	39.c	40.c	41.c	42.a	43.a	44.c	45.a	46.T	47.F	48.T
49.F	50.T	51.b	52.a	53.a	54.d	55.c					

CHAPTER 18

1.F	2.F	3.c	4.T	5.d	6.a	7.a	8.b	9.b	10.a	11.b	12.b
13.F	14.b	15.b	16.a	17.b	18.T	19.F	20.F	21.T	22.F	23.a	24.d
25.d	26.b	27.c	28.d	29.T	30.T	31.T	32.b	33.d	34.b	35.d	36.a
37.b	38.c	39.b	40.b	41.b	42.d	43.c	44.d	45.b	46.d	47.a	48.c